Interest Boxes

Natural Versus Synthetic (1.0)

Albert Einstein (1.1)

Diamond and Graphite: Substances that Contain Only Carbon Atoms (1.8)

Water—A Unique Compound (1.11)

Acid Rain (2.2)

Blood: A Buffered Solution (2.5)

Bad-Smelling Compounds (3.5)

Cholesterol and Heart Disease (3.13)

Clinical Treatment of High Cholesterol (3.13)

Pheromones (4.0)

A Few Words About Curved Arrows (4.7)

Pesticides: Natural and Synthetic (4.8)

Naturally Occurring Alkynes (5.6)

Trans Fats (5.12)

Designing a Polymer (5.16)

Recycling Polymers (5.16)

Ethyne Chemistry or the Forward Pass (5.16)

Kekulé's Dream (6.1)

Delocalized Electrons in Vision (6.6)

Ultraviolet Light and Sunscreens (6.10)

Anthocyanins: A Colorful Class of Compounds (6.12)

Measuring Toxicity (7.0)

Buckyballs (7.2)

Heme and Chlorophyll (7.3)

Porphyrin, Bilirubin, and Jaundice (7.3)

The Toxicity of Benzene (7.4)

Thyroxine (7.7)

The Enantiomers of Thalidomide (8.12)

Chiral Drugs (8.12)

Octane Number (9.0)

Fossil Fuels: A Problematic Energy Source (9.0)

Decaffeinated Coffee and the Cancer Scare (9.6)

Food Preservatives (9.6)

Survival Compounds (10.0)

Artificial Blood (10.0)

Environmental Adaptation (10.3)

Why Carbon Instead of Silicon? (10.3)

Investigating Naturally Occurring Halogen Containing Compounds (10.7)

Eradicating Termites (10.11)

S-Adenosylmethionine: A Natural Antidepressant (10.11)

Grain Alcohol and Wood Alcohol (11.2)

Biological Dehydrations (11.3)

Blood Alcohol Content (11.4)

Alkaloids (11.5)

Anesthetics (11.6)

An Unusual Antibiotic (11.7)

Mustard—A Chemical Warfare Agent (11.8)

Antidote to a Chemical Warfare Agent (11.8)

Benzo[a]Pyrene and Cancer (11.9)

Chimney Sweeps and Cancer (11.9)

The Discovery of Penicillin (12.4)

Dalmatians: Don't Try to Fool Mother Nature (12.4)

Nerve Impulses, Paralysis, and Insecticides (12.8)

Biodegradable Polymers (12.8)

Aspirin (12.9)

Nature's Sleeping Pill (12.11)

Penicillin and Drug Resistance (12.12)

Penicillins in Clinical Use (12.12)

Synthetic Polymers (12.13)

Dissolving Sutures (12.13)

Butanedione: An Unpleasant Compound (13.1)

Synthesizing Organic Compounds (13.4)

Semisynthetic Drugs (13.4)

Preserving Biological Specimens (13.7)

Antiancer Drugs (13.8)

Synthesis of Aspirin (14.4)

Nikola Tesla (15.8)

Magnetic Resonance Imaging (15.18)

Measuring the Blood Glucose Levels of Diabetics (16.5)

Glucose/Dextrose (16.7)

Lactose Intolerance (16.12)

Galactosemia (16.12)

Why the Dentist is Right (16.13)

Controlling Fleas (16.13)

Heparin (16.14)

Vitamin C (16.14)

Acceptable Daily Intake (16.16)

Proteins and Nutrition (17.1)

Amino Acids and Disease (17.2)

A Peptide Antibiotic (17.2)

Enkephalins (17.6)

Hair: Straight or Curly? (17.6)

Peptide Hormones (17.6)

Primary Structure and Evolution (17.7)

Nutrasweet (17.13)

Vitamin B_1 (18.4)

"Vitamine"—An Amine Required for Life (18.4)

Niacin (18.5)

Niacin Deficiency (18.5)

Heart Attacks: Assessing the Damage (18.9)

Cancer Chemotherapy (18.11)

The First Antibiotics (18.11)

Too Much Broccoli (18.12)

Differences in Metabolism (19.0)

The Nobel Prize (19.2)

Phenylketonuria: An Inborn Error of Metabolism (19.6)

Alcaptonuria (19.6)

Basal Metabolic Rate (19.9)

Omega Fatty Acids (20.1)

Olestra: Nonfat with Flavor (20.3)

Whales and Echolocation (20.3)

Making Soap (20.4)

Snake Venom (20.5)

Is Chocolate a Health Food? (20.5)

Multiple Sclerosis and the Myelin Sheath (20.5)

Lipoproteins (20.8)

The Structure of DNA: Watson, Crick, Franklin, and Wilkins (21.0)

Sickle-Cell Anemia (21.8)

Antibiotics that Act by Inhibiting Translation (21.8)

DNA Fingerprinting (21.11)

Resisting Herbicides (21.12)

Drug Safety (22.4)

Orphan Drugs (22.10)

Essential
Organic Chemistry

Essential Organic Chemistry

Paula Yurkanis Bruice

University of California, Santa Barbara

PEARSON

Prentice Hall

Upper Saddle River, NJ 07458

Library of Congress Cataloging-in-Publication Data

Bruice, Paula Yurkanis
   Essential Organic Chemistry/Paula Yurkanis Bruice.—1st ed.
      p. cm.
   Includes index.
   ISBN 0-13-149858-4
   1. Chemistry, Organic—Textbooks. I. Title.

QD251.3.B777 2006
547—dc22   2005048872

Executive Editor: Nicole Folchetti
Media Editor: Margaret Trombley
Project Manager: Kristen Kaiser
Art Director: Maureen Eide
Assistant Art Director: John Christiana
Assistant Managing Editor, Science Media: Nicole M. Jackson
Assistant Managing Editor, Science Supplements: Becca Richter
Executive Marketing Manager: Steve Sartori
Media Production Editor: Karen Bosch
Production Assistant: Nancy Bauer
Editorial Assistant: Timothy Murphy
Director, Creative Services: Paul Belfanti
Manufacturing Manager: Alexis Heydt-Long
Manufacturing Buyer: Alan Fischer
Senior Managing Editor, Art Production and Management:
   Patricia Burns
Manager, Production Technologies: Matthew Haas
Managing Editor, Art Management: Abigail Bass
Art Production Editor: Denise Keller
Manager, Art Production: Sean Hogan
Assistant Manager, Art Production: Ronda Whitson

Illustrations: ESM Art Production; Lead Illustrators:
   Scott Wieber and Nathan Storck
Contributing Art Studios: Imagineering and Wavefunction /
   Richard Johnson
Spectra: Reproduced by permission of Aldrich Chemical Co.
Interior Designer: Joseph Sengotta
Director, Image Resource Center: Melinda Reo
Manager, Rights and Permissions: Zina Arabia
Interior Image Specialist: Beth Boyd-Brenzel
Photo Editor: Cynthia Vincenti
Photo Researcher: Melinda Alexander
Cover Designer: John Christiana
Cover Researcher: Karen Sanatar
Cover Photos: *Slug*: © David Welling/Nature Picture Library;
   DNA: © Paul Morrell/Getty Images Inc.–Stone Allstock;
   Nudibranch: © David Fleetham/Getty Images Inc.–Taxi;
   Bottle of pills: © Steve Allen/Brand X Pictures/Getty Images Inc.;
   Water droplets: © PIER/The Image Bank/Getty Images Inc.;
   Killer frogs: © George Grall/National Geographic/
   Getty Images Inc.
Production Services/Composition: Prepare, Inc.

© 2006 by Pearson Education, Inc.
Pearson Prentice Hall
Pearson Education, Inc.
Upper Saddle River, NJ 07458

Pearson Prentice Hall™ is a trademark of Pearson Education, Inc.

Printed in the United States of America
10 9 8 7 6 5 4 3 2

ISBN 0-13-149858-4

Pearson Education Ltd., *London*
Pearson Education Australia Pty. Ltd., *Sydney*
Pearson Education Singapore, Pte. Ltd.
Pearson Education North Asia Ltd., *Hong Kong*
Pearson Education Canada, Inc., *Toronto*
Pearson Educacíon de Mexico, S.A. de C.V.
Pearson Education—Japan, *Tokyo*
Pearson Education Malaysia, Pte. Ltd.

To Meghan, Kenton, and Alec
with love and immense respect,
and to Tom, my best friend

Brief Contents

Contents

4 Alkenes: Structure, Nomenclature, Stability, and an Introduction to Reactivity 81

5 Reactions of Alkenes and Alkynes: An Introduction to Multistep Synthesis 103

12 Carbonyl Compounds I: Nucleophilic Acyl Substitution   288

13 Carbonyl Compounds II: Reactions of Aldehydes and Ketones • More Reactions of Carboxylic Acid Derivatives   320

17 Amino Acids, Peptides, and Proteins   434

18 Enzymes, Coenzymes, and Vitamins   462

Preface

To the Instructor

The guiding principle in writing this book was to create a text that will allow students to see organic chemistry as an exciting science and to understand why it is an important one. They should not think that studying organic chemistry involves simply memorizing molecules and reactions. Thus, the book revolves around shared features and unifying concepts, and it emphasizes principles that can be applied again and again. I want students to learn how to apply what they have learned to a new setting, reasoning their way to a solution, rather than memorizing a multitude of facts.

I hope that as students proceed through their study of organic chemistry, they see that it is a subject that unfolds and grows and allows them to use what they learn at the beginning of the course to predict what follows. I also want students to see that organic chemistry is integral to biology as well as to their daily lives. Consequently, there are about 100 interest boxes sprinkled throughout the book. These boxes are designed to show students the relevance of organic chemistry to medicine (e.g., dissolving sutures, cholesterol and heart disease, artificial blood), agriculture (e.g., resisting herbicides, acid rain, pesticides: natural and synthetic), nutrition (e.g., trans fats, basal metabolic rate, omega fatty acids), and to our shared life on this planet (e.g., fossil fuels, measuring toxicity, biodegradable polymers).

Some hard choices had to be made in deciding what was "essential" to this audience of students. In writing the book, these choices were made with three goals kept in mind: Students should understand how and why organic compounds react the way they do, they should experience the fun and challenge of designing simple syntheses, and they should learn in earlier chapters those reactions they will encounter again in later chapters that focus on bioorganic topics. When I wrote the chapter on spectroscopy, I did not want students to be overwhelmed by a topic that they may never revisit in their lives, but I did want them to enjoy being able to interpret some simple spectra.

I hope your students enjoy this book. I am always eager to hear your comments—positive comments are the most fun, but critical comments are the most useful.

Pedagogical Features

Problems, Solved Problems, and Problem-Solving Strategies

The book contains lots of problems. The answers to all the problems (along with explanations when needed) are in the accompanying *Study Guide and Solutions Manual*, which I authored for consistency in language with that of the text. The *Study Guide and Solutions Manual* also contains an exercise on "electron pushing." I have found this exercise (which should take a student less than 15 minutes to work through) to be successful in making my students comfortable with a topic that should be easy, but somehow perplexes even the best of them unless they have sufficient practice.

The problems within each chapter are primarily drill problems; one or more of these are at the end of most sections. These problems allow students to test themselves on material just covered before moving on to the next section. The solutions to selected problems are worked out in detail to provide insight into how to solve problems. Most chapters also contain at least one "Problem-Solving Strategy," which teaches students how to approach certain kinds of problems. Each "Problem-Solving Strategy" is followed by an exercise giving the student an opportunity to use the problem-solving skill just learned. Short answers to problems marked with a diamond are provided at the end of the book so students can quickly test their understanding.

The end-of-chapter problems vary in difficulty. The initial problems are drill problems that integrate material from the entire chapter. These problems provide a greater challenge by requiring the student to think in terms of all the material in the chapter, rather than material just from individual sections. The problems become more challenging as the student proceeds, often reinforcing concepts covered in previous chapters. The net effect is to progressively build both problem-solving ability and confidence.

Margin Notes and Boxed Material to Engage the Student

Margin notes and biographical sketches appear throughout the text. The margin notes remind students of important principles and succinctly recap key points to facilitate review. The biographical sketches give students some appreciation of the history of chemistry and the people who contributed to that history.

With the conviction that learning should be fun, I have put material in strategically placed interest boxes that cover intriguing asides: For example, these boxes explain why Dalmatians are the only mammals that excrete uric acid, why life is based on carbon instead of silicon, how a microorganism has learned to use industrial waste as a source of carbon, and why SAMe is a product prominently displayed in health food stores.

Summaries and Voice Boxes to Help the Student

Each chapter concludes with a "Summary" to help students synthesize the key points of the chapter. Chapters that cover reactions conclude with a "Summary of Reactions." Illustrations with annotations (voice boxes) are found throughout the book to help students focus on points being discussed.

Art Program: Rich in Three-Dimensional, Computer-Generated Structures

Energy-minimized, three-dimensional structures appear throughout the text, giving students an appreciation of the three-dimensional shapes of organic molecules. Color is used to highlight and organize the information, not simply for show. I have attempted to make the color consistent (e.g., mechanism arrows are always red), but there is no need for a student to memorize a color palette.

Companion Website with GradeTracker

WWW icons in the margins identify 3-D molecules, movies, and interactive animations on the Companion Website (*http://www.prenhall.com/bruice*) that are pertinent to the material being discussed. Each chapter on the Website has practice exercises and quizzes. Although I had never been a fan of this type of question, the quality of these questions has changed my mind. With GradeTracker, students can work problems and track their progress throughout the semester. Instructors can import these grades at any time during the semester.

List of Resources

For Students

Study Guide and Solutions Manual (0-13-149860-6) by Paula Yurkanis Bruice. This Study Guide and Solutions Manual contains complete and detailed explanations of the solutions to the problems in the text.

Companion Website (*http://www.prenhall.com/bruice*) Built to complement *Essential Organic Chemistry* as part of an integrated course package, the easy-to-use Companion Website features the following modules for each chapter:

- **Interactive Tutorial and Animation Galleries** highlight central concepts and illustrate key mechanisms. The Tutorials often allow the student to make incorrect choices and then explain why there is a better answer.
- **Molecule Galleries** feature hundreds of 3-D molecular models of compounds. Students can rotate and compare models, change their representation, and examine electrostatic potential map surfaces—a unique feature of learning organic chemistry on the Web.
- **Practice Exercises and Quizzes** offer new exercises to test the student's understanding of the material. Each question includes a hint with a cross-reference to a text reading and detailed feedback.

Molecular Modeling Workbook (0-13-141040-7) Features SpartanView™ and SpartanBuild™ software. This workbook includes a software tutorial and numerous challenging exercises students can tackle to solve problems involving structure building and analysis using the tools included in the two pieces of Spartan software. Available free when packaged with the text; contact your Prentice Hall representative for details.

ChemOffice Student CD This software includes ChemDraw LTD and Chem3D LTD. A free workbook is available on the Companion Website that includes a software tutorial and numerous exercises for each chapter of the text. The software is sold at the instructor's discretion and is available at a substantial discount if purchased with the textbook.

Prentice Hall Molecular Model Kit (0-205-508136-3) This best-selling model kit allows students to build space-filling and ball-and-stick models of common organic molecules. It allows accurate depiction of double and triple bonds, including hetero-atomic molecules (which some model kits cannot handle well).

Prentice Hall Framework Molecular Model Kit (0-13-330076-5) This model kit allows students to build scale models that show the mutual relations of atoms in organic molecules, including precise interatomic distances and bond angles. This is the most accurate model kit available.

For Instructors

Instructor Resource Center on CD/DVD (0-13-149861-4) This lecture resource provides a fully searchable and integrated collection of resources to help you make efficient and effective use of your lecture preparation time, as well as to enhance your classroom presentations and assessment efforts. This resource features almost all the art from the text, including tables, in JPEG, PDF and PowerPoint™ formats; two pre-built PowerPoint™ presentations; and all interactive and dynamic media objects from the Companion Website. This CD also features a search-engine tool that enables you to find relevant resources via key terms, learning objectives, figure numbers, and resource type. This CD/DVD set also contains TestGen test-generation software and a TestGen version of the Test Item File that enables professors to create and tailor exams to their needs, or to create online quizzes for delivery in WebCT, Blackboard, or CourseCompass.

Test Item File (0-13-149859-1) by Debbie Beard, Mississippi State University and Gary Hollis, Roanoke College. Includes a selection of more than 1,200 multiple choice, short answer, and essay test questions.

Acknowledgments

I am enormously grateful to the following reviewers who made this book a reality. The value of their work cannot be overstated.

Manuscript Reviewers

Ardeshir Azadnia, *Michigan State University*
Debbie Beard, *Mississippi State University*
J. Phillip Bowen, *University of North Carolina-Greensboro*
Tim Burch, *Milwaukee Area Technical College*
Dana Chatellier, *University of Delaware*
Michelle Chatellier, *University of Delaware*
Long Chiang, *University of Massachusetts-Lowell*
Jan Dekker, *Reedley College*
Olga Dolgounitcheva, *Kansas State University*
John Droske, *University of Wisconsin-Stevens Point*
Eric Enholm, *University of Florida*
Gregory Friestad, *University of Vermont*
Wesley Fritz, *College of Dupage*
Robert Gooden, *Southern University*
Michael Groziak, *California State University-Hayward*
Steve Holmgren, *Montana State University*
Robert Hudson, *University of Western Ontario*
Richard Johnson, *University of New Hampshire*
Alan Kennan, *Colorado State University*
Spencer Knapp, *Rutgers University*
Mike Nuckols, *North Carolina State University*
Ed Parish, *Auburn University*
Mark W. Peczuh, *University of Connecticut*
Suzanne Purrington, *North Carolina State University*
Charles Rose, *University of Nevada-Reno*
Preet Saluja, *Triton College*
Joseph Sloop, *United States Military Academy*
Robert Swindell, *University of Arkansas*
Amar Tung, *Lincoln University*
Kraig Wheeler, *Delaware State University*
Randy Winchester, *Grand Valley State University*
Mark Workentin, *University of Western Ontario*

Focus Group Attendees

Ardeshir Azadina, *Michigan State University*
Gregory L. Baker, *Michigan State University*
Jay Brown, *Southwest Minnesota State University*
Jerry Easdon, *College of Ozarks*
Nancy Gardner, *California State University-Long Beach*
Cyril Parkanyi, *Florida Atlantic University*
Bob Swindell, *University of Arkansas*
Kathleen Trahanovsky, *Iowa State University*

Accuracy Reviewer

Susan Schelble, *University of Colorado-Denver*

I am deeply grateful to my editor, Nicole Folchetti, who was always ready to do whatever was needed to make this book the best that it could be. Her creative talents are extraordinary. I also want to thank the other talented and dedicated people at Prentice Hall who played important roles in the development of the book. Kathleen

Schiaparelli, Executive Managing Editor, kept the project on track and managed an infinite number of critical details, Steve Sartori, Executive Marketing Manager, brought the book to the attention of the global community of organic chemistry instructors, Timothy Murphy, Editorial Assistant, assembled a talented panel of manuscript reviewers and kept dozens of balls in the air, Kristen Kaiser, Project Manager, produced the supplements, Margaret Trombley, Nicole Jackson, and Karen Bosch developed and managed the media program, Maureen Eide, Art Director, and Joseph Christiana created a striking cover, Denise Keller produced the new art, and Dave Theisen, National Sales Director, has been, as always, a champion, resource, and friend in the field. I am also grateful to the compositor, Rosaria Cassinese, who kept me on track—with more patience than I deserved—during the production process.

I particularly want to thank the many wonderful and talented students who have taught me more than they will ever know. And I want to thank my children, from whom I may have learned the most.

To make this book as user friendly as possible, I would appreciate any comments that will help me achieve this goal in future editions. If you find sections that should be clarified or expanded, please let me know. Finally, I am enormously grateful to Susan Schelbel of the University of Colorado, Denver, who painstakingly combed the book looking for errors. Any that remain are my responsibility; if you find any, please send me a quick e-mail so that they can be corrected in future printings.

Paula Yurkanis Bruice
University of California, Santa Barbara
pybruice@chem.ucsb.edu

To the Student

Welcome to organic chemistry! You are about to embark on an exciting journey. This book has been written with students like you in mind—those who are encountering the subject for the first time. The book's central goal is to make your journey both stimulating and enjoyable by helping you understand the guiding principles of the subject and applying these principles to your field of study. You should start by familiarizing yourself with the book. The material on the inside of the front and back covers contains information that you may want to refer to many times during the course. Chapter summaries and summaries of the reactions at the end of the chapters are useful tools to remind yourself of what you should have learned; they also provide a good mental checklist of what you should understand as you get to the end of the chapter. The glossary at the end of the book can be a very useful study aid. Take advantage of all of these features! Appendices are written to consolidate useful categories of information, so be sure to see what kind of information is provided there. The molecular models and electrostatic potential maps that you will find throughout the book are there to give you an appreciation of what molecules look like in three dimensions and how charge is distributed within a molecule. Think of the margin notes as the author's opportunity to provide personal reminders and place emphasis on important points. Be sure to read them.

Work all the problems within each chapter. These are drill problems that allow you to check whether you have mastered the material. Some of them are solved for you in the text. Others—those marked with a diamond—have short answers provided at the end of the book. Don't overlook the "Problem-Solving Strategies" sprinkled throughout the text; they provide practical suggestions on the best way to approach important types of problems.

Work as many end-of-chapter problems as you can. The more problems you work, the more comfortable you will be with the subject and the more prepared you will be for the material in subsequent chapters. Do not let any problem frustrate you. If you cannot figure out the answer in a reasonable amount of time, turn to the *Study Guide and Solutions Manual* to learn how you should have approached the problem. Later on, go back and try to work the problem again, this time on your own. Be sure to visit the Companion Website (*http://www.prenhall.com/bruice*) to try out the tutorials, study the 3-D molecules, and take the tests.

The most important thing for you to remember in organic chemistry is DO NOT GET BEHIND! Organic chemistry consists of a lot of simple steps, each by itself very easy to master. But the subject can quickly become overwhelming if you don't keep up.

Before many of its theories and mechanisms were worked out, organic chemistry was a discipline that could be mastered only through memorization. Fortunately, that is no longer true. You will find many common threads that allow you to use what you have learned in one situation to predict what will happen in other situations. So, as you read the book and study your notes, always try to understand why each thing happens. If the reasons behind the reactivity are understood, most reactions can be predicted. Approaching the class with the misconception that you must memorize hundreds of unrelated reactions could be your downfall. There is simply too much material to memorize! Reasoning, not memorization provides the necessary foundation on which to lay subsequent material. From time to time, however, some memorization will be required. Some fundamental rules have to be memorized, and you will have to memorize the common names of a number of organic compounds. But the latter should not be a problem; after all, your friends have common names that you have been able to learn.

Good luck in your study. I hope you enjoy your course in organic chemistry and learn to appreciate the logic of the discipline. If you have any comments about the book or any suggestions about how it can be improved for the students who will follow you, I would love to hear from you.

Paula Yurkanis Bruice
pybruice@chem.ucsb.edu

To the Student

The First Edition of *Essential Organic Chemistry*, by Paula Yurkanis Bruice conveys the science of organic chemistry as one that is exciting and particularly relevant for the students who take this course. Written in the hallmark style of the author, through outstanding clarity of explanation, *Essential Organic Chemistry* remains true to the text's main goal—encouraging students to understand the "why" of organic chemistry.

Approach

• **A strong bioorganic flavor** shows how organic chemistry is integral to biology as well as to our daily lives. There are over 100 special-interest boxes that show the relevance of the science to the fields of medicine, agriculture, nutrition, and our shared life on this planet. A strong bioorganic flavor throughout the text encourages students to recognize that organic chemistry and biochemistry are not separate entities, but two parts of a continuum of knowledge. This material is found in special interest boxes, specific chapter sections, and chapters that focus on bioorganic topics. A complete list of the special interest boxes is located opposite the inside front cover.

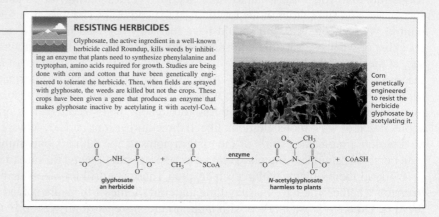

RESISTING HERBICIDES

Glyphosate, the active ingredient in a well-known herbicide called Roundup, kills weeds by inhibiting an enzyme that plants need to synthesize phenylalanine and tryptophan, amino acids required for growth. Studies are being done with corn and cotton that have been genetically engineered to tolerate the herbicide. Then, when fields are sprayed with glyphosate, the weeds are killed but not the crops. These crops have been given a gene that produces an enzyme that makes glyphosate inactive by acetylating it with acetyl-CoA.

Corn genetically engineered to resist the herbicide glyphosate by acetylating it.

glyphosate
an herbicide

N-acetylglyphosate
harmless to plants

Practice and Help with Problem Solving

• **Solved Problems and practice exercises** throughout the text carefully walk you through the steps involved in solving a particular type of problem.

PROBLEM 5 SOLVED

Using the pK_a values of the conjugate acids of the leaving groups (the pK_a of HBr is −9; the pK_a of H_2O is 15.7; the pK_a, of H_3O^+ is −1.7), explain the difference in reactivity of

a. CH_3Br and CH_3OH **b.** $CH_3\overset{+}{O}H_2$ and CH_3OH

SOLUTION TO 5a The conjugate acid of the leaving group of CH_3Br is HBr; its pK_a is = −9; the conjugate acid of the leaving group of CH_3OH is H_2O; its pK_a is = 15.5. Because HBr is a much stronger acid than H_2O, Br^- is a much weaker base than HO^-. (Recall the stronger the acid, the weaker is its conjugate base.) Therefore, Br^- is a much better leaving group than HO^- causing CH_3Br to be much more reactive than CH_3OH.

• **Problem-Solving Strategies** in each chapter demonstrate how to approach a variety of problems, organize your thoughts, and improve your problem-solving abilities. Every strategy is followed by an exercise that allows you to immediately practice the strategy just discussed.

PROBLEM-SOLVING STRATEGY

(*S*)-Alanine is a naturally occurring amino acid. Draw its structure using a perspective formula.

$$CH_3CHCOO^-$$
$$\overset{|}{\underset{}{^+NH_3}}$$
alanine

First draw the bonds about the asymmetric center. Remember that the two bonds in the plane of the paper must be adjacent to one another.

Put the group with the lowest priority on the hatched wedge. Put the group with the highest priority on any remaining bond.

Because you have been asked to draw the *S* enantiomer, draw an arrow counterclockwise from the group with the highest priority to the next available bond; put the group with the next highest priority on that bond.

Put the remaining substituent on the last available bond.

Now continue on to Problem 13.

• **End-of-chapter problems** focus on overarching principles and concepts that tend to be especially difficult for students. These problems progress from drill problems that incorporate the material learned in individual sections in the entire chapter, to more challenging ones that will challenge you to think in terms of the material learned throughout the entire chapter.

Enhanced Pedagogy

- **Biographical Sketches** give you an appreciation of the history of chemistry and the people who contributed to that history.

- **Margin Notes**, emphasize important principles and recap key points to facilitate review and remind students of important principles to help them grasp concepts in the text.

Emil Fischer (1852–1919) *was born in a village near Cologne, Germany. He became a chemist against the wishes of his father, a successful merchant, who wanted him to enter the family business. He was a professor of chemistry at the Universities of Erlangen, Würzburg, and Berlin. In 1902, he received the Nobel Prize in chemistry for his work on sugars. During World War I, he organized German chemical production. Two of his three sons died in that war.*

proper positions, so they cannot bind efficiently to the enzyme. For example, fumarase catalyzes the addition of water to fumarate (the trans isomer) but not to maleate (the cis isomer).

An achiral molecule reacts identically with both enantiomers. A sock, which is achiral, fits on either foot.

A chiral molecule reacts differently with each enantiomer. A shoe, which is chiral, fits on only one foot.

$+ \; H_2O$ $\xrightarrow{\text{fumarase}}$ no reaction

maleate

An enzyme's behavior can be likened to a right-handed glove, which fits only the right hand: It forms only one stereoisomer and it reacts with only one stereoisomer.

PROBLEM 27◆

a. What would be the product of the reaction of fumarate and H_2O if H^+ were used as a catalyst instead of fumarase?

b. What would be the product of the reaction of maleate and H_2O if H^+ were used as a catalyst instead of fumarase?

- **Summary of Reactions** sections list reactions covered in the chapter for review. Cross-references make it easy to locate the sections covering specific reaction types.

Summary of Reactions

1. Reaction of *carbonyl compounds* with a Grignard reagent (Section 13.4).
   a. Reaction of *formaldehyde* with a Grignard reagent forms a primary alcohol:

   $$\underset{H}{\overset{O}{\underset{\|}{C}}}_{H} \quad \xrightarrow[\text{2. } H_3O^+]{\text{1. } CH_3MgBr} \quad CH_3CH_2OH$$

   b. Reaction of an *aldehyde* (other than formaldehyde) with a Grignard reagent forms a secondary alcohol:

   $$\underset{R}{\overset{O}{\underset{\|}{C}}}_{H} \quad \xrightarrow[\text{2. } H_3O^+]{\text{1. } CH_3MgBr} \quad R\overset{OH}{\underset{CH_3}{\overset{|}{\underset{|}{C}}}}H$$

   c. Reaction of a *ketone* with a Grignard reagent forms a tertiary alcohol:

   $$\underset{R}{\overset{O}{\underset{\|}{C}}}_{R'} \quad \xrightarrow[\text{2. } H_3O^+]{\text{1. } CH_2MgBr} \quad R\overset{OH}{\underset{CH_3}{\overset{|}{\underset{|}{C}}}}R'$$

- **End-of-Chapter Summaries** review the major concepts of the chapter in a concise, narrative format.

Summary

A hydrogen bonded to an **α-carbon** of an aldehyde, ketone, or ester is sufficiently acidic to be removed by a strong base, because the base formed when the proton is removed is stabilized by delocalization of its negative change onto an oxygen. Aldehydes and ketones (p$K_a \sim 16$–20) are more acidic than esters (p$K_a \sim 25$). **β-Diketones** (p$K_a \sim 9$) and **β-keto esters** (p$K_a \sim 11$) are even more acidic. The interconversion of **keto** and **enol tautomers** is called tautomerization or enolization;

the carbonyl carbon of the other molecule. The product of an aldol addition can be dehydrated to give an **aldol condensation** product. In a **Claisen condensation**, the cnolate ion of an ester attacks the carbonyl carbon of a second molecule of ester, eliminating an ^-OR group to form a β-keto ester.

Carboxylic acids with a carbonyl group at the 3-position **decarboxylate** when they are heated. Carboxylic acids can be prepared by a **malonic ester synthesis**

Mechanisms

- **Accurate and Complete Mechanisms.** This text includes hundreds of complete mechanisms integrated into the text prose to promote true understanding, not just memorization.

mechanism for the conversion of an acyl chloride into an ester

the weaker base is expelled

formation of a tetrahedral intermediate

proton dissociation

weaker base

Visualization

One of the greatest challenges facing organic chemistry students lies in the often abstract nature of the subject. To help you better visualize important concepts, we have developed striking artwork on paper and on the Companion Website that that accompanies this book.

• **Voice Balloons**, will help you focus on important points being discussed

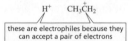

these are electrophiles because they can accept a pair of electrons

An electron-rich atom or molecule is called a **nucleophile**. A nucleophile has a pair of electrons it can share. Some nucleophiles are neutral and some are negatively charged. Because a nucleophile has electrons to share and an electrophile is seeking electrons, it should not be surprising that they attract each other. Thus, the preceding rule can be restated as *a nucleophile reacts with an electrophile*.

A nucleophile reacts with an electrophile.

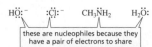

these are nucleophiles because they have a pair of electrons to share

We have seen that a π bond is weaker than a σ bond (Section 1.14). The π bond, therefore, is the bond that is most easily broken when an alkene undergoes a reaction. We also have seen that the π bond of an alkene consists of a cloud of electrons above and below the σ bond. This cloud of electrons causes an alkene to be an electron-rich molecule—it is a nucleophile. (Notice the relatively electron-rich pale orange area in

Electrostatic Potential Maps

• **Electrostatic potential maps** will help visualize the electronic structure of molecules and atoms and give you a greater understanding of why and how reactions occur. Use these to better understand why certain molecules and ions behave the way they do.

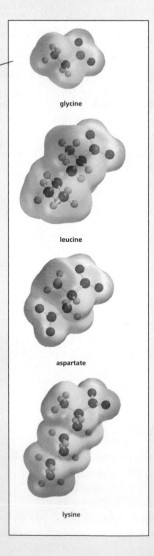

glycine

leucine

aspartate

lysine

Molecular Art

• Three-dimensional structures appear throughout the text to give a more accurate appreciation for the "shapes" of organic molecules.

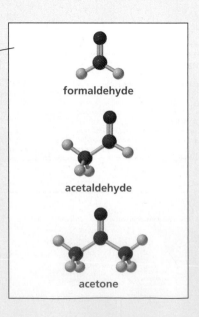

formaldehyde

acetaldehyde

acetone

Media Resources

Companion Website with GradeTracker
http://www.prenhall.com/bruice

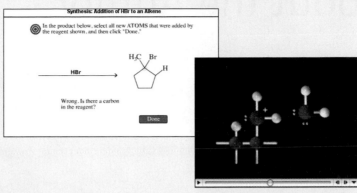

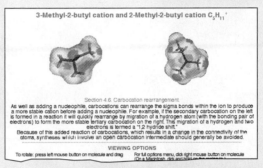

- **Interactive Tutorial and Animation Galleries** highlight central concepts in each chapter and illustrate key mechanisms. The Tutorials often allow students to make incorrect choices and then explain why there is a better answer.

- **Molecule Galleries** feature hundreds of 3-D molecular models of compounds noted in the chapter. You can rotate and compare models, change their representation, and examine electrostatic potential map surfaces—a unique feature of learning organic chemistry on the Web.

- **Practice Exercises and Quizzes** offer exercises to test your understanding of the material. Each question includes a hint with a cross-reference to a text reading, and detailed feedback. With GradeTracker, students can work problems and track their progress throughout the semester. Instructors can import these grades at any time during the semester.

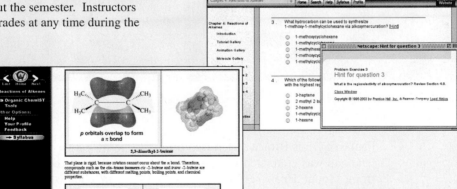

Instructor Resource Center on CD/DVD

- Hundreds of assets from the textbook and from the online resources are made available to instructors in one, simple CD-ROM/DVD, including images and tables, interactive tutorials, animations, and 3-D molecular models. Static art is provided in JPEG files for easy importing, ready-made in PowerPoint presentations (with and without lecture notes), or Adobe PDF format for high-resolution printing. Web-ready objects will run directly from the CD-ROM/DVD in your Web browser, even if you aren't connected to the Internet. The included MediaPortfolio browser allows you to browse through thumbnails or search for items by keyword, title, or description.

VISUALIZATION & MEDIA

About the Author

Paula Bruice and Zeus

Paula Yurkanis Bruice was raised primarily in Massachusetts, Germany, and Switzerland and was graduated from the Girls' Latin School in Boston. She received an A.B. from Mount Holyoke College and a Ph.D. in chemistry from the University of Virginia. She received an NIH postdoctoral fellowship for study in biochemistry at the University of Virginia Medical School, and she held a postdoctoral appointment in the Department of Pharmacology at Yale Medical School.

She is a member of the faculty at the University of California, Santa Barbara, where she has received the Associated Students Teacher of the Year Award, the Academic Senate Distinguished Teaching Award, and two Mortar Board Professor of the Year Awards. Her research interests concern the mechanism and catalysis of organic reactions, particularly those of biological significance. Paula has a daughter and a son who are physicians and a son who is a lawyer. Her main hobbies are reading mystery/suspense novels and her pets (two dogs, two cats, and a parrot).

1 Electronic Structure and Covalent Bonding

Ethane

Ethene

Ethyne

T o stay alive, early humans must have been able to tell the difference between two kinds of materials in their world. "You can live on roots and berries," they might have said, "but you can't live on dirt. You can stay warm by burning tree branches, but you can't burn rocks."

By the early eighteenth century, scientists thought they had grasped the nature of that difference. Compounds derived from living organisms were believed to contain an unmeasurable vital force—the essence of life. Because they came from organisms, they were called "organic" compounds. Compounds derived from minerals—those lacking that vital force—were "inorganic."

Because chemists could not create life in the laboratory, they assumed they could not create compounds with a vital force. With this mind-set, you can imagine how surprised chemists were in 1828 when Friedrich Wöhler produced urea—a compound known to be excreted by mammals—by heating ammonium cyanate, an inorganic mineral.

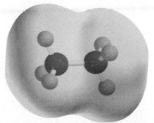

For the first time, an "organic" compound had been obtained from something other than a living organism and certainly without the aid of any kind of vital force. Clearly, chemists needed a new definition for "organic compounds." **Organic compounds** are now defined as *compounds that contain carbon*.

Why is an entire branch of chemistry devoted to the study of carbon-containing compounds? We study organic chemistry because just about all of the molecules that make life possible—proteins, enzymes, vitamins, lipids, carbohydrates, and nucleic acids—contain carbon; thus, the chemical reactions that take place in living systems, including our own bodies, are organic reactions. Most of the compounds found in nature—those we rely on for food, medicine, clothing (cotton, wool, silk), and

German chemist **Friedrich Wöhler (1800–1882)** *began his professional life as a physician and later became a professor of chemistry at the University of Göttingen. Wöhler codiscovered the fact that two different chemicals could have the same molecular formula. He also developed methods of purifying aluminum—at the time, the most expensive metal on Earth—and beryllium.*

energy (natural gas, petroleum)—are organic as well. Organic compounds are not, however, limited to the ones we find in nature. Chemists have learned to synthesize millions of organic compounds never found in nature, including synthetic fabrics, plastics, synthetic rubber, medicines, and even things like photographic film and Super glue. Many of these synthetic compounds prevent shortages of naturally occurring products. For example, it has been estimated that if synthetic materials were not available for clothing, all of the arable land in the United States would have to be used for the production of cotton and wool just to provide enough material to clothe us. Currently, there are about 16 million known organic compounds, and many more are possible.

What makes carbon so special? Why are there so many carbon-containing compounds? The answer lies in carbon's position in the periodic table. Carbon is in the center of the second row of elements. The atoms to the left of carbon have a tendency to give up electrons, whereas the atoms to the right have a tendency to accept electrons (Section 1.3).

the second row of the periodic table

Because carbon is in the middle, it neither readily gives up nor readily accepts electrons. Instead, it shares electrons. Carbon can share electrons with several different kinds of atoms, and it can also share electrons with other carbon atoms. Consequently, carbon is able to form millions of stable compounds with a wide range of chemical properties simply by sharing electrons.

When we study organic chemistry, we study how organic compounds react. When an organic compound reacts, some existing bonds break and some new bonds form. Bonds form when two atoms share electrons, and bonds break when two atoms no longer share electrons. How readily a bond forms and how easily it breaks depend on the particular electrons that are shared, which, in turn, depend on the atoms to which the electrons belong. So if we are going to start our study of organic chemistry at the beginning, we must start with an understanding of the structure of an atom—what electrons an atom has and where they are located.

NATURAL VERSUS SYNTHETIC

There are those who think that something that is natural—made in nature—is better than something that is synthetic—made in the laboratory. When a chemist synthesizes a compound, such as penicillin or estradiol, it is exactly the same in all respects as the compound synthesized in nature. Sometimes chemists can improve on nature. For example, chemists have synthesized analogs of morphine that have the painkilling effects of morphine, but, unlike morphine, are not habit forming. Chemists have synthesized analogs of penicillin that do not produce the allergic responses that a significant fraction of the population experiences from naturally produced penicillin, or do not have the bacterial resistance of the naturally produced antibiotic.

1.1 The Structure of an Atom

An atom consists of a tiny dense nucleus surrounded by electrons that are spread throughout a relatively large volume of space around the nucleus. The nucleus contains positively charged protons and neutral neutrons, so it is positively charged. The electrons are negatively charged. Because the amount of positive charge on a proton equals the amount of negative charge on an electron, a neutral atom has an equal number of protons and electrons. Atoms can gain electrons and thereby become negatively charged, or they can lose electrons and become positively charged. However, the number of protons in an atom never changes.

Protons and neutrons have approximately the same mass and are about 1800 times more massive than an electron. This means that most of the *mass* of an atom is in its nucleus. However, most of the *volume* of an atom is occupied by its electrons, and that is where our focus will be because it is the electrons that form chemical bonds.

The **atomic number** of an atom equals the number of protons in its nucleus. The atomic number is also the number of electrons that surround the nucleus of a neutral atom. For example, the atomic number of carbon is 6, which means that a neutral carbon atom has six protons and six electrons.

The **mass number** of an atom is the *sum* of its protons and neutrons. All carbon atoms have the same atomic number because they all have the same number of protons. They do not all have the same mass number because they do not all have the same number of neutrons. For example, 98.89% of naturally occurring carbon atoms have six neutrons—giving them a mass number of 12—and 1.11% have seven neutrons—giving them a mass number of 13. These two different kinds of carbon atoms (^{12}C and ^{13}C) are called isotopes. **Isotopes** have the same atomic number (i.e., the same number of protons), but different mass numbers because they have different numbers of neutrons.

^{14}C radioactive

Naturally occurring carbon also contains a trace amount of ^{14}C, which has six protons and eight neutrons. This isotope of carbon is radioactive, decaying with a half-life of 5730 years. (The half-life is the time it takes for one-half of the nuclei to decay.) As long as a plant or animal is alive, it takes in as much ^{14}C as it excretes or exhales. When it dies, it no longer takes in ^{14}C, so the ^{14}C in the organism slowly decreases. Therefore, the age of an organic substance can be determined by its ^{14}C content.

The **atomic weight** of a naturally occurring element is the average mass of its atoms. The **molecular weight** of a compound is the sum of the atomic weights of all the atoms in the molecule.

PROBLEM 1◆

Oxygen has three isotopes with mass numbers of 16, 17, and 18. The atomic number of oxygen is eight. How many protons and neutrons does each of the isotopes have?

1.2   The Distribution of Electrons in an Atom

The electrons in an atom can be thought of as occupying a set of shells that surround the nucleus. The way in which the electrons are distributed in these shells is based on a theory developed by Einstein. The first shell is the one closest to the nucleus. The second shell lies farther from the nucleus, and even farther out lie the third and higher numbered shells. Each shell contains subshells known as **atomic orbitals**. The first shell consists of only an *s* atomic orbital; the second shell consists of *s* and *p* atomic orbitals; and the third shell consists of *s*, *p*, and *d* atomic orbitals (Table 1.1).

Table 1.1 Distribution of Electrons in the First Three Shells That Surround the Nucleus			
	First shell	**Second shell**	**Third shell**
Atomic orbitals	*s*	*s, p*	*s, p, d*
Number of atomic orbitals	1	1, 3	1, 3, 5
Maximum number of electrons	2	8	18

Each shell contains one *s* orbital. The second and higher shells—in addition to their *s* orbital—each contain three *p* orbitals. The three *p* orbitals have the same energy. The third and higher shells—in addition to their *s* and *p* orbitals—also contain five *d* orbitals. Because an orbital can contain no more than two electrons (see below), the first shell, with only one atomic orbital, can contain no more than two electrons. The second shell, with four atomic orbitals—one *s* and three *p*—can have a total of eight electrons. Eighteen electrons can occupy the nine atomic orbitals—one *s*, three *p*, and five *d*—of the third shell.

An important point to remember is that *the closer the atomic orbital is to the nucleus, the lower is its energy*. Because a 1*s* orbital is closer to the nucleus than a 2*s* orbital, a 1*s* orbital is lower in energy. Comparing orbitals in the same shell, we see that an *s* orbital is lower in energy than a *p* orbital, and a *p* orbital is lower in energy than a *d* orbital.

The closer the orbital is to the nucleus, the lower is its energy.

Relative energies of atomic orbitals: $1s < 2s < 2p < 3s < 3p < 3d$

The **electronic configuration** of an atom describes what orbitals the electrons occupy. The following three rules are used to determine an atom's electronic configuration:

1. An electron always goes into the available orbital with the lowest energy.
2. No more than two electrons can occupy each orbital, and the two electrons must be of opposite spin. (Notice in Table 1.2 that spin in one direction is designated by ↑, and spin in the opposite direction by ↓.

From these first two rules, we can assign electrons to atomic orbitals for atoms that contain one, two, three, four, or five electrons. The single electron of a hydrogen atom occupies a 1*s* orbital, the second electron of a helium atom fills the 1*s* orbital, the third electron of a lithium atom occupies a 2*s* orbital, the fourth electron of a beryllium atom fills the 2*s* orbital, and the fifth electron of a boron atom occupies one of the 2*p* orbitals. (The subscripts *x*, *y*, and *z* distinguish the three 2*p* orbitals.) Because the three 2*p* orbitals have the same energy, the electron can be put into any one of them. Before we can continue to atoms containing six or more electrons, we need the third rule:

3. When there are two or more orbitals with the same energy, an electron will occupy an empty orbital before it will pair up with another electron.

Tutorial:
Electrons in orbitals

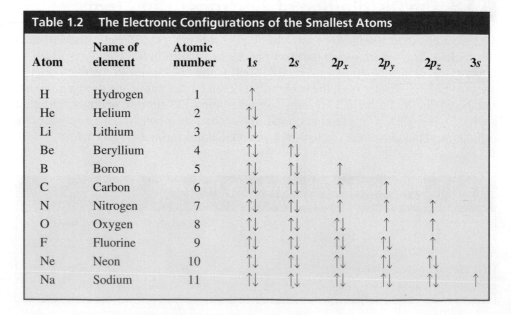

Table 1.2	The Electronic Configurations of the Smallest Atoms							
Atom	Name of element	Atomic number	1*s*	2*s*	2*p$_x$*	2*p$_y$*	2*p$_z$*	3*s*
H	Hydrogen	1	↑					
He	Helium	2	↑↓					
Li	Lithium	3	↑↓	↑				
Be	Beryllium	4	↑↓	↑↓				
B	Boron	5	↑↓	↑↓	↑			
C	Carbon	6	↑↓	↑↓	↑	↑		
N	Nitrogen	7	↑↓	↑↓	↑	↑	↑	
O	Oxygen	8	↑↓	↑↓	↑↓	↑	↑	
F	Fluorine	9	↑↓	↑↓	↑↓	↑↓	↑	
Ne	Neon	10	↑↓	↑↓	↑↓	↑↓	↑↓	
Na	Sodium	11	↑↓	↑↓	↑↓	↑↓	↑↓	↑

The sixth electron of a carbon atom, therefore, goes into an empty $2p$ orbital, rather than pairing up with the electron already occupying a $2p$ orbital (Table 1.2). There is one more empty $2p$ orbital, so that is where the seventh electron of a nitrogen atom goes. The eighth electron of an oxygen atom pairs up with an electron occupying a $2p$ orbital rather than going into a higher energy $3s$ orbital.

Electrons in inner shells (those below the outermost shell) are called **core electrons**. Electrons in the outermost shell are called **valence electrons**. Carbon, for example, has two core electrons and four valence electrons (Table 1.2).

Lithium and sodium each have one valence electron. Elements in the same column of the periodic table have the same number of valence electrons. Because the number of valence electrons is the major factor determining an element's chemical properties, elements in the same column of the periodic table have similar chemical properties. (You can find a periodic table inside the back cover of this book.) Thus, the chemical behavior of an element depends on its electronic configuration.

ALBERT EINSTEIN

Albert Einstein (1879–1955) is one of the most famous physicists of the twentieth century. Einstein was born in Germany. When he was in high school, his father's business failed and his family moved to Milan, Italy. Einstein had to stay behind because German law required compulsory military service after finishing high school. Einstein wanted to join his family in Italy. His high school mathematics teacher wrote a letter saying that Einstein could have a nervous breakdown without his family and also that there was nothing left to teach him. Eventually, Einstein was asked to leave the school because of his disruptive behavior. Popular folklore says he left because of poor grades in Latin and Greek, but his grades in those subjects were fine.

Einstein was visiting the United States when Hitler came to power, so he accepted a position at the Institute for Advanced Study in Princeton, becoming a U.S. citizen in 1940. Although a lifelong pacifist, he wrote a letter to President Roosevelt warning of ominous advances in German nuclear research. This led to the creation of the Manhattan Project, which developed the atomic bomb and tested it in New Mexico in 1945.

PROBLEM 2◆

How many valence electrons do the following atoms have?

a. carbon   **b.** nitrogen   **c.** oxygen   **d.** fluorine

PROBLEM 3◆

Table 1.2 shows that lithium and sodium each have one valence electron. Find potassium (K) in the periodic table and predict how many valence electrons it has.

PROBLEM 4◆

How many valence electrons do chlorine, bromine, and iodine have?

1.3 Ionic and Covalent Bonds

In trying to explain why atoms form bonds, G. N. Lewis proposed that *an atom is most stable if its outer shell is either filled or contains eight electrons and it has no electrons of higher energy*. According to Lewis's theory, an atom will give up, accept, or share electrons in order to achieve a filled outer shell or an outer shell that contains eight electrons. This theory has come to be called the **octet rule**.

Lithium (Li) has a single electron in its 2s orbital. If it loses this electron, the lithium atom ends up with a filled outer shell—a stable configuration. Lithium, therefore, loses an electron relatively easily. Sodium (Na) has a single electron in its 3s orbital, so it too loses an electron easily.

When we draw the electrons around an atom, as in the following equations, core electrons are not shown; only valence electrons are shown because only valence electrons are used in bonding. Each valence electron is shown as a dot. Notice that when the single valence electron of lithium or sodium is removed, the resulting atom—now called an ion—carries a positive charge.

$$\text{Li·} \longrightarrow \text{Li}^+ + e^-$$

$$\text{Na·} \longrightarrow \text{Na}^+ + e^-$$

Fluorine and chlorine each have seven valence electrons (Table 1.2 and Problem 4). Consequently, they readily acquire an electron in order to have an outer shell of eight electrons.

$$\text{:}\ddot{\text{F}}\text{·} + e^- \longrightarrow \text{:}\ddot{\text{F}}\text{:}^-$$

$$\text{:}\ddot{\text{Cl}}\text{·} + e^- \longrightarrow \text{:}\ddot{\text{Cl}}\text{:}^-$$

Ionic Bonds

Because sodium gives up an electron easily and chlorine acquires an electron readily, when sodium metal and chlorine gas are mixed, each sodium atom transfers an electron to a chlorine atom, and crystalline sodium chloride (table salt) is formed as a result. The positively charged sodium ions and negatively charged chloride ions are held together by the attraction of opposite charges (Figure 1.1). A **bond** is an attractive force between two atoms. A **bond** formed as a result of the attraction of opposite charges is called an **ionic bond**.

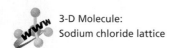
3-D Molecule:
Sodium chloride lattice

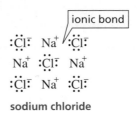

```
                          ionic bond

    :Cl:⁻  Na⁺  :Cl:⁻

    Na⁺  :Cl:⁻  Na⁺

    :Cl:⁻  Na⁺  :Cl:⁻
```

sodium chloride

a.

b.

▲ **Figure 1.1**
(a) Crystalline sodium chloride.
(b) The electron-rich chloride ions are red and the electron-poor sodium ions are blue. Each chloride ion is surrounded by six sodium ions, and each sodium ion is surrounded by six chloride ions. Ignore the "bonds" holding the balls together; they are there only to keep the model from falling apart.

Covalent Bonds

Instead of giving up or acquiring electrons, an atom can achieve a filled outer shell (or an outer shell of eight electrons) by sharing electrons. For example, two fluorine atoms can each attain a filled second shell by sharing their unpaired valence electrons. A bond formed as a result of *sharing electrons* is called a **covalent bond**.

$$:\ddot{F}\cdot \ + \ \cdot\ddot{F}: \ \longrightarrow \ :\ddot{F}\!:\!\ddot{F}:$$

a covalent bond

Two hydrogen atoms can form a covalent bond by sharing electrons. As a result of covalent bonding, each hydrogen acquires a filled first shell.

$$H\cdot \ | \ \cdot H \ \longrightarrow \ H\!:\!H$$

Similarly, hydrogen and chlorine can form a covalent bond by sharing electrons. In doing so, hydrogen fills its only shell and chlorine achieves an outer shell of eight electrons.

$$H\cdot \ + \ \cdot\ddot{C}\!l\!: \ \longrightarrow \ H\!:\!\ddot{C}\!l\!:$$

A hydrogen atom can achieve a completely empty shell by losing an electron. Loss of its sole electron results in a positively charged **hydrogen ion**. A positively charged hydrogen ion is called a **proton** because when a hydrogen atom loses its valence electron, only the hydrogen nucleus—which consists of a single proton—remains. A hydrogen atom can achieve a filled outer shell by gaining an electron, thereby forming a negatively charged hydrogen ion, called a **hydride ion**.

$$\underset{\text{a hydrogen atom}}{H\cdot} \ \longrightarrow \ \underset{\text{a proton}}{H^+} \ + \ e^-$$

$$\underset{\text{a hydrogen atom}}{H\cdot \ + \ e^-} \ \longrightarrow \ \underset{\text{a hydride ion}}{H\!:^-}$$

Because oxygen has six valence electrons, it needs to form two covalent bonds to achieve an outer shell of eight electrons (i.e., to complete its octet). Nitrogen, with five valence electrons, must form three covalent bonds, and carbon, with four valence electrons, must form four covalent bonds to complete their octets. Notice that all the atoms in water, ammonia, and methane have filled outer shells.

$$2\ H\cdot \ + \ \cdot\ddot{O}\!: \ \longrightarrow \ H\!:\!\underset{\underset{\textbf{water}}{\displaystyle H}}{\overset{\displaystyle ..}{\underset{\displaystyle ..}{O}}}\!:$$

$$3\ H\cdot \ + \ \cdot\dot{N}\cdot \ \longrightarrow \ H\!:\!\underset{\underset{\textbf{ammonia}}{\displaystyle H}}{N}\!:\!H$$

$$4\ H\cdot \ + \ \cdot\dot{C}\cdot \ \longrightarrow \ \underset{\underset{\textbf{methane}}{\displaystyle H}}{\overset{\displaystyle H}{H\!:\!C\!:\!H}}$$

Polar Covalent Bonds

The atoms that share the bonding electrons in the F—F or the H—H covalent bond are identical. Therefore, they share the electrons equally; that is, each electron spends as much time in the vicinity of one atom as in the other. Such a bond is called a **nonpolar covalent bond**.

Shown is a bronze sculpture of **Einstein** *on the grounds of the National Academy of Sciences in Washington, D.C. The statue measures 21 feet from the top of the head to the tip of the feet and weighs 7000 pounds. In his left hand, Einstein holds the mathematical equations that represent his three most important contributions to science: the photoelectric effect, the equivalency of energy and matter, and the theory of relativity. At his feet is a map of the sky.*

In contrast, the bonding electrons in hydrogen chloride, water, and ammonia are more attracted to one atom than to another because the atoms that share the electrons in these molecules are different and have different electronegativities. **Electronegativity** is the tendency of an atom to pull bonding electrons toward itself. The bonding electrons in hydrogen chloride, water, and ammonia are more attracted to the atom with the greater electronegativity. A **polar covalent bond** is a covalent bond between atoms of different electronegativities. The electronegativities of some of the elements are shown in Table 1.3. Notice that electronegativity increases as you go from left to right across a row of the periodic table or go up any of the columns.

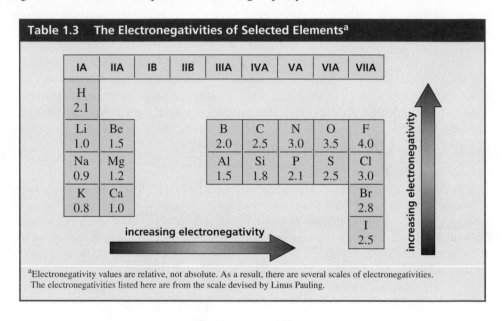

Table 1.3   The Electronegativities of Selected Elements[a]

IA	IIA	IB	IIB	IIIA	IVA	VA	VIA	VIIA
H 2.1								
Li 1.0	Be 1.5			B 2.0	C 2.5	N 3.0	O 3.5	F 4.0
Na 0.9	Mg 1.2			Al 1.5	Si 1.8	P 2.1	S 2.5	Cl 3.0
K 0.8	Ca 1.0							Br 2.8
								I 2.5

increasing electronegativity →

increasing electronegativity ↑

[a]Electronegativity values are relative, not absolute. As a result, there are several scales of electronegativities. The electronegativities listed here are from the scale devised by Linus Pauling.

A polar covalent bond has a slight positive charge on one end and a slight negative charge on the other. Polarity in a covalent bond is indicated by the symbols $\delta+$ and $\delta-$, which denote partial positive and partial negative charges, respectively. The negative end of the bond is the end that has the more electronegative atom. The greater the difference in electronegativity between the bonded atoms, the more polar the bond will be.

$$
\overset{\delta+\quad\delta-}{\text{H—Cl:}} \qquad \overset{\delta+\quad\delta-}{\underset{\underset{\delta+}{\text{H}}}{\text{H—O:}}} \qquad \overset{\delta+\quad\delta-\quad\delta+}{\underset{\underset{\delta+}{\text{H}}}{\text{H—N—H}}}
$$

You can think of ionic bonds and nonpolar covalent bonds as being at the opposite ends of a continuum of bond types. An ionic bond involves no sharing of electrons. A nonpolar covalent bond involves equal sharing. Polar covalent bonds fall somewhere in between, and the greater the difference in electronegativity between the atoms forming the bond, the closer the bond is to the ionic end of the continuum. C—H bonds are relatively nonpolar, because carbon and hydrogen have similar electronegativities (electronegativity difference = 0.4; see Table 1.3). N—H bonds are relatively polar (electronegativity difference = 0.9), but not as polar as O—H bonds (electronegativity difference = 1.4). The bond between sodium and chloride ions is closer to the ionic end of the continuum (electronegativity difference = 2.1), but sodium chloride is not as ionic as potassium fluoride (electronegativity difference = 3.2).

continuum of bond types

ionic bond	polar covalent bond	nonpolar covalent bond
K⁺F⁻ Na⁺Cl⁻	O—H N—H	C—H C—C

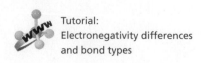

Tutorial:
Electronegativity differences
and bond types

PROBLEM 5◆

Which bond is more polar?

a. $H—CH_3$ or $Cl—CH_3$ **c.** $H—Cl$ or $H—F$

b. $H—OH$ or $H—H$ **d.** $Cl—Cl$ or $Cl—CH_3$

PROBLEM 6◆

Which of the following has

a. the most polar bond?   **b.** the least polar bond?

   NaI   LiBr Cl_2 KCl

Electrostatic potential maps (often simply called potential maps) are models that show how charge is distributed in the molecule under the map. The potential maps for LiH, H_2, and HF are shown below.

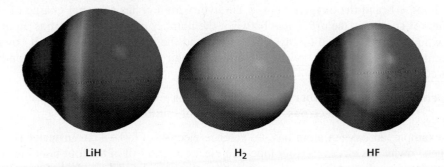

LiH H_2 HF

The colors indicate the distribution of charge in the molecule: Red signifies electron-rich areas (negative charge); blue signifies electron-deficient areas (positive charge); and green signifies no charge. For example, the potential map for LiH indicates that the hydrogen atom is more negatively charged than the lithium atom. By comparing the three maps, we can tell that the hydrogen in LiH is more negatively charged than a hydrogen in H_2, and the hydrogen in HF is more positively charged than a hydrogen in H_2.

3-D Molecules:
LiH; H_2; HF

red • orange • yellow • green • blue

most negative   most positive
electrostatic potential   electrostatic potential

PROBLEM 7◆

After examining the potential maps for LiH, HF, and H_2, answer the following questions:

a. Which compounds are polar?

b. Which compound has the most positively charged hydrogen?

PROBLEM 8◆

Use the symbols $\delta+$ and $\delta-$ to show the direction of polarity of the indicated bond in each

of the following compounds (for example, $H_3\overset{\delta+}{C}—\overset{\delta-}{O}H$).

a. $HO—H$ **b.** $H_3C—NH_2$ **c.** $HO—Br$ **d.** $I—Cl$

American chemist **Gilbert Newton Lewis (1875–1946)** *was born in Weymouth, Massachusetts, and received a Ph.D. from Harvard in 1899. He was the first person to prepare "heavy water," which has deuterium atoms in place of the usual hydrogen atoms (D_2O versus H_2O). Because heavy water can be used as a moderator of neutrons, it became important in the development of the atomic bomb. Lewis started his career as a professor at the Massachusetts Institute of Technology and joined the faculty at the University of California, Berkeley, in 1912.*

1.4 Representation of Structure

Lewis Structures

The chemical symbols we have been using, in which the valence electrons are represented as dots, are called **Lewis structures**. The Lewis structures for H_2O, H_3O^+, HO^-, and H_2O_2 are shown below.

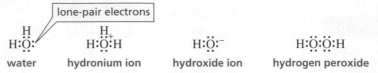

lone-pair electrons

water   hydronium ion   hydroxide ion   hydrogen peroxide

When you draw a Lewis structure, make sure that hydrogen atoms are surrounded by no more than two electrons and that C, O, N, and halogen (F, Cl, Br, I) atoms are surrounded by no more than eight electrons—they must obey the octet rule. Valence electrons not used in bonding are called **nonbonding electrons** or **lone-pair electrons**.

Once the atoms and the electrons are in place, each atom must be examined to see whether a charge should be assigned to it. A positive or a negative charge assigned to an atom is called a *formal charge*; the oxygen atom in the hydronium ion has a formal charge of +1, and the oxygen atom in the hydroxide ion has a formal charge of −1. A **formal charge** is the *difference* between the number of valence electrons an atom has when it is not bonded to any other atoms and the number of electrons it "owns" when it is bonded. An atom "owns" all of its lone-pair electrons and half of its bonding (shared) electrons.

formal charge = number of valence electrons − (number of lone-pair electrons + 1/2 number of bonding electrons)

For example, an oxygen atom has six valence electrons (Table 1.2). In water (H_2O), oxygen "owns" six electrons (four lone-pair electrons and half of the four bonding electrons). Because the number of electrons it "owns" is equal to the number of its valence electrons ($6 - 6 = 0$), the oxygen atom in water has no formal charge. The oxygen atom in the hydronium ion (H_3O^+) "owns" five electrons: two lone-pair electrons plus three (half of six) bonding electrons. Because the number of electrons it "owns" is one less than the number of its valence electrons ($6 - 5 = 1$), its formal charge is +1. The oxygen atom in hydroxide ion (HO^-) "owns" seven electrons: six lone-pair electrons plus one (half of two) bonding electron. Because it "owns" one more electron than the number of its valence electrons ($6 - 7 = -1$), its formal charge is −1.

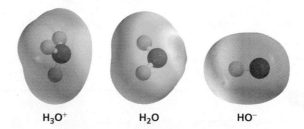

H_3O^+ H_2O HO^-

PROBLEM 9◆

The formal charge does not necessarily indicate that the atom has greater or less electron density than other atoms in the molecule without formal charges. You can see this by examining the potential maps for H_2O, H_3O^+, and HO^-.

a. Which atom bears the formal negative charge in the hydroxide ion? O

b. Which atom has the greater electron density in the hydroxide ion? O

c. Which atom bears the formal positive charge in the hydronium ion? O

d. Which atom has the least electron density in the hydronium ion? H

Knowing that nitrogen has five valence electrons (Table 1.2), convince yourself that the appropriate formal charges have been assigned to the nitrogen atoms in the following Lewis structures:

H:N̈:H
H
ammonia

H:N̈:H
H⁺
ammonium ion

H:N̈:⁻
H
amide anion

H:N̈:N̈:H
H H
hydrazine

Carbon has four valence electrons. Take a moment to understand why the carbon atoms in the following Lewis structures have the indicated formal charges:

H
H:C̈:H
H
methane

H
H:C⁺
H
methyl cation
a carbocation

H
H:C:⁻
H
methyl anion
a carbanion

H
H:C·
H
methyl radical

H H
H:C:C:H
H H
ethane

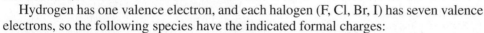

A species containing a positively charged carbon atom is called a **carbocation**, and a species containing a negatively charged carbon atom is called a **carbanion**. A species containing an atom with a single unpaired electron is called a **radical** (often called a **free radical**).

Hydrogen has one valence electron, and each halogen (F, Cl, Br, I) has seven valence electrons, so the following species have the indicated formal charges:

Movie:
Formal charge

H⁺
hydrogen
ion

H:⁻
hydride
ion

H·
hydrogen
radical

:B̈r:⁻
bromide
ion

:B̈r·
bromine
radical

:B̈r:B̈r:
bromine

:C̈l:C̈l:
chlorine

PROBLEM 10◆

Give each atom the appropriate formal charge:

a. CH₃—Ö—CH₃
  |
  H

b. H—C̈—H
  |
  H

c. CH₃—N—CH₃
   |
   CH₃
   (CH₃ above)

d. H—N—B—H
  | |
  H   H
  (H H above)

In studying the molecules in this section, notice that when the atoms don't bear a formal charge or an unpaired electron, hydrogen and the halogen atoms always have *one* covalent bond, oxygen always has *two* covalent bonds, nitrogen always has *three* covalent bonds, and carbon has *four* covalent bonds. Atoms that have more bonds or fewer bonds than the number required for a neutral atom will have either a formal charge or an unpaired electron. These numbers are very important to remember when you are first drawing structures of organic compounds because they provide a quick way to recognize when you have made a mistake.

H—
one bond

:F̈— :C̈l—
:Ï— :B̈r—
one bond

:Ö—
two bonds

—N̈—
three bonds

—C—
four bonds

In the following Lewis structures, notice that each atom has a filled outer shell. Also notice that since none of the molecules has a formal charge or an unpaired electron, C forms four bonds, N forms three bonds, O forms two bonds, and H and Br each form one bond.

H
H:C:B̈r:
H

H
H:C:Ö:H
H

H H
H:C:Ö:C:H
H H

H N
H:C:N:H
H H

H H
H:C:N:C:H
H H H

A pair of shared electrons can also be shown as a line between two atoms. Compare the preceding structures with the following ones:

$$
\text{H}-\overset{\overset{\displaystyle H}{|}}{\underset{\underset{\displaystyle H}{|}}{C}}-\ddot{\underset{..}{B}}\text{r}: \qquad
\text{H}-\overset{\overset{\displaystyle H}{|}}{\underset{\underset{\displaystyle H}{|}}{C}}-\overset{..}{\underset{..}{O}}-\text{H} \qquad
\text{H}-\overset{\overset{\displaystyle H}{|}}{\underset{\underset{\displaystyle H}{|}}{C}}-\overset{..}{\underset{..}{O}}-\overset{\overset{\displaystyle H}{|}}{\underset{\underset{\displaystyle H}{|}}{C}}-\text{H} \qquad
\text{H}-\overset{\overset{\displaystyle H}{|}}{\underset{\underset{\displaystyle H}{|}}{C}}-\overset{..}{\underset{\underset{\displaystyle H}{|}}{N}}-\text{H} \qquad
\text{H}-\overset{\overset{\displaystyle H}{|}}{\underset{\underset{\displaystyle H}{|}}{C}}-\overset{..}{\underset{\underset{\displaystyle H}{|}}{N}}-\overset{\overset{\displaystyle H}{|}}{\underset{\underset{\displaystyle H}{|}}{C}}-\text{H}
$$

PROBLEM 11◆

Draw the Lewis structure for each of the following:

a. $CH_3\overset{+}{N}H_3$ **b.** $^-C_2H_5$ **c.** NaOH   **d.** NH_4Cl

Kekulé Structures

In **Kekulé structures**, the two bonding electrons are drawn as lines and the lone-pair electrons are usually left out entirely, unless they are needed to draw attention to some chemical property of the molecule. (Although lone-pair electrons may not be shown, you should remember that neutral nitrogen, oxygen, and halogen atoms always have them: one pair in the case of nitrogen, two pairs in the case of oxygen, and three pairs in the case of a halogen.)

$$
\text{H}-\overset{\overset{\displaystyle H}{|}}{\underset{\underset{\displaystyle H}{|}}{C}}-\text{Br} \qquad
\text{H}-\overset{\overset{\displaystyle H}{|}}{\underset{\underset{\displaystyle H}{|}}{C}}-\text{O}-\text{H} \qquad
\text{H}-\overset{\overset{\displaystyle H}{|}}{\underset{\underset{\displaystyle H}{|}}{C}}-\text{O}-\overset{\overset{\displaystyle H}{|}}{\underset{\underset{\displaystyle H}{|}}{C}}-\text{H} \qquad
\text{H}-\overset{\overset{\displaystyle H}{|}}{\underset{\underset{\displaystyle H}{|}}{C}}-\overset{\underset{\displaystyle H}{|}}{N}-\text{H} \qquad
\text{H}-\overset{\overset{\displaystyle H}{|}}{\underset{\underset{\displaystyle H}{|}}{C}}-\overset{\underset{\displaystyle H}{|}}{N}-\overset{\overset{\displaystyle H}{|}}{\underset{\underset{\displaystyle H}{|}}{C}}-\text{H}
$$

Condensed Structures

Frequently, structures are simplified by omitting some (or all) of the covalent bonds and listing atoms bonded to a particular carbon (or nitrogen or oxygen) next to it with a subscript to indicate the number of such atoms. These kinds of structures are called **condensed structures**. Compare the preceding structures with the following ones:

$$
CH_3Br \qquad CH_3OH \qquad CH_3OCH_3 \qquad CH_3NH_2 \qquad CH_3NHCH_3
$$

You can find more examples of condensed structures and the conventions commonly used to create them in Table 1.4.

PROBLEM 12◆

Draw the lone-pair electrons that are not shown in the following structures:

a. $CH_3CH_2NH_2$ **c.** CH_3CH_2OH **e.** CH_3CH_2Cl

b. CH_3NHCH_3 **d.** CH_3OCH_3 **f.** $HONH_2$

PROBLEM 13◆

Draw condensed structures for the compounds represented by the following models (black = C, white = H, red = O, blue = N, green = Cl):

a.    **b.**    **c.** 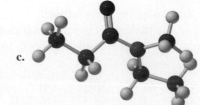   **d.**

Table 1.4   Kekulé and Condensed Structures

Kekulé structure	Condensed structures

Atoms bonded to a carbon are shown to the right of the carbon. Atoms other than H can be shown hanging from the carbon.

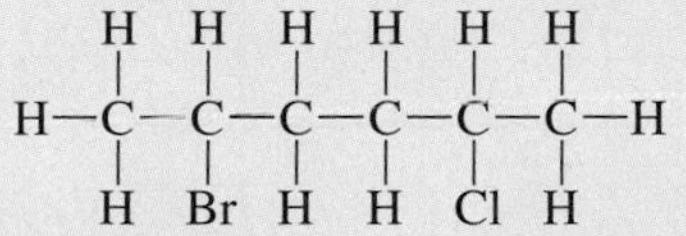

$$CH_3CHBrCH_2CH_2CHClCH_3 \quad \text{or} \quad CH_3CHCH_2CH_2CHCH_3$$
$$\qquad\qquad\qquad\qquad\qquad\qquad\qquad\quad | \qquad\qquad |$$
$$\qquad\qquad\qquad\qquad\qquad\qquad\qquad\ Br \qquad\quad Cl$$

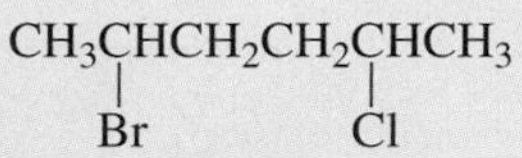

Repeating CH_2 groups can be shown in parentheses.

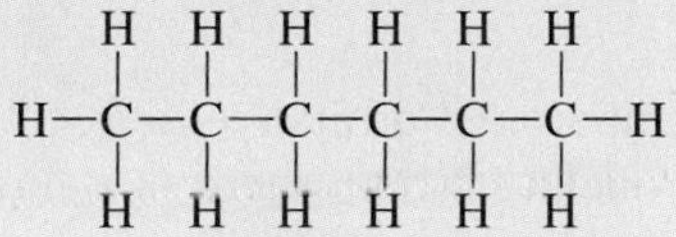

$$CH_3CH_2CH_2CH_2CH_2CH_3 \quad \text{or} \quad CH_3(CH_2)_4CH_3$$

Groups bonded to a carbon can be shown (in parentheses) to the right of the carbon, or hanging from the carbon.

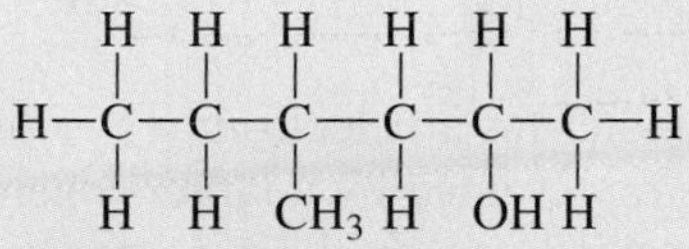

$$CH_3CH_2CH(CH_3)CH_2CH(OH)CH_3 \quad \text{or} \quad CH_3CH_2CHCH_2CHCH_3$$
$$\qquad\qquad\qquad\qquad\qquad\qquad\qquad\qquad\qquad\quad | \qquad\quad |$$
$$\qquad\qquad\qquad\qquad\qquad\qquad\qquad\qquad\qquad\ CH_3 \quad OH$$

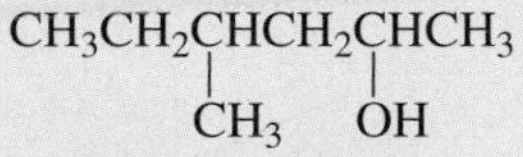

Groups bonded to the far-right carbon are not put in parentheses.

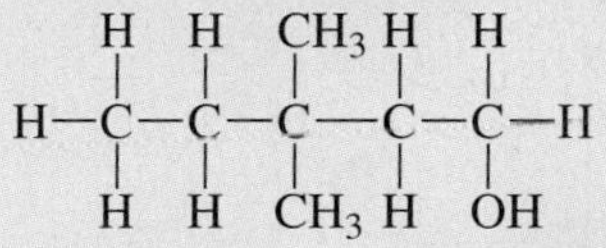

$$CH_3CH_2C(CH_3)_2CH_2CH_2OH \quad \text{or} \quad CH_3CH_2CCH_2CH_2OH$$

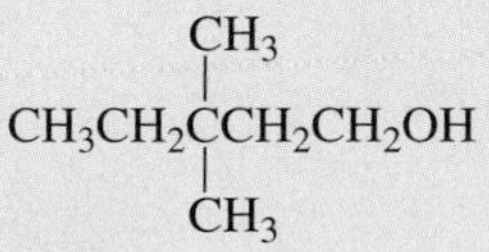

Two or more identical groups considered bonded to the "first" atom on the left can be shown (in parentheses) to the left of that atom, or hanging from the atom.

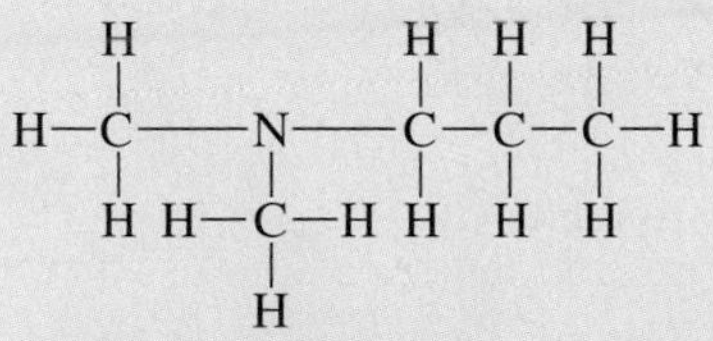

$$(CH_3)_2NCH_2CH_2CH_3 \quad \text{or} \quad CH_3NCH_2CH_2CH_3$$
$$\qquad\qquad\qquad\qquad\qquad\qquad\qquad\qquad\qquad | $$
$$\qquad\qquad\qquad\qquad\qquad\qquad\qquad\qquad CH_3$$

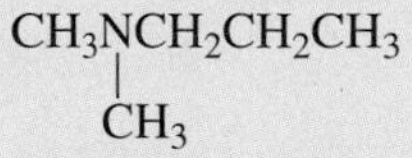

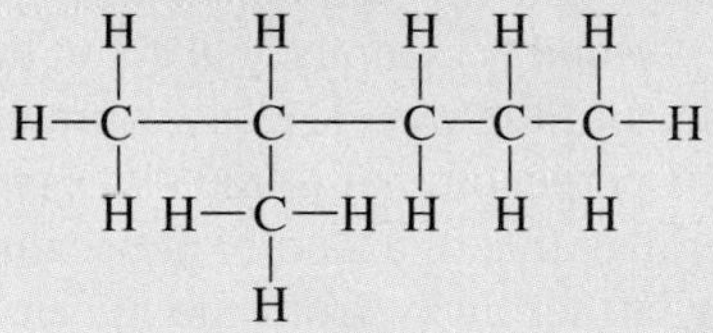

$$(CH_3)_2CHCH_2CH_2CH_3 \quad \text{or} \quad CH_3CHCH_2CH_2CH_3$$
$$\qquad\qquad\qquad\qquad\qquad\qquad\qquad\qquad\qquad | $$
$$\qquad\qquad\qquad\qquad\qquad\qquad\qquad\qquad CH_3$$

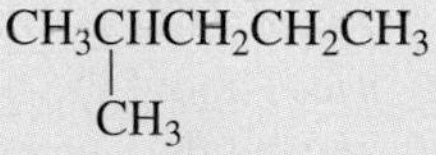

PROBLEM 14◆

Which of the atoms in the molecular models in Problem 13 have

a. three lone pairs?   **b.** two lone pairs?   **c.** one lone pair?   **d.** no lone pairs?

PROBLEM 15

a. Draw two Lewis structures for C_2H_6O.   **b.** Draw three Lewis structures for C_3H_8O.

(*Hint:* The two Lewis structures in part a are **constitutional isomers**; they have the same atoms, but differ in the way the atoms are connected; see page 46. The three Lewis structures in part b are also constitutional isomers.)

PROBLEM 16

Expand the following condensed structures to show the covalent bonds and lone-pair electrons:

a. $CH_3NH(CH_2)_2CH_3$ **c.** $(CH_3)_3COH$

b. $(CH_3)_2CHCl$ **d.** $(CH_3)_3C(CH_2)_3CH(CH_3)_2$

1.5 Atomic Orbitals

An orbital describes the volume of space around the nucleus where an electron is most likely to be found.

We have seen that electrons are distributed into different atomic orbitals (Table 1.2). An **orbital** is a three-dimensional region around the nucleus where there is a high probability of finding an electron. But what does an orbital look like? The *s* orbital is a sphere with the nucleus at its center. Thus, when we say that an electron occupies a 1*s* orbital, we mean that there is a greater than 90% probability that the electron is in the space defined by the sphere.

Because the average distance from the nucleus is greater for an electron in a 2*s* orbital than for an electron in a 1*s* orbital, a 2*s* orbital is represented by a larger sphere. Consequently, the average electron density in a 2*s* orbital is less than the average electron density in a 1*s* orbital.

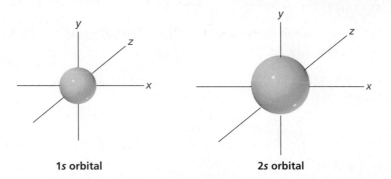

1s orbital   **2s orbital**

A *p* orbital has two lobes. In Section 1.2, you saw that the second and higher numbered shells each contains three *p* orbitals. The p_x orbital is symmetrical about the *x*-axis, the p_y orbital is symmetrical about the *y*-axis, and the p_z orbital is symmetrical about the *z*-axis. This means that each *p* orbital is perpendicular to the other two *p* orbitals. The energy of a 2*p* orbital is slightly greater than that of a 2*s* orbital because the average location of an electron in a 2*p* orbital is farther away from the nucleus. Generally, the lobes are depicted as teardrop-shaped, but computer-generated representations reveal that they are shaped more like doorknobs.

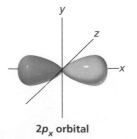
2p_x orbital

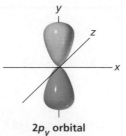
2p_y orbital

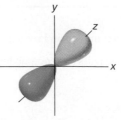
2p_z orbital

computer-generated 2p orbital

1.6   Covalent Bond Formation

How do atoms form covalent bonds in order to form molecules? Let's look first at the bonding in a hydrogen molecule (H_2). The covalent bond is formed when the $1s$ orbital of one hydrogen atom overlaps the $1s$ orbital of a second hydrogen atom. The covalent bond that is formed when the two orbitals overlap is called a **sigma (σ) bond**.

Movie:
H_2 bond formation

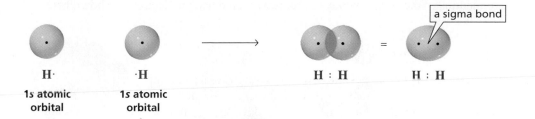

a sigma bond

H·   ·H   H : H   H : H
1s atomic   1s atomic
orbital   orbital

During bond formation, energy is released as the two orbitals start to overlap, because the electron in each atom not only is attracted to its own nucleus but also is attracted to the positively charged nucleus of the other atom (Figure 1.2). Thus, the attraction of the negatively charged electrons for the positively charged nuclei is what holds the atoms together. The more the orbitals overlap, the more energy is released until the atoms approach each other so closely that their positively charged nuclei start to repel each other. This repulsion causes a large increase in energy. We see that maximum stability (i.e., minimum energy) is achieved when the nuclei are a certain distance apart. This distance is the **bond length** of the new covalent bond. The length of the H—H bond is 0.74 Å.

Maximum stability corresponds to minimum energy.

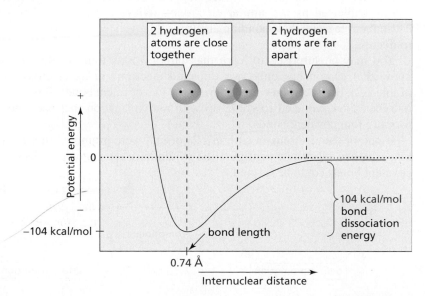

2 hydrogen atoms are close together

2 hydrogen atoms are far apart

104 kcal/mol bond dissociation energy

bond length

−104 kcal/mol

0.74 Å

Internuclear distance

◀ **Figure 1.2**
The change in energy that occurs as two $1s$ atomic orbitals approach each other. The internuclear distance at minimum energy is the length of the H—H covalent bond.

As Figure 1.2 shows, energy is released when a covalent bond forms. When the H—H bond forms, 104 kcal/mol or 435 kJ/mol of energy is released (1 kcal = 4.184 kJ)*. Breaking the bond requires precisely the same amount of energy. Thus, the **bond strength**—also called the **bond dissociation energy**—is the energy required to break the bond, or the energy released when the bond is formed. Every covalent bond has a characteristic bond length and bond strength.

*Joules are the Système International (SI) units for energy, although many chemists use calories. We will use both in this book.

Bonding in Methane

We will begin the discussion of bonding in organic compounds by looking at the bonding in methane, a compound with only one carbon atom. Methane (CH_4) has four covalent C—H bonds. Because all four bonds have the same length and all the bond angles are the same (109.5°), we can conclude that the four C—H bonds in methane are identical. Four different ways to represent a methane molecule are shown here.

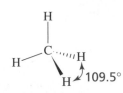

perspective formula
of methane

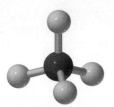

ball-and-stick model
of methane

space-filling model
of methane

electrostatic potential
map for methane

In a perspective formula, bonds in the plane of the paper are drawn as solid lines, bonds protruding out of the plane of the paper toward the viewer are drawn as solid wedges, and those protruding back from the plane of the paper away from the viewer are drawn as hatched wedges.

The potential map of methane shows that neither carbon nor hydrogen carries much of a charge: There are neither red areas, representing partially negatively charged atoms, nor blue areas, representing partially positively charged atoms. (Compare this map with the potential map for water on p. 24). The absence of partially charged atoms can be explained by the similar electronegativities of carbon and hydrogen, which cause them to share their bonding electrons relatively equally. Methane is a **nonpolar molecule**.

You may be surprised to learn that carbon forms four covalent bonds since you know that carbon has only two unpaired electrons in its electronic configuration (Table 1.2). But if carbon were to form only two covalent bonds, it would not complete its octet. Now we need to come up with an explanation that accounts for carbon's forming four covalent bonds.

If one of the electrons in carbon's $2s$ orbital were promoted into the empty $2p$ orbital, the new electronic configuration would have four unpaired electrons; thus, four covalent bonds could be formed.

Linus Carl Pauling (1901–1994)
was born in Portland, Oregon. A friend's home chemistry laboratory sparked Pauling's early interest in science. He received a Ph.D. from the California Institute of Technology and remained there for most of his academic career. He received the Nobel Prize in chemistry in 1954 for his work on molecular structure. Like Einstein, Pauling was a pacifist, winning the 1964 Nobel Peace Prize for his work on behalf of nuclear disarmament.

If carbon used an s orbital and three p orbitals to form these four bonds, the bond formed with the s orbital would be different from the three bonds formed with p orbitals. However, we know that the four C—H bonds in methane are identical. How can carbon form four identical bonds, using one s and three p orbitals? The answer is that carbon uses *hybrid orbitals*.

Hybrid orbitals are mixed orbitals—they result from combining atomic orbitals. The concept of combining orbitals was first proposed by Linus Pauling in 1931. If the one s and three p orbitals of the second shell are combined and then apportioned into four equal orbitals, each of the four resulting orbitals will be one part s and three parts p. This type of mixed orbital is called an sp^3 (stated "s-p-three" not "s-p-cubed") orbital.

(The superscript 3 means that three *p* orbitals were mixed with one *s* orbital to form the hybrid orbitals.) Each sp^3 orbital has 25% *s* character and 75% *p* character. Each of the four sp^3 orbitals has the same energy.

The superscript 3 means that three *p* orbitals were mixed

Like a *p* orbital, an sp^3 orbital has two lobes. Unlike the lobes of a *p* orbital, the two lobes of an sp^3 orbital are not the same size (Figure 1.3). The larger lobe of the sp^3 orbital is used in covalent bond formation.

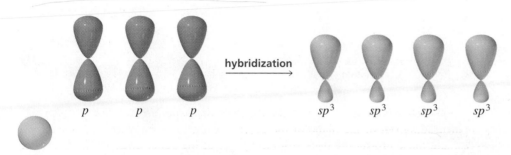

◀ **Figure 1.3**
An *s* orbital and three *p* orbitals hybridize to form four sp^3 orbitals. An sp^3 orbital is more stable than a *p* orbital, but not as stable as an *s* orbital.

The four sp^3 orbitals arrange themselves in space in a way that allows them to get as far away from each other as possible (Figure 1.4a). This occurs because electrons repel each other and getting as far from each other as possible minimizes the repulsion. When four orbitals spread themselves into space as far from each other as possible, they point toward the corners of a regular tetrahedron (a pyramid with four faces). Each of the four C—H bonds in methane is formed from overlap of an sp^3 orbital of carbon with the *s* orbital of a hydrogen (Figure 1.4b). This explains why the four C—H bonds are identical.

Electron pairs spread themselves into space as far from each other as possible.

a.   **b.**

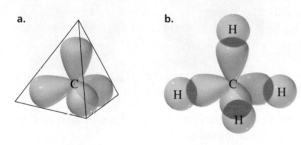

◀ **Figure 1.4**
(a) The four sp^3 orbitals are directed toward the corners of a tetrahedron, causing each bond angle to be 109.5°.
(b) An orbital picture of methane, showing the overlap of each sp^3 orbital of carbon with the *s* orbital of a hydrogen. (For clarity, the smaller lobes of the sp^3 orbitals are not shown.)

The angle formed between any two bonds of methane is 109.5°. This bond angle is called the **tetrahedral bond angle**. A carbon, such as the one in methane, that forms covalent bonds using four equivalent sp^3 orbitals is called a **tetrahedral carbon**.

The postulation of hybrid orbitals may appear to be a theory contrived just to make things fit—and that is exactly what it is. Nevertheless, it is a theory that gives us a very good picture of the bonding in organic compounds.

3-D Molecule:
Methane

Note to the student

It is important to understand what molecules look like in three dimensions. As you study each chapter, make sure to visit the website *www.prenhall.com/bruice* and look at the three-dimensional representations of molecules that can be found in the molecule gallery that accompanies the chapter.

Bonding in Ethane

The two carbon atoms in ethane are tetrahedral. Each carbon uses four sp^3 orbitals to form four covalent bonds:

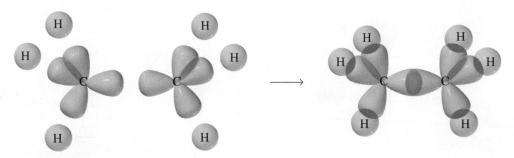

ethane

One sp^3 orbital of one carbon overlaps an sp^3 orbital of the other carbon to form the C—C bond. Each of the remaining three sp^3 orbitals of each carbon overlaps the s orbital of a hydrogen to form a C—H bond. Thus, the C—C bond is formed by sp^3–sp^3 overlap, and each C—H bond is formed by sp^3–s overlap (Figure 1.5).

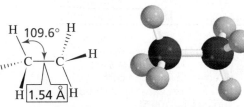 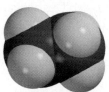 

▲ **Figure 1.5**
An orbital picture of ethane. The C—C bond is formed by sp^3–sp^3 overlap, and each C—H bond is formed by sp^3–s overlap. (The smaller lobes of the sp^3 orbitals are not shown.)

Each of the bond angles in ethane is nearly the tetrahedral bond angle of 109.5°, and the length of the C—C bond is 1.54 Å. Ethane, like methane, is a nonpolar molecule.

| perspective formula of ethane | ball-and-stick model of ethane | space-filling model of ethane | electrostatic potential map for ethane |

All the bonds in methane and ethane are sigma (σ) bonds. We will see that all **single bonds** found in organic compounds are sigma bonds.

PROBLEM 17♦

What orbitals are used to form the 10 covalent bonds in propane ($CH_3CH_2CH_3$)?

1.8 Bonding in Ethene: A Double Bond

Each of the carbon atoms in ethene (also called ethylene) forms four bonds, but each is bonded to only three atoms:

$$\text{H}\overset{\displaystyle}{\underset{\displaystyle\text{H}}{\text{C}}}=\overset{\displaystyle}{\underset{\displaystyle\text{H}}{\text{C}}}\text{H}$$

ethene
(ethylene)

To bond to three atoms, each carbon hybridizes three atomic orbitals. Because three orbitals (an *s* orbital and two of the *p* orbitals) are hybridized, three hybrid orbitals are obtained. These are called sp^2 orbitals. After hybridization, each carbon atom has three identical sp^2 orbitals and one unhybridized *p* orbital:

The axes of the three sp^2 orbitals lie in a plane. To minimize electron repulsion, the three orbitals need to get as far from each other as possible. This means that the bond angles are all close to 120°. The unhybridized *p* orbital is perpendicular to the plane defined by the axes of the sp^2 orbitals (Figure 1.6).

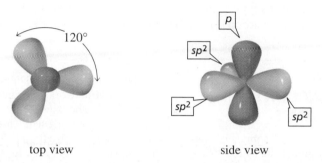

top view   side view

◀ **Figure 1.6**
An sp^2 hybridized carbon. The three sp^2 orbitals lie in a plane. The unhybridized *p* orbital is perpendicular to the plane. (The smaller lobes of the sp^2 orbitals are not shown.)

The carbons in ethene form two bonds with each other. This is called a **double bond**. The two carbon–carbon bonds in the double bond are not identical. One of the bonds results from the overlap of an sp^2 orbital of one carbon with an sp^2 orbital of the other carbon; this is a sigma (σ) bond. Each carbon uses its other two sp^2 orbitals to overlap the *s* orbital of a hydrogen to form the C—H bonds (Figure 1.7a). The second carbon–carbon bond results from side-to-side overlap of the two unhybridized *p* orbitals (Figure 1.7b). Side-to-side overlap of *p* orbitals forms a **pi (π) bond**. Thus, one of the bonds in a double bond is a σ bond and the other is a π bond. All the C—H bonds are σ bonds.

Side-to-side overlap of two *p* atomic orbitals forms a π bond. All other covalent bonds in organic molecules are σ bonds.

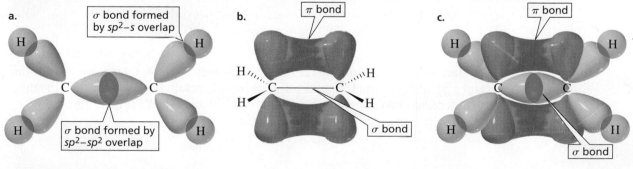

▲ **Figure 1.7**
(a) One C—C bond in ethene is a σ bond formed by sp^2–sp^2 overlap, and the C—H bonds are formed by sp^2–*s* overlap.
(b) The second C—C bond is a π bond formed by side-to-side overlap of a *p* orbital of one carbon with a *p* orbital of the other carbon.
(c) There is an accumulation of electron density above and below the plane containing the two carbons and four hydrogens.

The two *p* orbitals that overlap to form the π bond must be parallel to each other for maximum overlap to occur. This forces the triangle formed by one carbon and two hydrogens to lie in the same plane as the triangle formed by the other carbon and two hydrogens.

3-D Molecule:
Ethene

This means that all six atoms of ethene lie in the same plane, and the electrons in the *p* orbitals occupy a volume of space above and below the plane (Figure 1.7c). The potential map for ethene shows that it is a nonpolar molecule with a slight accumulation of negative charge (the pale orange area) above the two carbons. (If you could turn the potential map over, a similar accumulation of negative charge would be found on the other side.)

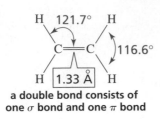

a double bond consists of
one σ bond and one π bond

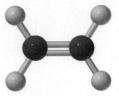

ball-and-stick model
of ethene

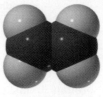

space-filling model
of ethene

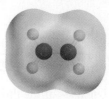

electrostatic potential map
for ethene

Four electrons hold the carbons together in a carbon–carbon double bond; only two electrons hold the carbons together in a carbon–carbon single bond. This means that a carbon–carbon double bond is stronger and shorter than a carbon–carbon single bond.

DIAMOND AND GRAPHITE: SUBSTANCES THAT CONTAIN ONLY CARBON ATOMS

Diamond is the hardest of all substances. Graphite, in contrast, is a slippery, soft solid most familiar to us as the "lead" in pencils. Both materials, in spite of their very different physical properties, contain only carbon atoms. The two substances differ solely in the nature of the bonds holding the carbon atoms together. Diamond consists of a rigid three-dimensional network of atoms, with each carbon bonded to four other carbons via sp^3 orbitals. The carbon atoms in graphite, on the other hand, are sp^2 hybridized, so each bonds to only three other carbons. This planar arrangement causes the atoms in graphite to lie in flat, layered sheets that can shear off of neighboring sheets. You experience this when you write with a pencil—sheets of carbon atoms shear off, leaving a thin trail of graphite.

1.9 Bonding in Ethyne: A Triple Bond

The carbon atoms in ethyne (also called acetylene) are each bonded to only two atoms—a hydrogen and another carbon:

$$H-C\equiv C-H$$
**ethyne
(acetylene)**

Because each carbon forms covalent bonds with two atoms, only two orbitals (an *s* and a *p*) are hybridized. Two identical *sp* orbitals result. Each carbon atom in ethyne, therefore, has two *sp* orbitals and two unhybridized *p* orbitals (Figure 1.8).

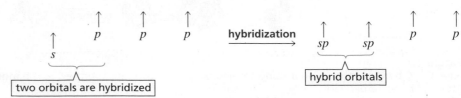

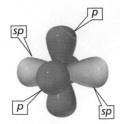

▲ **Figure 1.8**
An *sp* hybridized carbon. The two *sp* orbitals are oriented 180° away from each other, perpendicular to the two unhybridized *p* orbitals. (The smaller lobes of the *sp* orbitals are not shown.)

To get as far away from each other as possible, the two *sp* orbitals point in opposite directions (Figure 1.8).

The carbons in ethyne form three bonds with each other. This is called a **triple bond**. One of the *sp* orbitals of one carbon in ethyne overlaps an *sp* orbital of the other carbon to form a carbon–carbon σ bond. The other *sp* orbital of each carbon overlaps the *s* orbital of a hydrogen to form a C—H σ bond (Figure 1.9a). Because the two *sp* orbitals point in opposite directions, the bond angles are 180°. The two unhybridized *p* orbitals are perpendicular to each other, and both are perpendicular to the *sp* orbitals. Each of the unhybridized *p* orbitals engages in side-to-side overlap with a parallel *p* orbital on the other carbon, with the result that two π bonds are formed (Figure 1.9b).

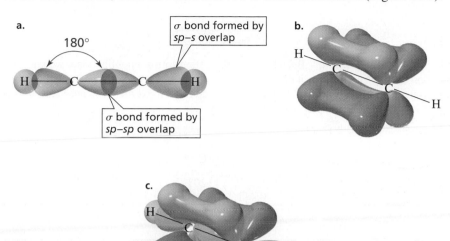

a.

180°

σ bond formed by *sp–s* overlap

σ bond formed by *sp–sp* overlap

b.

c.

◀ **Figure 1.9**
(a) The C—C σ bond in ethyne is formed by *sp–sp* overlap, and the C—H bonds are formed by *sp–s* overlap. The carbon atoms and the atoms bonded to them are in a straight line.
(b) The two carbon–carbon π bonds are formed by side-to-side overlap of the *p* orbitals of one carbon with the *p* orbitals of the other carbon.
(c) The triple bond has an electron-dense region above and below and in front of and in back of the internuclear axis of the molecule.

A **triple bond** consists of one σ bond and two π bonds. Because the two unhybridized *p* orbitals on each carbon are perpendicular to each other, there is a region of high electron density above and below, *and* in front of and in back of, the internuclear axis of the molecule (Figure 1.9c). The potential map for ethyne shows that negative charge accumulates in a cylinder that wraps around the egg-shaped molecule.

3-D Molecule: Ethyne

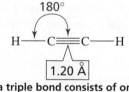

180°

H—C≡C—H

1.20 Å

a triple bond consists of one σ bond and two π bonds

ball-and-stick model of ethyne

space-filling model of ethyne

electrostatic potential map for ethyne

Because the two carbon atoms in a triple bond are held together by six electrons, a triple bond is stronger and shorter than a double bond.

PROBLEM 18 SOLVED

For each of the following species:

a. draw its Lewis structure.

b. describe the orbitals used by each carbon atom in bonding and indicate the approximate bond angles.

   1. HCOH   **2.** CCl₄   **3.** HCN

SOLUTION TO 18a Because HCOH is neutral, we know that each H forms one bond, the oxygen forms two bonds, and the carbon forms four bonds. Our first attempt at a Lewis structure (drawing the atoms in the order given by the Kekulé structure) shows that carbon is the only atom that does not form the needed number of bonds.

H—C—O—H

By forming a double bond between the carbon and the oxygen, all the atoms end up with the correct number of bonds. Lone-pair electrons are used to give each atom a filled outer shell. When we check to see if any atom needs to be assigned a formal charge, we find that none of the atoms has a formal charge.

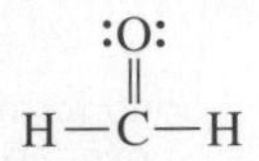

$$\text{:O:} \atop \underset{\displaystyle H-C-H}{\|}$$

SOLUTION TO 18b   Because the carbon atom forms a double bond, we know that carbon uses sp^2 orbitals (as it does in ethene) to bond to the two hydrogens and the oxygen. It uses its "leftover" p orbital to form the second bond to oxygen. Because carbon is sp^2 hybridized, the bond angles are approximately 120°.

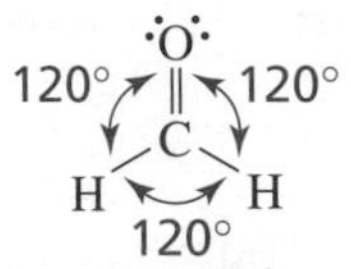

$$120° \nearrow \overset{\cdot\ddot{O}\cdot}{\|} \nwarrow 120°$$
$$\underset{120°}{H \; \overset{\displaystyle C}{\diagdown} \; H}$$

1.10 Bonding in the Methyl Cation, the Methyl Radical, and the Methyl Anion

Not all carbon atoms form four bonds. A carbon with a positive charge, a negative charge, or an unpaired electron forms only three bonds. Now we will see what orbitals carbon uses when it forms three bonds.

The Methyl Cation ($^+CH_3$)

The positively charged carbon in the methyl cation is bonded to three atoms, so it hybridizes three orbitals—an s orbital and two p orbitals. Therefore, it forms its three covalent bonds using sp^2 orbitals. Its unhybridized p orbital remains empty. The positively charged carbon and the three atoms bonded to it lie in a plane. The p orbital stands perpendicular to the plane.

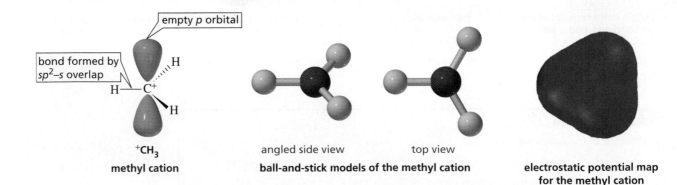

empty *p* orbital

bond formed by *sp²–s* overlap

$^+CH_3$
methyl cation

angled side view   top view
ball-and-stick models of the methyl cation

electrostatic potential map for the methyl cation

The Methyl Radical ($^\cdot CH_3$)

The carbon atom in the methyl radical is also sp^2 hybridized. The methyl radical differs by one unpaired electron from the methyl cation. That electron is in the p orbital. Notice the similarity in the ball-and-stick models of the methyl cation and the methyl radical. The potential maps, however, are quite different because of the additional electron in the methyl radical.

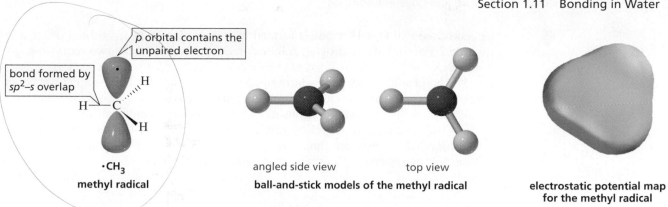

bond formed by sp^2–s overlap

p orbital contains the unpaired electron

•CH_3
methyl radical

ball-and-stick models of the methyl radical

angled side view   top view

electrostatic potential map for the methyl radical

The Methyl Anion ($\bar{:}CH_3$)

The negatively charged carbon in the methyl anion has three pairs of bonding electrons and one lone pair. The four pairs of electrons are farthest apart when the four orbitals containing the bonding and lone-pair electrons point toward the corners of a tetrahedron. In other words, a negatively charged carbon is sp^3 hybridized. In the methyl anion, three of carbon's sp^3 orbitals each overlaps the s orbital of a hydrogen, and the fourth sp^3 orbital holds the lone-pair electrons.

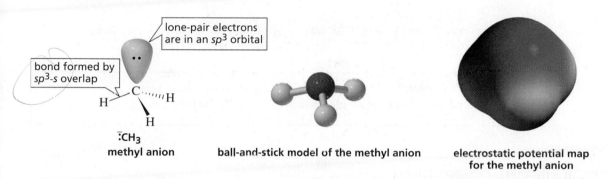

lone-pair electrons are in an sp^3 orbital

bond formed by sp^3-s overlap

$\bar{:}CH_3$
methyl anion

ball-and-stick model of the methyl anion

electrostatic potential map for the methyl anion

Take a moment to compare the potential maps for the methyl cation, the methyl radical, and the methyl anion.

1.11 Bonding in Water

The oxygen atom in water (H_2O) forms two covalent bonds. Because the electronic configuration of oxygen shows that it has two unpaired electrons (Table 1.2), oxygen does not need to promote an electron to form two covalent bonds. If we assume that oxygen uses p orbitals to form the two O—H bonds, as predicted by oxygen's electronic configuration, we would expect a bond angle of about 90° because the two p orbitals are at right angles to each other. However, the experimentally observed bond angle is 104.5°.

To explain the observed bond angle, oxygen must use hybrid orbitals to form covalent bonds—just as carbon does. The s orbital and the three p orbitals must hybridize to produce four identical sp^3 orbitals.

The bond angles in a molecule indicate which orbitals are used in bond formation.

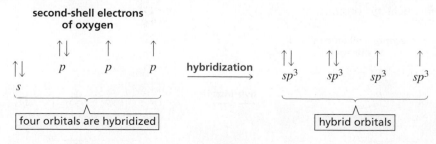

second-shell electrons of oxygen

hybridization

s p p p

sp^3 sp^3 sp^3 sp^3

four orbitals are hybridized

hybrid orbitals

Each of the two O—H bonds is formed by the overlap of an sp^3 orbital of oxygen with the s orbital of a hydrogen. A lone pair occupies each of the two remaining sp^3 orbitals.

The bond angle in water is a little smaller (104.5°) than the bond angle (109.5°) in methane, presumably because each of oxygen's lone pairs "feels" only one nucleus, which makes the lone-pair electrons more diffuse than a bonding pair that "feels" two nuclei and is therefore relatively confined between them. Consequently, there is more electron repulsion between lone-pair electrons, causing the O—H bonds to squeeze closer together, thereby decreasing the bond angle.

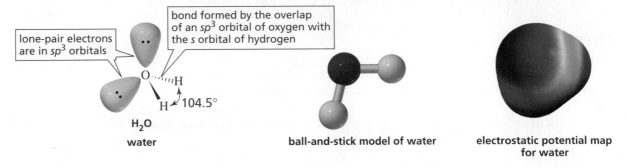

| | ball-and-stick model of water | electrostatic potential map for water |

lone-pair electrons are in sp^3 orbitals

bond formed by the overlap of an sp^3 orbital of oxygen with the s orbital of hydrogen

H_2O
water

3-D Molecule: Water

Compare the potential map for water with that for methane. Water is a polar molecule; methane is nonpolar.

PROBLEM 19◆

The bond angles in H_3O^+ are greater than _____ and less than _____.

WATER—A UNIQUE COMPOUND

Water is the most abundant compound found in living organisms. Its unique properties have allowed life to originate and evolve. Its high heat of fusion (the heat required to convert a solid to a liquid) protects organisms from freezing at low temperatures because a lot of heat must be removed from water to freeze it. Its high heat capacity (the heat required to raise the temperature of a substance a given amount) minimizes temperature changes in organisms, and its high heat of vaporization (the heat required to convert a liquid to a gas) allows animals to cool themselves with a minimal loss of body fluid. Because liquid water is denser than ice, ice formed on the surface of water floats and insulates the water below. That is why oceans and lakes don't freeze from the bottom up. It is also why plants and aquatic animals can survive when the ocean or lake they live in freezes.

1.12 Bonding in Ammonia and in the Ammonium Ion

The experimentally observed bond angles in NH_3 are 107.3°. The bond angles indicate that nitrogen also uses hybrid orbitals when it forms covalent bonds. Like carbon and oxygen, the one s and three p orbitals of the second shell of nitrogen hybridize to form four identical sp^3 orbitals:

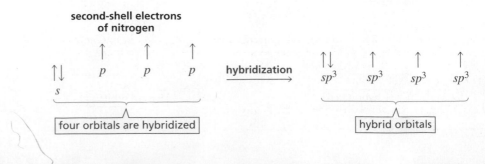

second-shell electrons of nitrogen

hybridization

four orbitals are hybridized

hybrid orbitals

Each of the N—H bonds in NH_3 is formed by the overlap of an sp^3 orbital of nitrogen with the *s* orbital of a hydrogen. The single lone pair occupies an sp^3 orbital. The bond angle (107.3°) is smaller than the tetrahedral bond angle (109.5°) because of the relatively diffuse lone pair. Notice that the bond angles in NH_3 (107.3°) are larger than the bond angles in H_2O (104.5°) because nitrogen has only one lone pair, whereas oxygen has two lone pairs.

3-D Molecule: Ammonia

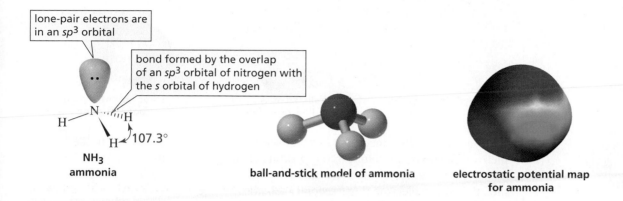

lone-pair electrons are in an sp^3 orbital

bond formed by the overlap of an sp^3 orbital of nitrogen with the *s* orbital of hydrogen

107.3°

NH_3
ammonia

ball-and-stick model of ammonia

electrostatic potential map for ammonia

Because the ammonium ion ($^+NH_4$) has four identical N—H bonds and no lone pairs, all the bond angles are 109.5°—just like the bond angels in methane.

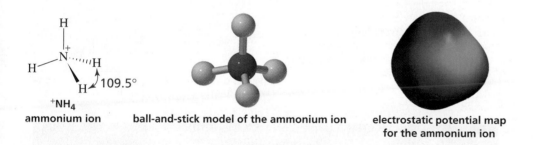

109.5°

$^+NH_4$
ammonium ion

ball-and-stick model of the ammonium ion

electrostatic potential map for the ammonium ion

PROBLEM 20◆

According to the potential map for the ammonium ion, which atom(s) has (have) the least electron density?

PROBLEM 21◆

Compare the potential maps for methane, ammonia, and water. Which is the most polar molecule? Which is the least polar?

electrostatic potential map for methane

electrostatic potential map for ammonia

electrostatic potential map for water

PROBLEM 22◆

Predict the approximate bond angles in the methyl carbanion.

hydrogen fluoride

hydrogen chloride

hydrogen bromide

hydrogen iodide

The greater the electron density in the region of orbital overlap, the stronger is the bond.

Fluorine, chlorine, bromine, and iodine are collectively known as the halogens. HF, HCl, HBr, and HI are called hydrogen halides. Bond angles will not help us determine the orbitals involved in a hydrogen halide bond, as they did with other molecules, because hydrogen halides have only one bond. We will assume that the halogens use *p* orbitals to form bonds.

hydrogen fluoride

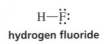

ball-and-stick model of hydrogen fluoride

electrostatic potential map for hydrogen fluoride

In the case of fluorine, the *p* orbital used in bond formation belongs to the second shell of electrons. In chlorine, the *p* orbital belongs to the third shell of electrons. Because the average distance from the nucleus is greater for an electron in the third shell than for an electron in the second shell, the average electron density is less in a $3p$ orbital than in a $2p$ orbital. This means that the electron density in the region where the *s* orbital of hydrogen overlaps the *p* orbital of the halogen decreases as the size of the halogen increases (Figure 1.10). Therefore, the hydrogen–halogen bond becomes longer and weaker as the size (atomic weight) of the halogen increases (Table 1.5).

Figure 1.10 ▶
There is greater electron density in the region of overlap of an *s* orbital with a $2p$ orbital than in the region of overlap of an *s* orbital with a $3p$ orbital.

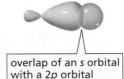
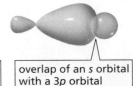

| overlap of an *s* orbital with a $2p$ orbital | overlap of an *s* orbital with a $3p$ orbital |

The shorter the bond, the stronger it is.

Table 1.5   Hydrogen–Halogen Bond Lengths and Bond Strengths

Hydrogen halide	Bond length (Å)	Bond strength kcal/mol	kJ/mol
H—F	0.917	136	571
H—Cl	1.2746	103	432
H—Br	1.4145	87	366
H—I	1.6090	71	298

PROBLEM 23◆

  a. Predict the relative lengths and strengths of the bonds in Cl_2 and Br_2.

  b. Predict the relative lengths and strengths of the bonds in HF, HCl, and HBr.

PROBLEM 24◆

  a. Which bond is longer?

  b. Which bond is stronger?

    1. C—Cl or C—Br   **2.** C—C or C—H   **3.** H—Cl or H—H

1.14   Summary: Orbital Hybridization, Bond Lengths, Bond Strengths, and Bond Angles

All single bonds are σ bonds. All double bonds are composed of one σ bond and one π bond. All triple bonds are composed of one σ bond and two π bonds. The easiest way to determine the hybridization of a carbon, oxygen, or nitrogen atom is to look at the number of π bonds it forms: If it forms no π bonds, it is sp^3 hybridized; if it forms one π bond, it is sp^2 hybridized; if it forms two π bonds, it is sp hybridized. The exceptions are carbocations and carbon radicals, which are sp^2 hybridized—not because they form a π bond, but because they have an empty or a half-filled p orbital (Section 1.10).

All single bonds found in organic compounds are sigma bonds.

The hybridization of a C, O, or N is $sp^{(3-\text{the number of }\pi\text{ bonds})}$.

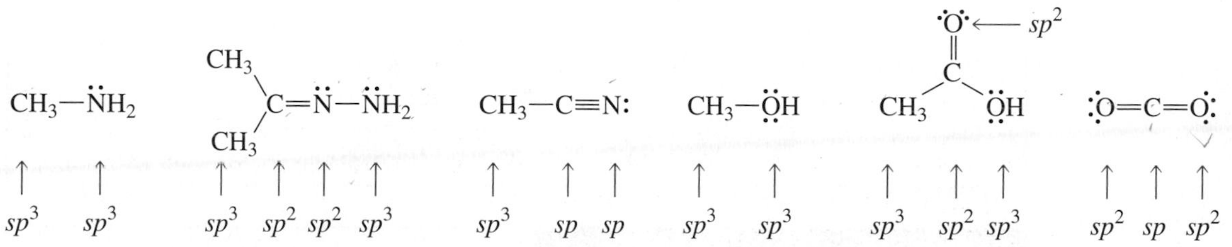

In comparing the lengths and strengths of carbon–carbon single, double, and triple bonds, we see that the more bonds holding two carbon atoms together, the shorter and stronger is the carbon–carbon bond (Table 1.6): Triple bonds are shorter and stronger than double bonds, which are shorter and stronger than single bonds.

A double bond (a σ bond plus a π bond) is stronger than a single bond (a σ bond), but it is not twice as strong. We can conclude, therefore, that a π bond is weaker than a σ bond.

A π bond is weaker than a σ bond.

You may wonder how an electron "knows" what orbital it should go into. In fact, electrons know nothing about orbitals. They simply arrange themselves around atoms in the most stable manner possible. It is chemists who use the concept of orbitals to explain this arrangement.

Table 1.6   Comparison of the Bond Angles and the Lengths and Strengths of the Carbon–Carbon Bonds in Ethane, Ethene, and Ethyne

Molecule	Hybridization of carbon	Bond angles	Length of C—C bond (Å)	Strength of C—C bond (kcal/mol)	(kJ/mol)
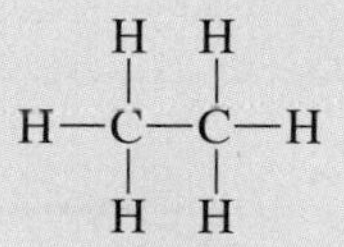 ethane	sp^3	109.5°	1.54	90	377
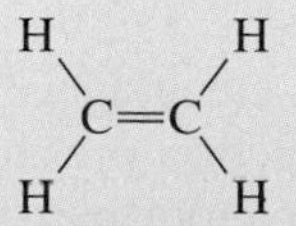 ethene	sp^2	120°	1.33	174	728
ethyne	sp	180°	1.20	231	967

PROBLEM 25◆

Which of the bonds of a carbon–carbon double bond has more effective orbital–orbital overlap, the σ bond or the π bond?

PROBLEM 26◆

a. What is the hybridization of each of the carbon atoms in the following compound?

$$CH_3CHCH=CHCH_2C\equiv CCH_3$$
$$|$$
$$CH_3$$

b. What is the hybridization of each of the carbon, oxygen, and nitrogen atoms in the following compounds?

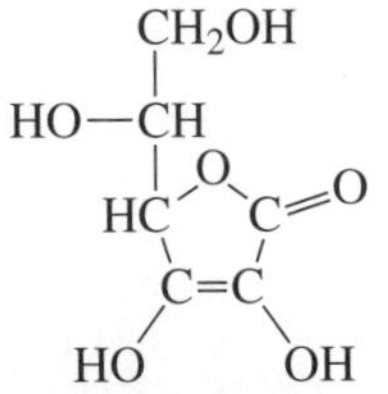

vitamin C 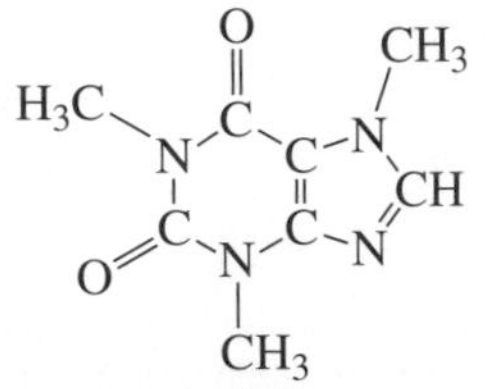   caffeine

PROBLEM SOLVING STRATEGY

Predict the approximate bond angle of the C—N—H bond in $(CH_3)_2NH$.

First we need to determine the hybridization of the central atom. Because the nitrogen atom forms only single bonds, we know it is sp^3 hybridized. Then we need to see if there are lone pairs that will affect the bond angle. A neutral nitrogen has one lone pair. Therefore, we can predict that the C—N—H bond angle will be about 107.3°, which is the same as the H—N—H bond angle in NH_3, another compound with a sp^3 hybridized N with one lone pair.

Now continue on to Problem 27.

PROBLEM 27◆

Predict the approximate bond angles:

a. the C—N—C bond angle in $(CH_3)_2\overset{+}{N}H_2$
b. the C—C—N bond angle in $CH_3CH_2NH_2$
c. the H—C—N bond angle in $(CH_3)_2NH$
d. the C—O—C bond angle in CH_3OCH_3

PROBLEM 28

Describe the orbitals used in bonding and the approximate bond angles in the following compounds. (*Hint:* see Table 1.6).

a. CH_3OH **b.** $HONH_2$ **c.** $HCOOH$ **d.** N_2

Summary

Organic compounds are compounds that contain carbon. The **atomic number** of an atom is the number of protons in its nucleus. The **mass number** of an atom is the sum of its protons and neutrons. **Isotopes** have the same atomic number, but different mass numbers.

An **atomic orbital** indicates where there is a high probability of finding an electron. The closer the atomic orbital is to the nucleus, the lower is its energy. Electrons are assigned to atomic orbitals according to three rules: an electron goes into the available orbital with the lowest energy; no more than two electrons can be in an orbital; and an electron will occupy an empty orbital with the same energy before pairing up.

The **octet rule** states that an atom will give up, accept, or share electrons in order to fill its outer shell or attain an outer shell with eight electrons. The **electronic configuration** of an atom describes the orbitals occupied by the atom's electrons. Electrons in inner shells are called **core electrons**; electrons in the outermost shell are called **valence electrons**. **Lone-pair electrons** are valence electrons that are not used in bonding. A **bond** formed as a result of the attraction of opposite charges is called an **ionic bond**; a bond formed as a result of sharing electrons is called a **covalent bond**. A **polar covalent bond** is a covalent bond between atoms with different **electronegativites**.

Lewis structures indicate which atoms are bonded together and show **lone-pair electrons** and **formal charges**. A **carbocation** has a positively charged carbon, a **carbanion** has a negatively charged carbon, and a **radical** has an unpaired electron.

Bond strength is measured by the **bond dissociation energy**. A σ bond is stronger than a π bond. All **single bonds** in organic compounds are **sigma (σ) bonds**. A **double bond** consists of one σ bond and one π bond, and a **triple bond** consists of one σ bond and two π bonds. Carbon-carbon triple bonds are shorter and stronger than carbon-carbon double bonds, which are shorter and stronger than carbon-carbon single bonds. To form four bonds, carbon promotes an electron from a $2s$ to a $2p$ orbital. C, N, and O form bonds using **hybrid orbitals**. The **hybridization** of C, N, or O depends on the number of π bonds the atom forms: No π bonds means that the atom is sp^3 **hybridized**, one π bond indicates that it is sp^2 **hybridized**, and two π bonds signifies that it is sp **hybridized**. Exceptions are carbocations and carbon radicals, which are sp^2 **hybridized**. Bonding and lone-pair electrons around an atom are positioned as far apart as possible.

Problems

29. Draw a Lewis structure for each of the following species:

   a. H_2CO_3 **b.** CO_3^{2-} **c.** H_2CO **d.** CH_3NH_2 **e.** CO_2 **f.** N_2H_4

30. How many valence electrons do the following atoms have?

   a. carbon and silicon   **b.** oxygen and sulfur   **c.** fluorine and bromine   **d.** magnesium and calcium

31. For each of the following species, give the hybridization of the central atom and indicate the bond angles:

   a. NH_3 **c.** $^-CH_3$ **e.** $^\cdot CH_3$ **g.** HCN

   b. $^+NH_4$ **d.** $C(CH_3)_4$ **f.** $^+CH_3$ **h.** H_3O^+

32. Use the symbols $\delta+$ and $\delta-$ to show the direction of polarity of the indicated bond in each of the following compounds:

   a. F—Br   **b.** H_3C—Cl   **c.** H_3C—MgBr   **d.** H_2N—OH

33. Draw the condensed structure of a compound that contains only carbon and hydrogen atoms and that has

   a. three sp^3 hybridized carbons.
   b. one sp^3 hybridized carbon and two sp^2 hybridized carbons.
   c. two sp^3 hybridized carbons and two sp hybridized carbons.

34. Predict the approximate bond angles:

   a. the H—C—O bond angle in CH_3OH **c.** the H—C—H bond angle in $H_2C{=}O$

   b. the C—O—H bond angle in CH_3OH **d.** the C—C—N bond angle in $CH_3C{\equiv}N$

35. Give each atom the appropriate formal charge:

   a. $H{:}\ddot{\underset{..}{O}}{:}$ **b.** $H{:}\ddot{\underset{..}{O}}{\cdot}$ **c.** $H{-}\overset{..}{N}{-}H$ **d.** $H{-}\ddot{C}{-}H$

36. Write the electronic configuration for the following species (carbon's electronic configuration is written as $1s^2\,2s^2\,2p^2$):

   a. Ca   **b.** Ca^{2+} **c.** Ar   **d.** Mg^{2+}

37. Only one of the following formulas describes a compound that exists. Fix the other formulas so they also describe compounds that exist.

   a. $CH_3CH_3CH_3$ **c.** $(CH_3)_2CCH_3$ **e.** $CH_3CH_2CH_2$

   b. CH_5 **d.** $(CH_3)_2CHCH_2CH_3$ **f.** $CH_3CHCH_2CH_3$

38. List the bonds in order of decreasing polarity (i.e., list the most polar bond first).

   a. C—O, C—F, C—N   **b.** C—Cl, C—I, C—Br   **c.** H—O, H—N, H—C   **d.** C—H, C—C, C—N

39. Write the Kekulé structure for each of the following compounds:

   a. CH_3CHO **c.** CH_3COOH **e.** $CH_3CH(OH)CH_2CN$

   b. CH_3OCH_3 **d.** $(CH_3)_3COH$ **f.** $(CH_3)_2CHCH(CH_3)CH_2C(CH_3)_3$

40. Assign the missing formal charges.

a.
$$\begin{array}{c} \text{H} \quad \text{H} \\ | \quad | \\ \text{H}-\overset{|}{\underset{|}{\text{C}}}-\overset{|}{\underset{|}{\text{C}}}\!: \\ \text{H} \quad \text{H} \end{array}$$

b.
$$\begin{array}{c} \text{H} \quad \text{H} \\ | \quad | \\ \text{H}-\overset{|}{\underset{|}{\text{C}}}-\overset{|}{\underset{|}{\text{C}}} \\ \text{H} \quad \text{H} \end{array}$$

c.
$$\begin{array}{c} \text{H} \quad \text{H} \\ | \quad | \\ \text{H}-\overset{|}{\underset{|}{\text{C}}}-\overset{|}{\underset{|}{\text{C}}}\!\cdot \\ \text{H} \quad \text{H} \end{array}$$

d.
$$\begin{array}{c} \text{H} \quad :\!\overset{..}{\text{O}}\!: \quad \text{H} \\ | \quad | \quad | \\ \text{H}-\overset{|}{\underset{|}{\text{C}}}-\overset{|}{\text{C}}-\overset{|}{\underset{|}{\text{C}}}-\text{H} \\ \text{H} \qquad \text{H} \end{array}$$

41. Draw the missing lone-pair electrons and assign the missing formal charges.

a.
$$\begin{array}{c} \text{H} \\ | \\ \text{H}-\overset{|}{\underset{|}{\text{C}}}-\text{O}-\text{H} \\ \text{H} \end{array}$$

b.
$$\begin{array}{c} \text{H} \\ | \\ \text{H}-\overset{|}{\underset{|}{\text{C}}}-\text{O}-\text{H} \\ \text{H} \quad \text{H} \end{array}$$

c.
$$\begin{array}{c} \text{H} \\ | \\ \text{H}-\overset{|}{\underset{|}{\text{C}}}-\text{O} \\ \text{H} \end{array}$$

d.
$$\begin{array}{c} \text{H} \\ | \\ \text{H}-\overset{|}{\underset{|}{\text{C}}}-\text{N}-\text{H} \\ \text{H} \quad \text{H} \end{array}$$

42. Account for the difference in the shape and color of the potential maps for ammonia and the ammonium ion in Section 1.12.

43. What is the hybridization of the indicated atom in each of the following molecules?

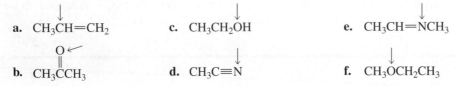

a. $CH_3\overset{\downarrow}{C}H{=}CH_2$

c. $CH_3CH_2\overset{\downarrow}{O}H$

e. $CH_3CH{=}\overset{\downarrow}{N}CH_3$

b. $CH_3\overset{\overset{\displaystyle O}{\|}}{C}CH_3$

d. $CH_3\overset{\downarrow}{C}{\equiv}N$

f. $CH_3\overset{\downarrow}{O}CH_2CH_3$

44. a. Which of the indicated bonds in each molecule is shorter?
   b. Indicate the hybridization of the C, O, and N atoms in each of the molecules.

1. $CH_3CH{=}\overset{\downarrow}{C}H\overset{\downarrow}{C}{\equiv}CH$

2. $CH_3\overset{\overset{\displaystyle O}{\|}}{C}\overset{\downarrow}{C}H_2{-}OH$

3. $CH_3NH{-}\overset{\downarrow}{C}H_2CH_2\overset{\downarrow}{N}{=}CHCH_3$

45. Which of the following molecules have a tetrahedral atom?

$$H_2O \quad H_3O^+ \quad {}^+CH_3 \quad NH_3 \quad {}^+NH_4 \quad {}^-CH_3$$

46. Do the two sp^2 hybridized carbons and the two indicated atoms lie in the same plane?

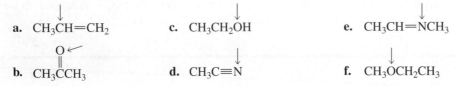

47. For each of the following molecules, indicate the hybridization of each carbon atom and give the approximate values of all the bond angles:
   a. $CH_3C{\equiv}CH$ b. $CH_3CH{=}CH_2$ c. $CH_3CH_2CH_3$ d. $CH_2{=}CH{-}CH{=}CH_2$

48. Sodium methoxide (CH_3ONa) has both ionic and covalent bonds. Which bond is ionic? How many covalent bonds does it have?

49. Explain why the following compound is not stable:

$$\begin{array}{c} \text{H} \quad \text{H} \\ \text{H}\diagdown \quad \diagup\text{C}\diagdown \\ \quad \text{C} \\ \text{H}-\text{C} \qquad \text{C} \\ | \qquad\qquad \||| \\ \text{H}-\text{C} \qquad \text{C} \\ \quad \diagup \quad \diagdown\text{C}\diagup \\ \text{H} \quad \text{H} \quad \text{H} \end{array}$$

50. Because a potential map roughly marks the "edge" of the molecule's electron cloud, the map tells us something about the relative size and shape of the molecule. A given kind of atom can have different sizes in different molecules. For example, each of the hydrogens in the three potential maps on p. 9 is a different size. Why does LiH have the largest hydrogen?

2 | Acids and Bases

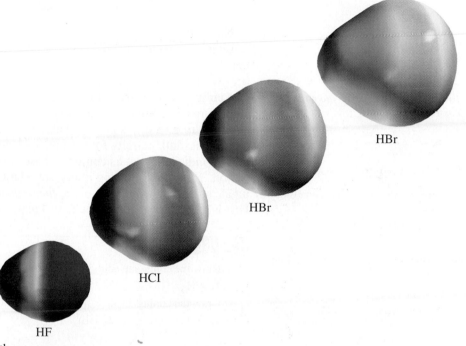

HBr

HBr

HCl

HF

Early chemists called any compound that tasted sour an acid (from *acidus*, Latin for "sour"). Some familiar acids are citric acid (found in lemons and other citrus fruits), acetic acid (found in vinegar), and hydrochloric acid (found in stomach acid—the sour taste associated with vomiting). Compounds that neutralize acids were called bases, or alkaline compounds. Glass cleaners and solutions designed to unclog drains are familiar alkaline solutions.

2.1 | Acid–Base Reactions

The terms "acid" and "base" have two definitions—the Brønsted–Lowry definitions and the Lewis definitions (Section 2.5). In the Brønsted–Lowry definitions, an **acid** is a species that donates a proton, and a **base** is a species that accepts a proton. (Remember that positively charged hydrogen ions are called protons.) In the following reaction, hydrogen chloride (HCl) is an acid because it donates a proton to water, and water is a base because it accepts a proton from HCl. Water can accept a proton because it has lone-pair electrons; either lone pair can form a covalent bond with a proton. In the reverse reaction, H_3O^+ is an acid because it donates a proton to Cl^-, and Cl^- is a base because it accepts a proton from H_3O^+. The reaction of an acid with a base is called an **acid–base reaction**. Notice that according to the Brønsted–Lowry definitions, *any species that has a hydrogen can potentially act as an acid, and any compound possessing lone-pair electrons can potentially act as a base.*

$$H\ddot{C}l\text{:} \ + \ H_2\ddot{O}\text{:} \ \rightleftharpoons \ \text{:}\ddot{C}l\text{:}^- \ + \ H_3\ddot{O}^+$$

an acid   a base   a base   an acid

When a compound loses a proton, the resulting species is called its **conjugate base**. Thus, in the above acid–base reaction, Cl^- is the conjugate base of HCl, and H_2O is

Born in Denmark, **Johannes Nicolaus Brønsted (1879–1947)** *studied engineering before he switched to chemistry. He was a professor of chemistry at the University of Copenhagen. During World War II, he became known for his anti-Nazi position, and in 1947 he was elected to the Danish parliament. He died before he could take his seat.*

Thomas M. Lowry (1874–1936) *was born in England, the son of an army chaplain. He earned a Ph.D. at Central Technical College, London (now Imperial College). He was head of chemistry at Westminster Training College and, later, at Guy's Hospital in London. In 1920, he became a professor of chemistry at Cambridge University.*

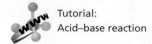

Tutorial:
Acid–base reaction

the conjugate base of H_3O^+. When a compound accepts a proton, the resulting species is called its **conjugate acid**. Thus, HCl is the conjugate acid of Cl^-, and H_3O^+ is the conjugate acid of H_2O.

In a reaction involving ammonia and water, ammonia (NH_3) is a base because it accepts a proton, and water is an acid because it donates a proton. Thus, HO^- is the conjugate base of H_2O, and ($^+NH_4$) is the conjugate acid of (NH_3). In the reverse reaction, ammonium ion ($^+NH_4$) is an acid because it donates a proton, and hydroxide ion (HO^-) is a base because it accepts a proton.

$$\ddot{N}H_3 \ + \ H_2\ddot{O}: \ \rightleftharpoons \ ^+NH_4 \ + \ H\ddot{O}:^-$$
$$\text{a base} \qquad \text{an acid} \qquad \qquad \text{an acid} \qquad \text{a base}$$

Notice that water can behave as either an acid or a base. It can behave as an acid because it has a proton that it can donate, but it can also behave as a base because it has a lone pair that can accept a proton. In Section 2.2, we will see how we know that water acts as a base in the reaction on page 31 and acts as an acid in the above reaction.

Acidity is a measure of the tendency of a compound to give up a proton. **Basicity** is a measure of a compound's affinity for a proton. A strong acid is one that has a strong tendency to give up its proton. This means that its conjugate base must be weak because it has little affinity for the proton. A weak acid has little tendency to give up its proton, indicating that its conjugate base is strong because it has a high affinity for the proton. Thus, the following important relationship exists between an acid and its conjugate base: *The stronger the acid, the weaker is its conjugate base.* For example, since HBr is a stronger acid than HCl, we know that Br^- is a weaker base than Cl^-.

> **The stronger the acid, the more readily it gives up a proton.**

> **The stronger the acid, the weaker is its conjugate base.**

PROBLEM 1◆

a. Draw the conjugate acid of each of the following:
   1. NH_3 **2.** Cl^- **3.** HO^- **4.** H_2O

b. Draw the conjugate base of each of the following:
   1. NH_3 **2.** HBr   **3.** HNO_3 **4.** H_2O

PROBLEM 2

a. Write an equation showing CH_3OH reacting as an acid with NH_3 and an equation showing it reacting as a base with HCl.

b. Write an equation showing NH_3 reacting as an acid with HO^- and an equation showing it reacting as a base with HBr.

2.2 Organic Acids and Bases; pK_a and pH

When a strong acid such as hydrogen chloride is dissolved in water, almost all the molecules dissociate (break into ions), which means that the *products* are favored at equilibrium—the equilibrium lies to the right. When a much weaker acid, such as acetic acid, is dissolved in water, very few molecules dissociate, so the *reactants* are favored at equilibrium—the equilibrium lies to the left. Two half-headed arrows are used to designate equilibrium reactions. A longer arrow is drawn toward the species favored at equilibrium.

$$H\ddot{C}l: \ + \ H_2\ddot{O}: \ \rightleftharpoons \ H_3\ddot{O}^+ \ + \ :\ddot{\underset{..}{C}}l:^-$$
$$\text{hydrogen}$$
$$\text{chloride}$$

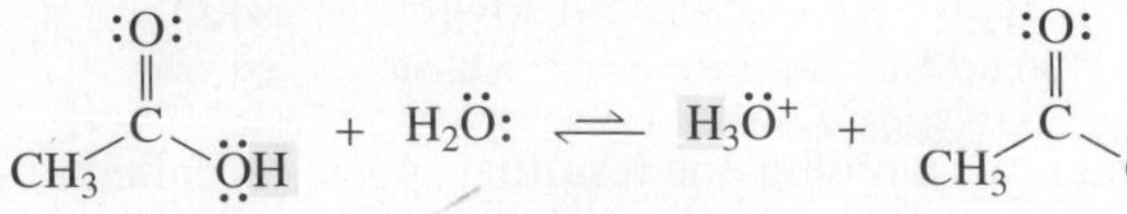

acetic acid

The degree to which an acid (HA) dissociates is indicated by the **acid dissociation constant, K_a**. Brackets are used to indicate the concentration in moles/liter [i.e., molarity (M)].

$$K_a = \frac{[H_3O^+][A^-]}{[HA]}$$

The larger the acid dissociation constant, the stronger the acid—that is, the greater is its tendency to give up a proton. Hydrogen chloride, with an acid dissociation constant of 10^7, is a stronger acid than acetic acid, with an acid dissociation constant of only 1.74×10^{-5}. For convenience, the strength of an acid is generally indicated by its **pK_a** value rather than its K_a value, where

$$pK_a = -\log K_a$$

The pK_a of hydrogen chloride is -7 and the pK_a of acetic acid, a much weaker acid, is 4.76. Notice that the smaller the pK_a, the stronger the acid.

very strong acids	p$K_a < 1$
moderately strong acids	p$K_a = 1-3$
weak acids	p$K_a = 3-5$
very weak acids	p$K_a = 5-15$
extremely weak acids	p$K_a > 15$

The concentration of positively charged hydrogen ions in the solution is indicated by **pH**. The concentration can be indicated as [H$^+$] or, because a hydrogen ion in water is solvated, as [H$_3$O$^+$]. The lower the pH, the more acidic is the solution.

$$pH = -\log[H_3O^+]$$

Acidic solutions have pH values less than 7; basic solutions have pH values greater than 7. The pH values of some commonly encountered solutions are shown in the margin. The pH of a solution can be changed simply by adding acid or base to the solution. Do not confuse pH and pK_a: The pH scale is used to describe the acidity of a *solution*; the pK_a is characteristic of a particular *compound*, much like a melting point or a boiling point—it indicates the tendency of the compound to give up its proton.

The stronger the acid, the smaller is its pK_a.

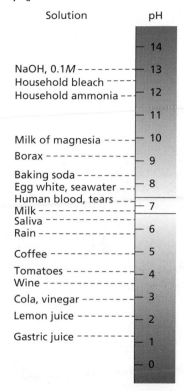

Solution	pH
	14
NaOH, 0.1M --------	13
Household bleach -----	
Household ammonia --	12
	11
Milk of magnesia -----	10
Borax -----------	9
Baking soda --------	
Egg white, seawater --	8
Human blood, tears ---	
Milk --------------	7
Saliva -------------	
Rain -------------	6
Coffee -----------	5
Tomatoes ---------	4
Wine -------------	
Cola, vinegar --------	3
Lemon juice --------	2
Gastric juice --------	1
	0

ACID RAIN

Rain is mildly acidic (pH = 5.5) because when the CO$_2$ in the air reacts with water, a weak acid—carbonic acid (pK_a = 6.4)—is formed.

$$CO_2 + H_2O \rightleftharpoons H_2CO_3$$
carbonic acid

In some parts of the world, rain has been found to be much more acidic—with pH values as low as 4.3. Acid rain is formed in areas where sulfur dioxide and nitrogen oxides are produced, because when these gases react with water, strong acids— sulfuric acid (pK_a = -5.0) and nitric acid (pK_a = -1.3)— are formed. Burning fossil fuels for the generation of electric power is the factor most responsible for forming these acid-producing gases.

Acid rain has many deleterious effects. It can cause lakes and streams to no longer be able to support aquatic life; it can make soil so acidic that crops cannot grow; and it can cause the deterioration of paint and building materials, including monuments and statues that are part of our cultural heritage. Marble—a form of calcium carbonate—decays because acid

reacts with CO$_3^{2-}$ to form carbonic acid, which decomposes to CO$_2$ and H$_2$O—the reverse of the reaction shown to the left.

$$CO_3^{2-} \underset{}{\overset{H^+}{\rightleftharpoons}} HCO_3^- \underset{}{\overset{H^+}{\rightleftharpoons}} H_2CO_3 \rightleftharpoons CO_2 + H_2O$$

photo taken in 1935

photo taken in 1994

Statue of George Washington in Washington Square Park in Greenwich Village, New York.

PROBLEM 3◆

a. Which is a stronger acid, one with a pK_a of 5.2 or one with a pK_a of 5.8?

b. Which is a stronger acid, one with an acid dissociation constant of 3.4×10^{-3} or one with an acid dissociation constant of 2.1×10^{-4}?

PROBLEM-SOLVING STRATEGY

Vitamin C has a pK_a value of 4.17. What is its K_a value?

You will need a calculator to answer this question. Remembering that $pK_a = -\log K_a$:

1. Enter the pK_a value on your calculator.

2. Multiply it by -1.

3. Determine the inverse log by pressing the key labeled 10^X.

You should find that vitamin C has a K_a value of 6.76×10^{-5}.

Now continue on to Problem 4.

PROBLEM 4◆

Butyric acid, the compound responsible for the unpleasant odor and taste of sour milk, has a pK_a value of 4.82. What is its K_a value? Is it a stronger or a weaker acid than vitamin C?

PROBLEM 5

Antacids are compounds that neutralize stomach acid. Write the equations that show how Milk of Magnesia, Alka-Seltzer, and Tums remove excess acid.

a. Milk of Magnesia: $Mg(OH)_2$ **c.** Tums: $CaCO_3$

b. Alka-Seltzer: $KHCO_3$ and $NaHCO_3$

PROBLEM 6◆

Are the following body fluids acidic or basic?

a. bile $(pH = 8.4)$ **b.** urine $(pH = 5.9)$ **c.** spinal fluid $(pH = 7.4)$

    The most common organic acids are carboxylic acids—compounds that have a COOH group. Acetic acid and formic acid are examples of carboxylic acids. Carboxylic acids have pK_a values ranging from about 3 to 5. (They are weak acids.) The pK_a values of a wide variety of organic compounds are given in Appendix II.

3-D Molecule:
Acetic acid

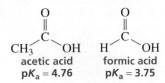

acetic acid
$pK_a = 4.76$ formic acid
$pK_a = 3.75$

    Alcohols are compounds that have an OH group. They are much weaker acids than carboxylic acids, with pK_a values close to 16. Methyl alcohol and ethyl alcohol are examples of alcohols.

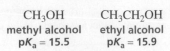

CH_3OH CH_3CH_2OH
methyl alcohol   ethyl alcohol
$pK_a = 15.5$ $pK_a = 15.9$

An alcohol can behave as an acid and donate a proton, or as a base and accept a proton.

$$CH_3OH + HO^- \rightleftharpoons CH_3O^- + H_2O$$
an acid

$$CH_3OH + H_3O^+ \rightleftharpoons CH_3\overset{+}{\underset{H}{O}}H + H_2O$$
a base

A carboxylic acid also can behave as an acid and donate a proton, or as a base and accept a proton.

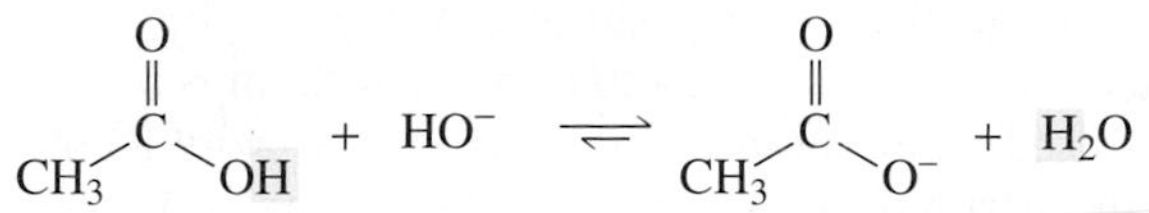

an acid

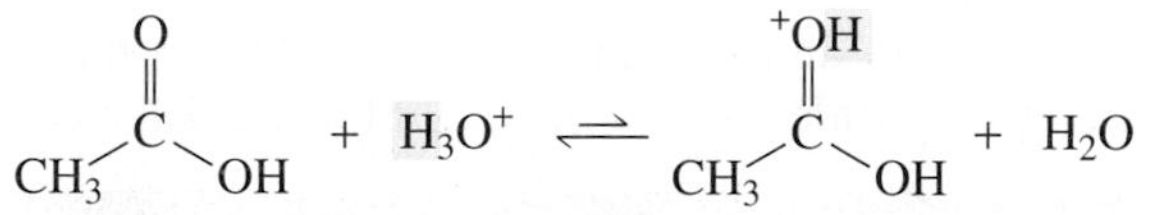

a base

A *protonated* compound is a compound that has gained an additional proton. Protonated alcohols and protonated acids are very strong acids. For example, protonated methyl alcohol has a pK_a of −2.5 and protonated acetic acid has a pK_a of −6.1. Notice that the sp^2 oxygen of the carboxylic acid is the one that is protonated. We will see why this is so in Section 12.9.

$$CH_3\overset{+}{\underset{H}{O}}H$$
protonated methyl alcohol
pK_a = −2.5

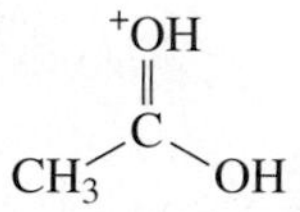

protonated acetic acid
pK_a = −6.1

Compounds with NH$_2$ groups are amines. An amine can behave as an acid and donate a proton, or as a base and accept a proton.

$$CH_3NH_2 + HO^- \rightleftharpoons CH_3\overset{-}{N}H + H_2O$$
an acid

$$CH_3NH_2 + H_3O^+ \rightleftharpoons CH_3\overset{+}{N}H_3 + H_2O$$
a base

Amines, however, have such high pK_a values that they rarely behave as acids. Ammonia also has a high pK_a value.

$$CH_3NH_2$$
methylamine
pK_a = 40

$$NH_3$$
ammonia
pK_a = 36

Amines are much more likely to act as bases. In fact, amines are the most common organic bases. Instead of talking about the strength of a base in terms of its pK_b value, it is easier to talk about the strength of its conjugate acid as indicated by its pK_a value, remembering that the stronger the acid, the weaker is its conjugate base. For example, protonated methylamine is a stronger acid than protonated ethylamine, which means that methylamine is a weaker base than ethylamine. Notice that the pK_a values of protonated amines are about 10 to 11.

$$CH_3\overset{+}{N}H_3$$
protonated methylamine
pK_a = 10.7

$$CH_3CH_2\overset{+}{N}H_3$$
protonated ethylamine
pK_a = 11.0

PROBLEM 7◆

a. Which is a stronger base, CH_3COO^- or $HCOO^-$? (The pK_a of CH_3COOH is 4.8; the pK_a of $HCOOH$ is 3.8.)

b. Which is a stronger base, HO^- or $^-NH_2$? (The pK_a of H_2O is 15.7; the pK_a of NH_3 is 36.)

c. Which is a stronger base, H_2O or CH_3OH? (The pK_a of H_3O^+ is -1.7; the pK_a of $CH_3OH_2^+$ is -2.5.)

It is important to know the approximate pK_a values of the various classes of compounds we have discussed. An easy way to remember them is in units of five, as shown in Table 2.1. (R is used when the particular carboxylic acid, alcohol, or amine is not specified.) Protonated alcohols, protonated carboxylic acids, and protonated water have pK_a values less than 0, carboxylic acids have pK_a values of about 5, protonated amines have pK_a values of about 10, and alcohols and water have pK_a values of about 15. These values are also listed inside the back cover of this book for easy reference.

Be sure to learn the approximate pK_a values given in Table 2.1.

Table 2.1 Approximate pK_a Values			
$pK_a < 0$	**$pK_a \sim 5$**	**$pK_a \sim 10$**	**$pK_a \sim 15$**
$R\overset{+}{O}H_2$ a protonated alcohol $R-\underset{\overset{\parallel}{^+OH}}{C}-OH$ a protonated carboxylic acid H_3O^+ protonated water	$R-\underset{\overset{\parallel}{O}}{C}-OH$ a carboxylic acid	$R\overset{+}{N}H_3$ a protonated amine	ROH an alcohol H_2O water

Now let's see how we knew that water acts as a base in the first reaction in Section 2.1 and as an acid in the second reaction. To determine which of the reactants will be the acid, we need to compare their pK_a values: The pK_a of hydrogen chloride is -7 and the pK_a of water is 15.7. Because hydrogen chloride is the stronger acid, it will donate a proton to water. Water, therefore, is a base in this reaction. When we compare the pK_a values of the two reactants of the second reaction, we see that the pK_a of ammonia is 36 and the pK_a of water is 15.7. In this case, water is the stronger acid, so it donates a proton to ammonia. Water, therefore, is an acid in this reaction.

To determine the position of equilibrium for an acid–base reaction (i.e., whether reactants or products are favored at equilibrium), we need to compare the pK_a value of the acid on the left of the arrow with the pK_a value of the acid on the right of the arrow. The equilibrium favors *reaction* of the strong acid and *formation* of the weak acid. In other words, *strong reacts to form weak*. Thus, the equilibrium lies away from the stronger acid and toward the weaker acid. Notice that the stronger acid has the weaker conjugate base.

Strong reacts to form weak.

$$CH_3-\underset{\overset{\parallel}{O}}{C}-OH \ + \ NH_3 \ \rightleftharpoons \ CH_3-\underset{\overset{\parallel}{O}}{C}-O^- \ + \ \overset{+}{N}H_4$$

stronger acid   stronger base   weaker base   weaker acid
$pK_a = 4.8$ $pK_a = 9.4$

$$CH_3CH_2OH \ + \ CH_3NH_2 \ \rightleftharpoons \ CH_3CH_2O^- \ + \ CH_3\overset{+}{N}H_3$$

weaker acid   weaker base   stronger base   stronger acid
$pK_a = 15.9$ $pK_a = 10.7$

PROBLEM 8

a. For each of the acid–base reactions in Section 2.2, compare the pK_a values of the acids on either side of the equilibrium arrows and convince yourself that the position of equilibrium is in the direction indicated. (The pK_a values you need can be found in Section 2.2 or in Problem 7.)

b. Do the same thing for the equilibria in Section 2.1. (The pK_a of $^+NH_4$ is 9.4.)

PROBLEM 9◆

Using the pK_a values in Section 2.2, rank the following species in order of decreasing base strength (i.e., list the strongest base first):

$$CH_3NH_2 \quad CH_3NH^- \quad CH_3OH \quad CH_3O^- \quad CH_3\overset{\overset{\displaystyle O}{\|}}{C}O^-$$

2.3 The Effect of Structure on pK_a

The strength of an acid is determined by the stability of the conjugate base that is formed when the acid gives up its proton. The more stable the base, the stronger is its conjugate acid. A stable base is a base that readily bears the electrons it formerly shared with a proton. In other words, stable bases are weak bases—they don't share their electrons well. So we can say, *the stronger the acid, the weaker is its conjugate base* or, *the stronger the acid, the more stable is its conjugate base.*

The stability of an acid's conjugate base is determined by two factors, its *size* and its *electronegativity*. The elements in the second row of the periodic table are all about the same size, but they have very different electronegativities. The electronegativities increase across the row from left to right. Of the atoms shown, carbon is the least electronegative and fluorine is the most electronegative.

> The weaker the base, the stronger is its conjugate acid.
>
> Stable bases are weak bases.
>
> The more stable the base, the stronger is its conjugate acid.

relative electronegativities: C < N < O < F
                                                              └─ most electronegative

If we look at the acids formed when hydrogens are attached to these elements, we see that the most acidic compound is the one that has its hydrogen attached to the most electronegative atom. Thus, HF is the strongest acid and methane is the weakest acid (Table 2.2).

relative acidities: CH_4 < NH_3 < H_2O < HF
                                                                     └─ strongest acid

If we look at the stabilities of the conjugate bases of these acids, we find that they too increase from left to right because the most electronegative atom is better able to bear its negative charge. Thus, the strongest acid has the most stable conjugate base.

relative stabilities: $^-CH_3$ < $^-NH_2$ < HO^- < F^-
                                                                      └─ most stable

> When atoms are similar in size, the strongest acid will have its hydrogen attached to the most electronegative atom.

We therefore can conclude that *when the atoms are similar in size, the strongest acid will have its hydrogen attached to the most electronegative atom.*

Table 2.2 The pK_a Values of Some Simple Acids			
CH_4	NH_3	H_2O	HF
pK_a = ~60	pK_a = 36	pK_a = 15.7	pK_a = 3.2
		H_2S	HCl
		pK_a = 7.0	pK_a = −7
			HBr
			pK_a = −9
			HI
			pK_a = −10

The effect that the electronegativity of the atom bonded to a hydrogen has on the acidity of that hydrogen can be appreciated when the pK_a values of alcohols and amines are compared. Because oxygen is more electronegative than nitrogen, an alcohol is more acidic than an amine.

CH_3OH CH_3NH_2
methyl alcohol   **methylamine**
pK_a = 15.5   **pK_a = 40**

Similarly, a protonated alcohol is more acidic than a protonated amine.

$CH_3\overset{+}{O}H_2$ $CH_3\overset{+}{N}H_3$
protonated methyl alcohol   **protonated methylamine**
pK_a = −2.5   **pK_a = 10.7**

In comparing atoms that are very different in size, the *size* of the atom is much more important than its *electronegativity* in determining how well it bears its negative charge. For example, as we proceed down a column in the periodic table, as the elements get larger, their stability increases, so the strength of their conjugate acid also *increases*. Thus, HI is the strongest acid of the hydrogen halides (even though iodine is the least electronegative of the halogens). Thus, *when atoms are very different in size, the strongest acid will have its hydrogen attached to the largest atom.*

When atoms are very different in size, the strongest acid will have its hydrogen attached to the largest atom.

relative electronegativities: F   >   Cl   >   Br   >   I

most electronegative   largest

relative stabilities: F^- < Cl^- < Br^- < I^-

most stable

relative acidities: HF   <   HCl   <   HBr   <   HI

strongest acid

Why does the size of an atom have such a significant effect on the stability of the base that it more than overcomes the difference in electronegativity? The valence electrons of F^- are in a $2p$ orbital, the valence electrons of Cl^- are in a $3p$ orbital, those of Br^- are in a $4p$ orbital, and those of I^- are in a $5p$ orbital. The volume of space occupied by a $3p$ orbital is significantly greater than the volume of space occupied by a $2p$ orbital because a $3p$ orbital extends out farther from the nucleus. Because its negative charge is spread over a larger volume of space, Cl^- is more stable than F^-.

Thus, as the halide ion increases in size, its stability increases because its negative charge is spread over a larger volume of space—its electron density decreases. Therefore, HI is the strongest acid of the hydrogen halides because I^- is the most stable halide ion, even though iodine is the least electronegative of the halogens (Table 2.2). The potential maps illustrate the large difference in size of the halide ions:

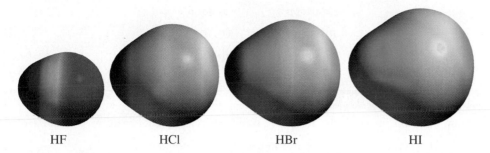

| HF | HCl | HBr | HI |

PROBLEM 10◆

For each of the following pairs, indicate which is the stronger acid:

a. HCl or HBr

b. $CH_3CH_2CH_2\overset{+}{N}H_3$ or $CH_3CH_2CH_2\overset{+}{O}H_2$

c.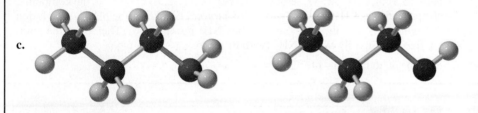

PROBLEM 11◆

a. Which of the halide ions (F^-, Cl^-, Br^-, I^-) is the strongest base?

b. Which is the weakest base?

PROBLEM 12◆

a. Which is more electronegative, oxygen or sulfur?

b. Which is a stronger acid, H_2O or H_2S?

c. Which is a stronger acid, CH_3OH or CH_3SH?

PROBLEM 13◆

For each of the following pairs, indicate which is the stronger base:

a. H_2O or HO^- **b.** H_2O or NH_3 **c.** $CH_3\overset{\overset{\displaystyle O}{\|}}{C}O^-$ or CH_3O^- **d.** CH_3O^- or CH_3S^-

---

2.4   The Effect of pH on the Structure of an Organic Compound

Whether an acid will lose a proton in an aqueous solution depends on both the pK_a of the acid and the pH of the solution. *A compound will exist primarily in its acidic form (with its proton) in solutions that are more acidic than the pK_a value of the group that*

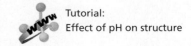
Tutorial:
Effect of pH on structure

> A compound will exist primarily in its acidic form if the pH of the solution is less than its pK_a.
>
> A compound will exist primarily in its basic form if the pH of the solution is greater than its pK_a.

undergoes dissociation and primarily in its basic form (without its proton) in solutions that are more basic than the pK_a value of the group that undergoes dissociation. When the pH of a solution equals the pK_a value of the group that undergoes dissociation, the concentration of the compound in its acidic form will equal the concentration of the compound in its basic form.

acidic form		basic form
$RCOOH$	$\rightleftharpoons$	$RCOO^- + H^+$
$R\overset{+}{N}H_3$	$\rightleftharpoons$	$RNH_2 + H^+$

PROBLEM SOLVING STRATEGY

Draw the form in which the following compounds will predominate in a solution of pH 5.5:

a. CH_3CH_2OH ($pK_a = 15.9$)

b. $CH_3CH_2\overset{+}{O}H_2$ ($pK_a = -2.5$)

c. $CH_3\overset{+}{N}H_3$ ($pK_a = 11.0$)

To answer this kind of question, you need to compare the pH of the solution with the pK_a value of the compound's dissociable proton. For part a, the pH of the solution is more acidic (5.5) than the pK_a value of the OH group (15.9). Therefore, the compound will exist primarily as CH_3CH_2OH (with its proton). For part b, the pH of the solution is more basic (5.5) than the pK_a value of the $^+OH_2$ group (-2.5). Therefore, the compound will exist primarily as CH_3CH_2OH (without its proton). For part c, the pH of the solution is more acidic (5.5) than the pK_a value of the $^+NH_3$ group (11.0). Therefore, the compound will exist primarily as $CH_3\overset{+}{N}H_3$ (with its proton).

Now continue on to Problem 14.

PROBLEM 14◆

For each of the following compounds, shown in their acidic forms, draw the form in which it will predominate in a solution of pH = 5.5:

a. CH_3COOH ($pK_a = 4.76$)   e. $^+NH_4$ ($pK_a = 9.4$)

b. $CH_3CH_2\overset{+}{N}H_3$ ($pK_a = 11.0$)   f. $HC\equiv N$ ($pK_a = 9.1$)

c. H_3O^+ ($pK_a = -1.7$)   g. HNO_2 ($pK_a = 3.4$)

d. HBr ($pK_a = -9$)   h. HNO_3 ($pK_a = -1.3$)

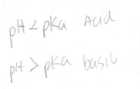

PROBLEM 15◆

a. Indicate whether a carboxylic acid (RCOOH) with a pK_a of 4.5 will have more charged molecules or more neutral molecules in a solution with the following pH value:

1. pH = 1   3. pH = 5   5. pH = 10

2. pH = 3   4. pH = 7   6. pH = 13

b. Answer the same question for a protonated amine ($R\overset{+}{N}H_3$) with a pK_a of 9.

c. Answer the same question for an alcohol (ROH) with a pK_a of 15.

2.5 Buffer Solutions

A solution of a weak acid (HA) and its conjugate base (A^-) is called a **buffer solution**. A buffer solution will maintain nearly constant pH when small amounts of acid or base are added to it, because the weak acid can donate a proton to any HO^- added to the solution, and its conjugate base can accept any H^+ that is added to the solution.

> can donate an H^+
> to HO^-

$$HA + HO^- \longrightarrow A^- + H_2O$$

$$A^- + H_3O^+ \longrightarrow HA + H_2O$$

> can accept an H^+ from H_3O^+

PROBLEM 16

Write the equation that shows how a buffer made by dissolving CH_3COOH and $CH_3COO^-Na^+$ in water prevents the pH of a solution from changing when

a. a small amount of H^+ is added to the solution.

b. a small amount of HO^- is added to the solution.

BLOOD: A BUFFERED SOLUTION

Blood is the fluid that transports oxygen to all the cells of the human body. The normal pH of human blood is 7.3 to 7.4. Death will result if this pH decreases to a value less than ~6.8 or increases to a value greater than ~8.0 for even a few seconds.

Oxygen is carried to cells by a protein in the blood called hemoglobin (HbH^+). When hemoglobin binds O_2, hemoglobin loses a proton, which would make the blood more acidic if it did not contain a buffer to maintain its pH.

$$HbH^+ + O_2 \rightleftharpoons HbO_2 + H^+$$

A carbonic acid/bicarbonate (H_2CO_3/HCO_3^-) buffer is used to control the pH of blood. An important feature of this buffer is that carbonic acid decomposes to CO_2 and H_2O:

$$\underset{\text{carbonic acid}}{CO_2 + H_2O \rightleftharpoons H_2CO_3} \rightleftharpoons \underset{\text{bicarbonate}}{HCO_3^- + H^+}$$

The increase in metabolism during periods of exercise produces large amounts of CO_2. The increased concentration of CO_2 shifts the equilibrium between carbonic acid and bicarbonate to the right, which increases the concentration of H^+. Significant amounts of lactic acid are also produced during exercise, and this further increases the concentration of H^+. Receptors in the brain respond to the increased concentration of H^+ and trigger a reflex that increases the rate of breathing. This causes hemoglobin to release more oxygen to the cells and increases the elimination of CO_2 by exhalation. Both processes decrease the concentration of H^+ in the blood by shifting both equilibria to the left.

Thus, any condition that decreases the rate and depth of ventilation, such as emphysema, will decrease the pH of the blood—a condition called acidosis. In contrast, any condition that increases the rate and depth of ventilation, such as hyperventilation due to anxiety, will increase the pH of blood—a condition called alkalosis.

2.6 Lewis Acids and Bases

In 1923, G. N. Lewis (see page 10) offered new definitions for the terms *acid* and *base*. He defined an acid as a species that *accepts a share in an electron pair* and a base as a species that *donates a share in an electron pair*. All proton-donating acids fit the Lewis definition because all proton-donating acids lose a proton and the proton accepts a share in an electron pair.

Lewis base: Have pair, will share.

Lewis acid: Need two from you.

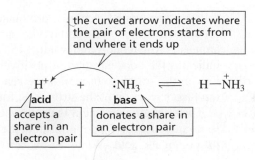

> the curved arrow indicates where the pair of electrons starts from and where it ends up

$$\underset{\text{acid}}{H^+} + \underset{\text{base}}{:NH_3} \rightleftharpoons H-\overset{+}{N}H_3$$

> accepts a share in an electron pair

> donates a share in an electron pair

Lewis acids, however, are not limited to compounds that donate protons. According to the Lewis definition, compounds such as aluminum chloride ($AlCl_3$), boron trifluoride (BF_3), and borane (BH_3) are acids because they have unfilled valence orbitals and thus can accept a share in an electron pair. These compounds react with a compound that has a lone pair, just as a proton reacts with ammonia. Thus, the Lewis definition of an acid includes all proton-donating acids and some additional acids that do not have protons. Throughout this text, the term *acid* is used to mean a proton-donating acid, and the term **Lewis acid** is used to refer to non-proton-donating acids such as $AlCl_3$ or BF_3.

All bases are **Lewis bases** because they all have a pair of electrons that they can share, either with an atom such as aluminum or boron or with a proton.

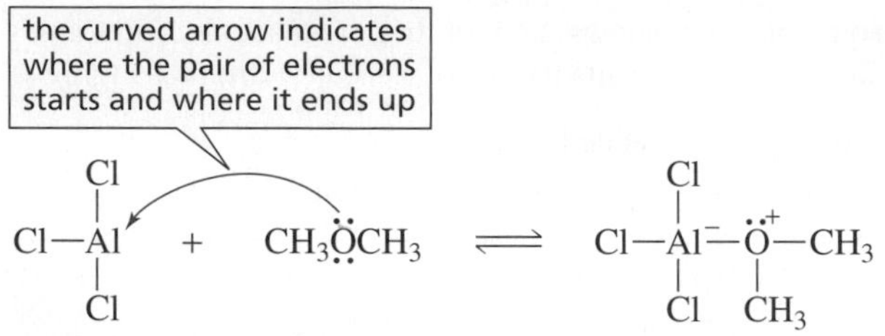

the curved arrow indicates where the pair of electrons starts and where it ends up

aluminum trichloride dimethyl ether
a Lewis acid   a Lewis base

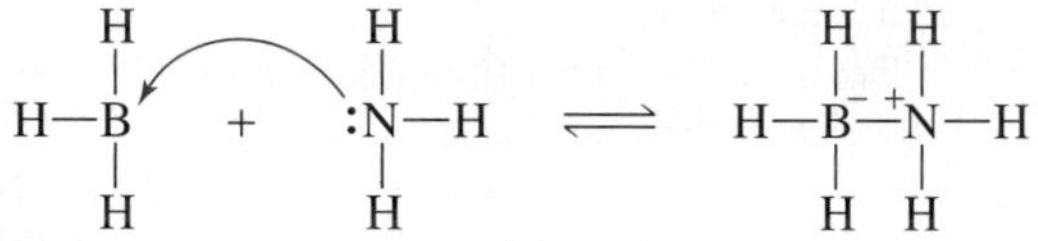

borane   ammonia
a Lewis acid   a Lewis base

PROBLEM 17

Give the products of the following reactions using arrows to show where the pair of electrons starts and where it ends up.

a. $ZnCl_2$ + $CH_3\ddot{O}H$ $\rightleftharpoons$

b. $FeBr_3$ + $:\ddot{B}r:^-$ $\rightleftharpoons$

c. $AlCl_3$ + $:\ddot{C}l:^-$ $\rightleftharpoons$

PROBLEM 18

Show how each of the following compounds reacts with HO^-:

a. CH_3OH **c.** $CH_3\overset{+}{N}H_3$ **e.** $^+CH_3$ **g.** $AlCl_3$

b. $^+NH_4$ **d.** BF_3 **f.** $FeBr_3$ **h.** CH_3COOH

Summary

An **acid** is a species that donates a proton, and a **base** is a species that accepts a proton. A **Lewis acid** is a species that accepts a share in an electron pair; a **Lewis base** is a species that donates a share in an electron pair.

Acidity is a measure of the tendency of a compound to give up a proton. **Basicity** is a measure of a compound's affinity for a proton. The stronger the acid, the weaker is its conjugate base. The strength of an acid is given by the **acid dissociation constant** (K_a). Approximate pK_a values are as

follows: protonated alcohols, protonated carboxylic acids, protonated water <0; carboxylic acids ~ 5; protonated amines ~ 10; alcohols and water ~ 15. The **pH** of a solution indicates the concentration of positively charged hydrogen ions in the solution. In **acid–base reactions**, the equilibrium favors reaction of the strong and formation of the weak.

The strength of an acid is determined by the stability of its conjugate base: The more stable the base, the stronger is its conjugate acid. When atoms are similar in size, the strongest acid has its hydrogen attached to the most electronegative atom. When atoms are very different in size, the strongest acid has its hydrogen attached to the largest atom.

A compound exists primarily in its acidic form in solutions more acidic than its pK_a value and primarily in its basic form in solutions more basic than its pK_a value. A **buffer solution** contains both an acid and its conjugate base.

Problems

19. For each of the following compounds, draw the form in which it will predominate at pH = 3, pH = 6, pH = 10, and pH = 14:

a. CH_3COOH
$pK_a = 4.8$

b. $CH_3CH_2\overset{+}{N}H_3$
$pK_a = 11.0$

c. CF_3CH_2OH
$pK_a = 12.4$

20. Give the products of the following acid–base reactions, and indicate whether reactants or products are favored at equilibrium (use the pK_a values that are given in Section 2.2):

a. $CH_3\overset{O}{\overset{\|}{C}}OH + CH_3O^- \rightleftharpoons$

c. $CH_3\overset{O}{\overset{\|}{C}}OH + CH_3NH_2 \rightleftharpoons$

b. $CH_3CH_2OH + {}^-NH_2 \rightleftharpoons$

d. $CH_3CH_2OH + HCl \rightleftharpoons$

21. a. Which of the following is the strongest acid?
b. Which is the weakest acid?
c. Which acid has the strongest conjugate base
   1. nitrous acid (HNO_2), $K_a = 4.0 \times 10^{-4}$
   2. nitric acid (HNO_3), $K_a = 22$
   3. bicarbonate (HCO_3^-), $K_a = 6.3 \times 10^{-11}$
   4. hydrogen cyanide (HCN), $K_a = 7.9 \times 10^{-10}$
   5. formic acid (HCOOH), $K_a = 2.0 \times 10^{-4}$

22. For each of the following pairs, indicate which is the stronger base:

a. HS^- or HO^- **b.** CH_3O^- or $CH_3\overset{-}{N}H$ **c.** CH_3OH or CH_3O^- **d.** Cl^- or Br^-

23. Locate the three nitrogen atoms in the electrostatic potential map of histamine, the compound that causes the symptoms associated with the common cold and allergic responses. Which of the nitrogen atoms is the most basic?

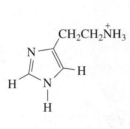
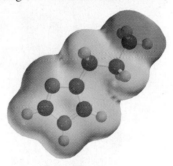

histamine

24. Using the table of pK_a values given in Appendix II, answer the following:
a. Which is the most acidic organic compound in the table?
b. Which is the least acidic organic compound in the table?
c. Which is the most acidic carboxylic acid in the table?

25. As long as the pH is greater than _____, more than 50% of a protonated amine with a pK_a of 10.4 will be in its neutral, nonprotonated form.

26. a. List the following carboxylic acids in order of decreasing acidity:

   1. $CH_3CH_2CH_2COOH$
      $K_a = 1.52 \times 10^{-5}$

   2. $CH_3CH_2\overset{\displaystyle|}{\underset{\displaystyle Cl}{CH}}COOH$
      $K_a = 1.39 \times 10^{-3}$

   3. $ClCH_2CH_2CH_2COOH$
      $K_a = 2.96 \times 10^{-5}$

   4. $CH_3\overset{\displaystyle|}{\underset{\displaystyle Cl}{CH}}CH_2COOH$
      $K_a = 8.9 \times 10^{-5}$

b. How does the presence of an electronegative substituent such as Cl affect the acidity of a carboxylic acid?
c. How does the location of the substituent affect the acidity of a carboxylic acid?
d. Why does the substituent affect the acidity of the carboxylic acid?

27. Explain the difference in the pK_a values of the following compounds:

$$CH_3 \overset{\overset{O}{\parallel}}{C} OH \qquad ICH_2 \overset{\overset{O}{\parallel}}{C} OH \qquad BrCH_2 \overset{\overset{O}{\parallel}}{C} OH \qquad ClCH_2 \overset{\overset{O}{\parallel}}{C} OH \qquad FCH_2 \overset{\overset{O}{\parallel}}{C} OH$$

$pK_a = 4.76$ $\qquad$ $pK_a = 3.15$ $\qquad$ $pK_a = 2.86$ $\qquad$ $pK_a = 2.81$ $\qquad$ $pK_a = 2.66$

28. Ethyne has a pK_a value of 25, water has a pK_a value of 15.7, and ammonia (NH_3) has a pK_a value of 36. Draw the equation, showing equilibrium arrows that indicate whether reactants or products are favored, for the reaction of ethyne with
   a. HO^- **b.** $^-NH_2$
   c. Which would be a better base to use if you wanted to remove a proton from ethyne, HO^- or $^-NH_2$?

29. For each of the following pairs of reactions, indicate which one has the more favorable equilibrium constant (that is, which one most favors products):

   a. $CH_3CH_2OH + NH_3 \rightleftharpoons CH_3CH_2O^- + \overset{+}{N}H_4$

   or

   $CH_3OH + NH_3 \rightleftharpoons CH_3O^- + \overset{+}{N}H_4$

   b. $CH_3CH_2OH + NH_3 \rightleftharpoons CH_3CH_2O^- + \overset{+}{N}H_4$

   or

   $CH_3CH_2OH + CH_3NH_2 \rightleftharpoons CH_3CH_2O^- + CH_3\overset{+}{N}H_3$

30. Carbonic acid has a pK_a of 6.1 at physiological temperature. Is the carbonic acid/bicarbonate buffer system that maintains the pH of the blood at 7.3 better at neutralizing excess acid or excess base?

31. Water and diethyl ether are immiscible liquids. Charged compounds dissolve in water, and uncharged compounds dissolve in ether (Section 3.7). $C_6H_{11}COOH$ has a pK_a of 4.8 and $C_6H_{11}\overset{+}{N}H_3$ has a pK_a of 10.7.

a. What pH would you make the water layer in order to cause both compounds to dissolve in it?

b. What pH would you make the water layer in order to cause the acid to dissolve in the water layer and the amine to dissolve in the ether layer?

c. What pH would you make the water layer in order to cause the acid to dissolve in the ether layer and the amine to dissolve in the water layer?

— ether
— water

For more information about acid–base chemistry see Special Topic I in the Study Guide and Solutions Manual.

32. How could you separate a mixture of the following compounds? The reagents available to you are water, ether, 1.0 M HCl, and 1.0 M NaOH. (*Hint:* See Problem 31.)

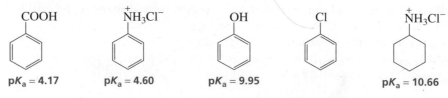

COOH $\qquad$ $\overset{+}{N}H_3Cl^-$ $\qquad$ OH $\qquad$ Cl $\qquad$ $\overset{+}{N}H_3Cl^-$

$pK_a = 4.17$ $\qquad$ $pK_a = 4.60$ $\qquad$ $pK_a = 9.95$ $\qquad\qquad\qquad$ $pK_a = 10.66$

3

An Introduction to Organic Compounds

Nomenclature, Physical Properties, and Representation of Structure

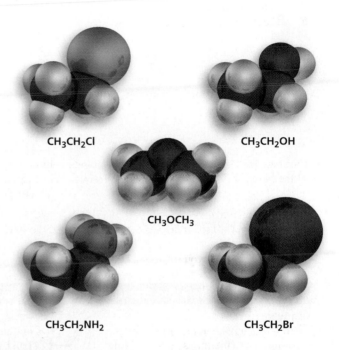

CH$_3$CH$_2$Cl

CH$_3$CH$_2$OH

CH$_3$OCH$_3$

CH$_3$CH$_2$NH$_2$

CH$_3$CH$_2$Br

If we are going to talk about organic compounds, we need to know how to name them. First, we will learn how *alkanes* are named because their names form the basis for the names of almost all organic compounds. **Alkanes** are composed of only carbon atoms and hydrogen atoms and contain only single bonds. Compounds that contain only carbon and hydrogen are called **hydrocarbons**, so an alkane is a hydrocarbon that has only single bonds. Alkanes in which the carbons form a continuous chain with no branches are called **straight-chain alkanes**. The names of several straight-chain alkanes are given in Table 3.1.

If you look at the relative numbers of carbon and hydrogen atoms in the alkanes listed in Table 3.1, you will see that the general molecular formula for an alkane is C$_n$H$_{2n+2}$, where n is any integer. So, if an alkane has one carbon atom, it must have four hydrogen atoms; if it has two carbon atoms, it must have six hydrogen atoms.

We have seen that carbon forms four covalent bonds and hydrogen forms only one covalent bond (Section 1.4). This means that there is only one possible structure for an alkane with molecular formula CH$_4$ (methane) and only one structure for an alkane with molecular formula C$_2$H$_6$ (ethane). We examined the structures of these compounds in Section 1.7. There is also only one possible structure for an alkane with molecular formula C$_3$H$_8$ (propane).

As the number of carbons in an alkane increases beyond three, the number of possible structures increases. There are two possible structures for an alkane with molecular formula C$_4$H$_{10}$. In addition to butane—a straight-chain alkane—there is a **branched-chain alkane** called isobutane. Both of these structures fulfill the requirement that each carbon forms four bonds and each hydrogen forms only one bond.

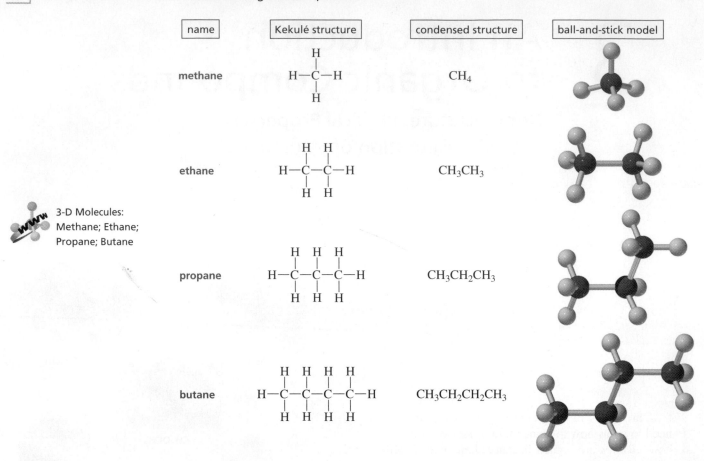

name	Kekulé structure	condensed structure	ball-and-stick model
methane		CH_4	
ethane		CH_3CH_3	
propane		$CH_3CH_2CH_3$	
butane		$CH_3CH_2CH_2CH_3$	

3-D Molecules:
Methane; Ethane;
Propane; Butane

Table 3.1   Nomenclature and Physical Properties of Some Straight-Chain Alkanes

Number of carbons	Molecular formula	Name	Condensed structure	Boiling point (°C)	Melting point (°C)	Density[a] (g/mL)
1	CH_4	methane	CH_4	−167.7	−182.5	
2	C_2H_6	ethane	CH_3CH_3	−88.6	−183.3	
3	C_3H_8	propane	$CH_3CH_2CH_3$	−42.1	−187.7	0.5005
4	C_4H_{10}	butane	$CH_3CH_2CH_2H_3$	−0.5	−138.3	0.5787
5	C_5H_{12}	pentane	$CH_3(CH_2)_3CH_3$	36.1	−129.8	0.5572
6	C_6H_{14}	hexane	$CH_3(CH_2)_4CH_3$	68.7	−95.3	0.6603
7	C_7H_{16}	heptane	$CH_3(CH_2)_5CH_3$	98.4	−90.6	0.6837
8	C_8H_{18}	octane	$CH_3(CH_2)_6CH_3$	125.7	−56.8	0.7026
9	C_9H_{20}	nonane	$CH_3(CH_2)_7CH_3$	150.8	−53.5	0.7177
10	$C_{10}H_{22}$	decane	$CH_3(CH_2)_8CH_3$	174.0	−29.7	0.7299

[a]Density is temperature dependent. The densities given are those determined at 20°C.

Compounds such as butane and isobutane, which have the same molecular formula but differ in the order in which the atoms are connected, are called **constitutional isomers**—their molecules have different constitutions. In fact, isobutane got its name because it is an "iso"mer of butane. The structural unit—a carbon bonded to a hydrogen and two CH_3 groups—that occurs in isobutane has come to be called "iso."

Thus, the name isobutane tells you that the compound is a four-carbon alkane with an iso structural unit.

$$CH_3CH_2CH_2CH_3$$
butane

$$CH_3\overset{\displaystyle|}{\underset{\displaystyle CH_3}{CH}}CH_3$$
isobutane

$$CH_3\overset{\displaystyle|}{\underset{\displaystyle CH_3}{CH}}-$$
an "iso" structural unit

There are three alkanes with molecular formula C_5H_{12}. We are able to name two of them. Pentane is the straight-chain alkane. Isopentane, as its name indicates, has an iso structural unit and five carbon atoms. We cannot name the other branched-chain alkane without defining a name for a new structural unit. (For now, ignore the names written in blue.)

$$CH_3CH_2CH_2CH_2CH_3$$
pentane

$$CH_3\overset{\displaystyle|}{\underset{\displaystyle CH_3}{CH}}CH_2CH_3$$
isopentane

$$CH_3\overset{\displaystyle CH_3}{\overset{\displaystyle|}{\underset{\displaystyle CH_3}{\underset{\displaystyle|}{C}}}}CH_3$$
2,2-dimethylpropane

There are five constitutional isomers with molecular formula C_6H_{14}. Again, we are able to name only two of them, unless we define new structural units.

common name: $CH_3CH_2CH_2CH_2CH_2CH_3$ $CH_3\overset{\displaystyle|}{\underset{\displaystyle CH_3}{CH}}CH_2CH_2CH_3$ $CH_3\overset{\displaystyle CH_3}{\overset{|}{\underset{\displaystyle CH_3}{\underset{|}{C}}}}CH_2CH_3$
systematic name: hexane   hexane

isohexane
2-methylpentane

2,2-dimethylbutane

$$CH_3CH_2\overset{\displaystyle|}{\underset{\displaystyle CH_3}{CH}}CH_2CH_3$$
3-methylpentane

$$CH_3\overset{\displaystyle|}{\underset{\displaystyle CH_3}{CH}}-\overset{\displaystyle|}{\underset{\displaystyle CH_3}{CH}}CH_3$$
2,3-dimethylbutane

The number of constitutional isomers increases rapidly as the number of carbons in an alkane increases. For example, there are 75 alkanes with molecular formula $C_{10}H_{22}$ and 4347 alkanes with molecular formula $C_{15}H_{32}$. To avoid having to memorize the names of thousands of structural units, chemists have devised rules that allow compounds to be named on the basis of their structures. That way, only the rules have to be learned. Because the name is based on the structure, these rules also make it possible to deduce the structure of a compound from its name.

This method of nomenclature is called **systematic nomenclature**. It is also called **IUPAC nomenclature** because it was designed by a commission of the International Union of Pure and Applied Chemistry (abbreviated IUPAC and pronounced "eye-you-pack"). Names such as isobutane—nonsystematic names—are called **common names** and are shown in red in this text. The systematic or IUPAC names are shown in blue. Before we can understand how a systematic name for an alkane is constructed, we must learn how to name alkyl substituents.

3.1   Nomenclature of Alkyl Substituents

Removing a hydrogen from an alkane results in an **alkyl substituent** (or an alkyl group). Alkyl substituents are named by replacing the "ane" ending of the alkane with "yl." The letter "R" is used to indicate any alkyl group.

$$CH_3-$$
a methyl group

$$CH_3CH_2-$$
an ethyl group

$$CH_3CH_2CH_2-$$
a propyl group

$$CH_3CH_2CH_2CH_2-$$
a butyl group

$$CH_3CH_2CH_2CH_2CH_2-$$
a pentyl group

$$R-$$
any alkyl group

If a hydrogen of an alkane is replaced by an OH, the compound becomes an **alcohol**; if it is replaced by an NH_2, the compound becomes an **amine**; if it is replaced by a halogen, the compound becomes an **alkyl halide**, and if it is replaced by an OR, the compound becomes an ether.

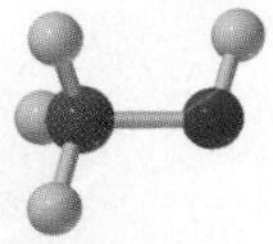

methyl alcohol

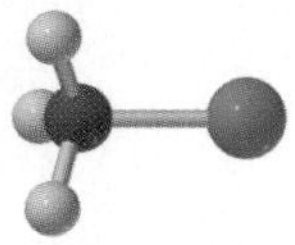

methyl chloride

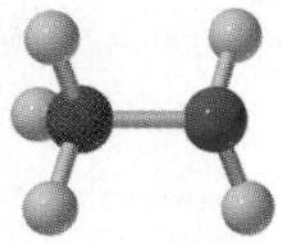

methylamine

R—OH R—NH$_2$ R—X | X = F, Cl, Br, or I | R—O—R
an alcohol an amine an alkyl halide an ether

The alkyl group name followed by the name of the class of the compound (alcohol, amine, etc.) yields the common name of the compound. The two alkyl groups in ethers are cited in alphabetical order. The following examples show how alkyl group names are used to build common names:

CH_3OH $CH_3CH_2NH_2$ $CH_3CH_2CH_2Br$ $CH_3CH_2CH_2CH_2Cl$
methyl alcohol ethylamine propyl bromide butyl chloride

CH_3I CH_3CH_2OH $CH_3CH_2CH_2NH_2$ $CH_3CH_2OCH_3$
methyl iodide ethyl alcohol propylamine ethyl methyl ether

Notice that there is a space between the name of the alkyl group and the name of the class of compound, except in the case of amines where the entire name is written as one word.

PROBLEM 1◆

Name the following compounds:

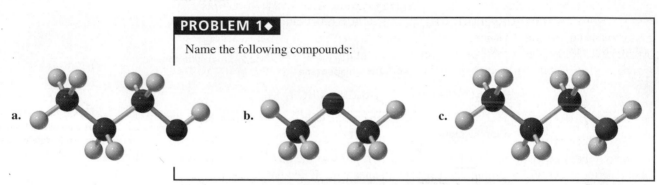

a. b. c.

Two alkyl groups—a propyl group and an isopropyl group—contain three carbon atoms. A propyl group is obtained when a hydrogen is removed from a *primary carbon* of propane. A **primary carbon** is a carbon that is bonded to only one other carbon. An isopropyl group is obtained when a hydrogen is removed from the *secondary carbon* of propane. A **secondary carbon** is a carbon that is bonded to two other carbons. Notice that an isopropyl group, as its name indicates, has its three carbon atoms arranged as an iso structural unit.

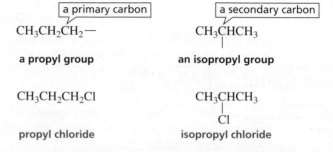

| a primary carbon | | a secondary carbon |
$CH_3CH_2CH_2-$ CH_3CHCH_3

a propyl group   **an isopropyl group**

$CH_3CH_2CH_2Cl$ CH_3CHCH_3
                      |
                      Cl

propyl chloride   **isopropyl chloride**

Molecular structures can be drawn in different ways. Isopropyl chloride, for example, is drawn here in two ways. Both represent the same compound. At first glance, the two-dimensional representations appear to be different: The methyl groups are across from one another in one structure and at right angles in the other. The structures are identical, however, because carbon is tetrahedral. The four groups bonded to the central carbon—a hydrogen, a chlorine, and two methyl groups—point to the corners of a tetrahedron. (If you visit the Molecule Gallery in Chapter 3 of the website (*www.prenhall.com/bruice*), you will be able to rotate isopropyl chloride to convince yourself that the two structures are identical.)

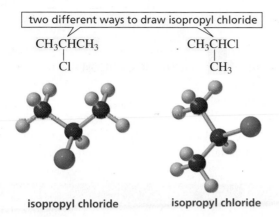

two different ways to draw isopropyl chloride

CH_3CHCH_3 CH_3CHCl
| |
Cl CH_3

isopropyl chloride   isopropyl chloride

Build models of the two representations of isopropyl chloride, and convince yourself that they represent the same compound.

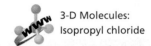
3-D Molecules:
Isopropyl chloride

There are four alkyl groups that contain four carbon atoms. The butyl and isobutyl groups have a hydrogen removed from a primary carbon. A *sec*-butyl group has a hydrogen removed from a secondary carbon (*sec*-, often abbreviated *s*-, stands for secondary), and a *tert*-butyl group has a hydrogen removed from a tertiary carbon (*tert*-, sometimes abbreviated *t*-, stands for tertiary). A **tertiary carbon** is a carbon that is bonded to three other carbons. Notice that the isobutyl group is the only group with an iso structural unit.

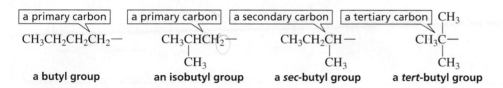

a primary carbon   a primary carbon   a secondary carbon   a tertiary carbon CH_3

$CH_3CH_2CH_2CH_2-$ CH_3CHCH_2- CH_3CH_2CH- CH_3C-
                               | | |
                               CH_3 CH_3 CH_3

a butyl group   an isobutyl group   a *sec*-butyl group   a *tert*-butyl group

A primary carbon is bonded to one carbon, a secondary carbon is bonded to two carbons, and a tertiary carbon is bonded to three carbons.

The name of a straight-chain alkyl group sometimes has the prefix "*n*" (for "normal"), to emphasize that its carbon atoms are in an unbranched chain. If the name does not have a prefix such as "*n*" or "iso," it is assumed that the carbons are in an unbranched chain.

$CH_3CH_2CH_2CH_2Br$ $CH_3CH_2CH_2CH_2CH_2F$

butyl bromide   **pentyl fluoride**
or   **or**
***n*-butyl bromide**   ***n*-pentyl fluoride**

Because a chemical name must specify only one compound, the only time you will see the prefix "*sec*" is in *sec*-butyl. The name "*sec*-pentyl" cannot be used because pentane has two different secondary carbon atoms. Therefore, there are two different alkyl groups that result from removing a hydrogen from a secondary carbon of pentane. Because the name "*sec*-pentyl chloride" would specify two different alkyl chlorides, it is not a correct name.

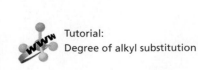
3-D Molecules:
n-Butyl alcohol; *sec*-Butyl alcohol; *tert*-Butyl alcohol

Tutorial:
Degree of alkyl substitution

Both alkyl halides have five carbon atoms with a chlorine attached to a secondary carbon, so both compounds would be named *sec*-pentyl chloride.

$CH_3CHCH_2CH_2CH_3$ $CH_3CH_2CHCH_2CH_3$
| |
Cl   Cl

If you examine the following structures, you will see that whenever the prefix "iso" is used, the iso structural unit will be at one end of the molecule and any group replacing a hydrogen will be at the other end:

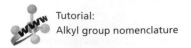

Tutorial:
Alkyl group nomenclature

$$CH_3CHCH_2CH_2OH$$
$$|$$
$$CH_3$$
isopentyl alcohol

$$CH_3CHCH_2CH_2CH_2Cl$$
$$|$$
$$CH_3$$
isohexyl chloride

$$CH_3CHCH_2NH_2$$
$$|$$
$$CH_3$$
isobutylamine

$$CH_3CHCH_2Br$$
$$|$$
$$CH_3$$
isobutyl bromide

$$CH_3CHCH_2CH_2OH$$
$$|$$
$$CH_3$$
isopentyl alcohol

$$CH_3CHBr$$
$$|$$
$$CH_3$$
isopropyl bromide

Alkyl group names are used so frequently that you should learn them. Some of the most common alkyl group names are compiled in Table 3.2 for your convenience.

Table 3.2   Names of Some Alkyl Groups

methyl	CH_3-	isobutyl	CH_3CHCH_2-
			    $|$
			    CH_3
				pentyl	$CH_3CH_2CH_2CH_2CH_2-$

ethyl	CH_3CH_2-			isopentyl	$CH_3CHCH_2CH_2-$
					    $|$
					    CH_3

propyl	$CH_3CH_2CH_2-$	*sec*-butyl	CH_3CH_2CH-
			     $|$
			     CH_3		

isopropyl	CH_3CH-
	    $|$
	    CH_3
				hexyl	$CH_3CH_2CH_2CH_2CH_2CH_2-$

				isohexyl	$CH_3CHCH_2CH_2CH_2-$
					    $|$
					    CH_3

butyl	$CH_3CH_2CH_2CH_2-$	*tert*-butyl	CH_3C-
			   $|$
			   CH_3		

PROBLEM 2

Draw the structures and name the four constitutional isomers with molecular formula C_4H_9Br.

PROBLEM 3◆

Write a structure for each of the following compounds:

a. isopropyl alcohol   **d.** *sec*-butyl iodide

b. isopentyl fluoride   **e.** *tert*-butylamine

c. ethyl propyl ether   **f.** *n*-octyl bromide

PROBLEM 4◆

Name the following compounds:

a. $CH_3OCH_2CH_3$

b. $CH_3OCH_2CH_2CH_3$

c. $CH_3CH_2CHNH_2$
          $|$
         CH_3

d. $CH_3CH_2CH_2CH_2OH$

e. CH_3CHCH_2Br
       $|$
      CH_3

f. CH_3CH_2CHCl
       $|$
      CH_3

> √ **PROBLEM 5♦**
>
> Draw the structure and give the systematic name of a compound with molecular formula C_5H_{12} that has
>
> **a.** no tertiary carbons.   **b.** no secondary or tertiary carbons.

3.2   Nomenclature of Alkanes

The systematic name of an alkane is obtained using the following rules:

1. Determine the number of carbons in the longest continuous carbon chain. This chain is called the **parent hydrocarbon**. The name that indicates the number of carbons in the parent hydrocarbon becomes the alkane's "last name." For example, a parent hydrocarbon with eight carbons would be called *octane*. The longest continuous chain is not always a straight chain; sometimes you have to "turn a corner" to obtain the longest continuous chain.

First, determine the number of carbons in the longest continuous chain.

$$\overset{8}{C}H_3\overset{7}{C}H_2\overset{6}{C}H_2\overset{5}{C}H_2\overset{4}{C}H\overset{3}{C}H_2\overset{2}{C}H_2\overset{1}{C}H_3 \qquad \overset{8}{C}H_3\overset{7}{C}H_2\overset{6}{C}H_2\overset{5}{C}H_2\overset{4}{C}HCH_2CH_3$$

| |
CH₃   CH₂CH₂CH₃
                                  3   2   1

4-methyloctane | two different alkanes with an eight-carbon parent hydrocarbon | **4-ethyloctane**

2. The name of any alkyl substituent that hangs off the parent hydrocarbon is cited before the name of the parent hydrocarbon, together with a number to designate the carbon to which the alkyl substituent is attached. The chain is numbered in the direction that gives the substituent as low a number as possible. The substituent's name and the name of the parent hydrocarbon are joined in one word, and there is a hyphen between the number and the substituent's name.

Number the chain so that the substituent gets the lowest possible number.

$$\overset{1}{C}H_3\overset{2}{C}H\overset{3}{C}H_2\overset{4}{C}H_2\overset{5}{C}H_3 \qquad \overset{6}{C}H_3\overset{5}{C}H_2\overset{4}{C}H_2\overset{3}{C}H\overset{2}{C}H_2\overset{1}{C}H_3$$

CH₃   CH₂CH₃

2-methylpentane   **3-ethylhexane**

Notice that only systematic names have numbers; common names never contain numbers.

Numbers are used only for systematic names, never for common names.

$$CH_3$$
$$|$$
$$CH_3CHCH_2CH_2CH_3$$

common name: **isohexane**
systematic name: **2-methylpentane**

3. If more than one substituent is attached to the parent hydrocarbon, the chain is numbered in the direction that will result in the lowest possible number in the name of the compound. The substituents are listed in alphabetical (not numerical) order, with each substituent getting the appropriate number. In the following example, the correct name (5-ethyl-3-methyloctane) contains a 3 as its lowest number, whereas the incorrect name (4-ethyl-6-methyloctane) contains a 4 as its lowest number:

Substituents are listed in alphabetical order.

$$CH_3CH_2CHCH_2CHCH_2CH_2CH_3$$
$$| \qquad |$$
$$CH_3 \quad CH_2CH_3$$

5-ethyl-3-methyloctane
not
4-ethyl-6-methyloctane
because 3 < 4

A number and a word are separated by a hyphen; numbers are separated by a comma.

di, tri, tetra, *sec*, and *tert* are ignored in alphabetizing.

iso and cyclo are not ignored in alphabetizing.

If two or more substituents are the same, the prefixes "di," "tri," and "tetra" are used to indicate how many identical substituents the compound has. The numbers indicating the locations of the identical substituents are listed together, separated by commas. Notice that there must be as many numbers in a name as there are substituents. The prefixes di, tri, tetra, *sec*, and *tert* are ignored in alphabetizing substituent groups, but the prefixes iso and cyclo are not ignored.

$$CH_3CH_2CHCH_2CHCH_3$$
$$\quad\quad|\quad\quad\quad|$$
$$\quad\quad CH_3\quad CH_3$$
2,4-dimethylhexane

$$\quad\quad\quad\quad CH_2CH_3$$
$$\quad\quad\quad\quad\;|$$
$$CH_3CH_2CCH_2CH_2CHCH_3$$
$$\quad\quad\quad|\quad\quad\quad\quad|$$
$$\quad\quad\quad CH_3\quad\quad CH_3$$
5-ethyl-2,5-dimethylheptane

$$\quad CH_2CH_3\quad CH_3$$
$$\quad\;|\quad\quad\quad\;|$$
$$CH_3CH_2CCH_2CH_2CHCHCH_2CH_2CH_3$$
$$\quad\quad\;|\quad\quad\quad\quad\quad|$$
$$\quad\quad CH_2CH_3\;\; CH_2CH_3$$
3,3,6-triethyl-7-methyldecane

$$\quad\quad\quad\quad\quad\quad\quad\quad CH_3$$
$$\quad\quad\quad\quad\quad\quad\quad\quad\;|$$
$$CH_3CH_2CH_2CHCH_2CH_2CHCH_3$$
$$\quad\quad\quad\quad|$$
$$\quad\quad\quad\quad CH_3CHCH_3$$
5-isopropyl-2-methyloctane

4. When both directions lead to the same lowest number for one of the substituents, the direction is chosen that gives the lowest possible number to one of the remaining substituents.

$$\quad\quad CH_3$$
$$\quad\quad\;|$$
$$CH_3CCH_2CHCH_3$$
$$\quad\;|\quad\quad|$$
$$\quad CH_3\;\; CH_3$$
2,2,4-trimethylpentane
not
2,4,4-trimethylpentane
because 2 < 4

$$\quad\quad\quad\quad CH_3\quad CH_2CH_3$$
$$\quad\quad\quad\quad\;|\quad\quad\;|$$
$$CH_3CH_2CHCHCH_2CHCH_2CH_3$$
$$\quad\quad\quad\quad\quad|$$
$$\quad\quad\quad\quad\quad CH_3$$
6-ethyl-3,4-dimethyloctane
not
3-ethyl-5,6-dimethyloctane
because 4 < 5

Only if the same set of numbers is obtained in both directions does the first group cited get the lower number.

5. If the same substituent numbers are obtained in both directions, the first group cited receives the lower number.

$$\quad\quad Cl$$
$$\quad\quad|$$
$$CH_3CHCHCH_3$$
$$\quad\quad\quad|$$
$$\quad\quad\quad Br$$
2-bromo-3-chlorobutane
not
3-bromo-2-chlorobutane

$$\quad\quad\quad CH_2CH_3$$
$$\quad\quad\quad\;|$$
$$CH_3CH_2CHCH_2CHCH_2CH_3$$
$$\quad\quad\quad\quad\quad\quad|$$
$$\quad\quad\quad\quad\quad\quad CH_3$$
3-ethyl-5-methylheptane
not
5-ethyl-3-methylheptane

In the case of two hydrocarbon chains with the same number of carbons, choose the one with the most substituents.

6. If a compound has two or more chains of the same length, the parent hydrocarbon is the chain with the greatest number of substituents.

$$\quad\quad\quad\;3\;\;\;4\;\;\;5\;\;\;6$$
$$CH_3CH_2CHCH_2CH_2CH_3$$
$$\quad\quad\quad\;|$$
$$\quad\quad\;2 CHCH_3$$
$$\quad\quad\quad\;|$$
$$\quad\quad\;1 CH_3$$
3-ethyl-2-methylhexane (two substituents)

$$\quad\;1\;\;\;2\;\;\;3\;\;\;4\;\;\;5\;\;\;6$$
$$CH_3CH_2CHCH_2CH_2CH_3$$
$$\quad\quad\quad\;|$$
$$\quad\quad\quad CHCH_3$$
$$\quad\quad\quad\;|$$
$$\quad\quad\quad CH_3$$
not
3-isopropylhexane (one substituent)

PROBLEM 6◆

Draw the structure for each of the following compounds:

a. 2,3-dimethylhexane

b. 4-isopropyl-2,4,5-trimethylheptane

c. 2,2-dimethyl-4-propyloctane

d. 4-isobutyl-2,5-dimethyloctane

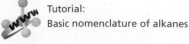

Tutorial:
Basic nomenclature of alkanes

PROBLEM 7 SOLVED

a. Draw the 18 isomeric octanes.

b. Give each isomer its systematic name.

c. Which isomers contain an isopropyl group?

d. Which isomers contain a *sec*-butyl group?

e. Which isomers contain a *tert*-butyl group?

SOLUTION TO 7a   Start with the isomer with an eight-carbon continuous chain. Then draw isomers with a seven-carbon continuous chain plus one methyl group. Next, draw isomers with a six-carbon continuous chain plus two methyl groups or one ethyl group. Then draw isomers with a five-carbon continuous chain plus three methyl groups or one methyl group and one ethyl group. Finally, draw a four-carbon continuous chain with four methyl groups. (You will be able to tell whether you have drawn duplicate structures by your answers to 7b because if two structures have the same systematic name, they are the same compound.)

PROBLEM 8◆

Give the systematic name for each of the following compounds:

$$\overset{\displaystyle CH_3 \quad CH_3}{\textbf{a.} \ CH_3CH_2CHCH_2\underset{\displaystyle CH_3}{C}CH_3}$$

$$\textbf{d.} \ \overset{\displaystyle CH_3}{CH_3CHCH_2CH_2\underset{\displaystyle CH_2CH_3}{C}HCH_3}$$

b. $CH_3CH_2C(CH_3)_3$

e. $CH_3CH_2CH_2CH_2\underset{\displaystyle CH(CH_3)_2}{C}HCH_2CH_2CH_3$

c. $CH_3CH_2C(CH_2CH_3)_2CH_2CH_2CH_3$

f. $CH_3C(CH_3)_2CH(CH_3)CH(CH_2CH_3)_2$

3.3 Nomenclature of Cycloalkanes

Cycloalkanes are alkanes with their carbon atoms arranged in a ring. Because of the ring, a cycloalkane has two fewer hydrogens than a noncyclic alkane with the same number of carbons. This means that the general molecular formula for a cycloalkane is C_nH_{2n}. Cycloalkanes are named by adding the prefix "cyclo" to the alkane name that signifies the number of carbon atoms in the ring.

cyclopropane   cyclobutane   cyclopentane   cyclohexane

Cycloalkanes are almost always written as **skeletal structures**. Skeletal structures show the carbon–carbon bonds as lines, but do not show the carbons or the hydrogens bonded to carbons. Atoms other than carbon are shown, and hydrogens bonded to atoms other than carbon are shown. Each vertex in a skeletal structure represents a carbon. It is understood that each carbon is bonded to the appropriate number of hydrogens to give the carbon four bonds.

cyclopropane   cyclobutane   cyclopentane   cyclohexane

Noncyclic molecules can also be represented by skeletal structures. In a skeletal structure of an noncyclic molecule, the carbon chains are represented by zigzag lines. Again, each vertex represents a carbon, and carbons are assumed to be present where a line begins or ends.

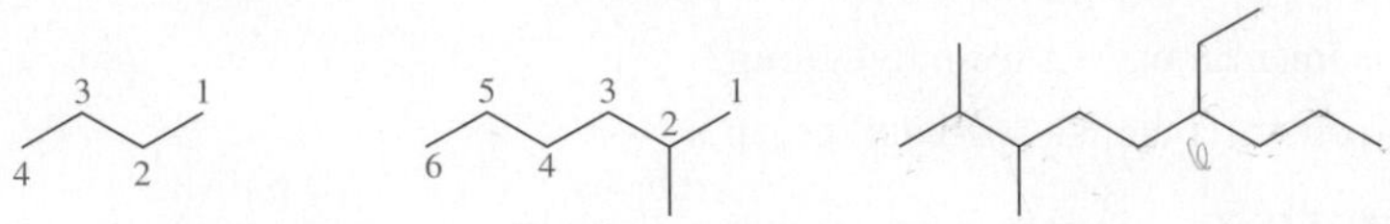

butane   2-methylhexane   6-ethyl-2,3-dimethylnonane

The rules for naming cycloalkanes resemble the rules for naming noncyclic alkanes:

If there is only one substituent on a ring, do not give that substituent a number.

1. In the case of a cycloalkane with an attached alkyl substituent, the ring is the parent hydrocarbon. There is no need to number the position of a single substituent on a ring.

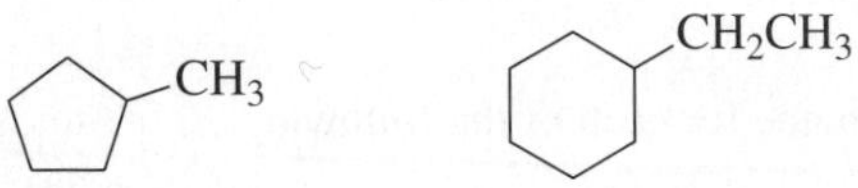

methylcyclopentane   ethylcyclohexane

Tutorial:
Advanced alkane
nomenclature

2. If the ring has two different substituents, they are cited in *alphabetical order* and the number 1 position is given to the substituent cited first.

1-methyl-2-propylcyclopentane   1,3-dimethylcyclohexane

PROBLEM-SOLVING STRATEGY

Indicate how many hydrogens are attached to each of the indicated carbon atoms in the following compound:

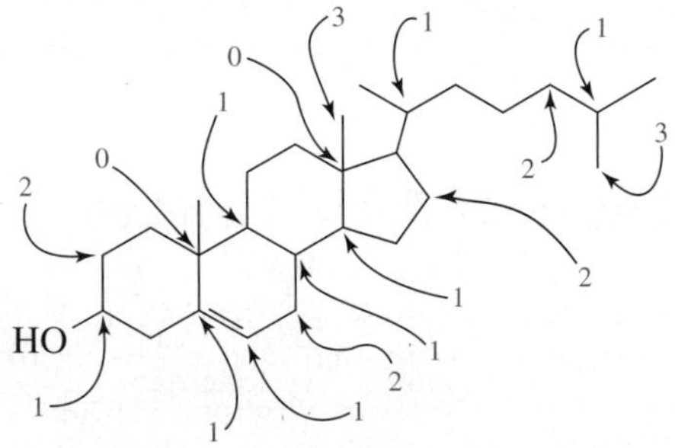

cholesterol

All the carbon atoms in the compound are neutral, so each needs to be bonded to four atoms. Thus, if the carbon has only one bond that is shown, it must be attached to three hydrogens that are not shown; if the carbon has two bonds that are shown, it must be attached to two hydrogens that are not shown, etc.

Now continue on to Problem 9.

PROBLEM 9

Indicate how many hydrogens are attached to each of the indicated carbon atoms in the following compound:

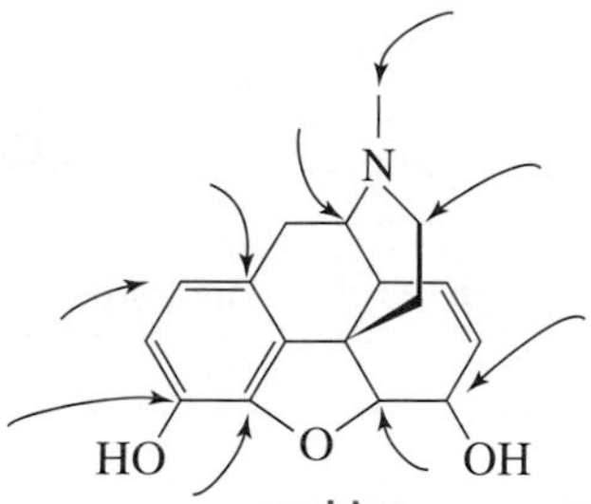

morphine

PROBLEM 10◆

Convert the following condensed structures into skeletal structures:

a. $CH_3CH_2CH_2CH_2CH_2CH_2OH$

$$\qquad CH_3$$
c. $CH_3CHCH_2CH_2CHCH_3$
$$\qquad\qquad\qquad Br$$

$$\quad CH_3 \quad CH_3$$
b. $CH_3CH_2CHCH_2CHCII_2CH_3$

d. $CH_3CH_2CH_2CH_2OCH_3$

PROBLEM 11◆

Give the systematic name for each of the following compounds:

a. 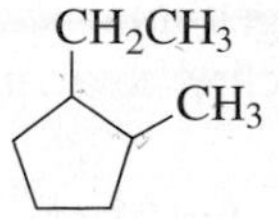
CH_2CH_3
CH_3

c.

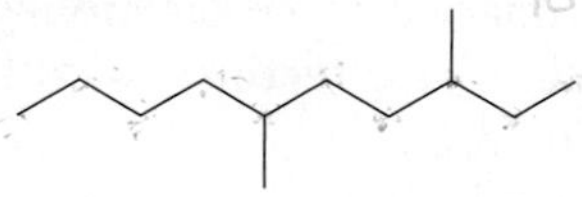

b. 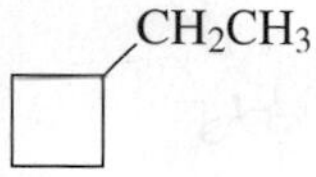
CH_2CH_3

d.

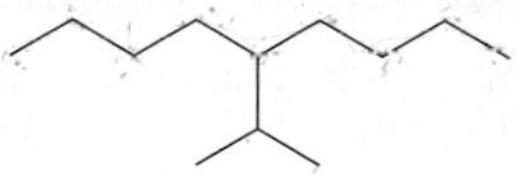

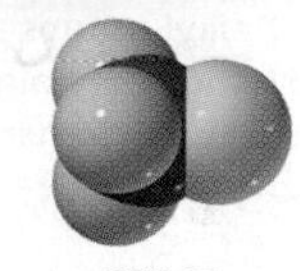

3.4 Nomenclature of Alkyl Halides

An alkyl halide is a compound in which a hydrogen of an alkane has been replaced by a halogen. The lone-pair electrons on the halogen are generally not shown unless they are needed to draw your attention to some chemical property of the atom. The common name of an alkyl halide consists of the name of the alkyl group, followed by the name of the halogen—with the "ine" ending of the halogen name replaced by "ide" (i.e., fluoride, chloride, bromide, iodide).

	CH_3Cl	CH_3CH_2F	CH_3CHI $\quad CH_3$	CH_3CH_2CHBr $\quad CH_3$
common name:	methyl chloride	ethyl fluoride	isopropyl iodide	sec-butyl bromide
systematic name:	chloromethane	fluoroethane	2-iodopropane	2-bromobutane

CH_3F
methyl fluoride

CH_3Cl
methyl chloride

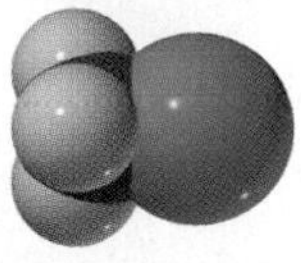

CH_3Br
methyl bromide

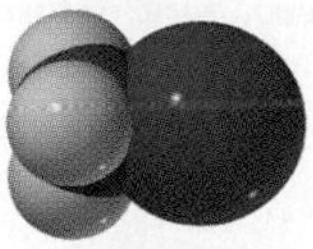

CH_3I
methyl iodide

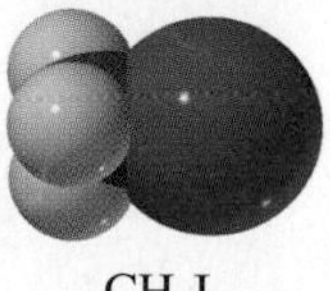

A compound can have more than one name, but a name must specify only one compound.

In the IUPAC system, alkyl halides are named as substituted alkanes. The substituent prefix names for the halogens end with "o" (i.e., "fluoro," "chloro," "bromo," "iodo"). Notice that each of the four alkyl halides just shown has two names: A compound can have more than one name, but a name must specify only one compound.

$$CH_3CH_2CHCH_2CH_2CHCH_3$$
$$Br$$
2-bromo-5-methylheptane

$$CH_3CCH_2CH_2CH_2CH_2Cl$$
$$CH_3$$
1-chloro-5,5-dimethylhexane

1-ethyl-2-iodocyclopentane

PROBLEM 12◆

Give two names for each of the following compounds:

a. $CH_3CH_2CHCH_3$
        Cl

b. $CH_3CHCH_2CH_2CH_2Cl$
    CH_3

c. Br

d. CH_3CHCH_3
       F

3.5 Classification of Alkyl Halides, Alcohols, and Amines

The number of alkyl groups attached to the carbon to which the halogen is bonded determines whether an alkyl halide is primary, secondary, or tertiary.

Alkyl halides are classified as *primary*, *secondary*, or *tertiary*, depending on the carbon to which the halogen is attached. **Primary alkyl halides** have the halogen attached to a primary carbon, **secondary alkyl halides** have the halogen attached to a secondary carbon, and **tertiary alkyl halides** have the halogen attached to a tertiary carbon (Section 3.1).

a primary carbon	a secondary carbon	a tertiary carbon
$R-CH_2-Br$	$R-CH-R$ $\quad\;\; Br$	$R-C-R$ (R, Br)
a primary alkyl halide	a secondary alkyl halide	a tertiary alkyl halide

The number of alkyl groups attached to the carbon to which the OH group is attached determines whether an alcohol is primary, secondary, or tertiary.

Alcohols are classified in the same way.

$R-CH_2-OH$ $R-CH-OH$ (R) $R-C-OH$ (R, R)

a primary alcohol   a secondary alcohol   a tertiary alcohol

The number of alkyl groups attached to the nitrogen determines whether an amine is primary, secondary, or tertiary.

There are also *primary*, *secondary*, or *tertiary amines*; but in the case of amines, the terms have different meanings. The classification refers to how many alkyl groups are bonded to the nitrogen. **Primary amines** have one alkyl group bonded to the nitrogen, **secondary amines** have two, and **tertiary amines** have three alkyl groups bonded to the nitrogen. The common name of an amine consists of the names of all the alkyl groups bonded to the nitrogen, in alphabetical order, followed by "amine."

$R-NH_2$ $R-NH$ (R) $R-N-R$ (R) $CH_3NCH_2CH_2CH_3$ (CH_2CH_3)

a primary amine   a secondary amine   a tertiary amine   ethylmethylpropylamine
a tertiary amine

BAD-SMELLING COMPOUNDS

Amines are characterized by their unpleasant odors. Amines with relatively small alkyl groups have a fishy smell. Fermented shark, for example, a traditional dish in Iceland, smells exactly like triethylamine. Putrecine and cadavarine are amines that are formed when amino acids are degraded. Because they are poisonous compounds, the body excretes them in the quickest way possible. Their odors are detected in urine and in bad breath. These compounds are also responsible for the odor of decaying flesh.

putrecine   cadavarine

PROBLEM 13◆

Tell whether the following compounds are primary, secondary, or tertiary:

         CH₃   CH₃   CH₃

a. $CH_3-\overset{\overset{\displaystyle CH_3}{|}}{\underset{\underset{\displaystyle CH_3}{|}}{C}}-Br$ **b.** $CH_3-\overset{\overset{\displaystyle CH_3}{|}}{\underset{\underset{\displaystyle CH_3}{|}}{C}}-OH$ **c.** $CH_3-\overset{\overset{\displaystyle CH_3}{|}}{\underset{\underset{\displaystyle CH_3}{|}}{C}}-NH_2$

PROBLEM 14◆

Name the following amines and tell whether they are primary, secondary, or tertiary:

a. $CH_3NHCH_2CH_2CH_3$ **c.** $CH_3CH_2NHCH_2CH_3$

         CH₃   CH₃

b. $CH_3\overset{\overset{\displaystyle CH_3}{|}}{N}CH_3$ **d.** $CH_3\overset{\overset{\displaystyle CH_3}{|}}{N}CH_2CH_2CH_2CH_3$

PROBLEM 15◆

Draw the structures for a–c by substituting a chlorine for a hydrogen of methylcyclohexane:

a. a primary alkyl halide   **c.** three secondary alkyl halides

b. a tertiary alkyl halide

3.6   Structures of Alkyl Halides, Alcohols, Ethers, and Amines

The C—X bond (where X denotes a halogen) of an alkyl halide is formed from the overlap of an sp^3 orbital of carbon with a p orbital of the halogen (Section 1.13). Fluorine uses a $2p$ orbital, chlorine a $3p$ orbital, bromine a $4p$ orbital, and iodine a $5p$ orbital. Because the electron density of the orbital decreases with increasing volume, the C—X bond becomes longer and weaker as the size of the halogen increases (Table 3.3). Notice that this is the same trend shown by the H—X bond (Table 1.5, page 26).

The oxygen of an alcohol has the same geometry it has in water (Section 1.11). In fact, an alcohol molecule can be thought of as a water molecule with an alkyl group in place of one of the hydrogens. The oxygen atom in an alcohol is sp^3 hybridized, as it is in water. One of the sp^3 orbitals of oxygen overlaps an sp^3 orbital of a carbon, one sp^3 orbital overlaps the s orbital of a hydrogen, and the other two sp^3 orbitals each contain a lone pair.

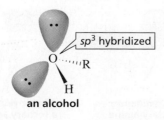

an alcohol

electrostatic potential
map for methyl alcohol

Table 3.3   Carbon–Halogen Bond Lengths and Bond Strengths

	Orbital interactions	Bond lengths	Bond strength kcal/mol	kJ/mol
H_3C—F	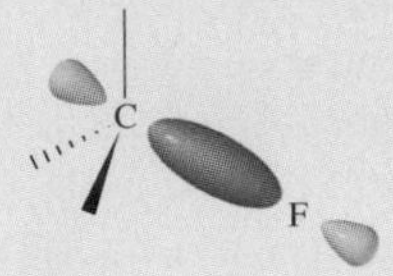	1.39 Å	108	451
H_3C—Cl	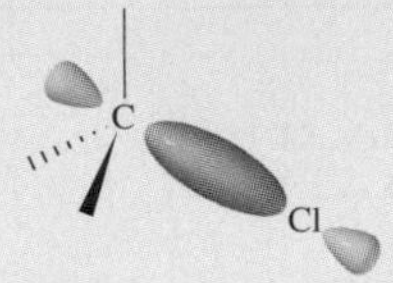	1.78 Å	84	350
H_3C—Br	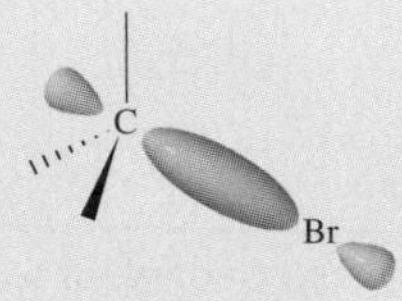	1.93 Å	70	294
H_3C—I	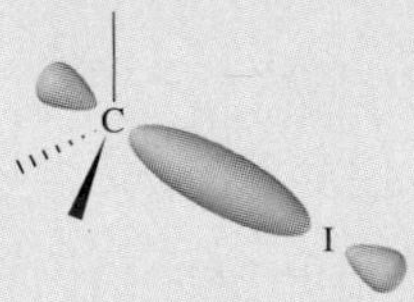	2.14 Å	57	239

The oxygen of an ether also has the same geometry it has in water. An ether molecule can be thought of as a water molecule with alkyl groups in place of both hydrogens.

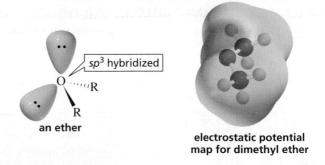

sp^3 hybridized

an ether

electrostatic potential map for dimethyl ether

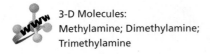

3-D Molecules:
Methylamine; Dimethylamine;
Trimethylamine

The nitrogen of an amine has the same geometry it has in ammonia (Section 1.12). One, two, or three hydrogens may be replaced by alkyl groups. Remember that the number of hydrogens replaced by alkyl groups determines whether the amine is primary, secondary, or tertiary (Section 3.5).

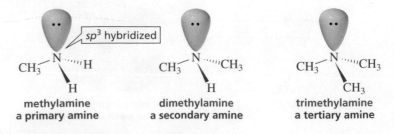

sp^3 hybridized

methylamine
a primary amine

dimethylamine
a secondary amine

trimethylamine
a tertiary amine

 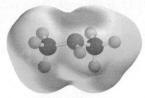 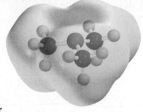

electrostatic potential maps for

methylamine   dimethylamine   trimethylamine

PROBLEM 16◆

Predict the approximate size of the following bond angles. (*Hint*: See Sections 1.11 and 1.12.)

a. the C—O—C bond angle in an ether
b. the C—N—C bond angle in a secondary amine
c. the C—O—H bond angle in an alcohol

3.7 Physical Properties of Alkanes, Alkyl Halides, Alcohols, Ethers, and Amines

Boiling Points

The **boiling point (bp)** of a compound is the temperature at which the liquid form of the compound becomes a gas (vaporizes). In order for a compound to vaporize, the forces that hold the individual molecules close to each other in the liquid must be overcome. This means that the boiling point of a compound depends on the strength of the attractive forces between the individual molecules. If the molecules are held together by strong forces, it will take a lot of energy to pull the molecules away from each other and the compound will have a high boiling point. In contrast, if the molecules are held together by weak forces, only a small amount of energy will be needed to pull the molecules away from each other and the compound will have a low boiling point.

Relatively weak forces hold alkane molecules together. Alkanes contain only carbon and hydrogen atoms. Because the electronegativities of carbon and hydrogen are similar, the bonds in alkanes are nonpolar. Consequently, there are no significant partial charges on any of the atoms in an alkane—alkanes are neutral molecules.

It is, however, only the average charge distribution over the alkane molecule that is neutral. Electrons are moving continuously, so at any instant the electron density on one side of the molecule can be slightly greater than that on the other side, causing the molecule to have a temporary dipole. A molecule with a dipole has a negative end and a positive end.

A temporary dipole in one molecule can induce a temporary dipole in a nearby molecule, as shown in Figure 3.1. Because the dipoles in the molecules are induced, the interactions between the molecules are called **induced-dipole–induced-dipole interactions**.

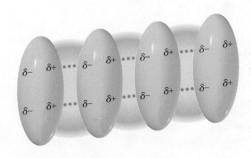

◀ **Figure 3.1**
Van der Waals forces are induced-dipole–induced-dipole interactions.

The molecules of an alkane are held together by these induced-dipole–induced-dipole interactions, which are known as **van der Waals forces**. Van der Waals forces are the weakest of all the intermolecular attractions.

In order for an alkane to boil, the van der Waals forces must be overcome. The magnitude of the van der Waals forces that hold alkane molecules together depends on the area of contact between the molecules. The greater the area of contact, the stronger are the van der Waals forces and the greater is the amount of energy needed to overcome those forces. If you look at the alkanes in Table 3.1, you will see that their boiling points increase as their size increases. This relationship holds because each additional methylene (CH_2) group increases the area of contact between the molecules. The four smallest alkanes have boiling points below room temperature (room temperature is about 25 °C), so they exist as gases at room temperature.

Because the strength of the van der Waals forces depends on the area of contact between the molecules, branching in a compound lowers its boiling point because it reduces the area of contact. If you think of unbranched pentane as a cigar and its most highly branched isomer as a tennis ball, you can see that branching decreases the area of contact between molecules: Two cigars make contact over a greater area than do two tennis balls. Thus, if two alkanes have the same molecular weight, the more highly branched alkane will have a lower boiling point.

Johannes Diderik van der Waals (1837–1923) *was a Dutch physicist. He was born in Leiden, the son of a carpenter, and was largely self-taught when he entered the University of Leiden, where he earned a Ph.D. He was a professor of physics at the University of Amsterdam from 1877 to 1903. He won the 1910 Nobel Prize for his research on the gaseous and liquid states of matter.*

$$CH_3CH_2CH_2CH_2CH_3$$

pentane
bp = 36.1 °C

$$CH_3CHCH_2CH_3$$
$$|$$
$$CH_3$$

2-methylbutane
bp = 27.9 °C

$$CH_3$$
$$|$$
$$CH_3CCH_3$$
$$|$$
$$CH_3$$

2,2-dimethylpropane
bp = 9.5 °C

PROBLEM 17◆

What is the smallest alkane that is a liquid at room temperature?

The boiling points of a series of ethers, alkyl halides, alcohols, or amines also increase with increasing molecular weight because of the increase in van der Waals forces. (See Appendix I.) The boiling points of these compounds, however, are also affected by the polar C—Z bond (where Z denotes N, O, F, Cl, or Br). The C—Z bond is polar because nitrogen, oxygen, and the halogens are more electronegative than the carbon to which they are attached.

$$R-\overset{|}{\underset{|}{C}}-\overset{\delta+ \ \delta-}{Z} \qquad Z = N, O, F, Cl, or Br$$

Molecules with polar bonds are attracted to one another because they can align themselves in such a way that the positive end of one molecule is adjacent to the negative end of another. These attractive forces, called **dipole–dipole interactions**, are stronger than van der Waals forces, but not as strong as ionic or covalent bonds.

More extensive tables of physical properties can be found in Appendix I.

Ethers generally have higher boiling points than alkanes of comparable molecular weight because both van der Waals forces and dipole–dipole interactions must be overcome for an ether to boil (Table 3.4).

cyclopentane
bp = 49.3 °C

tetrahydrofuran
bp = 65 °C

Table 3.4   Comparative Boiling Points (°C)

Alkanes	Ethers	Alcohols	Amines
$CH_3CH_2CH_3$	CH_3OCH_3	CH_3CH_2OH	$CH_3CH_2NH_2$
−42.1	−23.7	78	16.6
$CH_3CH_2CH_2CH_3$	$CH_3OCH_2CH_3$	$CH_3CH_2CH_2OH$	$CH_3CH_2CH_2NH_2$
−0.5	10.8	97.4	47.8
$CH_3CH_2CH_2CH_2CH_3$	$CH_3CH_2OCH_2CH_3$	$CH_3CH_2CH_2CH_2OH$	$CH_3CH_2CH_2CH_2NH_2$
36.1	34.5	117.3	77.8

As Table 3.4 shows, alcohols have much higher boiling points than alkanes or ethers of comparable molecular weight because, in addition to van der Waals forces and the dipole–dipole interactions of the C—O bond, alcohols can form **hydrogen bonds** and the hydrogen bonds have to be broken as well. A hydrogen bond is a special kind of dipole–dipole interaction that occurs between a hydrogen that is bonded to an oxygen, a nitrogen, or a fluorine and the lone-pair electrons of an oxygen, nitrogen, or fluorine in another molecule.

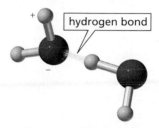

hydrogen bonding in water

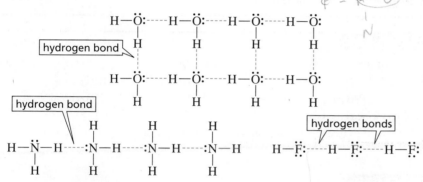

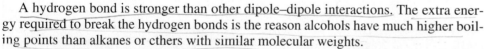

A hydrogen bond is stronger than other dipole–dipole interactions. The extra energy required to break the hydrogen bonds is the reason alcohols have much higher boiling points than alkanes or ethers with similar molecular weights.

The boiling point of water illustrates the dramatic effect that hydrogen bonding has on boiling points. Water has a molecular weight of 18 and a boiling point of 100 °C. The alkane nearest in size is methane, with a molecular weight of 16. Methane boils at −167.7 °C.

Primary and secondary amines also form hydrogen bonds, so these amines have higher boiling points than alkanes with similar molecular weights. Nitrogen is not as electronegative as oxygen, however, which means that the hydrogen bonds between amine molecules are weaker than the hydrogen bonds between alcohol molecules. An amine, therefore, has a lower boiling point than an alcohol with a similar molecular weight (Table 3.4).

Because primary amines have two N—H bonds, hydrogen bonding is more significant in primary amines than in secondary amines. Tertiary amines cannot form hydrogen bonds between their own molecules because they do not have a hydrogen attached to the nitrogen. Consequently, if you compare amines with the same molecular weight and similar structures, you will find that a primary amine has a higher boiling point than a secondary amine and a secondary amine has a higher boiling point than a tertiary amine.

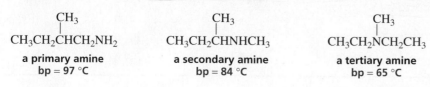

PROBLEM 18◆

a. Which is longer, an O—H hydrogen bond or an O—H covalent bond?
b. Which is stronger?

PROBLEM-SOLVING STRATEGY

a. Which of the following compounds will form hydrogen bonds between its molecules?
1. $CH_3CH_2CH_2OH$ **2.** $CH_3CH_2CH_2F$ **3.** $CH_3OCH_2CH_3$
b. Which of these compounds will form hydrogen bonds with a solvent such as ethanol?

In solving this type of question, start by defining the kind of compound that will do what is being asked.

a. A hydrogen bond forms when a hydrogen that is attached to an O, N, or F of one molecule interacts with a lone pair on an O, N, or F of another molecule. Therefore, a compound that will form hydrogen bonds with itself must have a hydrogen bonded to an O, N, or F. Only compound 1 will be able to form hydrogen bonds with itself.
b. A solvent such as ethanol has a hydrogen attached to an O, so it will be able to form hydrogen bonds with a compound that has a lone pair on an O, N, or F. All three compounds will be able to form hydrogen bonds with ethanol.

Now continue on to Problem 19.

PROBLEM 19◆

a. Which of the following compounds will form hydrogen bonds between its molecules?
1. $CH_3CH_2CH_2COOH$
2. $CH_3CH_2N(CH_3)_2$
3. $CH_3CH_2CH_2CH_2Br$
4. $CH_3CH_2CH_2NHCH_3$
5. $CH_3CH_2OCH_2CH_2OH$
6. $CH_3CH_2CH_2CH_2F$
b. Which of the preceding compounds will form hydrogen bonds with a solvent such as ethanol?

PROBLEM 20◆

List the following compounds in order of decreasing boiling point:

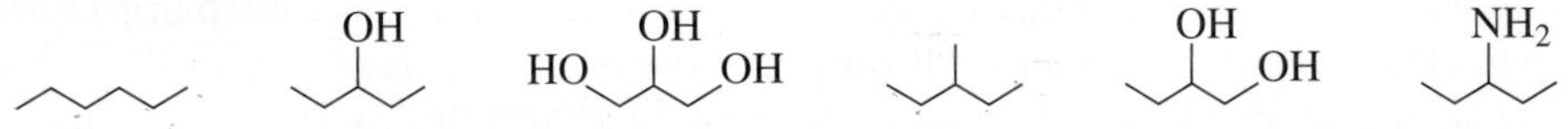

Both van der Waals forces and dipole–dipole interactions must be overcome in order for an alkyl halide to boil. As the halogen atom increases in size, the size of its electron cloud increases, and the larger the electron cloud, the stronger are the van der Waals interactions. Therefore, an alkyl fluoride has a lower boiling point than an alkyl chloride with the same alkyl group. Similarly, alkyl chlorides have lower boiling points than alkyl bromides, which have lower boiling points than alkyl iodides (Table 3.5).

Table 3.5 Comparative Boiling Points of Alkanes and Alkyl Halides (°C)

	Y				
	H	F	Cl	Br	I
$CH_3—Y$	−161.7	−78.4	−24.2	3.6	42.4
$CH_3CH_2—Y$	−88.6	−37.7	12.3	38.4	72.3
$CH_3CH_2CH_2—Y$	−42.1	−2.5	46.6	71.0	102.5
$CH_3CH_2CH_2CH_2—Y$	−0.5	32.5	78.4	101.6	130.5
$CH_3CH_2CH_2CH_2CH_2—Y$	36.1	62.8	107.8	129.6	157.0

PROBLEM 21◆

List the following compounds in order of decreasing boiling point:

a. $CH_3CH_2CH_2CH_2CH_2CH_2Br$ $CH_3CH_2CH_2CH_2Br$ $CH_3CH_2CH_2CH_2CH_2Br$

b. $CH_3CHCH_2CH_2CH_2CH_2CH_3$ $\underset{\underset{CH_3}{|}}{CH_3}C\overset{\overset{CH_3}{|}}{—}\overset{\overset{CH_3}{|}}{\underset{\underset{CH_3}{|}}{C}}CH_3$
     |
     CH_3

   $CH_3CH_2CH_2CH_2CH_2CH_2CH_2CH_3$ $CH_3CH_2CH_2CH_2CH_2CH_2CH_2CH_2CH_3$

c. $CH_3CH_2CH_2CH_2CH_3$ $CH_3CH_2CH_2CH_2OH$ $CH_3CH_2CH_2CH_2Cl$
   $CH_3CH_2CH_2CH_2CH_2OH$

Melting Points

The **melting point (mp)** is the temperature at which a solid is converted into a liquid. If you examine the melting points of the alkanes listed in Table 3.1, you will see that the melting points increase (with a few exceptions) as the molecular weight increases (Figure 3.2). The increase in melting point is less regular than the increase in boiling point because *packing* influences the melting point of a compound. **Packing** is a property that determines how well the individual molecules in a solid fit together in a crystal lattice. The tighter the fit, the more energy is required to break the lattice and melt the compound.

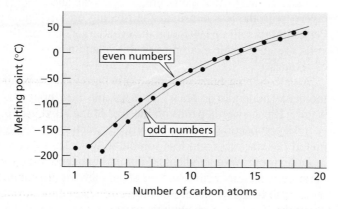

◀ **Figure 3.2**
Melting points of straight-chain alkanes.

PROBLEM 22◆

Use Figure 3.2 to answer this question. Which pack more tightly, the molecules of an alkane with an even number of carbons or with an odd number of carbons?

Solubility

The general rule that governs **solubility** is "like dissolves like." In other words, *polar compounds dissolve in polar solvents, and nonpolar compounds dissolve in nonpolar solvents.* This is because a polar solvent such as water has partial charges that can interact with the partial charges on a polar compound. The negative poles of the solvent molecules surround the positive pole of the polar *solute*, and the positive poles of the solvent molecules surround the negative pole of the polar *solute*. A **solute** is a molecule or an ion dissolved in a solvent. Clustering of the solvent molecules around the

"Like" dissolves "like."

solute molecules separates solute molecules from each other, which is what makes them dissolve. The interaction between solvent molecules and solute molecules is called **solvation**.

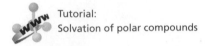

Tutorial:
Solvation of polar compounds

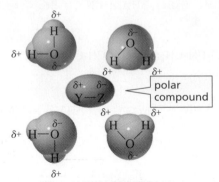

solvation of a polar compound by water

Because nonpolar compounds have no net charge, polar solvents are not attracted to them. In order for a nonpolar molecule to dissolve in a polar solvent such as water, the nonpolar molecule would have to push the water molecules apart, disrupting their hydrogen bonding. Hydrogen bonding is strong enough to exclude the nonpolar compound. In contrast, nonpolar solutes dissolve in nonpolar solvents because the van der Waals interactions between solvent and solute molecules are about the same as between solvent–solvent and solute–solute molecules.

Alkanes are nonpolar, which causes them to be soluble in nonpolar solvents and insoluble in polar solvents such as water. The densities of alkanes (Table 3.1) increase with increasing molecular weight, but even a 30-carbon alkane is less dense than water. This means that a mixture of an alkane and water will separate into two distinct layers, with the less dense alkane floating on top. The Alaskan oil spill of 1989, the Persian Gulf spill of 1991, and the even larger spill off the northwest coast of Spain in 2002 are large-scale examples of this phenomenon. (Crude oil is primarily a mixture of alkanes.)

An alcohol has both a nonpolar alkyl group and a polar OH group. So is an alcohol molecule nonpolar or polar? Is it soluble in a nonpolar solvent, or is it soluble in water? The answer depends on the size of the alkyl group. As the alkyl group increases in size, it becomes a more significant fraction of the alcohol molecule and the compound becomes less and less soluble in water. In other words, the molecule becomes more and more like an alkane. Four carbons tend to be the dividing line at room temperature. Alcohols with fewer than four carbons are soluble in water, but alcohols with more than four carbons are insoluble in water. In other words, an OH group can drag about three or four carbons into water.

The four-carbon dividing line is only an approximate guide because the solubility of an alcohol also depends on the structure of the alkyl group. Alcohols with branched alkyl groups are more soluble in water than alcohols with nonbranched alkyl groups with the same number of carbons, because branching minimizes the contact surface of the nonpolar portion of the molecule. So *tert*-butyl alcohol is more soluble than *n*-butyl alcohol in water.

Similarly, the oxygen atom of an ether can drag only about three carbons into water (Table 3.6). We have already seen (photo on page 44) that diethyl ether—an ether with four carbons—is not soluble in water.

Low-molecular-weight amines are soluble in water because amines can form hydrogen bonds with water. Comparing amines with the same number of carbons, we find that primary amines are more soluble than secondary amines because primary amines have two hydrogens that can engage in hydrogen bonding. Tertiary amines, like primary and secondary amines, have lone-pair electrons that can accept hydrogen bonds, but unlike primary and secondary amines, tertiary amines do not have hydrogens

Oil from a 70,000-ton oil spill in 1996 off the coast of Wales.

Table 3.6   Solubilities of Ethers in Water

2 C's	CH_3OCH_3	soluble
3 C's	$CH_3OCH_2CH_3$	soluble
4 C's	$CH_3CH_2OCH_2CH_3$	slightly soluble (10 g/100 g H_2O)
5 C's	$CH_3CH_2OCH_2CH_2CH_3$	minimally soluble (1.0 g/100 g H_2O)
6 C's	$CH_3CH_2CH_2OCH_2CH_2CH_3$	insoluble (0.25 g/100 g H_2O)

to donate for hydrogen bonds. Tertiary amines, therefore, are less soluble in water than are secondary amines with the same number of carbons.

Alkyl halides have some polar character, but only alkyl fluorides have an atom that can form a hydrogen bond with water. This means that alkyl fluorides are the most water soluble of the alkyl halides. The other alkyl halides are less soluble in water than ethers or alcohols with the same number of carbons (Table 3.7).

Table 3.7   Solubilities of Alkyl Halides in Water

CH_3F	CH_3Cl	CH_3Br	CH_3I
very soluble	soluble	slightly soluble	slightly soluble
CH_3CH_2F	CH_3CH_2Cl	CH_3CH_2Br	CH_3CH_2I
soluble	slightly soluble	slightly soluble	slightly soluble
$CH_3CH_2CH_2F$	$CH_3CH_2CH_2Cl$	$CH_3CH_2CH_2Br$	$CH_3CH_2CH_2I$
slightly soluble	slightly soluble	slightly soluble	slightly soluble
$CH_3CH_2CH_2CH_2F$	$CH_3CH_2CH_2CH_2Cl$	$CH_3CH_2CH_2CH_2Br$	$CH_3CH_2CH_2CH_2I$
insoluble	insoluble	insoluble	insoluble

PROBLEM 23◆

Rank the following groups of compounds in order of decreasing solubility in water:

a. $CH_3CH_2CH_2OH$ $CH_3CH_2CH_2CH_2Cl$ $CH_3CH_2CH_2CH_2OH$

   $HOCH_2CH_2CH_2OH$

b.

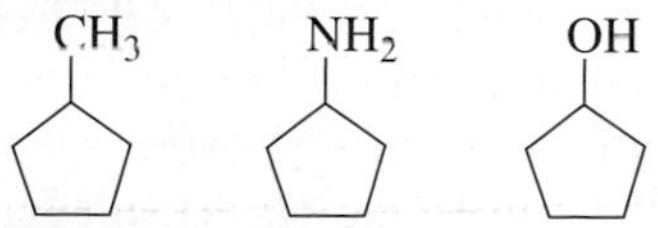

PROBLEM 24◆

In which of the following solvents would cyclohexane have the lowest solubility: 1-pentanol, diethyl ether, ethanol, or hexane?

PROBLEM 25◆

The effectiveness of a barbiturate as a sedative is related to its ability to penetrate the non-polar membrane of a cell. Which of the following barbiturates would you expect to be the more effective sedative?

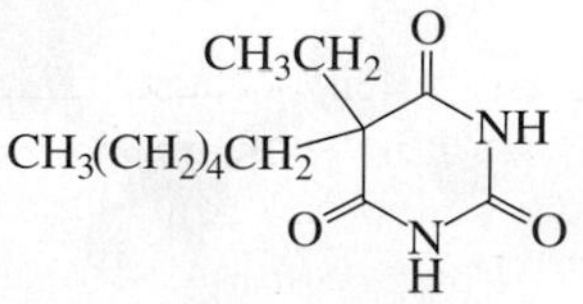

hexethal 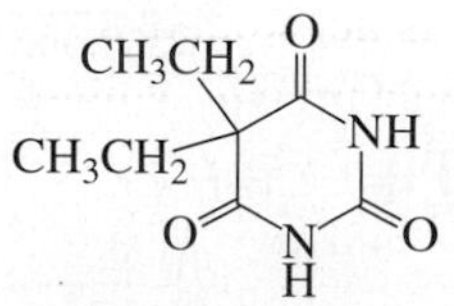   barbital

3.8 Conformations of Alkanes: Rotation About Carbon–Carbon Bonds

We have seen that a carbon–carbon single bond (a σ bond) is formed when an sp^3 orbital of one carbon overlaps an sp^3 orbital of a second carbon (Section 1.7). Figure 3.3 shows that rotation about a carbon–carbon single bond can occur without any change in the amount of orbital overlap. The different spatial arrangements of the atoms that result from rotation about a single bond are called **conformations**. A specific conformation is called a **conformer**.

When rotation occurs about the carbon–carbon bond of ethane, two extreme conformations can result—a *staggered conformation* and an *eclipsed conformation*. An infinite number of conformations between these two extremes are also possible.

Compounds are three dimensional, but we are limited to a two-dimensional sheet of paper when we show their structures. Chemists commonly use *Newman projections* to represent on paper the three-dimensional spatial arrangements of the atoms that result from rotation about a σ bond. In a **Newman projection**, you are looking down the length of a particular C—C bond. The carbon in front is represented by the point at which three bonds intersect, and the carbon in back is represented by a circle. The three lines emanating from each of the carbons represent its other three bonds.

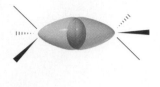

▲ **Figure 3.3**
Rotation about a carbon–carbon bond can occur without changing the amount of orbital overlap.

Melvin S. Newman (1908–1993)
was born in New York. He received a Ph.D. from Yale University in 1932 and was a professor of chemistry at Ohio State University from 1936 to 1973.

3-D Molecule:
Staggered and eclipsed
conformations of ethane

Newman projections

a staggered conformer for rotation about the carbon–carbon bond in ethane

$60°$

an eclipsed conformer for rotation about the carbon–carbon bond in ethane

A staggered conformer is more stable than an eclipsed conformer.

The electrons in a C—H bond will repel the electrons in another C—H bond if the bonds get too close to each other. **Torsional strain** is the name given to the repulsion felt by the bonding electrons of one substituent as they pass close to the bonding electrons of another substituent. The **staggered conformer**, therefore, is the most stable conformer of ethane because the C—H bonds are as far away from each other as possible; therefore, the compound has less torsional strain. The **eclipsed conformer** is the least stable conformer because in no other conformer are the C—H bonds as close to one another. Rotation about a carbon–carbon single bond is not completely free because of the energy difference between the staggered and eclipsed conformers. The eclipsed conformer is higher in energy, so an energy barrier must be overcome when rotation about the C—C bond occurs (Figure 3.4). However, the barrier in ethane is small enough (2.9 kcal/mol or 12 kJ/mol) to allow the conformers to interconvert millions of times per second at room temperature.

Figure 3.4 shows the potential energies of all the conformers of ethane obtained during one complete 360° rotation. Notice that the **staggered conformers** are at energy minima, whereas the **eclipsed conformers** are at energy maxima.

Butane has three carbon–carbon single bonds, and rotation can occur about each of them.

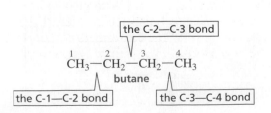

the C-2—C-3 bond

$$\overset{1}{C}H_3 - \overset{2}{C}H_2 - \overset{3}{C}H_2 - \overset{4}{C}H_3$$
butane

the C-1—C-2 bond

the C-3—C-4 bond

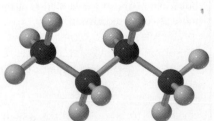

ball-and-stick model of butane

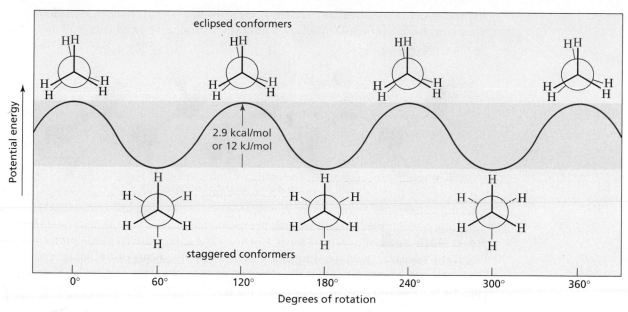

▲ **Figure 3.4**
Potential energy of ethane as a function of the angle of rotation about the carbon–carbon bond.

The staggered and eclipsed conformers are shown for one complete 360° rotation about the C-2—C-3 bond.

The three staggered conformers do not have the same energy. Conformer D, in which the two methyl groups are as far apart as possible, is more stable than the other two staggered conformers (B and F) because of steric strain. **Steric strain** is the strain (i.e., the extra energy) put on a molecule when atoms or groups are too close to one another, which results in repulsion between the electron clouds of these atoms or groups. In general, steric strain in molecules increases as the size of the group increases.

The eclipsed conformers resulting from rotation about the C-2—C-3 bond in butane also have different energies. The eclipsed conformer in which the two methyl groups are closest to each other (A) is less stable than the eclipsed conformers in which they are farther apart (C and E).

Because there is continuous rotation about all the carbon–carbon single bonds in a molecule, organic molecules with carbon–carbon single bonds are not static balls and sticks—they have many interconvertible conformers. The conformers cannot be separated, however, because the small difference in their energy allows them to interconvert rapidly.

The relative number of molecules in a particular conformation at any one time depends on the stability of the conformer: The more stable the conformer, the greater is the fraction of molecules that will be in that conformation. Most molecules, therefore,

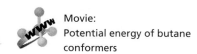

Movie:
Potential energy of butane conformers

are in staggered conformations. The tendency to assume a staggered conformation causes carbon chains to orient themselves in a zigzag fashion, as shown by the ball-and-stick model of decane.

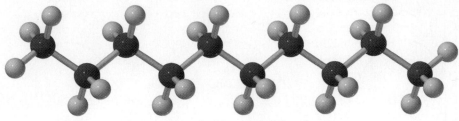

ball-and-stick model of decane

3-D Molecule: Decane

PROBLEM 26

a. Draw the three staggered conformers of butane for rotation about the C-1—C-2 bond. (The carbon in the foreground in a Newman projection should have the lower number.)

b. Do the three staggered conformers have the same energy?

c. Do the three eclipsed conformers have the same energy?

PROBLEM 27

a. Draw the most stable conformer of pentane for rotation about the C-2—C-3 bond.

b. Draw the least stable conformer of pentane for rotation about the C-2—C-3 bond.

cyclopropane

3.9 Cycloalkanes: Ring Strain

Cyclopropane is a planar compound with bond angles of 60°. The bond angles, therefore, are 49.5° less than the ideal sp^3 bond angle of 109.5°. This deviation of the bond angle from the ideal bond angle causes strain called **angle strain**. Figure 3.5 shows that angle strain results from less effective orbital overlap. The less effective orbital overlap causes the C—C bonds of cyclopropane to be weaker than normal C—C bonds.

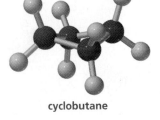

cyclobutane

Figure 3.5 ▶
(a) Overlap of sp^3 orbitals in a normal σ bond.
(b) Overlap of sp^3 orbitals in cyclopropane.

a.

**good overlap
strong bond**

b.

**poor overlap
weak bond**

The bond angles in planar cyclobutane would have to be compressed from 109.5° to 90°, the bond angle associated with a planar four-membered ring. Planar cyclobutane would then be expected to have less angle strain than cyclopropane because the bond angles in cyclobutane are only 19.5° away from the ideal bond angle.

Although planar cyclobutane would have less angle strain than cyclopropane, it would have more torsional strain because it has eight pairs of eclipsed hydrogens, compared with the six pairs of cyclopropane. So cyclobutane is not a planar molecule—it is a bent molecule. One of its CH_2 groups is bent away from the plane defined by the other three carbon atoms. This increases the angle strain, but the increase is more than compensated for by the decreased torsional strain as a result of the adjacent hydrogens not being as eclipsed as they would be in a planar ring.

If cyclopentane were planar, it would have essentially no angle strain (its bond angles would be 108°), but its 10 pairs of eclipsed hydrogens would be subject to considerable torsional strain. So cyclopentane puckers, allowing the hydrogens to become nearly staggered. In the process, however, the compound acquires some angle strain.

cyclopentane

cyclopentane

3-D Molecules:
Cyclopropane; Cyclobutane; Cyclopentane

3.10   Conformations of Cyclohexane

The cyclic compounds most commonly found in nature contain six-membered rings because such rings can exist in a conformation—called a *chair conformation*—that is almost completely free of strain. All the bond angles in a **chair conformer** are 111°, which is very close to the ideal tetrahedral bond angle of 109.5°, and all the adjacent bonds are staggered (Figure 3.6). The chair conformer is such an important conformer that you should learn how to draw it:

1. Draw two parallel lines of the same length, slanted upward. Both lines should start at the same height.

2. Connect the tops of the lines with a V; the left-hand side of the V should be slightly longer than the right-hand side. Connect the bottoms of the lines with an inverted V; the lines of the V and the inverted V should be parallel. This completes the framework of the six-membered ring.

3. Each carbon has an axial bond and an equatorial bond. The **axial bonds** (red lines) are vertical and alternate above and below the ring. The axial bond on one of the uppermost carbons is up, the next is down, the next is up, and so on.

axial
bonds

4. The **equatorial bonds** (red lines with blue balls) point outward from the ring. Because the bond angles are greater than 90°, the equatorial bonds are on a slant. If the axial bond points up, the equatorial bond on the same carbon is on a downward slant. If the axial bond points down, the equatorial bond on the same carbon is on an upward slant.

equatorial bond

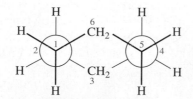

chair conformer of
cyclohexane

Newman projection of
the chair conformer

ball-and-stick model of the
chair conformer of cyclohexane

◀ **Figure 3.6**
The chair conformer of cyclohexane, a Newman projection of the chair conformer, and a ball-and-stick model showing that all the bonds are staggered.

Notice that each equatorial bond is parallel to two ring bonds (two carbons over).

Remember that cyclohexane is viewed on edge. The lower bonds of the ring are in front and the upper bonds of the ring are in back.

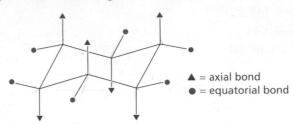

▲ = axial bond
● = equatorial bond

3-D Molecule:
Chair cyclohexane

Bonds that are equatorial in one chair conformer are axial in the other chair conformer.

PROBLEM 28

Draw 1,2,3,4,5,6-hexamethylcyclohexane with

a. all the methyl groups in axial positions.

b. all the methyl groups in equatorial positions.

Cyclohexane rapidly interconverts between two stable chair conformers because of the ease of rotation about its C—C bonds. This interconversion is called ring-flip; at room temperature, cyclohexane undergoes 10^5 ring flips per second. When the two chair conformers interconvert, bonds that are equatorial in one chair conformer become axial in the other chair conformer and vice versa (Figure 3.7).

Figure 3.7 ▶
The bonds that are axial in one chair conformer are equatorial in the other chair conformer. The bonds that are equatorial in one chair conformer are axial in the other chair conformer.

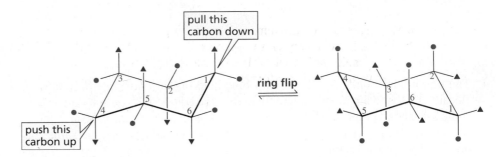

To convert from one chair conformer to the other, the bottommost carbon must be pushed up and what was previously the topmost carbon must be pulled down. The conformations that cyclohexane can assume when interconverting from one chair conformer to the other are shown in Figure 3.8. Because the chair conformers are the most

3-D Molecule:
Boat cyclohexane

Figure 3.8 ▶
The conformers of cyclohexane—and their relative energies—as one chair conformer interconverts to the other chair conformer.

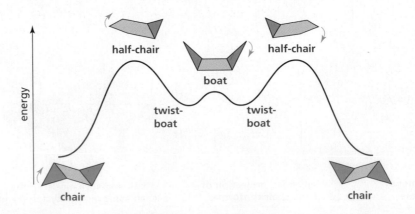

stable of the conformers, at any instant, more molecules of cyclohexane are in chair conformations than in any other conformation. It has been calculated that, for every thousand molecules of cyclohexane in a chair conformation, no more than two molecules are in the next most stable conformation—the twist-boat.

Build a model of cyclohexane, and convert it from one chair conformer to the other. To do this, pull the topmost carbon down and push the bottommost carbon up.

Go to the website for three-dimensional representations of the conformers of cyclohexane.

3.11   Conformations of Monosubstituted Cyclohexanes

Unlike cyclohexane, which has two equivalent chair conformers, the two chair conformers of a monosubstituted cyclohexane such as methylcyclohexane are not equivalent. The methyl substituent is in an equatorial position in one conformer and in an axial position in the other (Figure 3.9), because we have just seen that substituents that are equatorial in one chair conformer are axial in the other (Figure 3.7).

Build a model of methylcyclohexane, and convert it from one chair conformer to the other.

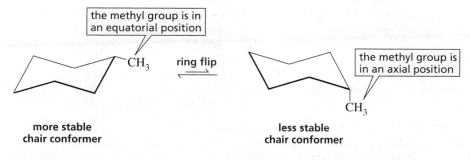

3-D Molecule:
Chair conformers of methylcyclohexane

▲ **Figure 3.9**
A substituent is in an equatorial position in one chair conformer and in an axial position in the other. The conformer with the substituent in the equatorial position is more stable.

Because the three axial bonds on the same side of the ring are parallel to each other, any axial substituent will be relatively close to the axial substituents on the other two carbons. Therefore, the chair conformer with the methyl substituent in an equatorial position is more stable than the chair conformer with the methyl substituent in an axial position, because a substituent has more room and, therefore, fewer steric interactions when it is in an equatorial position.

The larger the substituent on a cyclohexane ring, the more the equatorial-substituted conformer will be favored.

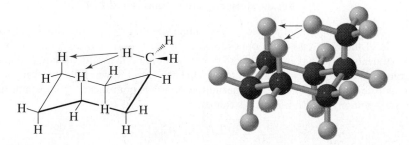

Because of the difference in stability of the two chair conformers of methylcyclohexane, at any one time, there will be more chair conformers with the substituent in the equatorial position than chair conformers with the substituent in the axial position.

PROBLEM 29◆

At any one time, will there be more conformers with the substituent in the equatorial position in a sample of ethylcyclohexane or in a sample of isopropylcyclohexane?

| 3.12 | **Conformations of Disubstituted Cyclohexanes** |

If a cyclohexane ring has two substituents, both substituents have to be taken into account when determining which of the two chair conformers is the more stable. Let's start by looking at 1,4-dimethylcyclohexane. First of all, note that there are two different dimethylcyclohexanes. One has both methyl substituents on the *same side* of the cyclohexane ring (they are both pointing downward); it is called the **cis isomer** (*cis* is Latin for "on this side"). The other has the two methyl substituents on *opposite sides* of the ring (one is pointing upward and one is pointing downward); it is called the **trans isomer** (*trans* is Latin for "across"). *cis*-1,4-Dimethylcyclohexane and *trans*-1,4-dimethylcyclohexane are called **geometric isomers** or **cis–trans isomers**: They have the same atoms, and the atoms are linked in the same order, but they differ in the spatial arrangement of the atoms. The cis and trans isomers are different compounds—they can be separated from one another.

The cis isomer has its substituents on the same side of the ring.

The trans isomer has its substituents on opposite sides of the ring.

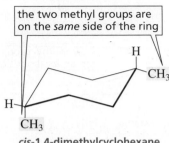

the two methyl groups are on the *same* side of the ring

cis-**1,4-dimethylcyclohexane**

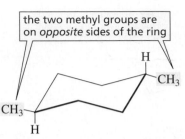

the two methyl groups are on *opposite* sides of the ring

trans-**1,4-dimethylcyclohexane**

PROBLEM-SOLVING STRATEGY

Is the conformer of 1,2-dimethylcyclohexane with one methyl group in an equatorial position and the other in an axial position the cis isomer or the trans isomer?

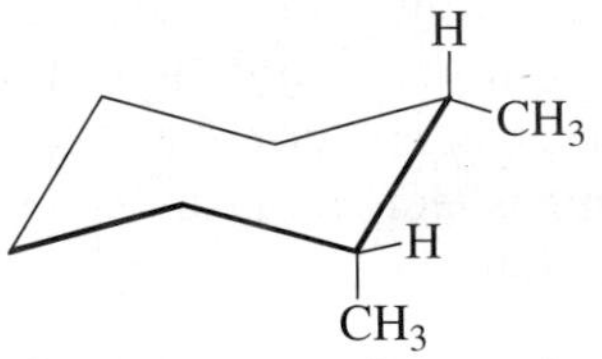

Is this the cis isomer or the trans isomer?

To solve this kind of problem we need to determine whether the two substituents are on the same side of the ring (cis) or on opposite sides of the ring (trans). If the bonds bearing the substituents are both pointing upward or both pointing downward, the compound is the cis isomer; if one bond is pointing upward and the other downward, the compound is the trans isomer. Because the conformer in question has both methyl groups attached to downward-pointing bonds, it is the cis isomer.

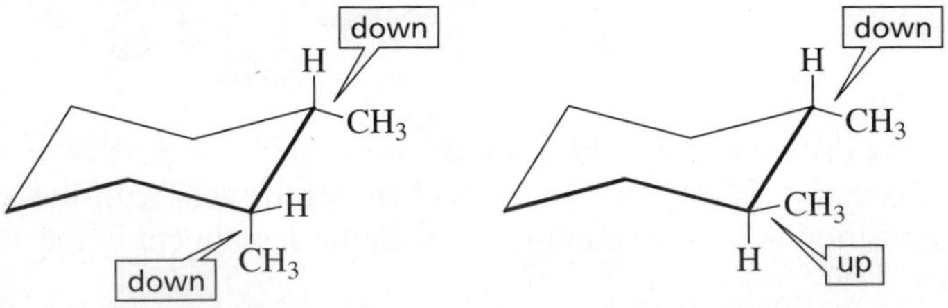

the cis isomer

the trans isomer

Now continue on to Problem 30.

PROBLEM 30◆

Determine whether each of the following is a cis isomer or a trans isomer:

a.

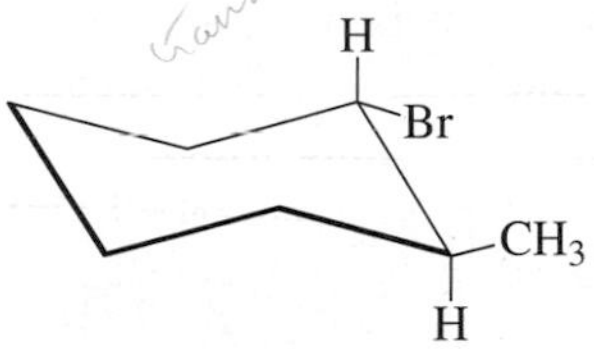

b.

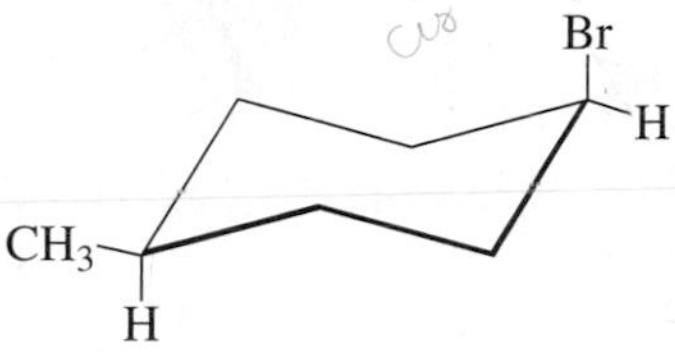

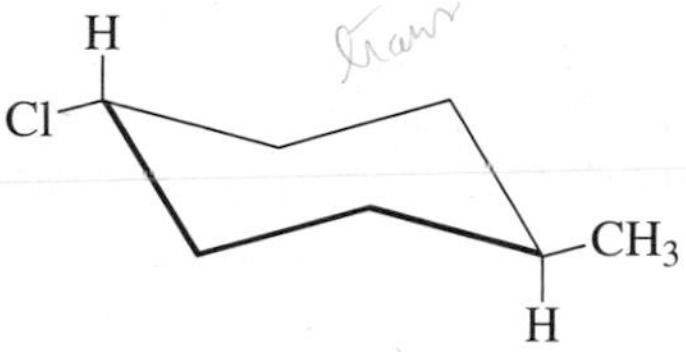

c.

d.

e.

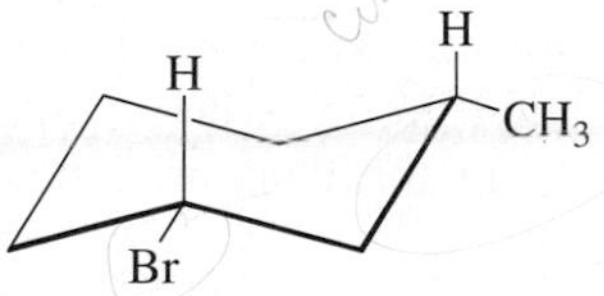

f.

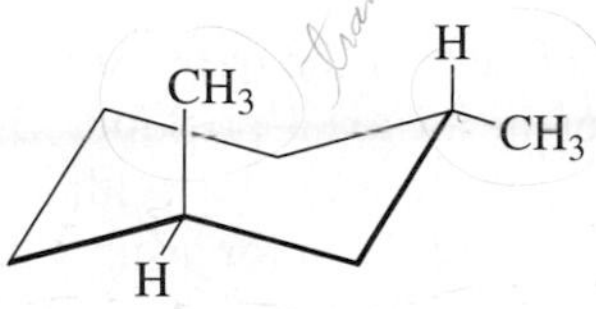

Every compound with a cyclohexane ring has two chair conformers. First we will determine which of the two chair conformers of *cis*-1,4-dimethylcyclohexane is more stable. One chair conformer has one methyl group in an equatorial position and one methyl group in an axial position. The other chair conformer also has one methyl group in an equatorial position and one methyl group in an axial position. Therefore, both chair conformers are equally stable.

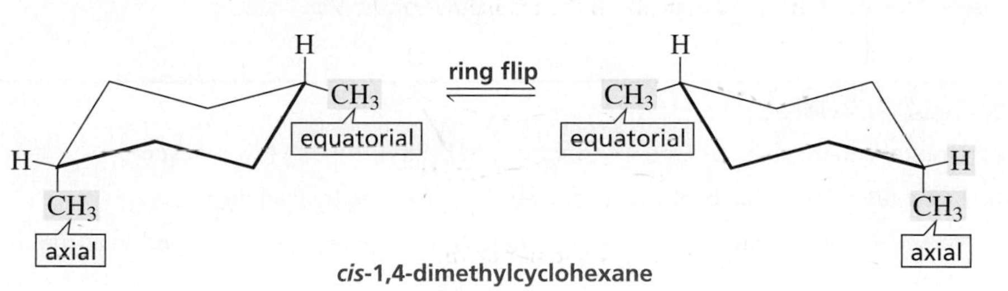

cis-**1,4-dimethylcyclohexane**

In contrast, the two chair conformers of *trans*-1,4-dimethylcyclohexane have different stabilities because one has both methyl substituents in equatorial positions and the other has both methyl groups in axial positions. The conformer with both methyl groups in equatorial positions is more stable.

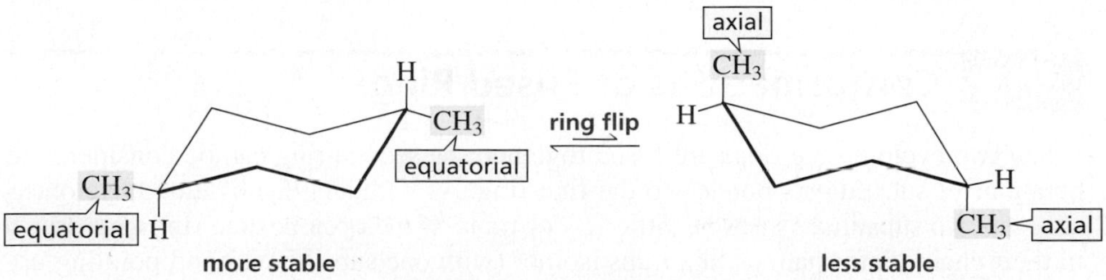

trans-**1,4-dimethylcyclohexane**

Now let's look at the geometric isomers of 1-*tert*-butyl-3-methylcyclohexane. Both substituents of the cis isomer are in equatorial positions in one conformer and in axial positions in the other conformer. The conformer with both substituents in equatorial positions is more stable.

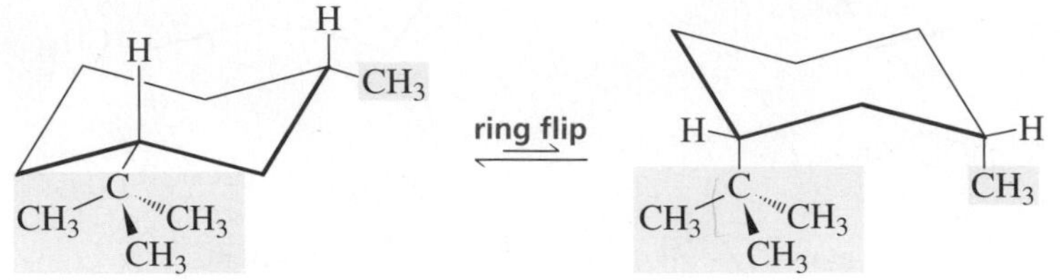

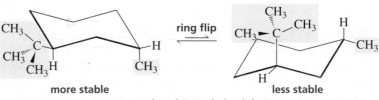

more stable — less stable

cis-**1**-*tert*-**butyl-3-methylcyclohexane**

Both conformers of the trans isomer have one substituent in an equatorial position and the other in an axial position. Because the *tert*-butyl group is larger than the methyl group, the conformer with the *tert*-butyl group in the equatorial position, where there is more room for a substituent, is more stable.

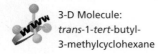

3-D Molecule:
trans-1-*tert*-butyl-
3-methylcyclohexane

more stable — less stable

trans-**1**-*tert*-**butyl-3-methylcyclohexane**

PROBLEM 31 | **SOLVED**

a. Draw the more stable chair conformer of *cis*-1-ethyl-2-methylcyclohexane.

b. Draw the more stable conformer of *trans*-1-ethyl-2-methylcyclohexane.

c. Which is more stable, *cis*-1-ethyl-2-methylcyclohexane or *trans*-1-ethyl-2-methyl-cyclohexane?

SOLUTION TO 31a   If the two substituents of a 1,2-disubstituted cyclohexane are to be on the same side of the ring, one must be in an equatorial position and the other must be in an axial position. The more stable chair conformer is the one in which the larger of the two substituents (the ethyl group) is in the equatorial position.

3.13 Conformations of Fused Rings

When two cyclohexane rings are fused together, the second ring can be considered to be a pair of substituents bonded to the first ring. As with any disubstituted cyclohexane, the two substituents can be either cis or trans. If the cyclohexane rings are drawn in their chair conformations, the trans isomer (with one substituent bond pointing upward and the other downward) will have both substituents in the equatorial position.

The cis isomer will have one substituent in the equatorial position and one substituent in the axial position. **Trans-fused** cyclohexane rings, therefore, are more stable than **cis-fused** cyclohexane rings.

equatorial   H

equatorial

equatorial   H

trans-fused rings
more stable

equatorial   H

H

axial

cis-fused rings
less stable

Hormones are chemical messengers—organic compounds synthesized in glands and delivered to target tissues in order to stimulate or inhibit some process. Many hormones are **steroids**. The four rings in steroids are designated A, B, C, and D. The B, C, and D rings are all trans fused, and in most naturally occurring steroids, the A and B rings are also trans fused.

C   D

A   B

the steroid ring system

A   B   C   D

all the rings are trans fused

Michael S. Brown *and* **Joseph Leonard Goldstein** *shared the 1985 Nobel Prize in physiology or medicine for their work on the regulation of cholesterol metabolism and the treatment of disease caused by elevated cholesterol levels in the blood. Brown was born in New York in 1941; Goldstein was born in South Carolina in 1940. They are both professors of medicine at the University of Texas Southwestern Medical Center.*

The most abundant member of the steroid family in animals is **cholesterol**, the precursor of all other steroids. Cholesterol is an important component of cell membranes. We will see that its ring structure makes it more rigid than other membrane components (Section 20.5).

H_3C

H_3C H

H   H

HO

cholesterol

Tutorial:
Steroids

CHOLESTEROL AND HEART DISEASE

Cholesterol is probably the best-known steroid because of the correlation between cholesterol levels in the blood and heart disease. It is synthesized in the liver and is also found in almost all body tissues. Cholesterol is also found in many foods, but we do not require it in our diet because the body can synthesize all we need. A diet high in cholesterol can lead to high levels of cholesterol in the bloodstream; the excess can accumulate on the walls of arteries, restricting the flow of blood. This disease of the circulatory system is known as *atherosclerosis* and is a primary cause of heart disease. Cholesterol travels through the bloodstream packaged in particles that are classified according to their density. LDL (low-density lipoprotein) particles transport cholesterol from the liver to other tissues. Receptors on the surfaces of cells bind LDL particles, allowing them to be brought into the cell so that it can use the cholesterol. HDL (high-density lipoprotein) is a cholesterol scavenger, removing cholesterol from the surfaces of membranes and delivering it back to the liver, where it is converted into bile acids. LDL is the so-called bad cholesterol, whereas HDL is the "good" cholesterol. The more cholesterol we eat, the less the body synthesizes. But this does not mean that dietary cholesterol has no effect on the total amount of cholesterol in the bloodstream, because dietary cholesterol also inhibits the synthesis of the LDL receptors. So the more cholesterol we eat, the less the body synthesizes, but also, the less the body can get rid of by bringing it into target cells.

CLINICAL TREATMENT OF HIGH CHOLESTEROL

Statins are the newest class of cholesterol-reducing drugs. Statins reduce serum cholesterol levels by inhibiting the enzyme that catalyzes the formation of a compound needed for the synthesis of cholesterol. As a consequence of diminished cholesterol synthesis in the liver, the liver expresses more LDL receptors—the receptors that help clear LDL from the bloodstream. Studies show that for every 10% that cholesterol is reduced, deaths from coronary heart disease are reduced by 15% and total death risk is reduced by 11%.

Compactin and lovastatin are natural statins used clinically under the trade names Zocor and Mevacor. Atorvastatin (Lipitor), a synthetic statin, is now the most popular statin. It has greater potency and a longer half-life than natural statins, because its metabolites are as active as the parent drug in reducing cholesterol levels. Therefore, smaller doses of the drug may be administered. In addition, Lipitor is less polar than compactin and lovastatin, so it has a greater tendency to remain in the endoplasmic reticulum of the liver cells, where it is needed.

lovastatin
Mevacor

simvastatin
Zocor

atorvastatin
Lipitor

Summary

Alkanes are **hydrocarbons** that contain only single bonds. Their general molecular formula is C_nH_{2n+2}. **Constitutional isomers** have the same molecular formula, but their atoms are linked differently. Alkanes are named by determining the number of carbons in their **parent hydrocarbon**—the longest continuous chain. **Substituents** are listed in alphabetical order, with a number to designate their position on the chain.

Systematic names can contain numbers; **common names** never do. A compound can have more than one name, but a name must specify only one compound. Whether an alkyl halide or an alcohol is **primary**, **secondary**, or **tertiary** depends on whether the X (halogen) or OH group is bonded to a primary, secondary, or tertiary carbon. A **primary carbon** is bonded to one carbon, a **secondary carbon** is bonded to two carbons, and a **tertiary carbon** is bonded to three carbons. Whether an amine is **primary**, **secondary**, or **tertiary** depends on the number of alkyl groups bonded to the nitrogen.

The oxygen of an alcohol or an ether has the same geometry it has in water; the nitrogen of an amine has the same geometry it has in ammonia. The greater the attractive forces between molecules—**van der Waals forces**, **dipole–dipole interactions**, **hydrogen bonds**—the higher is the **boiling point** of the compound. A **hydrogen bond** is an interaction between a hydrogen bonded to an O, N, or F and a lone pair of an O, N, or F in another molecule. The boiling point of straight-chain compounds increases with increasing molecular weight. Branching lowers the boiling point.

Polar compounds dissolve in **polar solvents**, and **nonpolar compounds** dissolve in **nonpolar solvents**. The interaction between a solvent and a molecule or an ion dissolved in that solvent is called **solvation**. The oxygen of an alcohol or an ether can drag about three or four carbons into water.

Rotation about a C—C bond results in two extreme **conformations** that rapidly interconvert: **staggered** and **eclipsed**. A **staggered conformer** is more stable than an **eclipsed conformer** because of **torsional strain**—repulsion between pairs of bonding electrons.

Five- and six-membered rings are more stable than three- and four-membered rings because of **angle strain** that results when bond angles deviate from the ideal bond angle of 109.5°. In a process called **ring flip**, cyclohexane rapidly interconverts between two stable chair conformations. **Bonds** that are **axial** in one chair conformer are **equatorial** in the other and vice versa. The chair conformer with a substituent in the equatorial position is more stable, because there is more room in an equatorial position; thus, the conformer has less **steric strain**. In the case of disubstituted cyclohexanes, the more stable conformer will have its larger substituent in the equatorial position. A **cis isomer** has its two substituents on the same side of the ring; a **trans isomer** has its substituents on opposite sides of the ring. Cis and trans isomers are called **geometric isomers** or **cis–trans isomers**. Cis and trans isomers are different compounds; conformers are different conformations of the same compound.

Problems

32. Write a structural formula for each of the following:
   a. *sec*-butyl *tert*-butyl ether
   b. isoheptyl alcohol
   c. *sec*-butylamine
   d. 4-*tert*-butylheptane
   e. 1,1-dimethylcyclohexane
   f. 4,5-diisopropylnonane
   g. triethylamine
   h. cyclopentylcyclohexane
   i. 3,4-dimethyloctane
   j. 5,5-dibromo-2-methyloctane

33. Give the systematic name for each of the following:

   a. 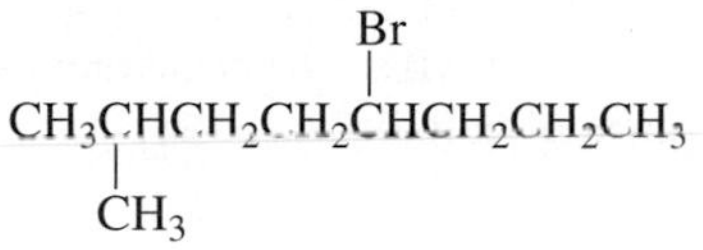 $\overset{\text{Br}}{\underset{\overset{|}{\text{CH}_3}}{\text{CH}_3\text{CHCH}_2\text{CH}_2\text{CHCH}_2\text{CH}_2\text{CH}_3}}$

   c. $\underset{\overset{|}{\text{CH}_3}\ \overset{|}{\text{CH}_3}}{\overset{\overset{\text{CH}_3}{|}}{\text{CH}_3\text{CHCH}_2\text{CHCHCH}_3}}$

   e.

   b. $(CH_3)_3CCH_2CH_2CH_2CH(CH_3)_2$

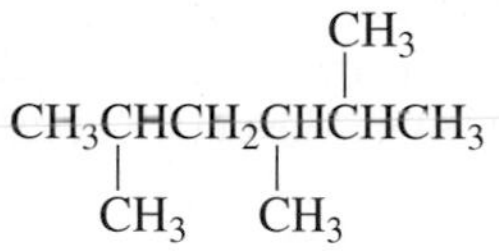

   d. $(CH_3CH_2)_4C$

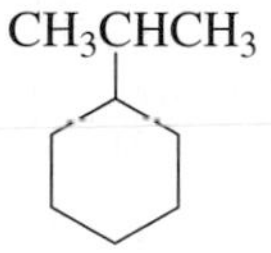

   f.

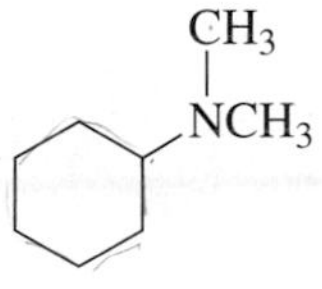

34. a. How many primary carbons does the following structure have?

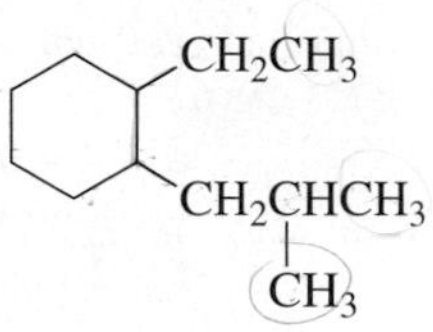

$$\text{CH}_2\text{CH}_3$$
$$\text{CH}_2\text{CHCH}_3$$
$$\text{CH}_3$$

   b. How many secondary carbons does the structure have?
   c. How many tertiary carbons does it have?

35. Draw the structure and give the systematic name of a compound with a molecular formula C_5H_{12} that has
   a. only one tertiary carbon
   b. no secondary carbons

36. Which of the following conformers of isobutyl chloride is the most stable?

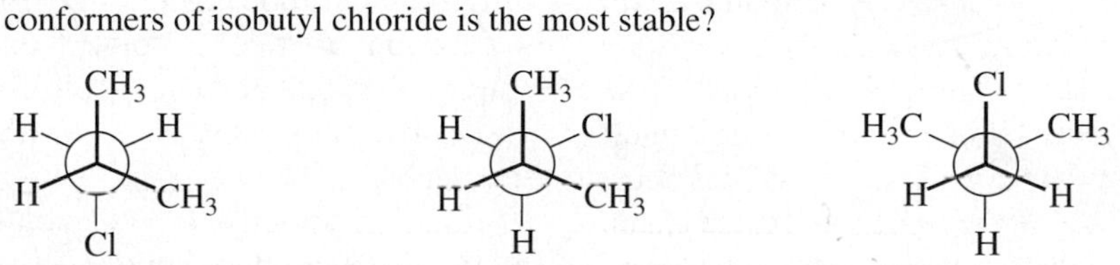

37. Name the following amines and tell whether they are primary, secondary, or tertiary:

   a. 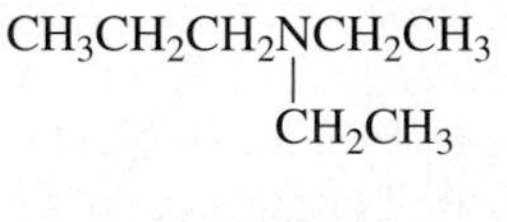 $\underset{\overset{|}{\text{CH}_2\text{CH}_3}}{\text{CH}_3\text{CH}_2\text{CH}_2\text{NCH}_2\text{CH}_3}$

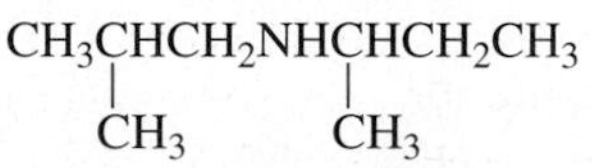

   c. 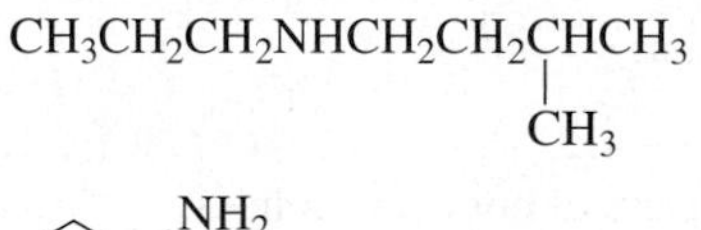 $\underset{\overset{|}{\text{CH}_3}}{\text{CH}_3\text{CH}_2\text{CH}_2\text{NHCH}_2\text{CH}_2\text{CHCH}_3}$

   b. $\underset{\overset{|}{\text{CH}_3}\ \overset{|}{\text{CH}_3}}{\text{CH}_3\text{CHCH}_2\text{NHCHCH}_2\text{CH}_3}$

   d.

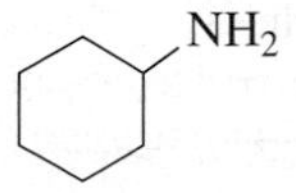

38. Draw the structural formula of an alkane that has
   a. six carbons, all secondary
   b. eight carbons and only primary hydrogens
   c. seven carbons with two isopropyl groups

39. Name each of the following:

   a. $CH_3CH_2CH_2OCH_2CH_3$

   c. 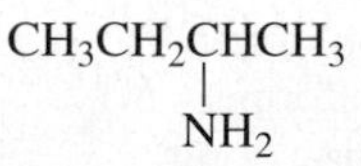 $\underset{\overset{|}{\text{NH}_2}}{\text{CH}_3\text{CH}_2\text{CHCH}_3}$

   e. 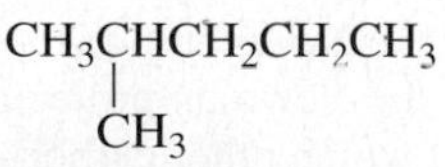 $\underset{\overset{|}{\text{CH}_3}}{\text{CH}_3\text{CHCH}_2\text{CH}_2\text{CH}_3}$

   b. $\underset{\overset{|}{\text{CH}_3}}{\text{CH}_3\text{CHCH}_2\text{CH}_2\text{CH}_2\text{OH}}$

   d. $\underset{\overset{|}{\text{Cl}}}{\text{CH}_3\text{CH}_2\text{CHCH}_3}$

   f. 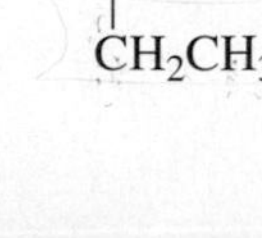 $\underset{\overset{|}{\text{CH}_2\text{CH}_3}}{\overset{\overset{\text{CH}_3}{|}}{\text{CH}_3\text{CBr}}}$

g. ⟨cyclohexane⟩-OH

h. ⟨cyclopentane⟩-Br

i. CH_3CHNH_2
      |
     CH_3

j. $CH_3CH_2CH(CH_3)NHCH_2CH_3$

40. Which of the following pairs of compounds has
  a. the higher boiling point: 1-bromopentane or 1-bromohexane?
  b. the higher boiling point: pentyl chloride or isopentyl chloride?
  c. the greater solubility in water: butyl alcohol or pentyl alcohol?
  d. the higher boiling point: hexyl alcohol or methyl pentyl ether?
  e. the higher melting point: hexane or isohexane?
  f. the higher boiling point: 1-chloropentane or pentyl alcohol?

41. Ansaid and Motrin belong to the group of drugs known as nonsteroidal anti-inflammatory drugs (NSAIDs). Both are only slightly soluble in water, but one is a little more soluble than the other. Which of the drugs has the greater solubility in water?

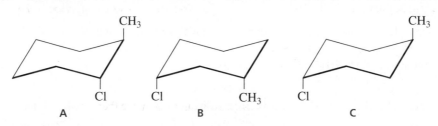

Ansaid®   Motrin®

42. Al Kane was given the structural formulas of several compounds and was asked to give them systematic names. How many did Al name correctly? Correct those that are misnamed.
  a. 3-isopropyloctane
  b. 2,2-dimethyl-4-ethylheptane
  c. isopentyl bromide
  d. 3,3-dichlorooctane
  e. 5-ethyl-2-methylhexane
  f. 2-methyl-2-isopropylheptane

43. Which of the following conformers has the highest energy?

A   B   C

44. Give systematic names for all the alkanes with molecular formula C_7H_{16} that do not have any secondary hydrogens.

45. Draw the skeletal structures of the following compounds:
  a. 5-ethyl-2-methyloctane
  b. 1,3-dimethylcyclohexane
  c. propylcyclopentane
  d. 2,3,3,4-tetramethylheptane

46. Which of the following pairs of compounds has
  a. the higher boiling point: 1-bromopentane or 1-chloropentane?
  b. the higher boiling point: diethyl ether or butyl alcohol?
  c. the greater density: heptane or octane?
  d. the higher boiling point: isopentyl alcohol or isopentylamine?
  e. the higher boiling point: hexylamine or dipropylamine?

47. Why are alcohols of lower molecular weight more soluble in water than those of higher molecular weight?

48. For rotation about the C-3—C-4 bond of 2-methylhexane:
  a. Draw the Newman projection of the most stable conformer.
  b. Draw the Newman projection of the least stable conformer.
  c. About which other carbon–carbon bonds may rotation occur?
  d. How many of the carbon–carbon bonds in the compound have staggered conformers that are all equally stable?

49. Which of the following structures represents a cis isomer?

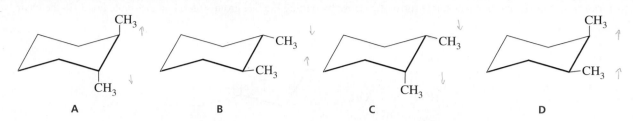

A   B   C   D

50. Draw all the isomers that have the molecular formula $C_5H_{11}Br$. (*Hint:* There are eight such isomers.)
  a. Give the systematic name for each of the isomers.
  b. How many of the isomers are primary alkyl halides?
  c. How many of the isomers are secondary alkyl halides?
  d. How many of the isomers are tertiary alkyl halides?

51. Give the systematic name for each of the following:
  a. 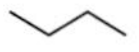
  c. 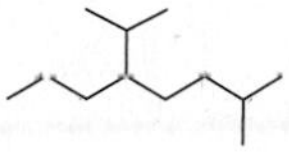
  e.

  b. OH
  d.
  Cl

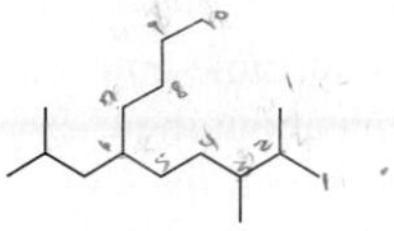

52. Draw the two chair conformers of each compound, and indicate which conformer is more stable:
  a. *cis*-1-ethyl-3-methylcyclohexane   **d.** *trans*-1-ethyl-3-methylcyclohexane
  b. *trans*-1-ethyl-2-isopropylcyclohexane   **e.** *cis*-1-ethyl-3-isopropylcyclohexane
  c. *trans*-1-ethyl-2-methylcyclohexane   **f.** *cis*-1-ethyl-4-isopropylcyclohexane

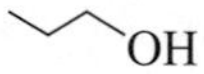

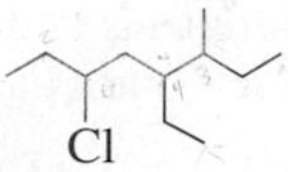

53. Explain why
  a. H_2O has a higher boiling point than CH_3OH (65 °C).
  b. H_2O has a higher boiling point than NH_3 (−33 °C).
  c. H_2O has a higher boiling point than HF (20 °C).

54. How many ethers have molecular formula $C_5H_{12}O$? Draw their structures and name them.

55. Draw the most stable conformer of the following molecule:

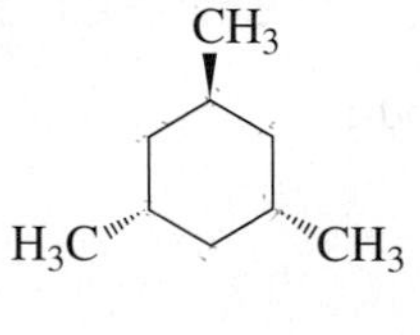

CH_3

H_3C⁗   ⁗CH_3

56. Give the systematic name for each of the following:
  a. 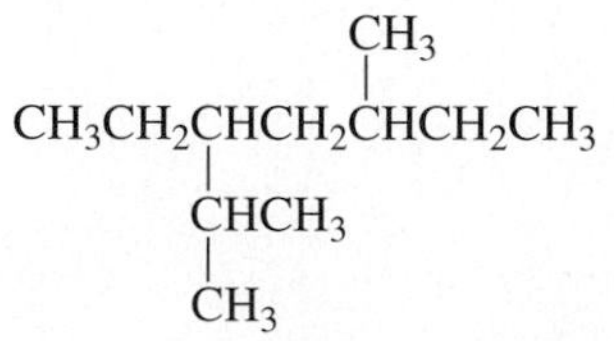
  CH_3
  $CH_3CH_2CHCH_2CHCH_2CH_3$
  $CHCH_3$
  CH_3

  b. 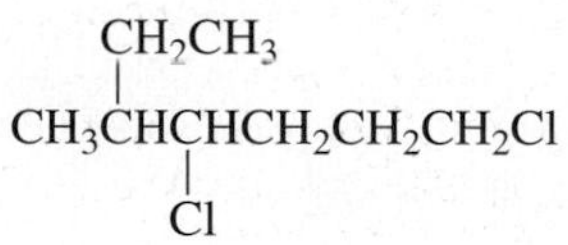
  CH_2CH_3
  $CH_3CHCHCH_2CH_2CH_2Cl$
  Cl

57. Which of the following can be used to verify that carbon is tetrahedral?
  a. Methyl bromide does not have constitutional isomers.
  b. Tetrachloromethane does not have a dipole.
  c. Dibromomethane does not have constitutional isomers.

58. The most stable form of glucose (blood sugar) is a six-membered ring in a chair conformation with its five substituents all in equatorial positions. Draw the most stable form of glucose by putting the OH groups on the appropriate bonds in the chair conformer.

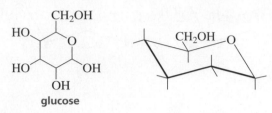

glucose

59. Draw the nine isomeric heptanes and name each isomer.

60. Draw the most stable conformer of 1,2,3,4,5,6-hexachlorocyclohexane.

61. a. Draw all the staggered and eclipsed conformers that result from rotation about the C-2—C-3 bond of pentane.
   b. Draw a potential-energy diagram for rotation about the C-2—C-3 bond of pentane through 360°, starting with the least stable conformer.

62. Using Newman projections, draw the most stable conformer for the following:
   a. 3-methylpentane, considering rotation about the C-2—C-3 bond
   b. 3-methylhexane, considering rotation about the C-3—C-4 bond

63. For each of the following disubstituted cyclohexanes, indicate whether the substituents in the two chair conformers would be both equatorial in one chair conformer and both axial in the other *or* one equatorial and one axial in each of the chair conformers:
   a. *cis*-1,2-   **c.** *cis*-1,3-   **e.** *cis*-1,4-
   b. *trans*-1,2-   **d.** *trans*-1,3-   **f.** *trans*-1,4-

64. Which will have a higher percentage of the diequatorial-substituted conformer, compared with the diaxial-substituted conformer: *trans*-1,4-dimethylcyclohexane or *cis*-1-*tert*-butyl-3-methylcyclohexane?

4 Alkenes

Structure, Nomenclature, Stability, and an Introduction to Reactivity

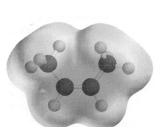

**E isomer
of 2-butene**

**Z isomer
of 2-butene**

I n Chapter 3, we saw that alkanes are hydrocarbons that contain only carbon–carbon *single* bonds. Hydrocarbons that contain a carbon–carbon *double* bond are called **alkenes**. Alkenes play many important roles in biology. Ethene, for example, is a plant hormone—a compound that controls the plant's growth and other changes in its tissues. Ethene affects seed germination, flower maturation, and fruit ripening. Many of the flavors and fragrances produced by certain plants also belong to the alkene family.

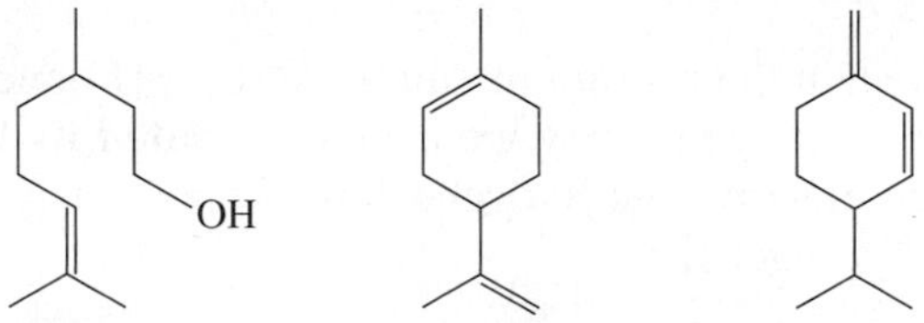

**citronellol
in rose and
geranium oils**

**limonene
in lemon and
orange oils**

**β-phellandrene
oil of eucalyptus**

Ethene is the hormone that causes tomatoes to ripen.

We will start our study of alkenes by looking at how they are named, their structures, and their relative stabilities. Then we will take a look at a reaction of an alkene, paying close attention to why alkenes react the way they do. We will also look at how chemists portray the step-by-step process by which a reaction occurs and the energy changes that occur during a reaction. You will see that some of this material is based on information with which you are familiar while some is new. Both will add to your foundation of knowledge that will be built on in subsequent chapters.

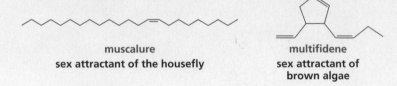

PHEROMONES

Insects communicate by releasing **pheromones**—chemical substances that other insects of the same species detect with their antennae. There are sex, alarm, and trail pheromones, and many of these are alkenes. Interfering with an insect's ability to send or receive chemical signals is an environmentally safe way to control insect populations. For example, traps with synthetic sex attractants have been used to capture such crop-destroying insects as the gypsy moth and the boll weevil.

muscalure
sex attractant of the housefly

multifidene
**sex attractant of
brown algae**

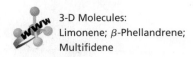

3-D Molecules:
Limonene; β-Phellandrene;
Multifidene

The general molecular formula for a hydrocarbon is C_nH_{2n+2}, minus two hydrogens for every π bond or ring present in the molecule.

Molecular Formulas

We have seen that the general molecular formula for a noncyclic alkane is C_nH_{2n+2} (Section 3.0). You also learned that the general molecular formula for a cyclic alkane is C_nH_{2n} because the cyclic structure reduces the number of hydrogens by two (Section 3.3).

The general molecular formula for a *noncyclic alkene* is also C_nH_{2n} because, as a result of the double bond, an alkene has two fewer hydrogens than an alkane with the same number of carbons. Thus, the general molecular formula for a *cyclic alkene* must be C_nH_{2n-2}. We can, therefore, make the following statement: *The general molecular formula for a hydrocarbon is C_nH_{2n+2}, minus two hydrogens for every π bond and/or ring in the molecule.*

$CH_3CH_2CH_2CH_2CH_3$	$CH_3CH_2CH_2CH{=}CH_2$		
an alkane	**an alkene**	**a cyclic alkane**	**a cyclic alkene**
C_5H_{12}	C_5H_{10}	C_5H_{10}	C_5H_8
C_nH_{2n+2}	C_nH_{2n}	C_nH_{2n}	C_nH_{2n-2}

Because alkanes contain the maximum number of C—H bonds possible—that is, they are saturated with hydrogen—they are called **saturated hydrocarbons**. In contrast, alkenes are called **unsaturated hydrocarbons** because they have fewer than the maximum number of hydrogens.

$CH_3CH_2CH_2CH_3$
a saturated hydrocarbon

$CH_3CH{=}CHCH_3$
an unsaturated hydrocarbon

PROBLEM 1◆ **SOLVED**

Determine the molecular formula for each of the following:

a. a 5-carbon hydrocarbon with one π bond and one ring

b. a 4-carbon hydrocarbon with two π bonds and no rings

c. a 10-carbon hydrocarbon with one π bond and two rings

SOLUTION TO 1a For a 5-carbon hydrocarbon with no π bonds and no rings, $C_nH_{2n+2} = C_5H_{12}$. A 5-carbon hydrocarbon with one π bond and one ring has four fewer hydrogens, because two hydrogens are subtracted for every π bond or ring present in the molecule. Its molecular formula, therefore, is C_5H_8.

PROBLEM 2◆ **SOLVED**

Determine the total number of double bonds and/or rings for the hydrocarbons with the following molecular formulas:

a. $C_{10}H_{16}$ **b.** $C_{20}H_{34}$ **c.** C_8H_{16}

SOLUTION TO 2a   For a 10-carbon hydrocarbon with no π bonds and no rings, $C_nH_{2n+2} = C_{10}H_{22}$. Thus, a 10-carbon compound with molecular formula $C_{10}H_{16}$ has six fewer hydrogens. Therefore, the total number of its double bonds and/or rings is three.

4.2   Nomenclature of Alkenes

The **functional group** is the center of reactivity in a molecule. In an alkene, the double bond is the functional group. The IUPAC system uses a suffix to denote certain functional groups. The systematic name of an alkene, for example, is obtained by replacing the *ane* at the end of the parent hydrocarbon's name with the suffix *ene*. Thus, a two-carbon alkene is called ethene and a three-carbon alkene is called propene. Ethene also is frequently called by its common name: ethylene.

| systematic name: | $H_2C{=}CH_2$ ethene | $CH_3CH{=}CH_2$ propene | cyclopentene | cyclohexene |
| common name: | ethylene | propylene | | |

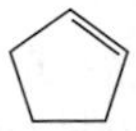

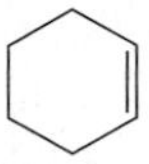

The following rules are used to name a compound that has a functional group suffix:

1. The longest continuous chain containing the functional group (in this case, the carbon–carbon double bond) is numbered in a direction that gives the functional group suffix the lowest possible number. The position of the double bond is indicated by the number immediately preceding the name of the alkene. For example, 1-butene signifies that the double bond is between the first and second carbons of butene; 2-hexene signifies that the double bond is between the second and third carbons of hexene. (The four alkene names shown above do not need a number, because there is no ambiguity.)

> Number the longest continuous chain containing the functional group in the direction that gives the functional group suffix the lowest possible number.

$$\overset{4}{C}H_3\overset{3}{C}H_2\overset{2}{C}H{=}\overset{1}{C}H_2$$
1-butene

$$\overset{1}{C}H_3\overset{2}{C}H{=}\overset{3}{C}H\overset{4}{C}H_3$$
2-butene

$$\overset{1}{C}H_3\overset{2}{C}H{=}\overset{3}{C}H\overset{4}{C}H_2\overset{5}{C}H_2\overset{6}{C}H_3$$
2-hexene

Notice that 1-butene does not have a common name. You might be tempted to call it "butylene," which is analogous to "propylene" for propene, but butylene is not an appropriate name. A name must be unambiguous, and "butylene" could signify either 1-butene or 2-butene.

2. The name of a substituent is cited before the name of the longest continuous chain that contains the functional group, together with a number to designate the carbon to which the substituent is attached. Notice that *if there is a functional group suffix and a substituent, the functional group suffix gets the lowest possible number.*

> When there is only a substituent, the substituent gets the lowest possible number.
>
> When there is only a functional group suffix, the functional group suffix gets the lowest possible number.
>
> When there is both a functional group suffix and a substituent, the functional group suffix gets the lowest possible number.

$$\overset{1}{C}H_3\overset{2}{C}H{=}\overset{3}{C}H\overset{4}{\underset{|}{C}}H\overset{5}{C}H_3$$
4-methyl-2-pentene

$$\overset{3}{C}H_3\overset{}{C}{=}\overset{4}{C}H\overset{5}{C}H_2\overset{6}{C}H_2\overset{7}{C}H_3$$
3-methyl-3-heptene

3. If a chain has more than one double bond, we first identify the chain that contains all the double bonds by its alkane name, replacing the "ne" ending with the appropriate suffix: *diene, triene, etc.* The chain is numbered in the direction that yields the lowest number in the name of the compound.

$$\overset{1}{CH_2}=\overset{2}{CH}-\overset{3}{CH_2}-\overset{4}{CH}=\overset{5}{CH_2}$$
1,4-pentadiene

$$\overset{1}{CH_3}\overset{2}{CH}=\overset{3}{CH}-\overset{4}{CH}=\overset{5}{CH}\overset{6}{CH_2}\overset{7}{CH_3}$$
2,4-heptadiene

$$\overset{5}{CH_3}\overset{4}{CH}=\overset{3}{CH}-\overset{2}{CH}=\overset{1}{CH_2}$$
1,3-pentadiene

Substituents are cited in alphabetical order.

4. If a chain has more than one substituent, the substituents are cited in alphabetical order, using the same rules for alphabetizing that you learned in Section 3.2. The appropriate number is then assigned to each substituent.

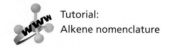
3,6-dimethyl-3-octene

$$\underset{7}{CH_3}\underset{6}{CH_2}\underset{5}{\overset{Br}{\underset{|}{CH}}}\underset{4}{\overset{Cl}{\underset{|}{CH}}}\underset{3}{CH_2}\underset{2}{CH}=\underset{1}{CH_2}$$
5-bromo-4-chloro-1-heptene

A substituent receives the lowest possible number only if there is no functional group suffix or if the same number for the functional group suffix is obtained in both directions.

5. If the same number for the alkene functional group suffix is obtained in both directions, the correct name is the name that contains the lowest substituent number.

$$CH_3CH_2CH_2C=CHCH_2CHCH_3$$
$$\overset{|}{CH_3} \qquad \overset{|}{CH_3}$$
2,5-dimethyl-4-octene
not
4,7-dimethyl-4-octene
because 2 < 4

$$CH_3CHCH=CCH_2CH_3$$
$$\overset{|}{Br} \quad \overset{|}{CH_3}$$
2-bromo-4-methyl-3-hexene
not
5-bromo-3-methyl-3-hexene
because 2 < 3

6. A number is not needed to denote the position of the double bond in a cyclic alkene because the ring is always numbered so that the double bond is between carbons 1 and 2. In determining a substituent number, you need to move around the ring in the direction (clockwise or counterclockwise) that puts the lowest number into the name.

Tutorial:
Alkene nomenclature

3-ethylcyclopentene   **4,5-dimethylcyclohexene**   **4-ethyl-3-methylcyclohexene**

7. Numbers are needed if the ring has more than one double bond.

1,3-cyclohexadiene   **1,4-cyclohexadiene**   **2-methyl-1,3-cyclopentadiene**

Remember that the name of a substituent is stated *before* the name of the parent hydrocarbon, and the functional group suffix is stated *after* the name of the parent hydrocarbon.

[substituent] [parent hydrocarbon] [functional group suffix]

The sp^2 carbons of an alkene are called **vinylic carbons**. An sp^3 carbon that is adjacent to a vinylic carbon is called an **allylic carbon**.

vinylic carbons
$$RCH_2-CH=CH-CH_2R$$
allylic carbons

Two groups containing a carbon–carbon double bond are used in common names—the **vinyl group** and the **allyl group**. The vinyl group is the smallest possible group that contains a vinylic carbon; the allyl group is the smallest possible group that contains an allylic carbon. When "allyl" is used in nomenclature, the substituent must be attached to the allylic carbon.

Tutorial:
Common names of
alkyl groups

$$H_2C=CH-$$
the vinyl group

$$H_2C=CHCH_2-$$
the allyl group

$$H_2C=CHCl$$

$$H_2C=CHCH_2Br$$

systematic name: chloroethene   3-bromopropene
common name: vinyl chloride   allyl bromide

PROBLEM 3

Draw the structure for each of the following compounds:

a. 3,3-dimethylcyclopentene

b. 6-bromo-2,3-dimethyl-2-hexene

c. ethyl vinyl ether

d. allyl alcohol

PROBLEM 4♦

Give the systematic name for each of the following compounds:

a. $CH_3CHCH=CHCH_3$
     |
     CH_3

b. $CH_3CH_2C=CCHCH_3$
                 $\overset{\displaystyle CH_3}{|}$ and $\overset{}{\underset{CH_3 \quad Cl}{|}}$

c. $BrCH_2CH_2CH=CCH_3$
                          |
                          CH_2CH_3

d.

4.3   The Structure of Alkenes

The structure of the smallest alkene (ethene) was described in Section 1.8. Other alkenes have similar structures. Each double-bonded carbon of an alkene has three sp^2 orbitals that lie in a plane with angles of 120°. Each of these sp^2 orbitals overlaps an orbital of another atom to form a σ bond. Thus, one of the carbon–carbon bonds in a double bond is a σ bond, formed by the overlap of an sp^2 orbital of one carbon with an sp^2 orbital of the other carbon. The second carbon–carbon bond in the double bond (the π bond) is formed from side-to-side overlap of the remaining p orbitals of the sp^2 carbons. Because three points determine a plane, each sp^2 carbon and the two atoms singly bonded to it lie in a plane. In order to achieve maximum orbital–orbital overlap, the two p orbitals must be parallel to each other. Therefore, all six atoms of the double-bond system are in the same plane.

$$
\begin{array}{ccc}
H_3C & & CH_3 \\
 & C=C & \\
H_3C & & CH_3
\end{array}
$$

**the six carbon atoms
are in the same plane**

It is important to remember that the π bond represents the cloud of electrons that is above and below the plane defined by the two sp^2 carbons and the four atoms bonded to them.

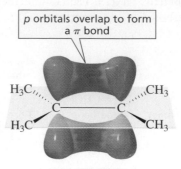

p orbitals overlap to form a π bond

3-D Molecule:
2,3-Dimethyl-2-butene

PROBLEM 5◆   SOLVED

For each of the following compounds, tell how many of its carbon atoms lie in the same plane:

a. ![structure a]   b. ![structure b]   c. ![structure c]   d. ![structure d]

SOLUTION TO 5a   The two sp^2 carbons (blue dots) and the carbons bonded to each of the sp^2 carbons (red dots) lie in the same plane. Therefore, five carbons lie in the same plane.

4.4   Cis–Trans Isomerism

We have just seen that the two *p* orbitals that form the π bond must be parallel to achieve maximum overlap. Therefore, rotation about a double bond does not readily occur. If rotation were to occur, the two *p* orbitals would cease to overlap—in other words, the π bond would break (Figure 4.1).

Because rotation about a double bond does not readily occur, an alkene such as 2-butene can exist in two distinct forms: The hydrogens bonded to the sp^2 carbons can be on the same side of the double bond or on opposite sides of the double bond. The

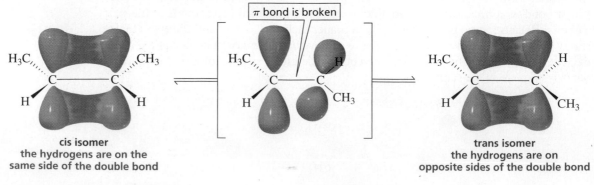

cis isomer
the hydrogens are on the
same side of the double bond

π bond is broken

trans isomer
the hydrogens are on
opposite sides of the double bond

▲ **Figure 4.1**
Rotation about the carbon–carbon double bond would break the π bond.

isomer with the hydrogens on the same side of the double bond is called the **cis isomer**, and the isomer with the hydrogens on opposite sides of the double bond is called the **trans isomer**. A pair of isomers such as *cis*-2-butene and *trans*-2-butene is called **cis–trans isomers** or **geometric isomers**. This should remind you of the cis–trans isomers of disubstituted cyclohexanes you encountered in Section 3.12— the cis isomer had its substituents on the same side of the ring, and the trans isomer had its substituents on opposite sides of the ring.

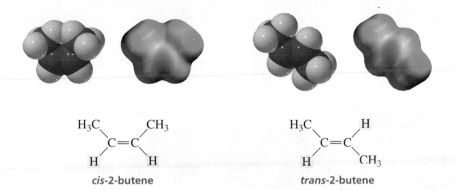

cis-2-butene   trans-2-butene

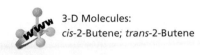

3-D Molecules:
cis-2-Butene; *trans*-2-Butene

If one of the sp^2 carbons of the double bond is attached to two identical substituents, there is only one possible structure for the alkene. In other words, cis and trans isomers are not possible for an alkene that has identical substituents attached to one of the double-bonded carbons.

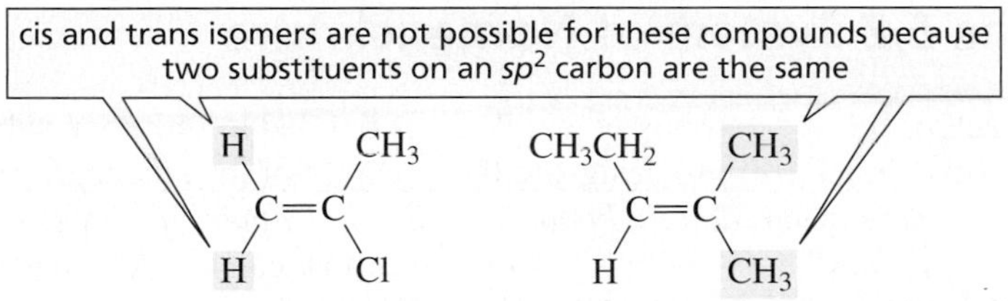

cis and trans isomers are not possible for these compounds because two substituents on an sp^2 carbon are the same

3-D Molecule:
2-Methyl-2-pentene

Cis and trans isomers can be interconverted only when the molecule absorbs sufficient heat or light energy to cause the π bond to break, because once the π bond is broken, rotation can occur easily about the remaining σ bond (Section 3.8).

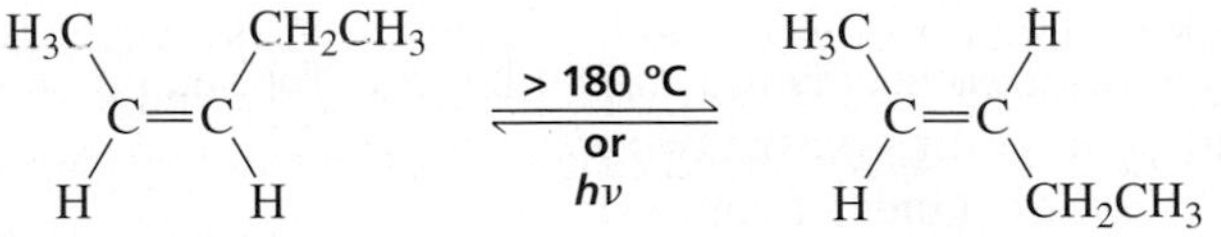

cis-2-pentene   trans-2-pentene

PROBLEM 6◆

a. Which of the following compounds can exist as cis–trans isomers?
b. For those compounds, draw and label the cis and trans isomers.

1. $CH_3CH=CHCH_2CH_3$ **3.** $CH_3CH=CHCH_3$

2. $CH_3CH=CCH_3$ **4.** $CH_3CH_2CH=CH_2$
         |
        CH_3

CIS–TRANS INTERCONVERSION IN VISION

When rhodopsin absorbs light, a double bond interconverts between the cis and trans forms. This process plays an important role in vision.

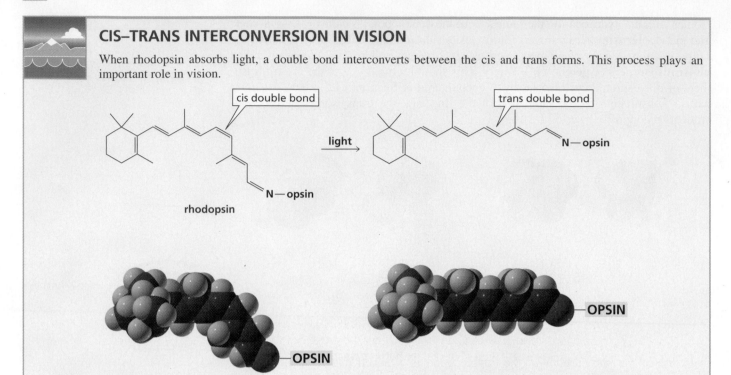

cis double bond

trans double bond

light

N—opsin

rhodopsin

N—opsin

OPSIN

cis form

OPSIN

trans form

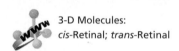

3-D Molecules:
cis-Retinal; *trans*-Retinal

4.5 The *E,Z* System of Nomenclature

As long as each of the sp^2 carbons of an alkene is bonded to only one substituent, we can use the terms *cis* and *trans* to designate the structure of the alkene. *If the hydrogens are on the same side of the double bond, it is the cis isomer; if they are on opposite sides of the double bond, it is the trans isomer.* But how can we designate the isomers of a compound such as 1-bromo-2-chloropropene?

Br   Cl   Br   CH₃
  C=C   C=C
H   CH₃   H   Cl

Which isomer is cis and which is trans?

The cis–trans system of nomenclature cannot be used for such a compound because there are four different groups on the two sp^2 carbons. The *E,Z* system of nomenclature was devised for these kinds of compounds.

To name an isomer by the *E,Z* system, we first determine the relative priorities of the two groups bonded to one of the sp^2 carbons and then the relative priorities of the two groups bonded to the other sp^2 carbon. (The rules for assigning relative priorities will soon be explained.) If the high-priority groups are on the same side of the double bond, the isomer has the *Z* configuration (*Z* is for *zusammen*, German for "together"). If the high-priority groups are on opposite sides of the double bond, the isomer has the *E* configuration (*E* is for *entgegen*, German for "opposite").

The *Z* isomer has the high-priority groups on the same side.

low priority   low priority
        C=C
high priority   high priority

the Z isomer

low priority   high priority
        C=C
high priority   low priority

the E isomer

The relative priorities of the two groups bonded to an sp^2 carbon are determined using the following rules:

- **Rule 1.** The relative priorities of the two groups depend on the atomic numbers of the atoms that are bonded directly to the sp^2 carbon. The greater the atomic number, the higher is the priority of the group.

  For example, in the following compounds, one of the sp^2 carbons is bonded to a Br and to an H:

The greater the atomic number of the atom bonded to an sp^2 carbon, the higher is the priority of the substituent.

high priority

Br	Cl
C=C
H	CH₃

the Z isomer

high priority

Br	CH₃
C=C
H	Cl

the E isomer

Br has a greater atomic number than H, so **Br** has a higher priority than **H**. The other sp^2 carbon is bonded to a Cl and to a C. Cl has the greater atomic number, so **Cl** has a higher priority than **C**. (Notice that you use the atomic number of C, not the mass of the CH₃ group, because the priorities are based on the atomic numbers of atoms, *not* on the masses of groups.) The isomer on the left has the high-priority groups (Br and Cl) on the same side of the double bond, so it is the **Z isomer**. (Zee groups are on Zee Zame Zide.) The isomer on the right has the high-priority groups on opposite sides of the double bond, so it is the **E isomer**.

- **Rule 2.** If the two substituents attached to the sp^2 carbon start with the same atom (there is a tie), you must move outward from the point of attachment and consider the atomic numbers of the atoms that are attached to the "tied" atoms.

  In the following compounds, one of the sp^2 carbons is bonded both to a Cl and to a C:

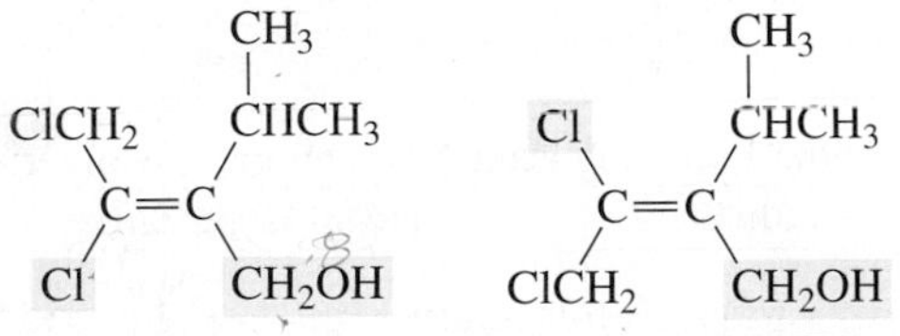

	CH₃	
ClCH₂	CHCH₃	
	C=C	
Cl	CH₂OH	

the Z isomer

	CH₃	
Cl	CHCH₃	
	C=C	
ClCH₂	CH₂OH	

the E isomer

Cl has a greater atomic number than C, so Cl has the higher priority. Both atoms bonded to the other sp^2 carbon are C's (from a CH₂OH group and a CH(CH₃)₂ group), so there is a tie at this point. The C of the CH₂OH group is bonded to **O, H**, and **H**, and the C of the CH(CH₃)₂ group is bonded to **C, C,** and **H**. Of these six atoms, O has the greatest atomic number, so **CH₂OH** has a higher priority than **CH(CH₃)₂**. (Note that you do not add the atomic numbers; you take the single atom with the greatest atomic number.) The *E* and *Z* isomers are as shown.

If the atoms attached to sp^2 carbons are the same, the atoms attached to the "tied" atoms are compared; the one with the greatest atomic number belongs to the group with the higher priority.

- **Rule 3.** If an atom is doubly bonded to another atom, the priority system treats it as if it were singly bonded to two of those atoms. If an atom is triply bonded to another atom, the priority system treats it as if it were singly bonded to three of those atoms.

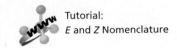

Tutorial:
E and *Z* Nomenclature

For example, one of the sp^2 carbons in the following pair of isomers is bonded to a CH_2OH group and to a $C\equiv CH$ group:

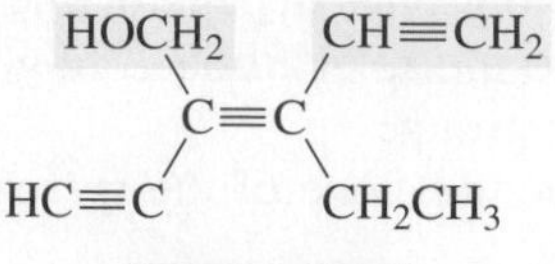

$$HOCH_2 \quad CH=CH_2$$
$$C=C$$
$$HC\equiv C \quad CH_2CH_3$$
the Z isomer

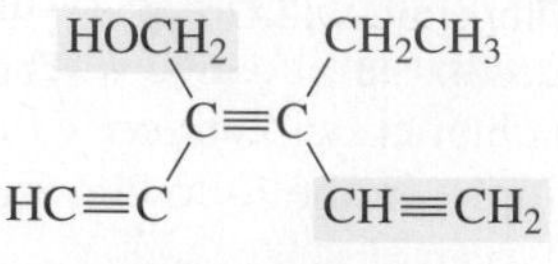

$$HOCH_2 \quad CH_2CH_3$$
$$C=C$$
$$HC\equiv C \quad CH=CH_2$$
the E isomer

If an atom is doubly bonded to another atom, treat it as if it were singly bonded to two of those atoms.

If an atom is triply bonded to another atom, treat it as if it were singly bonded to three of those atoms.

Cancel atoms that are identical in the two groups; use the remaining atoms to determine the group with the higher priority.

Because the atoms immediately bonded to the sp^2 carbon are both C's, there is a tie. The triple-bonded C is considered to be bonded to **C, C,** and **C**; the other C is bonded to **O, H,** and **H**. Of the six atoms, O has the greatest atomic number, so **CH₂OH** has a higher priority than **C≡CH**. Both atoms that are bonded to the other sp^2 carbon are C's, so there is a tie there as well. The first carbon of the CH_2CH_3 group is bonded to **C, H,** and **H**. The first carbon of the **CH=CH₂** group is bonded to an H and doubly bonded to a C. Therefore, it is considered to be bonded to **H, C,** and **C**. One C cancels in each of the two groups, leaving H and H in the CH_2CH_3 group and C and H in the CH=CH₂ group. C has a greater atomic number than H, so **CH=CH₂** has a higher priority than **CH₂CH₃**.

PROBLEM 7◆

Assign relative priorities to each set of substituents:

a. —Br, —I, —OH, —CH₃

b. —CH₂CH₂OH, —OH, —CH₂Cl, —CH=CH₂

PROBLEM 8

Draw and label the E and Z isomers for each of the following compounds:

a. CH₃CH₂CH=CHCH₃

b. 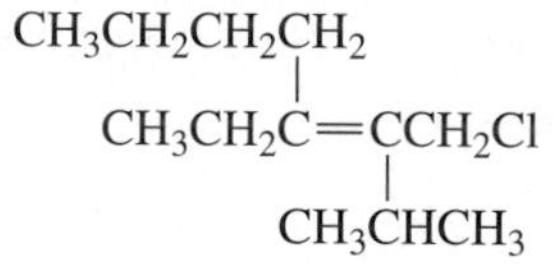 CH₃CH₂CH₂CH₂
$\qquad\qquad$ |
$\quad$ CH₃CH₂C=CCH₂Cl
$\qquad\qquad$ |
$\qquad\qquad$ CH₃CHCH₃

PROBLEM 9

Indicate whether each of the following is an E isomer or a Z isomer:

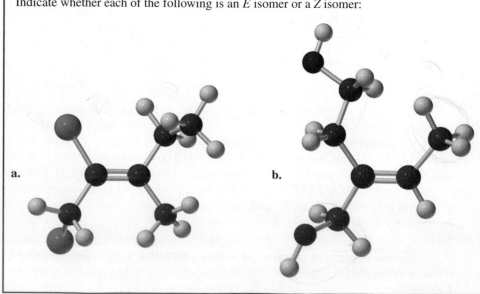

a.

b.

PROBLEM-SOLVING STRATEGY

Draw the structure of (*E*)-1-bromo-2-methyl-2-butene.

First draw the compound without specifying the isomer; this allows you to identify the substituents bonded to the sp^2 carbons. Then determine the relative priorities of the two substituents on each of the sp^2 carbons.

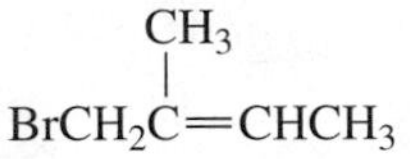

$$CH_3$$
$$|$$
$$BrCH_2C\!=\!CHCH_3$$

One sp^2 carbon is attached to a CH_3 and an H; CH_3 has the higher priority. The other sp^2 carbon is attached to a CH_3 and a CH_2Br; CH_2Br has the higher priority. To get the *E* isomer, draw the compound with the two high-priority substituents on opposite sides of the double bond.

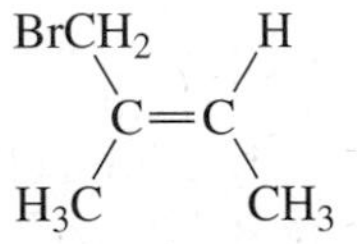

$$\begin{array}{ccc} BrCH_2 & & H \\ & \diagdown \quad \diagup & \\ & C\!=\!C & \\ & \diagup \quad \diagdown & \\ H_3C & & CH_3 \end{array}$$

Now continue on to Problem 10.

PROBLEM 10◆

Draw (*Z*)-3-isopropyl-2-heptene.

4.6   The Relative Stabilities of Alkenes

Alkyl substituents that are bonded to the sp^2 carbons of an alkene have a stabilizing effect on the alkene.

relative stabilities of alkyl-substituted alkenes

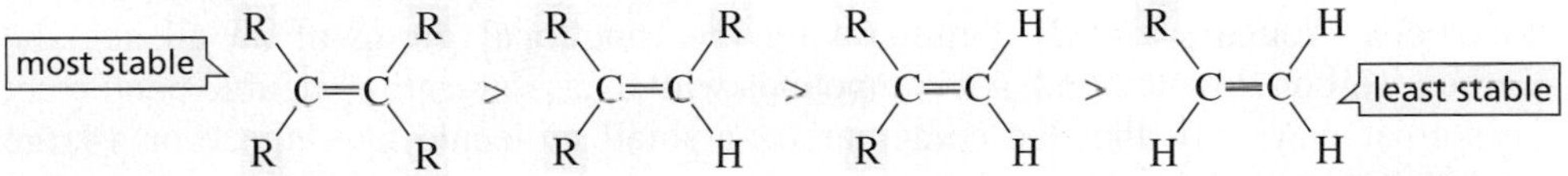

We can, therefore, make the following statement: *The more alkyl substituents bonded to the sp^2 carbons of an alkene, the greater is its stability.* (Some students find it easier to look at the number of hydrogens bonded to the sp^2 carbons. In terms of hydrogens, we can use the following rule: *The fewer hydrogens bonded to the sp^2 carbons of an alkene, the greater is its stability.*)

> The fewer hydrogens bonded to the sp^2 carbons of an alkene, the more stable it is.

PROBLEM 11

a. Which of the following compounds is the most stable?

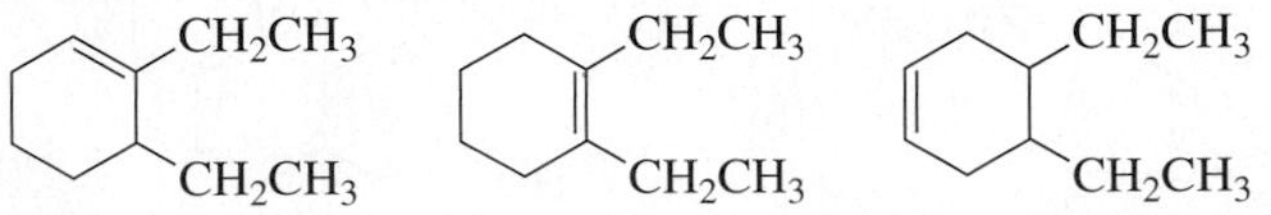

b. Which is the least stable?

Both *trans*-2-butene and *cis*-2-butene have two alkyl groups bonded to their sp^2 carbons. The trans isomer, in which the large substituents are farther apart, is more stable than the cis isomer, in which the large substituents are closer together.

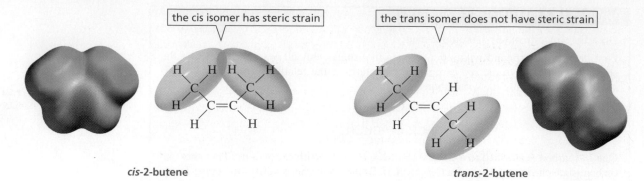

the cis isomer has steric strain

the trans isomer does not have steric strain

cis-2-butene

trans-2-butene

When the large substituents are on the same side of the molecule, their electron clouds can interfere with each other, causing strain in the molecule and making it less stable. This kind of strain is called *steric strain* (Section 3.8). When the large substituents are on opposite sides of the molecule, their electron clouds cannot interact; thus, the molecule has less steric strain and is more stable.

PROBLEM 12◆

Rank the following compounds in order of decreasing stability:

trans-3-hexene, *cis*-3-hexene, 1-hexene, *cis*-3,4-dimethyl-3-hexene

---

4.7 How Alkenes React • Curved Arrows

There are many millions of organic compounds. If you had to memorize how each of them reacts, studying organic chemistry would be a horrendous experience. Fortunately, organic compounds can be divided into families, and all the members of a family react in similar ways. What determines the family to which an organic compound belongs is its functional group. The **functional group** is the center of reactivity of a molecule. You will find a table of common functional groups inside the back cover of this book. You are already familiar with the functional group of an alkene: the carbon–carbon double bond. All compounds with a carbon–carbon double bond react in similar ways, whether the compound is a small molecule like ethene or a large molecule like cholesterol.

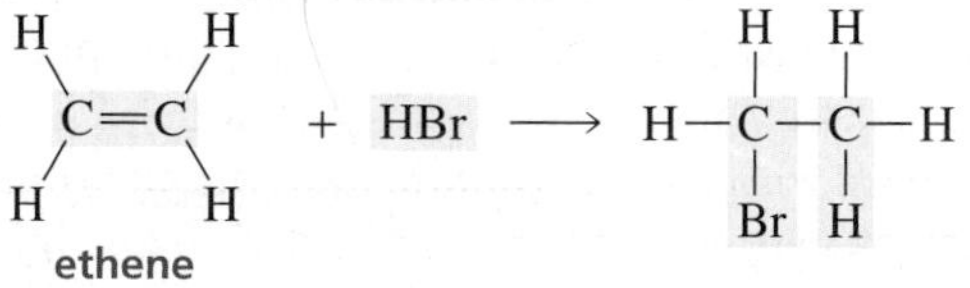

ethene

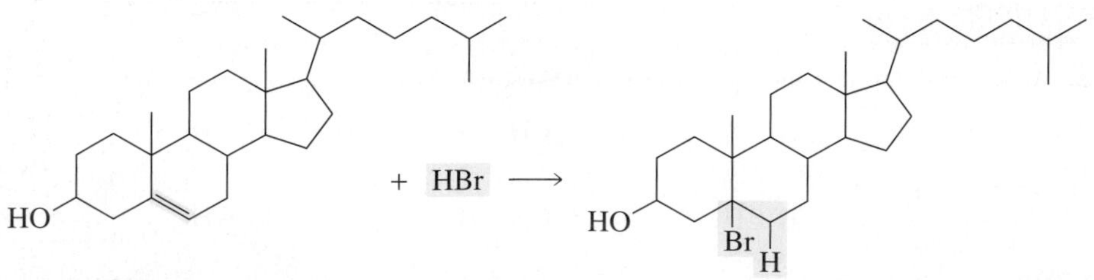

cholesterol

To further reduce the need to memorize, one needs to understand *why* a functional group reacts the way it does. It is not sufficient to know that a compound with a carbon–carbon double bond reacts with HBr to form a product in which the H and Br atoms have taken the place of the π bond; we need to understand *why* the compound reacts with HBr. In each chapter that discusses the reactivity of a particular functional group, we will see how the nature of the functional group allows us to predict the kind of reactions it will undergo. Then, when you are confronted with a reaction you have never seen before, knowledge of how the structure of the molecule affects its reactivity will help you predict the products of the reaction.

In essence, organic chemistry is about the interaction between electron-rich atoms or molecules and electron-deficient atoms or molecules. It is these forces of attraction that make chemical reactions happen. From this follows a very important rule that determines the reactivity of organic compounds: *Electron-rich atoms or molecules are attracted to electron-deficient atoms or molecules.* Each time you study a new functional group, remember that the reactions it undergoes can be explained by this very simple rule.

> Electron-rich atoms or molecules are attracted to electron-deficient atoms or molecules.

Therefore, to understand how a functional group reacts, you must first learn to recognize electron-deficient and electron-rich atoms and molecules. An electron-deficient atom or molecule is called an **electrophile**. An electrophile has an atom that can accept a pair of electrons. Thus, an electrophile looks for electrons. Literally, "electrophile" means "electron loving" (*phile* is the Greek suffix for "loving").

$$\text{H}^+ \qquad \text{CH}_3\overset{+}{\text{C}}\text{H}_2$$

these are electrophiles because they can accept a pair of electrons

An electron-rich atom or molecule is called a **nucleophile**. A nucleophile has a pair of electrons it can share. Some nucleophiles are neutral and some are negatively charged. Because a nucleophile has electrons to share and an electrophile is seeking electrons, it should not be surprising that they attract each other. Thus, the preceding rule can be restated as *a nucleophile reacts with an electrophile.*

> A nucleophile reacts with an electrophile.

$$\text{H}\ddot{\text{O}}:^- \qquad :\ddot{\text{Cl}}:^- \qquad \text{CH}_3\ddot{\text{N}}\text{H}_2 \qquad \text{H}_2\ddot{\text{O}}:$$

these are nucleophiles because they have a pair of electrons to share

We have seen that a π bond is weaker than a σ bond (Section 1.14). The π bond, therefore, is the bond that is most easily broken when an alkene undergoes a reaction. We also have seen that the π bond of an alkene consists of a cloud of electrons above and below the σ bond. This cloud of electrons causes an alkene to be an electron-rich molecule—it is a nucleophile. (Notice the relatively electron-rich pale orange area in the electrostatic potential maps for *cis*- and *trans*-2-butene in Section 4.3.) We can, therefore, predict that an alkene will react with an electrophile and, in the process, the π bond will break. So if a reagent such as hydrogen bromide is added to an alkene, the alkene (a nucleophile) will react with the partially positively charged hydrogen (an electrophile) of hydrogen bromide and a carbocation will be formed. In the second step of the reaction, the positively charged carbocation (an electrophile) will react with the negatively charged bromide ion (a nucleophile) to form an alkyl halide.

$$\text{CH}_3\text{CH}=\text{CHCH}_3 \;+\; \overset{\delta+}{\text{H}}-\overset{\delta-}{\text{Br}} \;\longrightarrow\; \underset{\underset{\text{H}}{|}}{\text{CH}_3\overset{+}{\text{C}}\text{H}-\text{CHCH}_3} \;+\; \text{Br}^- \;\longrightarrow\; \underset{\underset{\text{Br}\;\;\text{H}}{|\;\;\;|}}{\text{CH}_3\text{CH}-\text{CHCH}_3}$$

a carbocation   2-bromobutane
                                        an alkyl halide

The description of the step-by-step process by which reactants (e.g., alkene + HBr) are changed into products (e.g., alkyl halide) is called the **mechanism of the reaction**. To help us understand a mechanism, curved arrows are drawn to show how the electrons move as new covalent bonds are formed and existing covalent bonds are broken. In other

Curved arrows show the flow of electrons; they are drawn from an electron-rich center to an electron-deficient center.

words, the curved arrows show which bonds are formed and which are broken. Because the curved arrows show how the electrons flow, *they are drawn from an electron-rich center* (at the tail of the arrow) *to an electron-deficient center* (at the point of the arrow). Notice that these arrows represent the simultaneous movement of two electrons (an electron pair). These are called "curved" arrows to distinguish them from the "straight" arrows used to link reactants with products in chemical reactions.

$$CH_3CH{=}CHCH_3 \ + \ \overset{\delta+}{H}{-}\overset{\delta-}{\overset{..}{\underset{..}{Br}}}{:} \ \longrightarrow \ CH_3\overset{+}{CH}{-}CHCH_3 \ + \ :\overset{..}{\underset{..}{Br}}{:}^-$$

π bond has broken

new σ bond has formed

For the reaction of 2-butene with HBr, an arrow is drawn to show that the two electrons of the π bond of the alkene are attracted to the partially positively charged hydrogen of HBr. The hydrogen, however, is not free to accept this pair of electrons because it is already bonded to a bromine, and hydrogen can be bonded to only one atom at a time (Section 1.4). Therefore, as the π electrons of the alkene move toward the hydrogen, the H—Br bond breaks, with bromine keeping the bonding electrons. Notice that the π electrons are pulled away from one carbon, but remain attached to the other. Thus, the two electrons that formerly formed the π bond now form a σ bond between carbon and the hydrogen from HBr. The product of this first step in the reaction is a carbocation because the sp^2 carbon that did not form the new bond with hydrogen has lost a share in a pair of electrons (the electrons of the π bond). It is, therefore, positively charged.

In the second step of the reaction, a lone pair on the negatively charged bromide ion forms a bond with the positively charged carbon of the carbocation. Notice that both steps of the reaction involve *the reaction of an electrophile with a nucleophile*.

$$CH_3\overset{+}{CH}{-}CHCH_3 \ + \ :\overset{..}{\underset{..}{Br}}{:}^- \ \longrightarrow \ CH_3CH{-}CHCH_3$$

H   :Br: H

new σ bond

Solely from the knowledge that an electrophile reacts with a nucleophile and a π bond is the weakest bond in an alkene, we have been able to predict that the product of the reaction of 2-butene and HBr is 2-bromobutane. Overall, the reaction involves the *addition* of 1 mole of HBr to 1 mole of the alkene. The reaction, therefore, is called an **addition reaction**. Because the first step of the reaction involves the addition of an electrophile (H^+) to the alkene, the reaction is more precisely called an **electrophilic addition reaction**. Electrophilic addition reactions are the characteristic reactions of alkenes.

At this point, you may think that it would be easier just to memorize the fact that 2-bromobutane is the product of the reaction, without trying to understand the mechanism that explains why 2-bromobutane is the product. Keep in mind, however, that the number of reactions you encounter is going to increase substantially, and it will be impossible to memorize them all. If you strive to understand the mechanism of each reaction, the unifying principles of organic chemistry will become apparent, making mastery of the material much easier and a lot more fun.

It will be helpful to do the exercise on drawing curved arrows in the Study Guide/Solution Manual (Special Topic II).

PROBLEM 13◆

Which of the following are electrophiles, and which are nucleophiles?

$$H^- \qquad CH_3O^- \qquad CH_3C{\equiv}CH \qquad CH_3\overset{+}{CH}CH_3 \qquad NH_3$$

A FEW WORDS ABOUT CURVED ARROWS

1. Make certain that the curved arrows are drawn in the direction of the electron flow and never against the flow. This means that an arrow will always be drawn away from a negative charge and/or toward a positive charge.

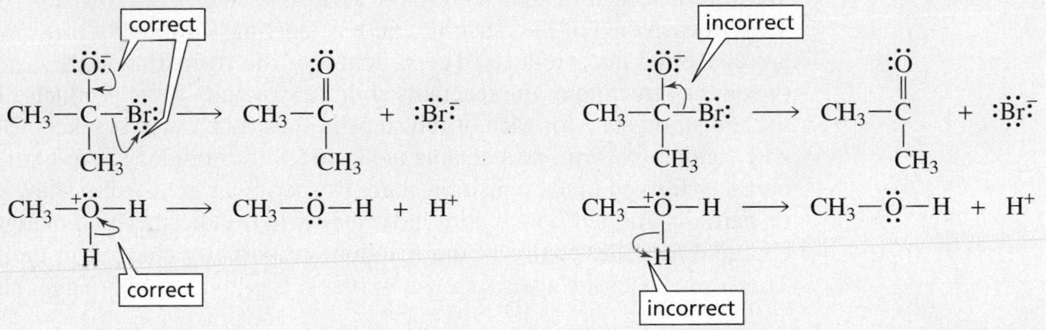

correct

$$CH_3-\overset{\overset{\displaystyle :\ddot{O}:^-}{|}}{\underset{\underset{\displaystyle CH_3}{|}}{C}}-\ddot{B}r: \longrightarrow CH_3-\overset{\overset{\displaystyle :\ddot{O}}{\|}}{\underset{\underset{\displaystyle CH_3}{|}}{C}} + :\ddot{B}r:^-$$

$$CH_3-\overset{+}{\ddot{O}}-H \longrightarrow CH_3-\ddot{O}-H + H^+$$
$$\underset{\displaystyle H}{|}$$

correct

incorrect

$$CH_3-\overset{\overset{\displaystyle :\ddot{O}:^-}{|}}{\underset{\underset{\displaystyle CH_3}{|}}{C}}-\ddot{B}r: \longrightarrow CH_3-\overset{\overset{\displaystyle :\ddot{O}}{\|}}{\underset{\underset{\displaystyle CH_3}{|}}{C}} + :\ddot{B}r:^-$$

$$CH_3-\overset{+}{\ddot{O}}-H \longrightarrow CH_3-\ddot{O}-H + H^+$$
$$\underset{\displaystyle H}{|}$$

incorrect

2. Curved arrows are drawn to indicate the movement of electrons. Never use a curved arrow to indicate the movement of an atom. For example, you can't use an arrow as a lasso to remove the proton, as shown here:

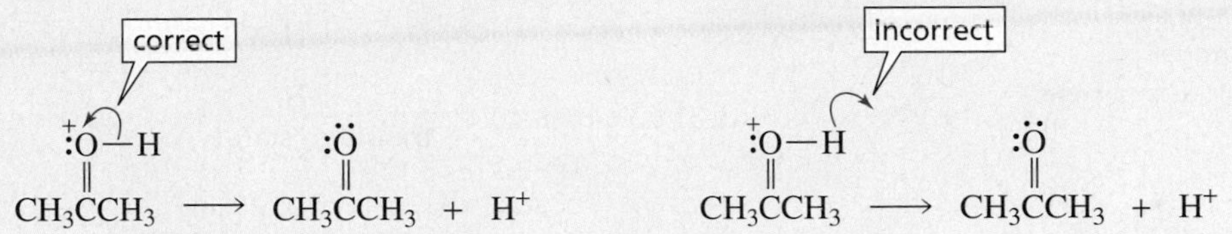

correct

$$\overset{+}{:}\!\ddot{O}-H$$
$$\underset{\displaystyle \underset{CH_3CCH_3}{\|}}{\overset{:\ddot{O}}{}} \longrightarrow CH_3CCH_3 + H^+$$

incorrect

$$\overset{+}{:}\!O-H$$
$$\underset{\displaystyle \underset{CH_3CCH_3}{\|}}{\overset{:\ddot{O}}{}} \longrightarrow CH_3CCH_3 + H^+$$

3. The arrow starts at the electron source. It does not start at an atom. In the following example, the arrow starts at the electrons of the π bond, not at a carbon atom:

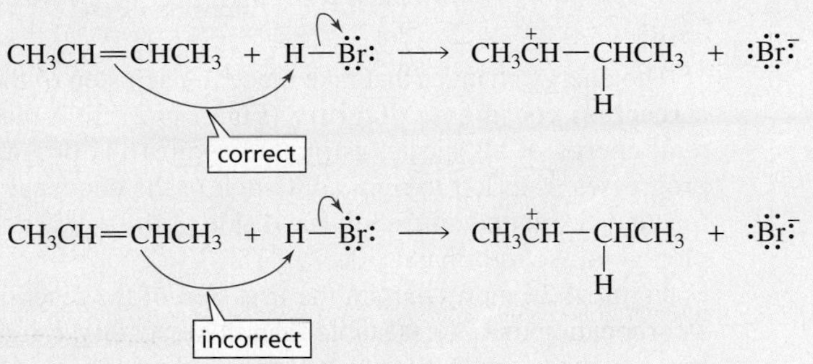

$$CH_3CH{=}CHCH_3 + H-\ddot{B}r: \longrightarrow CH_3\overset{+}{C}H-\underset{\underset{\displaystyle H}{|}}{C}HCH_3 + :\ddot{B}r:^-$$

correct

$$CH_3CH{=}CHCH_3 + H-\ddot{B}r: \longrightarrow CH_3\overset{+}{C}H-\underset{\underset{\displaystyle H}{|}}{C}HCH_3 + :\ddot{B}r:^-$$

incorrect

PROBLEM 14

Use curved arrows to show the movement of electrons in each of the following reaction steps:

a. 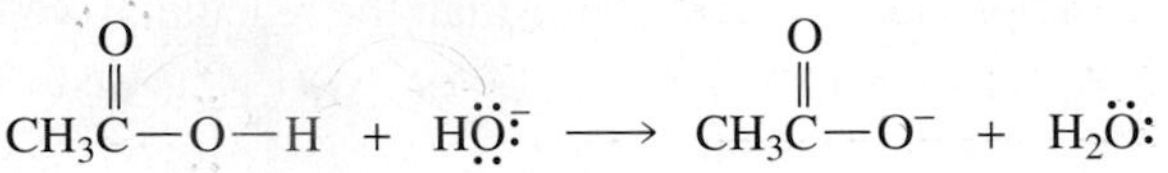 $CH_3\overset{\overset{\displaystyle O}{\|}}{C}-O-H + H\ddot{O}:^- \longrightarrow CH_3\overset{\overset{\displaystyle O}{\|}}{C}-O^- + H_2\ddot{O}:$

b. (hexene ring) $+ Br^+ \longrightarrow$ (cyclohexane ring with Br, +)

c. $CH_3\overset{\overset{\displaystyle \ddot{O}:}{\|}}{C}OH + H-\underset{\underset{\displaystyle H}{|}}{\ddot{O}}-H \longrightarrow CH_3\overset{\overset{\displaystyle \overset{+}{\ddot{O}}H}{\|}}{C}OH + H_2O$

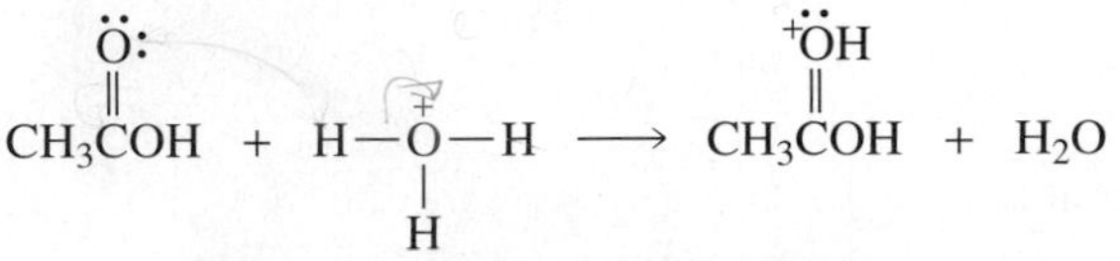

d. 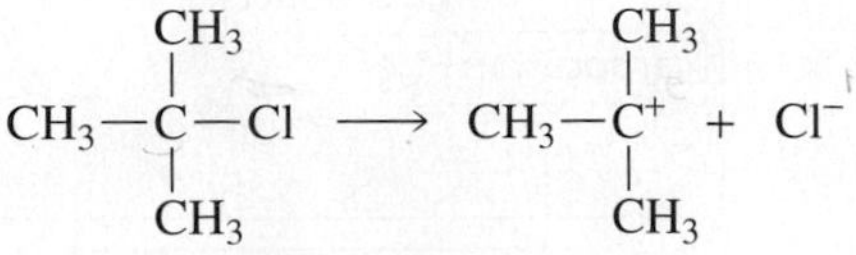 $CH_3-\overset{\overset{\displaystyle CH_3}{|}}{\underset{\underset{\displaystyle CH_3}{|}}{C}}-Cl \longrightarrow CH_3-\overset{\overset{\displaystyle CH_3}{|}}{\underset{\underset{\displaystyle CH_3}{|}}{\overset{+}{C}}} + Cl^-$

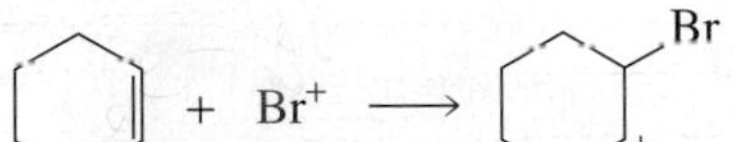

4.8 Using a Reaction Coordinate Diagram to Describe a Reaction

We have just seen that the addition of HBr to 2-butene is a two-step reaction (Section 4.7). In each step of the reaction, the reactants pass through a *transition state* as they are converted into products. The structure of the **transition state** lies somewhere between the structure of the reactants and the structure of the products. The structure of the transition state for each of the steps is shown below in brackets. Bonds that break and bonds that form, as reactants are converted to products, are partially broken and partially formed in the transition state. Dashed lines are used to show partially broken or partially formed bonds. Similarly, atoms that either become charged or lose their charge during the course of the reaction are partially charged in the transition state. Transition states are always shown in brackets with a double-dagger superscript.

Mechanistic Tutorial:
Addition of HBr to an alkene

$$CH_3CH=CHCH_3 \ + \ HBr \ \longrightarrow \left[\begin{array}{c} \overset{\delta+}{CH_3CH} \cdots CHCH_3 \\ \vdots \\ H \\ \vdots \\ \overset{\delta-}{Br} \end{array} \right]^{\ddagger} \longrightarrow CH_3\overset{+}{C}HCH_2CH_3 \ + \ Br^-$$
transition state

$$CH_3\overset{+}{C}HCH_2CH_3 \ + \ Br^- \ \longrightarrow \left[\begin{array}{c} \overset{\delta+}{CH_3CH}CH_2CH_3 \\ \vdots \\ \overset{\delta-}{Br} \end{array} \right]^{\ddagger} \longrightarrow \begin{array}{c} CH_3CHCH_2CH_3 \\ | \\ Br \end{array}$$
transition state

The energy changes that take place in each step of the reaction can be described by a **reaction coordinate diagram** (Figure 4.2). In a reaction coordinate diagram, the total energy of all species is plotted against the progress of the reaction. A reaction progresses from left to right as written in the chemical equation, so the energy of the reactants is plotted on the left-hand side of the *x*-axis and the energy of the products is plotted on the right-hand side.

Figure 4.2a shows that, in the first step of the reaction, the alkene is converted into a carbocation that is less stable than the reactants. Remember that *the more stable the species, the lower is its energy*. Because the product of the first step is less stable than the reactants, we know that this step consumes energy. We see that as the carbocation is formed, the reaction passes through the transition state. Notice that the transition state is a *maximum* energy state on the reaction coordinate diagram.

The carbocation reacts in the second step with the bromide ion to form the final product (Figure 4.2b). Because the product is more stable than the reactants, we know that this step releases energy.

The more stable the species, the lower is its energy.

Figure 4.2 ▶
Reaction coordinate diagrams for the two steps in the addition of HBr to 2-butene: (a) the first step; (b) the second step.

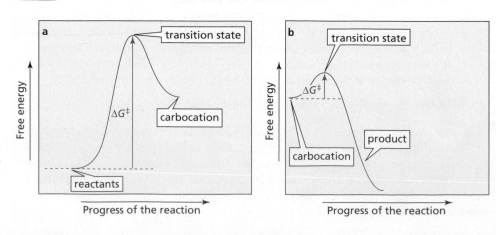

A chemical species that is the product of one step of a reaction and is the reactant for the next step is called an **intermediate**. Thus, the carbocation is an intermediate. Although the carbocation is more stable than either of the transition states, it is still too unstable to be isolated. Do not confuse transition states with intermediates: *Transition states have partially formed bonds, whereas intermediates have fully formed bonds.*

Because the product of the first step is the reactant in the second step, we can hook the two reaction coordinate diagrams together to obtain the reaction coordinate diagram for the overall reaction (Figure 4.3).

Transition states have partially formed bonds. Intermediates have fully formed bonds.

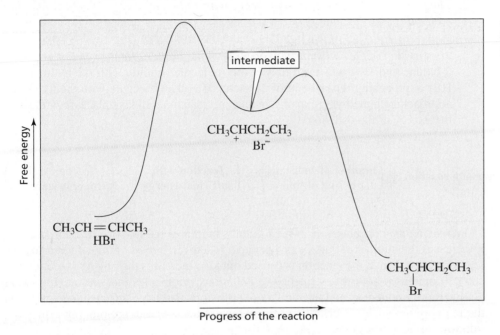

◀ **Figure 4.3**
Reaction coordinate diagram for the addition of HBr to 2-butene.

A reaction is over when the system reaches equilibrium. The relative concentrations of products and reactants at equilibrium depend on their relative stabilities: *The more stable the compound, the greater is its concentration at equilibrium.* Thus, if the products are more stable (have a lower free energy) than the reactants, there will be a higher concentration of products than reactants at equilibrium. On the other hand, if the reactants are more stable than the products, there will be a higher concentration of reactants than products at equilibrium. We see that the free energy of the final products in Figure 4.3 is lower than the free energy of the initial reactants. Therefore, we know that there will be more products than reactants when the reaction has reached equilibrium. A reaction that leads to a higher concentration of products compared with the concentration of reactants is called a **favorable reaction**.

The more stable the compound, the greater is its concentration at equilibrium.

How fast a reaction occurs is indicated by the energy "hill" that must be climbed for the reactants to be converted into products. The higher the energy barrier, the slower is the reaction. The energy barrier is called the **free energy of activation**. The free energy of activation for each step is indicated by $\Delta G^{\ddagger}$ in Figure 4.2. It is the difference between the free energy of the transition state and the free energy of the reactants:

$$\Delta G^{\ddagger} = \text{(free energy of the transition state)} - \text{(free energy of the reactants)}$$

We can see from the reaction coordinate diagram that the free energy of activation for the first step of the reaction is greater than the free energy of activation for the second step. In other words, the first step of the reaction is slower than the second step. This is what you would expect because the molecules in the first step of this reaction must collide with sufficient energy to break covalent bonds, whereas no bonds are broken in the second step.

The higher the energy barrier, the slower is the reaction.

The reaction step that has its transition state *at the highest point on the reaction coordinate* is called the **rate-determining step**. The rate-determining step controls the overall rate of the reaction because the overall rate cannot exceed the rate of the rate-determining step. In Figure 4.3, the rate-determining step is the first step—the addition of the electrophile (the proton) to the alkene.

What determines how fast a reaction occurs? The rate of a reaction depends on the following factors:

1. *The number of collisions that take place between the reacting molecules in a given period of time.* The greater the number of collisions, the faster is the reaction.

2. *The fraction of the collisions that occur with sufficient energy to get the reacting molecules over the energy barrier.* If the free energy of activation is small, more collisions will lead to reaction than if the free energy of activation is large.

3. *The fraction of the collisions that occur with the proper orientation.* For example, 2-butene and HBr will react only if the molecules collide with the hydrogen of HBr approaching the π bond of 2-butene. If collision occurs with the hydrogen approaching a methyl group of 2-butene, no reaction will take place, regardless of the energy of the collision.

$$\text{rate of a reaction} = \left(\begin{matrix}\textbf{number of collisions}\\\textbf{per unit of time}\end{matrix}\right) \times \left(\begin{matrix}\textbf{fraction with}\\\textbf{sufficient energy}\end{matrix}\right) \times \left(\begin{matrix}\textbf{fraction with}\\\textbf{proper orientation}\end{matrix}\right)$$

Increasing the concentration of the reactants increases the rate of a reaction because it increases the number of collisions that occur in a given period of time. Increasing the temperature at which the reaction is carried out also increases the rate of a reaction because it increases both the frequency of collisions (molecules that are moving faster collide more frequently) and the number of collisions that have sufficient energy to get the reacting molecules over the energy barrier (molecules that are moving faster collide with greater energy).

PROBLEM 15

Draw a reaction coordinate diagram for

a. a fast reaction with products that are more stable than the reactants.

b. a slow reaction with products that are more stable than the reactants.

c. a slow reaction with products that are less stable than the reactants.

PROBLEM 16◆

Given the following reaction coordinate diagram for the reaction of A to give D, answer the following questions:

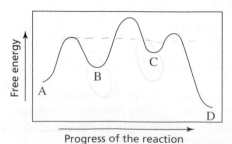

a. How many intermediates are there?

b. Which intermediate is the most stable?

c. How many transition states are there?

d. Which transition state is the most stable?

e. Which is more stable, reactants or products?

f. What is the fastest step in the reaction?

g. What is the reactant of the rate-determining step?

h. Is the overall reaction favorable?

PESTICIDES: NATURAL AND SYNTHETIC

Long before chemists learned how to create compounds that would protect plants from predators, plants were doing the job themselves. Plants had every incentive to synthesize pesticides—when you can't run, you need to find another way to protect yourself. But this poses a question: Which pesticides are more harmful—those synthesized by chemists or those synthesized by plants? Unfortunately, we don't know the answer because federal laws require that all human-made pesticides be tested for any cancer-causing effects, but do not require plant-made pesticides to be tested. In addition, risk evaluations of chemicals are usually done on rats, and a chemical that is carcinogenic in a rat may not be carcinogenic in a human. Some chemicals are harmful only at high doses, and rats are exposed to much greater concentrations of the chemical during testing than humans are exposed to during normal activities. For example, we all need sodium chloride for survival, but high concentrations are poisonous. We associate alfalfa sprouts with healthy eating, but monkeys fed very large amounts of alfalfa sprouts develop an immune system disorder.

Summary

Alkenes are hydrocarbons that contain a double bond. The double bond is the **functional group** or center of reactivity of the alkene. The **functional group suffix** of an alkene is *ene*. The general molecular formula for a hydrocarbon is C_nH_{2n+2}, minus two hydrogens for every π bond or ring in the molecule. Because alkenes contain fewer than the maximum number of hydrogens, they are called **unsaturated hydrocarbons**.

Because of restricted rotation about the double bond, an alkene can exist as **cis–trans isomers**. The **cis isomer** has its hydrogens on the same side of the double bond; the **trans isomer** has its hydrogens on opposite sides of the double bond. The **Z isomer** has the high-priority groups on the same side of the double bond; the **E isomer** has the high-priority groups on opposite sides of the double bond. The relative priorities depend on the atomic numbers of the atoms bonded directly to the sp^2 carbons. The more alkyl substituents bonded to the sp^2 carbons of an alkene, the greater is its stability. **Trans alkenes** are more stable than **cis alkenes** because of steric strain.

All compounds with a particular **functional group** react similarly. Due to the cloud of electrons above and below its π bond, an alkene is an electron-rich molecule (a **nucleophile**). Nucleophiles are attracted to electron-deficient atoms or molecules, called **electrophiles**. Alkenes undergo **electrophilic addition reactions**. The description of the step-by-step process by which reactants are changed into products is called the **mechanism of the reaction**. **Curved arrows** show which bonds are formed and which are broken and the direction of the electron flow that accompany these changes.

A **reaction coordinate diagram** shows the energy changes that take place in a reaction. The more stable the species, the lower is its energy. As reactants are converted into products, a reaction passes through a maximum energy **transition state**. An **intermediate** is the product of one step of a reaction and the reactant for the next step. Transition states have partially formed bonds; intermediates have fully formed bonds. The **rate-determining step** has its transition state at the highest point on the reaction coordinate.

The relative concentrations of products and reactants at equilibrium depend on their relative stabilities. The more stable the compound, the greater is its concentration at equilibrium. The **free energy of activation**, $\Delta G^{\ddagger}$, is the energy barrier of a reaction. It is the difference between the free energy of the reactants and the free energy of the transition state. The smaller the $\Delta G^{\ddagger}$, the faster is the reaction.

Problems

17. Give the systematic name for each of the following compounds:

a.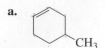

b.

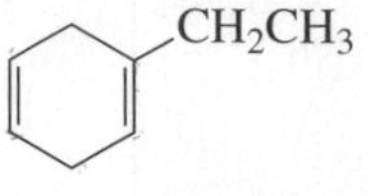

c.

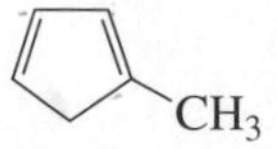

18. Squalene, a hydrocarbon with molecular formula $C_{30}H_{50}$, is obtained from shark liver. (*Squalus* is Latin for "shark.") If squalene is a noncyclic compound, how many π bonds does it have?

19. Draw and label the E and Z isomers for each of the following compounds:

a. 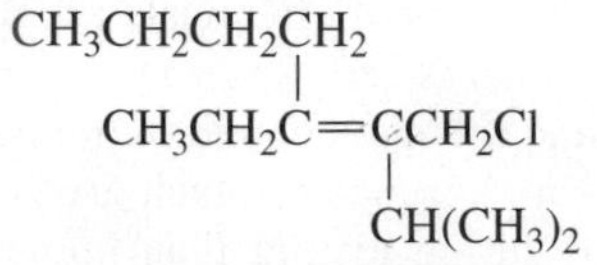 CH₃CH₂CH₂CH₂
           |
   CH₃CH₂C=CCH₂Cl   CH Cl
             |
         CH(CH₃)₂   CH CC

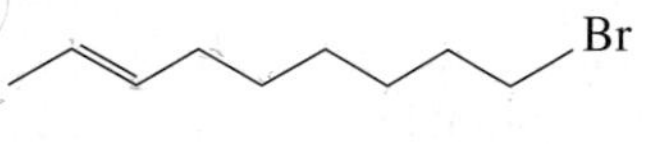

b. 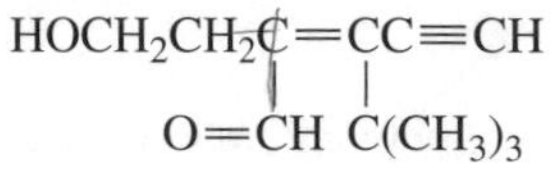 HOCH₂CH₂C=CC≡CH
            | |
       O=CH C(CH₃)₃

20. For each of the following pairs, indicate which member is more stable:

a. 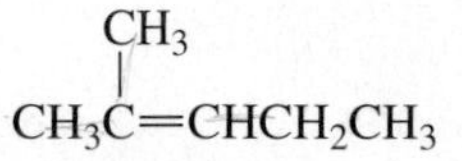
    CH₃
    |
  CH₃C=CHCH₂CH₃   or   CH₃CH=CHCHCH₃
                              |
                              CH₃

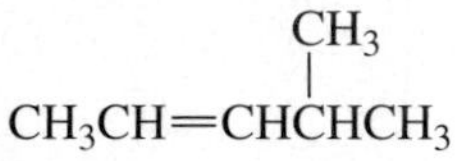

b. 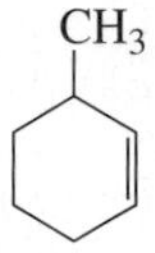 (cyclohexene with CH₃) or (cyclohexene with CH₃)

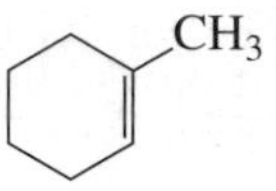

21. a. Give the structures and the systematic names for all alkenes with molecular formula C_4H_8, ignoring cis–trans isomers. (*Hint:* There are three.)

    b. Which of the compounds have E and Z isomers?

22. Draw the structure for each of the following:

    a. (Z)-1,3,5-tribromo-2-pentene
    b. (Z)-3-methyl-2-heptene
    c. (E)-1,2-dibromo-3-isopropyl-2-hexene

    d. vinyl bromide
    e. 1,2-dimethylcyclopentene
    f. diallylamine

23. Determine the total number of double bonds and/or rings for the hydrocarbons with the following molecular formulas:

    a. $C_{12}H_{20}$ **b.** $C_{40}H_{56}$

24. Name the following compounds:

a.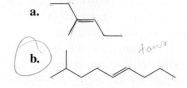

b.

c. 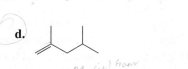 Br

d.

e.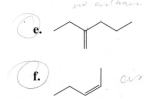

f.

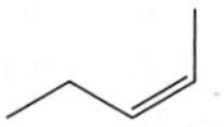

25. Draw curved arrows to show the flow of electrons responsible for the conversion of the reactants into the products:

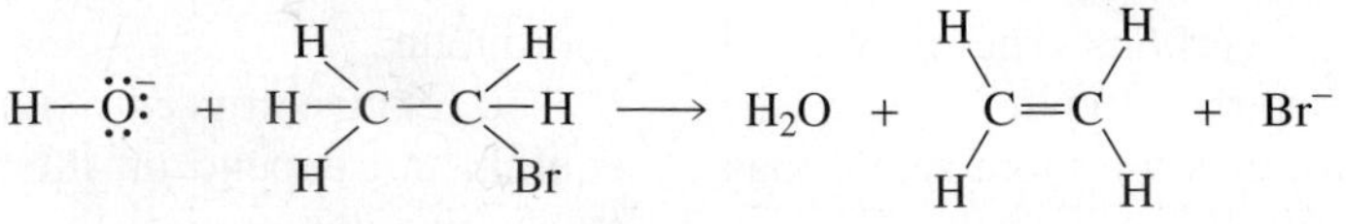

H—Ö:⁻ + H—C—C—H ⟶ H₂O + C=C + Br⁻

(with structural diagrams showing H, H on carbons, Br substituent, and the alkene product with H atoms)

26. Draw three alkenes with molecular formula C_5H_{10} that do not have cis–trans isomers.

27. Tell whether each of the following compounds has the *E* or the *Z* configuration:

a.

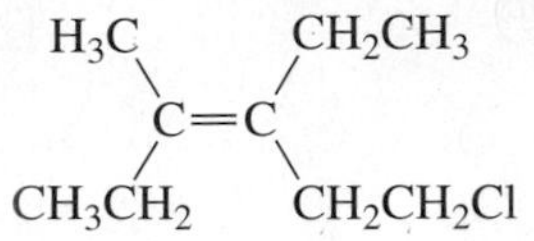

$$\underset{CH_3CH_2}{\overset{H_3C}{\diagdown}}C=C\underset{CH_2CH_2Cl}{\overset{CH_2CH_3}{\diagup}}$$

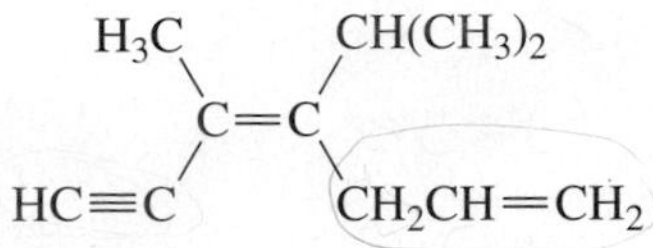

c.

$$\underset{Br}{\overset{H_3C}{\diagdown}}C=C\underset{CH_2CH_2CH_2CH_3}{\overset{CH_2Br}{\diagup}}$$

b.

$$\underset{HC\equiv C}{\overset{H_3C}{\diagdown}}C=C\underset{CH_2CH=CH_2}{\overset{CH(CH_3)_2}{\diagup}}$$

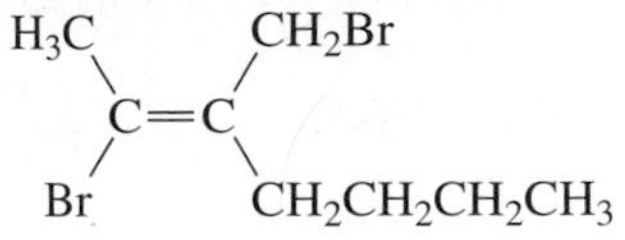

d.

$$\underset{HOCH_2}{\overset{\overset{\displaystyle O}{\parallel}}{CH_3C}}C=C\underset{CH_2CH_2Cl}{\overset{CH_2Br}{\diagup}}$$

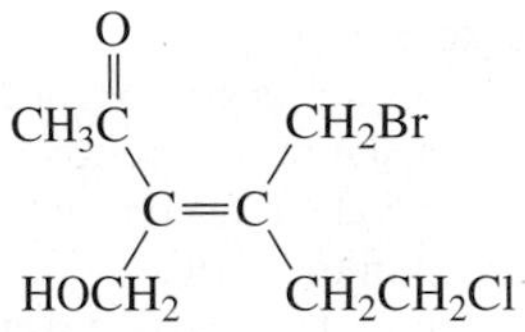

28. Which of the following compounds is the most stable? Which is the least stable?

3,4-dimethyl-2-hexene; 2,3-dimethyl-2-hexene; 4,5-dimethyl-2-hexene

29. Assign relative priorities to each set of substituents:
  a. —CH₂CH₂CH₃,   —CH(CH₃)₂,   —CH=CH₂,   —CH₃
  b. —CH₂NH₂,   —NH₂, —OH,   —CH₂OH
  c. —COCH₃,   —CH=CH₂,   —Cl,   —C≡N

30. Determine the molecular formula for each of the following:
  a. a 5-carbon hydrocarbon with two π bonds and no rings
  b. an 8-carbon hydrocarbon with three π bonds and one ring

31. Give the systematic name for each of the following compounds:

a. 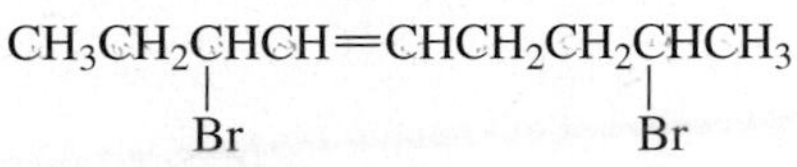 CH₃CH₂CHCH=CHCH₂CH₂CHCH₃
             | |
             Br   Br

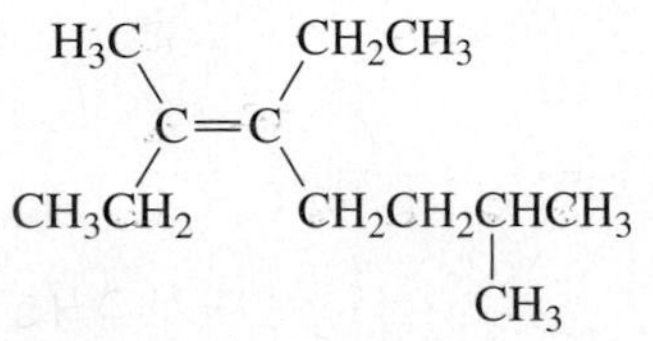

c.

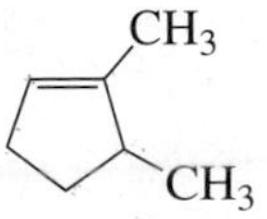

$$\text{(cyclopentene ring with CH}_3 \text{ at C1 and CH}_3 \text{ at C3)}$$

b.

$$\underset{CH_3CH_2}{\overset{H_3C}{\diagdown}}C=C\underset{CH_2CH_2CHCH_3}{\overset{CH_2CH_3}{\diagup}}$$
                                                           |
                                                          CH₃

d.

$$\underset{H_3C}{\overset{H_3C}{\diagdown}}C=C\underset{CH_2CH_2CH_2CH_3}{\overset{CH_2CH_3}{\diagup}}$$

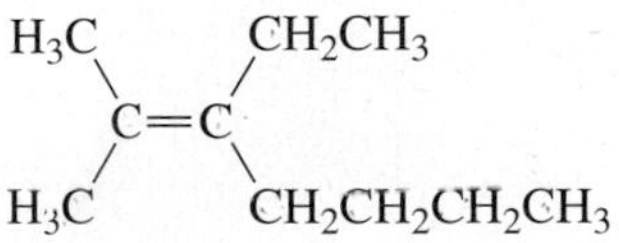

32. Draw a reaction coordinate diagram for a two-step reaction in which the products of the first step are less stable than the reactants, the reactants of the second step are less stable than the products of the second step, the final products are less stable than the initial reactants, and the second step is a the rate-determining step. Label the reactants, products, intermediates, and transition states.

33. Molly Kule was a lab technician who was asked by her supervisor to add names to the labels on a collection of alkenes that showed only structures on the labels. How many did Molly get right? Correct the incorrect names.
  a. 3-pentene
  b. 2-octene
  c. 2-vinylpentane
  d. 1-ethyl-1-pentene
  e. 5-ethylcyclohexene
  f. 5-chloro-3-hexene
  g. 5-bromo-2-pentene
  h. (*E*)-2-methyl-1-hexene
  i. 2-methylcyclopentene
  j. 2-ethyl-2-butene

34. Determine the number of double bonds and/or π bonds and then draw possible structures, for compounds with the following molecular formulas:
  a. C₃H₆   **b.** C₃H₄   **c.** C₄H₆

35. Draw a reaction coordinate diagram for the following reaction in which C is the most stable and B is the least stable of the three species and the transition state going from A to B is more stable than the transition state going from B to C:

$$A \underset{k_{-1}}{\overset{k_1}{\rightleftharpoons}} B \underset{k_{-2}}{\overset{k_2}{\rightleftharpoons}} C$$

   a. How many intermediates are there?
   b. How many transition states are there?
   c. Which step has the greater rate constant in the forward direction?
   d. Which step has the greater rate constant in the reverse direction?
   e. Of the four steps, which has the greatest rate constant?
   f. Which is the rate-determining step in the forward direction?
   g. Which is the rate-determining step in the reverse direction?

36. The rate constant for a reaction can be increased by _____ the stability of the reactant or by _____ the stability of the transition state.

37. α-Farnesene is a compound found in the waxy coating of apple skins. To complete its systematic name, include the E or Z designation after the number indicating the location of the double bond.

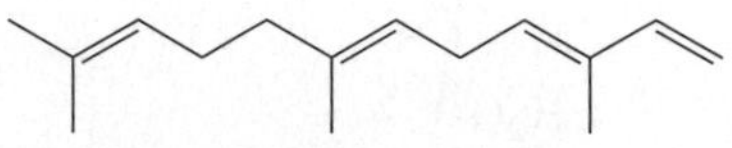

α-**farnesene**
3,7,11-trimethyl-(1,3?,6?,10)-dodcatetraene

38. Give the structures and the systematic names for all alkenes with molecular formula C_6H_{12}, ignoring cis–trans isomers. (*Hint:* There are 13.)
   a. Which of the compounds have E and Z isomers?
   b. Which of the compounds is the most stable?

5 Reactions of Alkenes and Alkynes

An Introduction to Multistep Synthesis

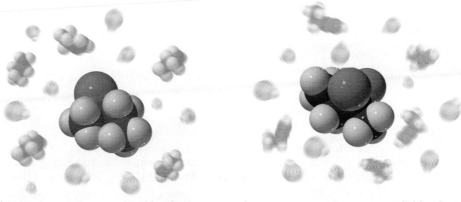

1-butene + HCl $\longrightarrow$ 2-chlorobutane   **1-butyne + 2HCl $\longrightarrow$ 2,2-dichlorobutane**

We have seen that an alkene such as 2-butene undergoes an **electrophilic addition reaction** with HBr (Section 4.7). The first step of the reaction is a relatively slow addition of the proton (an electrophile) to the alkene (a nucleophile) to form a carbocation intermediate. In the second step, the positively charged carbocation intermediate (an electrophile) reacts rapidly with the negatively charged bromide ion (a nucleophile).

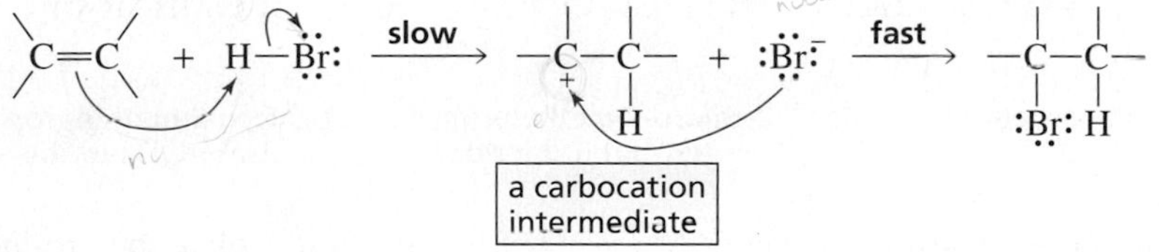

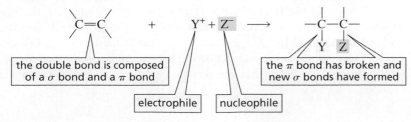

a carbocation intermediate

In this chapter, we will look at more reactions of alkenes. You will see that they all occur by similar mechanisms. As you study each reaction, notice the feature that all alkene reactions have in common: *The relatively loosely held π electrons of the carbon–carbon double bond are attracted to an electrophile. Thus, each reaction starts with the addition of an electrophile to one of the sp^2 carbons of the alkene and concludes with the addition of a nucleophile to the other sp^2 carbon.* The end result is that the π bond breaks and the sp^2 carbons form new σ bonds with the electrophile and the nucleophile.

the double bond is composed of a σ bond and a π bond

the π bond has broken and new σ bonds have formed

electrophile   nucleophile

This reactivity makes alkenes an important class of organic compounds because they can be used to synthesize a wide variety of other compounds. For example, we will see that alkyl halides, alcohols, ethers, and alkanes all can be synthesized from alkenes by electrophilic addition reactions. The particular product obtained depends only on the *electrophile* and the *nucleophile* used in the addition reaction.

5.1 Addition of a Hydrogen Halide to an Alkene

If the electrophilic reagent that adds to an alkene is a hydrogen halide (HF, HCl, HBr, or HI), the product of the reaction will be an alkyl halide:

$$CH_2\!=\!CH_2 + HCl \longrightarrow CH_3CH_2Cl$$

ethene   chloroethane
                                                   ethyl chloride

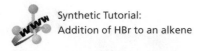 + HI $\longrightarrow$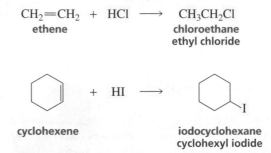

cyclohexene   iodocyclohexane
                               cyclohexyl iodide

Because the alkenes in the preceding reactions have the same substituents on both of the sp^2 carbons, it is easy to determine the product of the reaction: The electrophile (H^+) adds to one of the sp^2 carbons, and the nucleophile (X^-) adds to the other sp^2 carbon. It doesn't make any difference which sp^2 carbon the electrophile attaches to, because the same product will be obtained in either case.

But what happens if the alkene does not have the same substituents on both of the sp^2 carbons? Which sp^2 carbon gets the hydrogen? For example, does the addition of HCl to 2-methylpropene produce *tert*-butyl chloride or isobutyl chloride?

$$\underset{\text{2-methylpropene}}{\overset{\overset{\displaystyle CH_3}{|}}{CH_3C\!=\!CH_2}} + HCl \longrightarrow \underset{\substack{\text{2-chloro-2-methylpropane}\\ \textit{tert}\text{-butyl chloride}}}{\overset{\overset{\displaystyle CH_3}{|}}{\underset{\underset{\displaystyle Cl}{|}}{CH_3CCH_3}}} \quad\text{or}\quad \underset{\substack{\text{1-chloro-2-methylpropane}\\ \text{isobutyl chloride}}}{\overset{\overset{\displaystyle CH_3}{|}}{CH_3CHCH_2Cl}}$$

To answer this question, we need to carry out the reaction, isolate the products, and identify them. When we do, we find that the only product of the reaction is *tert*-butyl chloride. Now we need to find out why that compound is the product of the reaction so we can use this knowledge to predict the products of other alkene reactions. To do that, we need to look at the **mechanism of the reaction**.

Recall that the first step of the reaction—the addition of H^+ to an sp^2 carbon to form either the *tert*-butyl cation or the isobutyl cation—is the rate-determining step (Section 4.7). If there is any difference in the rate of formation of these two carbocations, the one that is formed faster will be the preferred product of the first step. Moreover, because carbocation formation is rate determining, the particular carbocation that is formed in the first step determines the final product of the reaction. That is, if the *tert*-butyl cation is formed, it will react rapidly with Cl^- to form *tert*-butyl chloride. On the other hand, if the isobutyl cation is formed, it will react rapidly with Cl^- to form isobutyl chloride. Knowing that the only product of the reaction is *tert*-butyl chloride, we know that the *tert*-butyl cation is formed faster than the isobutyl cation.

Synthetic Tutorial:
Addition of HBr to an alkene

$$
\begin{array}{c}
\text{CH}_3 \\
| \\
\text{CH}_3\overset{+}{\text{C}}\text{CH}_3 \\
\textit{tert-butyl cation}
\end{array}
\xrightarrow{\text{Cl}^-}
\begin{array}{c}
\text{CH}_3 \\
| \\
\text{CH}_3\text{CCH}_3 \\
| \\
\text{Cl} \\
\textit{tert-butyl chloride} \\
\textbf{only product formed}
\end{array}
$$

$$
\begin{array}{c}
\text{CH}_3 \\
| \\
\text{CH}_3\text{C}=\text{CH}_2
\end{array} + \text{HCl}
$$

$$
\xcancel{\longrightarrow}
\begin{array}{c}
\text{CH}_3 \\
| \\
\text{CH}_3\overset{+}{\text{C}}\text{HCH}_2 \\
\textbf{isobutyl cation}
\end{array}
\xrightarrow{\text{Cl}^-}
\begin{array}{c}
\text{CH}_3 \\
| \\
\text{CH}_3\text{CHCH}_2\text{Cl} \\
\textbf{isobutyl chloride} \\
\textbf{not formed}
\end{array}
$$

The sp^2 carbon that does *not* become attached to the proton is the carbon that is positively charged in the carbocation.

Why is the *tert*-butyl cation formed faster than the isobutyl cation? To answer this, we need to take a look at the factors that affect the stability of carbocations and, therefore, the ease with which they are formed.

5.2   Carbocation Stability

Carbocations are classified according to the number of alkyl substituents that are bonded to the positively charged carbon: A **primary carbocation** has one alkyl substituent, a **secondary carbocation** has two, and a **tertiary carbocation** has three. The stability of a carbocation increases as the number of alkyl substituents bonded to the positively charged carbon increases. Thus, tertiary carbocations are more stable than secondary carbocations, and secondary carbocations are more stable than primary carbocations.

Carbocation stability: 3° > 2° > 1°

relative stabilities of carbocations

$$
\boxed{\text{most stable}} >
\begin{array}{c}
\text{R} \\
| \\
\text{R}-\overset{+}{\text{C}} \\
| \\
\text{R} \\
\text{a tertiary} \\
\text{carbocation}
\end{array} >
\begin{array}{c}
\text{R} \\
| \\
\text{R}-\overset{+}{\text{C}} \\
| \\
\text{H} \\
\text{a secondary} \\
\text{carbocation}
\end{array} >
\begin{array}{c}
\text{H} \\
| \\
\text{R}-\overset{+}{\text{C}} \\
| \\
\text{H} \\
\text{a primary} \\
\text{carbocation}
\end{array} >
\begin{array}{c}
\text{H} \\
| \\
\text{H}-\overset{+}{\text{C}} \\
| \\
\text{H} \\
\text{methyl cation}
\end{array}
\boxed{\text{least stable}}
$$

Why does the stability of a carbocation increase as the number of alkyl substituents bonded to the positively charged carbon increases? Alkyl groups are able to donate electrons toward the positively charged carbon, which decreases the concentration of positive charge on the carbon—and decreasing the concentration of positive charge increases the stability of the carbocation. Notice that the blue—recall that blue represents electron-deficient areas (Section 1.3)—is most intense for the least stable methyl cation and is least intense for the most stable *tert*-butyl cation.

The greater the number of alkyl substituents bonded to the positively charged carbon, the more stable is the carbocation.

Alkyl substituents stabilize both alkenes *and* carbocations.

electrostatic potential map for the *tert*-butyl cation

electrostatic potential map for the isopropyl cation

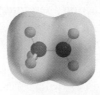
electrostatic potential map for the ethyl cation

electrostatic potential map for the methyl cation

PROBLEM 1◆

Which is more stable, a methyl cation or an ethyl cation?

PROBLEM 2◆

List the following carbocations in order of decreasing stability:

$$CH_3CH_2\overset{+}{\underset{|}{\overset{\displaystyle CH_3}{C}}}CH_3 \qquad CH_3CH_2\overset{+}{C}HCH_3 \qquad CH_3CH_2CH_2\overset{+}{C}H_2$$

Now we are prepared to understand why the *tert*-butyl cation is formed faster than the isobutyl cation when 2-methylpropene reacts with HCl. We know that the *tert*-butyl cation (a tertiary carbocation) is more stable than the isobutyl cation (a primary carbocation). The same factors that stabilize the positively charged carbocation stabilize the transition state for its formation because the transition state has a partial positive charge. Therefore, the transition state leading to the *tert*-butyl cation is more stable (i.e., lower in energy) than the transition state leading to the isobutyl cation (Figure 5.1). The more stable the transition state, the smaller is the free energy of activation, and therefore, the faster is the reaction (Section 4.8). Therefore, the *tert*-butyl cation will be formed faster than the isobutyl cation.

Figure 5.1 ▶
Reaction coordinate diagram for the addition of H⁺ to 2-methylpropene to form the primary isobutyl cation and the tertiary *tert*-butyl cation.

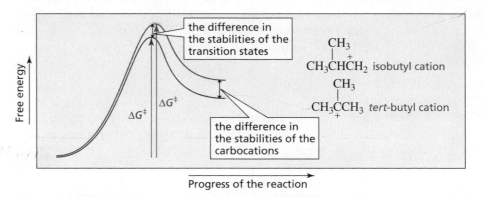

5.3 Regioselectivity of Electrophilic Addition Reactions

We have just seen that the major product of an electrophilic addition reaction is the one obtained by adding the electrophile (H^+) to the sp^2 carbon that results in the formation of the more stable carbocation. For example, when propene reacts with HCl, the proton can add to the number-1 carbon (C-1) to form a secondary carbocation, or it can add to the number-2 carbon (C-2) to form a primary carbocation. The secondary carbocation is formed more rapidly because it is more stable than the primary carbocation. (Primary carbocations are so unstable that they form only with great difficulty.) The product of the reaction, therefore, is 2-chloropropane.

The major product obtained from the addition of HI to 2-methyl-2-butene is 2-iodo-2-methylbutane; only a small amount of 2-iodo-3-methylbutane is obtained. The major product obtained from the addition of HBr to 1-methylcyclohexene is 1-bromo-1-methylcyclohexane. In both cases, the more stable tertiary carbocation is formed more rapidly than the less stable secondary carbocation, so the major product of each reaction is the one that results from forming the tertiary carbocation.

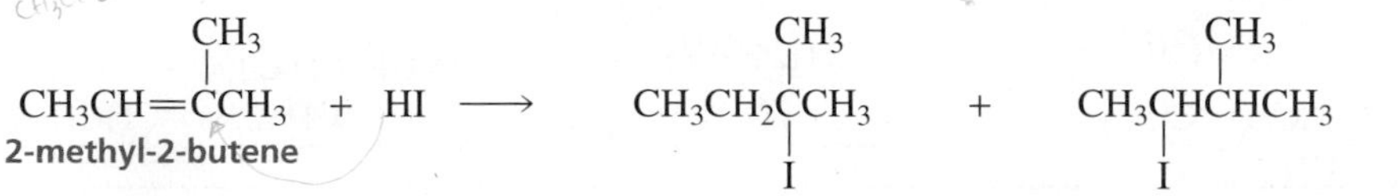

$$
\underset{\text{2-methyl-2-butene}}{CH_3CH=\overset{\overset{\displaystyle CH_3}{|}}{C}CH_3} + HI \longrightarrow \underset{\substack{\text{2-iodo-2-methylbutane}\\\text{major product}}}{CH_3CH_2\overset{\overset{\displaystyle CH_3}{|}}{\underset{\underset{\displaystyle I}{|}}{C}}CH_3} + \underset{\substack{\text{2-iodo-3-methylbutane}\\\text{minor product}}}{CH_3\overset{\overset{\displaystyle CH_3}{|}}{\underset{\underset{\displaystyle I}{|}}{C}H}CHCH_3}
$$

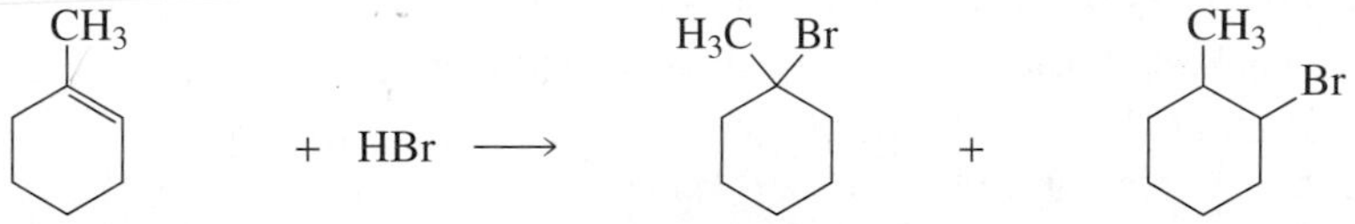

1-methylcyclohexene   + HBr ⟶   1-bromo-1-methyl-cyclohexane **major product**   +   1-bromo-2-methyl-cyclohexane **minor product**

The two products of each of these reactions are called *constitutional isomers.* **Constitutional isomers** have the same molecular formula, but differ in how their atoms are connected. A reaction (such as either of those just shown) in which two or more constitutional isomers could be obtained as products, but one of them predominates, is called a **regioselective reaction**.

The addition of HBr to 2-pentene is not regioselective. Because the addition of H^+ to either of the sp^2 carbons produces a secondary carbocation, both carbocation intermediates have the same stability, so both will be formed equally easily. Thus, approximately equal amounts of the two alkyl halides will be formed.

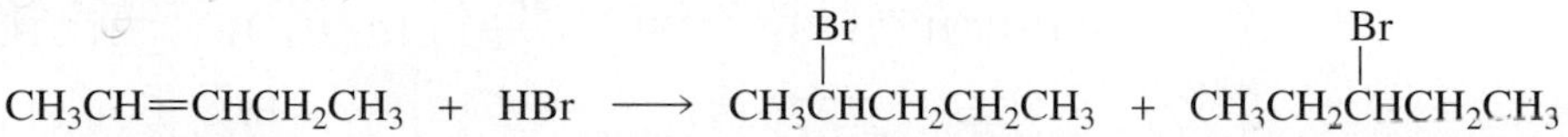

> **Regioselectivity is the preferential formation of one constitutional isomer over another.**

$$
\underset{\text{2-pentene}}{CH_3CH=CHCH_2CH_3} + HBr \longrightarrow \underset{\text{2-bromopentane}}{CH_3\overset{\overset{\displaystyle Br}{|}}{C}HCH_2CH_2CH_3} + \underset{\text{3-bromopentane}}{CH_3CH_2\overset{\overset{\displaystyle Br}{|}}{C}HCH_2CH_3}
$$

By examining the alkene reactions we have seen so far, we can devise a rule that applies to *all* alkene electrophilic addition reactions: *The* **electrophile** *adds to the* sp^2 *carbon that is bonded to the greater number of hydrogens.* Using this rule is simply a quick way to determine the relative stabilities of the intermediates that could be formed in the rate-determining step. You will get the same answer, whether you identify the major product of an electrophilic addition reaction by using the rule or whether you identify it by determining relative carbocation stabilities. In the following reaction for example, H^+ is the electrophile:

$$
\underset{}{CH_3CH_2\overset{2}{C}H=\overset{1}{C}H_2} + HCl \longrightarrow CH_3CH_2\overset{\overset{\displaystyle Cl}{|}}{C}HCH_3
$$

We can say that H^+ adds preferentially to C-1 because C-1 is bonded to two hydrogens, whereas C-2 is bonded to only one hydrogen. Or we can say that H^+ adds to C-1 because that results in the formation of a secondary carbocation, which is more stable than the primary carbocation that would have to be formed if H^+ added to C-2.

Vladimir Vasilevich Markovnikov (1837–1904) *was born in Russia, the son of an army officer. He was a professor of chemistry at Kazan, Odessa, and Moscow Universities. He was the first to recognize that in electrophilic addition reactions, the* H^+ *adds to the* sp^2 *carbon that is bonded to the greater number of hydrogens. Therefore, this is referred to as Markovnikov's rule.*

PROBLEM 3◆

What would be the major product obtained from the addition of HBr to each of the following compounds?

The electrophile adds to the sp^2 carbon that is bonded to the greater number of hydrogens.

a. $CH_3CH_2CH=CH_2$

c. (cyclopentene with CH_3)

e. (cyclohexane with $=CH_2$)

b. $CH_3CH=CCH_3$ with CH_3

d. $CH_2=CCH_2CH_2CH_3$ with CH_3

f. $CH_3CH=CHCH_3$

PROBLEM-SOLVING STRATEGY

a. What alkene should be used to synthesize 3-bromohexane?

$$? + HBr \longrightarrow CH_3CH_2CHCH_2CH_2CH_3$$
$$\overset{|}{Br}$$
3-bromohexane

The best way to answer this kind of question is to begin by listing all the alkenes that could be used. Because you want to synthesize an alkyl halide that has a bromo substituent at the C-3 position, the alkene should have an sp^2 carbon at that position. Two alkenes fit the description: 2-hexene and 3-hexene.

$$CH_3CH=CHCH_2CH_2CH_3 \qquad CH_3CH_2CH=CHCH_2CH_3$$
2-hexene **3-hexene**

Because there are two possibilities, we next need to determine whether there is any advantage to using one over the other. The addition of H^+ to 2-hexene can form two different carbocations. Because they are both secondary carbocations, they have the same stability; therefore, approximately equal amounts of each will be formed. As a result, half of the product will be 3-bromohexane and half will be 2-bromohexane.

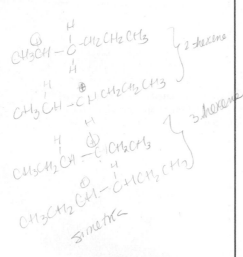

$$CH_3CH_2CHCH_2CH_2CH_3 \xrightarrow{Br^-} CH_3CH_2CHCH_2CH_2CH_3$$
secondary carbocation
$$\overset{|}{Br}$$
3-bromohexane

$$CH_3CH=CHCH_2CH_2CH_3$$
2-hexene

HBr

$$CH_3CHCH_2CH_2CH_2CH_3 \xrightarrow{Br^-} CH_3CHCH_2CH_2CH_2CH_3$$
secondary carbocation
$$\overset{|}{Br}$$
2-bromohexane

The addition of H^+ to either of the sp^2 carbons of 3-hexene, on the other hand, forms the same carbocation because the alkene is symmetrical. Therefore, all of the product will be the desired 3-bromohexane.

$$CH_3CH_2CH=CHCH_2CH_3 \xrightarrow{HBr} CH_3CH_2CHCH_2CH_2CH_3 \xrightarrow{Br^-} CH_3CH_2CHCH_2CH_2CH_3$$
only one carbocation is formed
$$\overset{|}{Br}$$
3-hexene **3-bromohexane**

Because all the alkyl halide formed from 3-hexene is 3-bromohexane, but only half the alkyl halide formed from 2-hexene is 3-bromohexane, 3-hexene is the best alkene to use to prepare 3-bromohexane.

b. What alkene should be used to synthesize 2-bromopentane?

$$? \quad + \quad HBr \quad \longrightarrow \quad CH_3CHCH_2CH_2CH_3$$
$$| \atop Br$$

2-bromopentane

Either 1-pentene or 2-pentene could be used because both have an sp^2 carbon at the C-2 position.

$$CH_2{=}CHCH_2CH_2CH_3 \qquad CH_3CH{=}CHCH_2CH_3$$

1-pentene   **2-pentene**

When H^+ adds to 1-pentene, one of the carbocations that could be formed is secondary and the other is primary. A secondary carbocation is more stable than a primary carbocation, which is so unstable that little, if any, will be formed. Thus, 2-bromopentane will be the only product of the reaction.

$$CH_2{=}CHCH_2CH_2CH_3 \quad \overset{HBr}{\underset{HBr}{\nearrow \atop \searrow}} \quad \begin{array}{l} CH_3\overset{+}{C}HCH_2CH_2CH_3 \xrightarrow{Br^-} CH_3CHCH_2CH_2CH_3 \\ \qquad\qquad\qquad\qquad\qquad\quad | \\ \qquad\qquad\qquad\qquad\qquad Br \\ \qquad\qquad\qquad\qquad\quad \textbf{2-bromopentane} \\[6pt] \overset{+}{C}H_2CH_2CH_2CH_2CH_3 \end{array}$$

1-pentene

When H^+ adds to 2-pentene, on the other hand, each of the two carbocations that can be formed is secondary. Both are equally stable, so they will be formed in approximately equal amounts. Thus, only about half of the product of the reaction will be 2-bromopentane. The other half will be 3-bromopentane.

$$CH_3CH{=}CHCH_2CH_3 \quad \overset{HBr}{\underset{HBr}{\nearrow \atop \searrow}} \quad \begin{array}{l} CH_3\overset{+}{C}HCH_2CH_2CH_3 \xrightarrow{Br^-} CH_3CHCH_2CH_2CH_3 \\ \qquad\qquad\qquad\qquad\qquad\quad | \\ \qquad\qquad\qquad\qquad\qquad Br \\ \qquad\qquad\qquad\qquad\quad \textbf{2-bromopentane} \\[6pt] CH_3CH_2\overset{+}{C}HCH_2CH_3 \xrightarrow{Br^-} CH_3CH_2CHCH_2CH_3 \\ \qquad\qquad\qquad\qquad\qquad\qquad | \\ \qquad\qquad\qquad\qquad\qquad\quad Br \\ \qquad\qquad\qquad\qquad\quad \textbf{3-bromopentane} \end{array}$$

2-pentene

Because all the alkyl halide formed from 1-pentene is 2-bromopentane, but only half the alkyl halide formed from 2-pentene is 2-bromopentane, 1-pentene is the best alkene to use to prepare 2-bromopentane.

Now continue on to answer the questions in Problem 4.

PROBLEM 4◆

What alkene should be used to synthesize each of the following alkyl bromides?

$$\textbf{a.} \quad CH_3\underset{\underset{Br}{|}}{\overset{\overset{CH_3}{|}}{C}}CH_3$$

b.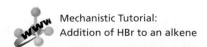

$$\textbf{c.} \quad \text{(cyclohexyl)}{-}\underset{\underset{Br}{|}}{\overset{\overset{CH_3}{|}}{C}}CH_3$$

$$\textbf{d.} \quad \text{(cyclohexyl)}\underset{Br}{\overset{CH_2CH_3}{<}}$$

Mechanistic Tutorial:
Addition of HBr to an alkene

5.4 Addition of Water to an Alkene

When water is added to an alkene, no reaction takes place, because there is no electrophile present to start a reaction by adding to the nucleophilic alkene. The O—H bonds of water are too strong—water is too weakly acidic—to allow the hydrogen to act as an electrophile for this reaction.

$$CH_3CH=CH_2 \ + \ H_2O \ \longrightarrow \ \text{no reaction}$$

If, however, an acid such as HCl or H_2SO_4 is added to the solution, a reaction will occur because the acid provides an electrophile. The product of the reaction is an alcohol. The addition of water to a molecule is called **hydration**, so we can say that an alkene will be *hydrated* in the presence of water and acid.

$$CH_3CH=CH_2 \ + \ H_2O \ \xrightarrow{\textbf{HCl}} \ CH_3CH-CH_2$$
$$\qquad\qquad\qquad\qquad\qquad\qquad\qquad | \quad\ | $$
$$\qquad\qquad\qquad\qquad\qquad\qquad\quad OH \ \ H$$

isopropyl alcohol

Mechanistic Tutorial: Addition of water to an alkene

The first two steps of the mechanism for the acid-catalyzed addition of water to an alkene are essentially the same as the two steps of the mechanism for the addition of a hydrogen halide to an alkene: The electrophile (H^+) adds to the sp^2 carbon that is bonded to the greater number of hydrogens, and the nucleophile (H_2O) adds to the other sp^2 carbon.

mechanism for the acid-catalyzed addition of water

protonated alcohol loses a proton

$$CH_3CH=CH_2 \ + \ H^+ \ \xrightarrow{\textbf{slow}} \ CH_3CHCH_3 \ + \ H_2\ddot{O}: \ \xrightarrow{\textbf{fast}} \ CH_3CHCH_3 \ \rightleftharpoons \ CH_3CHCH_3 \ + \ H^+$$

addition of the electrophile

addition of the nucleophile

a protonated alcohol

an alcohol

As we saw in Section 4.7, the addition of the electrophile to the alkene is relatively slow, and the subsequent addition of the nucleophile to the carbocation occurs rapidly. The reaction of the carbocation with a nucleophile is so fast that the carbocation combines with whatever nucleophile it collides with first. In this hydration reaction, there are two nucleophiles in solution: water and the counterion of the acid (e.g., Cl^-) that is used to start the reaction. (Notice that HO^- is not a nucleophile in this reaction because there is no appreciable concentration of HO^- in an acidic solution.)[1] Because the concentration of water is much greater than the concentration of the counterion, the carbocation is much more likely to collide with water. The product of the collision is a protonated alcohol. We have seen that protonated alcohols are very strong acids (Section 2.2). The protonated alcohol, therefore, loses a proton, and the final product of the addition reaction is an alcohol. A reaction coordinate diagram for the reaction is shown in Figure 5.2.

Synthetic Tutorial: Addition of water to an alkene

A proton adds to the alkene in the first step, but a proton is returned to the reaction mixture in the final step. Overall, a proton is not consumed. A species that increases the rate of a reaction and is not consumed during the course of the reaction is called a **catalyst**. Catalysts increase the rate of a reaction by decreasing the free energy of activation of the reaction (Section 4.8). Catalysts do *not* affect the equilibrium constant of the reaction. In other words, a catalyst increases the *rate* at which a product is formed, but does not affect the *amount* of product formed. The catalyst in the hydration of an alkene is an acid, so hydration is an **acid-catalyzed reaction**.

[1] At a pH of 4, for example, the concentration of HO^- is 1×10^{-10} M, whereas the concentration of water in a dilute aqueous solution is 55.5 M.

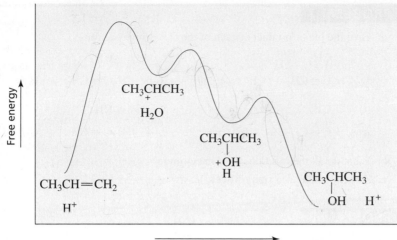

◀ **Figure 5.2**
A reaction coordinate diagram for the acid-catalyzed addition of water to an alkene.

PROBLEM 5◆

Use Figure 5.2 to answer the following questions about the acid-catalyzed hydration of an alkene:

a. How many transition states are there?
b. How many intermediates are there?
c. Which is more stable, the protonated alcohol or the neutral alcohol?
d. Of the six steps in the forward and reverse directions, which are the two fastest?

PROBLEM 6◆

Give the major product obtained from the acid-catalyzed hydration of each of the following alkenes:

a. $CH_3CH_2CH_2CH=CH_2$ **c.** $CH_3CH_2CH_2CH=CHCH_3$

b. (cyclohexene)   **d.** (cyclohexane ring) $=CH_2$

5.5   Addition of an Alcohol to an Alkene

Alcohols react with alkenes in the same way that water does. Like the addition of water, the addition of an alcohol requires an acid catalyst. The product of the reaction is an ether.

$$CH_3CH=CH_2 + CH_3OH \xrightarrow{\text{HCl}} CH_3CH-CH_2$$
$$\underset{\text{OCH}_3\ \text{H}}{\qquad\qquad}$$
isopropyl methyl ether

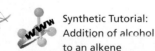 Synthetic Tutorial: Addition of alcohol to an alkene

Do not memorize the products of alkene addition reactions. Instead, for each reaction, ask yourself, "What is the electrophile?" and "What nucleophile is present in the greatest concentration?"

The mechanism for the acid-catalyzed addition of an alcohol is essentially the same as the mechanism for the acid-catalyzed addition of water—the only difference is the nucleophile is ROH instead of HOH.

$$CH_3CH=CH_2 + H^+ \xrightarrow{\text{slow}} CH_3CHCH_3 + CH_3\ddot{O}H \xrightarrow{\text{fast}} CH_3CHCH_3 \rightleftharpoons CH_3CHCH_3 + H^+$$
$$\underset{+\ :\!OH\ \ \ \ \ \ \ \ \ \ \ \ \ \ \ :\!OCH_3}{}$$
$$\underset{\ H\ \ \ \ \ \ \ \ \ \ \ \ \ \ \text{an ether}}{}$$

PROBLEM 7

a. Give the major product of each of the following reactions:

$$\overset{\overset{\displaystyle CH_3}{|}}{1.\ CH_3C}=CH_2\ +\ HCl\ \longrightarrow$$

$$\overset{\overset{\displaystyle CH_3}{|}}{3.\ CH_3C}=CH_2\ +\ H_2O\ \xrightarrow{HCl}$$

$$\overset{\overset{\displaystyle CH_3}{|}}{2.\ CH_3C}=CH_2\ +\ HBr\ \longrightarrow$$

$$\overset{\overset{\displaystyle CH_3}{|}}{4.\ CH_3C}=CH_2\ +\ CH_3OH\ \xrightarrow{HCl}$$

b. What do all the reactions have in common?

c. How do all the reactions differ?

PROBLEM 8

How could the following compounds be prepared, using an alkene as one of the starting materials?

a. 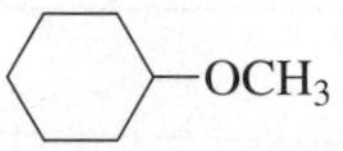 ⬡—OCH_3

c. $CH_3CH_2OCHCH_2CH_3$
    $\quad\quad\quad\quad\quad\ \ \underset{|}{CH_3}$

b. $CH_3O\underset{\underset{\displaystyle CH_3}{|}}{\overset{\overset{\displaystyle CH_3}{|}}{C}}CH_3$

[handwritten: CH₃, CH₃CClt₃, OCH₃]

d. $CH_3\underset{\underset{\displaystyle OH}{|}}{CH}CH_2CH_3$

PROBLEM 9◆

When chemists write reactions, they show reaction conditions, such as the solvent, the temperature, and any required catalyst above or below the arrow.

$$CH_2=CHCH_2CH_3\ +\ H_2O\ \xrightarrow{HCl}\ CH_3\underset{\underset{\displaystyle OH}{|}}{CH}CH_2CH_3$$

Sometimes reactions are written by placing only the organic (carbon-containing) reagent on the left-hand side of the arrow; the other reagents are written above or below the arrow.

$$CH_2=CHCH_2CH_3\ \xrightarrow[H_2O]{HCl}\ CH_3\underset{\underset{\displaystyle OH}{|}}{CH}CH_2CH_3$$

There are two nucleophiles in each of the following reactions. For each reaction, explain why there is a greater concentration of one nucleophile than the other. What will be the major product of each reaction?

a. $CH_3CH=CHCH_3\ +\ H_2O\ \xrightarrow{HCl}$

b. $CH_3CH=CHCH_3\ \xrightarrow[CH_3OH]{HBr}$

PROBLEM 10

Give the major product(s) obtained from the reaction of HBr with each of the following:

a. 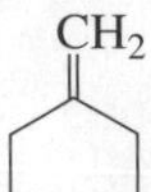 ⬡=CH_2

b. $CH_3\underset{\underset{\displaystyle CH_3}{|}}{CH}CH_2CH=CH_2$

c. 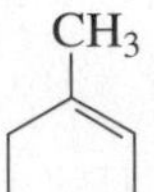 ⬡—CH_3

d. 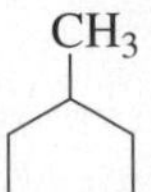 ⬡—CH_3

5.6 Introduction to Alkynes

An **alkyne** is a hydrocarbon that contains a carbon–carbon triple bond. Because of its triple bond, an alkyne has four fewer hydrogens than the corresponding alkane. Therefore, the general molecular formula for an noncyclic alkyne is C_nH_{2n-2}.

A few drugs contain alkyne functional groups. Those shown below are not naturally occurring compounds; they exist only because chemists have been able to synthesize them. Their trade names are shown in green. Trade names are always capitalized and can be used for commercial purposes only by the owner of the registered trademark (Section 22.1).

Parsal®
Sinovial®

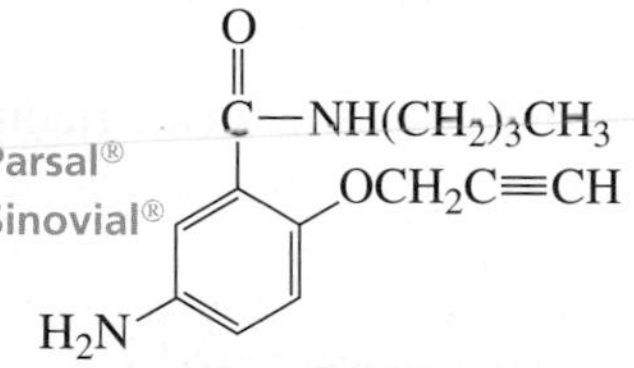

parsalmide
an analgesic

Eudatin®
Supirdyl®

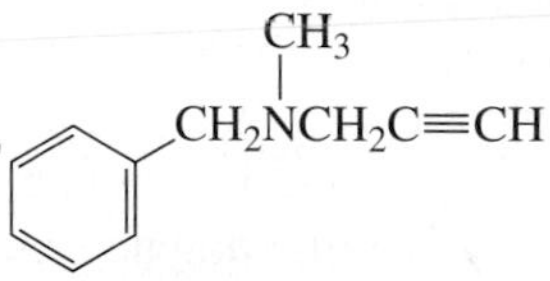

pargyline
an antihypertensive

Norquen®
Ovastol®

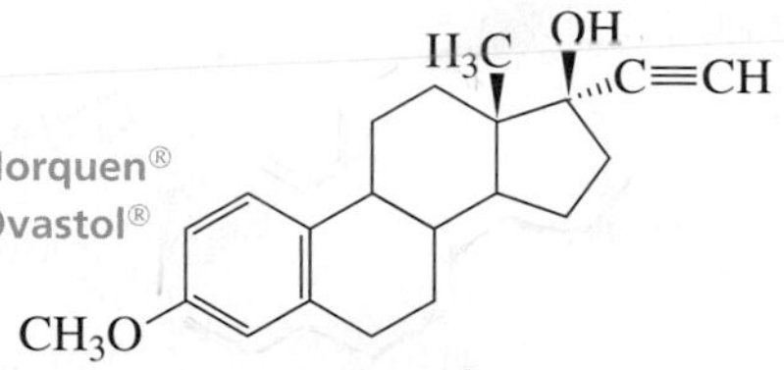

mestranol
a component in oral contraceptives

NATURALLY OCCURRING ALKYNES

There are only a few naturally occurring alkynes. Examples include capillin, which has fungicidal activity, and ichthyothereol, a convulsant used by the Amazon Indians for poisoned arrowheads. A class of naturally occurring compounds called enediynes has been found to have powerful antibiotic and anticancer properties. These compounds all have a nine- or ten-membered ring that contains two triple bonds separated by a double bond. Some enediynes are currently in clinical trials.

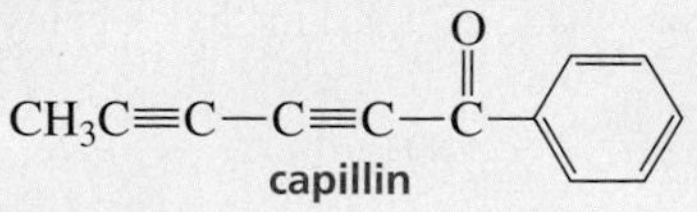

capillin

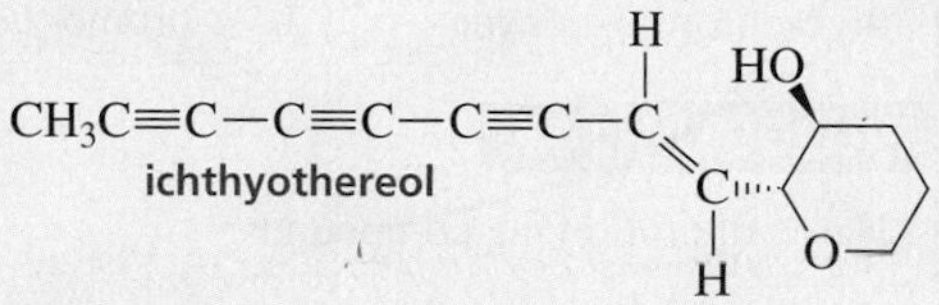

ichthyothereol

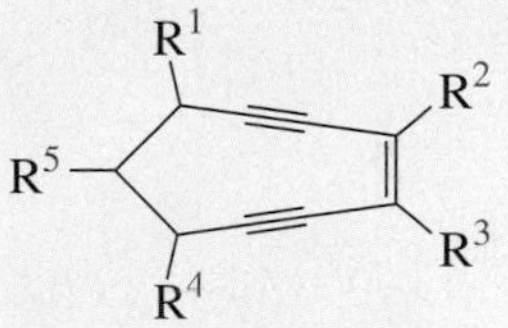

an enediyne

PROBLEM 11◆

What is the general molecular formula for a cyclic alkyne?

PROBLEM 12◆

What is the molecular formula for a cyclic hydrocarbon with 14 carbons and two triple bonds?

$C_{14}C_{30}$

5.7 Nomenclature of Alkynes

The systematic name of an alkyne is obtained by replacing the "ane" ending of the alkane name with "yne." Analogous to the way compounds with other functional groups are named, the longest continuous chain containing the carbon–carbon triple bond is numbered in the direction that gives the alkyne functional group suffix the lowest

1-hexyne
a terminal alkyne

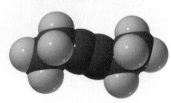

3-hexyne
an internal alkyne

3-D Molecules:
1-Hexyne; 3-Hexyne

possible number. If the triple bond is at the end of the chain, the alkyne is classified as a **terminal alkyne**. Alkynes with triple bonds located elsewhere along the chain are called **internal alkynes**. For example, 1-butyne is a terminal alkyne, whereas 2-pentyne is an internal alkyne.

$$\overset{5}{C}H_2\overset{6}{C}H_3$$

	HC≡CH	$\overset{4}{C}H_3\overset{3}{C}H_2\overset{2}{C}≡\overset{1}{C}H$	$\overset{1}{C}H_3\overset{2}{C}≡\overset{3}{C}\overset{4}{C}H_2\overset{5}{C}H_3$	$\overset{4}{C}H_3\overset{3}{C}H\overset{2}{C}≡\overset{1}{C}CH_3$
systematic:	ethyne	1-butyne	2-pentyne	4-methyl-2-hexyne
common:	acetylene	a terminal alkyne	an internal alkyne	

Acetylene (HC≡CH), the common name for the smallest alkyne, may be a familiar word because of the oxyacetylene torch used in welding. Acetylene is supplied to the torch from one high-pressure gas tank, and oxygen is supplied from another. Burning acetylene produces a high-temperature flame capable of melting and vaporizing iron and steel. It is an unfortunate common name for the smallest alkyne because its "ene" ending is characteristic of a double bond rather than a triple bond.

If the same number for the alkyne functional group suffix is obtained by counting from either direction along the carbon chain, the correct systematic name is the one that contains the lowest substituent number. If the compound contains more than one substituent, the substituents are listed in alphabetical order.

$$\overset{CH_3}{\underset{6\quad5\quad4\quad3\ 2\quad1}{CH_3CHC≡CCH_2CH_2Br}}$$

1-bromo-5-methyl-3-hexyne
not **6-bromo-2-methyl-3-hexyne**
because 1 < 2

$$\overset{Cl\ Br}{\underset{1\quad2\quad3\quad4\quad5\ 6\quad7\quad8}{CH_3CHCHC≡CCH_2CH_2CH_3}}$$

3-bromo-2-chloro-4-octyne
not **6-bromo-7-chloro-4-octyne**
because 2 < 6

PROBLEM 13◆

Draw the structure for each of the following compounds:

a. 1-chloro-3-hexyne   **b.** 4-bromo-2-pentyne   **c.** 4,4-dimethyl-1-pentyne

PROBLEM 14◆

Name the following compounds:

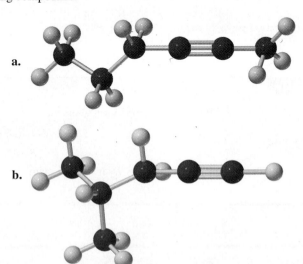

a.

b.

PROBLEM 15

Draw the structures and give the systematic names for the seven alkynes with molecular formula C_6H_{10}.

PROBLEM 16◆

Give the systematic name for each of the following compounds:

a. BrCH₂CH₂C≡CCH₃

c. CH₃CH₂CHC≡CCH₂CH₃
　　　　　|
　　　　　CH₃

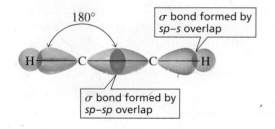

b. CH₃CH₂CHC≡CCH₂CHCH₃
　　　　|　　　　　　　|
　　　　Br　　　　　　Cl

d. CH₃CH₂CHC≡CH
　　　　|
　　　　CH₂CH₂CH₃

2sp y 2p

5.8　The Structure of Alkynes

The structure of ethyne was discussed in Section 1.9. We saw that each carbon is *sp* hybridized, so each has two *sp* orbitals and two *p* orbitals. One *sp* orbital overlaps the *s* orbital of a hydrogen, and the other overlaps an *sp* orbital of the other carbon. Because the *sp* orbitals are oriented as far from each other as possible to minimize electron repulsion, ethyne is a linear molecule with bond angles of 180°.

3-D Molecule:
Ethyne

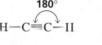

180°

σ bond formed by
sp–s overlap

180°

H—C≡C—H

σ bond formed by
sp–sp overlap

electrostatic potential map
for ethyne

　　The two remaining *p* orbitals on each carbon are oriented at right angles to one another and to the *sp* orbitals (Figure 5.3). Each of the two *p* orbitals on one carbon overlaps the parallel *p* orbital on the other carbon to form two π bonds. One pair of overlapping *p* orbitals results in a cloud of electrons above and below the σ bond, and the other pair results in a cloud of electrons in front of and behind the σ bond. The electrostatic potential map of ethyne shows that the end result can be thought of as a cylinder of electrons wrapped around the σ bond.

A triple bond is composed of a σ bond
and two π bonds.

a.

H
　C
　　C
　　　H

b.

H
　C
　　C
　　　H

◀ **Figure 5.3**
(a) Each of the two π bonds of a triple bond is formed by side-to-side overlap of a *p* orbital of one carbon with a parallel *p* orbital of the adjacent carbon.
(b) A triple bond consists of a σ bond formed by *sp–sp* overlap (yellow) and two π bonds formed by *p–p* overlap (blue and purple).

PROBLEM 17◆

What orbitals are used to form the carbon–carbon σ bond between the highlighted carbons?

a. CH₃CH=CHCH₃

d. CH₃C≡CCH₃

g. CH₃CH=CHCH₂CH₃

b. CH₃CH=CHCH₃

e. CH₃C≡CCH₃

h. CH₃C≡CCH₂CH₃

c. CH₃CH=C=CH₂

f. CH₂=CHCH=CH₂

i. CH₂=CHC≡CH

Tutorial:
Orbitals used to form
carbon–carbon single bonds

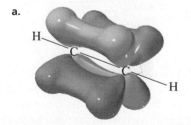
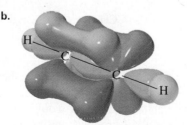

5.9 Physical Properties of Unsaturated Hydrocarbons

All hydrocarbons have similar physical properties. In other words, alkenes and alkynes have physical properties similar to those of alkanes (Section 3.7). All are insoluble in water and all are soluble in nonpolar solvents such as hexane. They are less dense than water and, like any other series of compounds, have boiling points that increase with increasing molecular weight (Table 3.1 on page 46). Alkynes are more linear than alkenes, causing alkynes to have stronger van der Waals interactions. As a result, an alkyne has a higher boiling point than an alkene containing the same number of carbon atoms (see Appendix I).

5.10 Addition of a Hydrogen Halide to an Alkyne

With a cloud of electrons completely surrounding the σ bond, an alkyne is an electron-rich molecule. In other words, it is a nucleophile and, consequently, it will react with electrophiles. For example, if a reagent such as HCl is added to an alkyne, the relatively weak π bond will break because the π electrons are attracted to the electrophilic proton. In the second step of the reaction, the positively charged carbocation intermediate reacts rapidly with the negatively charged chloride ion.

$$CH_3C\equiv CCH_3 + H-\ddot{C}l: \longrightarrow CH_3\overset{+}{C}=CHCH_3 + :\ddot{C}l:^- \longrightarrow CH_3\overset{Cl}{\underset{|}{C}}=CHCH_3$$

Thus, alkynes, like alkenes, undergo electrophilic addition reactions. We will see that the same electrophilic reagents that add to alkenes also add to alkynes. The addition reactions of alkynes, however, have a feature that alkenes do not have: Because the product of the addition of an electrophilic reagent to an alkyne is an alkene, a second electrophilic addition reaction can occur if excess hydrogen halide is present.

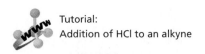

Tutorial:
Addition of HCl to an alkyne

a second electrophilic addition reaction occurs

$$CH_3C\equiv CCH_3 \xrightarrow{\text{HCl}} CH_3\overset{Cl}{\underset{|}{C}}=CHCH_3 \xrightarrow{\text{HCl}} CH_3\overset{Cl}{\underset{\underset{Cl}{|}}{\underset{|}{C}}}CH_2CH_3$$

If the alkyne is a *terminal* alkyne, the H$^+$ will add to the *sp* carbon bonded to the hydrogen, because the *secondary* vinylic cation that results is more stable than the *primary* vinylic cation that would be formed if the H$^+$ added to the other *sp* carbon. (Recall that alkyl groups stabilize positively charged carbon atoms; see Section 5.2.)

The electrophile adds to the *sp* carbon of a terminal alkyne that is bonded to the hydrogen.

$$CH_3CH_2C\equiv CH \xrightarrow{\text{HBr}} CH_3CH_2\overset{+}{C}=CH \longrightarrow CH_3CH_2\overset{Br}{\underset{|}{C}}=\overset{H}{\underset{|}{C}}H$$

1-butyne   Br$^-$ 2-bromo-1-butene
                                     a halo-substituted alkene

3-D Molecule:
Vinylic cation

$$CH_3CH_2\overset{+}{C}=CH_2 \qquad CH_3CH_2CH=\overset{+}{C}H$$

a secondary vinylic cation   a primary vinylic cation

A second addition reaction will take place if excess hydrogen halide is present. When the second equivalent of hydrogen halide adds to the double bond, the electrophile (H^+) adds to the sp^2 carbon that is bonded to the greater number of hydrogens—as predicted by the rule that governs electrophilic addition reactions (Section 5.3).

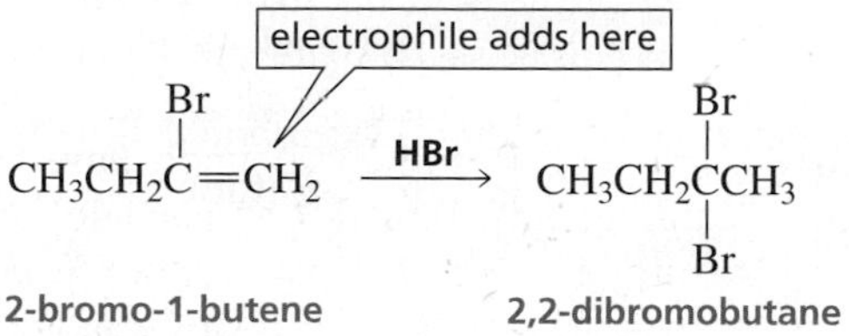

electrophile adds here

$$CH_3CH_2\overset{\overset{\displaystyle Br}{|}}{C}=CH_2 \xrightarrow{HBr} CH_3CH_2\overset{\overset{\displaystyle Br}{|}}{\underset{\underset{\displaystyle Br}{|}}{C}}CH_3$$

2-bromo-1-butene   2,2-dibromobutane

Addition of a hydrogen halide to an *internal* alkyne forms two products, because the initial addition of the proton can occur with equal ease to either of the *sp* carbons.

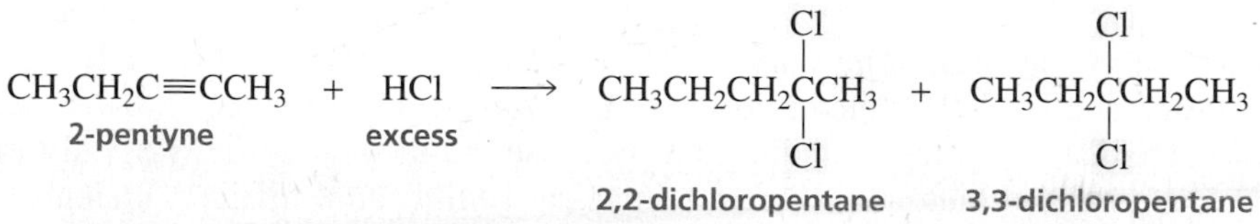

$$CH_3CH_2C \equiv CCH_3 \ + \ HCl \ \longrightarrow \ CH_3CH_2CH_2\overset{\overset{\displaystyle Cl}{|}}{\underset{\underset{\displaystyle Cl}{|}}{C}}CH_3 \ + \ CH_3CH_2\overset{\overset{\displaystyle Cl}{|}}{\underset{\underset{\displaystyle Cl}{|}}{C}}CH_2CH_3$$

2-pentyne   excess   2,2-dichloropentane   3,3-dichloropentane

If, however, the same group is attached to each of the *sp* carbons of the internal alkyne, only one product is obtained.

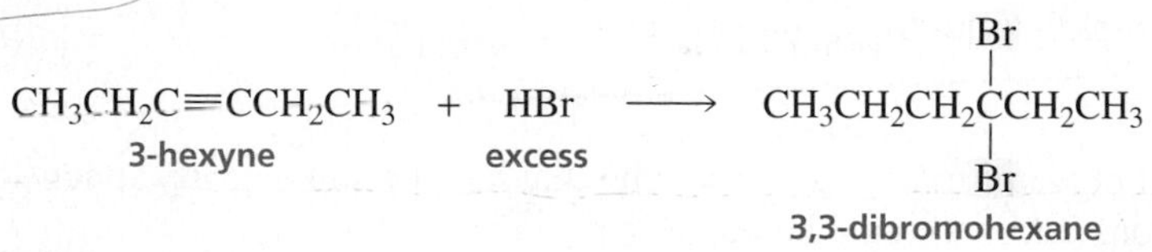

$$CH_3CH_2C \equiv CCH_2CH_3 \ + \ HBr \ \longrightarrow \ CH_3CH_2CH_2\overset{\overset{\displaystyle Br}{|}}{\underset{\underset{\displaystyle Br}{|}}{C}}CH_2CH_3$$

3-hexyne   excess   3,3-dibromohexane

PROBLEM 18◆

Give the major product of each of the following reactions:

a. $HC \equiv CCH_3 \xrightarrow{HBr}$

c. $CH_3C \equiv CCH_3 \xrightarrow[HBr]{excess}$

b. $HC \equiv CCH_3 \xrightarrow[HBr]{excess}$

d. $CH_3C \equiv CCH_2CH_3 \xrightarrow[HBr]{excess}$

5.11   Addition of Water to an Alkyne

In Section 5.4, we saw that alkenes undergo the acid-catalyzed addition of water. The product of the reaction is an alcohol.

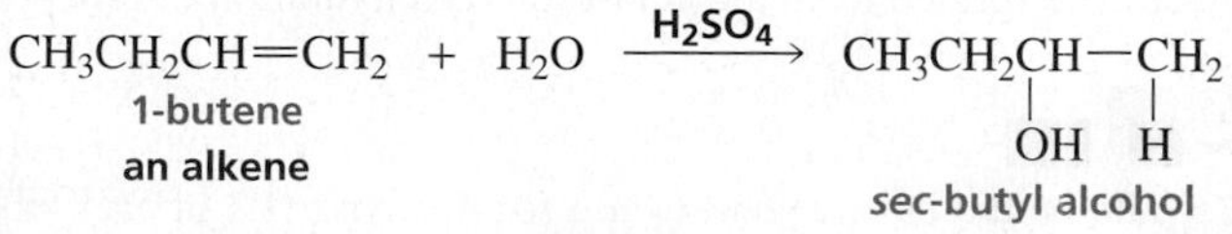

$$CH_3CH_2CH=CH_2 \ + \ H_2O \xrightarrow{H_2SO_4} CH_3CH_2\underset{\underset{\displaystyle OH}{|}}{C}H-\underset{\underset{\displaystyle H}{|}}{C}H_2$$

1-butene
an alkene   *sec*-butyl alcohol

Alkynes also undergo the acid-catalyzed addition of water. The initial product of the reaction is an *enol*. An **enol** has a carbon–carbon double bond and an OH group bonded to one of the sp^2 carbons. (The ending "ene" signifies the double bond, and "ol" the OH group. When the two endings are joined, the final e of "ene" is dropped to avoid two consecutive vowels, but it is pronounced as if the *e* were there, "ene-ol.")

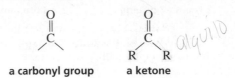

Addition of water to a terminal alkyne forms a ketone.

The enol immediately rearranges to a *ketone*. A carbon doubly bonded to an oxygen is called a **carbonyl ("car-bo-kneel") group**. A **ketone** is a compound that has two alkyl groups bonded to a carbonyl group.

$$\underset{\text{a carbonyl group}}{\overset{\displaystyle O}{\underset{\displaystyle \parallel}{C}}} \qquad \underset{\text{a ketone}}{\overset{\displaystyle O}{\underset{\displaystyle R \qquad R}{\overset{\displaystyle \parallel}{C}}}}$$

A ketone and an enol differ only in the location of a double bond and a hydrogen. The ketone and enol are called **keto–enol tautomers**. **Tautomers** ("taw-toe-mers") are isomers that are in rapid equilibrium. Because the keto tautomer is usually more stable than the enol tautomer, it predominates at equilibrium. Interconversion of the tautomers is called **tautomerization** or **enolization**.

$$\underset{\text{keto tautomer}}{RCH_2-\overset{\displaystyle O}{\overset{\displaystyle \parallel}{C}}-R} \;\rightleftharpoons\; \underset{\text{enol tautomer}}{RCH=\overset{\displaystyle OH}{\overset{\displaystyle |}{C}}-R}$$
$$\text{tautomerization}$$

Addition of water to an internal alkyne that has the same group attached to each of the *sp* carbons forms a single ketone as a product.

$$CH_3CH_2C\equiv CCH_2CH_3 + H_2O \xrightarrow{\;H_2SO_4\;} CH_3CH_2\overset{\displaystyle O}{\overset{\displaystyle \parallel}{C}}CH_2CH_2CH_3$$

If the two groups are not identical, two ketones are formed because the initial addition of the proton can happen to either of the *sp* carbons.

$$CH_3C\equiv CCH_2CH_3 + H_2O \xrightarrow{\;H_2SO_4\;} CH_3\overset{\displaystyle O}{\overset{\displaystyle \parallel}{C}}CH_2CH_2CH_3 + CH_3CH_2\overset{\displaystyle O}{\overset{\displaystyle \parallel}{C}}CH_2CH_3$$

Terminal alkynes are less reactive than internal alkynes toward the addition of water. Terminal alkynes will add water if mercuric ion (Hg^{2+}) is added to the acidic mixture. The mercuric ion is a catalyst—it increases the rate of the addition reaction.

$$CH_3CH_2C\equiv CH + H_2O \xrightarrow[HgSO_4]{\;H_2SO_4\;} \underset{\text{an enol}}{CH_3CH_2\overset{\displaystyle OH}{\overset{\displaystyle |}{C}}=CH_2} \;\rightleftharpoons\; \underset{\text{a ketone}}{CH_3CH_2\overset{\displaystyle O}{\overset{\displaystyle \parallel}{C}}-CH_3}$$

Tutorial:
Common terms in the reactions of alkynes

PROBLEM 19◆

What ketones would be formed from the acid-catalyzed hydration of 3-heptyne?

PROBLEM 20◆

Which alkyne would be the best reagent to use for the synthesis of each of the following ketones?

a. $CH_3\overset{\displaystyle O}{\overset{\displaystyle \parallel}{C}}CH_3$ 　　 b. $CH_3CH_2\overset{\displaystyle O}{\overset{\displaystyle \parallel}{C}}CH_2CH_2CH_3$ 　　 c. $CH_3\overset{\displaystyle O}{\overset{\displaystyle \parallel}{C}}\!-\!\bigcirc$

PROBLEM 21◆

Draw the enol tautomers for the following ketone.

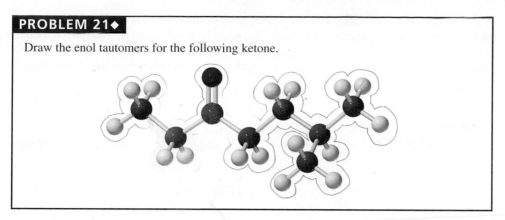

5.12   Addition of Hydrogen to Alkenes and Alkynes

In the presence of a metal catalyst such as platinum or palladium, hydrogen (H_2) adds to the double bond of an alkene to form an alkane. Without the catalyst, the energy barrier to the reaction would be enormous because the H—H bond is so strong. The catalyst decreases the energy of activation by breaking the H—H bond. Platinum and palladium are used in a finely divided state adsorbed on charcoal (Pt/C, Pd/C).

$$
\begin{array}{c}
CH_3 \\
| \\
CH_3C\!\!=\!\!CH_2 \quad + \quad H_2 \quad \xrightarrow{\textbf{Pd/C}} \quad CH_3CHCH_3 \\
\end{array}
$$

2-methylpropene   2-methylpropane

cyclohexene   + H_2 $\xrightarrow{\textbf{Pt/C}}$ cyclohexane

The addition of hydrogen is called **hydrogenation**. Because the preceding reactions require a catalyst, they are examples of **catalytic hydrogenation**. A reaction that increases the number of C—H bonds is called a **reduction reaction**. Thus, hydrogenation is a reduction reaction.

The details of the mechanism of catalytic hydrogenation are not completely understood. We know that hydrogen is adsorbed on the surface of the metal. Breaking the π bond of the alkene and the σ bond of H_2 and forming the C—H σ bonds all occur on the surface of the metal. The alkane product diffuses away from the metal surface as it is formed (Figure 5.4).

> **A reduction reaction increases the number of C—H bonds.**

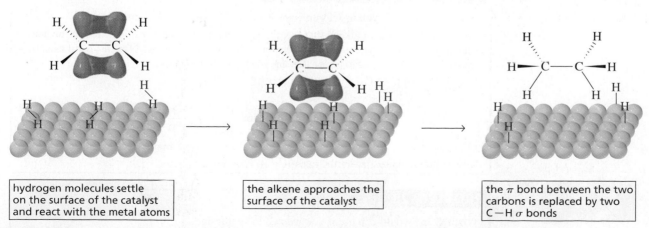

hydrogen molecules settle on the surface of the catalyst and react with the metal atoms

the alkene approaches the surface of the catalyst

the π bond between the two carbons is replaced by two C—H σ bonds

▲ **Figure 5.4**
Catalytic hydrogenation of an alkene.

Hydrogen adds to an alkyne in the presence of a metal catalyst such as palladium or platinum in the same manner that it adds to an alkene. It is difficult to stop the reaction at the alkene stage because hydrogen readily adds to alkenes in the presence of these efficient metal catalysts. The product of the hydrogenation reaction, therefore, is an alkane.

$$CH_3CH_2C\equiv CH \xrightarrow[\text{Pt/C}]{H_2} CH_3CH_2CH = CH_2 \xrightarrow[\text{Pt/C}]{H_2} CH_3CH_2CH_2CH_3$$

alkyne   alkene   alkane

The reaction can be stopped at the alkene stage if a partially deactivated metal catalyst is used. The most commonly used partially deactivated metal catalyst is **Lindlar catalyst**.

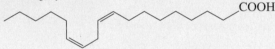

2-pentyne   cis-2-pentene

Because the alkyne sits on the surface of the metal catalyst and the hydrogens are delivered to the triple bond from the surface of the catalyst, both hydrogens are delivered to the same side of the triple bond. Therefore, the addition of hydrogen to an internal alkyne in the presence of Lindlar catalyst forms a **cis alkene**.

Herbert H. M. Lindlar *was born in Switzerland in 1909 and received a Ph.D. from the University of Bern. He worked at Hoffmann–La Roche and Co. in Basel, Switzerland, and he authored many patents. His last patent was a procedure for isolating the carbohydrate xylose from the waste produced in paper mills.*

TRANS FATS

Fats and oils contain carbon–carbon double bonds. Oils are liquids at room temperature because they contain more carbon–carbon double bonds than fats do: oils are polyunsaturated (Section 19.1).

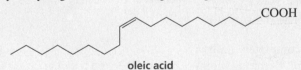

linoleic acid
an 18-carbon fatty acid with two cis double bonds

Some or all of the double bonds in oils can be reduced by catalytic hydrogenation. For example, margarine and shortening

are prepared by hydrogenating vegetable oils such as soybean oil and safflower oil until they have the desired consistency.

All the double bonds in naturally occurring fats and oils have the cis configuration. The heat used in the hydrogenation process breaks the π bond of the double bond. If, instead of being hydrogenated, the double bond reforms, a double bond with the trans configuration can be formed if the sigma bond rotates before the π bond forms (Section 4.4).

One reason that trans fats are of concern to our health is that they do not have the same shape as natural cis fats, but they are able to take their place in membranes. Thus, they can affect the ability of the membrane to correctly control the flow of molecules in and out of the cell.

oleic acid
an 18-carbon fatty acid with one cis double bond

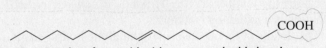

an 18-carbon fatty acid with one trans double bond

 Tutorial: Hydrogenation/Lindlar catalyst

PROBLEM-SOLVING STRATEGY

What alkene would you use if you wanted to synthesize methylcyclohexane?

You need to choose an alkene that has the same number of carbons, attached in the same way, as those in the desired product. Several alkenes could be used for this synthesis, because the double bond can be located anywhere in the molecule.

methylcyclohexane

Now continue on to answer the questions in Problem 22.

PROBLEM 22

What alkene would you use if you wanted to synthesize

a. pentane?   **b.** methylcyclopentane?

PROBLEM 23◆

What alkyne would you use if you wanted to synthesize

a. butane?   **b.** *cis*-2-butene?   **c.** 1-hexene?

5.13   Acidity of a Hydrogen Bonded to an *sp* Hybridized Carbon

Carbon forms nonpolar covalent bonds with hydrogen because carbon and hydrogen, having similar electronegativities, share their bonding electrons almost equally. However, all carbon atoms do not have the same electronegativity. An *sp* hybridized carbon is more electronegative than an sp^2 hybridized carbon, which is more electronegative than an sp^3 hybridized carbon.

relative electronegativities of carbon atoms

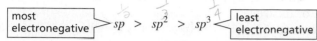

> **most electronegative** $sp > sp^2 > sp^3$ **least electronegative**

Because the electronegativity of carbon atoms follows the order $sp > sp^2 > sp^3$, ethyne is a stronger acid than ethene, and ethene is a stronger acid than ethane. (Don't forget, the stronger the acid, the lower is its pK_a.)

$$HC\equiv CH \qquad H_2C=CH_2 \qquad CH_3CH_3$$

ethyne	ethene	ethane
$pK_a = 25$	$pK_a = 44$	$pK_a > 60$

> *sp* hybridized carbons are more electronegative than sp^2 hybridized carbons, which are more electronegative than sp^3 hybridized carbons.

In order to remove a proton from an acid (in a reaction that strongly favors products), the base that removes the proton must be stronger than the base that is generated as a result of removing the proton (Section 2.2). In other words, you must start with a stronger base than the base that will be formed. Because NH_3 is a weaker acid ($pK_a = 36$) than a terminal alkyne ($pK_a = 25$), the amide ion ($^-NH_2$) is a stronger base than the carbanion—called an **acetylide ion**—that is formed when a hydrogen is removed from the *sp* carbon of a terminal alkyne. (Remember, the stronger the acid, the weaker is its conjugate base.) Therefore, the amide ion can be used to form an acetylide ion.

> **The stronger the acid, the weaker is its conjugate base.**

$$RC\equiv CH + {}^-NH_2 \rightleftharpoons RC\equiv C^- + NH_3$$

stronger acid	amide ion	acetylide ion		NH₃
	stronger base	**weaker base**		**weaker acid**

> **To remove a proton from an acid in a reaction that favors products, the base that removes the proton must be stronger than the base that is formed.**

The amide ion cannot remove a hydrogen bonded to an sp^2 or an sp^3 carbon. Only a hydrogen bonded to an *sp* carbon is sufficiently acidic to be removed by the amide ion. Consequently, a hydrogen bonded to an *sp* carbon sometimes is referred to as an "acidic" hydrogen. The "acidic" property of terminal alkynes is one way their reactivity differs from that of alkenes. Be careful not to misinterpret what is meant when we say that a hydrogen bonded to an *sp* carbon is "acidic." It is more acidic than most other carbon-bound hydrogens but it is much less acidic than a hydrogen of a water molecule, and we know that water is only a very weakly acidic compound.

relative acid strengths

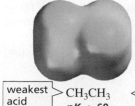   

weakest acid						strongest acid
CH_3CH_3	<	$H_2C=CH_2$	<	NH_3	<	$HC\equiv CH$
$pK_a > 60$		$pK_a = 44$		$pK_a = 36$		$pK_a = 25$

H_2O	<	HF
$pK_a = 15.7$		$pK_a = 3.2$

PROBLEM 24◆

Explain why sodium amide cannot be used to form a carbanion from an alkane in a reaction that favors products.

PROBLEM-SOLVING STRATEGY

a. List the following compounds in order of decreasing acidity:

$$CH_3CH_2\overset{+}{N}H_3 \qquad CH_3CH=\overset{+}{N}H_2 \qquad CH_3C\equiv\overset{+}{N}H$$

To compare the acidities of a group of compounds, first look at how they differ. These three compounds differ in the hybridization of the nitrogen to which the acidic hydrogen is attached. Now recall what you know about hybridization and acidity. You know that hybridization of an atom affects its electronegativity (sp is more electronegative than sp^2, and sp^2 is more electronegative than sp^3); and you know that the more electronegative the atom to which a hydrogen is attached, the more acidic is the hydrogen. Now you can answer the question.

relative acidities $\quad CH_3C\equiv\overset{+}{N}H \; > \; CH_3CH=\overset{+}{N}H_2 \; > \; CH_3CH_2\overset{+}{N}H_3$

b. Draw the conjugate bases and list them in order of decreasing basicity. First remove a proton from each acid to get the structures of the conjugate bases, and then recall that the stronger the acid, the weaker is its conjugate base.

relative basicities $\quad CH_3CH_2NH_2 \; > \; CH_3CH=NH \; > \; CH_3C\equiv N$

Now continue on to Problem 25.

PROBLEM 25◆

List the following species in order of decreasing basicity:

a. $CH_3CH_2CH=\overset{-}{C}H \qquad CH_3CH_2C\equiv C^- \qquad CH_3CH_2CH_2\overset{-}{C}H_2$

b. $CH_3CH_2O^- \qquad F^- \qquad CH_3C\equiv C^- \qquad ^-NH_2$

PROBLEM 26 **SOLVED**

Which carbocation in each of the following pairs is more stable?

a. $CH_3\overset{+}{C}H_2 \;$ or $\; H_2C=\overset{+}{C}H \qquad\qquad$ **b.** $H_2C=\overset{+}{C}H \;$ or $\; HC\equiv\overset{+}{C}$

SOLUTION TO 26a A double-bonded carbon is more electronegative than a single-bonded carbon. Being more electronegative, the double-bonded carbon would be less stable with a positive charge than would be a single-bonded carbon. Therefore, the ethyl carbocation is more stable.

5.14 Synthesis Using Acetylide Ions

Reactions that form carbon–carbon bonds are important in the synthesis of organic compounds because without such reactions, we could not convert molecules with small carbon skeletons into molecules with larger carbon skeletons. Instead, the product of a reaction would always have the same number of carbons as the starting material.

One reaction that forms a C—C bond is the reaction of an acetylide ion with an alkyl halide. Only primary alkyl halides or methyl halides should be used in this reaction.

$$CH_3CH_2C\equiv C^- \; + \; CH_3CH_2CH_2Br \longrightarrow CH_3CH_2C\equiv CCH_2CH_2CH_3 \; + \; Br^-$$
$$\text{an acetylide ion} \qquad\qquad\qquad\qquad\qquad\qquad \text{3-heptyne}$$

The mechanism of this reaction is well understood. Bromine is more electronegative than carbon, and as a result, the electrons in the C—Br bond are not shared equally by the two atoms. There is a partial positive charge on carbon and a partial negative charge on bromine. The negatively charged acetylide ion (a nucleophile) is attracted to the partially positively charged carbon (an electrophile) of the alkyl halide. As the electrons of the acetylide ion approach the carbon to form the new C—C bond, they push out the bromine and its bonding electrons because carbon can bond to no more than four atoms at a time.

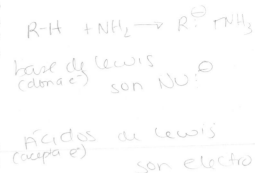

$$CH_3CH_2C\equiv\ddot{C}^- \;+\; CH_3CH_2\overset{(\delta)+}{C}H_2-\overset{(\delta)-}{Br} \;\longrightarrow\; CH_3CH_2C\equiv CCH_2CH_2CH_3 \;+\; Br^-$$

Simply by choosing an alkyl halide of the appropriate structure, terminal alkynes can be converted into internal alkynes of any desired chain length. (The numbers 1 and 2 in front of the reagents above and below the reaction arrow indicate two sequential reactions; the second reagent is not added until the reaction with the first reagent is completely over.)

$$CH_3CH_2CH_2C\equiv CH \quad\xrightarrow[\text{2. } CH_3CH_2CH_2CH_2CH_2Cl]{\text{1. } NaNH_2}\quad CH_3CH_2CH_2C\equiv CCH_2CH_2CH_2CH_2CH_3$$

1-pentyne   **4-decyne**

PROBLEM 27   SOLVED

A chemist wants to synthesize 4-decyne but cannot find any 1-pentyne, the starting material used in the synthesis just described. How else can 4-decyne be synthesized?

SOLUTION   The sp carbons of 4-decyne are bonded to a pentyl group and to a propyl group. Therefore, to obtain 4-decyne, the acetylide ion of 1-pentyne can react with a pentyl halide or the acetylide ion of 1-heptyne can react with a propyl halide. Since 1-pentyne is not available, the chemist should use 1-heptyne and a propyl halide.

$$CH_3CH_2CH_2CH_2CH_2C\equiv CH \quad\xrightarrow[\text{2. } CH_3CH_2CH_2Cl]{\text{1. } NaNH_2}\quad CH_3CH_2CH_2CH_2CH_2C\equiv CCH_2CH_2CH_3$$

1-heptyne   **4-decyne**

5.15   An Introduction to Multistep Synthesis

Synthetic chemists consider time, cost, and yield in designing syntheses. In the interest of time, a well-designed synthesis should require as few steps (sequential reactions) as possible, and those steps should each involve a reaction that is easy to carry out. If two chemists in a pharmaceutical company were each asked to prepare a new drug, and one synthesized the drug in three simple steps while the other used 20 difficult steps, which chemist would not get a raise? In addition, each step in the synthesis should provide the greatest possible yield of the desired product, and the cost of the starting materials must be considered—the more reactant needed to synthesize one gram of product, the more expensive it is to produce. Sometimes it is preferable to design a synthesis involving several steps if the starting materials are inexpensive, the reactions are easy to carry out, and the yield of each step is high. This would be better than designing a synthesis with fewer steps that require expensive starting materials and reactions that are more difficult or give lower yields. At this point, you don't know how much chemicals cost or how difficult it is to carry out certain reactions. So, for the time being, when you design a synthesis, just try to find the route with the fewest steps.

The following examples will give you an idea of the type of thinking required for the design of a successful synthesis.

Example 1. Starting with 1-butyne, how could you make the following ketone? You can use any organic and inorganic reagents.

$$CH_3CH_2C \equiv CH \xrightarrow{?} CH_3CH_2\overset{\overset{O}{\|}}{C}CH_2CH_2CH_3$$
1-butyne

Many chemists find that the easiest way to design a synthesis is to work backward. Instead of looking at the starting material and deciding how to do the first step of the synthesis, look at the product and decide how to do the last step. The product is a ketone. At this point, the only reaction you know that forms a ketone is the addition of water (in the presence of an acid catalyst) to an alkyne. If the alkyne used in the reaction has identical substituents on each of the *sp* carbons, only one ketone will be obtained. Thus, 3-hexyne is the best alkyne to use for the synthesis of the desired ketone.

$$CH_3CH_2C \equiv CCH_2CH_3 \xrightarrow[H_2SO_4]{H_2O} CH_3CH_2\overset{\overset{OH}{|}}{C}=CHCH_2CH_3 \rightleftharpoons CH_3CH_2\overset{\overset{O}{\|}}{C}CH_2CH_2CH_3$$
3-hexyne

3-Hexyne can be obtained from the starting material by removing the proton from its *sp* carbon, followed by alkylation. To obtain the desired product, a two-carbon alkyl halide must be used in the alkylation reaction.

$$CH_3CH_2C \equiv CH \xrightarrow[\text{2. CH}_3\text{CH}_2\text{Br}]{\text{1. NaNH}_2} CH_3CH_2C \equiv CCH_2CH_3$$
1-butyne   3-hexyne

Thus, the synthetic scheme for the synthesis of the desired ketone is given by

$$CH_3CH_2C \equiv CH \xrightarrow[\text{2. CH}_3\text{CH}_2\text{Br}]{\text{1. NaNH}_2} CH_3CH_2C \equiv CCH_2CH_3 \xrightarrow[\text{H}_2\text{SO}_4]{\text{H}_2\text{O}} CH_3CH_2\overset{\overset{O}{\|}}{C}CH_2CH_2CH_3$$

Example 2. Starting with ethyne, how could you make 2-bromopentane?

$$HC \equiv CH \xrightarrow{?} CH_3CH_2CH_2\overset{\overset{}{\underset{|}{C}}}{H}CH_3$$
ethyne   Br

2-bromopentane

The desired product can be prepared from 1-pentene, which can be prepared from 1-pentene. 1-Pentene can be prepared from ethyne and an alkyl halide with three carbons.

$$HC \equiv CH \xrightarrow[\text{2. CH}_3\text{CH}_2\text{CH}_2\text{Br}]{\text{1. NaNH}_2} CH_3CH_2CH_2C \equiv CH \xrightarrow[\substack{\text{Lindlar} \\ \text{catalyst}}]{\text{H}_2} CH_3CH_2CH_2CH = CH_2 \xrightarrow{\text{HBr}} CH_3CH_2CH_2\underset{\underset{Br}{|}}{C}HCH_3$$

Example 3. How could you prepare 3,3-dibromohexane from reagents that contain no more than two carbon atoms?

$$\text{reagents with no more than 2 carbon atoms} \xrightarrow{?} CH_3CH_2\underset{\underset{Br}{|}}{\overset{\overset{Br}{|}}{C}}CH_2CH_2CH_3$$

3,3-dibromohexane

3-D Molecules:
1-bromobutane;
3-octyne

The desired product can be prepared from an alkyne and excess HBr. 3-Hexyne is the alkyne that should be used, because it will form one dibromide, whereas 2-hexyne would form two different dibromides—3,3-dibromohexane and 2,2-dibromohexane. 3-Hexyne can be prepared from 1-butyne and ethyl bromide, and 1-butyne can be prepared from ethyne and ethyl bromide.

$$HC\equiv CH \xrightarrow[\text{2. CH}_3\text{CH}_2\text{Br}]{\text{1. NaNH}_2} CH_3CH_2C\equiv CH \xrightarrow[\text{2. CH}_3\text{CH}_2\text{Br}]{\text{1. NaNH}_2} CH_3CH_2C\equiv CCH_2CH_3 \xrightarrow{\text{excess HBr}} CH_3CH_2\overset{\displaystyle Br}{\underset{\displaystyle Br}{C}}CH_2CH_2CH_3$$

PROBLEM 28

Starting with acetylene, how could the following compounds be synthesized? $HC\equiv CH$

a. $CH_3CH_2CH_2C\equiv CH$

c. $CH_3CH=CH_2$

e. $\underset{H}{\overset{CH_3}{C}}=\underset{H}{\overset{CH_3}{C}}$

b. $CH_3CH_2CH_2\overset{O}{\overset{\|}{C}}CH_3$

d. $CH_3\underset{Br}{CH}CH_3$

f. $CH_3\underset{Cl}{\overset{Cl}{C}}CH_3$

5.16 Polymers

A **polymer** is a large molecule made by linking together repeating units of small molecules called **monomers**. The process of linking them together is called **polymerization**.

$$n\,M \xrightarrow{\text{polymerization}} -M-M-M-M-M-M-M-M-M-$$
monomer   polymer

ethylene monomers   polyethylene

Polymers can be divided into two broad groups: **synthetic polymers** and **biopolymers**. Synthetic polymers are synthesized by scientists, whereas biopolymers are synthesized by organisms. Examples of biopolymers are DNA, the storage molecule for genetic information—the molecule that determines whether a fertilized egg becomes a human or a honeybee; RNA and proteins, the molecules that induce biochemical transformations; and polysaccharides. The structures and properties of these biopolymers are presented in other chapters. Here, we will explore synthetic polymers.

Probably no group of synthetic compounds is more important to modern life than synthetic polymers. Some synthetic polymers resemble natural substances, but most are quite different from those found in nature. Such diverse items as photographic film, compact discs, food wrap, artificial joints, Super glue, toys, plastic bottles, weather stripping, automobile body parts, and shoe soles are made of synthetic polymers. Currently, there are approximately 30,000 patented polymers in the United States. More than 2.5×10^{13} kilograms of synthetic polymers are produced in the United States each year, and we can expect many more new materials to be developed by scientists in the years to come.

Synthetic polymers can be divided into two major classes, depending on their method of preparation. Here we will look at *chain-growth polymers*. *Step-growth polymers* are discussed in Section 12.13. **Chain-growth polymers** are made by **chain reactions**—the addition of monomers to the end of a growing chain. The monomers used most commonly in chain-growth polymerization are alkenes. Polystyrene—used for disposable food containers, insulation, and toothbrush handles, among other things—is an example of a chain-growth polymer. It is made by polymerizing an alkene called styrene. Polystyrene is pumped full of air to produce the material known as Styrofoam. Some of the many polymers synthesized by chain-growth polymerization are listed in Table 5.1.

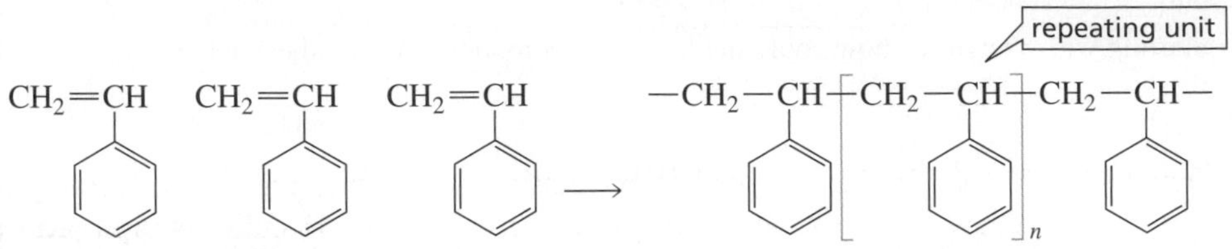

styrene

polystyrene
a chain-growth polymer

repeating unit

Table 5.1   Some Important Chain-Growth Polymers and Their Uses

Monomer	Repeating unit	Polymer name	Uses
$CH_2=CH_2$	$-CH_2-CH_2-$	polyethylene	film, toys, bottles, plastic bags
$CH_2=CH$ | Cl	$-CH_2-CH-$ | Cl	poly(vinyl chloride)	"squeeze" bottles, pipe, siding, flooring
$CH_2=CH-CH_3$	$-CH_2-CH-$ | CH_3	polypropylene	molded caps, margarine tubs, indoor/outdoor carpeting, upholstery
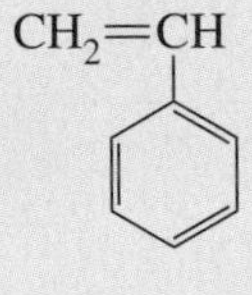 $CH_2=CH$ (phenyl)	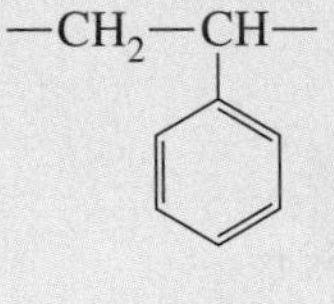 $-CH_2-CH-$ (phenyl)	polystyrene	packaging, toys, clear cups, egg cartons, hot drink cups
$CF_2=CF_2$	$-CF_2-CF_2-$	poly(tetrafluoroethylene) Teflon	nonsticking surfaces, liners, cable insulation
$CH_2=CH$ | $C\equiv N$	$-CH_2-CH-$ | $C\equiv N$	poly(acrylonitrile) Orlon, Acrilan	rugs, blankets, yarn, apparel, simulated fur
$CH_2=C-CH_3$ | $COCH_3$ || O	CH_3 | $-CH_2-C-$ | $COCH_3$ || O	poly(methyl methacrylate) Plexiglas, Lucite	lighting fixtures, signs, solar panels, skylights
$CH_2=CH$ | $OCCH_3$ || O	$-CH_2-CH-$ | $OCCH_3$ || O	poly(vinyl acetate)	latex paints, adhesives

The initiator for chain-growth polymerization can be an electrophile that adds to the alkene monomer, causing it to become a cation. This is the **initiation step**; it is the step that initiates the chain reaction. Here, boron trifluoride (BF_3) is used as the electrophile, because boron has an incomplete octet and can, therefore, accept a share in an electron pair.

initiation step

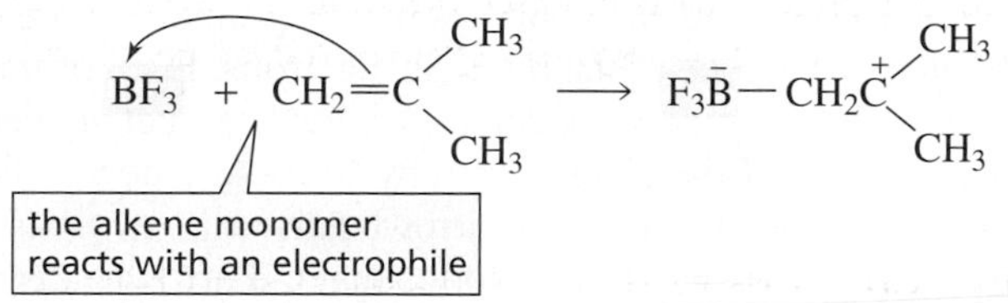

the alkene monomer
reacts with an electrophile

The cation formed in the initiation step reacts with a second monomer, forming a new cation that reacts in turn with a third monomer. These are called **propagation steps** because they propagate the chain reaction. The cation is now at the end of the unit that was most recently added to the end of the chain. This is called the **propagating site**.

propagation steps

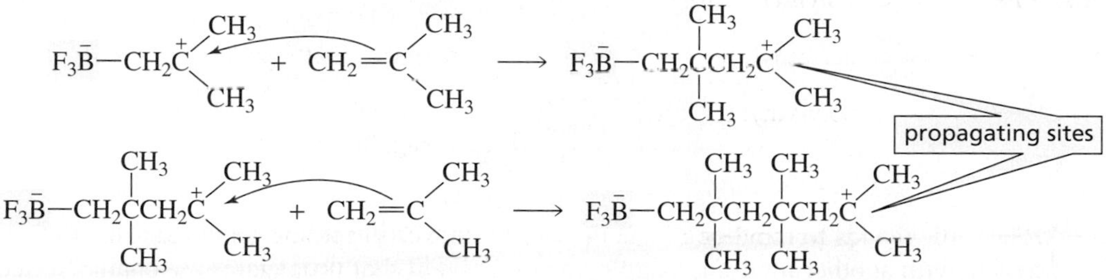

propagating sites

As each subsequent monomer adds to the chain, the positively charged propagating site always ends up on the last unit added. This process is repeated over and over. Hundreds or even thousands of alkene monomers can be added one at a time to the growing chain. Eventually, the chain reaction stops because the propagating sites are destroyed. The propagating step is destroyed when it reacts with a nucleophile. This is called a **termination step**.

termination step

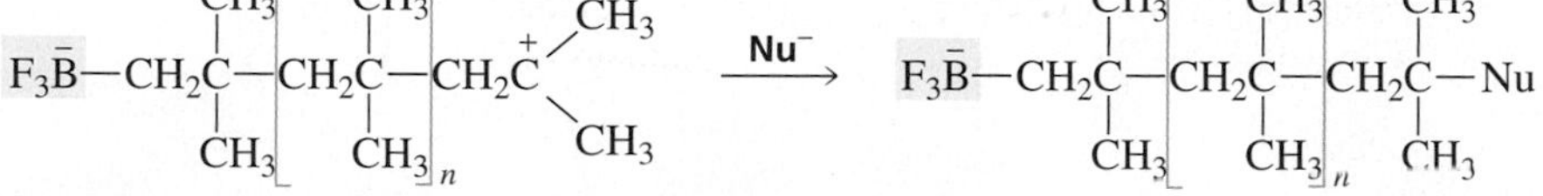

Chain reactions have initiation, propagation, and termination steps.

DESIGNING A POLYMER

A polymer used for making contact lenses must be sufficiently hydrophilic (water-loving) to allow lubrication of the eye. Such a polymer, therefore, has many OH groups.

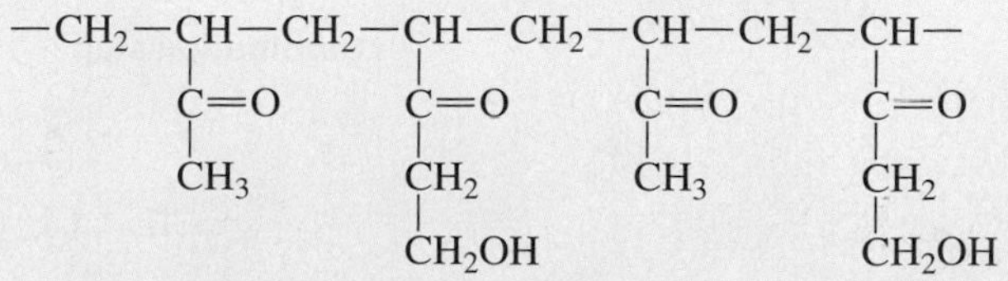

polymer used to make contact lenses

PROBLEM 29◆

Why is BF_3 used as the electrophile to initiate chain-growth polymerization, rather than an electrophile such as HCl?

The initiator can also be a species that breaks into radicals. Most initiators have an O—O bond because such a bond easily breaks in a way that allows each of the atoms that formed the bond to retain one of the bonding electrons. Each of the radicals that is formed seeks an electron to complete its octet. A radical can get an electron by adding to the electron-rich π bond of the alkene, thereby forming a new radical. The curved arrows that we have previously seen have arrowheads with two barbs because they represent the movement of two electrons. Notice that the arrowheads in the following mechanism have only one barb because they represent the movement of only one electron. There are two initiation steps; one creates radicals and the other forms the radical that propagates the chain reaction.

An arrowhead with two barbs signifies the movement of two electrons.

An arrowhead with one barb signifies the movement of one electron.

initiation steps

the alkene monomer reacts with a radical

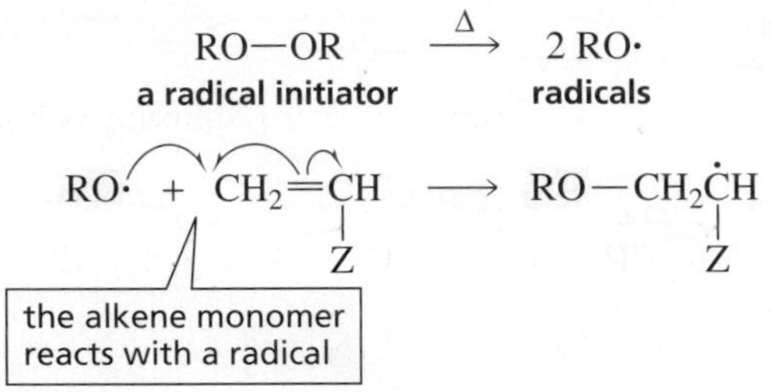

The radical adds to another alkene monomer, converting it into a radical. This radical reacts with another monomer, adding a new subunit that propagates the chain. Notice that when a radical is used to initiate polymerization, the propagating sites are also radicals.

propagation steps

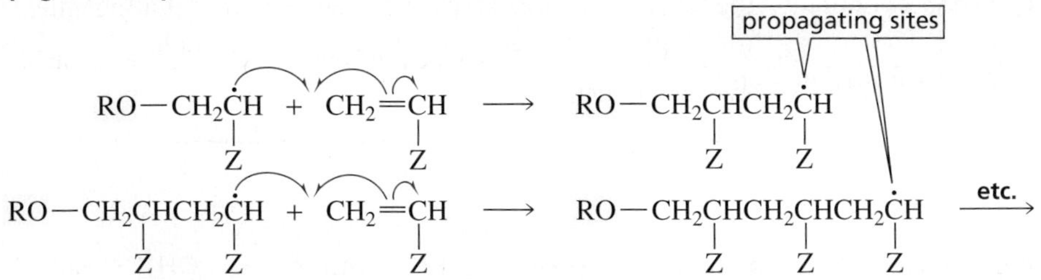

propagating sites

The chain reaction stops when the propagating site reacts with a species (XY) that can terminate the chain.

termination step

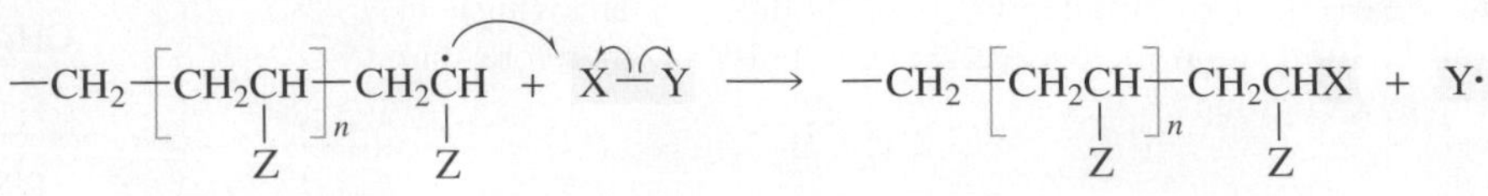

RECYCLING POLYMERS

When plastics are recycled, the various types must be separated from one another. To aid in the separation, many states require manufacturers to include a recycling symbol on their products to indicate the type of plastic. You are probably familiar with these symbols, which are found on the bottom of plastic containers. The symbols consist of three arrows around one of seven numbers; an abbreviation below the symbol indicates the type of polymer from which the container is made. The lower the number in the middle of the symbol, the greater is the ease with which the material can be recycled: 1 (PET) stands for poly(ethylene terephthalate), 2 (HDPE) for high-density polyethylene, 3 (V) for poly(vinyl chloride), 4 (LDPE) for low-density polyethylene, 5 (PP) for polypropylene, 6 (PS) for polystyrene, and 7 for all other plastics.

PROBLEM 30◆

What monomer would you use to form each of the following polymers?

a. 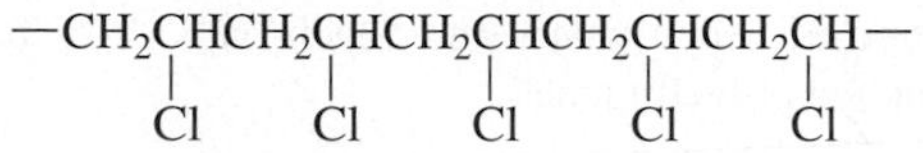 $-CH_2CHCH_2CHCH_2CHCH_2CHCH_2CH-$
          | | | | |
          Cl   Cl   Cl   Cl   Cl

b.

$$CH_3 \quad CH_3 \quad CH_3 \quad CH_3 \quad CH_3 \quad CH_3$$
$$-CH_2CCH_2CCH_2CCH_2CCH_2CCH_2C-$$
$$C=O \; C=O \; C=O \; C=O \; C=O \; C=O$$
$$O \quad\; O \quad\; O \quad\; O \quad\; O \quad\; O$$
$$CH_3 \quad CH_3 \quad CH_3 \quad CH_3 \quad CH_3 \quad CH_3$$

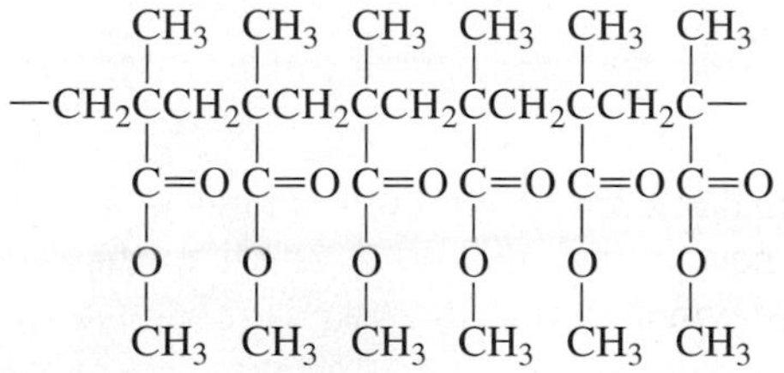

c. $-CF_2CF_2CF_2CF_2CF_2CF_2CF_2CF_2CF_2CF_2-$

PROBLEM 31

Show the mechanism for the formation of a segment of poly(vinyl chloride) containing three units of vinyl chloride and initiated by $HO\cdot$.

Branching of the Polymer Chain

If the propagating site removes a hydrogen atom from a chain, a branch can grow off the chain at that point.

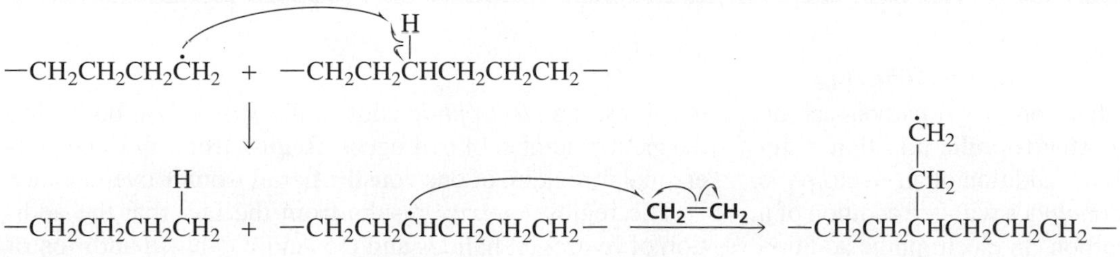

$$-CH_2CH_2CH_2\dot{C}H_2 \;+\; -CH_2CH_2\overset{H}{CHCH_2CH_2CH_2}-$$

$$\downarrow$$

$$-CH_2CH_2CH_2\overset{H}{CH_2} \;+\; -CH_2CH_2\dot{C}HCH_2CH_2CH_2- \;\xrightarrow{CH_2=CH_2}\; -CH_2CH_2\overset{\dot{C}H_2}{\underset{CH_2}{CHCH_2CH_2CH_2}}-$$

Removing a hydrogen atom from a carbon near the end of a chain leads to short branches, whereas removing a hydrogen atom from a carbon near the middle of a chain results in long branches. Short branches are more likely to be formed than long ones because the ends of the chain are more accessible.

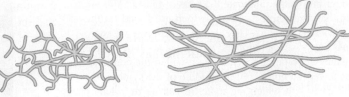

chain with short branches   chain with long branches

Branching greatly affects the physical properties of the polymer. Unbranched chains can pack together more closely than branched chains can. Consequently, linear polyethylene (known as high-density polyethylene) is a relatively hard plastic, used for the production of such things as artificial hip joints, whereas branched polyethylene (low-density polyethylene) is a much more flexible polymer, used for trash bags and dry-cleaning bags.

PROBLEM 32◆

Polyethylene can be used for the production of beach chairs and beach balls. Which of these items is made from more highly branched polyethylene?

ETHYNE CHEMISTRY OR THE FORWARD PASS?

Father Julius Arthur Nieuwland (1878–1936) did much of the early work that led to the synthesis of a polymer called neoprene, a synthetic rubber. He formed the required monomer by reacting vinylacetylene with HCl. Because of the vow of poverty he took as a priest, he refused to accept any royalties for his discovery.

$$HC\equiv CCH=CH_2 \xrightarrow{\textbf{HCl}} CH_2=\overset{\overset{\textstyle Cl}{\textstyle |}}{C}CH=CH_2 \longrightarrow$$

vinylacetylene   2-chloro-1,3-butadiene chloroprene

neoprene

Knute Rockne in his uniform during the year he was captain of the Notre Dame football team.

Father Nieuwland was born in Belgium and settled with his parents in South Bend, Indiana, two years later. He became a priest and a professor of botany and chemistry at the University of Notre Dame, where Knute Rockne—the inventor of the forward pass—worked for him as a research assistant. Rockne also taught chemistry at Notre Dame, but when he received an offer to coach the football team, he switched fields, in spite of Father Nieuwland's attempts to convince him to continue his work as a scientist.

Summary

An **alkene** is a hydrocarbon that contains a carbon–carbon double bond. Alkenes undergo **electrophilic addition reactions**. Each reaction starts with the addition of an electrophile to one of the sp^2 carbons and concludes with the addition of a nucleophile to the other sp^2 carbon. In electrophilic addition

reactions, the *electrophile* adds to the sp^2 carbon bonded to the greater number of hydrogens. **Regioselectivity** is the preferential formation of one **constitutional isomer** over another.

The regioselectivity results from the fact that the addition of hydrogen halides and the acid-catalyzed addition of water and alcohols form **carbocation intermediates**. **Tertiary carbocations** are more stable than **secondary carbocations**, which are more stable than **primary carbocations**. We have now seen that alkyl groups stabilize both alkenes and carbocations.

An **alkyne** is a hydrocarbon that contains a carbon–carbon triple bond. The functional group suffix of an alkyne is "yne." A **terminal alkyne** has the triple bond at the end of the chain; an **internal alkyne** has the triple bond located elsewhere along the chain.

Like alkenes, alkynes undergo electrophilic addition reactions. The same reagents that add to alkenes add to alkynes. Electrophilic addition to a *terminal* alkyne is regioselective; the *electrophile* adds to the *sp* carbon that is bonded to the hydrogen. If excess reagent is available, alkynes undergo a second addition reaction with hydrogen halides because the product of the first reaction is an alkene.

When an alkyne undergoes the acid-catalyzed addition of water, the product of the reaction is an enol, which immediately rearranges to a ketone. A **ketone** is a compound that has two alkyl groups bonded to a **carbonyl** ($C=O$) **group**. The ketone and enol are called **keto–enol tautomers**; they differ in the location of a double bond and a

hydrogen. Interconversion of the tautomers is called **tautomerization**. The keto tautomer predominates at equilibrium. Terminal alkynes add water if mercuric ion is added to the acidic mixture.

Hydrogen adds to alkenes and alkynes in the presence of a metal catalyst (Pd or Pt) to form an alkane. These are **reduction** reactions because they increase the number of $C—H$ bonds. Addition of hydrogen to an internal alkyne in the presence of a Lindlar catalyst forms a cis alkene. The addition of H_2 to a compound is called **hydrogenation**.

The electronegativities of carbon atoms decrease in the order: $sp > sp^2 > sp^3$. Ethyne is, therefore, a stronger acid than ethene, and ethene is a stronger acid than ethane. An amide ion can remove a hydrogen bonded to an *sp* carbon of a terminal alkyne because it is a stronger base than the **acetylide ion** that is formed. An acetylide ion can undergo a reaction with a methyl halide or a primary alkyl halide to form an internal alkyne.

Polymers are large molecules that are made by linking together many small molecules called **monomers**. **Chain-growth polymers** are formed by chain reactions with **initiation**, **propagation**, and **termination steps**. Polymerization of alkenes can be initiated by electrophiles and by radicals.

Summary of Reactions

1. Electrophilic addition reactions of alkenes
   a. Addition of hydrogen halides (Section 5.1)

$$RCH{=}CH_2 \ + \ HX \ \longrightarrow \ \underset{\underset{X}{|}}{RCHCH_3}$$

$$HX = HF, HCl, HBr, HI$$

   b. Acid-catalyzed addition of water and alcohols (Sections 5.4 and 5.5)

$$RCH{=}CH_2 \ + \ H_2O \ \xoverset{HCl}{\longrightarrow} \ \underset{\underset{OH}{|}}{RCHCH_3}$$

$$RCH{=}CH_2 \ + \ CH_3OH \ \xoverset{HCl}{\longrightarrow} \ \underset{\underset{OCH_3}{|}}{RCHCH_3}$$

2. Electrophilic addition reactions of alkynes
   a. Addition of hydrogen halides (Section 5.10)

$$RC{\equiv}CH \ \xoverset{HX}{\longrightarrow} \ \underset{\underset{X}{|}}{RC{=}CH_2} \ \xoverset{excess\ HX}{\longrightarrow} \ \underset{\underset{X}{|}}{\overset{\overset{X}{|}}{RC}}{-}CH_3$$

$$HX \ = \ HF, HCl, HBr, HI$$

b. Acid-catalyzed addition of water (Section 5.11)

$$RC\equiv CR' \quad \xrightarrow[\text{H}_2\text{SO}_4]{\text{H}_2\text{O}} \quad \underset{\text{an internal alkyne}}{} \qquad RCCH_2R' \; + \; RCH_2CR'$$

$$RC\equiv CH \quad \xrightarrow[\text{H}_2\text{SO}_4/\text{HgSO}_4]{\text{H}_2\text{O}} \quad RCCH_3$$

a terminal alkyne

3. Addition of hydrogen to alkenes and alkynes (Section 5.12)

$$RCH=CH_2 \; + \; H_2 \quad \xrightarrow{\textbf{Pd/C or Pt/C}} \quad RCH_2CH_3$$

$$RC\equiv CR' \; + \; 2\,H_2 \quad \xrightarrow{\textbf{Pd/C or Pt/C}} \quad RCH_2CH_2R'$$

$$R-C\equiv C-R' \; + \; H_2 \quad \xrightarrow[\textbf{catalyst}]{\textbf{Lindlar}} \quad \underset{R}{\overset{H}{\underset{\diagdown}{}}}C=C\underset{R'}{\overset{H}{\diagup}}$$

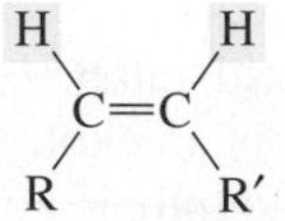

4. Removal of a proton from a terminal alkyne, followed by reaction with an alkyl halide (Sections 5.13 and 5.14)

$$RC\equiv CH \quad \xrightarrow{\textbf{NaNH}_2} \quad RC\equiv C^- \quad \xrightarrow{\textbf{R'CH}_2\textbf{Br}} \quad RC\equiv CCH_2R'$$

Problems

33-43 , 45 ,47-59 ,62-65

33. Identify the electrophile and the nucleophile in each of the following reaction steps. Then draw curved arrows to illustrate the bond-making and bond-breaking processes.

a. $CH_3\overset{+}{C}HCH_3 \; + \; :\!\overset{..}{\underset{..}{Cl}}\!:^- \quad \longrightarrow \quad CH_3CHCH_3$
   $:\!\overset{..}{\underset{..}{Cl}}\!:$

b. $CH_3\overset{CH_3}{\underset{CH_3}{\overset{|}{\underset{|}{C}}}}^+ \; + \; CH_3\overset{..}{\underset{..}{O}}H \quad \longrightarrow \quad CH_3\overset{CH_3}{\underset{CH_3}{\overset{|}{\underset{|}{C}}}}\!-\!{}^{+}\overset{..}{\underset{H}{O}}CH_3$

34. What will be the major product of the reaction of 2-methyl-2-butene with each of the following reagents?
   a. HBr
   b. HI
   c. $H_2/Pd/C$
   d. H_2O + trace HCl
   e. CH_3OH + trace HCl
   f. CH_3CH_2OH + trace HCl

35. Give the major product of each of the following reactions:

a. 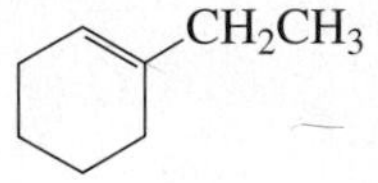 (cyclohexene with CH$_2$CH$_3$) + HBr $\longrightarrow$

c. 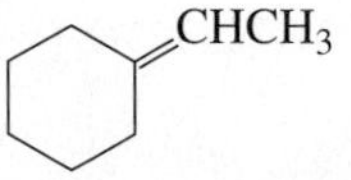 (cyclohexane ring with =CHCH$_3$) + HCl $\longrightarrow$

b. $CH_2=\overset{CH_3}{\overset{|}{C}}CH_2CH_3 \; + \; HBr \quad \longrightarrow$

d. $CH_3\overset{CH_3}{\overset{|}{C}}=CHCH_3 \; + \; HCl \quad \longrightarrow$

36. Draw curved arrows to show the flow of electrons responsible for the conversion of reactants into products.

a. $CH_3-\overset{:\overset{..}{O}:^-}{\underset{\underset{CH_3}{|}}{\overset{|}{C}}}-OCH_3 \quad \longrightarrow \quad CH_3-\overset{:\overset{..}{O}}{\overset{\|}{C}}-CH_3 \; + \; CH_3O^-$

b. CH₃C≡C—H + :N̈H₂ ⟶ CH₃C≡C⁻ + N̈H₃

c. CH₃CH₂—Br + CH₃Ö:⁻ ⟶ CH₃CH₂—ÖCH₃ + Br⁻

37. Give the reagents that would be required to carry out the following syntheses:

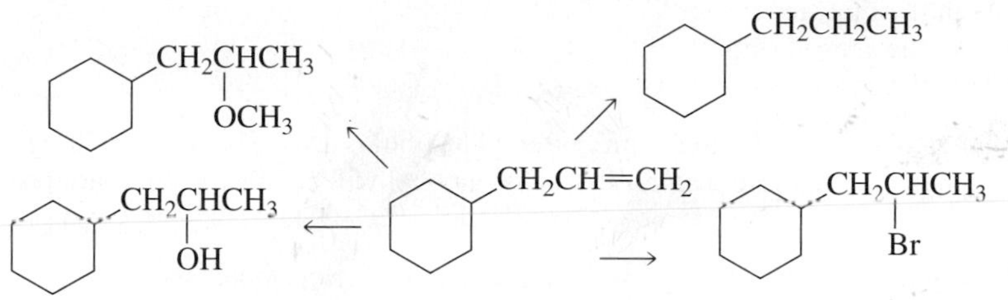

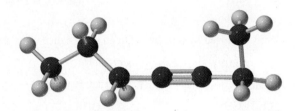

38. Draw all the enol tautomers for each of the ketones in Problem 20.

39. What ketones are formed when the following alkyne undergoes the acid-catalyzed addition of water?

40. Give the major product of each of the following reactions:

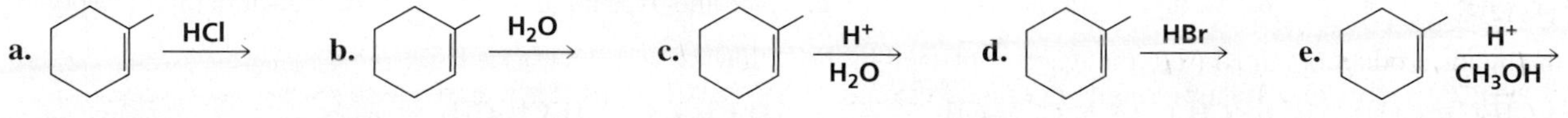

a. [cyclohexene] $\xrightarrow{\text{HCl}}$ **b.** [cyclohexene] $\xrightarrow{\text{H}_2\text{O}}$ **c.** [methylcyclohexene] $\xrightarrow[\text{H}_2\text{O}]{\text{H}^+}$ **d.** [methylcyclohexene] $\xrightarrow{\text{HBr}}$ **e.** [methylcyclohexene] $\xrightarrow[\text{CH}_3\text{OH}]{\text{H}^+}$

41. For each of the following pairs, indicate which member is the more stable:

a. CH₃ĊCH₃ (with CH₃ substituent) or CH₃ĊHCH₂CH₃   **b.** CH₃CH₂ĊH₂ or CH₃ĊHCH₃   **c.** CH₃ĊH₂ or CH₂=ĊH

42. Using any alkene and any other reagents, how would you prepare the following compounds?

a. [cyclohexane]   **b.** CH₃CH₂CH₂CHCH₃ with Cl   **c.** [cyclohexane]CH₂CHCH₃ with OH

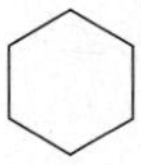

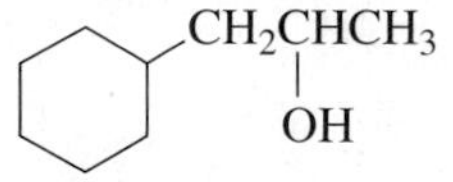

43. Identify the two alkenes that react with HBr to give 1-bromo-1-methylcyclohexane.

44. The second-order rate constant (in units of M⁻¹ s⁻¹) for acid-catalyzed hydration at 25 °C is given for each of the following alkenes:

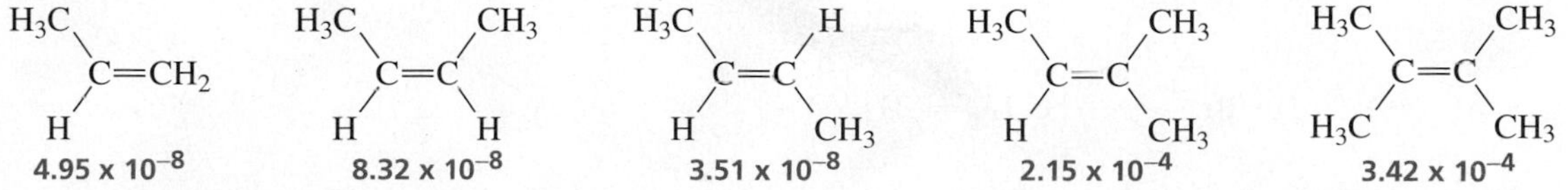

H₃C, H \ C=CH₂   4.95 x 10⁻⁸

H₃C, CH₃ \ C=C / H, H   8.32 x 10⁻⁸

H₃C, H \ C=C / H, CH₃   3.51 x 10⁻⁸

H₃C, CH₃ \ C=C / H, CH₃   2.15 x 10⁻⁴

H₃C, CH₃ \ C=C / H₃C, CH₃   3.42 x 10⁻⁴

a. Why does (Z)-2-butene react faster than (E)-2-butene?
b. Why does 2-methyl-2-butene react faster than (Z)-2-butene?
c. Why does 2,3-dimethyl-2-butene react faster than 2-methyl-2-butene?

45. a. Propose a mechanism for the following reaction (remember to use curved arrows when showing a mechanism):

$$CH_3CH_2CH{=}CH_2 \ + \ CH_3OH \ \xrightarrow{\ H^+\ } \ CH_3CH_2CHCH_3$$
$$\underset{\displaystyle OCH_3}{|}$$

b. Which step is the rate-determining step?
c. What is the electrophile in the first step?
d. What is the nucleophile in the first step?
e. What is the electrophile in the second step?
f. What is the nucleophile in the second step?

46. The pK_a of protonated ethyl alcohol is -2.4 and the pK_a of ethyl alcohol is 15.9. Therefore, as long as the pH of the solution is greater than _____ and less than _____, more than 50% of ethyl alcohol will be in its neutral, nonprotonated form. (*Hint:* See Section 2.4.)

47. a. How many alkenes could you treat with H_2/Pt in order to prepare methylcyclopentane?
b. Which of the alkenes is the most stable?

48. Starting with an alkene, indicate how each of the following compounds can be synthesized:

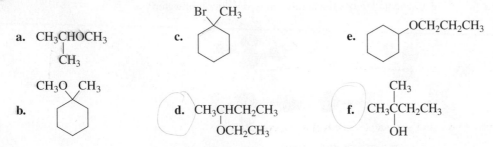

a. CH_3CHOCH_3
　　　$\underset{\displaystyle CH_3}{|}$

b.

c.

d. $CH_3CHCH_2CH_3$
　　　$\underset{\displaystyle OCH_2CH_3}{|}$

e.

f. $CH_3CCH_2CH_3$
　　　with CH_3 above and OH below

49. Draw a structure for each of the following:
a. 2-hexyne
b. 5-ethyl-3-octyne
c. 1-bromo-1-pentyne
d. 5,6-dimethyl-2-heptyne

50. Give the major product obtained from the reaction of each of the following with excess HCl:
a. $CH_3CH_2C{\equiv}CH$
b. $CH_3CH_2C{\equiv}CCH_2CH_3$
c. $CH_3CH_2C{\equiv}CCH_2CH_2CH_3$

51. Give the systematic name for each of the following compounds:

a. $CH_3C{\equiv}CCH_2CHCH_3$
　　　$\underset{\displaystyle Br}{|}$

b. $CH_3C{\equiv}CCH_2CHCH_3$
　　　$\underset{\displaystyle CH_2CH_2CH_3}{|}$

c. $CH_3C{\equiv}CCH_2CCH_3$
　　　with CH_3 above and CH_3 below

d. $CH_3CHCH_2C{\equiv}CCHCH_3$
　　　$\underset{\displaystyle Cl}{|}$ 　　$\underset{\displaystyle CH_3}{|}$

52. Identify the electrophile and the nucleophile in each of the following reaction steps. Then draw curved arrows to illustrate the bond-making and bond-breaking processes.

a. $CH_3CH_2\overset{+}{C}{=}CH_2 \ + \ :\ddot{\underset{..}{C}l}:^- \ \longrightarrow \ CH_3CH_2C{=}CH_2$
　　　$\underset{\displaystyle :\ddot{\underset{..}{C}l}:}{|}$

b. $CH_3C{\equiv}CH \ + \ H{-}Br \ \longrightarrow \ CH_3\overset{+}{C}{=}CH_2 \ + \ Br^-$

c. $CH_3C{\equiv}C{-}H \ + \ :\ddot{N}H_2^- \ \longrightarrow \ CH_3C{\equiv}C:^- \ + \ \ddot{N}H_3$

53. Al Kyne was given the structural formulas of several compounds and was asked to give them systematic names. How many did Al name correctly? Correct those that are misnamed.
a. 4-ethyl-2-pentyne
b. 1-bromo-4-heptyne
c. 2-methyl-3-hexyne
d. 3-pentyne

54. Draw the structures and give the common and systematic names for alkynes with molecular formula C_7H_{12}.

55. What reagents would you use for the following syntheses?
   a. (Z)-3-hexene from 3-hexyne
           b. hexane from 3-hexyne

56. What is the molecular formula of a hydrocarbon that has 1 triple bond, 2 double bonds, 1 ring, and 32 carbons?

57. What will be the major product of the reaction of 1 mol of propyne with each of the following reagents?
   a. HBr (1 mol)
   b. HBr (2 mol)
   c. aqueous H_2SO_4, $HgSO_4$
   d. excess H_2/Pt/C
   e. H_2/Lindlar catalyst
   f. sodium amide
   g. product of Problem 57f followed by 1-chloropentane

58. Answer Problem 57, using 2-butyne as the starting material instead of propyne. fg

59. What reagents could be used to carry out the following syntheses?

$$RCH_2CH_3$$
$$RCH=CH_2$$
$$\overset{Br}{\underset{|}{RCCH_3}}$$
$$\overset{|}{\underset{Br}{}}$$

$$RC=CH$$

$$\overset{RC=CH_2}{\underset{Br}{|}}$$
$$\overset{RCHCH_3}{\underset{Br}{|}}$$
$$\overset{O}{\underset{\parallel}{RCCH_3}}$$

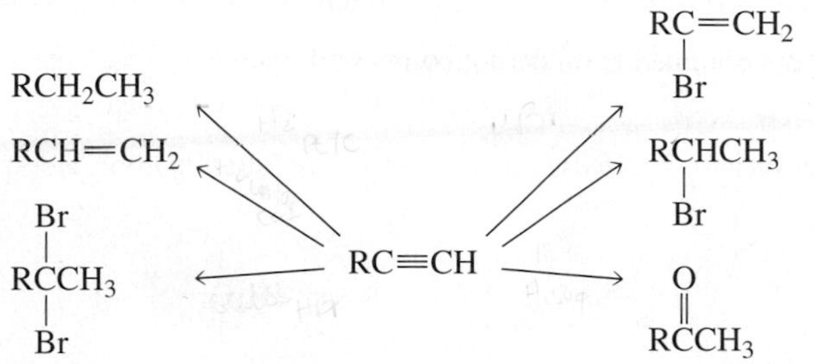

60. a. Starting with 5-methyl-2-hexyne, how could you prepare the following compound?

$$CH_3CH_2\overset{|}{\underset{OH}{C}}HCH_2\overset{|}{\underset{CH_3}{C}}HCH_3$$

   b. What other alcohol would also be obtained?

61. How many of the following names are correct? Correct the incorrect names.
   a. 4-heptyne
   b. 2-ethyl-3-hexyne
   c. 4-chloro-2-pentyne
   d. 2,3-dimethyl-5-octyne
   e. 4,4-dimethyl-2-pentyne
   f. 2,5-dimethyl-3-hexyne

62. Which of the following pairs are keto–enol tautomers?

   a. $CH_3\overset{|}{\underset{OH}{C}}HCH_3$ and $CH_3\overset{O}{\underset{\parallel}{C}}CH_3$

   c. $CH_3CH_2CH_2CH=CHOH$ and $CH_3CH_2CH_2\overset{O}{\underset{\parallel}{C}}CH_3$

   b. $CH_3CH_2CH_2\overset{OH}{\underset{|}{C}}=CH_2$ and $CH_3CH_2CH_2\overset{O}{\underset{\parallel}{C}}CH_3$

63. Using ethyne as the starting material, how can the following compounds be prepared?

   a. $CH_3\overset{O}{\underset{\parallel}{C}}CH_3$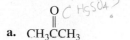

   b.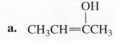

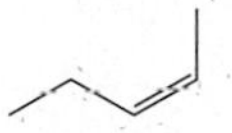

   c.

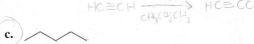

64. Draw the keto tautomer for each of the following:

   a. $CH_3CH=\overset{OH}{\underset{|}{C}}CH_3$
   b. $CH_3CH_2CH_2\overset{OH}{\underset{|}{C}}=CH_2$
   c. cyclohexene—OH
   d. 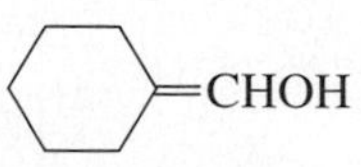 cyclohexane=CHOH

65. Show how each of the following compounds could be prepared using the given starting material, any necessary inorganic reagents, and any necessary organic compound that has no more than four carbon atoms:

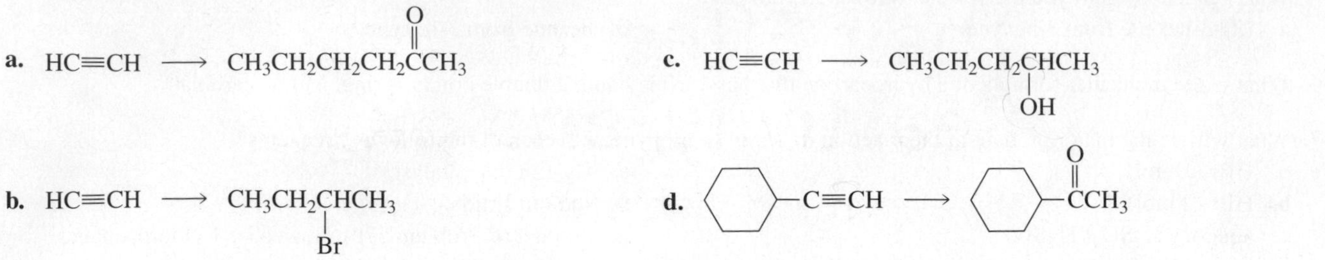

a. $HC\equiv CH \longrightarrow CH_3CH_2CH_2CH_2\overset{\overset{\displaystyle O}{\|}}{C}CH_3$

c. $HC\equiv CH \longrightarrow CH_3CH_2CH_2\underset{\underset{\displaystyle OH}{|}}{C}HCH_3$

b. $HC\equiv CH \longrightarrow CH_3CH_2\underset{\underset{\displaystyle Br}{|}}{C}HCH_3$

d. (cyclohexyl)$-C\equiv CH \longrightarrow$ (cyclohexyl)$-\overset{\overset{\displaystyle O}{\|}}{C}CH_3$

66. Any base whose conjugate acid has a pK_a greater than _____ can remove a proton from a terminal alkyne to form an acetylide ion in a reaction that favors products.

67. Dr. Polly Meher was planning to synthesize 3-octyne by adding 1-bromobutane to the product obtained from the reaction of 1-butyne with sodium amide. Unfortunately, however, she had forgotten to order 1-butyne. How else can she prepare 3-octyne?

68. Draw short segments of the polymers obtained from the following monomers:
a. $CH_2\!\!=\!\!CHF$
b. $CH_2\!\!=\!\!CHCO_2H$

69. Draw the structure of the monomer or monomers used to synthesize the following polymers:

a. 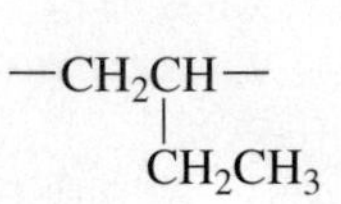 $-CH_2\underset{\underset{\displaystyle CH_2CH_3}{|}}{C}H-$

b. 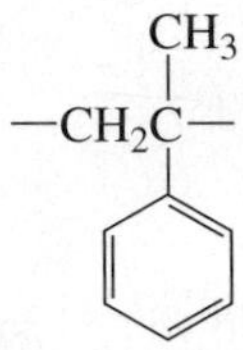 $-CH_2\underset{\underset{\displaystyle (phenyl)}{|}}{\overset{\overset{\displaystyle CH_3}{|}}{C}}-$

70. Draw short segments of the polymer obtained from 1-pentene, using BF_3 as an initiator.

6

Delocalized Electrons and Their Effect on Stability, Reactivity, and pK_a

Ultraviolet and Visible Spectroscopy

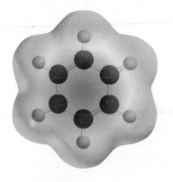

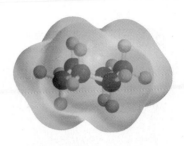

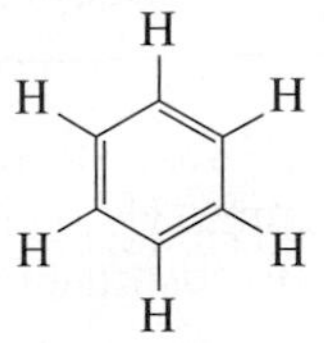

E lectrons that are restricted to a particular region are called **localized electrons**. Localized electrons either belong to a single atom or are confined to a bond between two atoms.

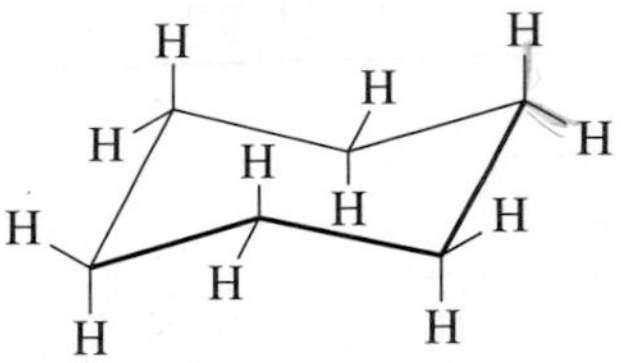

$$CH_3 \overset{\cdot\cdot}{-} NH_2$$
localized electrons

$$CH_3 \overset{}{-} CH = CH_2$$
localized electrons

benzene

cyclohexane

Not all electrons are confined to a single atom or bond. Many organic compounds contain *delocalized* electrons. **Delocalized electrons** neither belong to a single atom nor are confined to a bond between two atoms, but are shared by three or more atoms. The concept of having *delocalized electrons* is invoked frequently to explain the behavior of organic compounds.

6.1 Delocalized Electrons: The Structure of Benzene

To understand the concept of delocalized electrons, we will first look at the structure of benzene. Because early organic chemists didn't know about delocalized electrons, benzene's structure puzzled them. They knew that benzene had a molecular formula of C_6H_6; thus, the number of rings and p bonds in benzene must total four. They also knew that when one of the hydrogens is substituted with another atom, only one product is obtained and when two of the hydrogens are substituted with another atom, three products are obtained.

For every *two* hydrogens that are missing from the general molecular formula C_nH_{2n+2}, a hydrocarbon has either a π bond or a ring. Thus the number of rings and π bonds in benzene must total four.

$$C_6H_6 \xrightarrow[\text{with an X}]{\text{replace a hydrogen}} C_6H_5X \xrightarrow[\text{with an X}]{\text{replace a hydrogen}} C_6H_4X_2 + C_6H_4X_2 + C_6H_4X_2$$

one monosubstituted compound

three disubstituted compounds

A cyclic six-membered ring, with alternating single and double bonds, is a structure that gives the correct molecular formula and gives only one product when one of the hydrogens is substituted with another atom. However, when two of the hydrogens of this structure are substituted, four disubstituted products are obtained—a 1,3-disubstituted product, a 1,4-disubstituted product, and two 1,2-disubstituted products—because the two substituents can be placed either on two adjacent carbons joined by a single bond or on two adjacent carbons joined by a double bond.

3-D Molecules:
1,2-Difluorobenzene;
1,3-Difluorobenzene;
1,4-Difluorobenzene

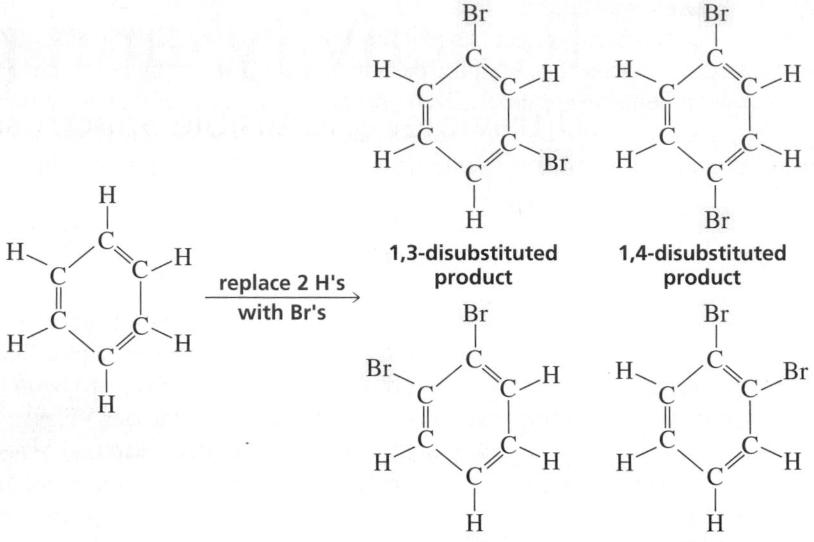

replace 2 H's with Br's →

1,3-disubstituted product

1,4-disubstituted product

1,2-disubstituted product

1,2-disubstituted product

In 1865, the German chemist Friedrich Kekulé suggested a way of resolving this dilemma. He proposed that there actually *are* four disubstituted products, but the two 1,2-disubstituted products interconvert too rapidly to be separated from each other.

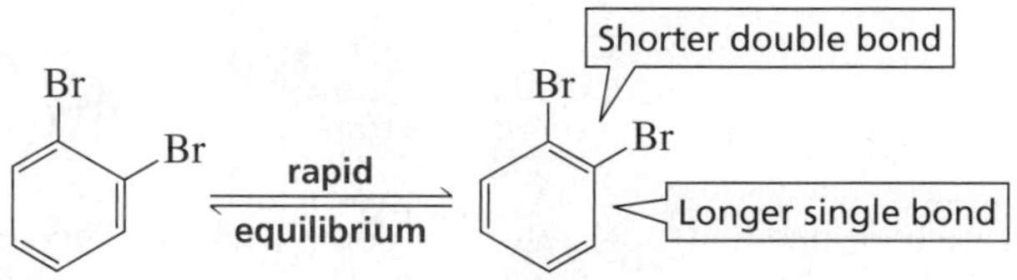

rapid equilibrium

Shorter double bond

Longer single bond

KEKULÉ'S DREAM

Friedrich August Kekulé von Stradonitz (1829–1896) was born in Germany. He entered the University of Giessen to study architecture, but switched to chemistry after taking a course in the subject. He was a professor of chemistry at the University of Heidelberg, at the University of Ghent in Belgium, and then at the University of Bonn. In 1890, he gave an extemporaneous speech at the twenty-fifth-anniversary celebration of his first paper on the cyclic structure of benzene. In this speech, he claimed that he had arrived at the Kekulé structures as a result of dozing off in front of a fire while working on a textbook. He dreamed of chains of carbon atoms twisting and turning in a snakelike motion, when suddenly the head of one snake seized hold of its own tail and formed a spinning ring. In 1895, Kekulé was made a nobleman by Emperor William II. This allowed him to add "von Stradonitz" to his name. Kekulé's students received three of the first five Nobel Prizes in chemistry.

Friedrich August Kekulé von Stradonitz

Controversy over the structure of benzene continued until the 1930s, when modern analytical techniques showed that benzene is a planar molecule and *the six carbon–carbon bonds have the same length*. The length of each carbon–carbon bond is 1.39 Å, which is shorter than a carbon–carbon single bond (1.54 Å) but longer than a carbon–carbon double bond (1.33 Å; Section 1.14). In other words, benzene does not have alternating single and double bonds.

If the carbon–carbon bonds in benzene all have the same length, they must also have the same number of electrons between the carbon atoms. This can be so, however, only if the π electrons of benzene are delocalized around the ring, rather than each pair of π electrons being localized between two carbon atoms. To better understand how electrons can be delocalized, we'll now take a close look at the bonding in benzene.

6.2   The Bonding in Benzene

Each of benzene's six carbon atoms is sp^2 hybridized. An sp^2 hybridized carbon has bond angles of 120°—identical to the size of the angles of a planar hexagon. Benzene, therefore, is a planar molecule. Each of its carbons uses two sp^2 orbitals to bond to two other carbons; the third sp^2 orbital overlaps the s orbital of a hydrogen (Figure 6.1a). Each carbon also has a p orbital at right angles to the sp^2 orbitals. Because benzene is planar, the six p orbitals are parallel (Figure 6.1b). The p orbitals are close enough for side-to-side overlap, so each p orbital overlaps the p orbitals on *both* adjacent carbons. As a result, the overlapping p orbitals form a continuous doughnut-shaped cloud of electrons above, and another doughnut-shaped cloud of electrons below, the plane of the benzene ring (Figure 6.1c). The electrostatic potential map (Figure 6.1d) shows that all the carbon–carbon bonds have the same electron density.

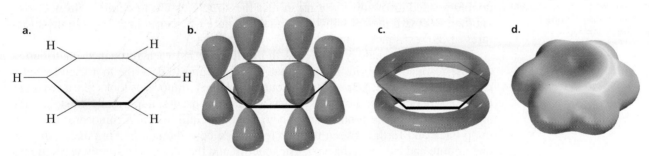

a.   b.   c.   d.

▲ **Figure 6.1**
(a) The carbon–carbon and carbon–hydrogen σ bonds in benzene.
(b) The p orbital on each carbon of benzene can overlap with two adjacent p orbitals.
(c) The clouds of π electrons above and below the plane of the benzene ring.
(d) The electrostatic potential map for benzene.

Each of the six π electrons, therefore, is localized neither on a single carbon nor in a bond between two carbons (as in an alkene). Instead, each π electron is shared by all six carbons. The six π electrons are delocalized—they roam freely within the doughnut-shaped clouds that lie over and under the ring of carbon atoms. Consequently, benzene can be represented by a hexagon containing either dashed lines or a circle, to symbolize the six delocalized π electrons.

3-D Molecule:
Benzene

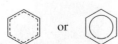

or

This type of representation makes it clear that there are no double bonds in benzene. We see now that Kekulé's structure for benzene was pretty close to the correct structure. The actual structure of benzene is a Kekulé structure with delocalized electrons.

6.3 Resonance Contributors and the Resonance Hybrid

A disadvantage to using dashed lines to represent delocalized electrons is that they do not tell us how many π electrons are present in the molecule. For example, the dashed lines inside the hexagon in the representation of benzene indicate that the π electrons are shared equally by all six carbons and that all the carbon–carbon bonds have the same length, but they do not show how many π electrons are in the ring. Consequently, chemists prefer to use structures with localized electrons to approximate the actual structure that has delocalized electrons. The *approximate* structure with localized electrons is called a **resonance contributor**. The *actual* structure with delocalized electrons is called a **resonance hybrid**. Notice that it is easy to see that there are six π electrons in the ring of each resonance contributor.

resonance contributor   resonance contributor

resonance hybrid

Resonance contributors are shown with a double-headed arrow between them. The double-headed arrow does *not* mean that the structures are in equilibrium with one another. Rather, it indicates that the actual structure lies somewhere between the structures of the resonance contributors.

The following analogy illustrates the difference between resonance contributors and the resonance hybrid. Imagine that you are trying to describe to a friend what a rhinoceros looks like. You might tell your friend that a rhinoceros looks like a cross between a unicorn and a dragon. The unicorn and the dragon don't really exist, so they are like the resonance contributors. They are not in equilibrium: A rhinoceros does not jump back and forth between the two resonance contributors, looking like a unicorn one minute and a dragon the next. The unicorn and the dragon are simply ways to represent what the actual structure—the rhinoceros—looks like. *Resonance contributors, like unicorns and dragons, are imaginary, not real. Only the resonance hybrid, like the rhinoceros, is real.*

> **Electron delocalization is shown by double-headed arrows (↔). Equilibrium is shown by two arrows pointing in opposite directions (⇌).**

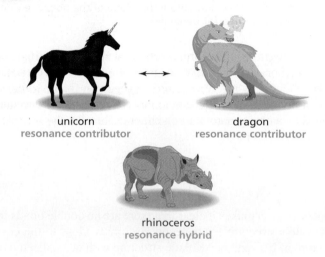

unicorn
resonance contributor

dragon
resonance contributor

rhinoceros
resonance hybrid

Electron delocalization occurs only if all the atoms sharing the delocalized electrons lie in or close to the same plane, so that their *p* orbitals can effectively overlap. For example, cyclooctatetraene is not planar; its *sp*² carbons have bond angles of 120° and a planar eight-member ring would have bond angles of 135°. Because it is not planar, the *p* orbitals cannot overlap; each pair of π electrons is *localized* between two carbons instead of being *delocalized* over the entire ring of eight carbons. Thus, cyclooctatetraene, unlike benzene, has alternating single and double bonds.

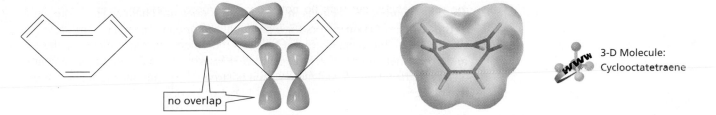

no overlap

cyclooctatetraene

3-D Molecule: Cyclooctatetraene

6.4   Drawing Resonance Contributors

We have seen that an organic compound with delocalized electrons is generally represented as a structure with localized electrons to let us know how many π electrons are present in the molecule. For example, nitroethane is represented as having a nitrogen–oxygen double bond and a nitrogen–oxygen single bond.

$$CH_3CH_2 \overset{+}{-}N \underset{:\overset{..}{O}:^-}{\overset{:\overset{..}{O}:}{\Big\langle}}$$

nitroethane

However, the two nitrogen–oxygen bonds in nitroethane are identical; they each have the same bond length. A more accurate description of the molecule's structure is obtained by drawing the two resonance contributors. Both resonance contributors show the compound with a nitrogen–oxygen double bond and a nitrogen–oxygen single bond, but to show that the electrons are delocalized, the double bond in one contributor is the single bond in the other.

$$CH_3CH_2 \overset{+}{-}N \underset{:\overset{..}{O}:^-}{\overset{:\overset{..}{O}:}{\Big\langle}} \qquad \longleftrightarrow \qquad CH_3CH_2 \overset{+}{-}N \underset{\overset{..}{O}:}{\overset{:\overset{..}{O}:^-}{\Big\langle}}$$

resonance contributor   **resonance contributor**

The resonance hybrid shows that the *p* orbital of nitrogen overlaps the *p* orbital of each oxygen. In other words, the two π electrons are shared by three atoms. The resonance hybrid also shows that the two nitrogen–oxygen bonds are identical and that the negative charge is shared by both oxygen atoms. Thus, we need to visualize and mentally average both resonance contributors to appreciate what the actual molecule—the resonance hybrid—looks like.

> Delocalized electrons result from a *p* orbital overlapping the *p* orbitals of more than one adjacent atom.

$$CH_3CH_2 \overset{+}{-}N \underset{\underset{\delta-}{\overset{..}{O}:}}{\overset{\overset{\delta-}{\overset{..}{O}:}}{\Big\langle}}$$

resonance hybrid

Rules for Drawing Resonance Contributors

To draw resonance contributors, the electrons in one resonance contributor are moved to generate the next resonance contributor. As you draw resonance contributors, keep in mind the following constraints:

1. Only electrons move. The nuclei of the atoms never move.

2. The only electrons that can move are π electrons (electrons in π bonds) and lone-pair electrons.

3. The total number of electrons in the molecule does not change. Therefore, each of the resonance contributors for a particular compound must have the same net charge. If one has a net charge of zero, all the others must also have net charges of zero. (A net charge of zero does not necessarily mean that there is no charge on any of the atoms: A molecule with a positive charge on one atom and a negative charge on another atom has a net charge of zero.)

To draw resonance contributors, move only π electrons or lone pairs toward an sp^2 carbon.

Tutorial:
Drawing resonance
contributors

Notice, as you study the following resonance contributors, that when resonance contributors are drawn, the electrons (π electrons or lone pairs) are always moved toward an sp^2 carbon. Remember that an sp^2 carbon is either a positively charged carbon (Section 1.10) or a double-bonded carbon. Electrons cannot be moved toward an sp^3 carbon, because an sp^3 carbon cannot accommodate any more electrons—it has a complete octet.

The following carbocation has delocalized electrons. To draw its resonance contributor, π electrons are moved toward a positive charge. The resonance hybrid shows that the positive charge is shared by two carbons.

$$CH_3CH{=}CH{-}\overset{+}{C}HCH_3 \quad \longleftrightarrow \quad CH_3\overset{+}{C}H{-}CH{=}CHCH_3$$
resonance contributors

$$CH_3\overset{\delta+}{CH}{=\!=\!=}CH{=\!=\!=}\overset{\delta+}{C}HCH_3$$
resonance hybrid

Tutorial:
Localized and delocalized
electrons

Now let's compare this carbocation with a similar compound in which all the electrons are localized. The π electrons in the following compound cannot move, because the carbon they would move to is an sp^3 carbon; *sp^3 carbons cannot accept electrons.*

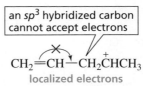

an sp^3 hybridized carbon
cannot accept electrons

$$CH_2{=}CH{-}CH_2\overset{+}{C}HCH_3$$
localized electrons

In the next example, π electrons again move toward a positive charge. The resonance hybrid shows that the positive charge is shared by three carbons.

$$CH_3CH{=}CH{-}CH{=}CH{-}\overset{+}{C}H_2 \quad \longleftrightarrow \quad CH_3CH{=}CH{-}\overset{+}{C}H{-}CH{=}CH_2 \quad \longleftrightarrow \quad CH_3\overset{+}{C}H{-}CH{=}CH{-}CH{=}CH_2$$
resonance contributors

$$CH_3\overset{\delta+}{CH}{=\!=\!=}CH{=\!=\!=}\overset{\delta+}{CH}{=\!=\!=}CH{=\!=\!=}\overset{\delta+}{C}H_2$$
resonance hybrid

The resonance contributor for the next compound is obtained by moving lone-pair electrons toward an sp^2 carbon. The sp^2 carbon can accommodate the new electrons by breaking a π bond.

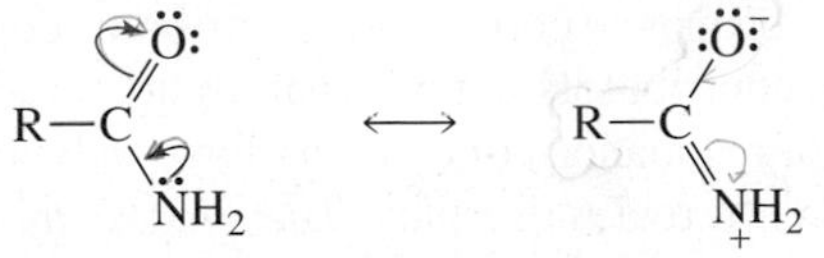

resonance contributors

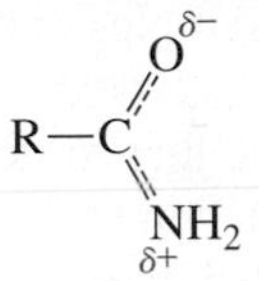

resonance hybrid

The following resonance contributors are obtained by moving π electrons toward an sp^2 carbon. Notice that the electrons move toward the most electronegative atom (the oxygen).

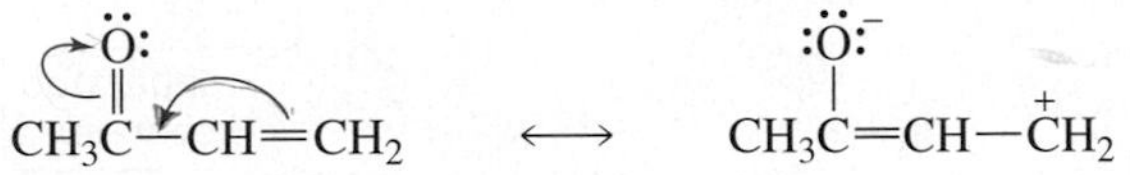

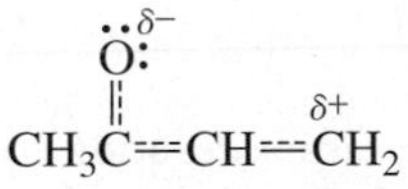

resonance hybrid

Radicals can also have delocalized electrons. Resonance contributors are obtained by moving the unpaired electron toward an sp^2 carbon.

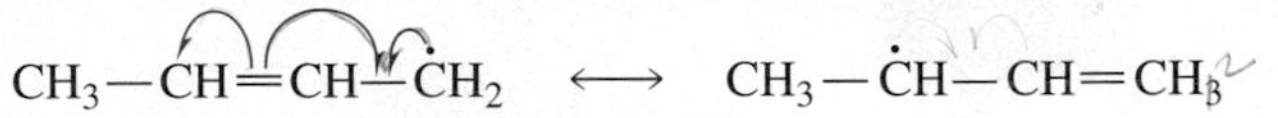

resonance contributors

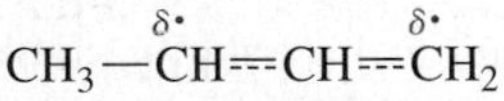

resonance hybrid

PROBLEM 1◆

a. Which compounds have delocalized electrons?

1. 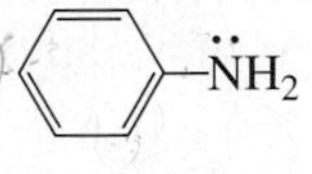

2. 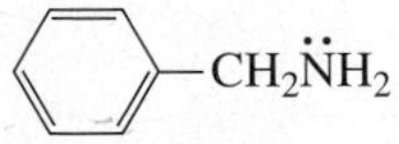

3. 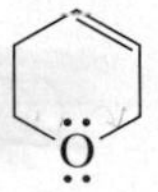

4. $CH_2{=}CHCH_2CH{=}CH_2$

5. 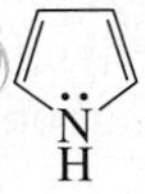

6. $CH_3CH{=}CHCH{=}CH\overset{+}{C}H_2$

7. $CH_3CH_2\ddot{N}HCH_2CH{=}CH_2$

8. $CH_2{=}CH-\dot{C}H-CH{=}CH_2$

b. Draw the contributing resonance structures for these compounds.

6.5 Predicted Stabilities of Resonance Contributors

The greater the predicted stability of the resonance contributor, the more it contributes to the structure of the resonance hybrid.

All resonance contributors do not necessarily contribute equally to the resonance hybrid. The degree to which each resonance contributor contributes depends on its predicted stability. Because resonance contributors are not real, their stabilities cannot be measured. Therefore, the stabilities of resonance contributors have to be predicted based on molecular features that are found in real molecules. *The greater the predicted stability of the resonance contributor, the more it contributes to the structure of resonance hybrid; and the more it contributes to the structure of resonance hybrid, the more similar the contributor is to the real molecule*. The examples that follow illustrate these points.

The two resonance contributors for a carboxylic acid are labeled **A** and **B**. Structure **B** has two features that make it less stable than structure **A**: One of its oxygen atoms has a positive charge—not a comfortable situation for an electronegative atom—and it has separated charges. A molecule with **separated charges** is a molecule with a positive charge and a negative charge that can be neutralized by the movement of electrons. Resonance contributors with separated charges are relatively unstable (are relatively high in energy) because it takes energy to keep opposite charges separated. Structure **A**, therefore, is predicted to be more stable than structure **B**. Consequently, structure **A** makes a greater contribution to the resonance hybrid; that is, the resonance hybrid looks more like **A** than like **B**.

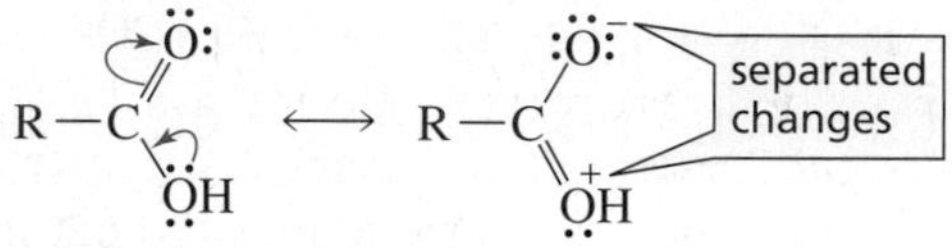

A B

a carboxylic acid

The two resonance contributors for a carboxylate ion are shown next.

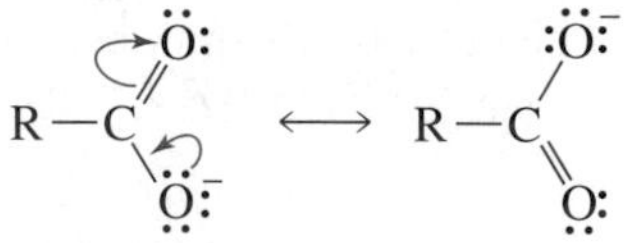

C D

a carboxylate ion

Structures **C** and **D** are predicted to be equally stable and therefore are expected to contribute equally to the resonance hybrid.

Let's now see which of the following resonance contributors has a greater predicted stability. Structure **E** has a negative charge on carbon, and structure **F** has a negative charge on oxygen. Oxygen is more electronegative than carbon, so oxygen can accommodate the negative charge better. Consequently, structure **F** is predicted to be more stable than structure **E**. The resonance hybrid, therefore, more closely resembles structure **F**; that is, the resonance hybrid has a greater concentration of negative charge on the oxygen atom than on the carbon atom.

3-D Molecule: An enolate ion

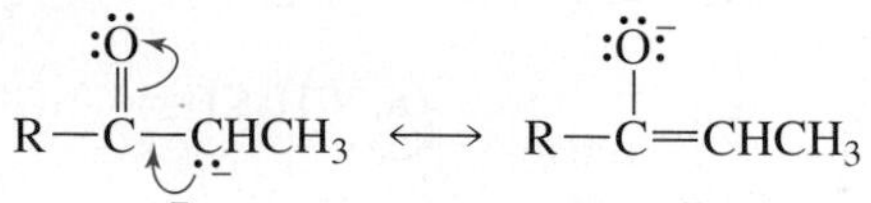

E F

PROBLEM 2 SOLVED

Draw contributing resonance structures for each of the following species, and rank the structures in order of decreasing contribution to the hybrid:

a. $CH_3\overset{+}{C}(CH_3)—CH{=}CHCH_3$ **b.** 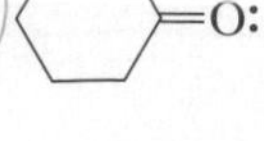 **c.** $CH_3\overset{+}{C}H—CH{=}CHCH_3$

SOLUTION TO 2a   Structure **A** is more stable than structure **B** because the positive charge is on a tertiary carbon in **A** and on a secondary carbon in **B**.

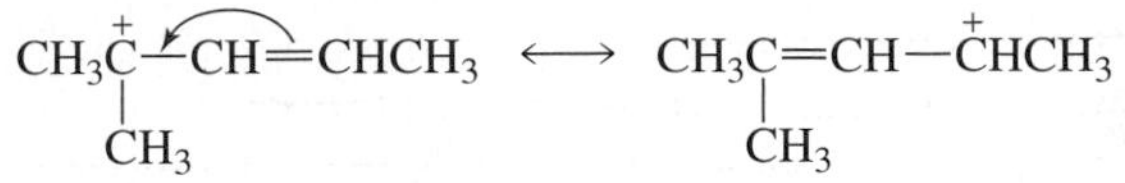

$$CH_3\overset{+}{C}\overset{\frown}{-}CH=CHCH_3 \quad \longleftrightarrow \quad CH_3C=CH-\overset{+}{C}HCH_3$$
$$\underset{CH_3}{|} \qquad\qquad\qquad \underset{CH_3}{|}$$
$$\textbf{A} \qquad\qquad\qquad\qquad \textbf{B}$$

PROBLEM 3◆

Draw the resonance hybrid for each of the species in Problem 2.

6.6   Resonance Stabilization

Delocalized electrons stabilize a compound. The extra stability a compound gains from having delocalized electrons is called **resonance stabilization** or **resonance energy**.

The amount of resonance stabilization associated with a compound that has delocalized electrons depends on the number *and* predicted stability of the resonance contributors: *The greater the number of relatively stable resonance contributors and the more nearly equivalent they are, the greater is the resonance stabilization.* For example, the resonance stabilization of a carboxylate ion with two relatively stable resonance contributors is significantly greater than the resonance stabilization of a carboxylic acid with only one relatively stable resonance contributor.

> The resonance stabilization is a measure of how much more stable a compound with delocalized electrons is than it would be if its electrons were localized.
>
> The greater the number of relatively stable resonance contributors, the greater is the resonance stabilization.
>
> The more nearly equivalent the resonance contributors, the greater is the resonance stabilization.

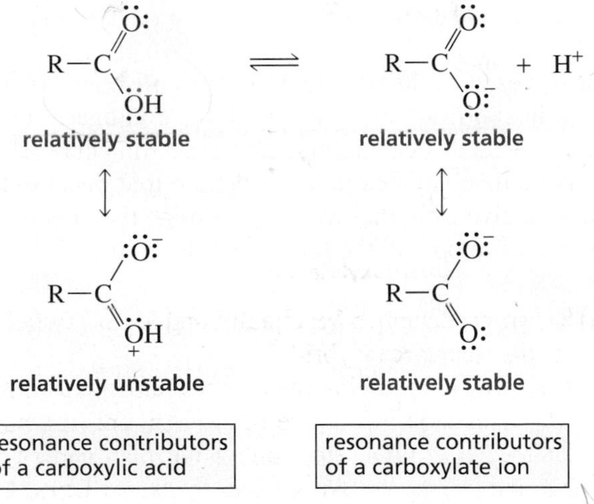

relatively stable   relatively stable   + H$^+$

relatively unstable   relatively stable

| resonance contributors of a carboxylic acid | resonance contributors of a carboxylate ion |

DELOCALIZED ELECTRONS IN VISION

We have seen that the interconversion of a double bond in rhodopsin between the cis and trans forms plays an important role in vision (Section 4.4). The first step in this process requires a π bond to break to form a pair of radicals. The energy required to break this bond depends on the stability of the radicals. Stabilization of the radicals by electron delocalization makes it easier to break the π bond.

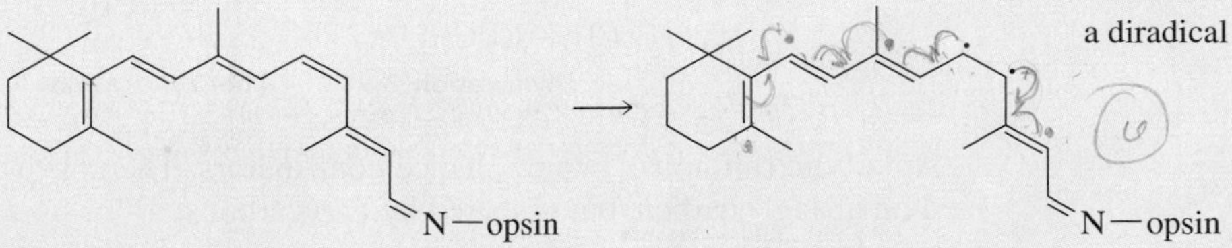

a diradical

rhodopsin   rhodopsin

PROBLEM 4◆

How many carbon atoms share the unpaired electrons in rhodopsin?

PROBLEM 5◆

Which compound has greatest resonance stabilization?

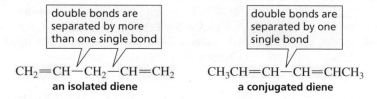

6.7 The Effect of Delocalized Electrons on Stability

Dienes are hydrocarbons with two double bonds. **Conjugated dienes** have conjugated double bonds; **conjugated double bonds** are separated by one single bond. **Isolated dienes** have isolated double bonds; **isolated double bonds** are separated by more than one single bond.

3-D Molecules:
2,3-Pentadiene;
1,4-Pentadiene;
1,3-Pentadiene

double bonds are
separated by more
than one single bond

double bonds are
separated by one
single bond

$CH_2{=}CH{-}CH_2{-}CH{=}CH_2$ $CH_3CH{=}CH{-}CH{=}CHCH_3$

an isolated diene   **a conjugated diene**

The π electrons in each of the double bonds of an isolated diene are *localized* between two carbons. In contrast, the π electrons in a conjugated diene are *delocalized*. We have seen that electron delocalization stabilizes a molecule. Therefore, conjugated dienes are more stable than isolated dienes. (Notice that because the compound does not have an electronegative atom that would determine the direction in which the electrons move, they can move both to the left and to the right.)

$\overset{..}{C}H_2{-}CH{=}CH{-}\overset{+}{C}H_2 \longleftrightarrow CH_2{=}CH{-}CH{=}CH_2 \longleftrightarrow \overset{+}{C}H_2{-}CH{=}CH{-}\overset{-}{C}H_2$

resonance contributors

delocalized
electrons

$CH_2{=\!=}CH{=\!=}CH{=\!=}CH_2$

resonance hybrid

An **allylic cation** has a positive charge on a carbon adjacent to an sp^2 carbon of an alkene. A **benzylic cation** has a positive charge on a carbon adjacent to an sp^2 carbon of a benzene ring.

$CH_2{=}CH\overset{+}{C}H_2$ [benzene ring with $\overset{+}{C}H_2$]

an allylic cation   **a benzylic cation**

An allylic cation has two resonance contributors. The positive charge is not localized on a single carbon, but is shared by two carbons.

Tutorial:
Common terms

$CH_2{=}CH{-}\overset{+}{C}H_2 \longleftrightarrow \overset{+}{C}H_2{-}CH{=}CH_2$

A benzylic cation has five resonance contributors. Notice that the positive charge is shared by four carbons.

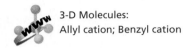

3-D Molecules:
Allyl cation; Benzyl cation

Because allylic and benzylic cations have delocalized electrons, they are more stable than other *primary* carbocations with localized electrons. (Indeed, they have about the same stability as secondary alkyl carbocations.) We can add allylic and benzylic cations to the group of carbocations whose relative stabilities were shown in Section 5.2.

relative stabilities of carbocations

most stable → a tertiary carbocation > a primary benzylic carbocation ≈ a primary allylic carbocation ≈ a secondary carbocation > a primary carbocation > methyl cation > vinyl cation ← least stable

Not all allylic and benzylic carbocations have the same stability. A tertiary allylic cation is more stable than a secondary allylic cation, which in turn is more stable than a primary allylic cation. Similarly, a tertiary benzylic cation is more stable than a secondary benzylic cation, which is more stable than a primary benzylic cation.

relative stabilities

most stable → $CH_2=CH-\overset{+}{\underset{R}{C}}-R$ > $CH_2=CH\overset{+}{C}H-R$ > $CH_2=CH\overset{+}{C}H_2$

tertiary allylic cation   secondary allylic cation   primary allylic cation

most stable → tertiary benzylic cation > secondary benzylic cation > primary benzylic cation

PROBLEM-SOLVING STRATEGY

Which carbocation is more stable?

$$CH_3CH=CH-\overset{+}{C}H_2 \quad \text{or} \quad CH_3\overset{CH_3}{\underset{}{C}}=CH-\overset{+}{C}H_2$$

Start by drawing the resonance contributors for each carbocation.

$$CH_3CH=CH-\overset{+}{C}H_2 \longleftrightarrow CH_3\overset{+}{C}H-CH=CH_2 \qquad CH_3\overset{CH_3}{\underset{}{C}}=CH-\overset{+}{C}H_2 \longleftrightarrow CH_3\underset{+}{\overset{CH_3}{C}}-CH=CH_2$$

Then compare the predicted stabilities of the set of resonance contributors for each carbocation.

Each carbocation has two resonance contributors. The positive charge of the carbocation on the left is shared by a primary allylic carbon and a secondary allylic carbon. The positive charge of the carbocation on the right is shared by a primary allylic carbon and a tertiary allylic carbon. Because a tertiary allylic carbon is more stable than a secondary allylic carbon, the carbocation on the right is more stable.

Now continue on to Problem 6.

PROBLEM 6◆

Which carbocation in each of the following pairs is more stable?

a. CH$_3$CH=CH$\overset{+}{C}$HCH$_3$ or   **b.** ⬡—$\overset{+}{C}$HCH$_3$ or   ⬡—$\overset{+}{C}$CH$_3$

CH$_3$CH=CH$\overset{+}{C}$H$_2$ CH$_3$

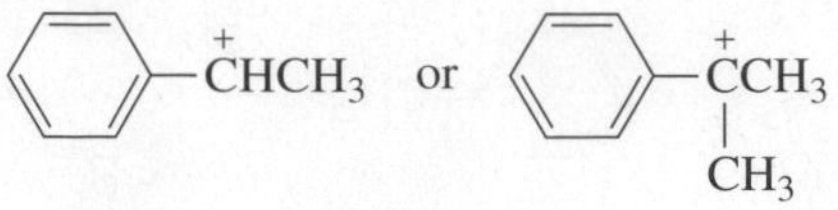

PROBLEM 7◆

Which species is more stable?

$\overset{+NH_2}{\overset{\|}{}}$ $\overset{+OH}{\overset{\|}{}}$

a. CH$_3$—C—NH$_2$ or   CH$_3$—C—NH$_2$

$\overset{O^-}{\overset{|}{}}$ $\overset{O^-}{\overset{|}{}}$

b. CH$_3$CHCH=CH$_2$ or   CH$_3$C=CHCH$_3$

6.8 The Effect of Delocalized Electrons on the Nature of the Product Formed in a Reaction

Our ability to predict the correct product of an organic reaction often depends upon recognizing when organic molecules have delocalized electrons. For example, in the following reaction, both sp^2 carbons of the alkene are bonded to the same number of hydrogens:

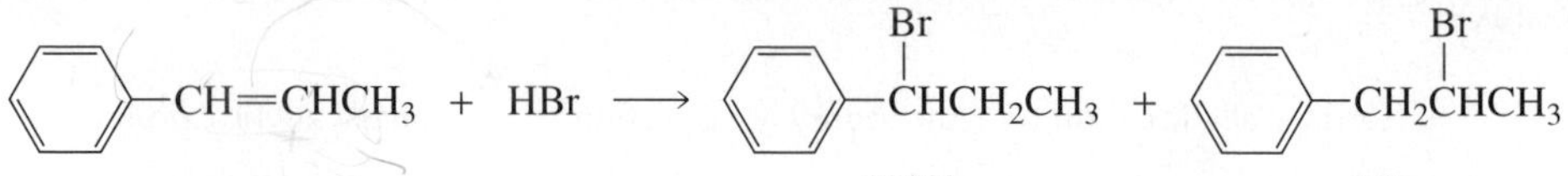

⬡—CH=CHCH$_3$ + HBr ⟶ ⬡—$\overset{Br}{\overset{|}{C}}HCH_2CH_3$ + ⬡—CH$_2$$\overset{Br}{\overset{|}{C}}HCH_3$

100%   **0%**

Therefore, the rule that tells us to add the electrophile to the sp^2 carbon bonded to the greater number of hydrogens predicts that approximately equal amounts of the two products will be formed. When the reaction is carried out, however, only one of the products is obtained.

The rule leads us to an incorrect prediction of the reaction product because it does not take electron delocalization into consideration. It presumes that both carbocation intermediates are equally stable since they are both secondary carbocations. The rule does not take into account the fact that one intermediate is a secondary carbocation and the other is a secondary benzylic cation. Because the secondary benzylic cation is stabilized by electron delocalization, it is formed more readily. The difference in the rates of formation of the two carbocations is large enough that only one product is obtained.

⬡—$\overset{+}{C}$HCH$_2$CH$_3$ ⬡—CH$_2$$\overset{+}{C}HCH_3$

a secondary benzylic cation   **a secondary carbocation**

PROBLEM 8 SOLVED

Predict the sites on each of the following compounds where protonation can occur:

a. CH$_3$CH=CHOCH$_3$ + H$^+$ **b.** 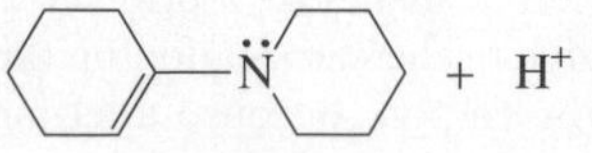 ⬡—N̈⬡ + H$^+$

SOLUTION TO 8a The contributing resonance structures show that there are two sites that can be protonated: the lone pair on oxygen and the lone pair on carbon.

sites of protonation

$$CH_3\overset{\frown}{CH}=CH\overset{\frown}{\underset{..}{O}}CH_3 \quad\longleftrightarrow\quad CH_3\overset{-}{\overset{..}{C}}H-CH=\overset{+}{\underset{..}{O}}CH_3 \qquad CH_3CH=CH\overset{..}{\underset{..}{O}}CH_3$$

resonance contributors

Let's now compare the products formed when isolated dienes (that have only localized electrons) undergo electrophilic addition reactions with the products formed when conjugated dienes (that have delocalized electrons) undergo the same reactions.

The reactions of *isolated dienes* are like the reactions of alkenes. If an excess of the electrophilic reagent is present, two independent addition reactions will occur, each following the rule that applies to all electrophilic addition reactions: *The electrophile adds to the sp^2 carbon that is bonded to the greater number of hydrogens.*

$$CH_2=CHCH_2CH_2CH=CH_2 \;+\; \boxed{HBr} \;\longrightarrow\; CH_3CHCH_2CH_2CHCH_3$$

1,5-hexadiene   excess $\underset{Br}{|}$ $\underset{Br}{|}$

The reaction proceeds exactly as we would predict from our knowledge of the mechanism for the reaction of alkenes with electrophilic reagents. The electrophile (H$^+$) adds to the electron-rich double bond in a manner that produces the more stable carbocation (Section 5.3). The bromide ion then adds to the carbocation. Because there is an excess of the electrophilic reagent, there is enough reagent to add to both double bonds.

mechanism for the reaction of 1,5-hexadiene with excess HBr

$$CH_2=CHCH_2CH_2CH=CH_2 \;+\; H-\overset{..}{\underset{..}{Br}}: \;\longrightarrow\; CH_3\overset{+}{C}HCH_2CH_2CH=CH_2 \;\longrightarrow\; CH_3CHCH_2CH_2CH=CH_2$$

$$+ :\overset{..}{\underset{..}{Br}}:^{-} \qquad\qquad \underset{Br}{|}$$

$$\downarrow H-\overset{..}{\underset{..}{Br}}:$$

$$CH_3CHCH_2CH_2CHCH_3 \;\longleftarrow\; CH_3CHCH_2CH_2\overset{+}{C}HCH_3$$

$$\underset{Br}{|} \qquad \underset{Br}{|} \qquad\qquad\qquad \underset{Br}{|} \qquad + :\overset{..}{\underset{..}{Br}}:^{-}$$

If there is only enough electrophilic reagent to add to one of the double bonds, it will add preferentially to the more reactive double bond. For example, in the reaction of 2-methyl-1,5-hexadiene with HCl, addition of HCl to the double bond on the left forms a secondary carbocation, whereas addition of HCl to the double bond on the right forms a tertiary carbocation. Because the transition state leading to formation of a tertiary carbocation is more stable than that leading to a secondary carbocation, the tertiary carbocation is formed faster (Section 5.3). Thus, in the presence of a limited amount of HCl, the major product of the reaction will be 5-chloro-5-methyl-1-hexene.

$$\overset{\overset{\displaystyle CH_3}{|}}{CH_2=CHCH_2CH_2C=CH_2} \;+\; HCl \;\longrightarrow\; \overset{\overset{\displaystyle CH_3}{|}}{\underset{\underset{\displaystyle Cl}{|}}{CH_2=CHCH_2CH_2CCH_3}}$$

2-methyl-1,5-hexadiene   **1 mol**   **5-chloro-5-methyl-1-hexene**
1 mol   **major product**

PROBLEM 9◆

Give the major product of each of the following reactions. (Equivalent amounts of reagents are used in each reaction.)

a. 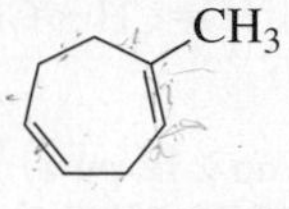 $\xrightarrow{\text{HCl}}$

b. $CH_2=CHCH_2CH_2CH=\overset{\overset{\displaystyle CH_3}{|}}{C}CH_3 \;\xrightarrow{\text{HBr}}$

> ### PROBLEM 10◆
>
> Which of the double bonds in zingiberene, the compound responsible for the odor of ginger, is the most reactive in an electrophilic addition reaction?
>
> **zingiberene**

When a *conjugated diene*, such as 1,3-butadiene, reacts with a limited amount of electrophilic reagent so that addition can occur at only one of the double bonds, two addition products are formed. One is a **1,2-addition** product, which is a result of addition at the 1- and 2-positions. The other is a **1,4-addition** product, the result of addition at the 1- and 4-positions.

$$CH_2=CH-CH=CH_2 + HBr \longrightarrow CH_3CH-CH=CH_2 + CH_3-CH=CH-CH_2$$

1,3-butadiene
1 mol   **1 mol**

3-bromo-1-butene   **1-bromo-2-butene**
1,2-addition product   **1,4-addition product**

Since we know how electrophilic reagents add to double bonds, we expect the 1,2-addition product to form. The fact that the 1,4-addition product also forms may be surprising because not only did the reagent not add to adjacent carbons, but a double bond has changed its position. The double bond in the 1,4-product is between the 2- and 3-positions, while the reactant had a single bond in this position.

When we talk about addition at the 1- and 2-positions or at the 1- and 4-positions, the numbers refer to the four carbons of the conjugated system. Thus, the carbon in the 1-position is one of the sp^2 carbons at the end of the conjugated system—it is not necessarily the first carbon in the molecule.

$$R-CH=CH-CH=CH-R$$

the conjugated system

To understand why both 1,2-addition and 1,4-addition products are obtained, we must look at the mechanism of the reaction. In the first step of the addition of HBr to 1,3-butadiene, the proton adds to C-1, forming an allylic cation. The π electrons of the allylic cation are delocalized—the positive charge is shared by two carbons. (Notice that because 1,3-butadiene is symmetrical, adding to C-1 is the same as adding to C-4.) The proton does not add to C-2 or C-3 because doing so would form a primary carbocation. The π electrons of a primary carbocation are localized; thus, it is not as stable as the delocalized allylic cation.

mechanism for the reaction of 1,3-butadiene with HBr

$$CH_2=CH-CH=CH_2 + H-\overset{..}{\underset{..}{Br}}: \longrightarrow CH_3-CH-CH=CH_2 \longleftrightarrow CH_3-CH=CH-CH_2$$

1,3-butadiene

$+ :\overset{..}{\underset{..}{Br}}:^-$ an allylic cation $+ :\overset{..}{\underset{..}{Br}}:^-$

$$CH_2-CH_2-CH=CH_2$$

a primary
carbocation

$$CH_3-CH-CH=CH_2 + CH_3-CH=CH-CH_2$$
|
Br   Br

3-bromo-1-butene   **1-bromo-2-butene**
1,2-addition product   **1,4-addition product**

The contributing resonance structures of the allylic cation show that the positive charge on the carbocation is not localized on C-2, but is shared by C-2 and C-4. Consequently, in the second step of the reaction, the bromide ion can attack either C-2 or C-4 to form the 1,2-addition product or the 1,4-addition product, respectively.

As we look at more examples, notice that the first step in all electrophilic additions to conjugated dienes is addition of the electrophile to one of the sp^2 carbons at the end of the conjugated system. This is the only way to obtain a carbocation that is stabilized by electron delocalization. If the electrophile were to add to one of the internal sp^2 carbons, the resulting carbocation would have only localized electrons.

PROBLEM 11♦

What diene would you expect to be more stable, 2,4-heptadiene or 2,5-heptadiene?

PROBLEM 12♦

Give the products of the following reactions (one equivalent of reagent is used in each case):

a. $CH_3CH{=}CH{-}CH{=}CHCH_3$ $\xrightarrow{HCl}$

b. $CH_3CH{=}\underset{\underset{CH_3}{|}}{\overset{\overset{CH_3}{|}}{C}}{-}C{=}CHCH_3$ $\xrightarrow{HBr}$

6.9   The Effect of Delocalized Electrons on pK_a

We have seen that a carboxylic acid is a much stronger acid than an alcohol. For example, the pK_a of acetic acid is 4.76, whereas the pK_a of ethanol is 15.9 (Section 2.2). We know, then, that the conjugate base of a carboxylic acid is considerably weaker and, therefore, more stable than the conjugate base of an alcohol. (Recall that the stronger the acid, the more stable is its conjugate base.)

$$\overset{\overset{\displaystyle O}{\|}}{CH_3C}OH \qquad CH_3CH_2OH$$

acetic acid   ethanol
pK_a = 4.76   pK_a = 15.9

The factor most responsible for the increased stability of the carboxylate ion is its *greater resonance stabilization* relative to that of its conjugate acid. The carboxylate ion has greater resonance stabilization, because the ion has two equivalent resonance contributors that are predicted to be relatively stable, whereas the carboxylic acid has only one (Section 6.5). Therefore, loss of a proton from a carboxylic acid is accompanied by an increase in resonance stabilization.

An increase in resonance stabilization means an increase in stability.

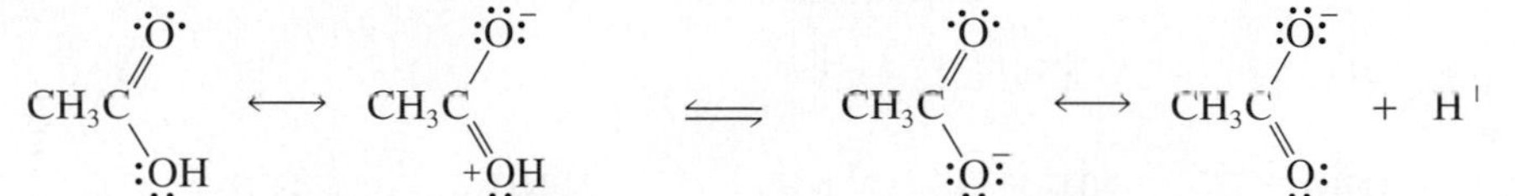

$CH_3C\overset{\displaystyle \cdot\ddot{O}}{\diagdown}\underset{\ddot{O}H}{}$ ⟷ $CH_3C\overset{\displaystyle :\ddot{O}:^-}{\diagup}\underset{+\ddot{O}H}{}$ ⇌ $CH_3C\overset{\displaystyle \cdot\ddot{O}}{\diagdown}\underset{:\ddot{O}:^-}{}$ ⟷ $CH_3C\overset{\displaystyle :\ddot{O}:^-}{\diagup}\underset{\ddot{O}:}{}$ + H$^+$

relatively stable   relatively unstable   relatively stable   relatively stable

resonance contributors of a carboxylic acid   resonance contributors of a carboxylate ion

In contrast, all the electrons in an alcohol—such as ethanol—and its conjugate base are localized, so loss of a proton from an alcohol is not accompanied by an increase in resonance stabilization.

$$CH_3CH_2\ddot{O}H \rightleftharpoons CH_3CH_2\ddot{O}:^- + H^+$$

ethanol

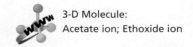

3-D Molecule:
Acetate ion; Ethoxide ion

Phenol, a compound in which an OH group is bonded to a benzene ring, is a stronger acid than an alcohol such as ethanol or cyclohexanol.

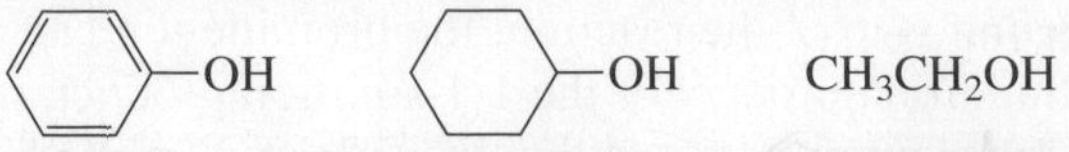

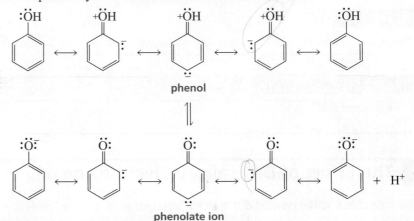

phenol
pK_a = 10

cyclohexanol
pK_a = 16

ethanol
pK_a = 16

While both phenol and the phenolate ion have delocalized electrons, the resonance stabilization of the phenolate ion is greater than that of phenol because three of phenol's resonance contributors have separated charges. The loss of a proton from phenol, therefore, is accompanied by an increase in resonance stabilization. In contrast, neither cyclohexanol nor its conjugate base has delocalized electrons, so loss of a proton is not accompanied by an increase in resonance stabilization.

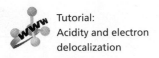
Tutorial:
Acidity and electron
delocalization

phenol

phenolate ion

Protonated aniline is a much stronger acid than protonated cyclohexylamine.

protonated aniline
pK_a = 4.60

protonated cyclohexylamine
pK_a = 11.2

The nitrogen atom of protonated aniline lacks a lone pair that can be delocalized. When it loses a proton, however, the lone pair that formerly held the proton can be delocalized. Loss of a proton, therefore, is accompanied by an increase in resonance stabilization.

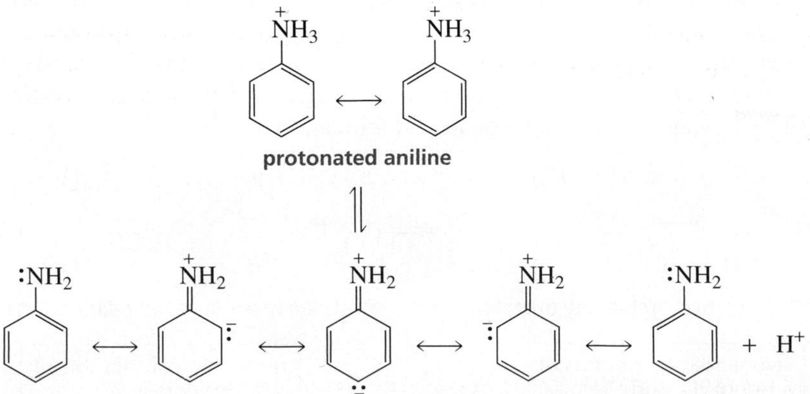

protonated aniline

aniline

An amine such as cyclohexylamine has no delocalized electrons either in the protonated form or in the unprotonated form, so proton loss is not associated with a change in resonance stabilization.

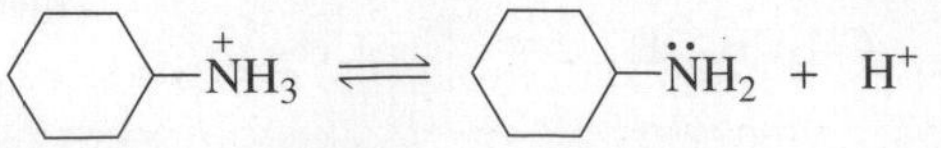

We can now add phenol and protonated aniline to the classes of organic compounds whose approximate pK_a values you should know (Table 6.1). They are also listed inside the back cover for easy reference.

Table 6.1 Approximate pK_a Values

pK_a < 0	pK_a ≈ 5	pK_a ≈ 10	pK_a ≈ 15
$\overset{+}{R\overset{H}{O}H}$	$\overset{O}{\underset{\parallel}{R}}COH$	$R\overset{+}{N}H_3$	ROH
$\overset{+OH}{\underset{RCOH}{\parallel}}$	⬡—$\overset{+}{N}H_3$	⬡—OII	H_2O
H_3O^+			

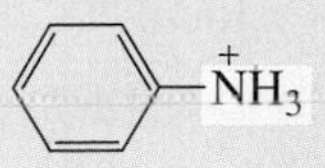

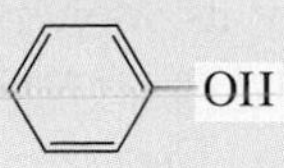

PROBLEM 13♦ SOLVED

Which is a stronger acid?

a. $CH_3CH_2CH_2OH$ or $CH_3CH{=}CHOH$

b. $\overset{O}{\underset{\parallel}{H}}CCH_2OH$ or $\overset{O}{\underset{\parallel}{CH_3}}COH$

c. $CH_3CH{=}CHCH_2OH$ or $CH_3CH{=}CHOH$

d. $CH_3CH_2CH_2\overset{+}{N}H_3$ or $CH_3CH{=}CH\overset{+}{N}H_3$

SOLUTION TO 13a The conjugate base of only one of the compounds has delocalized electrons. Because delocalized electrons stabilize the base, its conjugate acid is stronger.

$$CH_3CH_2CH_2\ddot{\overset{..}{O}}{:}^- \qquad CH_3CH{=}CH{-}\ddot{\overset{..}{O}}{:}^- \longleftrightarrow CH_3\overset{-}{\overset{..}{C}}H{-}CH{=}\ddot{O}{:}$$

only localized electrons   **delocalized electrons**

PROBLEM 14♦

Which is a stronger base?

a. ethylamine or aniline   **c.** phenolate ion or ethoxide ion

b. ethylamine or ethoxide ion ($CH_3CH_2O^-$)

PROBLEM 15♦

Rank the following compounds in order of decreasing acid strength:

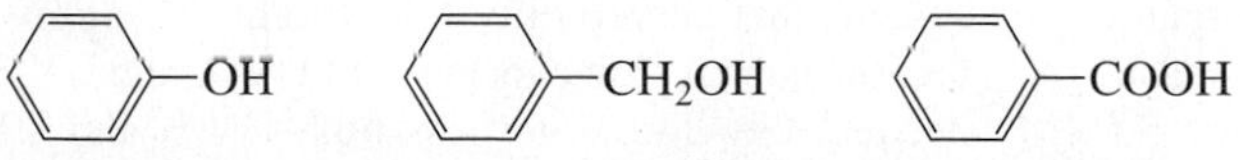

⬡—OH   ⬡—CH_2OH ⬡—COOH

6.10 Ultraviolet and Visible Spectroscopy

UV/Vis spectroscopy provides information about compounds with conjugated double bonds. Ultraviolet (UV) light has wavelengths ranging from 180 to 400 nm (nanometers); visible (Vis) light has wavelengths ranging from 400 to 780 nm. The shorter the wavelength, the greater is the energy of the radiation. Ultraviolet light, therefore, has greater energy than visible light. If a compound absorbs **ultraviolet light**, a UV spectrum is obtained; if it absorbs **visible light**, a visible spectrum is obtained.

The shorter the wavelength, the greater is the energy of the radiation.

The UV spectrum of acetone is shown in Figure 6.2. The $\boldsymbol{\lambda_{max}}$ (stated as "lambda max") is the wavelength corresponding to the highest point of the absorption band. The UV spectrum shows that acetone has a $\lambda_{max} = 195$ nm.

Figure 6.2 ▶
The UV spectrum of acetone.

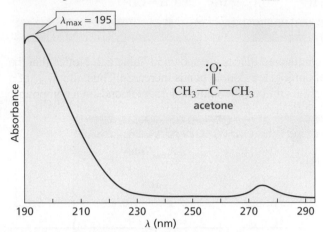

ULTRAVIOLET LIGHT AND SUNSCREENS

Exposure to ultraviolet light stimulates specialized cells in the skin to produce a black pigment known as melanin, which causes the skin to look tan. Melanin absorbs UV light, so it protects our bodies from the harmful effects of the sun. If more UV light reaches the skin than the melanin can absorb, the light will burn the skin and cause photochemical reactions that can result in skin cancer. UV-A is the lowest-energy UV light (315 to 400 nm) and does the least biological damage. Fortunately, most of the more dangerous, higher-energy UV light, UV-B (290 to 315 nm) and UV-C (180 to 290 nm), is filtered out by the ozone layer in the stratosphere. That is why there is such great concern about the apparent thinning of the ozone layer (Section 9.7).

para-aminobenzoic acid
PABA

2-ethylhexyl 4-(dimethylamino)benzoate
Padimate O

2-ethylhexyl (*E*)-3-(4-methoxyphenyl)-2-propenoate
Giv Tan F

Applying a sunscreen can protect skin against UV light. The amount of protection provided by a particular sunscreen is indicated by its SPF (sun protection factor). The higher the SPF, the greater the protection. Some sunscreens contain an inorganic component, such as zinc oxide, that reflects the light as it reaches the skin. Others contain a compound that absorbs UV light. PABA was the first commercially available UV-absorbing sunscreen. PABA absorbs UV-B light, but is not very soluble in oily skin lotions. Less polar compounds, such as Padimate O, are now commonly used. Recent research has shown that sunscreens that absorb only UV-B light do not give adequate protection against skin cancer; both UV-A and UV-B protection are needed. Giv Tan F absorbs both UV-B and UV-A light, so it gives better protection.

6.11 Effect of Conjugation on λ_{max}

The more conjugated double bonds there are in a compound, the longer is the wavelength at which the λ_{max} occurs. For example, methyl vinyl ketone has two conjugated double bonds, whereas acetone has one; hence, the λ_{max} for methyl vinyl ketone is at 219 nm—a longer wavelength than the λ_{max} for acetone.

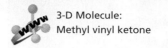

3-D Molecule:
Methyl vinyl ketone

$$:O:$$
$$\underset{H_3C \qquad CH_3}{C}$$
acetone
λ_{max} = **195 nm**

$$:O:$$
$$\underset{H_3C \qquad CH=CH_2}{C}$$
methyl vinyl ketone
λ_{max} = **219 nm**

The λ_{max} values for several conjugated dienes are shown in Table 6.2. Notice that the λ_{max} increases as the number of conjugated double bonds increases. Thus, the λ_{max} of a compound can be used to predict the number of conjugated double bonds in a compound.

The λ_{max} increases as the number of conjugated double bonds increases.

Table 6.2	Values for Ethene and Conjugated Dienes
Compound	λ_{max} **(nm)**
$H_2C=CH_2$	165
	217
	256
	290
	334
	364

If a compound has enough conjugated double bonds, it will absorb visible light $(\lambda_{max} > 400$ nm$)$ and the compound will be colored. β-Carotene, a precursor of vitamin A, is an orange substance found in carrots, apricots, and sweet potatoes. Lycopene is red and is found in tomatoes, watermelon, and pink grapefruit.

β-carotene
λ_{max} = **455 nm**

lycopene
λ_{max} = **474 nm**

PROBLEM 16◆

Rank the compounds in order of decreasing λ_{max}:

⬡   ⬡—⬡   ⬡—CH=CH—⬡   ⬡—CH=CH₂

6.12   The Visible Spectrum and Color

White light is a mixture of all wavelengths of visible light. If any color is removed from white light, the reflected light appears colored. So if a compound absorbs visible light, the compound will appear colored. Its color depends on the color of the light transmitted to the eye. In other words, it depends on the color produced from the wavelengths of light that are *not* absorbed.

Lycopene, β-carotene, and anthocyanins are found in the leaves of trees, but their characteristic colors are usually obscured by the green color of chlorophyll (Section 7.3). When chlorophyll degrades in the fall, the colors become apparent.

The relationship between the wavelengths of the absorbed light and the color observed is shown in Table 6.3. Notice that two absorption bands are necessary to produce green. Most colored compounds have fairly broad absorption bands; vivid colors have narrow absorption bands. The human eye is able to distinguish more than a million different shades of color!

Table 6.3 Dependence of the Color Observed on the Wavelength of Light Absorbed

Wavelengths absorbed (nm)	Observed color
380–460	yellow
380–500	orange
440–560	red
480–610	purple
540–650	blue
380–420 and 610–700	green

Azobenzenes (benzene rings connected by an N=N bond) have an extended conjugated system that causes them to absorb light from the visible region of the spectrum. Some substituted azobenzenes are used commercially as dyes. Varying the extent of conjugation and the substituents attached to the conjugated system provides a large number of different colors. Notice that the only difference between butter yellow and methyl orange is an SO₃⁻Na⁺ group. Methyl orange is a commonly used acid–base indicator. When margarine was first produced, it was colored with butter yellow to make it look more like butter. (White margarine would not have been very appetizing.) This dye was abandoned after it was found to be carcinogenic. β-Carotene is now used to color margarine (page 155).

methyl orange
an azobenzene

butter yellow
an azobenzene

PROBLEM 17◆

a. At pH = 7, one of the following ions is purple and the other is blue. Which is which?

b. What would be the difference in the colors of the compounds at pH = 3?

ANTHOCYANINS: A COLORFUL CLASS OF COMPOUNDS

A class of highly conjugated compounds called anthocyanins is responsible for the red, purple, and blue colors of many flowers (poppies, peonies, cornflowers), fruits (cranberries, rhubarb, strawberries, blueberries), and vegetables (beets, radishes, red cabbage).

In a neutral or basic solution, the ring on the right-hand side of the anthocyanin is not conjugated with the rest of the molecule. Thus, the anthocyanin does not absorb visible light and is,

therefore, a colorless compound. In an acidic environment, the OH group becomes protonated and water is eliminated. Loss of water results in the right-hand ring's becoming conjugated with the rest of the molecule. As a result of the number of conjugated double bonds, the anthocyanin absorbs visible light with wavelengths between 480 and 550 nm. The exact wavelength of light absorbed depends on the substituents (R and R′) of the anthocyanin. Thus, the flower, fruit, or vegetable appears red, purple, or blue, depending on what the R groups are. You can see this color change if you change the pH of cranberry juice so that it is no longer acidic.

anthocyanin
(three rings are conjugated)
red, blue, or purple

+ H_2O

(conjugation is disrupted)
colorless

(conjugation is disrupted)
colorless

R = H, OH, or OCH_3
R' = H, OH, or OCH_3

Summary

Localized electrons belong to a single atom or are confined to a bond between two atoms. **Delocalized electrons** are shared by more than two atoms; they result when a p orbital overlaps the p orbitals of more than one adjacent atom. Electron delocalization occurs only if all the atoms sharing the delocalized electrons lie in or close to the same plane.

Each of benzene's six carbon atoms is sp^2 hybridized, with bond angles of 120°. A p orbital of each carbon overlaps the p orbitals of both adjacent carbons. The six π electrons are shared by all six carbons. Thus, benzene is a planar molecule with six delocalized π electrons.

Chemists use **resonance contributors**—structures with localized electrons—to approximate the actual structure of a compound that has delocalized electrons: the **resonance hybrid**. To draw resonance contributors, move only π electrons or lone pairs toward an sp^2 carbon. The total number of electrons and the numbers of paired and unpaired electrons do not change.

The greater the **predicted stability** of the resonance contributor, the more it contributes to the hybrid and the more similar its structure is to the real molecule. The extra stability a compound gains from having delocalized electrons is called **resonance stabilization**. It tells us how much more stable a compound with delocalized electrons is than it would be if its electrons were localized. The greater the number of relatively stable resonance contributors and the more nearly equivalent they are, the greater is the resonance stabilization of the compound.

Electron delocalization can affect the stability of a compound, the nature of the product formed in a reaction, and the pK_a of a compound. Allylic and benzylic cations have delocalized electrons, so they are more stable than similarly substituted carbocations with localized electrons. A carboxylic acid and a phenol are more acidic than an alcohol such as ethanol, and a protonated aniline is more acidic than a protonated amine because the loss of a proton is accompanied by an increase in resonance stabilization.

Conjugated double bonds are separated by one single bond. **Isolated double bonds** are separated by more than one single bond. Because dienes with conjugated double bonds have delocalized electrons, they are more stable than dienes with isolated double bonds.

An isolated diene, like an alkene, undergoes only 1,2-addition. If there is only enough electrophilic reagent to add to one of the double bonds, it will add preferentially to the more reactive bond. A conjugated diene reacts with a limited amount of electrophilic reagent to form a **1,2-addition product** and a **1,4-addition product**. The first step is addition of the electrophile to one of the sp^2 carbons at the end of the conjugated system.

Ultraviolet and **visible (UV/Vis) spectroscopy** provide information about compounds with conjugated double bonds. UV light is higher in energy than visible light; the shorter the wavelength, the greater is its energy. The more conjugated double bonds there are in a compound, the longer is its λ_{max}.

Summary of Reactions

1. If there is excess electrophilic reagent, both double bonds of a *diene with isolated double bonds* will undergo electrophilic addition.

$$CH_2{=}CHCH_2CH_2\overset{\overset{\displaystyle CH_3}{|}}{C}{=}CH_2 \; + \; \underset{\textbf{excess}}{HBr} \; \longrightarrow \; CH_3\underset{\underset{\displaystyle Br}{|}}{CH}CH_2CH_2\overset{\overset{\displaystyle CH_3}{|}}{\underset{\underset{\displaystyle Br}{|}}{C}}CH_3$$

If there is only one equivalent of an electrophilic reagent, only the most reactive of the isolated double bonds will undergo electrophilic addition (Section 6.8).

$$CH_2{=}CHCH_2CH_2\overset{\overset{\displaystyle CH_3}{|}}{C}{=}CH_2 \; + \; HBr \; \longrightarrow \; CH_2{=}CHCH_2CH_2\overset{\overset{\displaystyle CH_3}{|}}{\underset{\underset{\displaystyle Br}{|}}{C}}CH_3$$

2. *Dienes with conjugated double bonds* undergo 1,2- and 1,4-addition with one equivalent of an electrophilic reagent (Section 6.8).

$$RCH{=}CHCH{=}CHR \; + \; HBr \; \longrightarrow \; RCH_2\underset{\underset{\displaystyle Br}{|}}{CH}CH{=}CHR \; + \; RCH_2CH{=}CH\underset{\underset{\displaystyle Br}{|}}{CH}R$$

$$\textbf{1,2-addition product} \qquad \textbf{1,4-addition product}$$

Problems

18. Which of the following compounds have delocalized electrons?

a. $CH_2{=}CH\overset{\overset{\displaystyle O}{\|}}{C}CH_3$

b.  (benzene ring)

c. 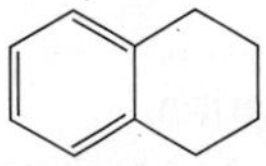 (bicyclic ring structure)

d. (cyclohexene ring)

e. $CH_2{=}CHCH_2CH{=}CH_2$

f. $CH_3\overset{\overset{\displaystyle CH_3}{|}}{\underset{+}{C}}CH_2CH{=}CH_2$

g. $CH_3CH_2NHCH_2CH{=}CHCH_3$

h. $CH_3CH_2N\overset{..}{H}CH{=}CHCH_3$

i. $CH_3CH_2\overset{}{C}H\overset{+}{C}H{=}CH_2$

j. $CH_3CH{=}CH\overset{..}{O}CH_2CH_3$

19. Draw resonance contributors for the following ions:

a. 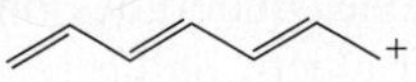 (structure) $+$

b. 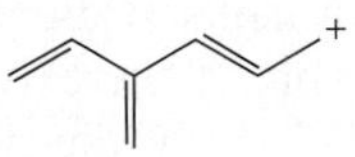 (structure) $+$

20. Are the following pairs of structures resonance contributors or different compounds?

a. 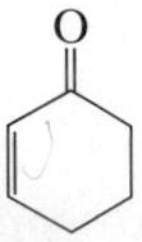 (cyclohexanone) and (cyclohexenone)

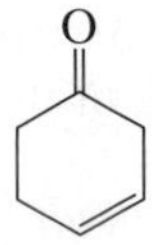

b. 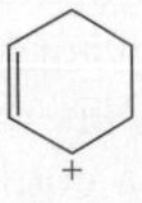 (cyclohexene with + charge) and (cyclohexenyl cation)

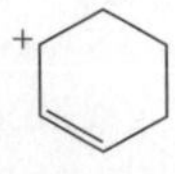

c. $CH_3\overset{\overset{\displaystyle O}{\|}}{C}CH_2CH_3$ and $CH_3\overset{\overset{\displaystyle OH}{|}}{C}{=}CHCH_3$

d. $CH_3\overset{}{C}HCH{=}CHCH_3$ and $CH_3CH{=}CHCH_2\overset{+}{C}H_2$

21. a. Draw resonance contributors for the following species. Indicate which are major contributors and which are minor contributors to the resonance hybrid.

1. $CH_3CH=CHOCH_3$

2. (benzene ring)$-CH_2\ddot{N}H_2$

3. $CH_3\ddot{C}HC\equiv N$

4. $CH_3CH=CH\overset{+}{C}H_2$

5. (cyclopentadienyl cation, ring with +)

6. (benzene ring)$-\overset{\cdot\cdot}{\underset{\cdot\cdot}{O}}CH_3$

b. Do any of the species have resonance contributors that all contribute equally to the resonance hybrid?

22. Which resonance contributor makes the greater contribution to the resonance hybrid?

a. $CH_3\overset{+}{C}HCH=CH_2$ or $CH_3CH=CH\overset{+}{C}H_2$

b. (cyclopentene ring with CH_3) or (cyclopentene ring with CH_3 and +)

c. (cyclohexadienone with $\overset{\cdot\cdot}{O}$) or (benzene ring with $:\overset{\cdot\cdot}{O}\overline{}$)

23. a. Which oxygen atom has the greater electron density?

$$CH_3\overset{\overset{\displaystyle O}{\|}}{C}OCH_3$$

b. Which compound has the greater electron density on its nitrogen atom?

(pyrrole ring $\underset{H}{N}$) or (pyrrole ring with O's $\underset{H}{N}$)

c. Which compound has the greater electron density on its oxygen atom?

(cyclohexane ring)$-NH\overset{\overset{\displaystyle O}{\|}}{C}CH_3$ or (benzene ring)$-NH\overset{\overset{\displaystyle O}{\|}}{C}CH_3$

24. Which can lose a proton more readily, a methyl group bonded to cyclohexane or a methyl group bonded to benzene? Explain your answer.

(cyclohexane ring)$-CH_3$ (benzene ring)$-CH_3$

25. a. Draw the resonance structures for $CO_3{}^{2-}$.
b. Predict the relative bond lengths of the three carbon–oxygen bonds.
c. What would you expect the charge to be on each oxygen atom?

26. Rank the following compounds in order of decreasing acidity:

(three skeletal structures)

27. Which species is more stable?

a. $CH_3CH_2O^-$ or $CH_3\overset{\overset{\displaystyle O}{\|}}{C}O^-$

b. $CH_3\overset{\overset{\displaystyle O}{\|}}{\underset{\cdot\cdot}{C}}HCH_2\overset{\overset{\displaystyle O}{\|}}{C}H$ or $CH_3\overset{\overset{\displaystyle O}{\|}}{C}\overset{-}{C}H\overset{\overset{\displaystyle O}{\|}}{C}CH_3$

c. $CH_3\overline{C}HCH_2\overset{\overset{\displaystyle O}{\|}}{C}CH_3$ or $CH_3CH_2\overset{\cdot\cdot}{C}H\overset{\overset{\displaystyle O}{\|}}{C}CH_3$

d. $CH_3\overset{\overset{\displaystyle NH_2}{|}}{C}HCH_3$ or $CH_3\overset{\overset{\displaystyle NH}{\|}}{C}NH_2$

e. $CH_3\overline{C}-\overset{\overset{\displaystyle O}{\|}}{\underset{\underset{\displaystyle CH_3}{|}}{C}}H$ or $CH_3\overline{C}-\overset{\overset{\displaystyle CH_2}{\|}}{\underset{\underset{\displaystyle CH_3}{|}}{C}}H$

f. (ring with two O's) $N:^-$ or (ring with one O) $N:^-$

28. Which species in each of the pairs in Problem 27 is the stronger base?

29. Draw resonance contributors for the following species. Do not include structures that are so unstable that their contributions to the resonance hybrid would be negligible. Indicate which species are major contributors to the resonance hybrid and which are minor contributors.

a. $CH_3-\overset{+}{N}\overset{O}{\underset{O^-}{\diagdown}}$

c. $H\overset{O}{\overset{||}{C}}CH=CH\overset{..}{C}H_2$

e. benzene ring with $CH=CH_2$

b. $H\overset{O}{\overset{||}{C}}NHCH_3$

d. $CH_3\overset{..}{C}H-\overset{+}{N}\overset{O}{\underset{O^-}{\diagdown}}$

f. $CH_3\overset{O}{\overset{||}{C}}\overset{..}{C}H\overset{O}{\overset{||}{C}}CH_3$

30. Rank the following compounds in order of decreasing acidity of the indicated hydrogen:

$$CH_3\overset{O}{\overset{||}{C}}CH_2CH_2\overset{O}{\overset{||}{C}}CH_3 \qquad CH_3\overset{O}{\overset{||}{C}}CH_2CH_2CH_2\overset{O}{\overset{||}{C}}CH_3 \qquad CH_3\overset{O}{\overset{||}{C}}CH_2\overset{O}{\overset{||}{C}}CH_3$$

31. Draw contributing resonance structures for each of the following species, and rank the structures in order of decreasing contribution to the hybrid:

a. $CH_3\overset{O}{\overset{||}{C}}OCH_3$

b. cyclohexene ring with $=O$

c. $CH_3-\overset{+OH}{\overset{||}{C}}-NHCH_3$

32. Draw the resonance hybrid for each of the species in Problem 31.

33. Name the following dienes and rank them in order of increasing stability. (Alkyl groups stabilize dienes in the same way that they stabilize alkenes; see Section 4.6.)

$$CH_3CH=CHCH=CHCH_3 \quad CH_2=CHCH_2CH=CH_2 \quad CH_3\overset{CH_3}{\overset{|}{C}}=CHCH=\overset{CH_3}{\overset{|}{C}}CH_3 \quad CH_3CH=CHCH=CH_2$$

34. Which carbocation in each of the following pairs is more stable?

a. $CH_3\overset{+}{O}CH_2$ or $CH_3\overset{+}{N}HCH_2$

b. $CH_3O\overset{+}{C}H_2CH_2$ or $CH_3O\overset{+}{C}H_2$

c. cyclohexene with $\overset{+}{C}HCH_3$ or cyclohexene with $\overset{+}{C}HCH_3$

35. a. How many linear dienes are there with molecular formula C_6H_{10}? (Ignore cis–trans isomers.)
b. How many of the linear dienes in part a are conjugated dienes?
c. How many are isolated dienes?

36. What products would be obtained from the reaction of 1,3,5-hexatriene with one equivalent of HBr?

37. Give the major product of each of the following reactions (one equivalent of each reagent is used):

a. cycloheptatriene with $-CH_3$ + HBr $\longrightarrow$

b. cyclohexene ring with $CH=CH_2$ and CH_3 + HBr $\longrightarrow$

38. How could you use UV spectroscopy to distinguish between the compounds in each of the following pairs?

a. (diene structure) and (diene structure)

b. (benzene-CH$_2$-C(=O)- structure) and (benzene-C(=O)- structure)

39. Give all the products of the following reaction:

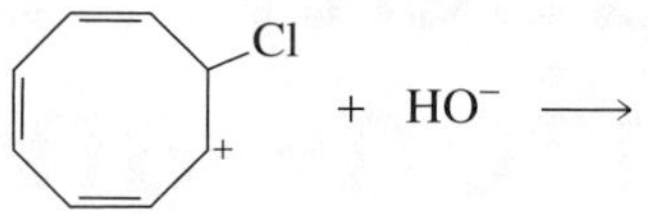

+ HO⁻ ⟶

40. When you sign some credit card slips, an imprint of your signature is made on the bottom copy. The carbonless paper contains tiny capsules that are filled with the colorless compound whose structure is shown here.

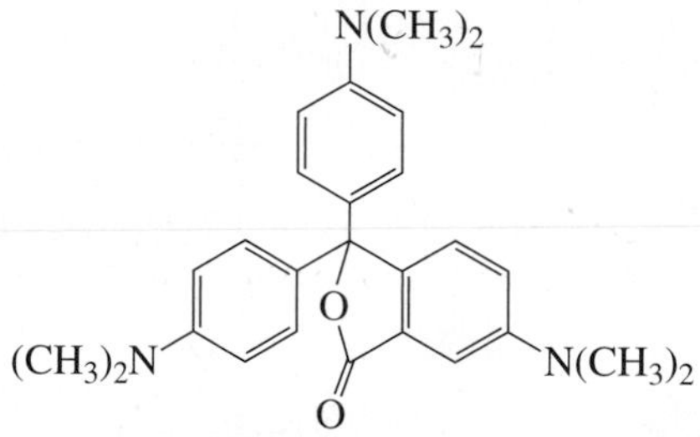

When you press on the paper, the capsules burst and the colorless compound comes into contact with the acid-treated paper, forming a highly colored compound. What is the structure of the colored compound?

41. a. Methyl orange (whose structure is given in Section 6.12) is an acid–base indicator. In solutions of pH < 4, it is red; in solutions of pH > 4, it is yellow. Account for the change in color.
  b. Phenolphthalein is also an indicator, but it exhibits a much more dramatic color change. In solutions of pH < 8.5, it is colorless; in solutions of pH > 8.5, it is deep red-purple. Account for the change in color.

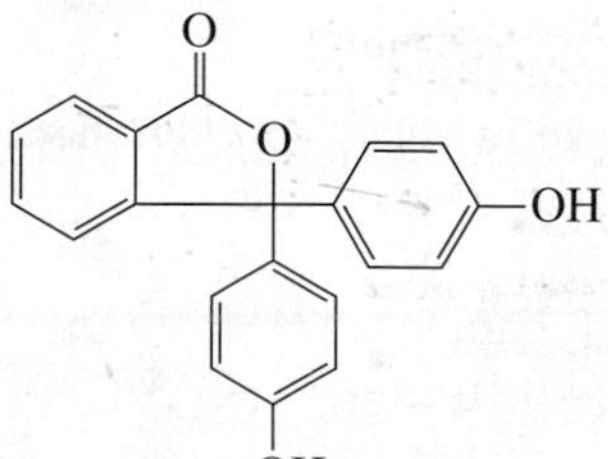

phenolphthalein

42. a. How could each of the following compounds be prepared from a hydrocarbon in a single step?

1. 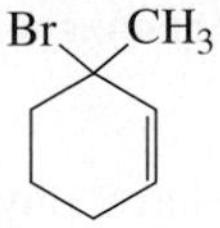   **2.**

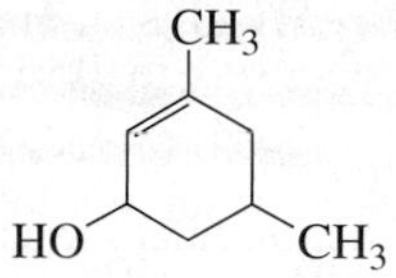

  b. What other organic compound would be obtained from each synthesis?

43. Give the major products obtained from the reaction of one equivalent of HCl with the following:
  a. 2,3-dimethyl-1,3-pentadiene   **b.** 2,4-dimethyl-1,3-pentadiene

7 Aromaticity • Reactions of Benzene and Substituted Benzenes

Michael Faraday (1791–1867)
was born in England, a son of a blacksmith. At the age of 14, he was apprenticed to a bookbinder and educated himself by reading the books that he bound. He became an assistant to Sir Humphry Davy in 1812 and taught himself chemistry. In 1825, he became the director of a laboratory at the Royal Institution; in 1833, he became a professor of chemistry there. He is best known for his work on electricity and magnetism.

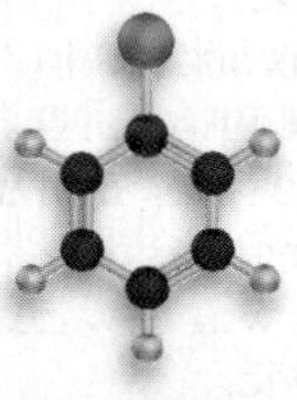

The compound we know as benzene was first isolated in 1825 by Michael Faraday. He extracted it from the liquid residue obtained after heating whale oil under pressure to produce the gas that was being used to illuminate buildings in London.

Many substituted benzenes are found in nature. A few that have physiological activity are shown here:

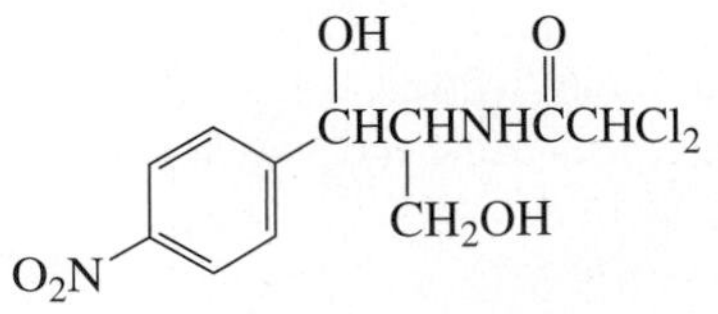

Chlorobenzene

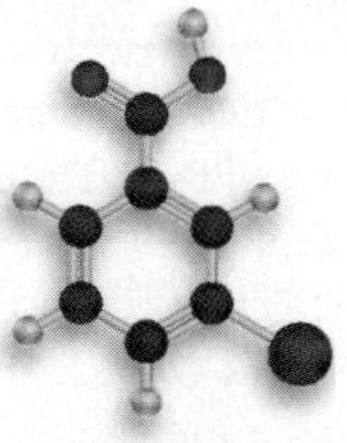

***meta*-Bromobenzoic acid**

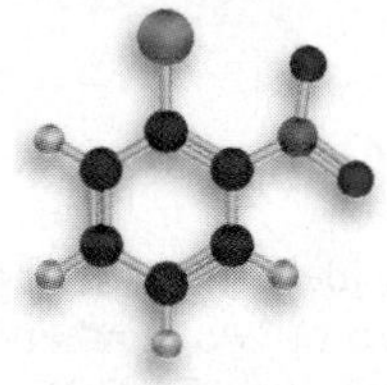

***ortho*-Chloronitrobenzene**

***para*-Iodobenzenesulfonic acid**

chloramphenicol
an antibiotic that is particularly effective against typhoid fever

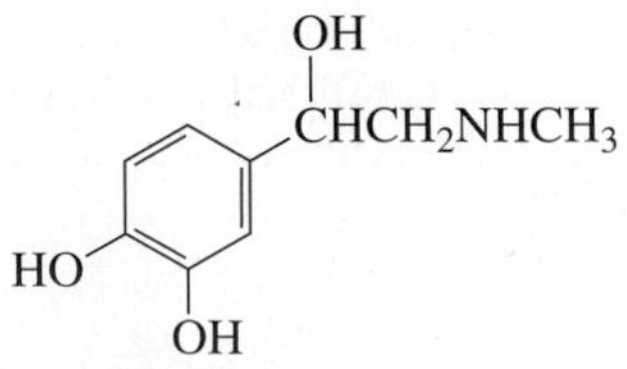

adrenaline
epinephrine

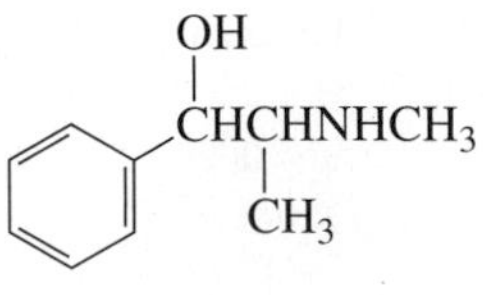

ephedrine
a bronchodilator

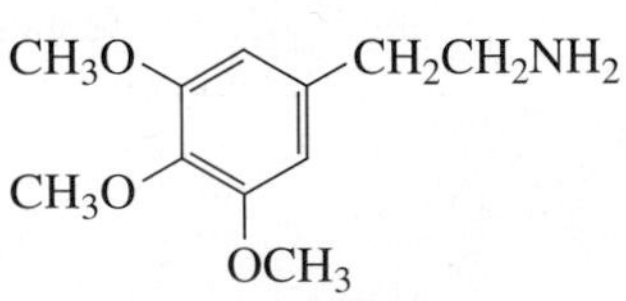

mescaline
active ingredient of the peyote cactus

Many physiologically active substituted benzenes are not found in nature, but exist because chemists have synthesized them. The now-banned diet drug "fen-phen" is a mixture of two synthetic substituted benzenes: fenfluramine and phentermine. Because of the known physiological activities of adrenaline and mescaline, chemists have synthesized compounds with similar structures. One such compound is amphetamine, a central nervous system stimulant. Amphetamine and a close relative, methamphetamine, are used clinically as appetite suppressants. Methamphetamine is the street drug known as "speed" because of its rapid and intense psychological effects. These compounds represent just a few of the many substituted benzenes that have been synthesized for commercial use by the chemical and pharmaceutical industries.

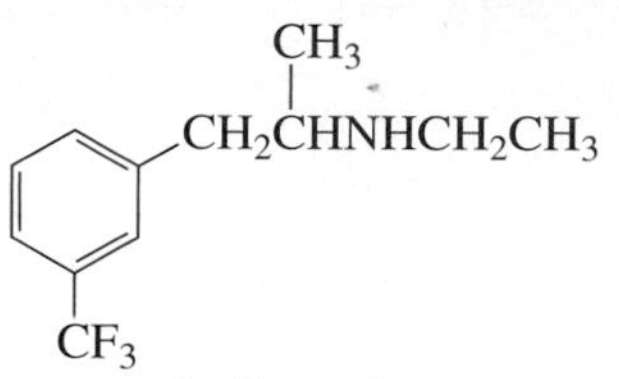

fenfluramine

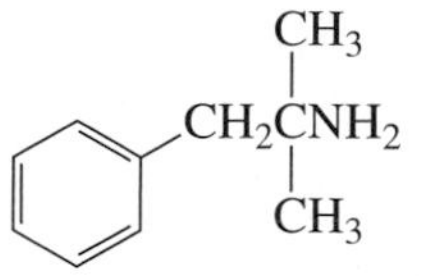

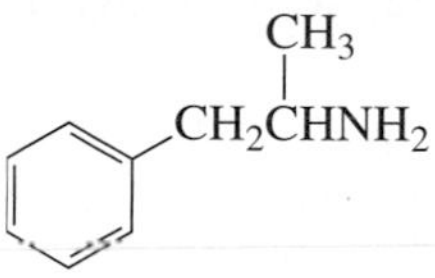

amphetamine

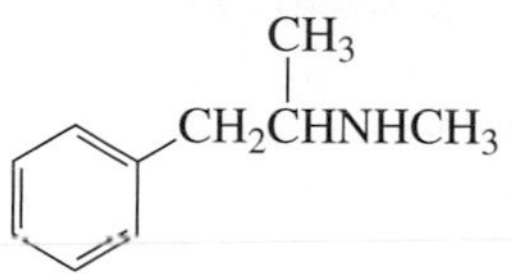

methamphetamine
"speed"

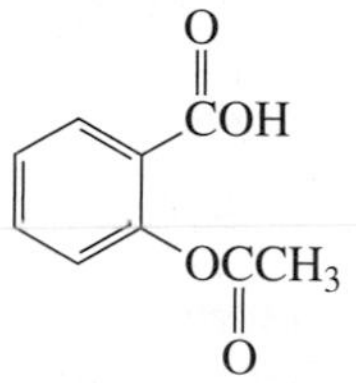

acetylsalicylic acid
aspirin

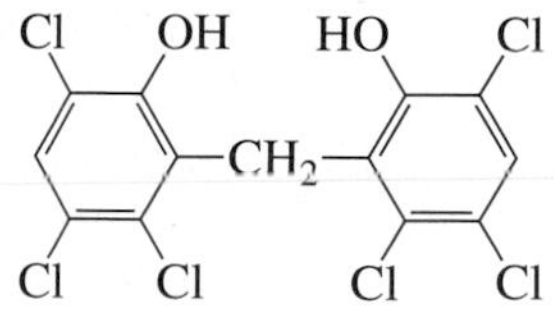

hexachlorophene
a disinfectant

phentermine

Compounds like benzene, that have relatively few hydrogens in relation to the number of carbons, are typically found in oils produced by trees and other plants. Early chemists called such compounds **aromatic compounds** because of their pleasing fragrances. In this way, they were distinguished from **aliphatic compounds**, with higher hydrogen-to-carbon ratios. The chemical meaning of the word "aromatic" now signifies certain kinds of chemical structures. We will now examine the criteria that a compound must satisfy to be classified as aromatic.

MEASURING TOXICITY

Agent Orange, a defoliant widely used during the Vietnam War, is a mixture of two synthetic substituted benzenes: 2,4-D and 2,4,5-T. Dioxin (TCDD), a contaminant formed during the manufacture of Agent Orange, has been implicated as the causative agent of the various symptoms suffered by those exposed to Agent Orange during the war.

The toxicity of a compound is indicated by its LD_{50} value—the quantity needed to kill 50% of the test animals exposed to the compound. Dioxin, with an LD_{50} value of 0.0006 mg/kg for guinea pigs, is an extremely toxic compound. Compare this with the LD_{50} values of some well-known poisons that are far less toxic: 0.96 mg/kg for strychnine and 15 mg/kg for both arsenic trioxide and sodium cyanide. One of the most toxic agents known is the botulism toxin, with an LD_{50} value of about 1×10^{-8} mg/kg.

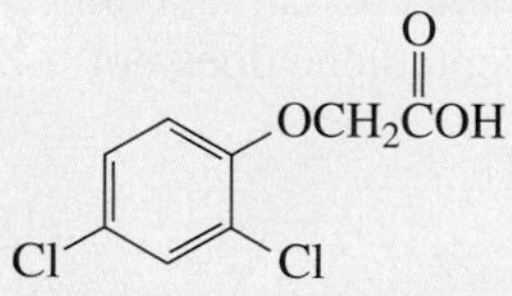

2,4-dichlorophenoxyacetic acid
2,4-D

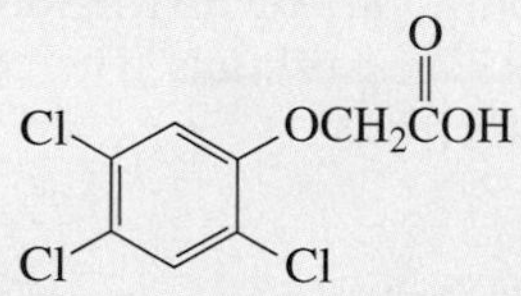

2,4,5-trichlorophenoxyacetic acid
2,4,5-T

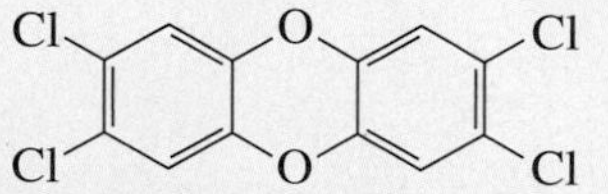

2,3,7,8-tetrachlorodibenzo[b,e][1,4]dioxin
TCDD

7.1   Criteria for Aromaticity

In Chapter 6, we saw that benzene is a planar, cyclic compound with a cyclic cloud of delocalized electrons above and below the plane of the ring. (See Figure 6.1 on page 139). Because its π electrons are delocalized, all the C—C bonds in benzene have the same length.

Aromatic compounds are particularly stable.

Benzene is a particularly stable compound because its **resonance stabilization**—the extra stability it gains from having delocalized electrons—is unusually large. Compounds with an unusually large amount of resonance stabilization are called **aromatic compounds**. How can we tell whether a compound is aromatic by looking at its structure? In other words, what structural features do aromatic compounds have in common?

To be classified as aromatic, a compound must meet both of the following criteria:

1. *It must have an uninterrupted cyclic cloud of π electrons* (called a π cloud) *above and below the plane of the molecule.* Let's look a little more closely at what this means:

    For the π cloud to be cyclic, *the molecule must be cyclic.*

    For the π cloud to be uninterrupted, *every atom in the ring must have a* p *orbital.*

    For the π cloud to form, each p orbital must overlap with the p orbitals on either side of it. Therefore, *the molecule must be planar.*

2. *The π cloud must contain an odd number of pairs of π electrons.*

For a compound to be aromatic, it must be cyclic and planar, and have an uninterrupted cloud of π electrons. The π cloud must contain an odd number of pairs of π electrons.

Benzene is an aromatic compound because it is cyclic and planar, every carbon in the ring has a p orbital, and the π cloud contains *three* pairs of π electrons.

7.2 Aromatic Hydrocarbons

Cyclobutadiene has two pairs of π electrons, and cyclooctatetraene has four pairs of π electrons. Unlike benzene, these compounds are *not* aromatic because they have an *even* number of pairs of π electrons. There is an additional reason why cyclooctatetraene is not aromatic—it is not planar (Section 6.3). Because cyclobutadiene and cyclooctatetraene are not aromatic, they do not have the unusual stability of aromatic compounds.

3-D Molecules:
Cyclobutadiene;
Benzene;
Cyclooctatetraene

cyclobutadiene [4]-annulene   benzene [6]-annulene   cyclooctatetraene [8]-annulene

Now let's look at some other compounds and determine whether they are aromatic. Cyclopentadiene is not aromatic because it does not have an uninterrupted ring of p orbital-bearing atoms. One of its ring atoms is sp^3 hybridized, and only sp^2 and sp hybridized carbons have p orbitals. Therefore, cyclopentadiene does not fulfill the first criterion for aromaticity.

Tutorial:
Aromaticity

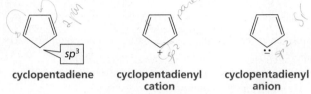

cyclopentadiene   cyclopentadienyl cation   cyclopentadienyl anion

The cyclopentadienyl cation is also not aromatic because, although it has an uninterrupted ring of p orbital-bearing atoms, its π cloud has *two* (an even number) pairs of π electrons. The cyclopentadienyl anion is aromatic: It has an uninterrupted ring of p orbital-bearing atoms, and its π cloud contains *three* (an odd number) pairs of delocalized π electrons.

Notice that the negatively charged carbon in the cyclopentadienyl anion is sp^2 hybridized because if it were sp^3 hybridized, the ion would not be aromatic. The resonance hybrid shows that all the carbons in the cyclopentadienyl anion are equivalent.

Each carbon has exactly one-fifth of the negative charge associated with the anion.

When drawing resonance contributors, remember that only electrons move, atoms never move.

resonance contributors of the cyclopentadienyl anion

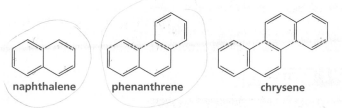

resonance hybrid

The criteria for determining whether a monocyclic hydrocarbon compound is aromatic can also be used to determine whether a polycyclic hydrocarbon compound is aromatic. Naphthalene (five pairs of π electrons), phenanthrene (seven pairs of π electrons), and chrysene (nine pairs of π electrons) are aromatic.

naphthalene   phenanthrene   chrysene

3-D Molecules:
Phenanthrene;
Naphthalene

BUCKYBALLS

A third form of pure carbon, in addition to diamond and graphite (Section 1.8), was discovered while scientists were conducting experiments designed to understand how long-chain molecules are formed in outer space. R. E. Smalley, R. F. Curl, Jr., and H. W. Kroto, the discoverers of this new form of carbon, shared the 1996 Nobel Prize in chemistry for their discovery. They named this new form buckminsterfullerene (often shortened to fullerene) because it reminded them of the geodesic domes popularized by R. Buckminster Fuller, an American architect and philosopher. The substance is nicknamed "buckyball." Consisting of a hollow cluster of 60 carbons, fullerene is the most symmetrical large molecule known. Like graphite, fullerene has only sp^2 hybridized carbons, but instead of being arranged in layers, the carbons are arranged in rings, forming a hollow cluster of 60 carbons that fit together like the seams of a soccer ball. Each molecule has 32 interlocking rings (20 hexagons and 12 pentagons). At first glance, fullerene would appear to be aromatic because of its benzene-like rings. However, it is not aromatic. The curvature of the ball prevents the molecule from fulfilling the first criterion for aromaticity—that it must be planar.

Buckyballs have extraordinary chemical and physical properties. They are exceedingly rugged and are capable of surviving the extreme temperatures of outer space. Because they are essentially hollow cages, they can be manipulated to make materials never before known. For example, when a buckyball is "doped" by inserting potassium or cesium into its cavity, it becomes an excellent organic superconductor. These molecules are presently being studied for use in many other applications, such as new polymers and catalysts and new drug delivery systems. The discovery of buckyballs is a strong reminder of the technological advances that can be achieved as a result of conducting basic research.

A geodesic dome

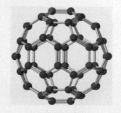

C_{60}
buckminsterfullerene
"buckyball"

PROBLEM 1

a. Draw arrows to show the movement of electrons in going from one resonance contributor to the next in the cyclopentadienyl anion.

b. How many ring atoms share the negative charge?

Richard E. Smalley *was born in 1943 in Akron, Ohio. He received a B.S. from the University of Michigan and a Ph.D. from Princeton University. He is a professor of chemistry at Rice University.*

Robert F. Curl, Jr., *was born in Texas in 1933. He received a B.A. from Rice University and a Ph.D. from the University of California, Berkeley. He is a professor of chemistry at Rice University.*

Sir Harold W. Kroto *was born in 1939 in England and is a professor of chemistry at the University of Sussex.*

PROBLEM 2◆

Which of the following are aromatic? Explain your choice.

cycloheptatriene   cycloheptatrienyl cation   cycloheptatrienyl anion

PROBLEM 3◆

Which of the following compounds are aromatic?

a.

c.

b.

d. CH₂=CHCH=CHCH=CH₂

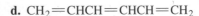

PROBLEM 4 | **SOLVED**

a. How many monobromonaphthalenes are there?

b. How many monobromophenanthrenes are there?

SOLUTION TO 4a There are two monobromonaphthalenes. Substitution cannot occur at either of the carbons shared by both rings, because those carbons are not bonded to a hydrogen. Naphthalene is a flat molecule, so substitution for a hydrogen at any other carbon will result in one of the compounds shown.

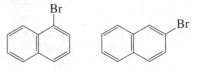

3-D Molecules:
1-Chloronaphthalene;
2-Chloronaphthalene

benzene

pyridine

pyrrole

7.3   Aromatic Heterocyclic Compounds

A compound does not have to be a hydrocarbon to be aromatic. Many *heterocyclic compounds* are aromatic. A **heterocyclic compound** is a cyclic compound in which one (or more) of the ring atoms is an atom other than carbon. A ring atom that is not carbon is called a **heteroatom.** The name comes from the Greek word *heteros,* which means "different." The most common heteroatoms found in heterocyclic compounds are N, O, and S.

heterocyclic compounds

        furan   thiophene

pyridine   pyrrole   furan   thiophene

Pyridine is an aromatic heterocyclic compound. Each of the six ring atoms of pyridine is sp^2 hybridized, which means that each has a p orbital; and the molecule contains three pairs of π electrons. Don't be confused by the lone-pair electrons on the nitrogen; they are not π electrons. Because nitrogen is sp^2 hybridized, it has three sp^2 orbitals and a p orbital. The p orbital is used to form the π bond. Two of nitrogen's sp^2 orbitals overlap the sp^2 orbitals of adjacent carbon atoms, and nitrogen's third sp^2 orbital contains the lone pair.

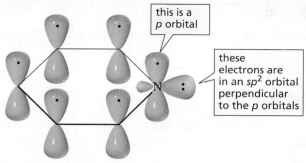

orbital structure of pyridine

It is not immediately apparent that the electrons represented as lone-pair electrons on the nitrogen atom of pyrrole are π electrons. The resonance contributors, however, show that the nitrogen atom is sp^2 hybridized and uses its three sp^2 orbitals to bond to two carbons and one hydrogen. The lone-pair electrons are in a p orbital that overlaps the p orbitals on adjacent carbons, forming a π bond—thus, they are π electrons. Pyrrole, therefore, has three pairs of π electrons and is aromatic.

resonance contributors of pyrrole

orbital structure of pyrrole   **orbital structure of furan**

Similarly, furan and thiophene are stable aromatic compounds. Both the oxygen in the former and the sulfur in the latter are sp^2 hybridized and have one lone pair in an sp^2 orbital. The second lone pair of each compound is in a p orbital that overlaps the p orbitals of adjacent carbons, forming a π bond. Thus, they are π electrons.

resonance contributors of furan

Quinoline, indole, imidazole, purine, and pyrimidine are other examples of heterocyclic aromatic compounds.

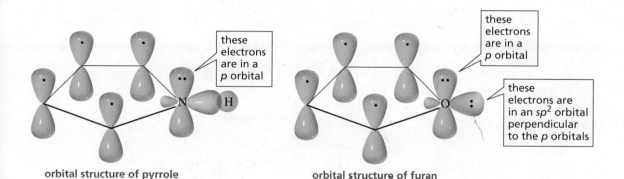

quinoline   **indole**   **imidazole**   **purine**   **pyrimidine**

PROBLEM 5

a. Draw arrows to show the movement of electrons in going from one resonance contributor to the next in pyrrole.

b. How many ring atoms share the negative charge in pyrrole?

PROBLEM 6◆

Answer the following questions by examining the electrostatic potential maps on page 166:

a. Why is the bottom part of the electrostatic potential map of pyrrole blue?

b. Why is the bottom part of the electrostatic potential map of pyridine red?

c. Why is the center of the electrostatic potential map of benzene more red than the center of the electrostatic potential map of pyridine?

HEME AND CHLOROPHYLL

Hemoglobin has four polypeptide chains and four heme groups (Section 17.11). Heme contains an iron ion (Fe^{2+}) ligated by the four nitrogens of a **porphyrin ring system**, which consists of four pyrrole rings joined by one-carbon bridges. **Ligation** is the sharing of lone-pair electrons with a metal ion.

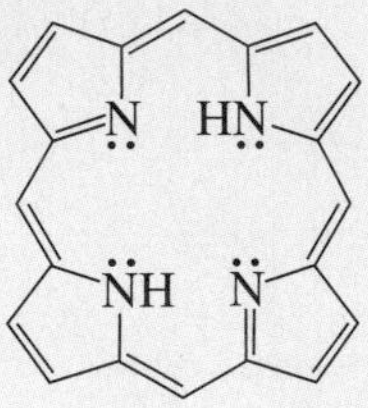

a porphyrin ring system

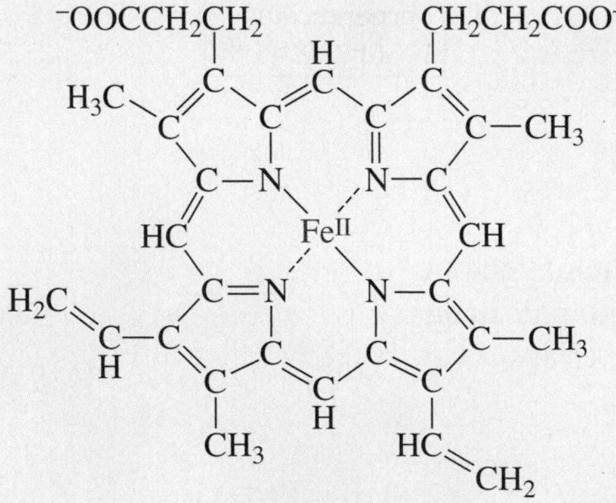

heme

The iron atom in hemoglobin, in addition to being ligated to the four nitrogens, is also ligated to a side chain of its peptide component (globin), and the sixth ligand is oxygen or carbon dioxide. Carbon monoxide is about the same size and shape as O_2, but CO binds more tightly than O_2 to Fe^{2+}. Consequently, breathing carbon monoxide can be fatal because it prevents the transport of oxygen in the bloodstream. The extensive conjugated system of porphyrin gives blood its characteristic red color. Concentrations as low as 1×10^{-8} M can be detected by UV spectroscopy (Section 6.12).

The ring system in chlorophyll *a*, the substance responsible for the green color of plants, is similar to porphyrin. The metal atom in chlorophyll *a* is magnesium (Mg^{2+}). Vitamin B_{12} also has a ring system similar to porphyrin (Section 18.10).

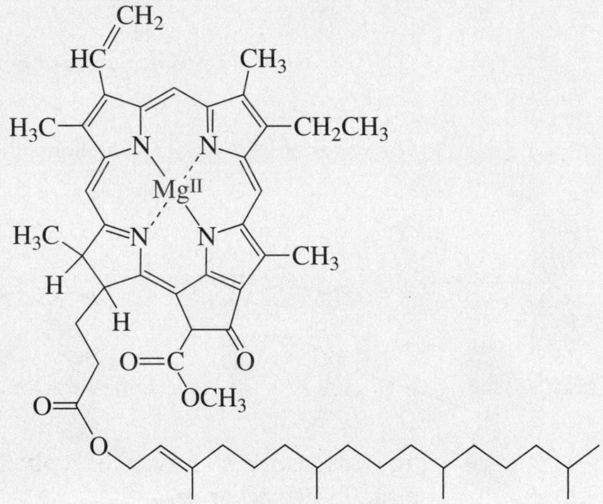

chlorophyll *a*

Chlorophyll *a* is the pigment that makes plants look green. This highly conjugated compound absorbs nongreen light. Therefore, plants reflect green light.

7.4   Nomenclature of Monosubstituted Benzenes

Some monosubstituted benzenes are named simply by stating the name of the substituent, followed by the word "benzene."

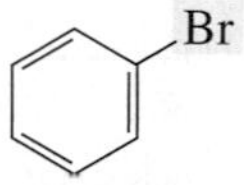

bromobenzene 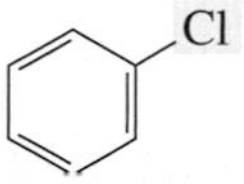   chlorobenzene 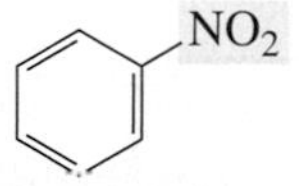   nitrobenzene   ethylbenzene
                                             used as a solvent
                                             in shoe polish

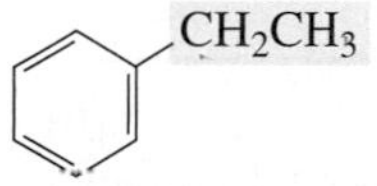

Some monosubstituted benzenes have names that incorporate the name of the substituent.

toluene   phenol   aniline   benzenesulfonic acid

3-D Molecules:
Toluene; Bromobenzene

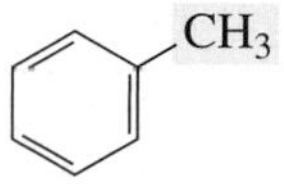

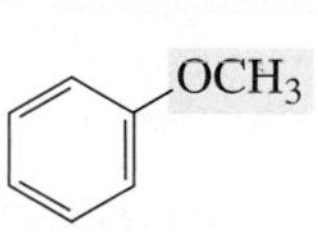

anisole 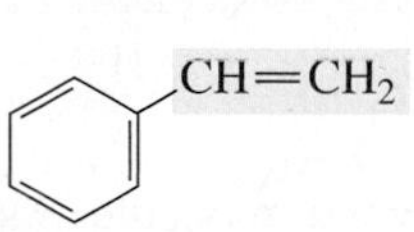   styrene 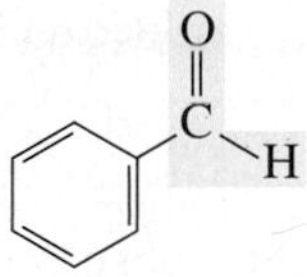   benzaldehyde   benzoic acid

When a benzene ring is a substituent, it is called a **phenyl group.** A benzene ring with a methylene group is called a **benzyl group.**

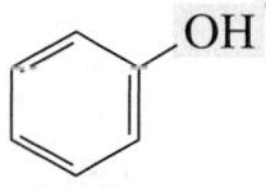

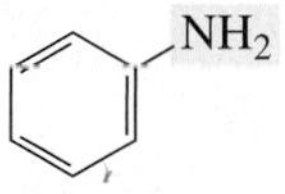

$-CH_2-$

a phenyl group   a benzyl group

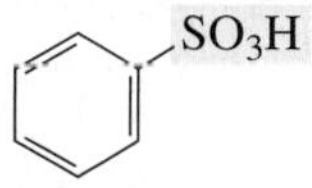

CH_2Cl

chloromethylbenzene   diphenyl ether $-CH_2OCH_2-$
benzyl chloride   dibenzyl ether

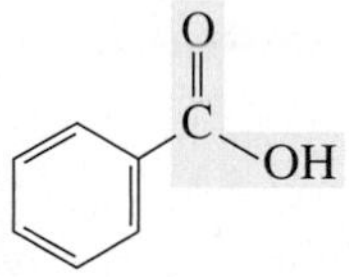

An **aryl group** (Ar) is the general term for either a phenyl group or a substituted phenyl group, just as an alkyl group (R) is the general term for a group derived from an alkane. In other words, ArOH could be used to designate any of the following phenols:

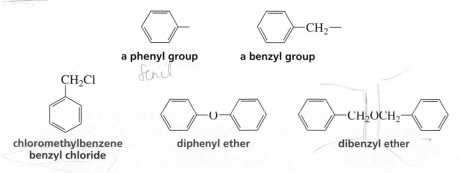

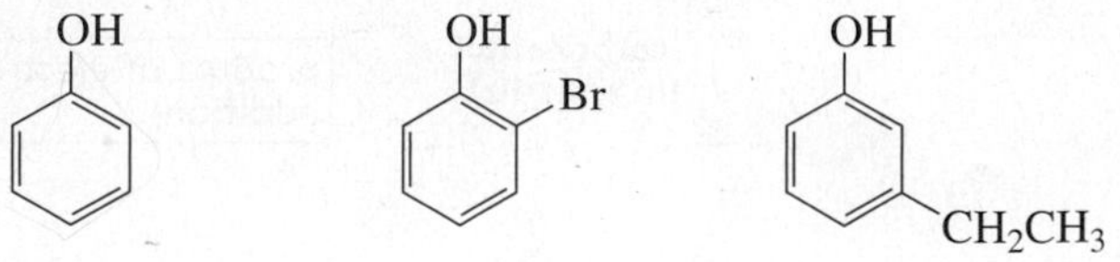

THE TOXICITY OF BENZENE

Although benzene has been widely used in chemical synthesis and has been frequently used as a solvent, it is toxic. Its major toxic effect is on the central nervous system and on bone marrow. Chronic exposure to benzene causes leukemia and aplastic anemia. A higher-than-average incidence of leukemia has been found in industrial workers with long-term exposure to as little as 1 ppm benzene in the atmosphere. Toluene has replaced benzene as a solvent because, although it is a central nervous system depressant like benzene, it does not cause leukemia or aplastic anemia. "Glue sniffers" seek the narcotic central nervous system effects of solvents such as toluene. This can be a highly dangerous activity.

PROBLEM 7◆

Draw the structure of each of the following compounds:

a. 2-phenylhexane　　　**b.** benzyl alcohol　　　**c.** 3-benzylpentane

7.5 How Benzene Reacts

Aromatic compounds (such as benzene) undergo **electrophilic aromatic substitution reactions**—an electrophile substitutes for one of the hydrogens attached to the benzene ring.

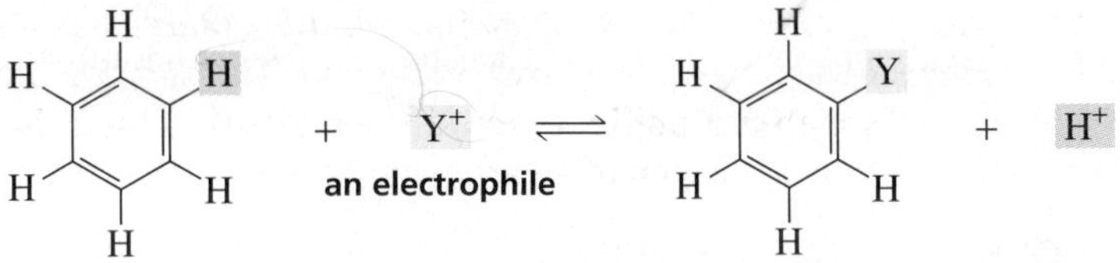

Now let's look at why this substitution reaction occurs. As a consequence of the π electrons above and below the plane of its ring, benzene is a nucleophile. It will, therefore, react with an electrophile (Y^+). When an electrophile attaches itself to a benzene ring, a carbocation intermediate is formed.

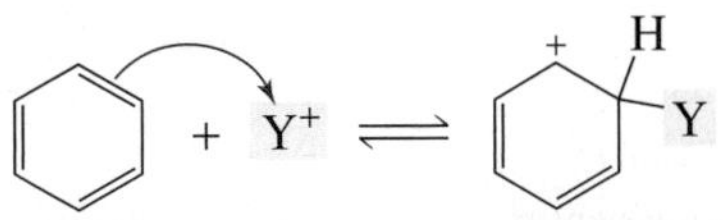

carbocation intermediate

This should remind you of the first step in an electrophilic addition reaction of an alkene: The alkene reacts with an electrophile and forms a carbocation intermediate (Section 4.7). In the second step of an electrophilic addition reaction, the carbocation reacts with a nucleophile (Z^-) to form an addition product.

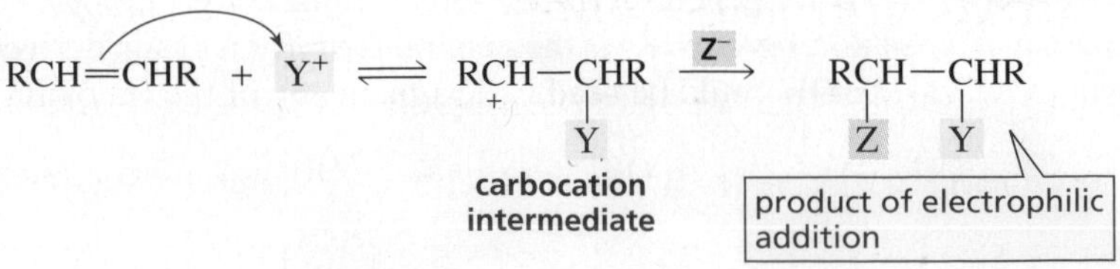

$$RCH=CHR + Y^+ \rightleftharpoons RCH-\underset{+}{CHR} \xrightarrow{Z^-} RCH-CHR$$

carbocation intermediate

product of electrophilic addition

If the carbocation intermediate formed from the reaction of benzene with an electrophile were to react similarly with a nucleophile (depicted as path *b* in Figure 7.1), the addition product would not be aromatic. If, however, the carbocation loses a proton from the site of electrophilic attack (depicted as path *a* in Figure 7.1), the aromaticity of the benzene ring is restored.

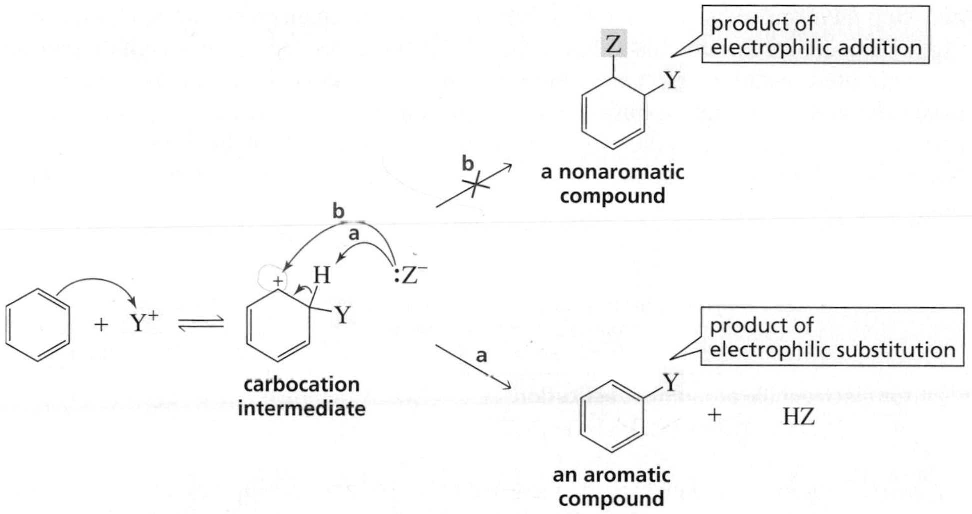

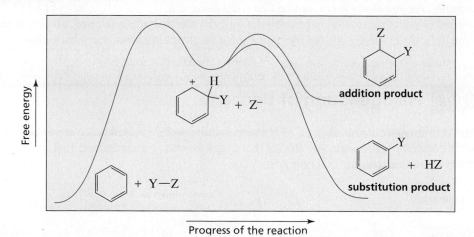

◄ Figure 7.1
Reaction of benzene with an electrophile. Because the aromatic product is more stable, the reaction proceeds as (a) an electrophilic substitution reaction rather than (b) an electrophilic addition reaction.

Because the aromatic substitution product is much more stable than the nonaromatic addition product (Figure 7.2), benzene undergoes *electrophilic substitution reactions* that preserve aromaticity rather than *electrophilic addition reactions*—the reactions characteristic of alkenes—that would destroy aromaticity. The substitution reaction is more precisely called an **electrophilic aromatic substitution reaction** since the electrophile substitutes for a hydrogen of an aromatic compound.

◄ Figure 7.2
Reaction coordinate diagrams for electrophilic substitution of benzene and electrophilic addition to benzene.

7.6 General Mechanism for Electrophilic Aromatic Substitution Reactions

The following are the five most common electrophilic aromatic substitution reactions:

1. **Halogenation:** A bromine (Br) or a chlorine (Cl) substitutes for a hydrogen.

2. **Nitration:** A nitro (NO_2) group substitutes for a hydrogen.

3. **Sulfonation:** A sulfonic acid (SO_3H) group substitutes for a hydrogen.

4. **Friedel–Crafts acylation:** An acyl ($RC{=}O$) group substitutes for a hydrogen.

5. **Friedel–Crafts alkylation:** An alkyl (R) group substitutes for a hydrogen.

All of these electrophilic aromatic substitution reactions take place by the same two-step mechanism. In the first step, benzene reacts with an electrophile (Y^+), forming a carbocation intermediate. The structure of the carbocation intermediate can be approximated by three resonance contributors. In the second step of the reaction, a base (:B) in the reaction mixture pulls off a proton from the carbocation intermediate, and the electrons that held the proton move into the ring to reestablish its aromaticity. Notice that *the proton is always removed from the carbon that has formed the new bond with the electrophile.*

general mechanism for electrophilic aromatic substitution

the proton is removed from the carbon that has formed the new bond with the electrophile

a base in the reaction mixture

$$\text{benzene} + Y^+ \xrightleftharpoons{\text{slow}} \left[\text{carbocation intermediate (three resonance contributors)} \right] \xrightarrow{\text{fast}} \text{product} + HB^+$$

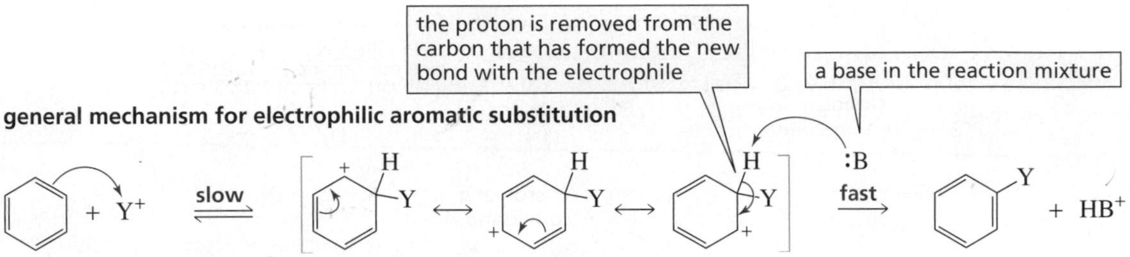

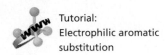

Tutorial:
Electrophilic aromatic substitution

In an electrophilic aromatic substitution reaction, an electrophile (Y^+) is put on a ring carbon, and the H^+ comes off the same ring carbon.

The reaction coordinate diagram in Figure 7.2 shows that the first step is relatively slow and consumes energy because an aromatic compound is being converted into a much less stable nonaromatic intermediate. The second step is fast and releases energy because this step restores the stability-enhancing aromaticity.

We will look at each of these five electrophilic aromatic substitution reactions individually. As you study them, notice that they differ only in how the electrophile (Y^+) needed to start the reaction is generated. Once the electrophile is formed, all five reactions follow the same two-step mechanism for electrophilic aromatic substitution.

7.7 Halogenation of Benzene

The bromination or chlorination of benzene requires a Lewis acid catalyst such as ferric bromide or ferric chloride. Recall that a *Lewis acid* is a compound that accepts a share in an electron pair (Section 2.6).

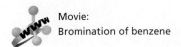

Movie:
Bromination of benzene

bromination

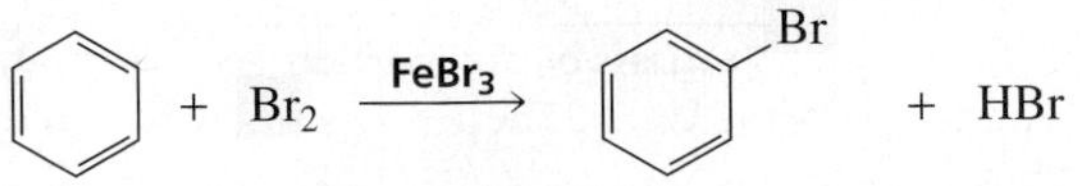

$$\text{benzene} + Br_2 \xrightarrow{FeBr_3} \text{bromobenzene} + HBr$$

bromobenzene

chlorination

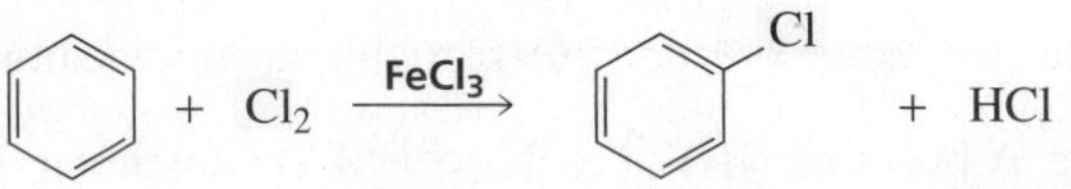

$$\text{benzene} + Cl_2 \xrightarrow{FeCl_3} \text{chlorobenzene} + HCl$$

chlorobenzene

Why does the reaction of benzene with Br_2 or Cl_2 require a catalyst when the reaction of an alkene with these reagents does not require a catalyst? Benzene, because of its stability, is much less reactive than an alkene; therefore, it requires a better electrophile. Donating a lone pair to the Lewis acid weakens the Br—Br (or Cl—Cl) bond, which makes Br_2 (or Cl_2) a better electrophile.

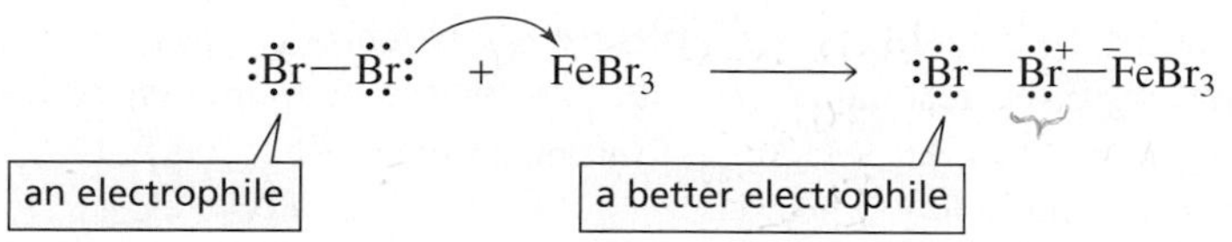

| an electrophile | | a better electrophile |

To make the mechanisms easier to understand, only one of the three resonance contributors of the carbocation intermediate is shown in this and subsequent mechanisms. Bear in mind, however, that each carbocation intermediate actually has the three resonance contributors shown in Section 7.6. In the last step of the reaction, a base (:B) from the reaction mixture (e.g., ⁻FeBr₄, solvent) removes a proton from the carbocation intermediate.

mechanism for bromination

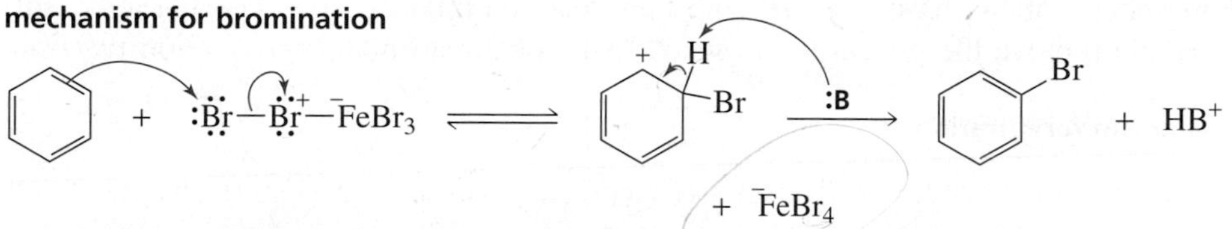

Chlorination of benzene occurs by the same mechanism as bromination.

mechanism for chlorination

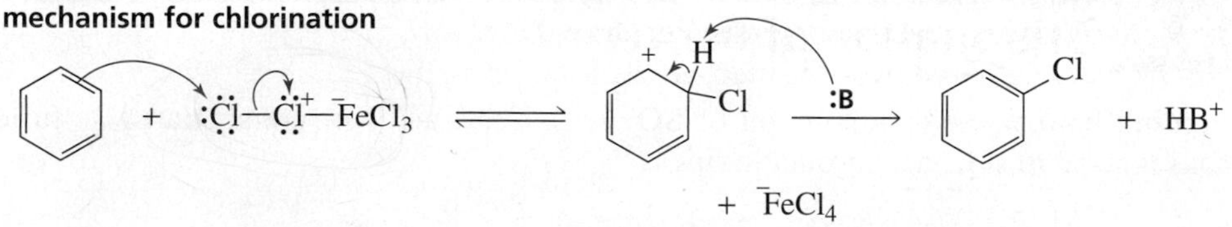

THYROXINE

Thyroxine is a hormone that regulates the metabolic rate, causing an increase in the rate at which fats, carbohydrates, and proteins are metabolized. Humans obtain thyroxine from tyrosine (an amino acid) and iodine. We get iodine primarily from the iodized salt in our diet. An enzyme called iodoperoxidase converts the I^- we ingest to I^+, the electrophile needed to place an iodo substituent on the benzene ring. Low thyroxine levels can be corrected by hormone supplements. Chronically low levels of thyroxine cause enlargement of the thyroid gland, a condition known as goiter.

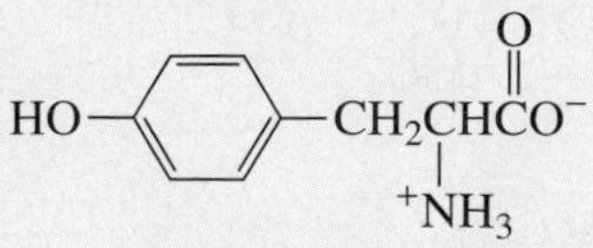

tyrosine

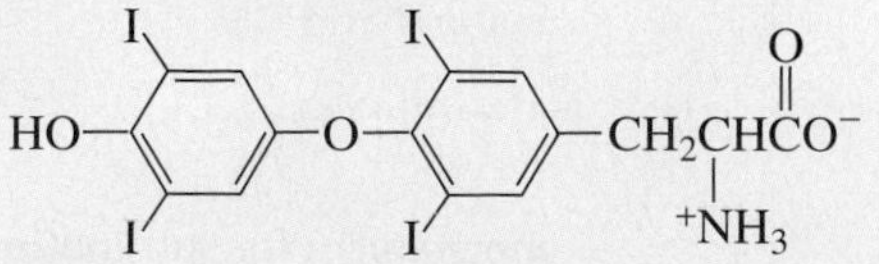

thyroxine

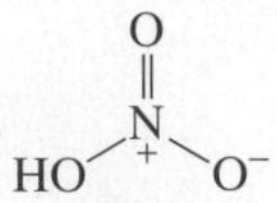

nitric acid

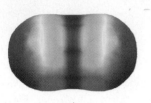

$O=\overset{+}{N}=O$
nitronium ion

7.8 Nitration of Benzene

Nitration of benzene with nitric acid requires sulfuric acid as a catalyst.

nitration

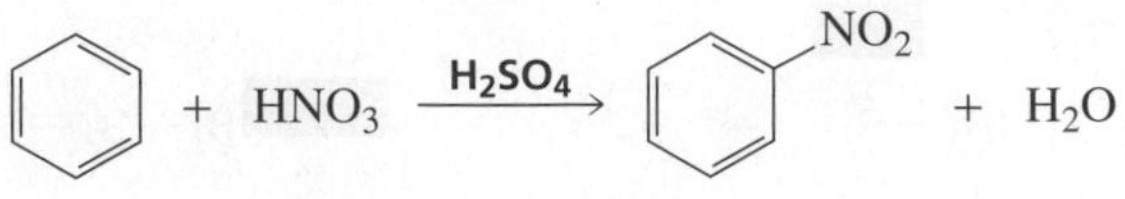

To generate the necessary electrophile, sulfuric acid protonates nitric acid. Loss of water from protonated nitric acid forms a nitronium ion, the electrophile required for nitration.

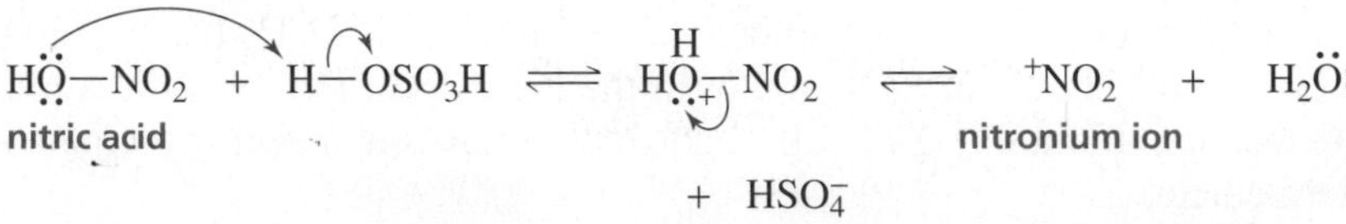

Remember that any base (:B) present in the reaction mixture (e.g., H_2O, HSO_4^-, solvent) can remove the proton in the second step of the aromatic substitution reaction.

mechanism for nitration

3-D Molecule:
Nitrobenzene

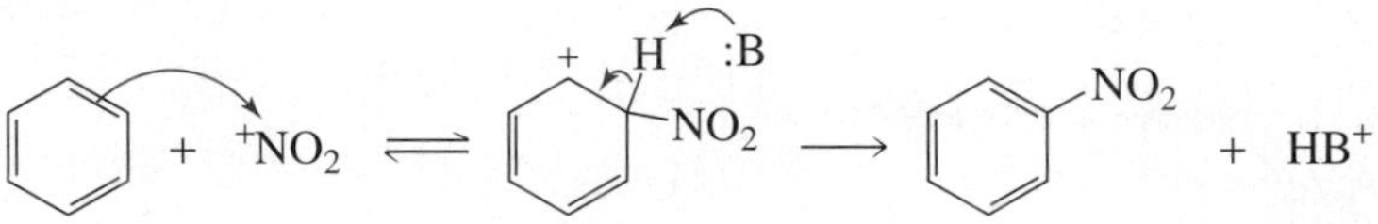

7.9 Sulfonation of Benzene

Fuming sulfuric acid (a solution of SO_3 in sulfuric acid) or concentrated sulfuric acid is used to sulfonate aromatic rings.

sulfonation

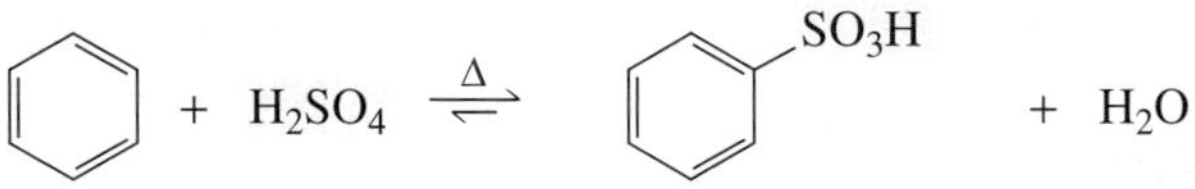

Take a minute to note the similarities in the mechanisms for forming the $^+SO_3H$ electrophile for sulfonation and the $^+NO_2$ electrophile for nitration.

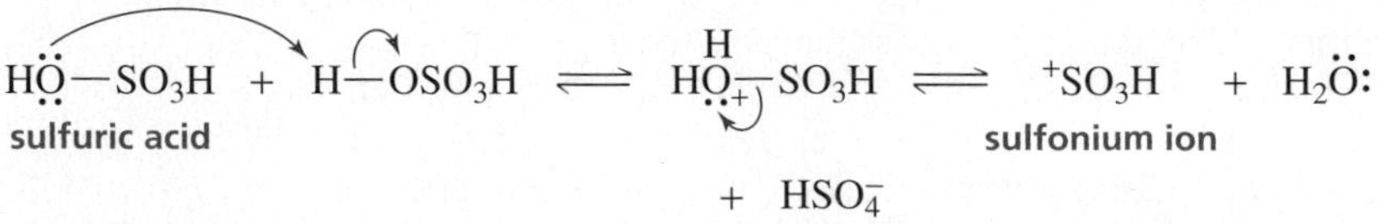

mechanism for sulfonation

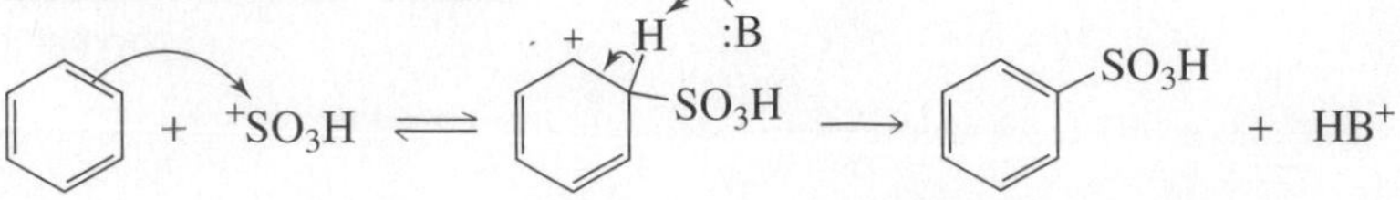

7.10   Friedel–Crafts Acylation of Benzene

Two electrophilic substitution reactions bear the names of chemists Charles Friedel and James Crafts. **Friedel–Crafts acylation** places an acyl group on a benzene ring.

Friedel–Crafts acylation

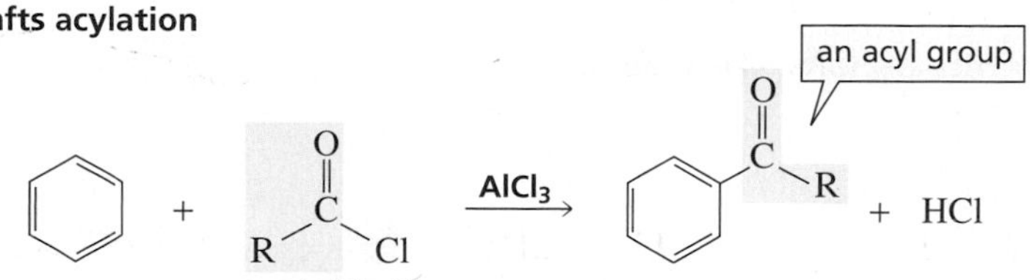

Charles Friedel (1832–1899) *was born in Strasbourg, France. He was a professor of chemistry and director of research at the Sorbonne. At one point, his interest in mineralogy led him to attempt to make synthetic diamonds. He met James Crafts when they both were doing research at L'Ecole de Médicine in Paris. They collaborated scientifically for most of their lives, discovering the Friedel–Crafts reactions in Friedel's laboratory in 1877.*

The electrophile (an acylium ion) required for a Friedel–Crafts acylation is formed by the reaction of an acyl chloride with $AlCl_3$, a Lewis acid. An **acyl chloride** has a chlorine in place of the OH group of a carboxylic acid.

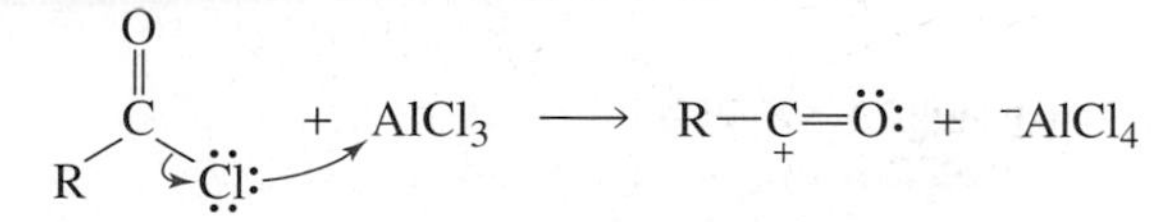

mechanism for Friedel–Crafts acylation

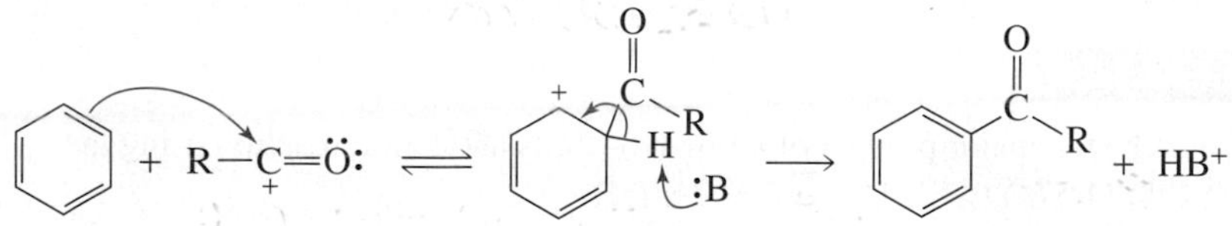

PROBLEM 8

Write the mechanism for the following reaction:

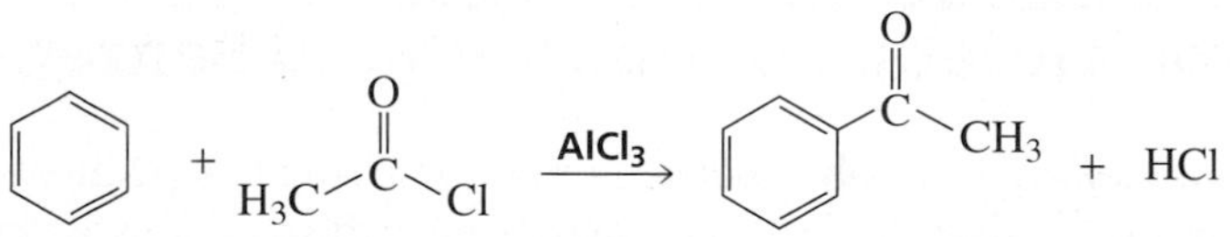

James Mason Crafts (1839–1917) *was born in Boston, the son of a woolen-goods manufacturer. He graduated from Harvard in 1858 and was a professor of chemistry at Cornell University and the Massachusetts Institute of Technology (MIT). He was president of MIT from 1897 to 1900, when he was forced to retire because of poor health.*

7.11   Friedel–Crafts Alkylation of Benzene

The Friedel–Crafts alkylation places an alkyl group on a benzene ring.

Friedel–Crafts alkylation

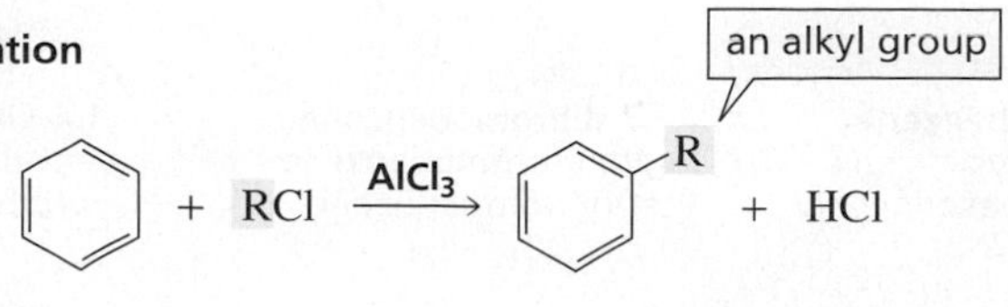

The electrophile in this reaction is a carbocation that is formed from the reaction of an alkyl halide with $AlCl_3$. Alkyl fluorides, alkyl chlorides, alkyl bromides, and alkyl iodides can all be used.

$$R\!-\!\ddot{C}\!l\!: + \ AlCl_3 \longrightarrow R^+ + \ ^-AlCl_4$$

an alkyl halide   a carbocation

mechanism for Friedel–Crafts alkylation

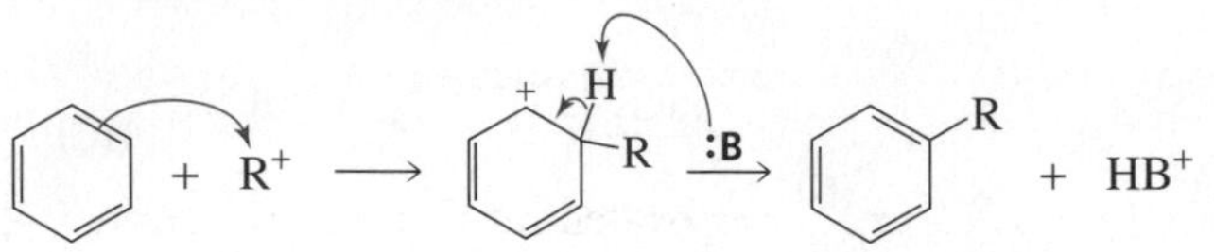

In addition to reacting with carbocations generated from alkyl halides, benzene can react with carbocations generated from the reaction of an alkene (Section 4.6).

alkylation of benzene by an alkene

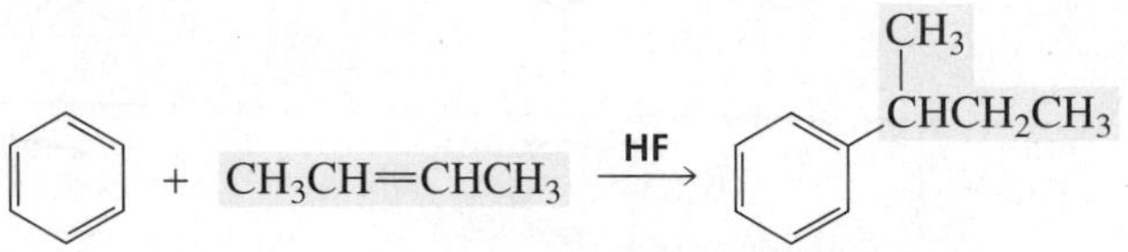

sec-butylbenzene

PROBLEM 9

Write the mechanism for the alkylation of benzene by 2-butene + HF.

PROBLEM 10◆

What would be the major product of a Friedel–Crafts alkylation reaction using the following alkyl chlorides?

a. CH_3CH_2Cl 　　　　**b.** $CH_3CH_2CHCH_3$ 　　　　**c.** $CH_2\!=\!CHCH_2Cl$
　　　　　　　　　　　　　　　　　　 $|$
　　　　　　　　　　　　　　　　　Cl

7.12 Nomenclature of Disubstituted Benzenes

The relative positions of two substituents on a benzene ring can be indicated either by numbers or by the prefixes *ortho, meta,* and *para.* Adjacent substituents are called *ortho,* substituents separated by one carbon are called *meta,* and substituents located opposite one another are designated *para.* Often, only their abbreviations (*o, m, p*) are used in naming compounds.

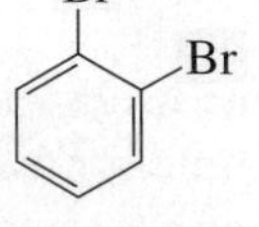

1,2-dibromobenzene
ortho-dibromobenzene
o-dibromobenzene

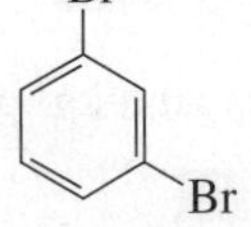

1,3-dibromobenzene
meta-dibromobenzene
m-dibromobenzene

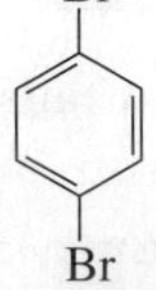

1,4-dibromobenzene
para-dibromobenzene
p-dibromobenzene

If the two substituents are different, they are listed in alphabetical order. The first stated substituent is given the 1-position, and the ring is numbered in the direction that gives the second substituent the lowest possible number.

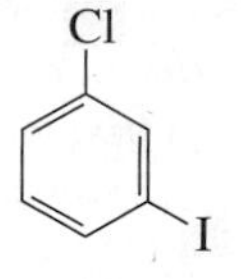

1-chloro-3-iodobenzene
meta-chloroiodobenzene

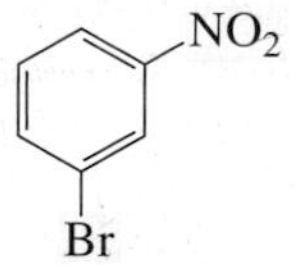

1-bromo-3-nitrobenzene
meta-bromonitrobenzene

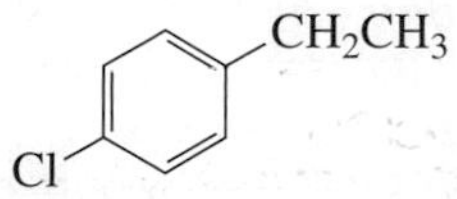

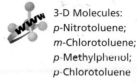

1-chloro-4-ethylbenzene
para-chloroethylbenzene

If one of the substituents can be incorporated into a name (Section 7.4), that name is used and the incorporated substituent is given the 1-position.

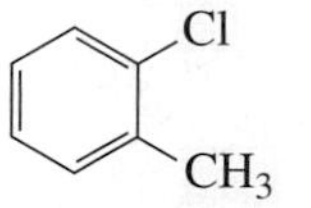

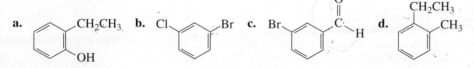

2-chlorotoluene	**4-nitroaniline**	**2-ethylphenol**
ortho-chlorotoluene	*para*-nitroaniline	*ortho*-ethylphenol
not	not	not
ortho-chloromethylbenzene	*para*-aminonitrobenzene	*ortho*-ethylhydroxybenzene

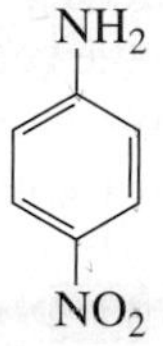

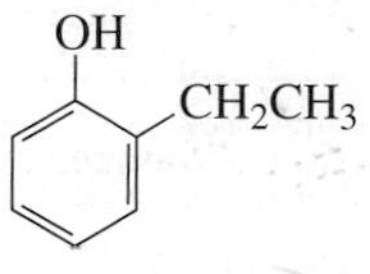

3-D Molecules:
p-Nitrotoluene;
m-Chlorotoluene;
p-Methylphenol;
p-Chlorotoluene

PROBLEM 11◆

Name the following compounds:

a.   b.   c.   d.

PROBLEM 12◆

Draw the structure of each of the following compounds:

a. *p*-bromophenol
b. *o*-nitroaniline

c. 2-bromo-4-iodophenol
d. 2,5-dinitrobenzaldehyde

PROBLEM 13◆

Correct the following incorrect names:

a. 2,4,6-tribromobenzene
b. 3-hydroxynitrobenzene

c. *para*-methylbromobenzene
d. 1,6-dichlorobenzene

7.13   The Effect of Substituents on Reactivity

Like benzene, substituted benzenes undergo the five electrophilic aromatic substitution reactions listed in Section 7.6: halogenation, nitration, sulfonation, acylation, and alkylation. Now we need to find out whether a substituted benzene is more reactive or less reactive than benzene itself. The answer depends on the substituent. Some substituents make the ring more reactive toward electrophilic aromatic substitution, and some make it less reactive.

The slow step of an electrophilic aromatic substitution reaction is the addition of an electrophile to the nucleophilic aromatic ring to form a carbocation intermediate (Section 7.6). *Substituents that are capable of donating electrons into the benzene ring stabilize both the carbocation and the transition state leading toward its formation, therefore, they increase the rate of electrophilic aromatic substitution. In contrast, substituents that withdraw electrons from the benzene ring destabilize the carbocation and the transition state leading toward its formation; therefore, they decrease the rate of electrophilic aromatic substitution.*

Electron-donating substituents increase the reactivity of the benzene ring toward electrophilic aromatic substitution.

Electron-withdrawing substituents decrease the reactivity of the benzene ring toward electrophilic aromatic substitution.

relative rates of electrophilic substitution

There are two ways substituents can donate electrons—*inductively* or by *resonance*. Substituents can also withdraw electrons *inductively* or by *resonance*.

Donating and Withdrawing Electrons Inductively

If a substituent that is bonded to a benzene ring is *less electron withdrawing than a hydrogen*, the electrons in the σ bond that attaches the substituent to the benzene ring will move toward the ring more readily than will those in the σ bond that attaches the hydrogen to the ring. Such a substituent donates electrons inductively compared with a hydrogen. Donation of electrons through a σ bond is called **inductive electron donation.** We have seen that alkyl substituents (such as CH_3) donate electrons inductively, compared with a hydrogen (Section 5.2).

If a substituent is *more electron withdrawing than a hydrogen*, it will withdraw the σ electrons away from the benzene ring more strongly than will a hydrogen. Withdrawal of electrons through a σ bond is called **inductive electron withdrawal**. The $^{+}NH_3$ group is an example of a substituent that withdraws electrons inductively: $^{+}NH_3$ is more electronegative than a hydrogen.

Donating and Withdrawing Electrons by Resonance

If a substituent has a lone pair on the atom that is directly attached to the benzene ring, the lone pair can be delocalized into the ring; these substituents are said to **donate electrons by resonance**. Substituents such as NH_2, OH, OR, and Cl donate electrons by resonance. These substituents also withdraw electrons inductively because the atom attached to the benzene ring is more electronegative than a hydrogen.

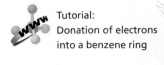

Tutorial:
Donation of electrons
into a benzene ring

donation of electrons into a benzene ring by resonance

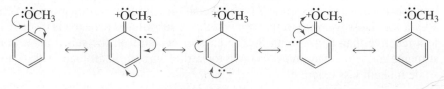

anisole

If a substituent is attached to the benzene ring by an atom that is doubly or triply bonded to a more electronegative atom, the π electrons of the ring can be delocalized onto the substituent; these substituents are said to **withdraw electrons by resonance**. Substituents such as $C{=}O$, $C{\equiv}N$, and NO_2 withdraw electrons by resonance. These substituents also withdraw electrons inductively because the atom attached to the benzene ring has a full or partial positive charge and, therefore, is more electronegative than a hydrogen.

Tutorial:
Withdrawal of electrons
from a benzene ring

withdrawal of electrons from a benzene ring by resonance

nitrobenzene

PROBLEM 14◆

For each of the following substituents, indicate whether it donates electrons inductively, withdraws electrons inductively, donates electrons by resonance, or withdraws electrons by resonance (inductive effects should be compared with a hydrogen; remember that many substituents can be characterized in more than one way):

a. $NHCH_3$

b. CH_2CH_3

c. $\overset{O}{\overset{\|}{C}}CH_3$

d. Br

e. OCH_3

f. $\overset{+}{N}(CH_3)_3$

Relative Reactivity of Substituted Benzenes

The substituents shown in Table 7.1 are listed according to how they affect the reactivity of the benzene ring toward electrophilic aromatic substitution compared with benzene—in which the substituent is a hydrogen. *The **activating substituents** make the benzene ring more reactive toward electrophilic substitution; the **deactivating substituents** make the benzene ring less reactive toward electrophilic substitution.* Remember that activating substituents donate electrons into the ring and deactivating substituents withdraw electrons from the ring.

All the *activating substituents* (except for alkyl substituents) donate electrons into the ring by resonance and withdraw electrons from the ring inductively. The fact that they have been found experimentally to be activators indicates that electron donation into the ring by resonance is more significant than inductive electron withdrawal from the ring. We have seen that an alkyl substituent, compared with a hydrogen, donates electrons inductively.

The halogens are *weakly deactivating substituents*; they also donate electrons into the ring by resonance and withdraw electrons from the ring inductively. Because the halogens have been found experimentally to be deactivators, we can conclude that they withdraw electrons inductively more strongly than they donate electrons by resonance.

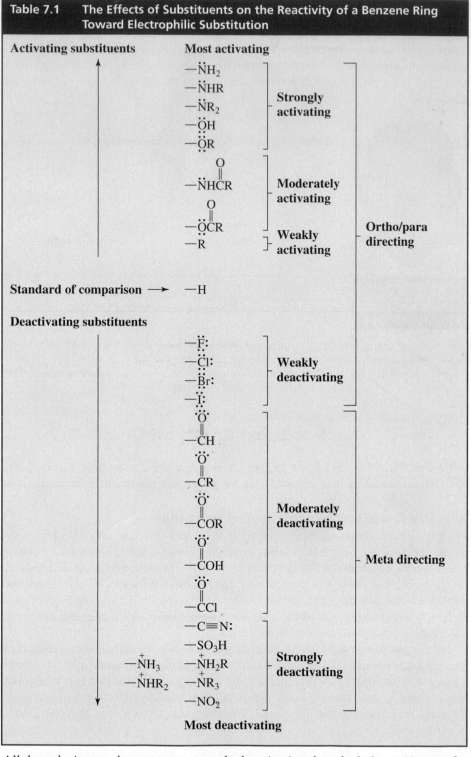

Table 7.1 The Effects of Substituents on the Reactivity of a Benzene Ring Toward Electrophilic Substitution

Activating substituents — Most activating

- $-\ddot{N}H_2$
- $-\ddot{N}HR$ — **Strongly activating**
- $-\ddot{N}R_2$
- $-\ddot{O}H$
- $-\ddot{O}R$

$$-\ddot{N}HCR \ (C{=}O)$$ — **Moderately activating**

$$-\ddot{O}CR \ (C{=}O)$$
$-R$ — **Weakly activating**

Standard of comparison → $-H$

Deactivating substituents

- $-\ddot{F}:$
- $-\ddot{C}l:$ — **Weakly deactivating**
- $-\ddot{B}r:$
- $-\ddot{I}:$

- $-CH \ (C{=}\ddot{O})$
- $-CR \ (C{=}\ddot{O})$
- $-COR \ (C{=}\ddot{O})$ — **Moderately deactivating**
- $-COH \ (C{=}\ddot{O})$
- $-CCl \ (C{=}\ddot{O})$

- $-C{\equiv}N:$
- $-SO_3H$
- $-\overset{+}{N}H_2R$ — **Strongly deactivating**
- $-\overset{+}{N}R_3$
- $-NO_2$

$-\overset{+}{N}H_3$
$-\overset{+}{N}HR_2$

Most deactivating

Ortho/para directing (for activating substituents and halogens)

Meta directing (for moderately and strongly deactivating substituents)

All the *substituents that are more strongly deactivating than the halogens* (except for the ammonium ions $^+NH_3$, $^+NH_2R$, $^+NHR_2$, and $^+NR_3$) withdraw electrons both inductively and by resonance. The ammonium ions have no resonance effect, but the positive charge on the nitrogen atom causes them to strongly withdraw electrons inductively.

Take a minute to compare the electrostatic potential maps for anisole, benzene, and nitrobenzene. Notice that an electron-donating substituent (OCH_3) makes the ring more red (more negative), whereas an electron-withdrawing substituent (NO_2) makes the ring less red (less negative).

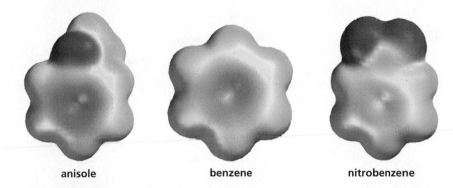

anisole   benzene   nitrobenzene

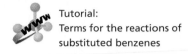

Tutorial:
Terms for the reactions of substituted benzenes

PROBLEM 15◆

List the following compounds in order of decreasing reactivity toward electrophilic aromatic substitution:

a. benzene, phenol, toluene, nitrobenzene, bromobenzene
b. dichloromethylbenzene, difluoromethylbenzene, toluene, chloromethylbenzene

---

7.14   The Effect of Substituents on Orientation

When a substituted benzene undergoes an electrophilic substitution reaction, where does the new substituent attach itself? Is the product of the reaction the ortho isomer, the meta isomer, or the para isomer?

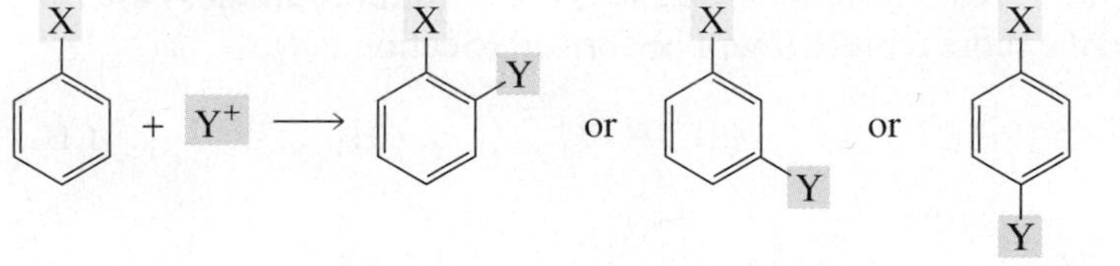

ortho isomer   meta isomer   para isomer

The substituent already attached to the benzene ring determines the location of the new substituent. There are two possibilities: A substituent will direct an incoming substituent either to the *ortho* and *para* positions, or it will direct an incoming substituent to the *meta* position. All activating substituents and the weakly deactivating halogens are **ortho–para directors**, and all substituents that are more deactivating than the halogens are **meta directors**. Thus, the substituents can be divided into three groups:

1. All activating substituents direct an incoming electrophile to the ortho and para positions.

All activating substituents are ortho–para directors.

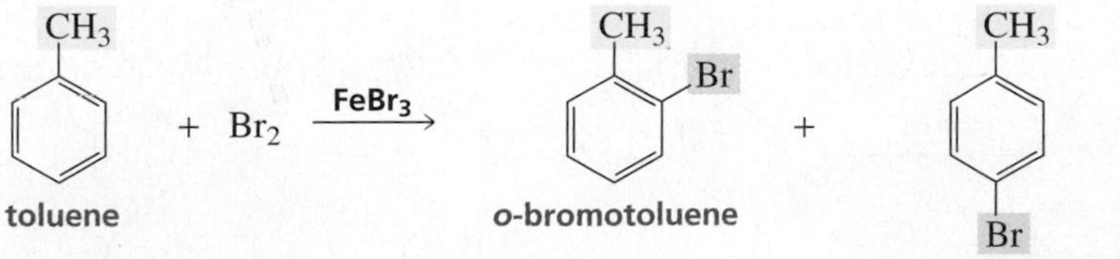

toluene   o-bromotoluene   p-bromotoluene

2. The weakly deactivating halogens also direct an incoming electrophile to the ortho and para positions.

The weakly deactivating halogens are ortho–para directors.

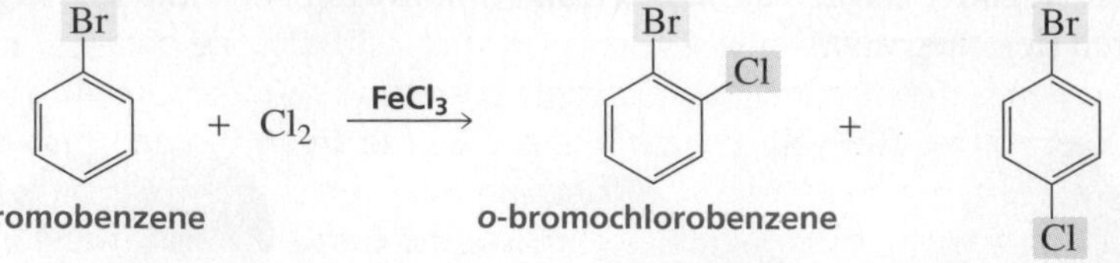

bromobenzene + Cl₂ → (FeCl₃) → o-bromochlorobenzene + p-bromochlorobenzene

3. All moderately deactivating and strongly deactivating substituents direct an incoming electrophile to the meta position.

All deactivating substituents (except the halogens) are meta directors.

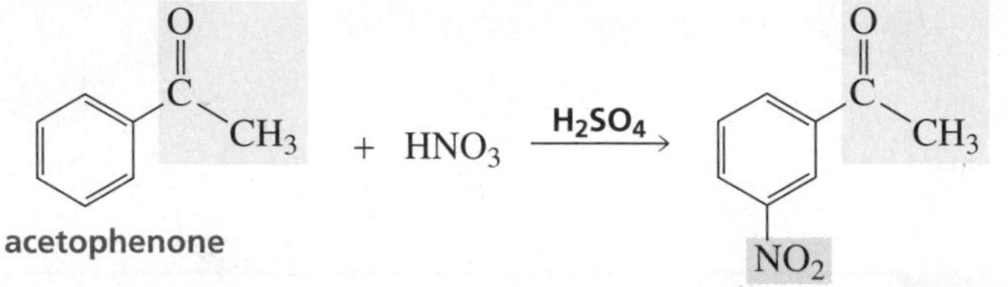

acetophenone + HNO₃ → (H₂SO₄) → m-nitroacetophenone

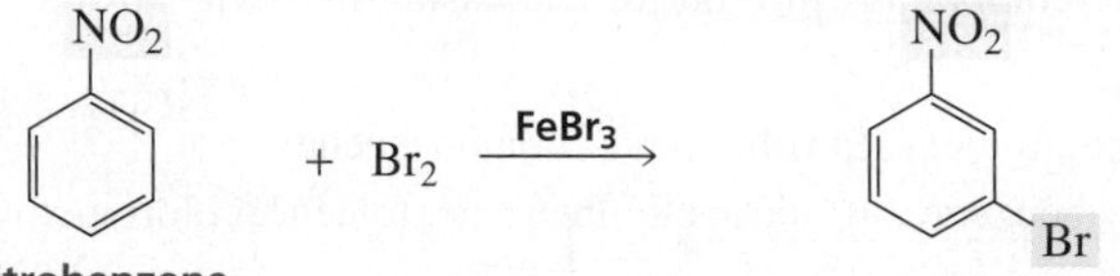

nitrobenzene + Br₂ → (FeBr₃) → m-bromonitrobenzene

To understand why a substituent directs an incoming electrophile to a particular position, we must look at the stability of the carbocation intermediate because Figure 7.2 shows that formation of the carbocation is the rate-determining step (Section 4.8). When a substituted benzene undergoes an electrophilic substitution reaction, three different carbocation intermediates can be formed: an *ortho*-substituted carbocation, a *meta*-substituted carbocation, and a *para*-substituted carbocation (Figure 7.3). The relative stabilities of the three carbocations enable us to determine the preferred pathway of the reaction because the more stable the carbocation, the less energy required to make it and the more rapidly it will be formed (Section 5.2).

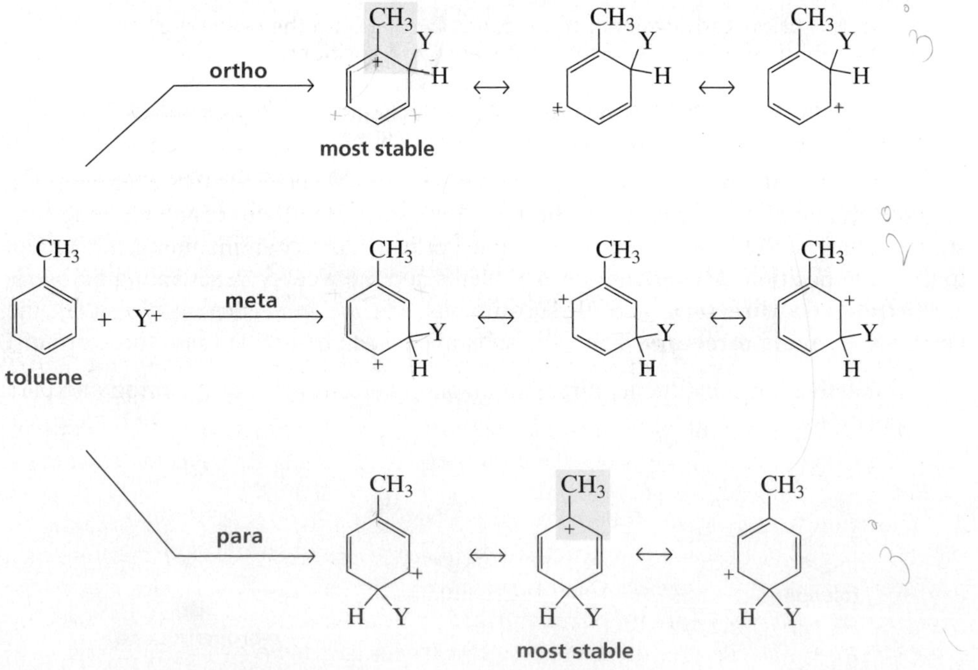

Figure 7.3 ▶
The structures of the carbocation intermediates formed from the reaction of an electrophile with toluene at the ortho, meta, and para positions.

If a substituent *donates electrons inductively*—a methyl group, for example—the indicated resonance contributors in Figure 7.3 are the most stable; the substituent is attached directly to the positively charged carbon, which the substituent can stabilize by inductive electron donation. These relatively stable resonance contributors are obtained only when the incoming group is directed to an ortho or para position. Therefore, the most stable carbocation is obtained by directing the incoming group to the ortho and para positions. Thus, *any substituent that donates electrons inductively is an ortho–para director.*

If a substituent *donates electrons by resonance*, the carbocations formed by putting the incoming electrophile on the ortho and para positions have a fourth resonance contributor (Figure 7.4). This is an especially stable resonance contributor because it is the only one whose atoms (except for hydrogen) all have complete octets (i.e., all have outer shells that contain eight electrons; Section 1.3). Therefore, *all substituents that donate electrons by resonance are ortho–para directors.*

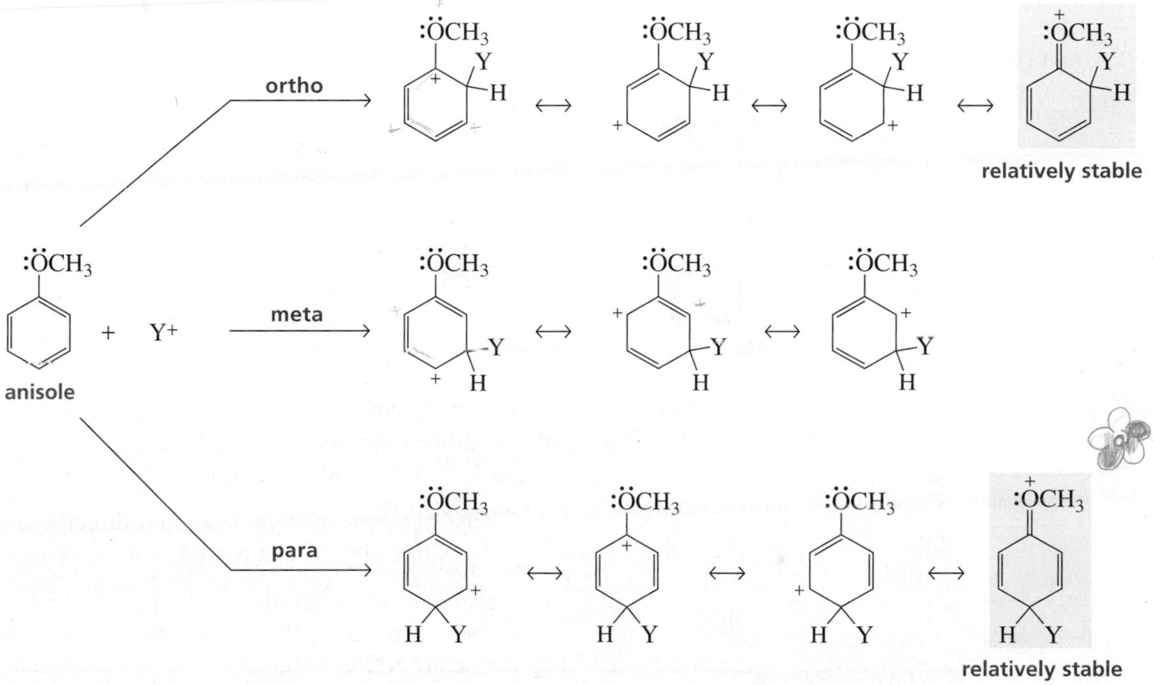

▲ **Figure 7.4**
The structures of the carbocation intermediates formed from the reaction of an electrophile with anisole at the ortho, meta, and para positions.

Substituents with a positive charge or a partial positive charge on the atom attached to the benzene ring, *withdraw electrons inductively* from the benzene ring, and most *withdraw electrons by resonance* as well. For all such substituents, the indicated resonance contributors in Figure 7.5 are the *least* stable because they have a positive charge on each of two adjacent atoms, so the most stable carbocation is formed when the incoming electrophile is directed to the meta position. Thus, *all substituents that withdraw electrons (except for the halogens, which are ortho–para directors because they donate electrons by resonance) are meta directors.*

Notice that the three possible carbocation intermediates in Figures 7.3 and 7.5 are the same, except for the substituent. The nature of the substituent determines whether the resonance contributors with the substituent directly attached to the positively charged carbon are the most stable (electron-donating substituents) or the least stable (electron-withdrawing substituents).

In summary, as shown in Table 7.1, all the activating substituents and the halogens are ortho–para directors. All substituents more deactivating than the halogens are meta directors. Put another way, *all substituents that donate electrons into the ring inductively or by resonance are ortho–para directors, and all substituents that cannot donate electrons into the ring inductively or by resonance are meta directors.*

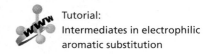

Tutorial:
Intermediates in electrophilic aromatic substitution

All substituents that donate electrons into the ring either inductively or by resonance are ortho–para directors.

All substituents that cannot donate electrons into the ring either inductively or by resonance are meta directors.

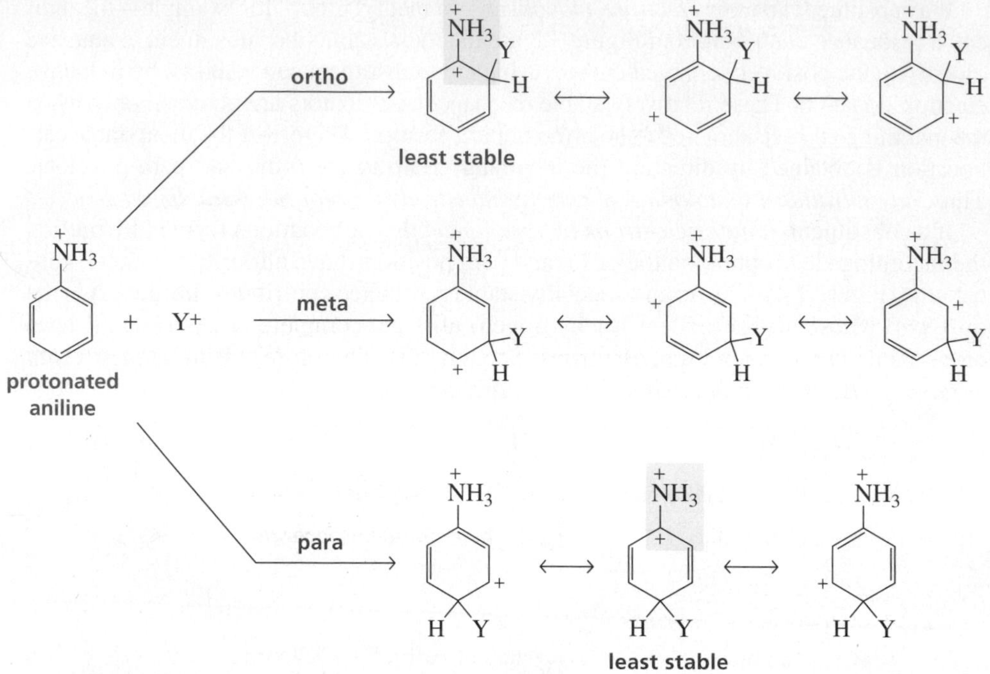

▲ **Figure 7.5**
The structures of the carbocation intermediates formed from the reaction of an electrophile with protonated aniline at the ortho, meta, and para positions.

You don't need to resort to memorization to determine whether a substituent is an ortho–para director or a meta director. It is easy to tell them apart: All ortho–para directors, except for alkyl groups, have at least one lone pair on the atom directly attached to the ring; all meta directors have a positive charge or a partial positive charge on the atom attached to the ring. Take a few minutes to examine the substituents listed in Table 7.1 to convince yourself that this is true.

PROBLEM 16

What product(s) would result from nitration of each of the following compounds?

a. propylbenzene

b. bromobenzene

c. benzaldehyde

d. benzenesulfonic acid

e. cyclohexylbenzene

f. benzoic acid

PROBLEM-SOLVING STRATEGY

In designing the synthesis of disubstituted benzenes, you must carefully consider the order in which the substituents are to be placed on the ring. For example, if you want to synthesize *meta*-bromobenzenesulfonic acid, the sulfonic acid group has to be placed on the ring first because that group will direct the bromo substituent to the desired meta position.

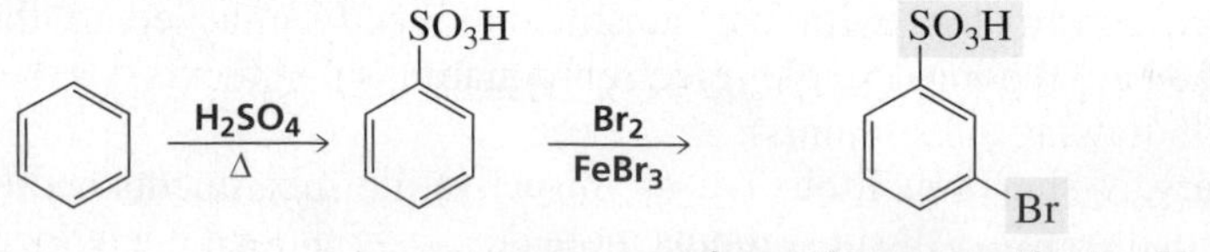

m-bromobenzenesulfonic acid

However, if the desired product is *para*-bromobenzenesulfonic acid, the order of the two re-actions must be reversed because only the bromo substituent is an ortho–para director.

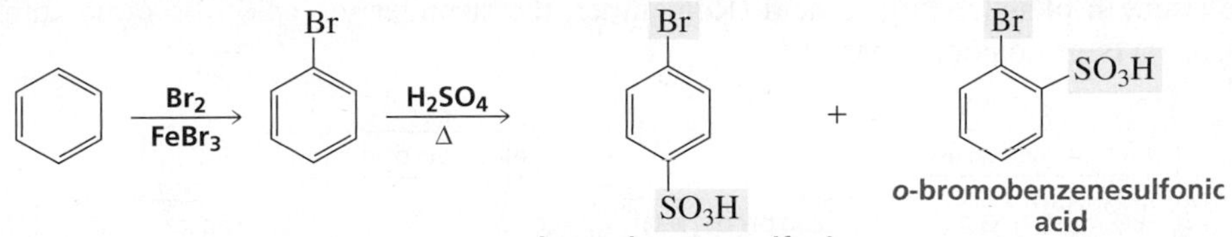

p-bromobenzenesulfonic
acid

o-bromobenzenesulfonic
acid

Now continue on to Problem 17.

PROBLEM 17◆

Show how the following compounds could be synthesized from benzene:

a. *m*-chloronitrobenzene   **b.** *p*-chloronitrobenzene

PROBLEM 18   SOLVED

Give the product(s) obtained from the reaction of each of the following compounds with one equivalent of Br_2 and a $FeBr_3$ catalyst:

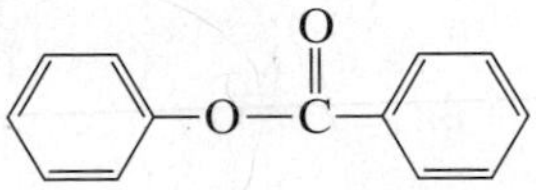

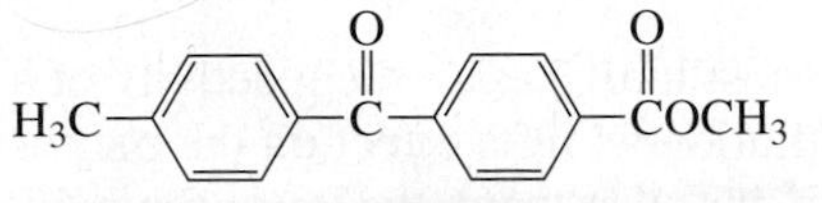

SOLUTION TO 18a   The left-hand ring is attached to a substituent that activates the ring by donating electrons into the ring by resonance. The right-hand ring is attached to a substituent that deactivates the ring by withdrawing electrons from the ring by resonance.

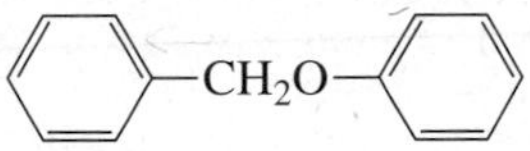

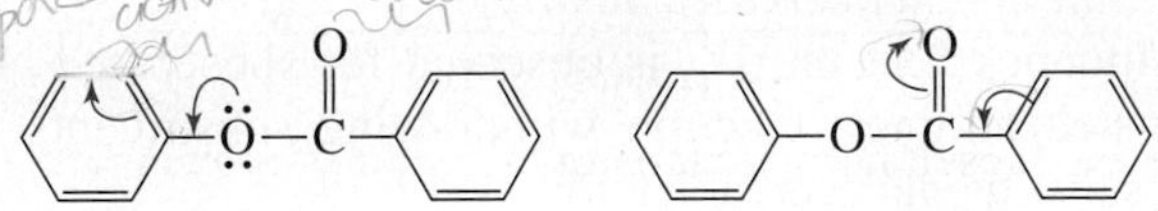

Thus, the left-hand ring is more reactive toward electrophilic aromatic substitution. The ac-tivating substituent will direct the bromine to the positions that are ortho and para to the substituent on the left-hand ring, giving the following products.

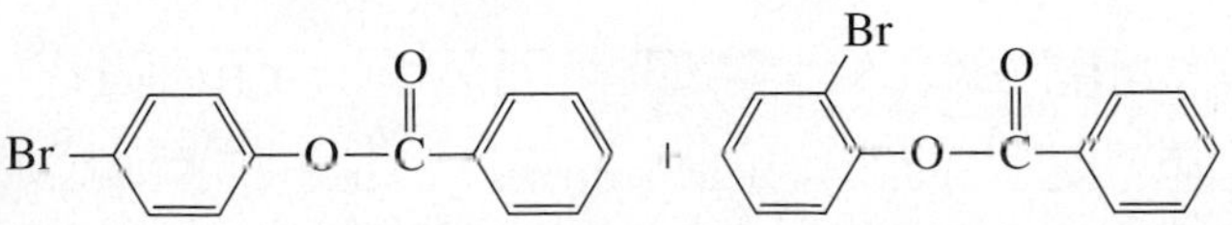

7.15   The Effect of Substituents on pK_a

When a substituent either withdraws electrons from or donates electrons into a ben-zene ring, the pK_a values of substituted phenols, benzoic acids, and protonated ani-lines will reflect this withdrawal or donation.

Electron-withdrawing groups stabilize a base and, therefore, increase the strength of its conjugate acid. Electron-donating groups destabilize a base, thereby decreasing the strength of its conjugate acid (Remember, the stronger the acid, the more stable (weaker) is its conjugate base.).

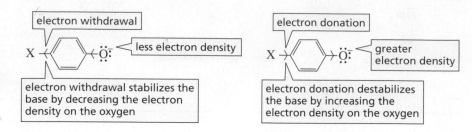

For example, the pK_a of phenol in H_2O at 25 °C is 9.95. The pK_a of *para*-nitrophenol is lower (7.14) because the nitro substituent withdraws electrons from the ring, whereas the pK_a of *para*-methyl phenol is higher (10.19) because the methyl substituent donates electrons into the ring.

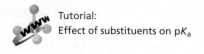
Tutorial:
Effect of substituents on pK_a

OH	OH	OH	OH	OH	OH
OCH₃	CH₃		Cl	HC=O	NO₂
$pK_a = 10.20$	$pK_a = 10.19$	$pK_a = 9.95$	$pK_a = 9.38$	$pK_a = 7.66$	$pK_a = 7.14$
		phenol			

The more deactivating (electron withdrawing) the substituent, the more it increases the acidity of a COOH, an OH, or an ⁺NH₃ group attached to a benzene ring.

The more activating (electron donating) the substituent, the more it decreases the acidity of a COOH, an OH, or an ⁺NH₃ group attached to a benzene ring.

Take a minute to compare the influence a substituent has on the reactivity of a benzene ring toward electrophilic aromatic substitution with its effect on the pK_a of phenol. Notice that the more strongly deactivating the substituent, the lower the pK_a of the phenol; and the more strongly activating the substituent, the higher the pK_a of the phenol. In other words, *electron withdrawal decreases reactivity toward electrophilic substitution and increases acidity, whereas electron donation increases reactivity toward electrophilic substitution and decreases acidity.*

A similar substituent effect on pK_a is observed for substituted benzoic acids and substituted protonated anilines: electron withdrawing substituents increase acidity; electron donating substituents decrease acidity.

COOH	COOH	COOH	COOH	COOH	COOH
OCH₃	CH₃		Br	CH₃C=O	NO₂
$pK_a = 4.47$	$pK_a = 4.34$	$pK_a = 4.20$	$pK_a = 4.00$	$pK_a = 3.70$	$pK_a = 3.44$

⁺NH₃	⁺NH₃	⁺NH₃	⁺NH₃	⁺NH₃	⁺NH₃
OCH₃	CH₃		Br	HC=O	NO₂
$pK_a = 5.29$	$pK_a = 5.07$	$pK_a = 4.58$	$pK_a = 3.91$	$pK_a = 1.76$	$pK_a = 0.98$

PROBLEM 19◆

Which of the compounds in each of the following pairs is more acidic?

a. $CH_3\overset{O}{\overset{\|}{C}}OH$ or $ClCH_2\overset{O}{\overset{\|}{C}}OH$

c. $FCH_2\overset{O}{\overset{\|}{C}}OH$ or $ClCH_2\overset{O}{\overset{\|}{C}}OH$

b. $CH_3CH_2\overset{O}{\overset{\|}{C}}OH$ or $H_3\overset{+}{N}CH_2\overset{O}{\overset{\|}{C}}OH$

d. $H\overset{O}{\overset{\|}{C}}OH$ or $CH_3\overset{O}{\overset{\|}{C}}OH$

Summary

To be classified as **aromatic**, a compound must have an un-interrupted cyclic cloud of π electrons that contains an *odd number of pairs* of π electrons.

A **heterocyclic compound** is a cyclic compound in which one or more of the ring atoms is a **heteroatom**—an atom other than carbon. Pyridine, pyrrole, furan, and thiophene are aromatic heterocyclic compounds.

Benzene's aromaticity causes it to undergo **electrophilic aromatic substitution reactions.** The electrophilic addition reactions characteristic of alkenes would lead to much less stable nonaromatic addition products. The most common electrophilic aromatic substitution reactions are halogenation, nitration, sulfonation, and Friedel–Crafts acylation and alkylation. Once the electrophile is generated, all electrophilic aromatic substitution reactions take place by the same two-step mechanism: (1) the aromatic compound reacts with an electrophile, forming a carbocation intermediate; and (2) a base pulls off a proton from the carbon that formed the bond with the electrophile.

Some monosubstituted benzenes are named as substituted benzenes (e.g., bromobenzene, nitrobenzene); some have names that incorporate the name of the substituent (e.g., toluene, phenol, aniline, anisole). The relative positions of two substituents on a benzene ring are indicated either by numbers or by the prefixes *ortho*, *meta*, and *para*.

Bromination or **chlorination** requires a Lewis acid catalyst. **Sulfonation** with sulfuric acid places an SO_3H group on the ring. **Nitration** with nitric acid requires sulfuric acid as a catalyst. An acyl chloride is used for a **Friedel–Crafts acylation**, a reaction that places an acyl group on a benzene ring; an alkyl halide is used for a **Friedel–Crafts** alkylation, a reaction that places an alkyl group on a benzene ring.

The nature of the substituent affects both the reactivity of the benzene ring and the placement of an incoming substituent: The rate of electrophilic aromatic substitution is increased by electron-donating substituents and decreased by electron-withdrawing substituents. Substituents can donate or withdraw electrons **inductively** or by **resonance.**

The stability of the carbocation intermediate determines the position to which the substituent directs an incoming electrophile. All activating substituents and the weakly deactivating halogens are **ortho–para directors**; all substituents more deactivating than the halogens are **meta directors.**

The acidity of substituted benzoic acids, phenols, and anilinium ions is increased by electron-withdrawing substituents and decreased by electron-donating substituents.

Summary of Reactions

Electrophilic aromatic substitution reactions:

a. Halogenation (Section 7.7)

b. Nitration and sulfonation (Sections 7.8 and 7.9)

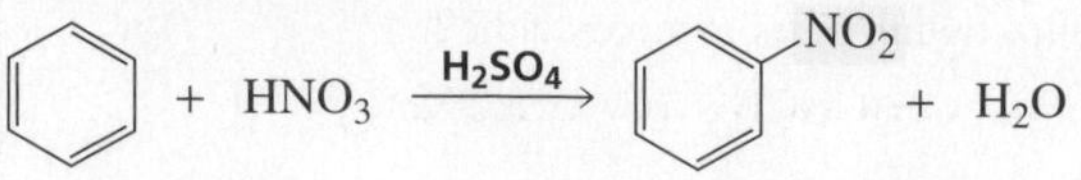

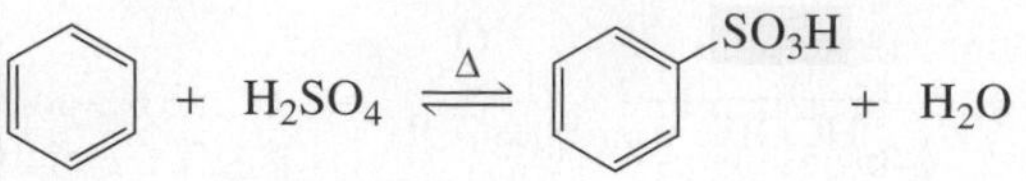

c. Friedel–Crafts acylation and alkylation (Sections 7.10 and 7.11)

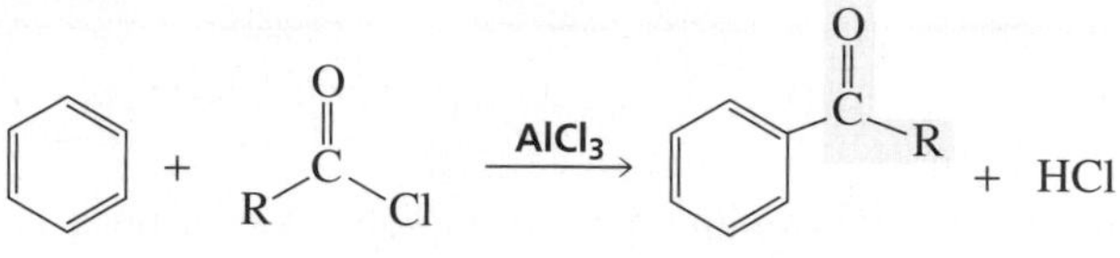

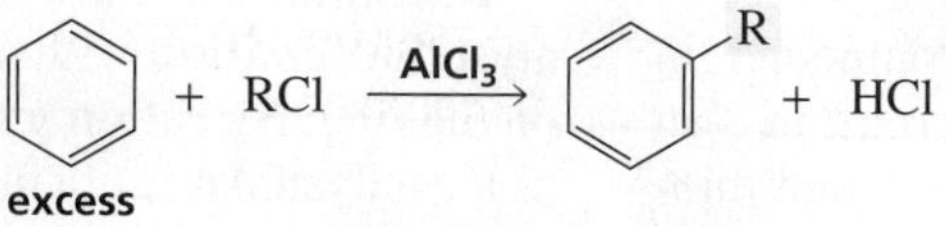

Problems

20. Which of the following compounds are aromatic?

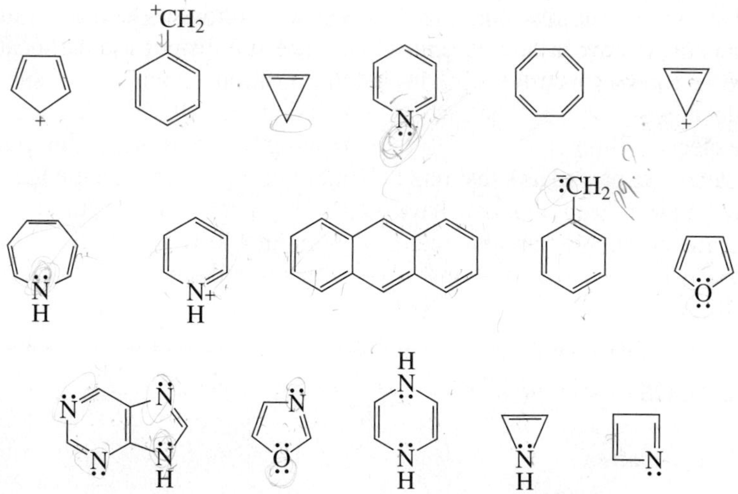

21. Give the product of the reaction of benzene with each of the following reagents:

a. $CH_3CHCH_3 + AlCl_3$
          |
          Cl

b. $CH_3CH=CH_2 + HF$

c. $CH_3CCl + AlCl_3$
              ‖
              O

22. Which ion in each of the following pairs is more stable? Explain your choice.

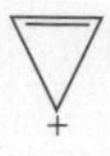

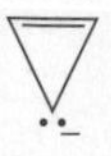

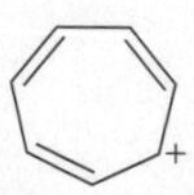

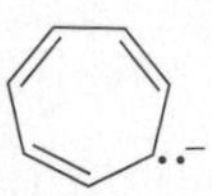

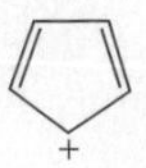

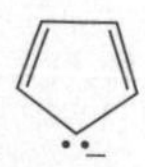

23. Which compound in each of the following pairs is a stronger base? Why?

a. 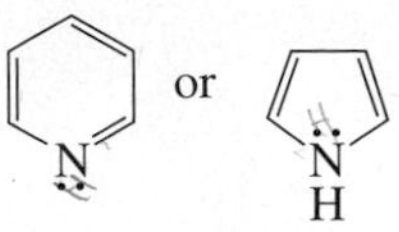 [pyridine structure] or [pyrrole structure N-H]

b. 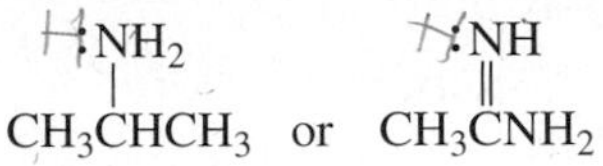 $CH_3\overset{\overset{\displaystyle \overset{+}{N}H_2}{|}}{C}HCH_3$ or $CH_3\overset{\overset{\displaystyle \overset{+}{N}H}{\|}}{C}NH_2$

24. Draw the structure of each of the following compounds:
a. *m*-ethylphenol
b. *p*-nitrobenzenesulfonic acid
c. *o*-bromoaniline
d. *p*-bromobenzoic acid
e. (*E*)-2-phenyl-2-pentene
f. 2,4-dichlorotoluene

25. Predict the relative pK_a values of cyclopentane and cyclopentadiene.

26. Name the following compounds:

a. 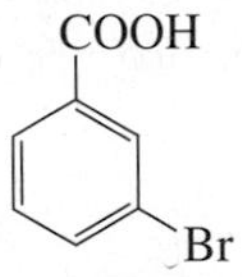 [benzene with COOH and Br substituents]

b. 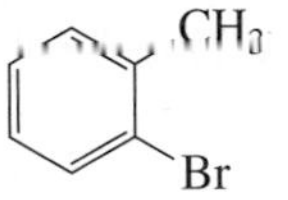 [benzene with CH₃ and Br substituents]

c. 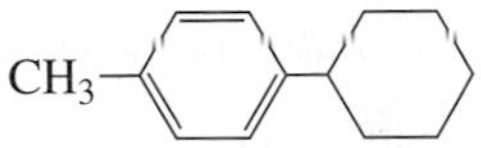 CH_3—[biphenyl/cyclohexyl structure]

27. Give the product(s) of each of the following reactions:
a. benzoic acid + HNO_3/H_2SO_4
b. isopropylbenzene + cyclohexene + HF
c. cyclohexylbenzene + $Br_2/FeBr_3$
d. phenol + H_2SO_4 + Δ
e. ethylbenzene + $Br_2/FeBr_3$

28. Rank the following anions in order of decreasing basicity:

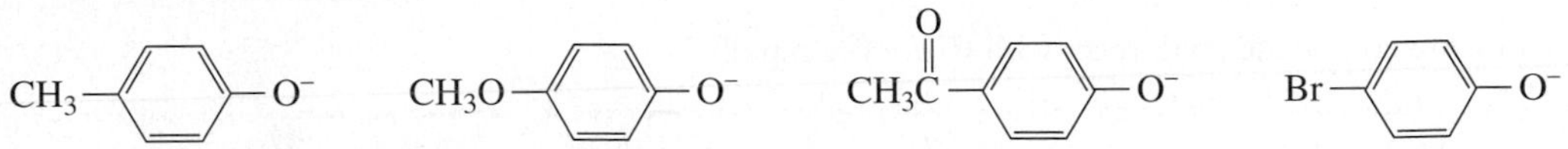

CH_3—⟨benzene⟩—O^- CH_3O—⟨benzene⟩—O^- $CH_3\overset{\overset{\displaystyle O}{\|}}{C}$—⟨benzene⟩—$O^-$ Br—⟨benzene⟩—O^-

29. Show how the following compounds could be synthesized from benzene:
a. *p*-nitrotoluene
b. *m*-chlorobenzenesulfonic acid
c. 3-phenylpentane

30. For each of the groups of substituted benzenes, indicate
a. the one that would be the most reactive in an electrophilic aromatic substitution reaction
b. the one that would be the least reactive in an electrophilic aromatic substitution reaction
c. the one that would yield the highest percentage of meta product

1. 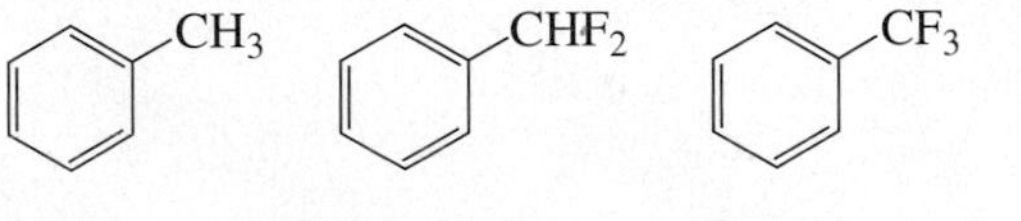 [benzene]—CH_3 [benzene]—CHF_2 [benzene]—CF_3

2. [benzene]—$\overset{+}{N}(CH_3)_3$ [benzene]—$CH_2\overset{+}{N}(CH_3)_3$ [benzene]—$CH_2CH_2\overset{+}{N}(CH_3)_3$

3. [benzene]—OCH_2CH_3 [benzene]—CH_2OCH_3 [benzene]—$\overset{\overset{\displaystyle O}{\|}}{C}OCH_3$

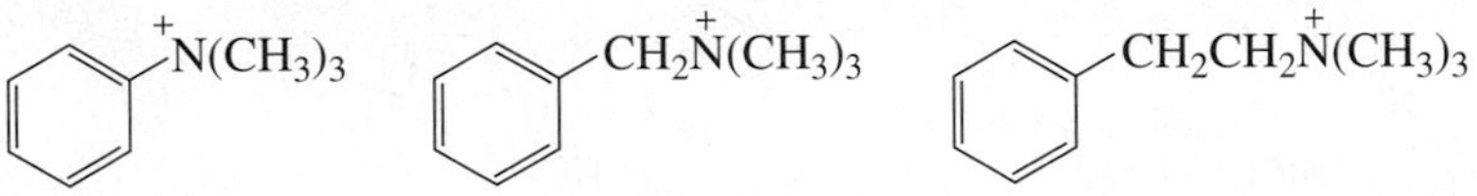

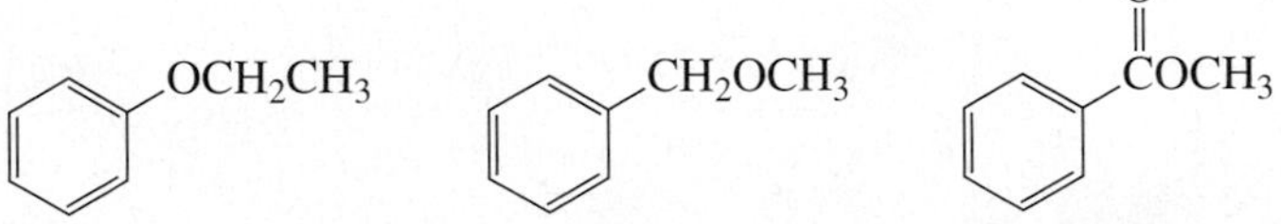

31. Draw the structures of the following groups of compounds and rank them in order of decreasing reactivity toward electrophilic aromatic substitution:
a. benzene, ethylbenzene, chlorobenzene, nitrobenzene
b. 1-chloro-2,4-dinitrobenzene, 2,4-dinitrophenol, 2,4-dinitrotoluene
c. benzene, benzoic acid, phenol, propylbenzene
d. *p*-nitrotoluene, 2-chloro-4-nitrotoluene, 2,4-dinitrotoluene, *p*-chlorotoluene

32. Give the products of the following reactions:

a. 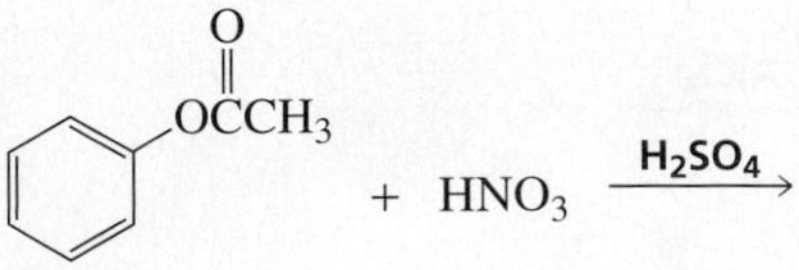 [benzene]—$\overset{\overset{\displaystyle O}{\|}}{O}CCH_3$ + HNO_3 $\xrightarrow{H_2SO_4}$

b. 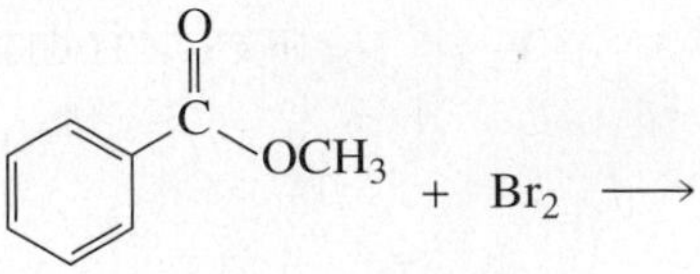 [benzene]—$\overset{\overset{\displaystyle O}{\|}}{C}$—$OCH_3$ + Br_2 $\longrightarrow$

c. [benzene with CF_3] + Cl_2 $\xrightarrow{FeCl_3}$

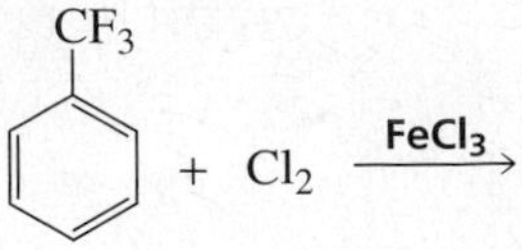

33. For each of the statements in Column I, choose a substituent from Column II that fits the description for the compound at the right:

Column I

a. Z donates electrons inductively, but does not donate or withdraw electrons by resonance.

b. Z withdraws electrons inductively and withdraws electrons by resonance.

c. Z deactivates the ring and directs ortho–para.

d. Z withdraws electrons inductively, donates electrons by resonance, and activates the ring.

e. Z withdraws electrons inductively, but does not donate or withdraw electrons by resonance.

Column II

OH
Br
$^+NH_3$
CH_2CH_3
NO_2

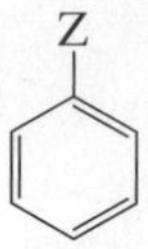

34. List the following compounds in order of decreasing reactivity toward electrophilic aromatic substitution:

dichloromethylbenzene, difluoromethylbenzene, toluene, chloromethylbenzene

35. Which of the compounds in each of the following pairs is more acidic?

a. 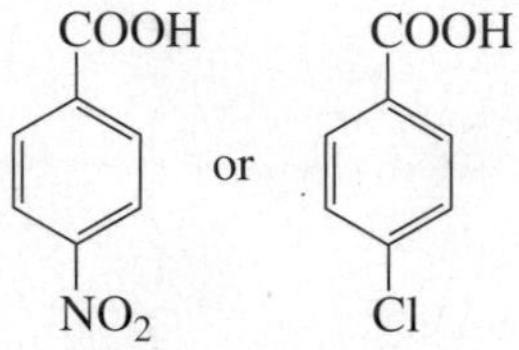 [structure: COOH-benzene-NO₂] or [structure: COOH-benzene-Cl]

b. 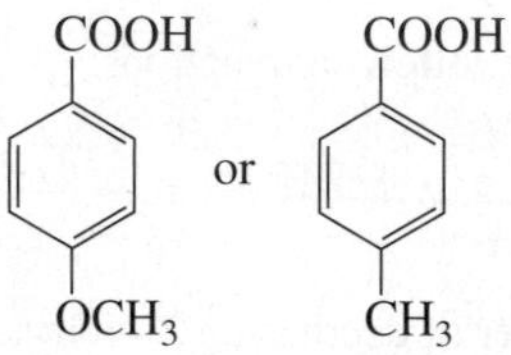 [structure: COOH-benzene-OCH₃] or [structure: COOH-benzene-CH₃]

36. Which of the following compounds will react with HBr more rapidly?

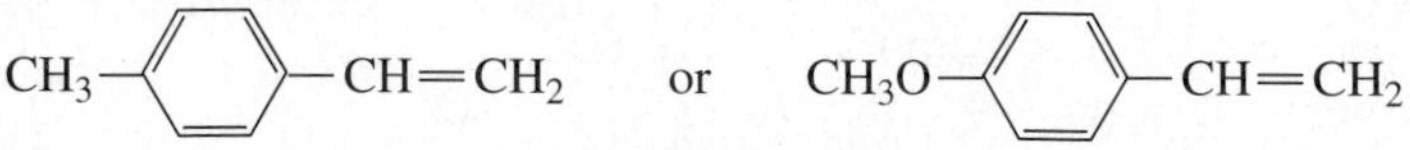

CH_3—[benzene]—$CH=CH_2$ or CH_3O—[benzene]—$CH=CH_2$

37. Show two ways that the following compound could be synthesized:

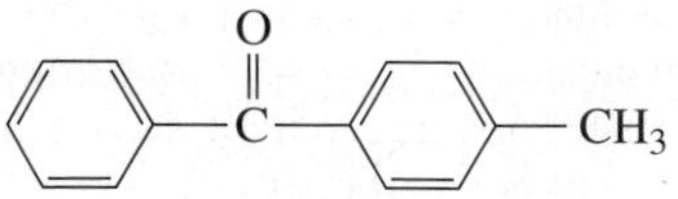

[structure: benzene–C(=O)–benzene–CH₃]

38. Give the product of each of the following reactions:

a. 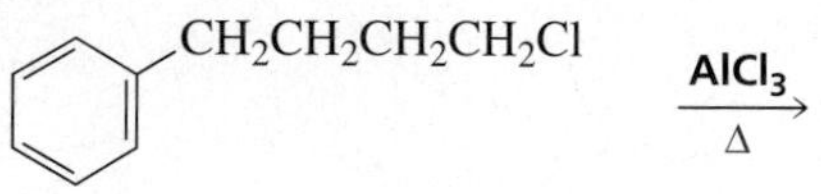 [benzene]—$CH_2CH_2CH_2CH_2Cl$ $\xrightarrow[\Delta]{AlCl_3}$

b. 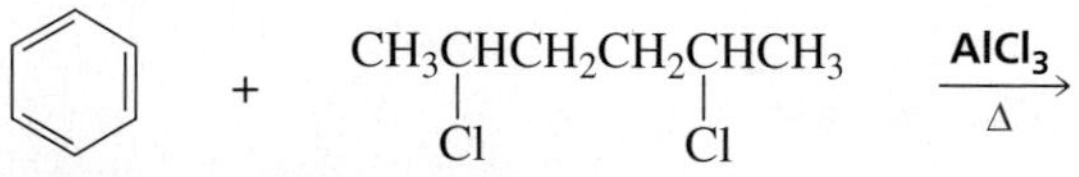 [benzene] + $CH_3CHCH_2CH_2CHCH_3$ (with Cl, Cl) $\xrightarrow[\Delta]{AlCl_3}$

39. Propose a mechanism for the following reaction:

a. 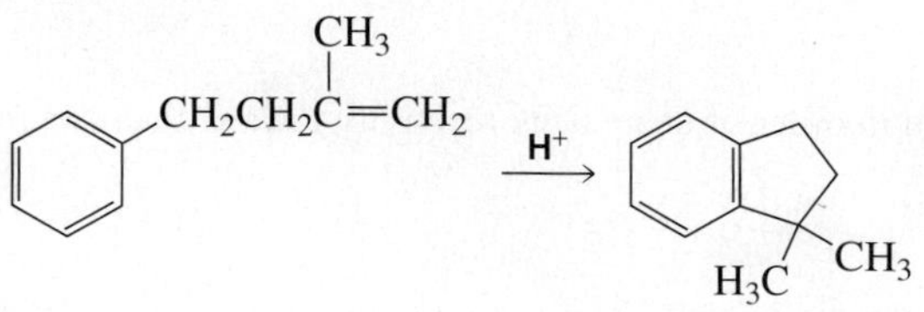 [structure with CH₃, CH₂CH₂C=CH₂ on benzene] $\xrightarrow{H^+}$ [indane structure with H₃C CH₃]

40. Give the products of the following reactions:

a. 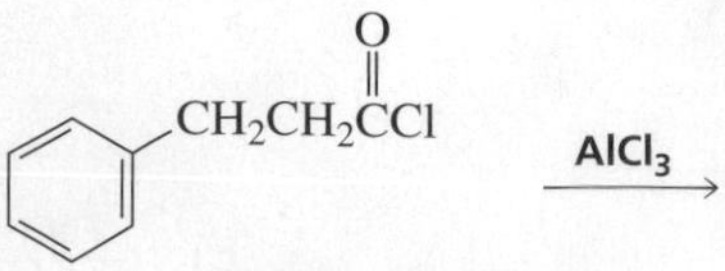 [benzene]—CH_2CH_2CCl (C=O) $\xrightarrow{AlCl_3}$

b. 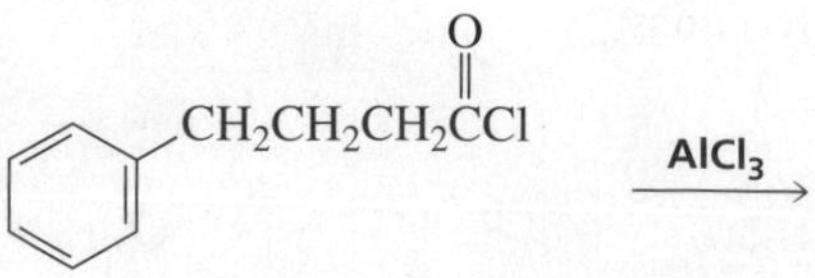 [benzene]—$CH_2CH_2CH_2CCl$ (C=O) $\xrightarrow{AlCl_3}$

8 Isomers and Stereochemistry

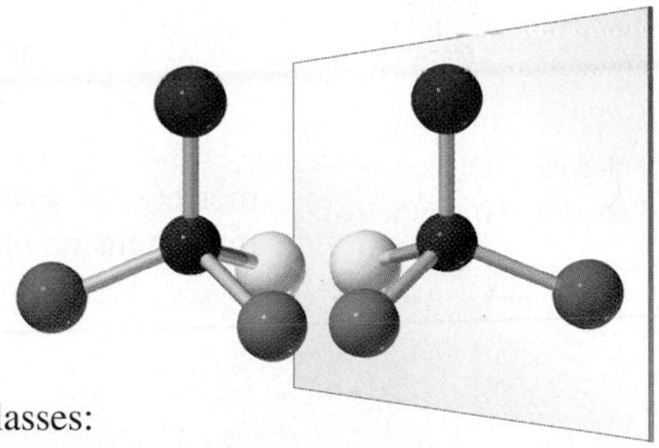

nonsuperimposable mirror images

Compounds that have the same molecular formula but are not identical are called **isomers**. Isomers fall into two main classes: *constitutional isomers* and *stereoisomers*. **Constitutional isomers** differ in the way their atoms are connected (Section 3.0). For example, ethyl alcohol and dimethyl ether are constitutional isomers because they have the same molecular formula, C_2H_6O, but the atoms in each compound are connected differently. The oxygen in ethyl alcohol is bonded to a carbon and to a hydrogen, whereas the oxygen in dimethyl ether is bonded to two carbons.

constitutional isomers

CH₃CH₂OH   and   CH₃OCH₃
ethyl alcohol   **dimethyl ether**

$$CH_3CH_2CH_2CH_2Cl \quad \text{and} \quad CH_3CH_2\overset{\text{Cl}}{\underset{|}{C}}HCH_3$$

1-chlorobutane   **2-chlorobutane**

Unlike the atoms in constitutional isomers, the atoms in *stereoisomers* are connected in the same way. **Stereoisomers** differ in the way their atoms are arranged in space. There are two kinds of stereoisomers: **cis–trans isomers** and isomers that contain **asymmetric centers**.

Movie:
Isomerism

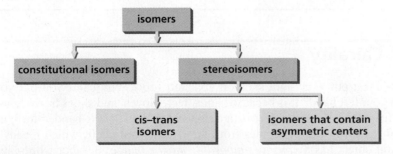

PROBLEM 1◆

a. Draw three constitutional isomers with molecular formula C_3H_8O.

b. How many constitutional isomers can you draw for $C_4H_{10}O$?

8.1   Cis–Trans Isomers

Cis–trans isomers result from restricted rotation (Section 3.12). Restricted rotation can be caused either by a *double bond* or by a *cyclic structure*. As a result of the restricted rotation about a carbon–carbon double bond, an alkene such as 2-pentene can exist as cis and trans isomers. The **cis isomer** has the hydrogens on the *same side* of the double bond, whereas the **trans isomer** has the hydrogens on *opposite sides* of the double bond.

***cis*-2-pentene**

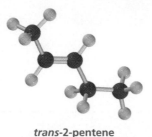

***trans*-2-pentene**

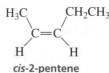

***cis*-2-pentene**

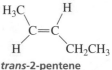

***trans*-2-pentene**

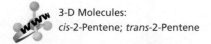

3-D Molecules:
cis-2-Pentene; *trans*-2-Pentene

Cyclic compounds can also have cis and trans isomers (Section 3.12). The cis isomer has the hydrogens on the same side of the ring, whereas the trans isomer has the hydrogens on opposite sides of the ring.

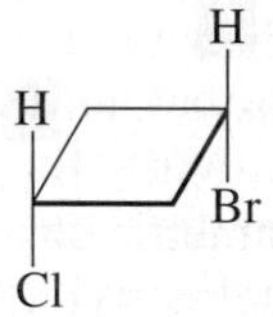

***cis*-1-bromo-3-chlorocyclobutane**

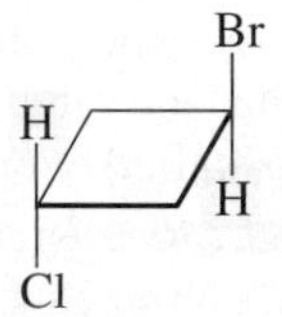

***trans*-1-bromo-3-chlorocyclobutane**

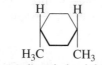

***cis*-1,4-dimethylcyclohexane**

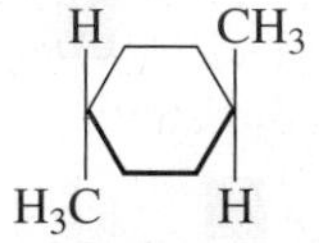

***trans*-1,4-dimethylcyclohexane**

PROBLEM 2

Draw the cis and trans isomers for the following compounds:

a. 1-ethyl-3-methylcyclobutane   **c.** 1-bromo-4-chlorocyclohexane

b. 3-hexene   **d.** 2-methyl-3-heptene

8.2   Chirality

Why can't you put your right shoe on your left foot? Why can't you put your right glove on your left hand? It is because hands, feet, gloves, and shoes have right-handed and left-handed forms. An object with a right-handed and a left-handed form is said to be **chiral** (ky-ral). "Chiral" comes from the Greek word *cheir*, which means "hand."

A chiral object has a *nonsuperimposable mirror image*. In other words, its mirror image is not the same as itself. A hand is chiral because if you look at your right hand in a mirror, you do not see your right hand; you see your left hand (Figure 8.1). In contrast, a chair is *not* chiral—it looks the same in the mirror. Objects that are not chiral are said to be **achiral**. An achiral object has a *superimposable mirror image*. Some other achiral objects would be a table, a fork, and a glass.

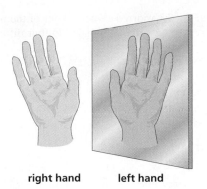 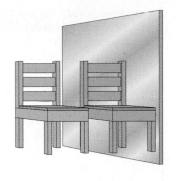

right hand **left hand**

◀ **Figure 8.1**
Using a mirror to test for chirality.
A chiral object is not the same as
its mirror image—they are
nonsuperimposable. An achiral
object is the same as its mirror
image—they are superimposable.

PROBLEM 3◆

a. Name five capital letters that are chiral.

b. Name five capital letters that are achiral.

Objects are not the only things that can be chiral. Molecules can be chiral, too. The feature that most often is the cause of chirality in a molecule is an *asymmetric center*.

An **asymmetric center** is an atom that is bonded to four different groups. The asymmetric center in each of the following compounds is the carbon indicated by a star. For example, the starred carbon in 4-octanol is an asymmetric center because it is bonded to four different groups (H, OH, $CH_2CH_2CH_3$, and $CH_2CH_2CH_2CH_3$). Notice that the difference in the groups bonded to the asymmetric center is not necessarily right next to the asymmetric center. For example, the propyl and butyl groups are different even though the point at which they differ is somewhat removed from the asymmetric center. The starred carbon in 2,4-dimethylhexane is an asymmetric center because it is bonded to four different groups—methyl, ethyl, isobutyl, and hydrogen.

A molecule with one asymmetric center is chiral.

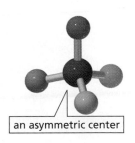

an asymmetric center

Tutorial:
Identification of asymmetric centers

$$CH_3CH_2CH_2\overset{*}{C}HCH_2CH_2CH_2CH_3$$
$$\underset{OH}{|}$$
4-octanol

$$CH_3\overset{*}{C}HCH_2CH_3$$
$$\underset{Br}{|}$$
2-bromobutane

$$\overset{CH_3}{\underset{}{|}}$$
$$CH_3CHCH_2\overset{*}{C}HCH_2CH_3$$
$$\underset{CH_3}{|}$$
2,4-dimethylhexane

PROBLEM 4◆

Which of the following compounds have an asymmetric center?

a. $CH_3CH_2CHCH_3$
$\underset{Cl}{|}$

b. $CH_3CH_2CHCH_3$
$\underset{CH_3}{|}$

c. $CH_3CH_2\overset{CH_3}{\underset{Br}{C}}CH_2CH_2CH_3$

d. CH_3CH_2OH

e. $CH_3CH_2CHCH_2CH_3$
$\underset{Br}{|}$

f. $CH_2{=}CHCHCH_3$
$\underset{NH_2}{|}$

PROBLEM 5 **SOLVED**

Tetracycline is called a broad-spectrum antibiotic because it is active against a wide variety of bacteria. How many asymmetric centers does tetracycline have?

SOLUTION The only carbons that can be asymmetric centers are sp^3 hybridized carbons, because an asymmetric center must have four different groups attached to it. Therefore, we start by locating all the sp^3 hybridized carbons in tetracycline. (They are numbered in red.) Tetracycline has nine sp^3 hybridized carbons. Four of them (#1, #2, #5, and #8) are not asymmetric centers because they are not bonded to four different groups. Tetracycline, therefore, has five asymmetric centers.

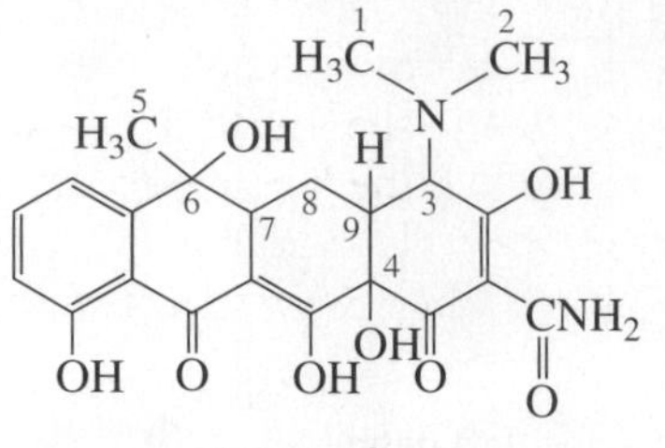

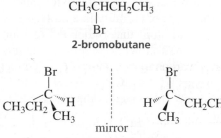

tetracycline

8.4 Isomers with One Asymmetric Center

A compound with one asymmetric center, such as 2-bromobutane, can exist as two different stereoisomers. The two isomers are analogous to a left and a right hand. Imagine a mirror between the two isomers; notice how they are mirror images of each other. The two stereoisomers are nonsuperimposable mirror images—they are different molecules.

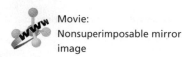

$$CH_3\overset{*}{C}HCH_2CH_3$$
$$|$$
$$Br$$

2-bromobutane

the two isomers of 2-bromobutane
enantiomers

Convince yourself that the two 2-bromobutane isomers are not identical by building ball-and-stick models; use four different-colored balls to represent the four different groups bonded to the asymmetric center. Try to superimpose them.

Nonsuperimposable mirror-image molecules are called **enantiomers**. The two stereoisomers of 2-bromobutane are enantiomers. A molecule that has a *nonsuperimposable* mirror image, like an object that has a nonsuperimposable mirror image, is **chiral**. Therefore, each of the enantiomers is chiral. Notice that chirality is a property of an entire object or an entire molecule.

A molecule that has a *superimposable* mirror image, like an object that has a superimposable mirror image, is **achiral**. To see that the achiral molecule is superimposable on its mirror image (i.e., they are identical molecules), mentally rotate it clockwise.

A chiral molecule has a nonsuperimposable mirror image.

An achiral molecule has a superimposable mirror image.

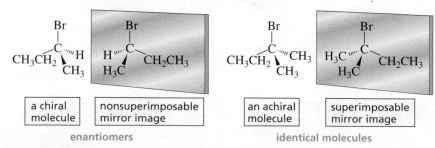

| a chiral molecule | nonsuperimposable mirror image | | an achiral molecule | superimposable mirror image |

enantiomers   *identical molecules*

PROBLEM 6◆

Which of the compounds in Problem 4 can exist as enantiomers?

8.5 Drawing Enantiomers

Chemists draw enantiomers using *perspective formulas*. **Perspective formulas** show two of the bonds to the asymmetric center in the plane of the paper, one bond as a solid wedge protruding out of the paper, and the fourth bond as a hatched wedge extending behind the paper. The solid wedge and the hatched wedge must be adjacent. The first enantiomer is drawn by putting the four groups bonded to the asymmetric center in any order. The second enantiomer is the mirror image of the first enantiomer.

A solid wedge represents a bond that points out of the plane of the paper toward the viewer.

A hatched wedge represents a bond that points back from the plane of the paper away from the viewer.

When you draw a perspective formula, make certain that the two bonds in the plane of the paper are adjacent to one another; neither the solid wedge nor the hatched wedge should be drawn between them.

$$
\begin{array}{cc}
\underset{\overset{|}{\underset{CH_2CH_3}{C}}}{Br} & \underset{\overset{|}{\underset{CH_3CH_2}{C}}}{Br} \\
H-C^{\prime\prime\prime\prime\prime}CH_3 & H_3C^{\prime\prime\prime\prime\prime}C-H
\end{array}
$$

perspective formulas of the enantiomers of 2-bromobutane

PROBLEM 7♦

Draw the enantiomers of each of the following compounds using perspective formulas:

a. $\underset{\overset{|}{\underset{}{Br}}}{CH_3CHCH_2OH}$ b. $\underset{\overset{|}{\underset{}{CH_3}}}{ClCH_2CH_2CHCH_2CH_3}$ c. $\underset{\overset{|}{\underset{}{OH}}}{CH_3\overset{\overset{CH_3}{|}}{C}HCHCH_3}$

8.6 Naming Enantiomers: The *R,S* System of Nomenclature

We need a way to name the individual stereoisomers of a compound such as 2-bromobutane so that we know which stereoisomer we are talking about. In other words, we need a system of nomenclature that indicates the **configuration** (arrangement) of the atoms or groups about the asymmetric center. Chemists use the letters *R* and *S* to indicate the configuration about an asymmetric center. For any pair of enantiomers with one asymmetric center, one will have the **R configuration** and the other will have the **S configuration**.

Let's first look at how we can determine the configuration of a compound if we have a three-dimensional model of the compound.

1. *Rank the groups (or atoms) bonded to the asymmetric center in order of priority.* The atomic numbers of the atoms directly attached to the asymmetric center determine the relative priorities. The higher the atomic number, the higher the priority. (This should remind you of the way relative priorities are determined for the *E,Z* system of nomenclature because the system of priorities was originally devised for the *R,S* system of nomenclature and was later borrowed for the *E,Z* system; Section 4.5).

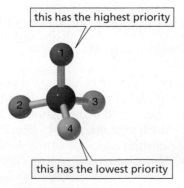

this has the highest priority

this has the lowest priority

The molecule is oriented so the group with the lowest priority points away from the viewer. If an arrow drawn from the highest priority group to the next highest priority group points clockwise, the molecule has the *R* configuration.

2. *Orient the molecule so that the group (or atom) with the lowest priority (4) is directed away from you. Then draw an imaginary arrow from the group (or atom) with the highest priority (1) to the group (or atom) with the next highest priority (2). If the arrow points clockwise, the asymmetric center has the R configuration (R is for rectus, which is Latin for "right"). If the arrow points counterclockwise, the asymmetric center has the S configuration (S is for sinister, which is Latin for "left").*

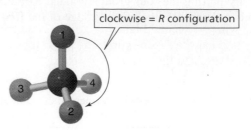

clockwise = *R* configuration

If you forget which is which, imagine driving a car and turning the steering wheel clockwise to make a right turn or counterclockwise to make a left turn.

If you are able to easily visualize spatial relationships, the above two rules are all you need to determine whether the asymmetric center of a molecule written on a two-dimensional piece of paper has the *R* or the *S* configuration. Just mentally rotate the molecule so that the group (or atom) with the lowest priority (4) is directed away from you, then draw an imaginary arrow from the group (or atom) with the highest priority to the group (or atom) with the next highest priority.

left turn

right turn

PROBLEM 8◆

Which of the following molecular models are identical?

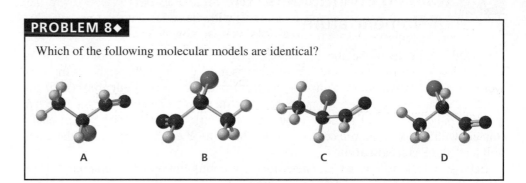

A   B   C   D

If you have trouble visualizing spatial relationships and you don't have access to a model, the following will allow you to determine the configuration about an asymmetric center without having to mentally rotate the molecule. As an example, we will determine which of the enantiomers of 2-bromobutane has the *R* configuration and which has the *S* configuration.

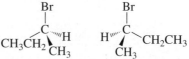

the enantiomers of 2-bromobutane

1. Rank the groups (or atoms) that are bonded to the asymmetric center in order of priority. In the following pair of enantiomers, bromine has the highest priority (1), the ethyl group has the second highest priority (2), the methyl group is next (3), and hydrogen has the lowest priority (4). (Revisit Section 4.5 if you don't understand how these priorities are assigned.)

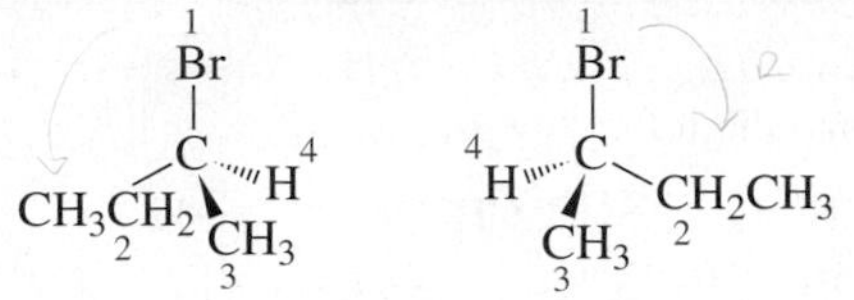

(diagram of two 2-bromobutane structures with Br, CH₃CH₂, CH₃, H, CH₂CH₃ labels and priority numbers)

2. If the group (or atom) with the lowest priority is bonded by a hatched wedge, draw an arrow from the group (or atom) with the highest priority (1) to the group (or atom) with the second highest priority (2). If the arrow points clockwise, the compound has the *R* configuration, and if it points counterclockwise, the compound has the *S* configuration.

Clockwise specifies *R* if the lowest priority substituent is attached to a hatched wedge.

Counterclockwise specifies *S* if the lowest priority substituent is attached to a hatched wedge.

the group with the lowest priority is bonded by a hatched wedge

(diagrams of two 2-bromobutane structures)

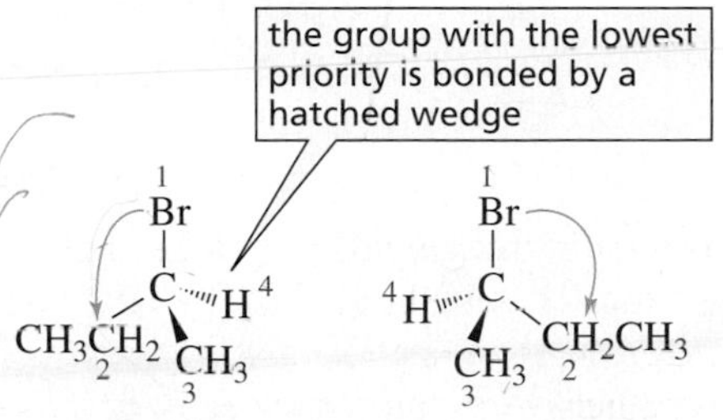

(S)-2-bromobutane   (R)-2-bromobutane

3-D Molecules:
(*R*)-2-Bromobutane;
(*S*)-2-Bromobutane

3. If the group with the lowest priority (4) is NOT bonded by a hatched wedge, then switch two groups so group 4 is bonded by a hatched wedge. Then proceed as in step #2 (above): Draw an arrow from the group (or atom) with the highest priority (1) to the group (or atom) with the second highest priority (2). Because you have switched two groups, you are now determining the configuration of the *enantiomer* of the original molecule. So if the arrow points clockwise, the enantiomer (with the switched groups) has the *R* configuration, which means the original molecule has the *S* configuration. On the other hand, if the arrow points counterclockwise, the enantiomer (with the switched groups) has the *S* configuration, which means the original molecule has the *R* configuration.

(diagram of molecule with CH₂CH₃, C, CH₃, HO, H labels)

switch
CH₃ and H

(diagram of molecule with CH₂CH₃, C, H, HO, CH₃ labels)

what is its configuration?

this molecule has the *R* configuration; therefore, the molecule had the *S* configuration before the groups were switched

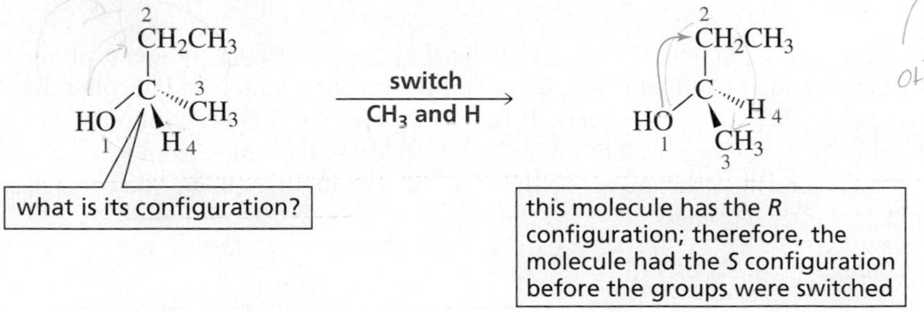

4. In drawing the arrow from group 1 to group 2, you can draw past the group (or atom) with the lowest priority (4), but never draw past the group (or atom) with the next lowest priority (3).

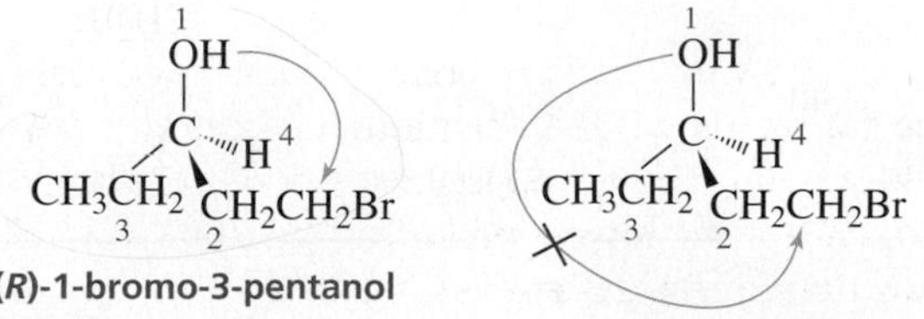

(diagram of two 1-bromo-3-pentanol structures with OH, C, H, CH₃CH₂, CH₂CH₂Br labels)

(*R*)-1-bromo-3-pentanol

PROBLEM 9◆

Assign relative priorities to the following groups:

a. $-CH_2OH$ $-CH_3$ $-CH_2CH_2OH$ $-H$

b. $-CH=O$ $-OH$ $-CH_3$ $-CH_2OH$

c. $-CH(CH_3)_2$ $-CH_2CH_2Br$ $-Cl$ $-CH_2CH_2CH_2Br$

d. $-CH=CH_2$ $-CH_2CH_3$ $-C\equiv CH$ $-CH_3$

PROBLEM 10◆

Indicate whether each of the following structures has the R or the S configuration:

a.

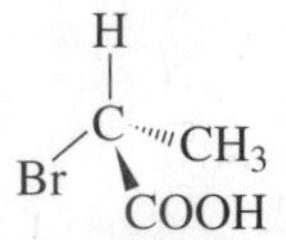

b.

PROBLEM 11◆

Name the following compounds:

a.

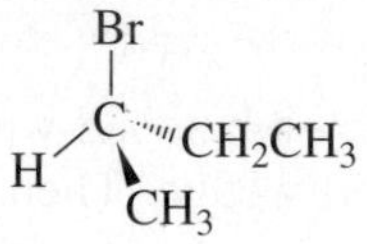

b.

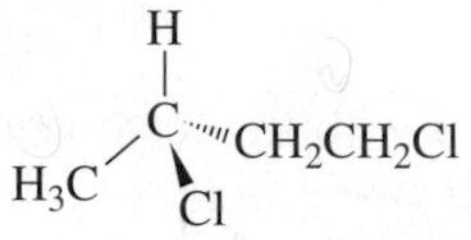

PROBLEM-SOLVING STRATEGY

Do the following structures represent identical molecules or a pair of enantiomers?

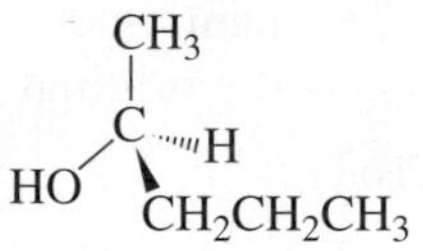

and

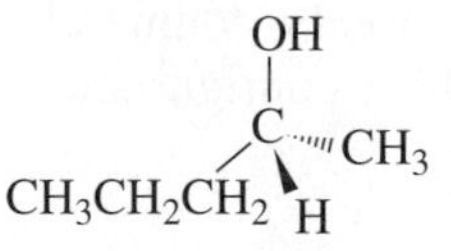

The easiest way to find out whether two molecules are enantiomers or identical molecules is to determine their configurations. If one has the R configuration and the other has the S configuration, they are enantiomers. If they both have the R configuration or both have the S configuration, they are identical molecules. Because the structure on the left has the S configuration and the structure on the right has the R configuration, we know that they represent a pair of enantiomers.

Now continue on to Problem 12.

PROBLEM 12◆

Do the following structures represent identical molecules or a pair of enantiomers?

a. 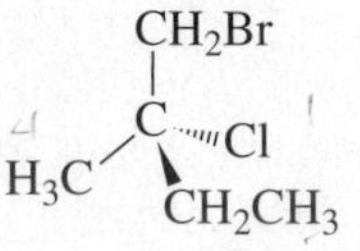 and

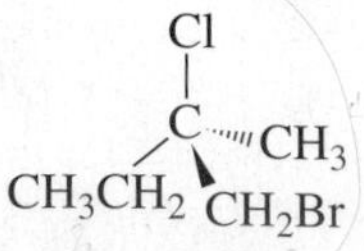

b. 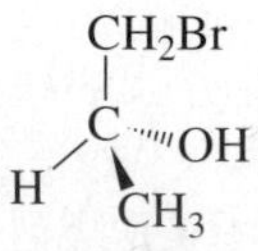 and

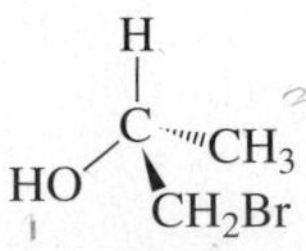

PROBLEM-SOLVING STRATEGY

(*S*)-Alanine is a naturally occurring amino acid. Draw its structure using a perspective formula.

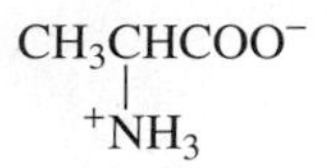

$$CH_3CHCOO^-$$
$$|$$
$$^+NH_3$$

alanine

First draw the bonds about the asymmetric center. Remember that the two bonds in the plane of the paper must be adjacent to one another.

Put the group with the lowest priority on the hatched wedge. Put the group with the highest priority on any remaining bond.

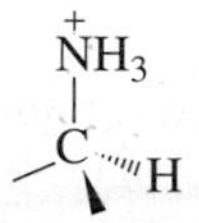

Because you have been asked to draw the *S* enantiomer, draw an arrow counterclockwise from the group with the highest priority to the next available bond; put the group with the next highest priority on that bond.

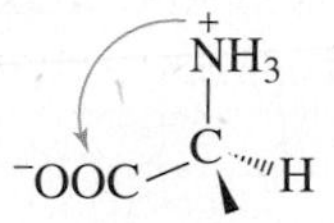

Put the remaining substituent on the last available bond.

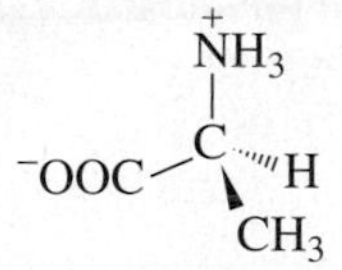

Now continue on to Problem 13.

PROBLEM 13

Draw perspective formulas for the following compounds:

a. (*S*)-2-chlorobutane   **b.** (*R*)-1,2-dibromobutane

8.7   Optical Activity

Enantiomers share many of the same properties—they have the same boiling points, the same melting points, and the same solubilities. In fact, all the physical properties of enantiomers are the same except the way they interact with *plane-polarized light*.

What is plane-polarized light? Normal light oscillates in all directions. **Plane-polarized light** (or simply polarized light) oscillates only in a single plane. Plane-polarized light is produced by passing normal light through a polarizer.

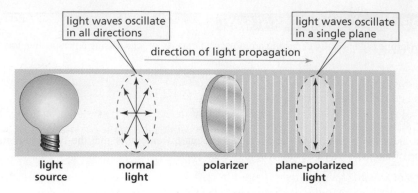

Born in France, **Jean-Baptiste Biot** **(1774–1862)** *was imprisoned for taking part in a street riot during the French Revolution. He became a professor of mathematics at the University of Beauvais and later a professor of physics at the Collège de France. He was awarded the Legion of Honor by Louis XVIII. (See page 208.)*

You experience the effect of a polarized lens with polarized sunglasses. Polarized sunglasses allow only light oscillating in a single plane to pass through them, so they block reflections (glare) more effectively than nonpolarized sunglasses.

In 1815, Jean-Baptiste Biot discovered that certain naturally occurring organic substances such as camphor and oil of turpentine are able to rotate the plane of polarization. He noted that some compounds rotated the plane clockwise and others counterclockwise, while some did not rotate the plane of polarization at all. He predicted that the ability to rotate the plane of polarization was attributable to some asymmetry in the molecules. It was later determined that the molecular asymmetry was associated with compounds having one or more asymmetric centers.

When polarized light passes through a solution of achiral molecules, the light emerges from the solution with its plane of polarization unchanged. *An achiral compound does not rotate the plane of polarization.*

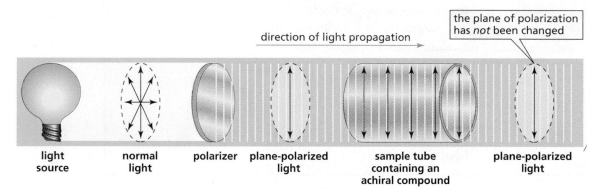

However, when polarized light passes through a solution of a chiral compound, the light emerges with its plane of polarization changed. Thus, *a chiral compound rotates the plane of polarization.* A chiral compound can rotate the plane of polarization clockwise or counterclockwise. If one enantiomer rotates the plane of polarization clockwise, its mirror image will rotate the plane of polarization exactly the same amount counterclockwise.

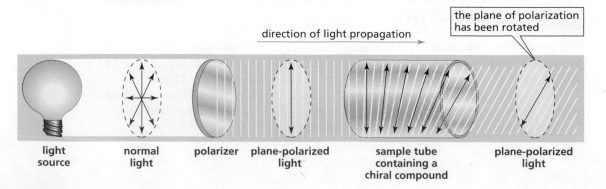

Movie:
Optical activity

A compound that rotates the plane of polarization is said to be **optically active**. In other words, chiral compounds are optically active and achiral compounds are **optically inactive**.

If an optically active compound rotates the plane of polarization clockwise, it is called **dextrorotatory**, indicated by (+). If it rotates the plane of polarization counterclockwise, it is called **levorotatory**, indicated by (−).

Do not confuse (+) and (−) with R and S. The (+) and (−) symbols indicate the direction in which an optically active compound rotates the plane of polarization, whereas R and S indicate the arrangement of the groups about an asymmetric center. Some compounds with the R configuration are (+) and some are (−). We can tell by looking at the structure of a compound whether it has the R or the S configuration, but the only way we can tell whether a compound is dextrorotatory (+) or levorotatory (−) is to put the compound in a polarimeter. For example, (S)-lactic acid and (S)-sodium lactate have the same configuration (they are both S), but (S)-lactic acid is dextrorotatory whereas (S)-sodium lactate is levorotatory. When we know the direction an optically active compound rotates the plane of polarization, we can incorporate (+) or (−) into its name.

Some molecules with the *R* configuration are (+) and some molecules with the *R* configuration are (−).

$$
\begin{array}{cc}
\text{CH}_3 & \text{CH}_3 \\
\text{C}\text{-----}\text{H} & \text{C}\text{-----}\text{H} \\
\text{HO} \quad \text{COOH} & \text{HO} \quad \text{COO}^-\text{Na}^+ \\
\textbf{(S)-(+)-lactic acid} & \textbf{(S)-(–)-sodium lactate}
\end{array}
$$

PROBLEM 14◆

a. Is (R)-lactic acid dextrorotatory or levorotatory?

b. Is (R)-sodium lactate dextrorotatory or levorotatory?

PROBLEM 15◆   SOLVED

What is the configuration of the following compounds?

a. (−)-glyceraldehyde   **c.** (+)-isoserine

b. (−)-glyceric acid   **d.** (+)-lactic acid

$$
\begin{array}{cccc}
\text{HC}=\text{O} & \text{COOH} & \text{COOH} & \text{COOH} \\
\text{C}\text{-----}\text{H} & \text{C}\text{-----}\text{H} & \text{C}\text{-----}\text{H} & \text{C}\text{-----}\text{H} \\
\text{HO} \quad \text{CH}_2\text{OH} & \text{HO} \quad \text{CH}_2\text{OH} & \text{HO} \quad \text{CH}_2\text{NH}_3^+ & \text{HO} \quad \text{CH}_3 \\
\textbf{(+)-glyceraldehyde} & \textbf{(–)-glyceric acid} & \textbf{(+)-isoserine} & \textbf{(–)-lactic acid}
\end{array}
$$

SOLUTION TO 15a   We know that (+)-glyceraldehyde has the R configuration because the arrow drawn from the OH group to the HC=O group is clockwise. Therefore, (−)-glyceraldehyde has the S configuration.

8.8   Specific Rotation

The degree to which an optically active compound rotates the plane of polarization can be measured with an instrument called a **polarimeter** (Figure 8.2). Light passes through the polarizer of a polarimeter and emerges as polarized light. The polarized light passes through an empty sample tube and emerges with its plane of polarization unchanged. The light then passes through an analyzer. The analyzer is a second polarizer mounted on an eyepiece with a dial marked in degrees. When using a polarimeter, the analyzer is rotated until the user's eye sees total darkness. At this point, the analyzer is at a right angle to the first polarizer, so no light passes through. This analyzer setting corresponds to zero rotation.

When light is filtered through two polarized lenses at a 90° angle to one another, no light is transmitted through them.

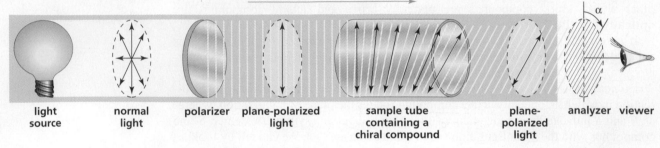

direction of light propagation

light source · normal light · polarizer · plane-polarized light · sample tube containing a chiral compound · plane-polarized light · analyzer · viewer

▲ **Figure 8.2**
Schematic of a polarimeter.

The sample to be measured is then placed in the sample tube. If the sample is optically active, it will rotate the plane of polarization. The analyzer will no longer block all the light, so light reaches the user's eye. The user rotates the analyzer until no light passes through. The degree to which the analyzer is rotated can be read from the dial. This is called the **observed rotation** (α); it is measured in degrees.

Each optically active compound has a characteristic specific rotation. The **specific rotation** can be calculated from the observed rotation using the following formula:

$$[\alpha]_{\lambda}^{T} = \frac{\alpha}{l \times c}$$

where $[\alpha]$ is the specific rotation; T is temperature in °C; λ is the wavelength of the incident light (when the sodium D-line is used, λ is indicated as D); α is the observed rotation; l is the length of the sample tube in decimeters; and c is the concentration of the sample in grams per milliliter of solution.

If one enantiomer has a specific rotation of +5.75°, the specific rotation of the other enantiomer must be −5.75° because the mirror image rotates the plane of polarization the same amount but in the opposite direction. The specific rotations of some common compounds are shown in Table 8.1.

$$[\alpha]_{D}^{20\,°C} = +5.75°$$ $$[\alpha]_{D}^{20\,°C} = -5.75°$$

Table 8.1	Specific Rotation of Some Naturally Occurring Compounds
Cholesterol	−31.5°
Cocaine	−16°
Codeine	−136°
Morphine	−132°
Penicillin V	+233°
Progesterone (female sex hormone)	+172°
Sucrose (table sugar)	+66.5°
Testosterone (male sex hormone)	+109°

A mixture of equal amounts of two enantiomers—such as (R)-(−)-lactic acid and (S)-(+)-lactic acid—is called a **racemic mixture**. Racemic mixtures are optically inactive because for every molecule in a racemic mixture that rotates the plane of polarization in one direction, there is a mirror-image molecule that rotates the plane in the opposite direction. As a result, the light emerges from a racemic mixture with its plane of polarization unchanged. The symbol ($\pm$) is used to specify a racemic mixture. Thus, ($\pm$)-2-bromobutane indicates a mixture of (+)-2-bromobutane and an equal amount of ($-$)-2-bromobutane.

PROBLEM 16◆

The observed rotation of 2.0 g of a compound in 50 mL of solution in a polarimeter tube 20-cm long is +13.4°. What is the specific rotation of the compound?

PROBLEM 17◆

(S)-(+)-Monosodium glutamate (MSG) is a flavor enhancer used in many foods. Some people have an allergic reaction to MSG (headache, chest pain, and an overall feeling of weakness). "Fast food" often contains substantial amounts of MSG, and it is widely used in Chinese food as well. MSG has a specific rotation of +24°.

$$COO^- Na^+$$
$$C\text{''''}H$$
$$^-OOCCH_2CH_2 \quad NH_3$$
$$+$$

(S)-(+)-monosodium glutamate

a. What is the specific rotation of (R)-($-$)-monosodium glutamate?
b. What is the specific rotation of a racemic mixture of MSG?

PROBLEM 18◆

Naproxen, a nonsteroidal anti-inflammatory drug, is the active ingredient in Aleve. Naproxen has a specific rotation of +66°. Does naproxen have the R or the S configuration?

8.9   Isomers with More than One Asymmetric Center

Many organic compounds have more than one asymmetric center. The more asymmetric centers a compound has, the more stereoisomers are possible for the compound. If we know how many asymmetric centers a compound has, we can calculate the maximum number of stereoisomers for that compound: *a compound can have a maximum of 2^n stereoisomers, where* n *equals the number of asymmetric centers*. For example, the amino acid threonine has two asymmetric centers. Therefore, it can have a maximum of four ($2^2 = 4$) stereoisomers.

$$CH_3\overset{*}{C}H - \overset{*}{C}HCOO^-$$
$$OH \quad ^+NH_3$$
threonine

The four stereoisomers of threonine consist of two pairs of enantiomers. Stereoisomers **1** and **2** are nonsuperimposable mirror images. They, therefore, are enantiomers. Stereoisomers **3** and **4** are also enantiomers. Stereoisomers **1** and **3** are not identical, and they are not mirror images. Such stereoisomers are called *diastereomers*. **Diastereomers** are stereoisomers that are not enantiomers. Stereoisomers **1** and **4**, **2** and **3**, and **2** and **4** are also diastereomers. Notice that in diastereomers the configuration of one of the asymmetric centers is the same in both stereoisomers but the configuration of the other asymmetric center is different in the two stereoisomers.

Diastereomers are stereoisomers that are not enantiomers.

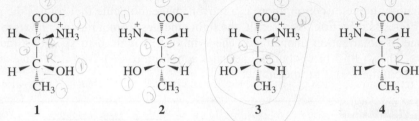

1   2   3   4

Enantiomers have *identical physical properties* (except for the way they interact with polarized light) and *identical chemical properties*—they react at the same rate with a given achiral reagent. Diastereomers have *different physical properties* (different melting points, different boiling points, different solubilities, different specific rotations, and so on) and *different chemical properties*—they react with the same achiral reagent at different rates.

PROBLEM 19◆

a. Stereoisomers with two asymmetric centers are called _____ if the configuration of both asymmetric centers in one isomer is the opposite of the configuration of the asymmetric centers in the other isomer.

b. Stereoisomers with two asymmetric centers are called _____ if the configuration of both asymmetric centers in one isomer is the same as the configuration of the asymmetric centers in the other isomer.

c. Stereoisomers with two asymmetric centers are called _____ if one of the asymmetric centers has the same configuration in both isomers and the other asymmetric center has the opposite configuration in the two isomers.

PROBLEM 20◆

a. How many asymmetric centers does cholesterol have?

b. What is the maximum number of stereoisomers that cholesterol can have? (Only one of these is found in nature.)

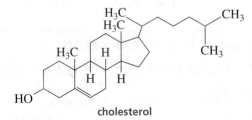

cholesterol

PROBLEM 21

Draw the stereoisomers of the following amino acids. Indicate pairs of enantiomers and pairs of diastereomers.

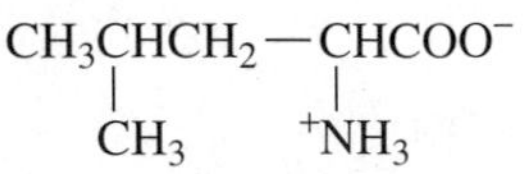

$$CH_3CHCH_2—CHCOO^-$$
$$\qquad CH_3 \qquad ^+NH_3$$
leucine

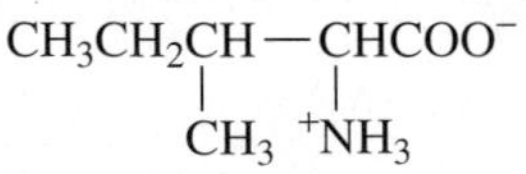

$$CH_3CH_2CH—CHCOO^-$$
$$\qquad CH_3 \ ^+NH_3$$
isoleucine

PROBLEM 22◆

Draw a diastereomer of the following compound:

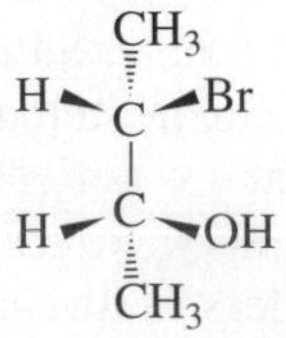

$$CH_3$$
$$H—C—Br$$
$$H—C—OH$$
$$CH_3$$

8.10 **Meso Compounds**

In the examples we have just seen, compounds with two asymmetric centers have four stereoisomers. However, some compounds with two asymmetric centers have only three stereoisomers. This is why we emphasized in Section 8.9 that the *maximum* number of stereoisomers a compound with n asymmetric centers can have is 2^n, instead of stating that a compound with n asymmetric centers has 2^n stereoisomers.

An example of a compound with two asymmetric centers that has only three stereoisomers is 2,3-dibromobutane.

$$CH_3CHCHCH_3$$
$$\quad | \quad |$$
$$\quad Br \ Br$$

2,3-dibromobutane

The "missing" isomer is the mirror image of **1**, because **1** and its mirror image are the same molecule. You can see that **1** and its mirror image are identical if you rotate the mirror image 180°.

superimposable mirror image

Stereoisomer **1** is called a *meso compound*. Even though a **meso** (mee-zo) **compound** has asymmetric centers, it is an achiral molecule—it does not rotate plane-polarized light—because it is superimposable on its mirror image.

A meso compound can be recognized by the fact that it has two (or more) asymmetric centers and a plane of symmetry. A plane of symmetry cuts the molecule in half, and one-half is the mirror image of the other half. Stereoisomer **1** has a **plane of symmetry**, which means that it does not have an enantiomer. Compare this with stereoisomer **2**, which does not have a plane of symmetry. Since it does not have a plane of symmetry, stereoisomer **2** has an enantiomer.

plane of symmetry — stereoisomer 1

It is easy to recognize when a compound with two asymmetric centers has a stereoisomer that is a meso compound—the four atoms or groups bonded to one asymmetric center are identical to the four atoms or groups bonded to the other asymmetric center. *A compound with the same four atoms or groups bonded to two different asymmetric centers will have three stereoisomers: One will be a meso compound, and the other two will be enantiomers.*

A meso compound is achiral.

Movie:
Plane of symmetry

A meso compound has two or more asymmetric centers and a plane of symmetry.

If a compound has a plane of symmetry, it is achiral.

If a compound with two asymmetric centers has the same four groups bonded to each of the asymmetric centers, one of its stereoisomers will be a meso compound.

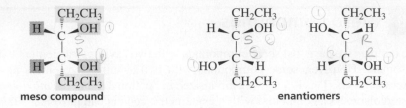

meso compound   enantiomers

Tutorial:
Identification of stereoisomers
with multiple asymmetric
centers

Tartaric acid has three stereoisomers because each of its two asymmetric centers has the same set of four substituents.

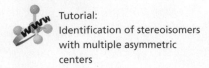

meso compound   enantiomers

The physical properties of the three stereoisomers of tartaric acid are listed in Table 8.2. The meso compound and either one of the enantiomers are diastereomers. Notice that the physical properties of the enantiomers are the same, whereas the physical properties of the diastereomers are different.

Table 8.2 Physical Properties of the Stereoisomers of Tartaric Acid			
	Melting point, °C	$[\alpha]_D^{25°C}$	**Solubility, g/100 g H_2O at 15°C**
(2R,3R)-(−)-Tartaric acid	170	+11.98°	139
(2S,3S)-(−)-Tartaric acid	170	−11.98°	139
(2R,3S)-Tartaric acid	140	0°	125
(±)-Tartaric acid	206	0°	139

PROBLEM-SOLVING STRATEGY

Which of the following compounds has a stereoisomer that is a meso compound?

    A = 2,3-dimethylbutane   C = 2-bromo-3-methylpentane

    B = 3,4-dimethylhexane   D = 3,4-diethylhexane

Check each compound to see if it has the necessary requirements to have a stereoisomer that is a meso compound. That is, does it have two asymmetric centers with the same four substituents attached to each of them?

Compounds A and D do *not* have a stereoisomer that is a meso compound because they don't have any asymmetric centers.

$$CH_3CHCHCH_3$$ with CH₃ groups above and below (A)

$$CH_3CH_2CHCHCH_2CH_3$$ with CH₂CH₃ groups above and below (D)

A   D

Compound C has two asymmetric centers. It does *not* have a stereoisomer that is a meso compound because each of the asymmetric centers is *not* bonded to the same four substituents.

$$Br$$
$$CH_3CHCHCH_2CH_3$$
$$CH_3$$
C

Compound B has two asymmetric centers and each asymmetric center is bonded to the same four atoms or groups. Therefore, compound B has a stereoisomer that is a meso compound.

$$CH_3$$
$$CH_3CH_2CHCHCH_2CH_3$$
$$CH_3$$
B

The isomer that is the meso compound is the one with a plane of symmetry.

$$CH_2CH_3$$
$$H \sim C \sim CH_3$$
$$H \sim C \sim CH_3$$
$$CH_2CH_3$$

Now continue on to Problem 23.

PROBLEM 23

Which of the following compounds has a stereoisomer that is a meso compound?

A 2,4-dibromohexane   **B** 2,4-dibromopentane   **C** 2,4-dimethylpentane

PROBLEM 24

Draw all the stereoisomers for each of the following compounds:

a. 1-bromo-2-methylbutane   **c.** 2,3-dichloropentane

b. 3-chloro-3-methylpentane   **d.** 1,3-dichloropentane

8.11   Separating Enantiomers   *NO → racémica*

Enantiomers cannot be separated by the usual separation techniques such as distillation or crystallization because their identical boiling points and solubilities cause them to distill or crystallize simultaneously. Louis Pasteur was the first to separate a pair of enantiomers successfully. While working with crystals of sodium ammonium tartrate, he noted that the crystals were not identical—some of the crystals were "right-handed" and some were "left-handed." He painstakingly separated the two kinds of crystals with a pair of tweezers. He found that a solution of the right-handed crystals rotated the plane of polarization clockwise, whereas a solution of the left-handed crystals rotated the plane of polarization counterclockwise.

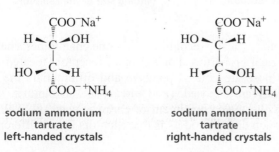

sodium ammonium
tartrate
left-handed crystals

sodium ammonium
tartrate
right-handed crystals

The French chemist and microbiologist **Louis Pasteur (1822–1895)** *was the first to demonstrate that microbes cause specific diseases. Asked by the French wine industry to find out why wine often went sour while aging, he showed that microorganisms cause grape juice to ferment, producing wine, and cause wine to slowly become sour. Gently heating the wine after fermentation, a process called pasteurization, kills the organisms so they cannot sour the wine.*

Eilhardt Mitscherlich (1794–1863), *a German chemist, first studied medicine so he could travel to Asia—a way to satisfy his interest in Oriental languages. He later became fascinated by chemistry. He was a professor of chemistry at the University of Berlin and wrote a successful chemistry textbook that was published in 1829.*

Pasteur was only 26 years old at the time and was unknown in scientific circles. He was concerned about the accuracy of his observations because a few years earlier, the well-known German organic chemist Eilhardt Mitscherlich had reported that crystals of the same salt were all identical. Pasteur immediately reported his findings to Jean-Baptiste Biot (Section 8.7) and repeated the experiment with Biot present. Biot was convinced that Pasteur had successfully separated the enantiomers of sodium ammonium tartrate. Pasteur's experiment also created a new chemical term. Tartaric acid is obtained from grapes, so it was also called racemic acid (*racemus* is Latin for "a bunch of grapes"). That is how a mixture of equal amounts of two enantiomers came to be known as a **racemic mixture**. (Section 8.8)

Later, chemists recognized how lucky Pasteur had been. Sodium ammonium tartrate forms asymmetric crystals only under certain conditions—precisely the conditions that Pasteur had employed. Under other conditions, the symmetrical crystals that had fooled Mitscherlich are formed. But to quote Pasteur, "Chance favors the prepared mind."

Separating enantiomers by hand, as Pasteur did, is not a universally useful method to separate enantiomers because few compounds form asymmetric crystals. A technique called **chromatography** is a more commonly used method. In this method, the mixture to be separated is dissolved in a solvent and the solution is passed through a column packed with material that tends to absorb organic compounds. If the chromatographic column is packed with *chiral* material, the two enantiomers can be expected to move through the column at different rates because they will have different affinities for the chiral material—just as a right hand prefers a right-hand glove—so one enantiomer will emerge from the column before the other.

8.12 Receptors

A **receptor** is a protein that binds a particular molecule. Because a receptor is chiral, it will bind one enantiomer better than the other. In Figure 8.3, the receptor binds the *R* enantiomer but it does not bind the *S* enantiomer.

Figure 8.3 ▶
Schematic diagram showing why only one enantiomer is bound by a receptor. One enantiomer fits into the binding site and one does not.

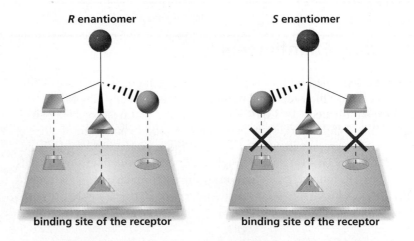

Because a receptor typically recognizes only one enantiomer, enantiomers can have different physiological properties. For example, receptors located on the exterior of nerve cells in the nose are able to perceive and differentiate the estimated 10,000 smells to which they are exposed. (*R*)-(−)-carvone is found in spearmint oil, and (*S*)-(+)-carvone is the main constituent of caraway seed oil. The reason these two enantiomers have such different odors is that each fits into a different receptor.

(R)-(−)-carvone
spearmint oil

(S)-(+)-carvone
caraway seed oil

$[\alpha]_D^{20\,°C} = -62.5°$

$[\alpha]_D^{20\,°C} = +62.5°$

Many drugs exert their physiological activity by binding to cellular receptors. If the drug has an asymmetric center, the receptor can preferentially bind one of the enantiomers. Enantiomers of a drug can have the same physiological activities, different degrees of the same activity, or very different activities, depending on the drug.

THE ENANTIOMERS OF THALIDOMIDE

Thalidomide was developed in West Germany. In 1957, it was first marketed for insomnia and morning sickness. Although it was available in more than 40 countries, it was not approved for use in the United States because the Federal Drug Administration had requested additional tests.

The dextrorotatory isomer has stronger sedative properties but the commercial drug was a racemic mixture. It was not recognized that the levorotatory isomer was highly teratogenic—it causes horrible birth defects—until it was noticed that women who were given the drug during the first three months of pregnancy gave birth to babies with a wide variety of defects—deformed limbs being the most common. About 10,000 children were damaged by the drug.

It was eventually determined that the dextrorotatory isomer also has mild teratogenic activity and that each of the enantiomers can racemize (interconvert) in vivo. Thus, it is not clear whether giving those women only the dextrorotatory isomer would have decreased the severity of the birth defects. Thalidomide recently has been approved—with restrictions—to treat leprosy as well as melanomas.

asymmetric center

thalidomide

CHIRAL DRUGS

Until relatively recently, most drugs have been marketed as racemic mixtures because of the high cost of separating the enantiomers. In 1992, the Food and Drug Administration (FDA) issued a policy statement encouraging drug companies to use recent advances in separation techniques to develop single enantiomer drugs. Now, one-third of all drugs sold are single enantiomers.

If a drug is sold as a racemate, the FDA requires that both enantiomers be tested, since the enantiomers can have similar or very different properties. The S isomer of Prozac, an antidepressant, is better at blocking serotonin; but it is used up faster than the R isomer. Testing has shown that the anesthetic (S)-(+)-ketamine is four times more potent than (R)-(−)-ketamine and the disturbing side effects appear to be associated only with the (R)-(−)-enantiomer. Only the S isomer of the beta-blocker propanolol shows activity; the R

isomer is inactive. The activity of ibuprofen, the popular analgesic marketed as Advil, Nuprin, and Motrin, resides primarily in the (S)-(+)-enantiomer. Heroin addicts can be maintained with (−)-α-acetylmethadol for a 72-hour period compared to 24 hours with racemic methadone. This means less frequent visits to the clinic, and a single dose can get an addict through an entire weekend.

Prescribing a single enantiomer prevents the patient from having to metabolize the less potent enantiomer and decreases the chance of unwanted drug interactions. Drugs that could not be given as racemates (because of the toxicity of one of the enantiomers) can now be used. For example, (S)-Penicillamine can be used to treat Wilson's disease even though (R)-Penicillamine causes blindness. Another reason for the increase in single-enantiomeric drugs is that drug companies may be able to extend their patents by developing a drug as a single enantiomer that was marketed previously as a racemate. (Section 22.10).

| 8.13 | **Stereochemistry of Reactions** |

In Chapter 5, when we looked at the electrophilic addition reactions of alkenes, we examined the step-by-step process by which each reaction occurs (the mechanism of the reaction), and we determined what products are formed. However, we did not consider the stereochemistry of the reactions.

Stereochemistry is the field of chemistry that deals with the structures of molecules in three dimensions. When we study the stereochemistry of a reaction, we are concerned with determining the stereoisomers that are formed in a reaction—does a reaction produce a single stereoisomer or all possible stereoisomers?

Stereochemistry of Electrophilic Addition Reactions of Alkenes

We have seen that when an alkene reacts with an electrophilic reagent such as HBr, the major product of the addition reaction is the one obtained by adding the electrophile (H^+) to the sp^2 carbon bonded to the greater number of hydrogens and adding the nucleophile (Br^-) to the other sp^2 carbon (Section 5.3). For example, the major product obtained from the reaction of propene with HBr is 2-bromopropane. This particular product does not have stereoisomers because it does not have an asymmetric center. Therefore, we do not have to be concerned with the stereochemistry of this reaction.

$$CH_3CH = CH_2 \xrightarrow{\ HBr\ } CH_3\overset{+}{C}HCH_3 \quad Br^- \longrightarrow CH_3CHCH_3$$

propene

2-bromopropane
major product

Tutorial:
Review addition of HBr
tutorial in Chapter 5

If, however, the reaction creates a product with an asymmetric center, we need to know which stereoisomers are formed. For example, the reaction of HBr with 1-butene forms 2-bromobutane, a compound with an asymmetric center. What stereoisomers of the product are obtained? Do we get the *R* enantiomer, the *S* enantiomer, or both?

asymmetric center

$$CH_3CH_2CH = CH_2 \xrightarrow{\ HBr\ } CH_3CH_2\overset{+}{C}HCH_3 \quad Br^- \longrightarrow CH_3CH_2CHCH_3$$

1-butene

2-bromobutane

When a reactant that does not have an asymmetric center undergoes a reaction that forms a product with *one* asymmetric center, the product will be a racemic mixture. For example, the reaction of 1-butene with HBr forms identical amounts of (*R*)-2-bromobutane and (*S*)-2-bromobutane.

You can understand why a racemic mixture is obtained if you examine the structure of the carbocation formed in the first step of the reaction. The positively charged carbon is sp^2 hybridized, so the three atoms to which it is bonded lie in a plane (Section 1.10). When the bromide ion approaches the positively charged carbon from above the plane, one enantiomer is formed, but when it approaches from below the plane, the other enantiomer is formed. Because the bromide ion has equal access to both sides of the plane, identical amounts of the *R* and *S* enantiomers are obtained from the reaction.

When a reactant that does not have an asymmetric center undergoes a reaction that forms a product with *one* asymmetric center, the product will be a racemic mixture.

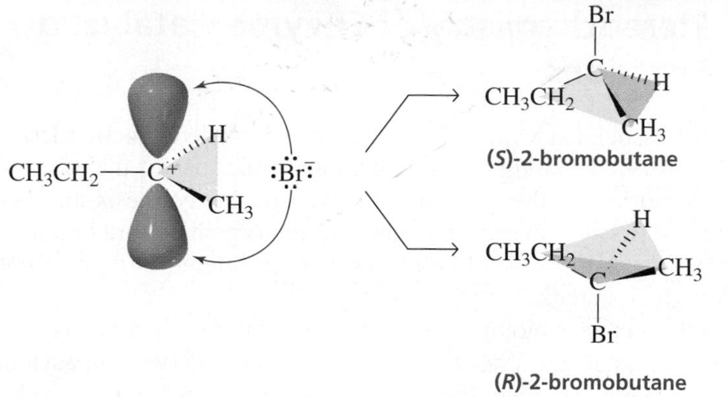

(S)-2-bromobutane

(R)-2-bromobutane

PROBLEM 25

What stereoisomers are obtained from each of the following reactions?

a. $CH_3CH_2CH_2CH=CH_2$ $\xrightarrow{HCl}$

b. 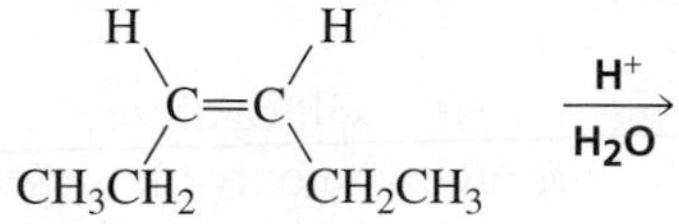 $\xrightarrow[H_2O]{H^+}$

c. $\xrightarrow{H_2 \atop Pt/C}$

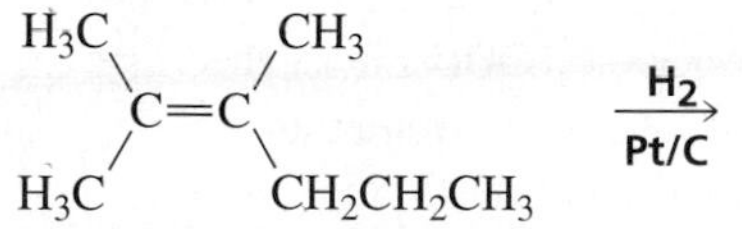

d. 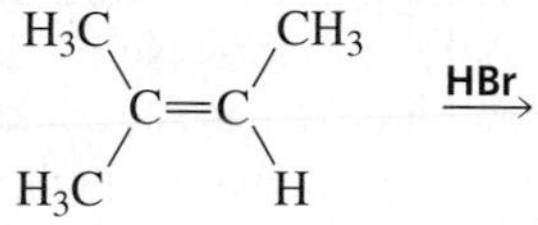 $\xrightarrow{HBr}$

Stereochemistry of Hydrogen Addition

We have seen that in catalytic hydrogenation, both hydrogens add to the same side of the double bond (Section 5.5). Thus, when the reactant is an alkyne, the product is a cis alkene.

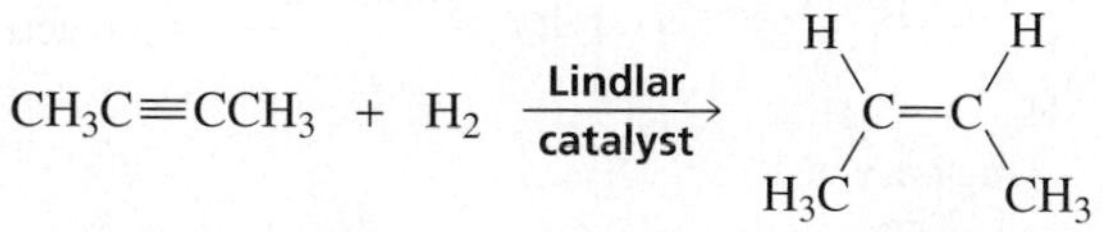

$$CH_3C\equiv CCH_3 + H_2 \xrightarrow[\text{catalyst}]{\text{Lindlar}}$$

cis-2-butene

When the reactant is a cyclic alkene, both hydrogens add to the same side of the ring, so only the cis isomer is formed.

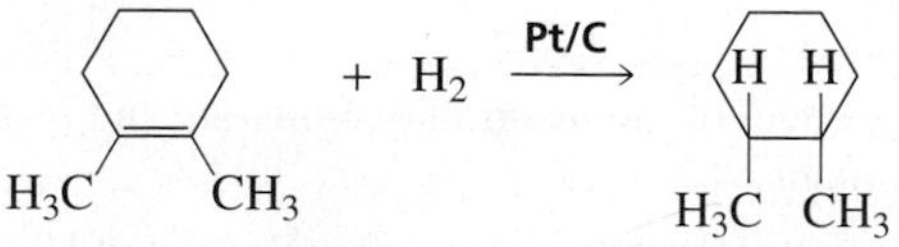

PROBLEM 26◆

Give the configuration of the products obtained from the following reactions:

a. *trans*-2-butene + H_2O + H^+

b. *cis*-3-hexene + HBr

c. 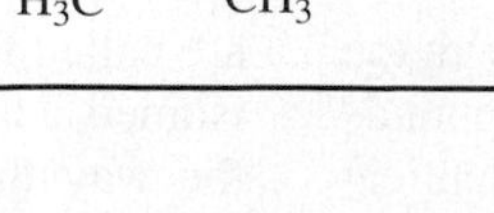 $\xrightarrow{H_2 \atop Pt/C}$

8.14 Stereochemistry of Enzyme-Catalyzed Reactions

*For studies on the stereochemistry of enzyme-catalyzed reactions, **Sir John Cornforth** received the Nobel Prize in chemistry in 1975. Born in Australia in 1917, he studied at the University of Sydney and received a Ph.D. from Oxford. His major research was carried out in laboratories at Britain's Medical Research Council and at Shell Research Ltd. He was knighted in 1977.*

*Fundamental work on the stereochemistry of enzyme-catalyzed reactions was done by **Frank H. Westheimer**. Westheimer was born in Baltimore in 1912 and received his graduate training at Harvard University. He served on the faculty of the University of Chicago and subsequently returned to Harvard as a professor of chemistry.*

The chemistry associated with living organisms is called **biochemistry**. When you study biochemistry, you study the structures and functions of the molecules found in the biological world and the reactions involved in the synthesis and degradation of these molecules. Because the compounds in living organisms are organic compounds, it is not surprising that many of the reactions encountered in organic chemistry also occur in biological systems.

Reactions that occur in biological systems are catalyzed by proteins called **enzymes**. When an enzyme catalyzes a reaction that forms a product with an asymmetric center, only one stereoisomer is formed. For example, the enzyme fumarase, which catalyzes the addition of water to fumarate, forms only (*S*)-malate—the *R* enantiomer is not formed.

fumarate + H_2O $\xrightarrow{\text{fumarase}}$ (*S*)-malate

An enzyme-catalyzed reaction forms only one stereoisomer because an enzyme is chiral. Its chiral binding site restricts delivery of reagents to only one side of the reactant. Consequently, only one stereoisomer is formed.

If the reactant can exist as stereoisomers, an enzyme typically will catalyze the reaction of only one of the stereoisomers, because the enzyme will bind only the stereoisomer whose substituents are in the correct positions to interact with substituents in the chiral binding site (Figure 8.3). Other stereoisomers do not have substituents in the proper positions, so they cannot bind efficiently to the enzyme. For example, fumarase catalyzes the addition of water to fumarate (the trans isomer) but not to maleate (the cis isomer).

maleate + H_2O $\xrightarrow{\text{fumarase}}$ no reaction

An achiral molecule reacts identically with both enantiomers. A sock, which is achiral, fits on either foot.

A chiral molecule reacts differently with each enantiomer. A shoe, which is chiral, fits on only one foot.

An enzyme's behavior can be likened to a right-handed glove, which fits only the right hand: It forms only one stereoisomer and it reacts with only one stereoisomer.

PROBLEM 27◆

a. What would be the product of the reaction of fumarate and H_2O if H^+ were used as a catalyst instead of fumarase?

b. What would be the product of the reaction of maleate and H_2O if H^+ were used as a catalyst instead of fumarase?

Summary

Stereochemistry is the field of chemistry that deals with the structures of molecules in three dimensions. Compounds that have the same molecular formula but are not identical are called **isomers**; they fall into two classes: constitutional isomers and stereoisomers. **Constitutional isomers** differ in the way their atoms are connected. **Stereoisomers** differ in the way their atoms are arranged in space. There are two kinds of stereoisomers: **cis–trans isomers** and isomers that contain **asymmetric centers**.

A **chiral** molecule has a nonsuperimposable mirror image. An **achiral** molecule has a superimposable mirror image. The feature that is most often the cause of chirality is an asymmetric center. An **asymmetric center** is a carbon bonded to four different atoms or groups. Nonsuperimposable mirror-image molecules are called **enantiomers**. **Diastereomers** are stereoisomers that are not enantiomers. Enantiomers have identical physical and chemical properties; diastereomers have different physical and chemical properties. An achiral molecule reacts identically with both enantiomers; a chiral molecule reacts differently with each enantiomer. A mixture of equal amounts of two enantiomers is called a **racemic mixture**.

The letters **R** and **S** indicate the **configuration** about an asymmetric center. If one stereoisomer has the R and the other has the S configuration, they are enantiomers; if they both have the R or both have the S configuration, they are identical.

Chiral compounds are **optically active**—they rotate the plane of polarized light; achiral compounds are **optically inactive**. If one enantiomer rotates the plane of polarization clockwise (+), its mirror image will rotate it the same amount counterclockwise (−). Each optically active compound has a characteristic **specific rotation**. A **racemic mixture** consists of equal amounts of two enantiomers; it is optically inactive. A **meso compound** has two or more asymmetric centers and a plane of symmetry; it is an achiral molecule. A compound with the same four groups bonded to two different asymmetric centers will have three stereoisomers, a meso compound, and a pair of enantiomers.

When a reactant that does not have an asymmetric center forms a product with one asymmetric center, the product will be a racemic mixture. Catalytic hydrogenation adds both hydrogens to the same side of the reactant. An enzyme-catalyzed reaction forms only one stereoisomer; an enzyme typically catalyzes the reaction of only one stereoisomer.

Problems

28. Neglecting stereoisomers, give the structures of all compounds with molecular formula C_5H_{10}. Which ones can exist as stereoisomers?

29. Draw all possible stereoisomers for each of the following compounds. State if no stereoisomers are possible.
  a. 2-bromo-4-methylpentane
  b. 2-bromo-4-chloropentane
  c. 3-heptene
  d. 1-bromo-4-methylcyclohexane
  e. 1-bromo-3-chlorocyclobutane
  f. 2-iodopentane
  g. 3,3-dimethylpentane
  h. 3-chloro-1-butene

30. Name the following compounds using *R,S* and *E,Z* (Section 4.5) designations where necessary:

a.
$$\underset{Cl}{\overset{F}{}}C=C\underset{Br}{\overset{I}{}}$$

b.
$$\underset{H_3C}{\overset{CH(CH_3)_2}{}}C\overset{CH_2CH_3}{\underset{H}{}}$$

c.
$$\underset{Cl}{\overset{F}{}}C=C\underset{Br}{\overset{H}{}}$$

d.
$$\underset{CH_3CH_2}{\overset{CH_3}{}}C\overset{CH_2CH_2Cl}{\underset{H}{}}$$

31. Mevacor is used clinically to lower serum cholesterol levels. How many asymmetric centers does Mevacor have?

Mevacor®

32. Indicate whether each of the following pairs of compounds are identical or are enantiomers, diastereomers, or constitutional isomers:

a. 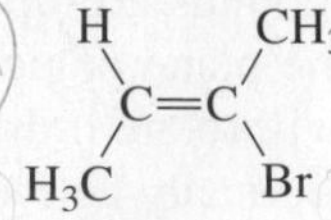 and 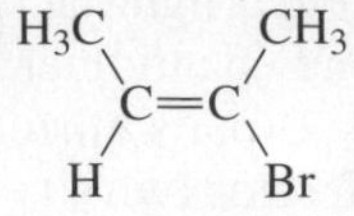

b. 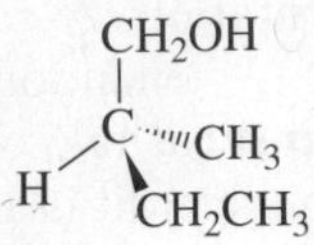 and 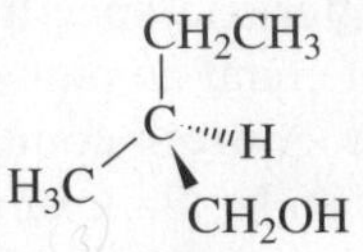

c. 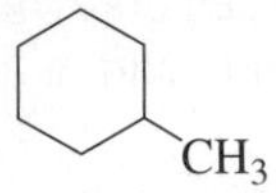 and 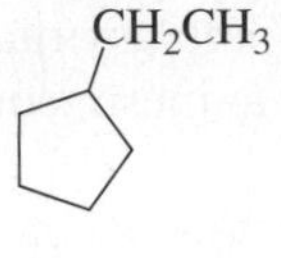

d. 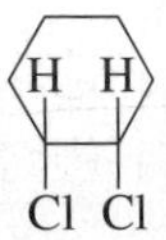 and 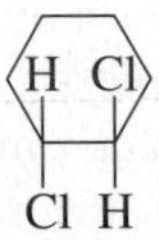

e. 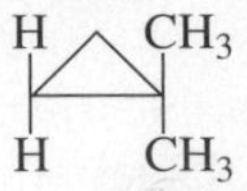 and 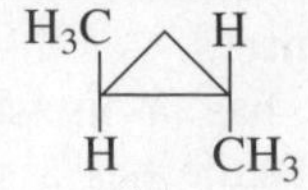

f. 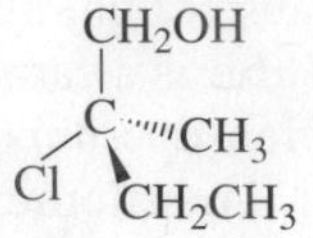 and 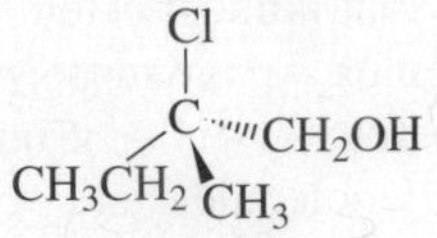

g. 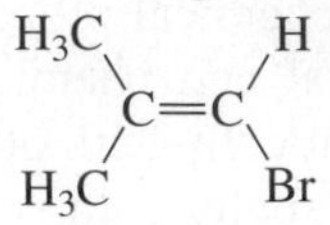 and 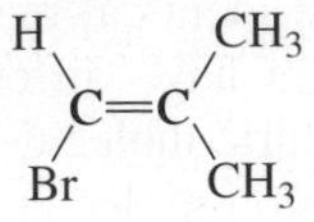

h. 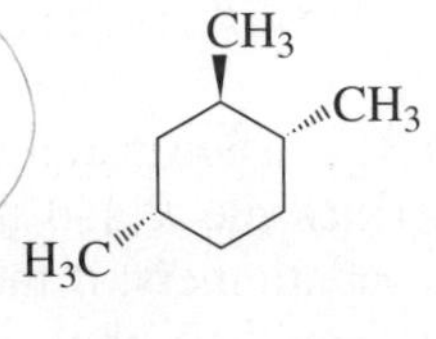 and 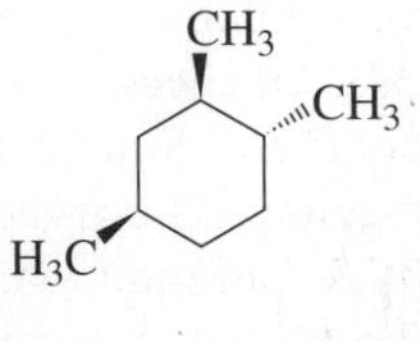

33. a. Give the product(s) that would be obtained from the reaction of *cis*-2-butene with each of the following reagents. If the products can exist as stereoisomers, show which stereoisomers are obtained.

1. HCl 2. HCl + H_2O 3. $H_2/Pt/C$ 4. HCl + CH_3OH

b. How would the product differ if the reactant had been *trans*-2-butene?

34. Which of the following compounds have an achiral stereoisomer?

a. 2,3-dichlorobutane **b.** 2,3-dichloropentane **c.** 2,4-dibromopentane **d.** 2,3-dibromopentane

35. Draw the stereoisomers of 2,4-dichlorohexane. Indicate pairs of enantiomers and pairs of diastereomers.

36. Citrate synthase, one of the enzymes in the series of enzyme-catalyzed reactions known as the Krebs cycle, catalyzes the synthesis of citric acid from oxaloacetic acid and acetyl-CoA. If the synthesis is carried out with acetyl-CoA that has radioactive carbon (^{14}C) in the indicated position, the isomer shown here is obtained. (Note: ^{14}C has a higher priority than ^{12}C.)

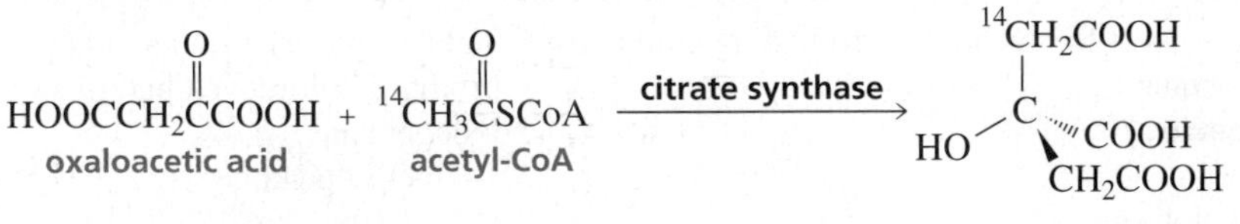
HOOCCH2CCOOH
oxaloacetic acid
14CH3CSCoA
acetyl-CoA
citrate synthase
14CH2COOH
HO
COOH
CH2COOH
citric acid

a. Which stereoisomer of citric acid is synthesized, *R* or *S*?
b. Why is the other stereoisomer not obtained?
c. If the acetyl-CoA used in the synthesis does not contain ^{14}C, will the product of the reaction be chiral or achiral?

37. Give the products of the following reactions. If the products can exist as stereoisomers, show which stereoisomers are obtained.

a. 1-butene + HCl **b.** *cis*-2-pentene + HCl **c.** 1-ethylcyclohexene + H_3O^+

38. A solution of an unknown compound (3.0 g of the compound in 20 mL of solution), when placed in a polarimeter tube 2.0 dm long, was found to rotate the plane of polarized light 1.8° in a counterclockwise direction. What is the specific rotation of the compound?

39. Butaclamol is a potent antipsychotic that has been used clinically in the treatment of schizophrenia. How many asymmetric centers does Butaclamol have?

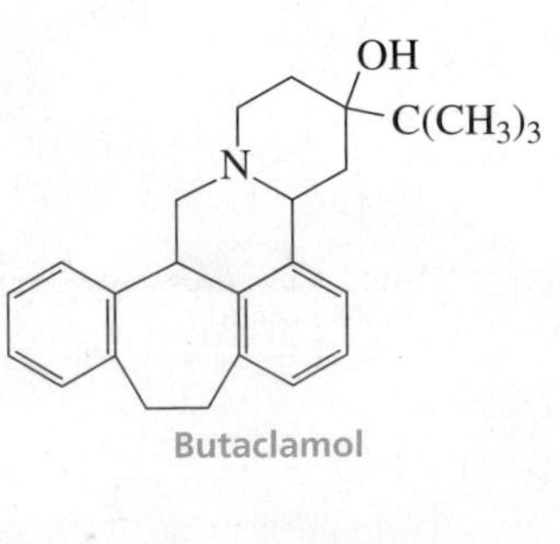
OH
C(CH3)3
N
Butaclamol

40. Which of the following objects are chiral?
  a. a mug with DAD written on one side
  b. a mug with MOM written on one side
  c. a mug with DAD written opposite the handle
  d. a mug with MOM written opposite the handle
  e. an automobile
  f. a wheelbarrow
  g. a nail
  h. a screw

41. Explain how R and S are related to $(+)$ and $(-)$.

42. Draw structures for each of the following:
  a. (S)-1-bromo-1-chlorobutane
  b. two achiral isomers of 3,4,5-trimethylheptane

43. Draw all the stereoisomers for each of the following compounds:
  a. 1-chloro-3-methylpentane
  b. 2,4-dichloroheptane
  c. 3,4-dichlorohexane

44. Indicate the configuration of the asymmetric centers in the following molecules:

  a.

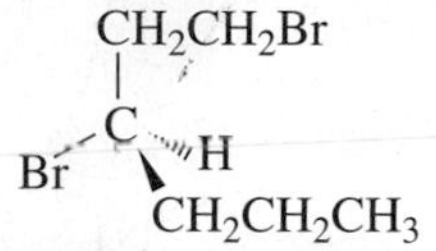

$$\begin{array}{c} CH_2CH_2Br \\ | \\ C \\ Br \quad \text{''''H} \\ CH_2CH_2CH_3 \end{array}$$

  b.

$$\begin{array}{c} H_3C \qquad CH_2CH_3 \\ \diagdown \qquad \diagup \\ C - C \text{—H} \\ Br \text{''''} \diagup \quad \diagdown \\ H \qquad Br \end{array}$$

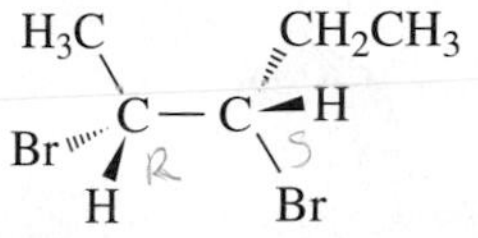

45. Two stereoisomers are obtained from the reaction of HBr with (S)-4-bromo-1-pentene. One of the stereoisomers is optically active, and the other is not. Give the structures of the stereoisomers, indicating their configurations, and explain the difference in the optical properties.

46. Chloramphenicol is a broad-spectrum antibiotic that is particularly useful against typhoid fever. What is the configuration of each asymmetric center in chloramphenicol?

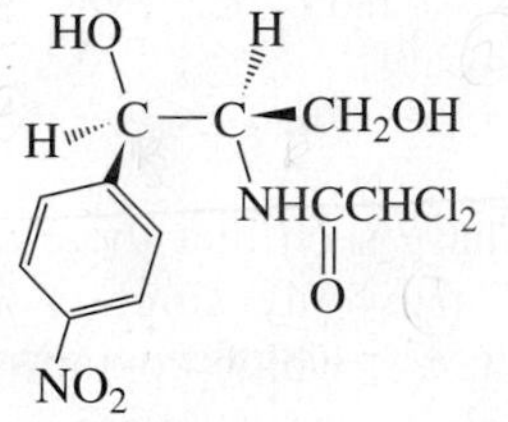

$$\begin{array}{c} HO \qquad H \\ | \qquad | \\ H\text{''''}C - C\text{—}CH_2OH \\ \\ \qquad NHCCHCl_2 \\ \qquad \| \\ \qquad O \\ \\ NO_2 \end{array}$$

chloramphenicol

47. For many centuries, the Chinese have used extracts from a group of herbs known as ephedra to treat asthma. Chemists have been able to isolate a compound from these herbs, which they named ephedrine, a potent dilator of air passages in the lungs.

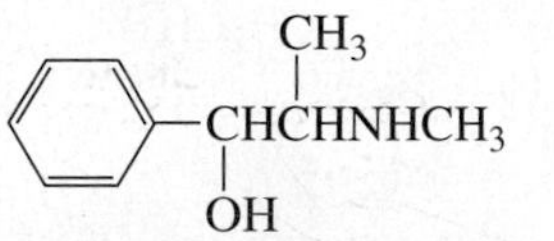

$$\begin{array}{c} CH_3 \\ | \\ -CHCHNHCH_3 \\ | \\ OH \end{array}$$

ephedrine

  a. How many stereoisomers are possible for ephedrine?
  b. The stereoisomer shown here is the one that is pharmacologically active. What is the configuration of each of the asymmetric centers?

$$\begin{array}{c} H \\ | \\ C - C\text{—}NHCH_3 \\ HO\text{''''} \diagup \quad \diagdown \\ H \qquad CH_3 \end{array}$$

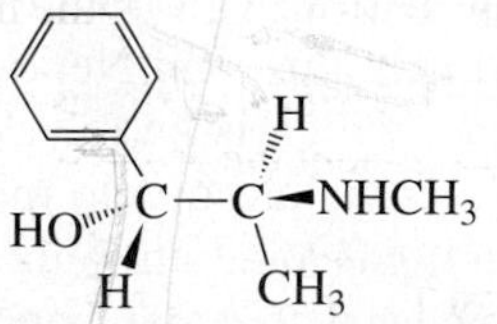

48. Indicate whether each of the following pairs of compounds are identical or are enantiomers:

  a.

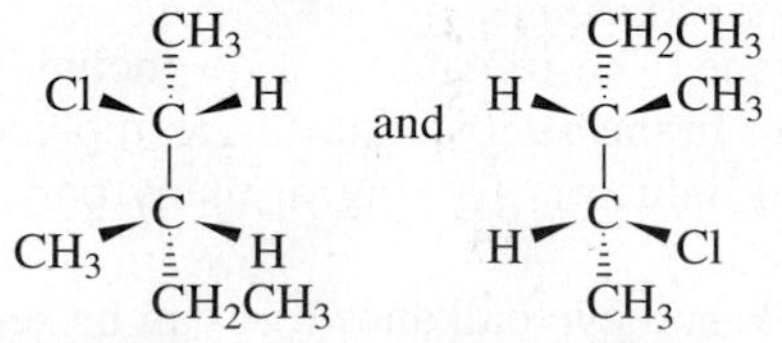

$$\begin{array}{c} CH_3 \\ | \\ Cl\diagdown C \diagup H \\ | \\ CH_3 \diagup C \diagdown H \\ CH_2CH_3 \end{array} \quad \text{and} \quad \begin{array}{c} CH_2CH_3 \\ | \\ H\diagdown C \diagup CH_3 \\ | \\ H \diagup C \diagdown Cl \\ CH_3 \end{array}$$

  b.

$$\begin{array}{c} CH_3 \\ | \\ HO\diagdown C \diagup H \\ | \\ H \diagup C \diagdown Cl \\ CH_3 \end{array} \quad \text{and} \quad \begin{array}{c} CH_3 \\ | \\ H\diagdown C \diagup OH \\ | \\ Cl \diagup C \diagdown H \\ CH_3 \end{array}$$

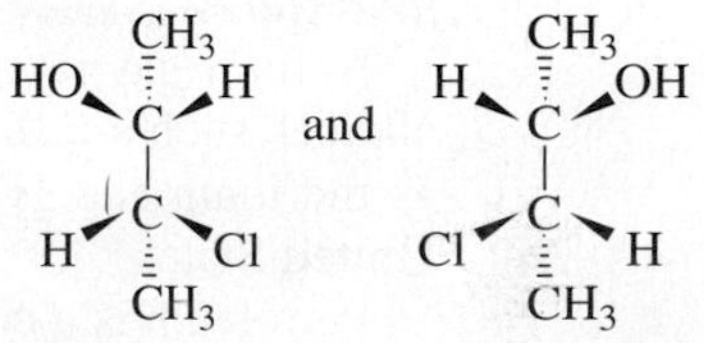

49. The following compound has only one asymmetric center. Why, then, does it have four stereoisomers?

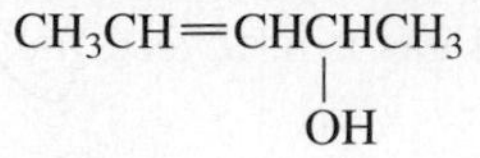

$$\begin{array}{c} CH_3CH \!=\! CHCHCH_3 \\ | \\ OH \end{array}$$

9 Reactions of Alkanes • Radicals

vitamin C   vitamin E

We have seen that there are three classes of hydrocarbons: *alkanes*, which contain only single bonds; *alkenes*, which contain double bonds; and *alkynes*, which contain triple bonds. Because **alkanes** do not contain any double or triple bonds, they are called **saturated hydrocarbons**—they are saturated with hydrogen. A few examples of alkanes are shown here.

CH₃CH₂CH₂CH₃
butane

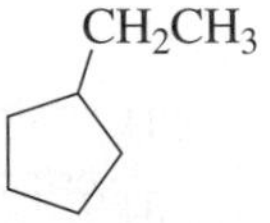

CH₂CH₃

ethylcyclopentane

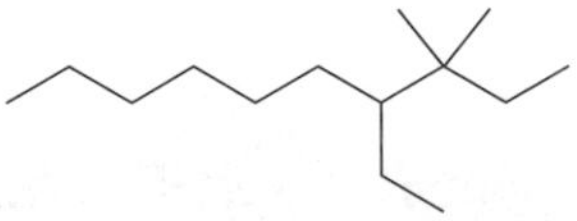

4-ethyl-3,3-dimethyldecane

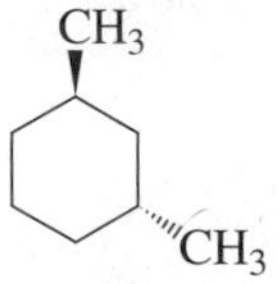

CH₃

CH₃

trans-1,3-dimethyl-cyclohexane

Alkanes are widespread both on Earth and on other planets. The atmospheres of Jupiter, Saturn, Uranus, and Neptune contain large quantities of methane (CH_4), the smallest alkane, which is an odorless and flammable gas. In fact, the blue colors of Uranus and Neptune are due to methane in their atmospheres. Alkanes on Earth are found in natural gas and petroleum, which are formed by the decomposition of plant and animal material that has been buried for long periods in the Earth's crust, an environment with little oxygen. Natural gas and petroleum, therefore, are known as *fossil fuels*.

Natural gas is approximately 75% methane. The remaining 25% includes small alkanes such as ethane, propane, and butane. In the 1950s, natural gas replaced coal as the main energy source for domestic and industrial heating in many parts of the United States.

Petroleum is a complex mixture of alkanes and cycloalkanes that can be separated into fractions by distillation. The fraction that boils at the lowest temperature (hydrocarbons containing three and four carbons) is a gas that can be liquefied under pressure. This gas is used as a fuel for cigarette lighters, camp stoves, and barbecues.

The fraction that boils at somewhat higher temperatures (hydrocarbons containing 5 to 11 carbons) is gasoline; the next fraction (9 to 16 carbons) includes kerosene and jet fuel. The fraction with 15 to 25 carbons is used for heating oil and diesel oil, and the highest boiling fraction is used for lubricants and greases. The nonpolar nature of these compounds gives them their oily feel. After distillation, a nonvolatile residue called asphalt or tar is left behind.

The 5- to 11-carbon fraction that is used for gasoline is, in fact, a poor fuel for internal combustion engines; it requires a process known as catalytic cracking to become a high-performance gasoline. Catalytic cracking converts straight-chain hydrocarbons that are poor fuels into branched-chain compounds that are high-performance fuels. Originally, cracking involved heating gasoline to very high temperatures in order to obtain hydrocarbons with three to five carbons. Modern cracking methods use catalysts to accomplish the same result at much lower temperatures.

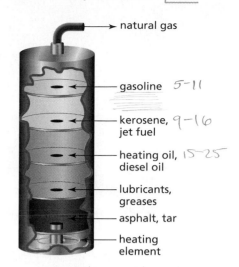

natural gas

gasoline 5–11

kerosene, 9–16
jet fuel

heating oil, 15–25
diesel oil

lubricants,
greases

asphalt, tar

heating
element

OCTANE NUMBER

When poor fuels are used in an engine, combustion can be initiated before the spark plug fires. A pinging or knocking may then be heard in the running engine. As the quality of the fuel improves, the engine is less likely to knock. The quality of a fuel is indicated by its octane number. Straight-chain hydrocarbons have low octane numbers and make poor fuels. Heptane, for example, with an arbitrarily assigned octane number of 0, causes engines to knock badly. Branched-chain fuels burn more slowly—thereby reducing knocking—because they have more hydrogens on primary carbons and these are the C—H bonds that require the most energy to break (Section 9.2). Consequently, branched-chain alkanes have high octane numbers. 2,2,4-Trimethylpentane, for example, does not cause knocking and has arbitrarily been assigned an octane number of 100.

$$CH_3CH_2CH_2CH_2CH_2CH_2CH_3$$

heptane
octane number = 0

$$\begin{array}{cc} & CH_3 \quad CH_3 \\ & | \qquad | \\ CH_3CCH_2CHCH_3 \\ & | \\ & CH_3 \end{array}$$

2,2,4-trimethylpentane
octane number = 100

The octane number of a gasoline is determined by comparing its knocking with the knocking of mixtures of heptane and 2,2,4-trimethylpentane. The octane number given to the gasoline corresponds to the percent of 2,2,4-trimethylpentane in the matching mixture. The term "octane number" originated from the fact that 2,2,4-trimethylpentane contains eight carbons.

FOSSIL FUELS: A PROBLEMATIC ENERGY SOURCE

We face three major problems as a consequence of our dependence on fossil fuels for energy. First, they are a nonrenewable resource and the world's supply is continually decreasing. Second, a group of Middle Eastern and South American countries controls a large portion of the world's supply of petroleum. These countries have formed a cartel known as the *Organization of Petroleum Exporting Countries* (*OPEC*), which controls both the supply and the price of crude oil. Political instability in any OPEC country can seriously affect the world oil supply. Third, burning fossil fuels increases the concentrations of CO_2 and SO_2 in the atmosphere. Scientists have established experimentally that atmospheric SO_2 causes "acid rain," a threat to the Earth's plants and, therefore, to our food and oxygen supplies (Section 2.2).

Since 1958, the concentration of atmospheric CO_2 at Mauna Loa, Hawaii has been carefully measured periodically. The concentration has increased 20% since the first measurements were taken, causing scientists to predict an increase in the Earth's temperature as a result of the absorption of infrared radiation by CO_2 (the *greenhouse effect*). A steady increase in the temperature of the Earth would have devastating consequences, including the formation of new deserts, massive crop failure, and the melting of polar ice caps with a concomitant rise in sea level. Clearly, what we need is a renewable, nonpolitical, nonpolluting, and economically affordable source of energy.

9.1 The Low Reactivity of Alkanes

The double and triple bonds of alkenes and alkynes are composed of strong σ bonds and weaker π bonds. We have seen that the reactivity of *alkenes* and *alkynes* is the result of an electrophile being attracted to the cloud of electrons that constitutes the π bond.

Alkanes are very unreactive compounds. They have only strong σ bonds. In addition, the electrons in the C—H and C—C σ bonds are shared equally by the bonding atoms; thus, none of the atoms in an alkane has any significant charge. This means that neither nucleophiles nor electrophiles are attracted to them. The failure of alkanes to undergo reactions prompted early organic chemists to call them **paraffins**, from the Latin *parum affinis*, which means "little affinity" (for other compounds).

9.2 Chlorination and Bromination of Alkanes

Alkanes do react with chlorine (Cl_2) or bromine (Br_2) to form alkyl chlorides or alkyl bromides. These **halogenation reactions** take place only at high temperatures or in the presence of light. (Irradiation with light is symbolized by $h\nu$.) They are the only reactions that alkanes undergo—with the exception of **combustion**, a reaction with oxygen that takes place at high temperatures and converts alkanes to carbon dioxide and water.

$$CH_4 \ + \ Cl_2 \ \xrightarrow[\substack{\text{or} \\ h\nu}]{\Delta} \ \underset{\text{methyl chloride}}{CH_3Cl} \ + \ HCl$$

$$CH_3CH_3 \ + \ Br_2 \ \xrightarrow[\substack{\text{or} \\ h\nu}]{\Delta} \ \underset{\text{ethyl bromide}}{CH_3CH_2Br} \ + \ HBr$$

The mechanism for the halogenation of an alkane is well understood. The high temperature (or light) supplies the energy required to break the Cl—Cl or Br—Br bond. When the bond breaks, each of the atoms retains one of the electrons. Notice that an arrowhead with one barb signifies the movement of one electron. Thus, chlorine (or bromine) *radicals* are formed. A **radical** (often called a **free radical**) is a species containing an atom with an unpaired electron.

$$:\!\ddot{C}l\!-\!\ddot{C}l\!: \ \xrightarrow[\substack{\text{or} \\ h\nu}]{\Delta} \ 2 \ :\!\ddot{C}l\!\cdot$$

$$:\!\ddot{B}r\!-\!\ddot{B}r\!: \ \xrightarrow[\substack{\text{or} \\ h\nu}]{\Delta} \ 2 \ :\!\ddot{B}r\!\cdot$$

Forming the chlorine (or bromine) radicals is the **initiation step** of the halogenation reaction. A radical is highly reactive because acquiring an electron will complete its octet. In the mechanism for the monochlorination of methane, the chlorine radical removes a hydrogen atom from methane, forming HCl and a methyl radical. The methyl radical removes a chlorine atom from Cl_2, forming methyl chloride and another chlorine radical, which can remove a hydrogen atom from another molecule of methane. These two steps are called **propagation steps** because the radical created in the first propagation step reacts in the second propagation step to produce a radical that can repeat the first propagation step. Thus, the two propagation steps are repeated over and over. The first propagation step is the rate-determining step of the overall reaction. Because the reaction has radical intermediates and repeating propagation steps, it is called a **radical chain reaction**.

mechanism for the monochlorination of methane

$$:\!\ddot{C}l\!-\!\ddot{C}l\!: \ \xrightarrow[\substack{\text{or} \\ h\nu}]{\Delta} \ 2 \ :\!\ddot{C}l\!\cdot \quad \boxed{\text{initiation step}}$$

The radical created in the second propagation step can repeat the first propagation step.

$$:\!\ddot{C}l\!\cdot \ + \ H\!-\!CH_3 \ \longrightarrow \ H\ddot{C}l\!: \ + \ \underset{\text{a methyl radical}}{\cdot CH_3}$$

$$\cdot CH_3 \ + \ :\!\ddot{C}l\!-\!\ddot{C}l\!: \ \longrightarrow \ CH_3Cl \ + \ :\!\ddot{C}l\!\cdot$$

$\left.\begin{array}{c} \\ \\ \end{array}\right\}$ $\boxed{\text{propagation steps}}$

$$:\overset{..}{\underset{..}{Cl}}\cdot \; + \; :\overset{..}{\underset{..}{Cl}}\cdot \longrightarrow \; Cl_2$$

$$\cdot CH_3 \; + \; \cdot CH_3 \longrightarrow \; CH_3CH_3 \left.\right\} \boxed{\text{termination steps}}$$

$$:\overset{..}{\underset{..}{Cl}}\cdot \; + \; \cdot CH_3 \longrightarrow \; CH_3Cl$$

3-D Molecule:
Methyl radical

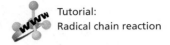

Tutorial:
Radical chain reaction

Any two radicals in the reaction mixture can combine to form a molecule in which all the electrons are paired. The combination of two radicals is called a **termination step** because it helps bring the reaction to an end by decreasing the number of radicals available to propagate the reaction. The radical chlorination of alkanes other than methane follows the same mechanism. A radical chain reaction, with its characteristic initiation, propagation, and termination steps, was first described in Section 5.16.

The reaction of an alkane with chlorine or bromine to form an alkyl halide is called a **radical substitution reaction** because *radicals* are involved as intermediates and the end result is the *substitution* of a halogen for one of the hydrogens of the alkane.

In order to maximize the amount of monohalogenated product obtained, a radical substitution reaction should be carried out in the presence of excess alkane. Excess alkane in the reaction mixture increases the probability that the halogen radical will collide with a molecule of alkane rather than with a molecule of alkyl halide—even toward the end of the reaction, by which time a considerable amount of alkyl halide will have been formed. If the halogen radical removes a hydrogen from a molecule of alkyl halide rather than from a molecule of alkane, a dihalogenated product will be obtained.

$$Cl\cdot \; + \; CH_3Cl \longrightarrow \; \cdot CH_2Cl \; + \; HCl$$

$$\cdot CH_2Cl \; + \; Cl_2 \longrightarrow \; CH_2Cl_2 \; + \; Cl\cdot$$

a dihalogenated compound

Bromination of alkanes follows the same mechanism as chlorination.

mechanism for the monobromination of ethane

$$Br\!-\!Br \; \xrightarrow[\substack{\text{or} \\ h\nu}]{\Delta} \; 2\,Br\cdot \qquad \boxed{\text{initiation step}}$$

$$Br\cdot \; + \; H\!-\!CH_2CH_3 \longrightarrow \; CH_3\dot{C}H_2 \; + \; HBr$$

$$CH_3\dot{C}H_2 \; + \; Br\!-\!Br \longrightarrow \; CH_3CH_2Br \; + \; Br\cdot \left.\right\} \boxed{\text{propagation steps}}$$

3-D Molecule:
Ethyl radical

$$Br\cdot \; + \; Br\cdot \longrightarrow \; Br_2$$

$$CH_3\dot{C}H_2 \; + \; CH_3\dot{C}H_2 \longrightarrow \; CH_3CH_2CH_2CH_3 \left.\right\} \boxed{\text{termination steps}}$$

$$CH_3\dot{C}H_2 \; + \; Br\cdot \longrightarrow \; CH_3CH_2Br$$

PROBLEM 1

Show the initiation, propagation, and termination steps for the monochlorination of cyclohexane.

PROBLEM 2

Write the mechanism for the formation of carbon tetrachloride, CCl_4, from the reaction of methane with $Cl_2 + h\nu$.

9.3 Factors that Determine Product Distribution

Two different alkyl halides are obtained from the monochlorination of butane. Substitution of a hydrogen bonded to one of the terminal carbons produces 1-chlorobutane, whereas substitution of a hydrogen bonded to one of the internal carbons forms 2-chlorobutane.

$$CH_3CH_2CH_2CH_3 + Cl_2 \xrightarrow{h\nu} CH_3CH_2CH_2CH_2Cl + CH_3CH_2\overset{\displaystyle Cl}{\overset{\displaystyle |}{C}}HCH_3 + HCl$$

butane   1-chlorobutane   2-chlorobutane
expected = 60%   expected = 40%
experimental = 29%   experimental = 71%

The expected (statistical) distribution of products is 60% 1-chlorobutane and 40% 2-chlorobutane because six of butane's 10 hydrogens can be substituted to form 1-chlorobutane, whereas only four can be substituted to form 2-chlorobutane. This assumes, however, that all of the C—H bonds in butane are equally easy to break. Then, the relative amounts of the two products would depend only on the probability of a chlorine radical colliding with a hydrogen on a primary carbon, compared with the probability of its colliding with a hydrogen on a secondary carbon. When we carry out the reaction in the laboratory and analyze the product, however, we find that it is 29% 1-chlorobutane and 71% 2-chlorobutane. Therefore, probability alone does not explain the product distribution. Because more 2-chlorobutane is obtained than expected and *the rate-determining step of the overall reaction is removing the hydrogen atom*, we conclude that it must be easier to remove a hydrogen atom from a secondary carbon than from a primary carbon.

Radicals, like alkenes and carbocations, are stabilized by alkyl groups (Sections 4.6 and 5.2). Thus, a tertiary radical—one with the unpaired electron on a tertiary carbocation—is more stable than a secondary radical which, in turn, is more stable than a primary radical.

relative stabilities of alkyl radicals

> **Alkyl groups stabilize radicals, carbocations, and alkenes.**

| most stable | tertiary radical | > | secondary radical | > | primary radical | > | methyl radical | least stable |

The more stable the radical, the more easily it is formed because the stability of the radical is reflected in the stability of the transition state leading to its formation (Section 5.2). Consequently, it is easier to remove a hydrogen atom from a secondary carbon to form a secondary radical than it is to remove a hydrogen atom from a primary carbon to form a primary radical.

> **Ease of radical formation: tertiary > secondary > primary.**

When a chlorine radical reacts with butane, it can remove a hydrogen atom from an internal carbon, thereby forming a secondary alkyl radical, or it can remove a hydrogen atom from a terminal carbon, thereby forming a primary alkyl radical. Because it is easier to form the more stable secondary alkyl radical, 2-chlorobutane is formed faster than 1-chlorobutane.

$$CH_3CH_2CH_2CH_3$$

$\xrightarrow{Cl\cdot}$ CH$_3$CH$_2$ĊHCH$_3$ $\xrightarrow{Cl_2}$ CH$_3$CH$_2$CHCH$_3$ + Cl·
a secondary alkyl radical + HCl   2-chlorobutane

$\xrightarrow{Cl\cdot}$ CH$_3$CH$_2$CH$_2$ĊH$_2$ $\xrightarrow{Cl_2}$ CH$_3$CH$_2$CH$_2$CH$_2$Cl + Cl·
a primary alkyl radical + HCl   1-chlorobutane

After experimentally determining the amount of each chlorination product obtained from various hydrocarbons, chemists were able to conclude that *at room temperature* it is 5.0 times easier for a chlorine radical to remove a hydrogen atom from a tertiary carbon than from a primary carbon, and it is 3.8 times easier to remove a hydrogen atom from a secondary carbon than from a primary carbon. (See Problem 12.) The precise ratios differ at different temperatures.

relative rates of alkyl radical formation by a chlorine radical at room temperature

tertiary > secondary > primary
5.0   3.8   1.0

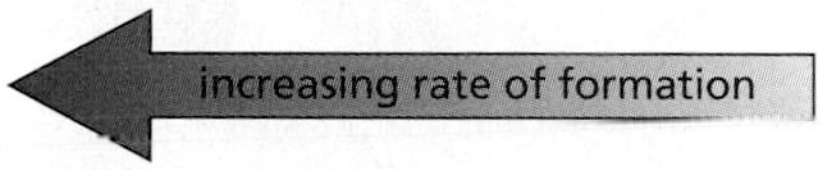

◀ increasing rate of formation

To determine the relative amounts of products obtained from radical chlorination of an alkane, both *probability* (the number of hydrogens that can be removed that will lead to the formation of the particular product) and *reactivity* (the relative rate at which a particular hydrogen is removed) must be taken into account. When both factors are considered, the calculated amounts of 1-chlorobutane and 2-chlorobutane agree with the amounts obtained experimentally.

relative amount of 1-chlorobutane

number of hydrogens × reactivity
$6 \times 1.0 = 6.0$

percent yield $= \dfrac{6.0}{21} = 29\%$

relative amount of 2-chlorobutane

number of hydrogens × reactivity
$4 \times 3.8 = 15$

percent yield $= \dfrac{15}{21} = 71\%$

The percent yield of each alkyl chloride is calculated by dividing the relative amount of the particular product by the sum of the relative amounts of all the alkyl chloride products (6 + 15 = 21).

Radical monochlorination of 2,2,5-trimethylhexane results in the formation of five monochlorination products. Because the relative amounts of the five alkyl chlorides total 35 (9.0 + 7.6 + 7.6 + 5.0 + 6.0 = 35), the percent yield of each product can be calculated as follows:

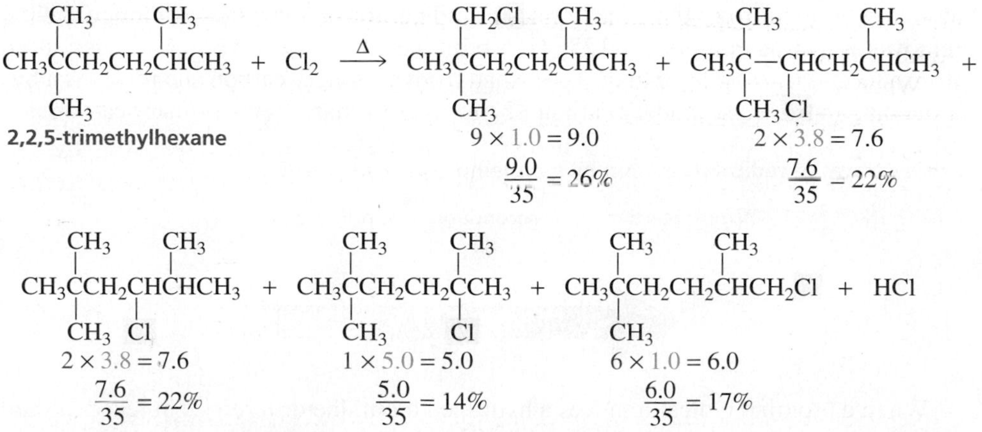

$$CH_3CCH_2CH_2CHCH_3 \;+\; Cl_2 \;\xrightarrow{\Delta}\; CH_3CCH_2CH_2CHCH_3 \;+\; CH_3C{-}CHCH_2CHCH_3 \;+$$

2,2,5-trimethylhexane

$9 \times 1.0 = 9.0$
$\dfrac{9.0}{35} = 26\%$

$2 \times 3.8 = 7.6$
$\dfrac{7.6}{35} = 22\%$

$$CH_3CCH_2CHCHCH_3 \;+\; CH_3CCH_2CH_2CCH_3 \;+\; CH_3CCH_2CH_2CHCH_2Cl \;+\; HCl$$

$2 \times 3.8 = 7.6$
$\dfrac{7.6}{35} = 22\%$

$1 \times 5.0 = 5.0$
$\dfrac{5.0}{35} = 14\%$

$6 \times 1.0 = 6.0$
$\dfrac{6.0}{35} = 17\%$

Tutorial:
Radicals: Common terms

PROBLEM 3◆

a. Which of the hydrogens in the following compound is the easiest to remove by a chlorine radical?

b. How many secondary hydrogens does the compound have?

PROBLEM 4◆

How many alkyl chlorides can be obtained from monochlorination of the following alkanes? Neglect stereoisomers.

a. $CH_3CH_2CH_2CH_2CH_3$

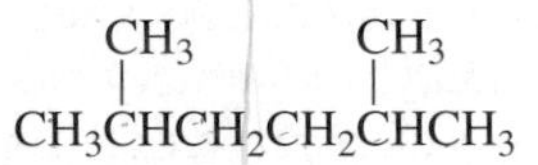

c. 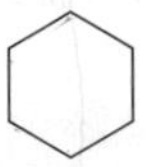 (cyclohexane)

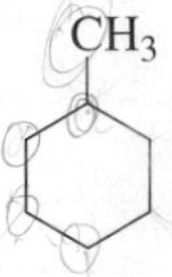

e. 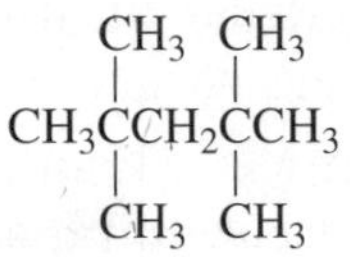
$$CH_3\underset{\underset{CH_3}{|}}{\overset{\overset{CH_3}{|}}{C}}CH_2\underset{\underset{CH_3}{|}}{\overset{\overset{CH_3}{|}}{C}}CH_3$$

b.
$$CH_3\underset{\underset{CH_3}{|}}{CH}CH_2CH_2\underset{\underset{CH_3}{|}}{CH}CH_3$$

d. (methylcyclohexane)

f. 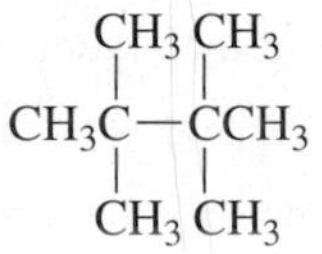
$$CH_3\underset{\underset{CH_3}{|}}{\overset{\overset{CH_3}{|}}{C}}—\underset{\underset{CH_3}{|}}{\overset{\overset{CH_3}{|}}{C}}CH_3$$

PROBLEM 5

Calculate the percent yield of each product obtained in Problems 4a and 4b if chlorination is carried out in the presence of light at room temperature.

9.4 The Reactivity–Selectivity Principle

The relative rates of radical formation when a bromine radical removes a hydrogen atom are different from the relative rates of radical formation when a chlorine radical removes a hydrogen atom. At 125 °C, a bromine radical removes a hydrogen atom from a tertiary carbon 1600 times faster than from a primary carbon and removes a hydrogen atom from a secondary carbon 82 times faster than from a primary carbon.

relative rates of radical formation by a bromine radical at 125 °C

tertiary	>	secondary	>	primary
1600		82		1

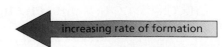
increasing rate of formation

When a bromine radical removes a hydrogen atom, the differences in reactivity are so great that the reactivity factor is vastly more important than the probability factor. For example, radical bromination of butane gives a 98% yield of 2-bromobutane, compared with the 71% yield of 2-chlorobutane obtained when butane is chlorinated (Section 9.3). In other words, bromination is more highly selective than chlorination.

Tutorial:
Radical bromination

$$CH_3CH_2CH_2CH_3 \ + \ Br_2 \ \xrightarrow{h\nu} \ CH_3CH_2CH_2CH_2Br \ + \ CH_3CH_2\overset{Br}{\overset{|}{C}}HCH_3 \ + \ HBr$$

    1-bromobutane   2-bromobutane
        2%   98%

Similarly, bromination of 2,2,5-trimethylhexane gives an 82% yield of the product in which bromine replaces the tertiary hydrogen. Chlorination of the same alkane results in a 14% yield of the tertiary alkyl chloride (Section 9.3).

$$CH_3\overset{\overset{\displaystyle CH_3}{|}}{\underset{\underset{\displaystyle CH_3}{|}}{C}}CH_2CH_2\overset{\overset{\displaystyle CH_3}{|}}{C}HCH_3 \ + \ Br_2 \ \xrightarrow{h\nu} \ CH_3\overset{\overset{\displaystyle CH_3}{|}}{\underset{\underset{\displaystyle CH_3}{|}}{C}}CH_2CH_2\overset{\overset{\displaystyle CH_3}{|}}{\underset{\underset{\displaystyle Br}{|}}{C}}CH_3 \ + \ HBr$$

    2,2,5-trimethylhexane   2-bromo-2,5,5-trimethylhexane
                                           82%

PROBLEM 6◆

Carry out the calculations that predict that

a. 2-bromobutane will be obtained in 98% yield.
b. 2-bromo-2,5,5,-trimethylhexane will be obtained in 82% yield.

Why are the relative amounts of the substitution products so different when a bromine radical rather than a chlorine radical is used to remove a hydrogen atom? It is because a chloride radical is much more reactive than a bromine radical. As a result of its greater reactivity, a chlorine radical makes primary, secondary, and tertiary radicals with almost equal ease. The less reactive bromine radical has a clear preference for the formation of the easiest-to-form tertiary radical. In other words, because a bromine radical is relatively unreactive, it is highly selective about which hydrogen atom it removes. The much more reactive chlorine radical is considerably less selective. These observations illustrate the **reactivity–selectivity principle**, which states that *the greater the reactivity of a species, the less selective it will be.*

> A chlorine radical is more reactive than a bromine radical.
>
> The more reactive a species is, the less selective it will be.

PROBLEM-SOLVING STRATEGY

Would chlorination or bromination of methylcyclohexane produce a greater yield of 1-halo-1-methylcyclohexane?
To solve this kind of problem, first draw the structures of the compounds being discussed.

$$\underset{}{\text{(methylcyclohexane)}} \ \xrightarrow[h\nu]{X_2} \ \underset{}{\text{(X–methylcyclohexane)}}$$

1-Halo-1-methylcyclohexane is a tertiary alkyl halide, so the question becomes, "Will bromination or chlorination produce a greater yield of a tertiary alkyl halide?" Because bromination is more selective, it will produce a greater yield of the desired compound. Chlorination will form some of the tertiary alkyl halide, but it will also form significant amounts of primary and secondary alkyl halides.

Now continue on to Problem 7.

PROBLEM 7◆

a. Would chlorination or bromination produce a greater yield of 1-halo-2,3-dimethylbutane?
b. Would chlorination or bromination produce a greater yield of 2-halo-2,3-dimethylbutane?
c. Would chlorination or bromination be a better way to make 1-halo-2,2-dimethylpropane?

9.5 Stereochemistry of Radical Substitution Reactions

If a reactant does not have an asymmetric center and a radical substitution reaction forms a product with an asymmetric center, a racemic mixture will be obtained.

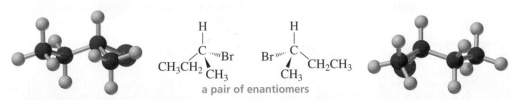

an asymmetric center

$$CH_3CH_2CH_2CH_3 + Br_2 \xrightarrow{h\nu} CH_3CH_2\overset{|}{\underset{Br}{C}}HCH_3 + HBr$$

configuration of the products

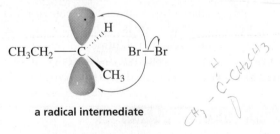

a pair of enantiomers

To understand why both enantiomers are formed, we must look at the propagation steps of the radical substitution reaction. In the first propagation step, the bromine radical removes a hydrogen atom from the alkane, creating a radical intermediate. The carbon bearing the unpaired electron is sp^2 hybridized; therefore, the three atoms to which it is bonded lie in a plane. In the second propagation step, the incoming halogen has equal access to both sides of the plane. As a result, identical amounts of the R and S enantiomers are formed.

$$CH_3CH_2 \!-\! \overset{\cdot}{\underset{CH_3}{C}} \quad Br\!-\!Br$$

a radical intermediate

PROBLEM 8◆

a. What hydrocarbon with molecular formula C_4H_{10} forms only two monochlorinated products? Both products are achiral.

b. What hydrocarbon with the same molecular formula as in part a forms only three monochlorinated products? One is achiral and two are chiral.

9.6 Radical Reactions in Biological Systems

For a long time, scientists assumed that radical reactions were not important in biological systems because a large amount of energy—heat or light—is required to initiate a radical reaction and it is difficult to control the propagation steps of the chain reaction once initiation occurs. However, it is now widely recognized that there are many biological reactions that involve radicals. Instead of being generated by heat or light, the radicals are formed by the interaction of organic molecules with metal ions. These radical reactions take place in the active site of an enzyme (Section 18.1). Containing the reaction in a specific site allows the reaction to be controlled.

Water-soluble compounds are readily eliminated by the body. In contrast, water-insoluble compounds are not readily eliminated; rather, they accumulate in the nonpolar components of cells. In order for cells to avoid becoming "toxic dumps," water insoluble compounds that are ingested (drugs, foods, environmental pollutants) must be converted into polar compounds that can be excreted.

A biological reaction, carried out in the liver, converts toxic water-insoluble hydrocarbons to less toxic water-soluble alcohols by substituting an H of the hydrocarbon with an OH. The reaction involves radicals and is catalyzed by an iron-containing enzyme called cytochrome P_{450}. A radical intermediate is created when Fe^VO removes a hydrogen atom from an alkane. In the next step, $Fe^{IV}OH$ dissociates homolytically into Fe^{III} and $HO\cdot$, and the $HO\cdot$, immediately combines with the radical intermediate to form the alcohol.

$$Fe^V{=}O \; + \; H{-}\overset{|}{\underset{|}{C}}{-} \;\longrightarrow\; Fe^{IV}{-}OH \; + \; \cdot\overset{|}{\underset{|}{C}}{-} \;\longrightarrow\; Fe^{III} \; + \; HO{-}\overset{|}{\underset{|}{C}}{-}$$

an alkane   **a radical**   **an alcohol**
                    intermediate

This reaction can also have the opposite toxicological effect. That is, substituting an OH for an H in some compounds causes a nontoxic compound to become toxic. For example, studies done on animals showed that substituting an OH for an H caused methylene chloride (CH_2Cl_2) to become a carcinogen when it is inhaled. This shows that compounds that are nontoxic *in vitro* (in a test tube) are not necessarily nontoxic *in vivo* (in an organism).

DECAFFEINATED COFFEE AND THE CANCER SCARE

Animal studies showing that methylene chloride becomes a carcinogen when inhaled caused some concern because methylene chloride was the solvent used to extract caffeine from coffee beans in the manufacture of decaffeinated coffee. However, when methylene chloride was added to drinking water fed to laboratory rats and mice, researchers found no toxic effects. They observed no toxicological responses of any kind either in rats that had consumed an amount of methylene chloride equivalent to the amount that would be ingested by drinking 120,000 cups of decaffeinated coffee per day or in mice that had consumed an amount equivalent to drinking 4.4 million cups of decaffeinated coffee per day. In addition, no increased risk of cancer was found in a study of thousands of workers exposed daily to inhaled methylene chloride. (This shows that studies done on humans do not always agree with those done on laboratory animals.) Because of the initial concern, however, researchers sought alternative methods for extracting caffeine from coffee beans. Extraction by liquid CO_2 at supercritical temperatures and pressures was found to be a better method since it extracts caffeine without simultaneously extracting some of the flavor compounds that are removed when methylene chloride is used.

Fats and oils react with radicals to form compounds that have strong odors (Section 20.3). These compounds are responsible for the unpleasant taste and smell associated with sour milk and rancid butter. The molecules that form cell membranes can undergo this same radical reaction (Section 20.5). Radical reactions in biological systems also have been implicated in the aging process.

Clearly, unwanted radicals in biological systems must be destroyed before they have an opportunity to damage cells. Unwanted radical reactions are prevented by **radical inhibitors**—compounds that destroy reactive radicals either by creating unreactive radicals or by creating compounds with only paired electrons.

Two examples of radical inhibitors that are present in biological systems are vitamin C and vitamin E. Both of these compounds form relatively stable (unreactive) radicals. Vitamin C (also called ascorbic acid) is a water-soluble compound that traps radicals formed in the aqueous environment of the cell and in blood plasma. Vitamin E is a water-insoluble (therefore, fat-soluble) compound that traps radicals formed in nonpolar membranes. Why one vitamin functions in aqueous environments and the other in nonaqueous environments is apparent from their structures and electrostatic potential maps (page 226), which show that vitamin C is a relatively polar compound, whereas vitamin E is nonpolar.

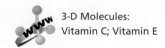
3-D Molecules:
Vitamin C; Vitamin E

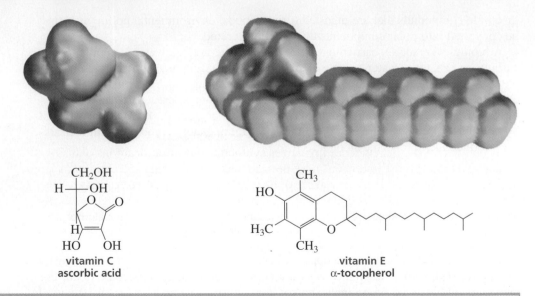

vitamin C
ascorbic acid

vitamin E
α-tocopherol

FOOD PRESERVATIVES

Radical inhibitors that are present in food are known as *preservatives* or *antioxidants*. They preserve food by preventing unwanted radical reactions. Vitamin E is a naturally occurring preservative found in vegetable oil. BHA and BHT are synthetic preservatives that are added to many packaged foods.

butylated hydroxyanisole
BHA

butylated hydroxytoluene
BHT

food preservatives

9.7 Radicals and Stratospheric Ozone

In 1995, the Nobel Prize in chemistry was awarded to **Sherwood Rowland**, **Mario Molina**, *and* **Paul Crutzen** *for their pioneering work in explaining the chemical processes responsible for the depletion of the ozone layer in the stratosphere. Their work demonstrated that human activities could interfere with global processes that support life. This was the first time that a Nobel Prize was presented for work in the environmental sciences.*

F. Sherwood Rowland *was born in Ohio in 1927. He received a B.A. from Ohio Weslyan University and a Ph.D. from the University of Chicago. He is a professor of chemistry at the University of California, Irvine.*

Ozone (O_3) a major constituent of smog, is a health hazard at ground level. In the stratosphere, however, a layer of ozone shields the Earth from harmful solar radiation. The greatest concentration of ozone occurs between 12 and 15 miles above the Earth's surface. The ozone layer is thinnest at the equator and densest toward the poles. Ozone is formed in the atmosphere from the interaction of molecular oxygen with very short wavelength ultraviolet light.

$$O_2 \xrightarrow{h\nu} O + O$$
$$O + O_2 \longrightarrow O_3$$

ozone

The stratospheric ozone layer acts as a filter for biologically harmful ultraviolet radiation that otherwise would reach the surface of the Earth. Among other effects, high-energy short-wavelength ultraviolet light can damage DNA in skin cells, causing mutations that trigger skin cancer. We owe our very existence to this protective ozone layer. According to current theories of evolution, life could not have developed on land in the absence of this ozone layer. Instead, life would have had to remain in the ocean, where water screens out the harmful ultraviolet radiation.

Since about 1985, scientists have noted a precipitous drop in stratospheric ozone over Antarctica. This area of ozone depletion, known as the "ozone hole," is unprecedented in the history of ozone observations. Scientists subsequently noted a similar decrease in ozone over Arctic regions; then, in 1988, they detected a depletion of ozone over the United States for the first time. Three years later, scientists determined that the rate of ozone depletion was two to three times faster than originally anticipated. Many in the scientific community blame recently observed increases in cataracts and skin cancer as well as diminished plant growth on the ultraviolet radiation that has penetrated the reduced ozone layer. It has been predicted that erosion of the protective ozone layer will cause an additional 200,000 deaths from skin cancer over the next 50 years.

Strong circumstantial evidence implicates synthetic chlorofluorocarbons (CFCs)—alkanes in which all the hydrogens have been replaced by fluorine and chlorine, such as $CFCl_3$ and CF_2Cl_2—as a major cause of ozone depletion. These gases, known commercially as Freons, have been used extensively as cooling fluids in refrigerators and air conditioners. They were also once widely used as propellants in aerosol spray cans (deodorant, hair spray, etc.) because of their odorless, nontoxic, and nonflammable properties and, being chemically inert, they do not react with the contents of the can. Such use now, however, has been banned.

Chlorofluorocarbons remain very stable in the atmosphere until they reach the stratosphere. There they encounter wavelengths of ultraviolet light that cause the carbon-chlorine bond to break, generating chlorine radicals.

$$
\begin{array}{ccc}
& \text{Cl} & & & \text{Cl} \\
& | & & & | \\
\text{F}-&\text{C}-\text{Cl} & \xrightarrow{h\nu} & \text{F}-&\text{C}\cdot & + & \text{Cl}\cdot \\
& | & & & | \\
& \text{F} & & & \text{F}
\end{array}
$$

The chlorine radicals are the ozone-removing agents. They react with ozone to form chlorine monoxide radicals. The chlorine monoxide radicals then react with ozone to form chlorine dioxide which dissociates to regenerate a chlorine radical. These three steps are repeated over and over, destroying a molecule of ozone in two of the steps. It has been calculated that each chlorine atom destroys 100,000 ozone molecules!

$$Cl\cdot + O_3 \longrightarrow ClO\cdot + O_2$$

$$ClO\cdot + O_3 \longrightarrow ClO_2 + O_2$$

$$ClO_2 \longrightarrow Cl\cdot + O_2$$

Mario Molina *was born in Mexico in 1943 and subsequently became a U.S. citizen. He received a Ph.D. from the University of California, Berkeley, and then became a postdoctoral fellow in Rowland's laboratory. He is currently a professor of earth, atmospheric, and planetary sciences at the Massachusetts Institute of Technology.*

Paul Crutzen *was born in Amsterdam in 1933. He was trained as a meteorologist and became interested in stratospheric chemistry and stratospheric ozone in particular. He is a professor at the Max Planck Institute for Chemistry in Mainz, Germany.*

Polar stratospheric clouds increase the rate of ozone destruction. These clouds form over Antarctica during the cold winter months. Ozone depletion in the Arctic is less severe because it generally does not get cold enough for the polar stratospheric clouds to form.

Movie: Chlorofluorocarbons and ozone

Growth of the Antarctic ozone hole, located mostly over the continent of Antarctica, since 1979. The images were made from data supplied by total ozone-mapping spectrometers (TOMS). The color scale depicts the total ozone values in Dobson units. The lowest ozone densities are represented by dark blue.

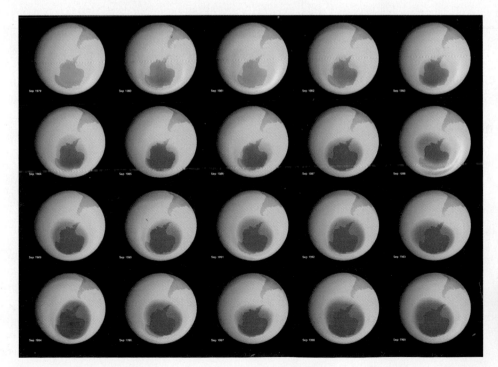

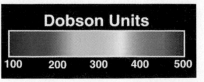

Dobson Units

100   200   300   400   500

Summary

Alkanes are called **saturated hydrocarbons** because they do not contain any double or triple bonds. Since they also have only strong σ bonds and atoms with no partial charges, alkanes are very unreactive. Alkanes do undergo **radical substitution reactions** with chlorine (Cl_2) or bromine (Br_2) at high temperatures or in the presence of light, to form alkyl chlorides or alkyl bromides. The substitution reaction is a **radical chain reaction** with **initiation**, **propagation**, and **termination steps**. If a reactant does not have an asymmetric center and a radical substitution reaction forms a product with an asymmetric center, a racemic mixture will be obtained.

The rate-determining step of the radical substitution reaction is formation of an alkyl **radical** by removing a hydrogen atom. The relative rates of alkyl radical formation are $3° > 2° > 1° >$ methyl. To determine the relative amounts of products obtained from the radical halogenation of an alkane, both probability and the relative rate at which a particular hydrogen is abstracted must be taken into account.

Some biological reactions involve radicals formed by the interaction of organic molecules with metal ions. The reactions take place in the active site of an enzyme. Unwanted radical reactions are prevented by **radical inhibitors**—compounds that destroy reactive radicals by creating relatively stable radicals or compounds with only paired electrons.

Strong circumstantial evidence implicates synthetic chlorofluorocarbons as being responsible for the diminishing ozone layer. The interaction of these compounds with UV light generates chlorine radicals, which are the ozone-removing agents.

Summary of Reactions

Alkanes undergo radical substitution reactions with Cl_2 or Br_2 in the presence of heat or light (Sections 9.2, 9.3, and 9.4).

$$CH_3CH_3 + Cl_2 \xrightarrow{\Delta \text{ or } h\nu} CH_3CH_2Cl + HCl$$
excess

$$CH_3CH_3 + Br_2 \xrightarrow{\Delta \text{ or } h\nu} CH_3CH_2Br + HBr$$
excess
bromination is more selective than chlorination

Problems

9. Give the product(s) of each of the following reactions, ignoring stereoisomers:

a.
$$\overset{\overset{\displaystyle CH_3}{|}}{CH_3CH_2CHCH_2CH_2CH_3} + Br_2 \xrightarrow{h\nu}$$

b. 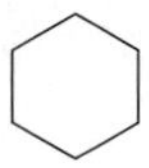 [hexagon] $+ Cl_2 \xrightarrow{h\nu}$

c. 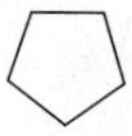 [pentagon] $+ Cl_2 \longrightarrow$

d. 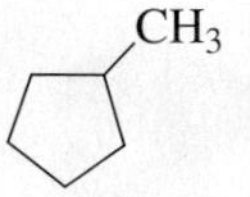 [pentagon with CH_3] $+ Cl_2 \xrightarrow{h\nu}$

10. Give the major product of each of the following reactions:

a.
$$\overset{\overset{\displaystyle CH_3}{|}}{CH_3CHCH_3} + Cl_2 \xrightarrow{h\nu}$$

b.
$$\overset{\overset{\displaystyle CH_3}{|}}{CH_3CHCH_3} + Br_2 \xrightarrow{h\nu}$$

11. a. What alkane with molecular formula C_5H_{12} forms only one monochlorinated product when heated with Cl_2.
   b. What alkane with molecular formula C_7H_{16} forms seven monochlorinated products (ignore stereoisomers) when heated with Cl_2.

12. When 2-methylpropane is monochlorinated in the presence of light at room temperature, 36% of the product is 2-chloro-2-methylpropane and 64% is 1-chloro-2-methylpropane. From these data, calculate how much easier it is to remove a hydrogen atom from a tertiary carbon than from a primary carbon under these conditions.

13. **a.** How many monochlorination products would be obtained from the radical chlorination of methylcyclohexane? Ignore stereoisomers.
   b. Which product would be obtained in greatest yield?
   c. Which product would be obtained in lowest yield?
   d. How many monochlorination products would be obtained if all stereoisomers are included?

14. Answer the questions in Problem 13 for the radical bromination of methylcyclohexane.

15. For each of the following compounds, give the major product that would be obtained from treating an excess of the compound with Cl_2 in the presence of light at room temperature. (Ignore Stereoisomers)

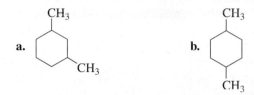

16. Answer the questions in Problem 15 if the compounds were treated with Br_2 at 125°C.

17. Dr. Al Cahall wanted to determine experimentally the relative ease of removal of a hydrogen atom from a tertiary, a secondary, and a primary carbon by a chlorine radical. He allowed 2-methylbutane to undergo chlorination at 300 °C and obtained as products 36% 1-chloro-2-methylbutane, 18% 2-chloro-2-methylbutane, 28% 2-chloro-3-methylbutane, and 18% 1-chloro-3-methylbutane. What values did he obtain for the relative ease of removing a hydrogen from tertiary, secondary, and primary carbons by a chlorine radical under the conditions of his experiment?

18. At 600 °C, the ratio of the relative rates of formation of a tertiary, a secondary, and a primary radical by a chlorine radical is 2.6 : 2.1 : 1. Explain the change in the degree of regioselectivity compared with what Dr. Al Cahall found in Problem 17.

10 | Substitution and Elimination Reactions of Alkyl Halides

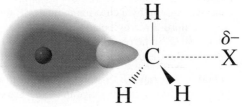

Organic compounds that have an electronegative atom or an electron-withdrawing group bonded to an sp^3 hybridized carbon undergo *substitution reactions* and/or *elimination reactions*.

In a **substitution reaction**, the electronegative atom or electron-withdrawing group is replaced by another atom or group. In an **elimination reaction**, the electronegative atom or electron-withdrawing group is eliminated, along with a hydrogen from an adjacent carbon. The atom or group that is *substituted* or *eliminated* in these reactions is called a **leaving group**. The substitution reaction is more precisely called a **nucleophilic substitution reaction** because it is a nucleophile that replaces the leaving group.

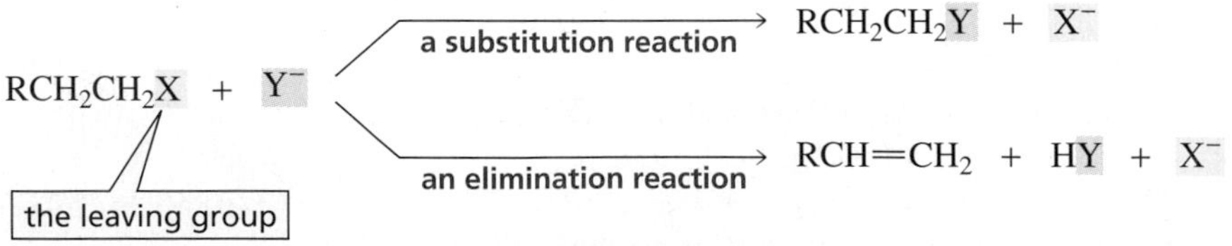

$$RCH_2CH_2X \ + \ Y^- \quad \xrightarrow{\text{a substitution reaction}} \quad RCH_2CH_2Y \ + \ X^-$$

$$\xrightarrow{\text{an elimination reaction}} \quad RCH=CH_2 \ + \ HY \ + \ X^-$$

the leaving group

Alkyl halides—compounds in which the leaving group is a halide ion—are a good family of compounds with which to start our study of substitution and elimination reactions because they have relatively good leaving groups; that is, the halide ions are easily displaced. We will then be prepared to discuss, in Chapter 11, the substitution and elimination reactions of compounds with poor leaving groups—those that are more difficult to displace.

alkyl halides

R—F	R—Cl	R—Br	R—I
an alkyl fluoride	an alkyl chloride	an alkyl bromide	an alkyl iodide

Substitution reactions are important in organic chemistry because they make it possible to convert readily available alkyl halides into a wide variety of other compounds. Substitution reactions are also important in the cells of plants and animals. Because alkyl halides are insoluble in water and cells exist in predominantly aqueous environments, we will see that biological systems use compounds in which the group that is replaced is more polar than a halogen and therefore more soluble in water.

SURVIVAL COMPOUNDS

Several marine organisms, including sponges, corals, and algae, synthesize alkyl halides that they use to deter predators. For example, red algae synthesize a toxic, foul-tasting alkyl halide that keeps predators from eating them. One predator, however, that is not deterred is a mollusk called a sea hare. After consuming red algae, a sea hare converts the original alkyl halide into a structurally similar compound it uses for its own defense. Unlike other mollusks, a sea hare does not have a shell. Its method of defense is to surround itself with a slimy material that contains the alkyl halide, thereby protecting itself from carnivorous fish.

synthesized by red algae   **synthesized by the sea hare**

a sea hare

ARTIFICIAL BLOOD

Clinical trials are currently underway to test the use of perfluorocarbons—alkanes in which all the hydrogens have been replaced by fluorines— as a substitute for blood. One compound under study has been found to be more effective than hemoglobin in carrying oxygen to cells and in transporting carbon dioxide to the lungs. Artificial blood has several advantages: It is safe from disease, it can be administered to any blood type, its availability is not dependent on blood donors, and it can be stored longer than whole blood (which is good for only about 40 days).

10.1 How Alkyl Halides React

A halogen is more electronegative than a carbon. Consequently, the two atoms do not share their bonding electrons equally. Because the more electronegative halogen has a larger share of the electrons, it has a partial negative charge (δ^-) and the carbon to which it is bonded has a partial positive charge (δ^+).

$$\overset{\delta+}{R CH_2} \!-\! \overset{\delta-}{X} \qquad X = F, Cl, Br, I$$

a polar bond

It is the polar carbon–halogen bond that causes alkyl halides to undergo substitution and elimination reactions. We will see that there are two mechanisms for the substitution reaction and two mechanisms for the elimination reaction. First, we will look at the substitution reaction.

10.2 The Mechanism of an S$_N$2 Reaction

How is the mechanism of a reaction determined? We can learn a great deal about the mechanism of a reaction by studying the factors that affect the rate of the reaction. The rate of a nucleophilic substitution reaction such as the reaction of methyl bromide with hydroxide ion depends on the concentrations of *both* reagents. If the concentration of methyl bromide in the reaction mixture is *doubled*, the rate of the nucleophilic substitution reaction *doubles*. If the concentration of the nucleophile (in this case hydroxide ion) is *doubled*, the rate of the reaction also *doubles*. If the concentrations of both reactants are doubled, the rate of the reaction quadruples.

$$CH_3Br \ + \ HO^- \ \longrightarrow \ CH_3OH \ + \ Br^-$$
methyl bromide \qquad\qquad methyl alcohol

When you know the relationship between the rate of a reaction and the concentration of the reactants, you can write a **rate law** for the reaction. Because the rate of the reaction of methyl bromide with hydroxide ion is dependent on the concentration of both reactants, the rate law for the reaction is

$$\text{rate} \propto [\textbf{alkyl halide}][\textbf{nucleophile}]$$

The proportionality sign ($\propto$) can be replaced by an equals sign and a proportionality constant. The proportionality constant, in this case k, is called the **rate constant**. The magnitude of the rate constant describes how difficult it is to overcome the energy barrier of the reaction—how hard it is to reach the transition state. The larger the rate constant, the lower is the energy barrier and, therefore, the easier it is to reach the transition state (see Figure 10.2).

$$\text{rate} \ = \ k[\textbf{alkyl halide}][\textbf{nucleophile}]$$

The rate law tells us which molecules are involved in the rate-determining step of the reaction. From the rate law for the reaction of methyl bromide with hydroxide ion, we know that both methyl bromide and hydroxide ion are involved in the rate-determining step.

PROBLEM 1◆

a. How is the rate of the reaction of methyl bromide with hydroxide ion affected if the concentration of hydroxide ion is tripled?

b. How is the rate of the reaction affected if the concentration of methyl bromide is changed from 1.00 M to 0.05 M?

Edward Davies Hughes
(1906–1963) *was born in North Wales. He earned two doctoral degrees: a Ph.D. from the University of Wales and a D.Sc. from the University of London, working with Sir Christopher Ingold. He was a professor of chemistry at University College, London.*

Sir Christopher Ingold
(1893–1970) *was born in Ilford, England. In addition to determining the mechanism of the S$_N$2 reaction, he was a member of the group that developed the R,S system of nomenclature for enantiomers. He also participated in developing the theory of resonance.*

The reaction of methyl bromide with hydroxide ion is an example of an **S$_N$2 reaction**, where "S" stands for substitution, "N" for nucleophilic, and "2" for bimolecular. **Bimolecular** means that two molecules are involved in the rate-determining step. In 1937, Edward Hughes and Christopher Ingold proposed a mechanism for an S$_N$2 reaction. Remember that a mechanism describes the step-by-step process by which reactants are converted into products. It is a theory that fits the experimental evidence that has been accumulated concerning the reaction. Hughes and Ingold based their mechanism for an S$_N$2 reaction on the following three pieces of experimental evidence:

1. The rate of the reaction depends on the concentration of the alkyl halide *and* on the concentration of the nucleophile. This means that both reactants are involved in the rate-determining step.

2. When the hydrogens of methyl bromide are successively replaced with methyl groups, the rate of the reaction with a given nucleophile becomes progressively slower (Table 10.1).

Table 10.1   Relative Rates of S$_N$2 Reactions for Several Alkyl Halides

$$R-Br \ + \ Cl^- \ \xrightarrow{\ S_N2\ } \ R-Cl \ + \ Br^-$$

Alkyl halide	Class of alkyl halide	Relative rate
CH$_3$—Br	methyl	1200
CH$_3$CH$_2$—Br	primary	40
CH$_3$CH$_2$CH$_2$—Br	primary	16

CH$_3$CH—Br
     |
     CH$_3$
	secondary	1

CH$_3$
     |
 CH$_3$C—Br
     |
     CH$_3$
	tertiary	too slow to measure

3. The reaction of an alkyl halide in which the halogen is bonded to an asymmetric center leads to the formation of only one stereoisomer, and the configuration of the asymmetric center in the product is inverted relative to its configuration in the reacting alkyl halide.

Hughes and Ingold proposed that an S$_N$2 reaction takes place in a single step. The nucleophile attacks the back side of the carbon that is bonded to the leaving group and displaces the leaving group.

The nucleophile attacks the back side of the carbon that is bonded to the leaving group.

mechanism of the S$_N$2 reaction

leaving group

Tutorial:
S$_N$2

How does Hughes and Ingold's mechanism account for the three observed pieces of experimental evidence? The mechanism shows the alkyl halide and the nucleophile coming together in the transition state of the one-step reaction. Therefore, increasing the concentration of either of them makes their collision more probable. This agrees with the observation that the reaction is dependent on the concentration of both reactants.

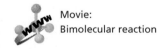

transition state

Movie:
Bimolecular reaction

Because the nucleophile attacks the back side of the carbon that is bonded to the halogen, bulky substituents attached to this carbon will make it harder for the nucleophile to have access to the back side and will therefore decrease the rate of the reaction (Figure 10.1). This explains why substituting methyl groups for the hydrogens in methyl bromide progressively slows the rate of the substitution reaction (Table 10.1).

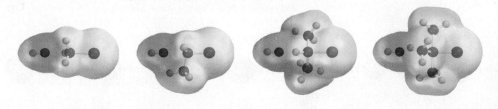

◀ **Figure 10.1**
The approach of HO$^-$ to a methyl halide, a primary alkyl halide, a secondary alkyl halide, and a tertiary alkyl halide. Increasing the bulk of the substituents bonded to the carbon that is undergoing nucleophilic attack decreases access to the back side of the carbon, thereby decreasing the rate of the S$_N$2 reaction.

Steric effects are caused by groups occupying a certain volume of space. A steric effect that decreases reactivity is called **steric hindrance**. Steric hindrance results when groups are in the way at the reaction site. Steric hindrance causes alkyl halides to have the following relative reactivities in an S_N2 reaction because, *generally*, primary alkyl halides are less sterically hindered than secondary alkyl halides, which, in turn, are less hindered than tertiary alkyl halides:

relative reactivities of alkyl halides in an S_N2 reaction

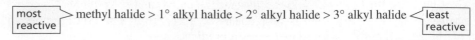

most reactive > methyl halide > 1° alkyl halide > 2° alkyl halide > 3° alkyl halide < least reactive

> Steric hindrance causes primary alkyl halides to be the most reactive in S_N2 reactions.

3-D Molecules:
Methyl chloride;
t-Butyl chloride

> Tertiary alkyl halides cannot undergo S_N2 reactions.

The three alkyl groups of a tertiary alkyl halide make it impossible for the nucleophile to come within bonding distance of the tertiary carbon, so tertiary alkyl halides are unable to undergo S_N2 reactions. The reaction coordinate diagrams for the S_N2 reactions of *unhindered* methyl bromide and a *sterically hindered* secondary alkyl bromide show that steric hindrance raises the energy of the transition state, slowing the reaction (Figure 10.2).

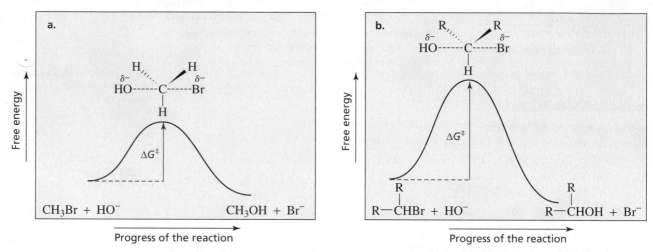

▲ **Figure 10.2**
Reaction coordinate diagrams for (a) the S_N2 reaction of methyl bromide with hydroxide ion; (b) an S_N2 reaction of a sterically hindered secondary alkyl bromide with hydroxide ion.

Figure 10.3 shows that as the nucleophile approaches the back side of the carbon of methyl bromide, the C—H bonds begin to move away from the nucleophile and its attacking electrons. By the time the transition state is reached, the C—H bonds are all in the same plane. As the nucleophile gets closer to the carbon and the bromine moves farther away from it, the C—H bonds continue to move in the same direction. Eventually, the bond between the carbon and the nucleophile is fully formed, and the bond between the carbon and bromine is completely broken, so the carbon is once

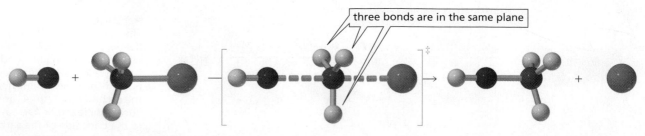

three bonds are in the same plane

▲ **Figure 10.3**
An S_N2 reaction between hydroxide ion and methyl bromide.

again tetrahedral. The carbon at which substitution occurs has *inverted its configuration* during the course of the reaction, just as an umbrella has a tendency to invert in a windstorm.

Because an S_N2 reaction takes place with **inversion of configuration,** only one substitution product is formed when an alkyl halide whose halogen atom is bonded to an asymmetric center undergoes an S_N2 reaction. The configuration of that product is inverted relative to the configuration of the alkyl halide. For example, the substitution product obtained from the reaction of hydroxide ion with (R)-2-bromopentane is (S)-2-pentanol. (The "ol" suffix indicates that the compound is an alcohol, and the "2" tells us that the OH group is bonded to the #2 carbon.) The configuration of the product is what the proposed mechanism predicts.

the configuration of the product is inverted relative to the configuration of the reactant

To draw the inverted product, draw the mirror image of the *reactant* and replace the halogen with the nucleophile.

(R)-2-bromobutane   (S)-2-butanol

PROBLEM 2◆

Arrange the following alkyl bromides in order of decreasing reactivity in an S_N2 reaction: 1-bromo-2-methylbutane, 1-bromo-3-methylbutane, 2-bromo-2-methylbutane, and 1-bromopentane.

PROBLEM 3◆ **SOLVED**

What product would be formed from the S_N2 reaction of

a. 2-bromobutane and hydroxide ion?   **c.** (S)-3-chlorohexane and hydroxide ion?

b. (R)-2-bromobutane and hydroxide ion?   **d.** 3-iodopentane and hydroxide ion?

SOLUTION TO 3a The product is 2-butanol. Because the reaction is an S_N2 reaction, we know that the configuration of the product is inverted relative to the configuration of the reactant. The configuration of the reactant is not specified, however, so we cannot specify the configuration of the product.

the configuration is not specified

$$CH_3CHCH_2CH_3 \ + \ \longrightarrow \ CH_3CHCH_2CH_3 \ + \ Br^-$$
$$\ \ |\qquad\qquad\qquad\qquad\qquad\qquad\quad |$$
$$\ \ Br \qquad\qquad\qquad\qquad\qquad\qquad OH$$

Viktor Meyer (1848–1897) *was born in Germany. To prevent him from becoming an actor, his parents persuaded him to enter the University of Heidelberg, where he earned a Ph.D. in 1867 at the age of 18. He was a professor of chemistry at the Universities of Stuttgart and Heidelberg. He coined the term "stereochemistry" for the study of molecular shapes and was the first to describe the effect of steric hindrance on a reaction.*

10.3 Factors Affecting S_N2 Reactions

The Leaving Group

If an alkyl iodide, an alkyl bromide, an alkyl chloride, and an alkyl fluoride (all with the same alkyl group) were allowed to react with the same nucleophile under the same conditions, we would find that the alkyl iodide is the most reactive and the alkyl fluoride is the least reactive.

relative rates of reaction

$HO^- + RCH_2I \longrightarrow RCH_2OH + I^-$		30,000
$HO^- + RCH_2Br \longrightarrow RCH_2OH + Br^-$		10,000
$HO^- + RCH_2Cl \longrightarrow RCH_2OH + Cl^-$		200
$HO^- + RCH_2F \longrightarrow RCH_2OH + F^-$		1

The only difference among these four reactions is the nature of the leaving group. From the relative reaction rates, we can see that the iodide ion is the best leaving group and the fluoride ion is the worst. This brings us to an important rule in organic chemistry: *The weaker the basicity of a group, the better is its leaving ability.* The reason leaving ability depends on basicity is because *weak bases are stable bases*—they readily bear the electrons they formerly shared with a proton (Section 2.3). Because weak bases don't share their electrons well, a weak base is not bonded as strongly to the carbon as a strong base would be, and a weaker bond is more easily broken (Section 2.3).

The iodide ion is the weakest base of the halide ions and the fluoride ion is the strongest. Therefore, alkyl iodides are the most reactive of the alkyl halides, and alkyl fluorides are the least reactive.

The weaker the base, the better it is as a leaving group.

Stable bases are weak bases.

relative reactivities of alkyl halides in an S_N2 reaction

$$\boxed{\text{most reactive}} \quad RI > RBr > RCl > RF \quad \boxed{\text{least reactive}}$$

The Nucleophile

When we talk about atoms or molecules that have lone-pair electrons, sometimes we call them bases and sometimes we call them nucleophiles. What is the difference between a base and a nucleophile?

Basicity is a measure of how well a compound (a **base**) shares its lone pair with a proton. The stronger the base, the better it shares its electrons. **Nucleophilicity** is a measure of how readily a compound (a **nucleophile**) is able to attack an electron-deficient atom. In the case of an S_N2 reaction, nucleophilicity is a measure of how readily the nucleophile attacks an sp^3 hybridized carbon bonded to a leaving group.

In general, *stronger bases are better nucleophiles*. For example, a species with a negative charge is a stronger base *and* a better nucleophile than a species with the same attacking atom that is neutral. Thus, HO^- is a stronger base and a better nucleophile than H_2O.

stronger base, better nucleophile		weaker base, poorer nucleophile
HO^-	>	H_2O
CH_3O^-	>	CH_3OH
$^-NH_2$	>	NH_3
$CH_3CH_2NH^-$	>	$CH_3CH_2NH_2$

If hydrogens are attached to the second-row elements, the resulting compounds have the following relative acidities (Section 2.3):

relative acid strengths

$$\boxed{\text{weakest acid}} \quad NH_3 < H_2O < HF$$

Because the weakest acid has the strongest conjugate base (Section 2.2), the conjugate bases have the following relative *base strengths* and relative *nucleophilicities*:

relative base strengths and relative nucleophilicities

$$\boxed{\text{strongest base}} \quad ^-NH_2 > HO^- > F^-$$

$$\boxed{\text{best nucleophile}}$$

Note that the amide anion is the strongest base, as well as the best nucleophile.

PROBLEM 4 ▐ SOLVED ▌

List the following species in order of decreasing nucleophilicity:

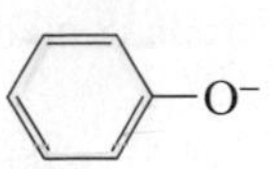

(handwritten: menos ácido → mas / mas básico → menos)

$$\text{C}_6\text{H}_5-\text{O}^-\quad CH_3OH\quad HO^-\quad CH_3\overset{\displaystyle O}{\overset{\|}{C}}O^-$$

SOLUTION   Let's first divide the nucleophiles into groups. There are three nucleophiles with negatively charged oxygens and one with a neutral oxygen. We know that the poorest nucleophile is the one with the neutral oxygen atom. So we need to rank the three nucleophiles with negatively charged oxygens in order of the pK_a's of their conjugate acids. A carboxylic acid is a stronger acid than phenol, which is a stronger acid than water (Section 6.9). Because water is the weakest acid, its conjugate base is the strongest base and the best nucleophile. Thus, the relative nucleophilicities are

$$HO^- \;>\; \text{C}_6\text{H}_5-\text{O}^- \;>\; CH_3\overset{\displaystyle O}{\overset{\|}{C}}O^- \;>\; CH_3OH$$

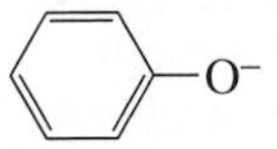

PROBLEM 5◆

For each of the following pairs of S_N2 reactions, indicate which reaction occurs faster:

a. $CH_3CH_2Br + H_2O$ or $CH_3CH_2Br + HO^-$

b. $CH_3\underset{\underset{\displaystyle CH_3}{|}}{C}HCH_2Br + HO^-$ or $CH_3CH_2\underset{\underset{\displaystyle CH_3}{|}}{C}HBr + HO^-$

c. $CH_3CH_2Cl + I^-$ or $CH_3CH_2Br + I^-$

d. $CH_3CH_2CH_2I + HO^-$ or $CH_3CH_2CH_2Br + HO^-$

Many different kinds of nucleophiles can react with alkyl halides. Therefore, a wide variety of organic compounds can be synthesized by means of S_N2 reactions.

$$CH_3CH_2Cl + HO^- \longrightarrow \underset{\text{an alcohol}}{CH_3CH_2OH} + Cl^-$$

$$CH_3CH_2Br + HS^- \longrightarrow \underset{\text{a thiol}}{CH_3CH_2SH} + Br^-$$

$$CH_3CH_2I + RO^- \longrightarrow \underset{\text{an ether}}{CH_3CH_2OR} + I^-$$

$$CH_3CH_2Br + RS^- \longrightarrow \underset{\text{a thioether}}{CH_3CH_2SR} + Br^-$$

$$CH_3CH_2Cl + {}^-NH_2 \longrightarrow \underset{\text{a primary amine}}{CH_3CH_2NH_2} + Cl^-$$

$$CH_3CH_2Br + {}^-C{\equiv}CR \longrightarrow \underset{\text{an alkyne}}{CH_3CH_2C{\equiv}CR} + Br^-$$

$$CH_3CH_2I + {}^-C{\equiv}N \longrightarrow \underset{\text{a nitrile}}{CH_3CH_2C{\equiv}N} + I^-$$

PROBLEM 6◆

What is the product of the reaction of ethyl bromide with each of the following nucleophiles?

a. $CH_3CH_2CH_2O^-$ **b.** $CH_3C{\equiv}C^-$ **c.** $(CH_3)_3N$ **d.** $CH_3CH_2S^-$

ENVIRONMENTAL ADAPTATION

The microorganism *Xanthobacter* has learned to use the alkyl halides that reach the ground as industrial pollutants as a source of carbon. The microorganism synthesizes an enzyme that uses the alkyl halide as a starting material to produce other carbon-containing compounds that it needs via S_N2 reactions.

WHY CARBON INSTEAD OF SILICON?

There are two reasons living organisms are composed primarily of carbon, oxygen, hydrogen, and nitrogen: the *fitness* of these elements for specific roles in life processes and their *availability* in the environment. Of the two reasons, fitness was probably more important than availability, because carbon, rather than silicon, became the fundamental building block of living organisms, even though silicon is more than 140 times more abundant than carbon in the Earth's crust and is just below carbon in the periodic table.

Abundance (atoms/100 atoms)

Element	In living organisms	In Earth's crust
H	49	0.22
C	25	0.19
O	25	47
N	0.3	0.1
Si	0.03	28

Why are hydrogen, carbon, oxygen, and nitrogen so fit for the roles they play in living organisms? First and foremost, they are among the smallest atoms that form covalent bonds, and carbon, oxygen, and nitrogen can also form multiple bonds. Because the atoms are small and can form multiple bonds, they form strong bonds that give rise to stable molecules. The compounds that make up living organisms must be stable (i.e., slow to react) if the organisms are to survive.

Silicon has almost twice the diameter of carbon, so silicon forms longer and weaker bonds. Consequently, an S_N2 reaction at silicon would occur much more rapidly than an S_N2 reaction at carbon. Moreover, silicon has another problem. The end product of carbon metabolism is CO_2. The analogous product of silicon metabolism would be SiO_2. Silicon dioxide molecules polymerize to form quartz (sea sand). It is hard to imagine that life could exist, much less proliferate, if animals exhaled sand instead of CO_2!

10.4 The Mechanism of an S_N1 Reaction

Given our understanding of S_N2 reactions, we would expect the rate of reaction of *tert*-butyl bromide with water to be very slow because water is a poor nucleophile and *tert*-butyl bromide is sterically hindered to attack by a nucleophile. It turns out, however, that the reaction is surprisingly fast. In fact, it is over one million times faster than the reaction of methyl bromide—a compound with no steric hindrance—with water (Table 10.2). Clearly, the reaction must be taking place by a mechanism different from that of an S_N2 reaction.

$$
\underset{\textbf{\textit{tert}-butyl bromide}}{CH_3-\overset{\overset{\displaystyle CH_3}{|}}{\underset{\underset{\displaystyle CH_3}{|}}{C}}-Br} + H_2O \longrightarrow \underset{\textbf{\textit{tert}-butyl alcohol}}{CH_3-\overset{\overset{\displaystyle CH_3}{|}}{\underset{\underset{\displaystyle CH_3}{|}}{C}}-OH} + HBr
$$

We have seen that in order to determine the mechanism of a reaction, we need to investigate the factors that affect the rate of the reaction and determine the configuration of the products of the reaction. Finding that doubling the concentration of the alkyl halide doubles the rate of the reaction but changing the concentration of the nucleophile has no effect on the rate of the reaction, allows us to write the rate law for the reaction:

$$\textbf{rate} = k[\textbf{alkyl halide}]$$

The rate law for the reaction of *tert*-butyl bromide with water differs from the rate law for the reaction of methyl bromide with hydroxide ion (Section 10.2), so the two reactions must have different mechanisms. We have seen that the reaction between methyl bromide and hydroxide ion is an S$_N$2 reaction. The reaction between *tert*-butyl bromide and water is an **S$_N$1** reaction, where "S" stands for substitution, "N" stands for nucleophilic, and "1" stands for unimolecular. **Unimolecular** means that only one molecule is involved in the rate-determining step. The mechanism of an S$_N$1 reaction is based on the following experimental evidence:

1. The rate law shows that the rate of the reaction depends only on the concentration of the alkyl halide. This means that the rate-determining step of the reaction involves only the alkyl halide.

2. When the methyl groups of *tert*-butyl bromide are successively replaced by hydrogens, the rate of the S$_N$1 reaction decreases progressively (Table 10.2). This is opposite to the order of reactivity exhibited by alkyl halides in S$_N$2 reactions (Table 10.1).

3. The reaction of an alkyl halide in which the halogen is bonded to an asymmetric center forms two stereoisomers: one with the same relative configuration at the asymmetric center as the reacting alkyl halide, the other with the inverted configuration.

Unlike an S$_N$2 reaction, where the leaving group departs and the nucleophile approaches *at the same time*, the leaving group in an S$_N$1 reaction departs *before* the nucleophile approaches. In the first step of an S$_N$1 reaction of an alkyl halide, the carbon–halogen bond breaks such that both of its electrons stay with the halogen. As a result, a carbocation intermediate is formed. In the second step, the nucleophile reacts rapidly with the carbocation to form a protonated alcohol. Because a protonated alcohol is a strong acid, it will lose a proton so the final product is an alcohol.

3-D Molecules:
t-Butyl bromide;
t-Butyl cation;
protonated *t*-Butyl alcohol;
t-Butyl alcohol

mechanism of the S$_N$1 reaction

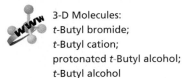

Tutorial:
S$_N$1

Table 10.2	Relative Rates of S$_N$1 Reactions for Several Alkyl Bromides (solvent is H$_2$O, nucleophile is H$_2$O)		
Alkyl bromide	**Class of alkyl bromide**		**Relative rate**
CH$_3$ | CH$_3$C—Br | CH$_3$	tertiary		1,200,000
CH$_3$CH—Br | CH$_3$	secondary		11.6
CH$_3$CH$_2$—Br	primary		1.00*
CH$_3$—Br	methyl		1.05*

*Although the rate of the S$_N$1 reaction of this compound with water is 0, a small rate is observed as a result of an S$_N$2 reaction.

Because the rate of an S_N1 reaction depends only on the concentration of the alkyl halide, the first step must be the rate-determining step. The nucleophile, therefore, is not involved in the rate-determining step, so its concentration has no effect on the rate of the reaction. If you look at the reaction coordinate diagram in Figure 10.4, you will see why increasing the rate of the second step will not make an S_N1 reaction go any faster.

Figure 10.4 ▶
Reaction coordinate diagram for an S_N1 reaction.

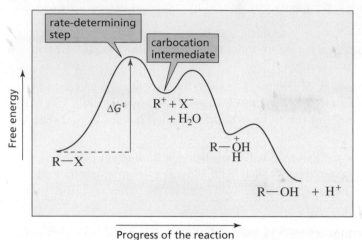

How does the mechanism for an S_N1 reaction account for the three pieces of experimental evidence? First, because the alkyl halide is the only species that participates in the rate-determining step, the mechanism agrees with the observation that the rate of the reaction depends on the concentration of the alkyl halide and does not depend on the concentration of the nucleophile.

Second, the mechanism shows that a carbocation is formed in the rate-determining step. We know that a tertiary carbocation is more stable, and is therefore easier to form, than a secondary carbocation, which in turn is more stable and easier to form than a primary carbocation (Section 5.2). Tertiary alkyl halides, therefore, are more reactive than secondary alkyl halides, which are more reactive than primary alkyl halides. This relative order of reactivity agrees with the observation that the rate of an S_N1 reaction decreases as the methyl groups of *tert*-butyl bromide are successively replaced by hydrogens (Table 10.2).

Carbocation stability: 3° > 2° > 1°.

relative reactivities of alkyl halides in an S_N1 reaction

most reactive > 3° alkyl halide > 2° alkyl halide > 1° alkyl halide < least reactive

Now let's see if the mechanism accounts for the formation of two stereoisomers. The positively charged carbon of the carbocation intermediate is sp^2 hybridized, and the three bonds connected to an sp^2 hybridized carbon are in the same plane. In the second step of the S_N1 reaction, the nucleophile can approach the carbocation from either side of the plane.

If the nucleophile attacks the side of the carbon from which the leaving group departed, the product will have the same relative configuration as that of the reacting alkyl halide. If, however, the nucleophile attacks the opposite side of the carbon, the product will have the inverted configuration relative to the configuration of the alkyl halide. Thus, the mechanism accounts for the observation that the reaction of an alkyl halide in which the leaving group is attached to an asymmetric center forms two stereoisomers—attack of the nucleophile on one side of the planar carbocation forms one stereoisomer and attack on the other side produces the other stereoisomer. The product of the reaction, therefore, is a pair of enantiomers.

if the leaving group in an S$_N$1 reaction is attached to an asymmetric center, a pair of enantiomers will be formed as products

PROBLEM 7◆

Arrange the following alkyl bromides in order of decreasing reactivity in an S$_N$1 reaction: isopropyl bromide, propyl bromide, *tert*-butyl bromide, methyl bromide.

10.5   Factors Affecting S$_N$1 Reactions

The Leaving Group

Because the rate-determining step of an S$_N$1 reaction is the dissociation of the alkyl halide to form a carbocation, two factors affect the rate of an S$_N$1 reaction: the ease with which the leaving group dissociates from the carbon and the stability of the carbocation that is formed. In the preceding section, we saw that tertiary alkyl halides are more reactive than secondary alkyl halides, which are more reactive than primary alkyl halides. This is because the more substituted the carbocation is, the more stable it is and therefore the easier it is to form. But how do we rank the relative reactivity of a series of alkyl halides with different leaving groups that dissociate to form the same carbocation?

As in the case of the S$_N$2 reaction, there is a direct relationship between basicity and leaving ability in the S$_N$1 reaction: The weaker the base, the less tightly it is bonded to the carbon and the easier it is to break the carbon–halogen bond. As a result, an alkyl iodide is the most reactive and an alkyl fluoride is the least reactive of the alkyl halides in both S$_N$1 and S$_N$2 reactions.

relative reactivities of alkyl halides in an S$_N$1 reaction

most reactive $>$ RI $>$ RBr $>$ RCl $>$ RF $<$ least reactive

The Nucleophile

We have seen that the rate-determining step of an S$_N$1 reaction is formation of the carbocation. Because the nucleophile comes into play *after* the rate-determining step, the reactivity of the nucleophile has no effect on the rate of an S$_N$1 reaction (Figure 10.4).

PROBLEM 8◆

Arrange the following alkyl halides in order of decreasing reactivity in an S$_N$1 reaction: 2-bromopentane, 2-chloropentane, 1-chloropentane, 3-bromo-3-methylpentane.

10.6 Comparison of the S$_N$2 and S$_N$1 Reactions

The characteristics of S$_N$2 and S$_N$1 reactions are summarized in Table 10.3. Remember that the "2" in "S$_N$2" and the "1" in "S$_N$1" refer to the number of molecules in the rate-determining step and not to the number of steps in the mechanism. In fact, the opposite is true: An S$_N$2 reaction proceeds by a *one*-step mechanism, and an S$_N$1 reaction proceeds by a *two*-step mechanism with a carbocation intermediate.

Table 10.3 Comparison of S$_N$2 and S$_N$1 Reactions	
S$_N$2	**S$_N$1**
A one-step mechanism	A two-step mechanism
A bimolecular rate-determining step	A unimolecular rate-determining step
Product has inverted configuration relative to the reactant	Products have both retained and inverted configurations relative to the reactant
Reactivity order: methyl $>1° > 2° > 3°$	Reactivity order: $3° > 2° > 1° >$ methyl

If the halogen is attached to an asymmetric center, the product of an S$_N$2 reaction will have a configuration that is inverted relative to that of the reactant.

the configuration is inverted relative to that of the reactant

(S)-2-bromobutane + HO⁻ →(S$_N$2 conditions)→ (R)-2-butanol + Br⁻

If the leaving group is attached to an asymmetric center, an S$_N$2 reaction forms the isomer with the inverted configuration.

If the leaving group is attached to an asymmetric center, an S$_N$1 reaction forms a pair of enantiomers.

If the same reaction is carried out under S$_N$1 conditions, two substitution products are obtained—one with the same relative configuration as the reactant and one with the inverted configuration.

product with inverted configuration   product with retained configuration

(S)-2-bromobutane + H$_2$O →(S$_N$1 conditions)→ (R)-2-butanol + (S)-2-butanol + HBr

Movies:
S$_N$1 inversion;
S$_N$1 retention

The difference between the products obtained from an S$_N$1 reaction and from an S$_N$2 reaction is a little easier to visualize in the case of cyclic compounds when they are drawn as planar rings. For example, when *cis*-1-bromo-4-methylcyclohexane undergoes an S$_N$2 reaction, only the trans product is obtained because the carbon bonded to the leaving group is attacked by the nucleophile only on its back side.

cis-1-bromo-4-methylcyclohexane + HO⁻ →(S$_N$2 conditions)→ *trans*-4-methylcyclohexanol + Br⁻

However, when *cis*-1-bromo-4-methylcyclohexane undergoes an S_N1 reaction, both the cis and the trans products are formed because the nucleophile can approach the carbocation intermediate from either side.

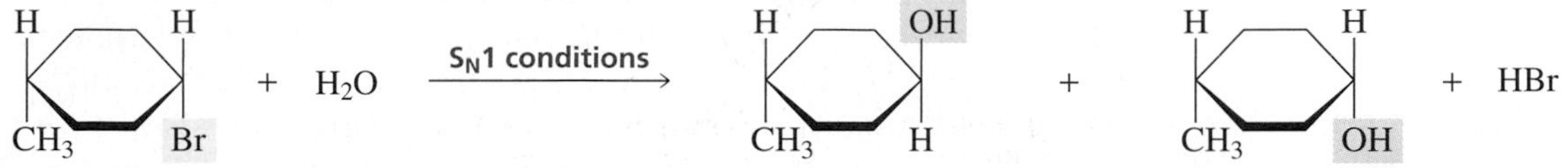

cis-1-bromo-4-methylcyclohexane $+ H_2O$ $\xrightarrow{S_N1 \text{ conditions}}$ trans-4-methylcyclohexanol $+$ cis-4-methylcyclohexanol $+ HBr$

PROBLEM 9

Give the substitution products that will be obtained from the following reactions if

a. the reaction is carried out under conditions that favor an S_N2 reaction.

b. the reaction is carried out under conditions that favor an S_N1 reaction.

    1. *trans*-1-iodo-4-methylcyclohexane + sodium methoxide/methanol

    2. *cis*-1-chloro-3-methylcyclobutane + sodium hydroxide/water

---

10.7 Elimination Reactions of Alkyl Halides

In addition to undergoing substitution reactions, alkyl halides can also undergo elimination reactions. In an elimination reaction, the halogen (X) is removed from one carbon and a hydrogen is removed from an *adjacent* carbon. A double bond is formed between the two carbons from which the atoms are eliminated. Therefore, the product of an elimination reaction is an alkene.

$CH_3CH_2CH_2X + Y^-$

    substitution $\longrightarrow$ $CH_3CH_2CH_2Y + X^-$

    elimination $\longrightarrow$ $CH_3CH=CH_2 + HY + X^-$

new double bond

The E2 Reaction

Just as there are two important nucleophilic substitution reactions—S_N1 and S_N2—there and two important elimination reactions—E1 and E2. The reaction of *tert*-butyl bromide with hydroxide ion is an example of an **E2 reaction**; "E" stands for *elimination* and "2" stands for *bimolecular*.

$CH_3-\underset{\underset{Br}{|}}{\overset{\overset{CH_3}{|}}{C}}-CH_3 + HO^- \longrightarrow CH_2=\underset{}{\overset{\overset{CH_3}{|}}{C}}-CH_3 + H_2O + Br^-$

tert-butyl bromide   2-methylpropene

3-D Molecule: 2-methylpropene

The rate of the E2 reaction depends on the concentrations of both *tert*-butyl bromide and hydroxide ion.

$$\text{rate} = k[\text{alkyl halide}][\text{base}]$$

The following mechanism agrees with the observation that both *tert*-butyl bromide and hydroxide ion are involved in the rate-determining step of the reaction:

mechanism of the E2 reaction

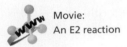
Movie:
An E2 reaction

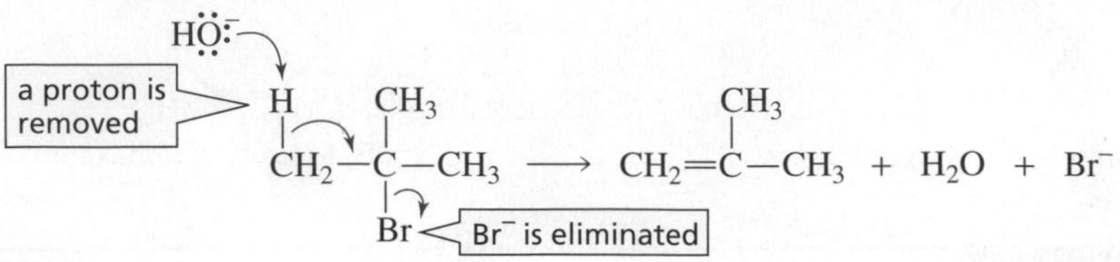

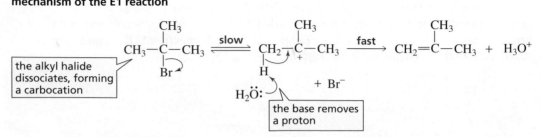

The mechanism shows that an E2 reaction is a one-step reaction: the base removes a proton from a carbon that is adjacent to the carbon bonded to the halogen. As the proton is removed, the electrons that the hydrogen shared with carbon move toward the carbon bonded to the halogen. As these electrons move toward the carbon, the halogen leaves, taking its bonding electrons with it. The electrons that were bonded to the hydrogen in the reactant have formed a π bond in the product.

The E1 Reaction

The reaction of *tert*-butyl bromide with water to form 2-methylpropene is an example of an **E1 reaction**.

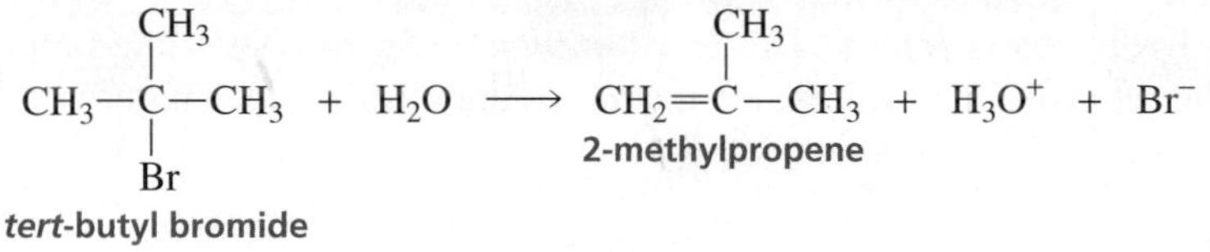

$$CH_3-\overset{\overset{\displaystyle CH_3}{|}}{\underset{\underset{\displaystyle Br}{|}}{C}}-CH_3 \ + \ H_2O \ \longrightarrow \ CH_2{=}\overset{\overset{\displaystyle CH_3}{|}}{C}-CH_3 \ + \ H_3O^+ \ + \ Br^-$$

tert-butyl bromide   **2-methylpropene**

The rate of the reaction depends only on the concentration of the alkyl halide. Thus, "E" stands for *elimination* and "1" stands for *unimolecular*.

rate = k[alkyl halide]

We know, then, that only the alkyl halide is involved in the rate-determining step of the reaction. Therefore, there must be at least two steps in the reaction.

mechanism of the E1 reaction

Movie:
An E1 reaction

The foregoing mechanism shows that an E1 reaction has two steps. In the first step, the alkyl halide dissociates, forming a carbocation. In the second step, the base forms the elimination product by removing a proton from a carbon that is adjacent to the positively charged carbon. Because the first step of the reaction is the rate-determining step, increasing the concentration of the base—which comes into play only in the second step of the reaction—has no effect on the rate of the reaction.

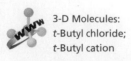
3-D Molecules:
t-Butyl chloride;
t-Butyl cation

Alkyl iodides are the most reactive and alkyl fluorides the least reactive in both E2 and E1 reactions, because weaker bases are better leaving groups (Section 10.3).

The weaker the base, the better it is as a leaving group.

relative reactivities of alkyl halides in E2 and E1 reactions

most reactive $\rightarrow$ RI > RBr > RCl > RF $\leftarrow$ least reactive

increasing reactivity

INVESTIGATING NATURALLY OCCURRING HALOGEN-CONTAINING COMPOUNDS

Complex halogen-containing compounds isolated from marine organisms have been found to have interesting and potent biological activity. Compounds produced in nature are called *natural products*. *Cyclocinamide A* is a natural product that comes from a marine sponge. This compound, as well as a host of analogs, has unique antitumor properties that are currently being exploited in the development of new anticancer drugs.

Jasplankinolide, also found in a sponge, modulates the formation and depolymerization of actin microtubules. Microtubules are found in all cells and are used for motile events, such as transportation of vesicles, migration, and cell division. Jasplankinolide is being used to further our understanding of these processes. Notice that *cyclocinamide A* has four asymmetric centers and *jasplankinolide* has six asymmetric centers. Because of the great diversity of marine life, the ocean probably contains many compounds with useful medicinal properties waiting to be discovered by scientists.

cyclocinamide A

jasplankinolide

PROBLEM 10◆

a. Which of the alkyl halides is more reactive in an E2 reaction?

$$CH_3CHCl \quad \text{or} \quad CH_3CHBr$$
$$\qquad | \qquad\qquad\qquad |$$
$$\quad CH_3 \qquad\qquad\quad CH_3$$

b. Which of the above alkyl halides is more reactive in an E1 reaction?

10.8   Products of Elimination Reactions

In an elimination reaction, the hydrogen is removed from a β-carbon. (The α-carbon is the carbon bonded to the halogen; the β-carbon is the carbon adjacent to the α-carbon.) An alkyl halide such as 2-bromopropane has two β-carbons from which a hydrogen can be removed in an E2 reaction. Because the two β-carbons are identical, the only product of this elimination reaction is propene.

β-carbons

$$CH_3CHCH_3 + CH_3O^- \longrightarrow CH_3CH{=}CH_2 + CH_3OH + Br^- + H_3$$
$$\text{propene}$$

α-carbon | Br

2-bromopropane

In contrast, 2-bromobutane has two structurally different β-carbons from which a hydrogen can be removed. So when 2-bromobutane reacts with a base, two elimination products are formed: 2-butene and 1-butene.

Tutorial:
E2 Elimination
regiochemistry

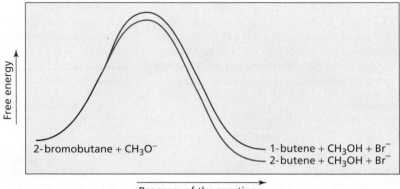

Which of the alkenes is obtained in greater yield? The answer is just what we would predict: The more stable alkene is formed in greater yield, because it has the more stable transition state leading to its formation and therefore is formed faster (Figure 10.5).

Figure 10.5 ▶
Reaction coordinate diagram for the E2 reaction of 2-bromobutane and methoxide ion.

We know that the stability of an alkene depends on the number of alkyl substituents bonded to its sp^2 carbons: The greater the number of substituents, the more stable is the alkene (Section 4.6). Therefore, 2-butene, with a total of two methyl substituents bonded to its sp^2 carbons, is more stable than 1-butene, with one ethyl substituent. Thus, 2-butene is formed faster than 1-butene. *Notice that the more substituted alkene is obtained when a hydrogen is removed from the β-carbon that is bonded to the fewest hydrogens.*

The most stable alkene is generally the most substituted alkene.

Because elimination from a tertiary alkyl halide typically leads to a more substituted alkene than does elimination from a secondary alkyl halide, and elimination from a secondary alkyl halide generally leads to a more substituted alkene than does elimination from a primary alkyl halide, the relative reactivities of alkyl halides in an E2 reaction are as follows:

relative reactivities of alkyl halides in an E2 reaction

tertiary alkyl halide > secondary alkyl halide > primary alkyl halide

The major product of an E1 reaction, such as the reaction of 2-chloro-2-methylbutane with water, is also the more stable alkene. 2-Methyl-2-butene (with a total of three alkyl substituents) is more stable than 2-methyl-1-butene (with only two alkyl substituents).

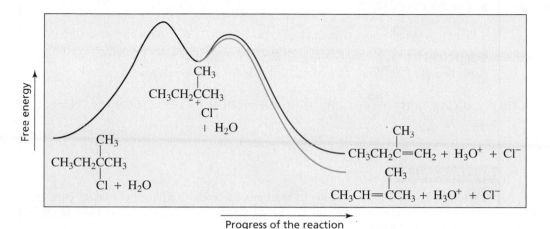

2-chloro-2-methylbutane

We see again that the more stable alkene is formed in greater yield, because the more stable alkene has the more stable transition state leading to its formation (Figure 10.6).

3-D Molecule:
2-Chloro-2-methylbutane

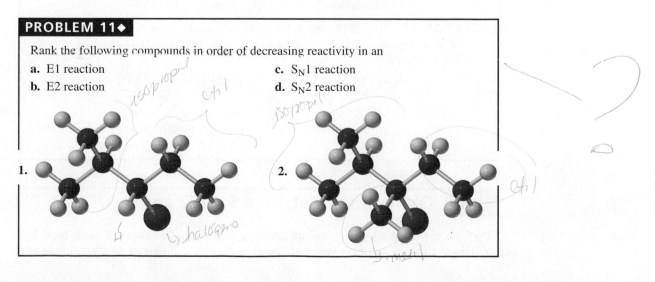

▲ **Figure 10.6**
Reaction coordinate diagram for the E1 reaction of 2-chloro-2-methylbutane. The major product is the more substituted alkene because its greater stability causes the transition state leading to its formation to be more stable.

3-D Molecule:
2-Methyl-2-butene

Because the rate-limiting step of an E1 reaction is the formation of the carbocation intermediate, alkyl halides have following relative reactivities since a tertiary carbocation is more stable, and is therefore easier to form, than a secondary carbocation, which in turn is more stable and easier to form than a primary carbocation (Section 5.2).

relative reactivities of alkyl halides in an E1 reaction

tertiary alkyl halide > secondary alkyl halide > primary alkyl halide

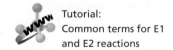
Tutorial:
Common terms for E1
and E2 reactions

Thus, for *both* E2 and E1 reactions, tertiary alkyl halides are the most reactive and primary alkyl halides are the least reactive, and the major product is the most stable alkene.

The major product of an E2 or E1 reaction is the most stable alkene.

PROBLEM 11◆

Rank the following compounds in order of decreasing reactivity in an

a. E1 reaction

c. S_N1 reaction

b. E2 reaction

d. S_N2 reaction

1.   2.

PROBLEM 12◆ | **SOLVED**

For each of the following alkyl halides, determine the major product that is formed when the alkyl halide reacts with hydroxide ion in an elimination reaction:

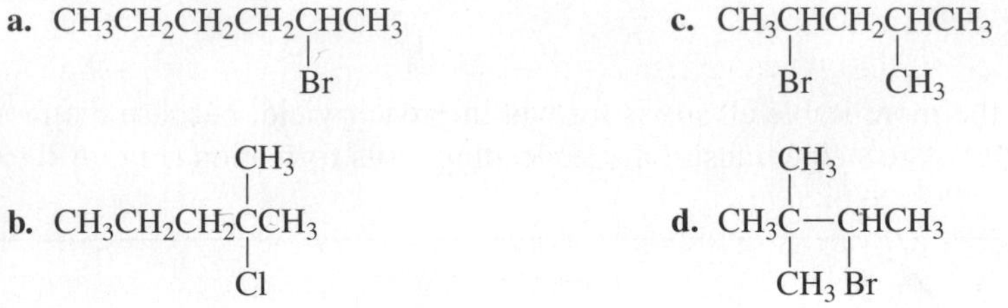

a. $CH_3CH_2CH_2CH_2CHCH_3$
$\quad\quad\quad\quad\quad\quad\quad\quad\quad\,|$
$\quad\quad\quad\quad\quad\quad\quad\quad\quad Br$

c. $CH_3CHCH_2CHCH_3$
$\quad\quad\quad\quad\quad\,|\quad\quad\,\,|$
$\quad\quad\quad\quad\quad Br\quad CH_3$

$\quad\quad\quad\quad\quad\quad\quad\quad\quad CH_3$
$\quad\quad\quad\quad\quad\quad\quad\quad\quad |$
b. $CH_3CH_2CH_2CCH_3$
$\quad\quad\quad\quad\quad\quad\quad\quad\quad |$
$\quad\quad\quad\quad\quad\quad\quad\quad\quad Cl$

$\quad\quad\quad\quad\quad\quad\quad\quad\quad CH_3$
$\quad\quad\quad\quad\quad\quad\quad\quad\quad |$
d. $CH_3C—CHCH_3$
$\quad\quad\quad\quad\quad\quad\quad\quad\quad |\quad\,|$
$\quad\quad\quad\quad\quad\quad\quad\quad\quad CH_3\,Br$

SOLUTION TO 12a More 2-hexene will be formed than 1-hexene: 2-Hexene is more stable because it has more alkyl substituents bonded to its sp^2 carbons.

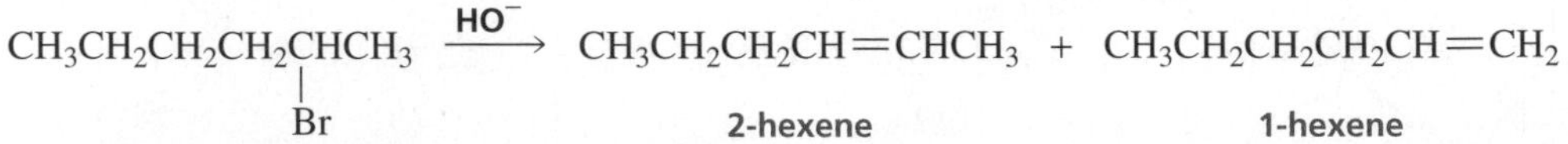

$$CH_3CH_2CH_2CH_2CHCH_3 \xrightarrow{\;HO^-\;} CH_3CH_2CH_2CH{=}CHCH_3 \;+\; CH_3CH_2CH_2CH_2CH{=}CH_2$$
$$\quad\quad\quad\quad\quad\quad\quad\quad\quad |$$
$$\quad\quad\quad\quad\quad\quad\quad\quad\quad Br$$

$\quad\quad\quad\quad\quad\quad\quad\quad\quad\quad\quad\quad\quad\quad\quad\quad\quad$ **2-hexene** $\quad\quad\quad\quad\quad\quad$ **1-hexene**
$\quad\quad\quad\quad\quad\quad\quad\quad\quad\quad\quad\quad\quad\quad\quad\quad$ **major product**

PROBLEM 13◆

Three alkenes are formed from the E1 reaction of 3-bromo-2,3-dimethylpentane. Give the structures of the alkenes, and rank them according to the amount that would be formed. (Ignore stereoisomers.)

We determined that 2-butene is the major elimination product when 2-bromobutane reacts with hydroxide ion (page 246). 2-Butene, however, has two stereoisomers, (E)-2-butene and (Z)-2-butene. Which of the stereoisomers is obtained in greater yield? Again, we find that the more stable product is the one formed in greater yield. Recall that the more stable alkene is the one with the *bulkiest groups on opposite sides of the double bond* (Section 4.6). Therefore, more (E)-2-butene is formed than (Z)-2-butene.

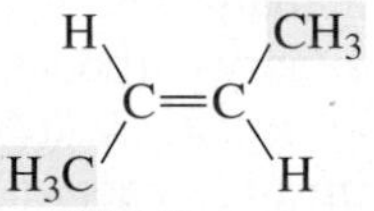

> **The major product of an E2 or E1 reaction is the alkene with the bulkiest substituents on opposite sides of the double bond.**

Tutorial:
E2 Stereochemistry

$\quad\quad\quad\quad$ H$\quad\quad$ CH$_3$ $\quad\quad\quad\quad\quad\quad$ H$\quad\quad\quad$ H
$\quad\quad\quad\quad\quad$C$=$C $\quad\quad\quad\quad\quad\quad\quad$ C$=$C
$\quad\quad$ H$_3$C$\quad\quad\quad$ H $\quad\quad\quad\quad\quad$ H$_3$C$\quad\quad\quad$ CH$_3$

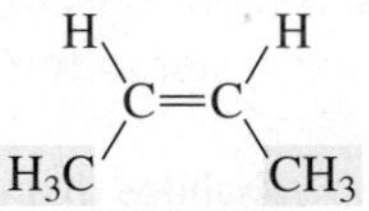

$\quad\quad\quad$ **(E)-2-butene** $\quad\quad\quad\quad\quad\quad\quad$ **(Z)-2-butene**
$\quad\quad\quad\quad$ **more stable** $\quad\quad\quad\quad\quad\quad\quad$ **less stable**

PROBLEM 14 | **SOLVED**

For each of the major elimination products determined in Problem 12 that can exist as stereoisomers, which stereoisomer is obtained in greater yield?

SOLUTION TO 14a 2-Hexene has two stereoisomers. More (E)-2-hexene will be formed than (Z)-2-hexene because (E)-2-hexene is more stable, since the bulkiest substituents are on opposite sides of the double bond.

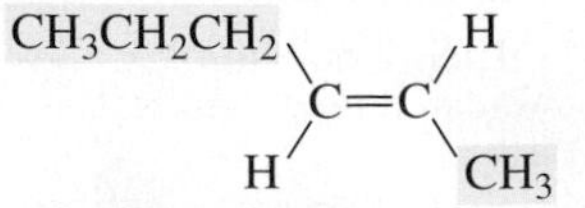

$\quad\quad$ CH$_3$CH$_2$CH$_2$ $\quad\quad\quad$ H $\quad\quad\quad$ CH$_3$CH$_2$CH$_2$ $\quad\quad$ CH$_3$
$\quad\quad\quad\quad\quad\quad\quad$C$=$C $\quad\quad\quad\quad\quad\quad\quad\quad\quad$ C$=$C
$\quad\quad\quad\quad$ H $\quad\quad\quad$ CH$_3$ $\quad\quad\quad\quad\quad\quad$ H $\quad\quad\quad$ H

$\quad\quad\quad\quad$ **(E)-2-hexene** $\quad\quad\quad\quad\quad\quad\quad\quad$ **(Z)-2-hexene**
$\quad\quad\quad\quad$ **major product**

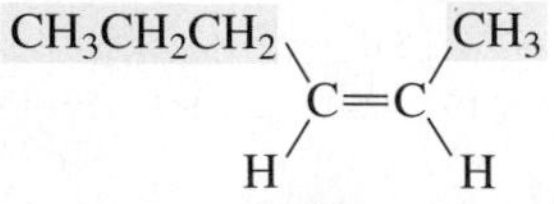

10.9 Competition Between S_N2/E2 and S_N1/E1

You have seen that alkyl halides can undergo four types of reactions: S_N2, S_N1, E2, and E1. At this point, it may seem a bit overwhelming to be given an alkyl halide and a nucleophile/base and be asked to predict the products of the reaction. Therefore, we need to organize what we know about the reactions of alkyl halides to make it a little easier to predict the products of any given reaction. Notice that HO^- is called a nucleophile in substitution reactions (because it attacks a carbon) and a base in elimination reactions (because it removes a proton).

First, we must determine whether the alkyl halide will undergo predominantly $S_N2/E2$ reactions or $S_N1/E1$ reactions. Two factors determine whether $S_N2/E2$ or $S_N1/E1$ reactions predominate: (1) the *concentration* of the nucleophile/base and (2) the *reactivity* of the nucleophile/base.

To understand how the concentration and the reactivity of the nucleophile/base affect which set of reactions predominates, we must look at the overall rate law for the reaction. The overall rate law is the sum of the individual rate laws for the S_N1, S_N2, E1, and E2 reactions. (Subscripts have been added to the rate constants to indicate that they have different values.)

$$\text{rate} = k_1[\text{alkyl halide}] + k_2[\text{alkyl halide}][\text{nucleophile}] + k_3[\text{alkyl halide}] + k_4[\text{alkyl halide}][\text{base}]$$

contribution to the rate by an S_N1 reaction	contribution to the rate by an S_N2 reaction	contribution to the rate by an E1 reaction	contribution to the rate by an E2 reaction

From the overall rate law, you can see that increasing the *concentration* of the nucleophile/base has no effect on the rate of the S_N1 and E1 reactions, because the concentration of the nucleophile/base is not in their rate laws. In contrast, increasing the *concentration* of the nucleophile/base increases the rate of the S_N2 and E2 reactions, because the concentration of the nucleophile/base is in their rate laws. Similarly, increasing the *reactivity* of the nucleophile/base increases the rate of the S_N2 and E2 reactions by increasing the value of the rate constants (k_2 and k_4), because more reactive nucleophiles/bases are better able to displace the leaving group. Increasing the reactivity of the nucleophile/base has no effect on the rate of S_N1 and E1 reactions, because the slow step in these reactions does not involve the nucleophile/base. In sum,

- S_N2 and E2 reactions are favored by a high concentration of a good nucleophile/strong base.
- S_N1 and E1 reactions are favored by a poor nucleophile/weak base, because a poor nucleophile/weak base disfavors S_N2 and E2 reactions.

Look back at the S_N1 and E1 reactions in previous sections and notice that they all have poor nucleophiles/weak bases (H_2O, CH_3OH), whereas the S_N2 and E2 reactions have good nucleophiles/strong bases (HO^-, CH_3O^-). In other words, a good nucleophile/strong base is used to encourage an $S_N2/E2$ reactions, and a poor nucleophile/weak base is used to encourage an $S_N1/E1$ reactions by discouraging an $S_N2/E2$ reactions.

An S_N2 reaction of an alkyl halide is favored by a high concentration of a good nucleophile/strong base.

An S_N1 reaction of an alkyl halide is favored by a poor nucleophile/weak base.

PROBLEM-SOLVING STRATEGY

This problem will help you determine whether an alkyl halide will undergo $S_N1/E1$ reactions or $S_N2/E2$ reactions.

Give the configuration of the substitution product(s) that will be obtained from the reaction of each of the following alkyl halides with the indicated nucleophile:

a.

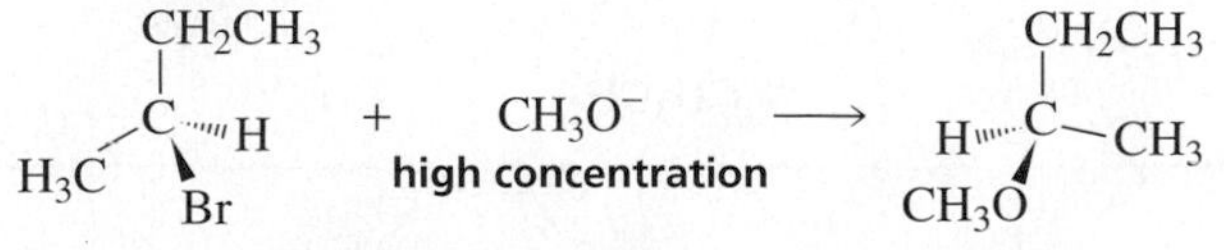

Because a high concentration of a good nucleophile is used, we can predict that the alkyl halide will undergo $S_N2/E2$ reactions. Therefore, the substitution product will have the inverted configuration relative to the configuration of the reactant. (An easy way to draw the inverted product is to draw the mirror image of the reacting alkyl halide and then put the nucleophile in the same location as the leaving group.)

b. 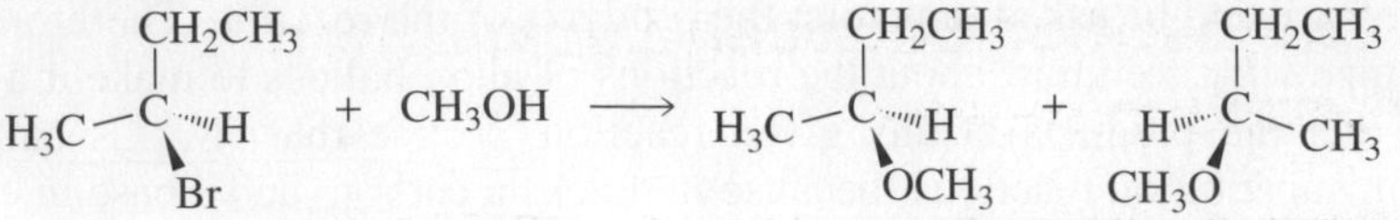 H$_3$C—C(CH$_2$CH$_3$)(—H)—Br + CH$_3$OH ⟶ H$_3$C—C(CH$_2$CH$_3$)(—H)—OCH$_3$ + H—C(CH$_2$CH$_3$)(CH$_3$O)—CH$_3$

Because a poor nucleophile is used, we can predict that the alkyl halide will undergo S$_N$1/E1 reactions. Therefore, two substitution products will be formed: one with the retained configuration and one with the inverted configuration, relative to the configuration of the reactant.

c. 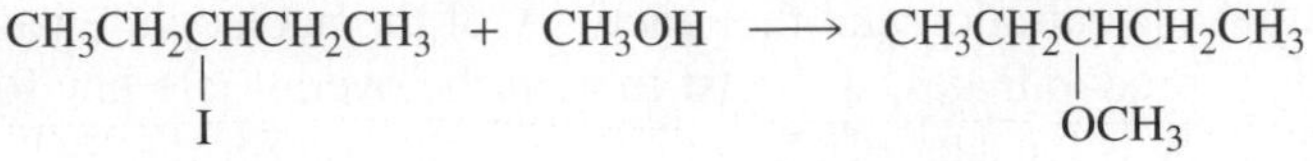 CH$_3$CH$_2$CHCH$_2$CH$_3$ + CH$_3$OH ⟶ CH$_3$CH$_2$CHCH$_2$CH$_3$
          | |
          I   OCH$_3$

The poor nucleophile allows us to predict that the alkyl halide will undergo S$_N$1/E1 reactions. However, the product does not have an asymmetric center, so it does not have stereoisomers. Therefore, only one substitution product will be formed.

Now continue on to Problem 15.

PROBLEM 15

Give the configuration of the substitution product(s) that will be obtained from the reaction of the following alkyl halides with the indicated nucleophile:

a. 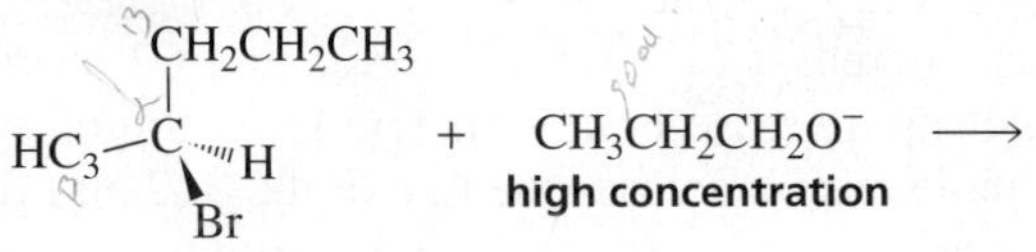 HC$_3$—C(CH$_2$CH$_2$CH$_3$)(—H)—Br + CH$_3$CH$_2$CH$_2$O$^-$ ⟶
                                              high concentration

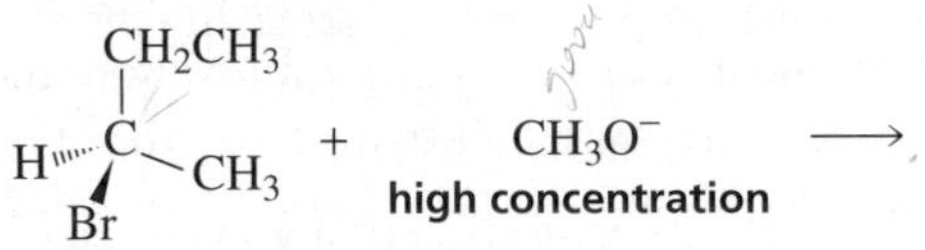

b. H—C(CH$_2$CH$_3$)(—CH$_3$)—Br + CH$_3$O$^-$ ⟶
                                        high concentration

c. H—C(CH$_2$CH$_3$)(—CH$_3$)—Br + CH$_3$OH ⟶

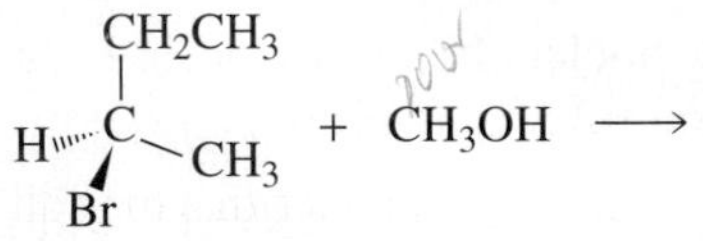

PROBLEM 16 SOLVED

What is the major elimination product that would be obtained from the reactions shown in Problem 15?

SOLUTION TO 16a   In determining the major elimination product, we do not need to be concerned whether it is an E1 or an E2 reaction, because both form the same major product. The major elimination product is 2-pentene. (Remember, the major elimination product is obtained by removing a hydrogen from the β-carbon bonded to the fewest hydrogens.) 2-Pentene has two stereoisomers. More (*E*)-2-pentene will be formed than (*Z*)-2-pentene because (*E*)-2-pentene is more stable, since the largest substituents are on opposite sides of the double bond.

CH$_3$CH$_2$ H   CH$_3$CH$_2$ CH$_3$
       \   /   \   /
        C=C   C=C
       /   \   /   \
      H   CH$_3$ H   H

(*E*)-2-pentene   (*Z*)-2-pentene
major product

10.10   Competition Between Substitution and Elimination

Having decided whether the reaction conditions favor S_N2/E2 reactions or S_N1/E1 reactions, we must next decide whether the reaction will form the substitution product, the elimination product, or both substitution and elimination products. *The relative amounts of substitution and elimination products depend on whether the alkyl halide is primary, secondary, or tertiary.*

S_N2/E2 Conditions

Let's first consider conditions that lead to S_N2/E2 reactions (a high concentration of a good nucleophile/strong base). The negatively charged species can act as a nucleophile and attack the back side of the α-carbon to form the substitution product, or it can act as a base and remove a proton from a β-carbon to form the elimination product. Thus, the two reactions compete with each other. Notice that both reactions occur for the same reason: The electron-withdrawing halogen gives the carbon to which it is bonded a partial positive charge.

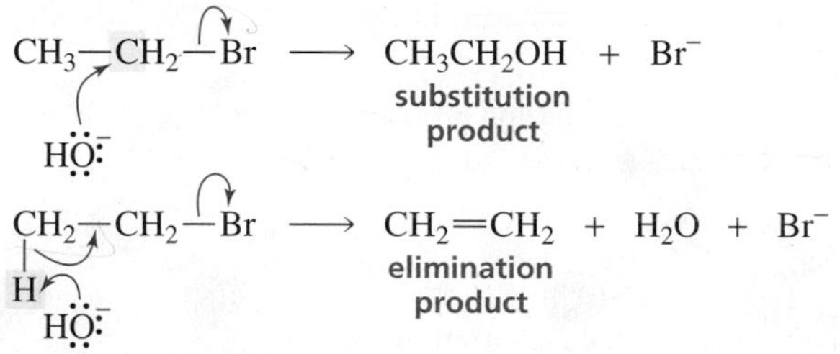

$$CH_3\!-\!CH_2\!-\!Br \longrightarrow CH_3CH_2OH + Br^-$$

$$H\ddot{O}:^-$$

substitution product

$$CH_2\!-\!CH_2\!-\!Br \longrightarrow CH_2\!=\!CH_2 + H_2O + Br^-$$

$$H\quad\ddot{O}:^-$$
$$H\ddot{O}:^-$$

elimination product

The relative reactivities of alkyl halides in S_N2 and E2 reactions are shown in Table 10.4. Because a *primary* alkyl halide is the most reactive in an S_N2 reaction and the least reactive in an E2 reaction, a primary alkyl halide forms principally the substitution product in a reaction carried out under conditions that favor S_N2/E2 reactions. In other words, substitution wins the competition.

> Primary alkyl halides undergo primarily substitution under S_N2/E2 conditions.

a primary alkyl halide

$$CH_3CH_2CH_2Br + CH_3O^- \xrightarrow{CH_3OH} CH_3CH_2CH_2OCH_3 + CH_3CH\!=\!CH_2 + CH_3OH + Br^-$$

propyl bromide   methyl propyl ether   propene
90%   10%

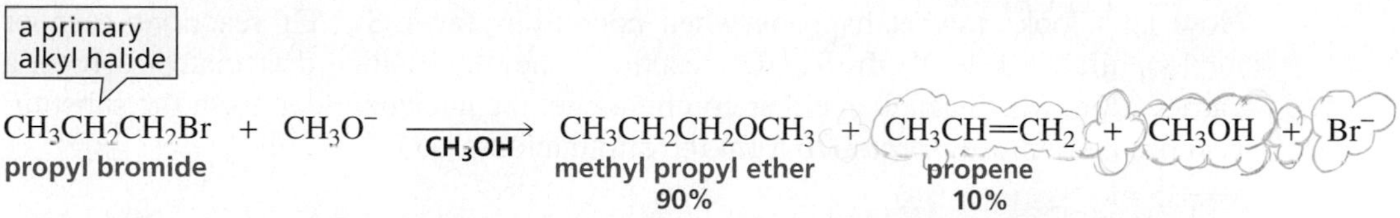

Table 10.4	Relative Reactivities of Alkyl Halides		
In an S_N2 reaction:	$1° > 2° > 3°$	In an S_N1 reaction:	$3° > 2° > 1°$
In an E2 reaction:	$3° > 2° > 1°$	In an E1 reaction:	$3° > 2° > 1°$

A *secondary* alkyl halide can form both substitution and elimination products under S_N2/E2 conditions.

a secondary alkyl halide

$$\underset{\text{2-chloropropane}}{\overset{\overset{\displaystyle Cl}{|}}{CH_3CHCH_3}} + CH_3O^- \xrightarrow{CH_3OH} \underset{\substack{\text{ethyl isopropyl ether}\\25\%}}{\overset{\overset{\displaystyle OCH_2CH_3}{|}}{CH_3CHCH_3}} + \underset{\substack{\text{propene}\\75\%}}{CH_3CH\!=\!CH_2} + CH_3OH + Cl^-$$

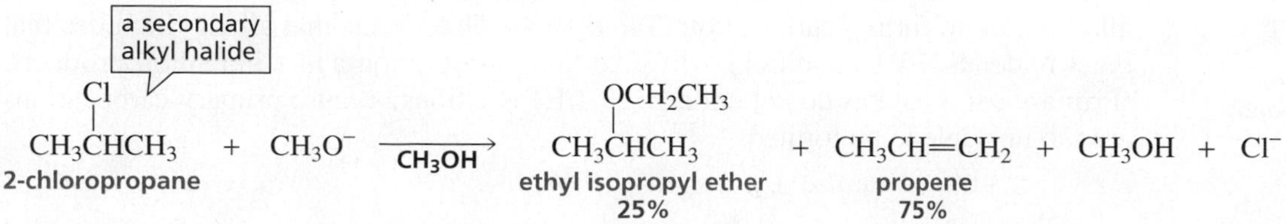

Tertiary alkyl halides undergo only elimination under S_N2/E2 conditions.

A *tertiary* alkyl halide is the least reactive of the alkyl halides in an S_N2 reaction and the most reactive in an E2 reaction (Table 10.4). Consequently, a tertiary alkyl halide forms *only* the elimination product in a reaction carried out under conditions that favor S_N2/E2 reactions.

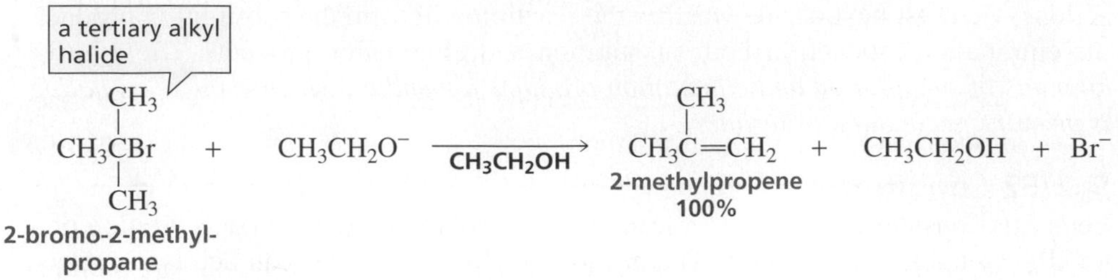

a tertiary alkyl halide

$$CH_3\underset{\underset{CH_3}{|}}{\overset{\overset{CH_3}{|}}{C}}Br \quad + \quad CH_3CH_2O^- \xrightarrow{\text{CH}_3\text{CH}_2\text{OH}} CH_3\underset{}{\overset{\overset{CH_3}{|}}{C}}{=}CH_2 \quad + \quad CH_3CH_2OH \quad + \quad Br^-$$

2-bromo-2-methyl-propane

2-methylpropene
100%

PROBLEM 17

Draw the stereoisomers that would be obtained in greatest yield from the reaction of the following alkyl chlorides with hydroxide ion:

a.

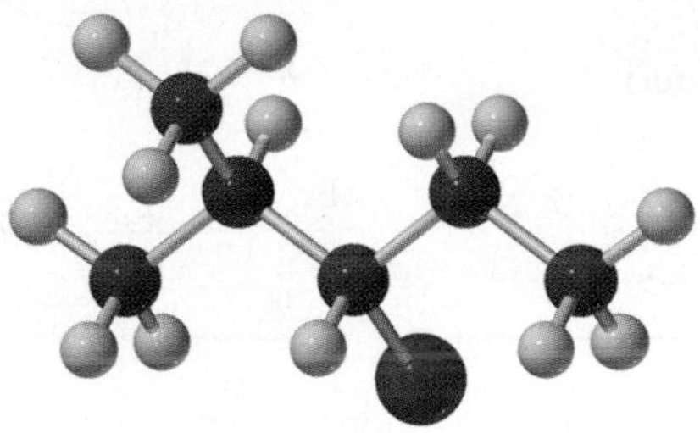

b.

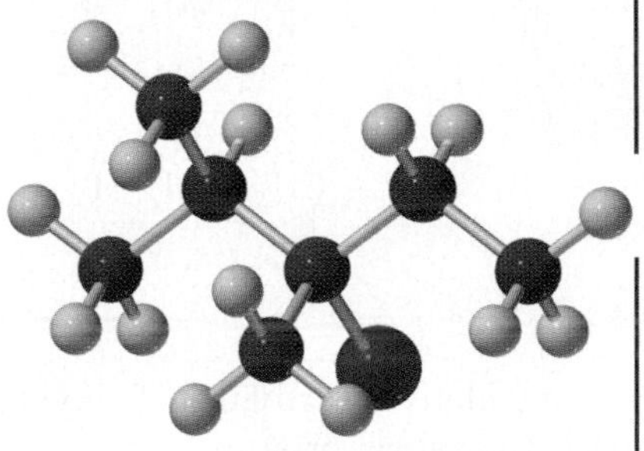

S_N1/E1 Conditions

Now let's look at what happens when conditions favor S_N1/E1 reactions (a poor nucleophile/weak base). In S_N1/E1 reactions, the alkyl halide dissociates to form a carbocation, which can then either combine with the nucleophile to form the substitution product or lose a proton to form the elimination product.

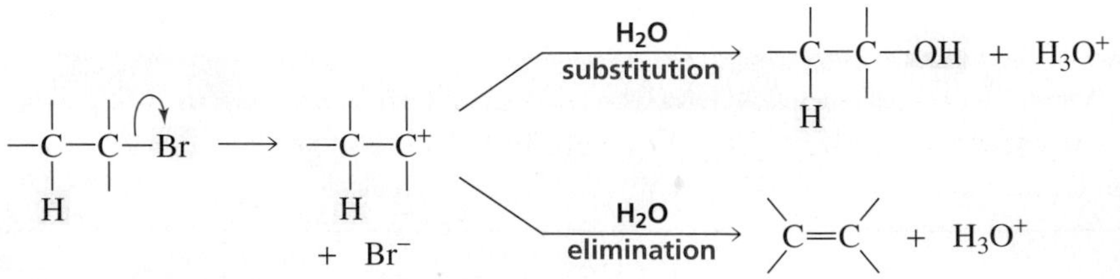

$$-\overset{|}{\underset{|}{C}}-\overset{|}{\underset{\underset{H}{|}}{C}}-Br \longrightarrow -\overset{|}{\underset{|}{C}}-\overset{|}{\underset{\underset{H}{|}}{C}}{}^+ + Br^-$$

substitution $\xrightarrow{\text{H}_2\text{O}}$ $-\overset{|}{\underset{\underset{H}{|}}{C}}-\overset{|}{\underset{|}{C}}-OH + H_3O^+$

elimination $\xrightarrow{\text{H}_2\text{O}}$ $\overset{}{C}{=}\overset{}{C} + H_3O^+$

Primary alkyl halides do not form carbocations; therefore they cannot undergo S_N1 and E1 reactions.

Alkyl halides have the same order of reactivity in S_N1 reactions as they do in E1 reactions because both reactions have the same rate-determining step: dissociation of the alkyl halide to form a carbocation (Table 10.4). This means that all alkyl halides that react under S_N1/E1 conditions will give both substitution and elimination products. (Primary alkyl halides do not undergo S_N1/E1 reactions because primary carbocations are too unstable to be formed.)

Table 10.5 summarizes the products obtained when alkyl halides react with nucleophiles/bases under $S_N2/E2$ and $S_N1/E1$ conditions.

Table 10.5 Summary of the Products Expected in Substitution and Elimination Reactions		
Class of alkyl halide	**Products under $S_N2/E2$ Conditions**	**Products under $S_N1/E1$ Conditions**
Primary alkyl halide	Primarily substitution	Cannot undergo $S_N1/E1$ reactions
Secondary alkyl halide	Both substitution and elimination	Both substitution and elimination
Tertiary alkyl halide	Only elimination	Both substitution and elimination

The stereoisomers obtained from substitution and elimination reactions are summarized in Table 10.6.

Table 10.6 Stereochemistry of Substitution and Elimination Reactions	
Reaction	**Products**
S_N1	Both stereoisomers (R and S) are formed.
E1	Both E and Z stereoisomers are formed (more of the stereoisomer with the bulkiest groups on opposite sides of the double bond).
S_N2	Only the inverted product is formed.
E2	Both E and Z stereoisomers are formed (more of the stereoisomer with the bulkiest groups on opposite sides of the double bond).

PROBLEM 18◆

Indicate whether the alkyl halides listed will give both substitution and elimination products, primarily substitution products, only elimination products, or no products when they react with the following:

a. methanol under $S_N1/E1$ conditions
b. sodium methoxide under $S_N2/E2$ conditions

   1. 1-bromobutane   **3.** 2-bromobutane
   2. 1-bromo-2-methylpropane   **4.** 2-bromo-2-methylpropane

Tutorial:
S_N2 promoting factors

Tutorial:
E2 Promoting factors

PROBLEM 19◆

a. Which reacts faster in an S_N2 reaction?

   $CH_3CH_2CH_2Br$ or $CH_3CH_2CHCH_3$
                                    |
                                    Br

c. Which reacts faster in an S_N1 reaction?

                CH_3 CH_3
                | |
   $CH_3CHCH_2CHCH_3$ or $CH_3CH_2CH_2CCH_3$
              | |
           Br   Br

b. Which reacts faster in an E1 reaction?

d. Which reacts faster in an E2 reaction?

               CH_3 CH_3
              | |
   CH_3CCH_2Cl or $CH_3CCH_2CH_2Cl$
        | |
        CH_3 CH_3

PROBLEM 20

Which of the following reactions will go faster if the concentration of the nucleophile is increased?

a. 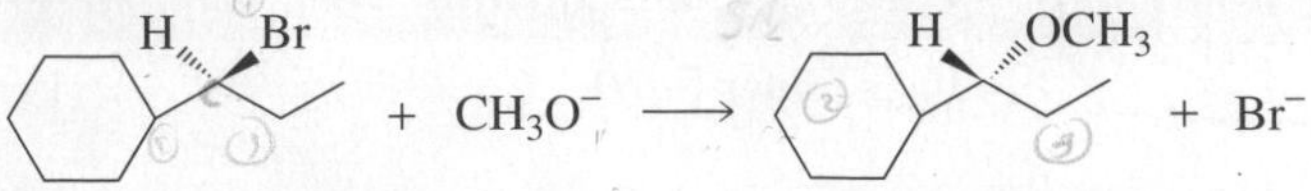
(cyclohexyl with H and Br) + CH_3O^- ⟶ (cyclohexyl with H and OCH_3) + Br^-

b. 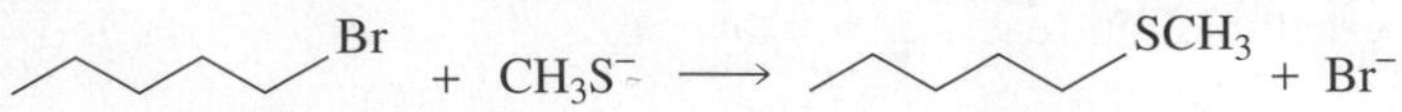
(pentyl-Br) + CH_3S^- ⟶ (pentyl-SCH_3) + Br^-

c. 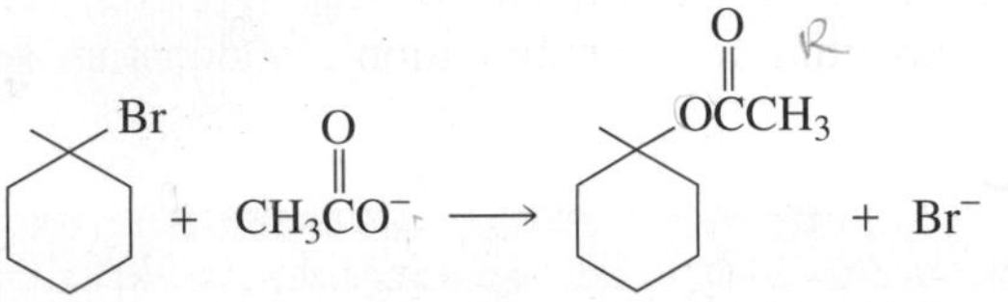
(methylcyclohexyl-Br) + $CH_3\overset{O}{\overset{\|}{C}}O^-$ ⟶ (methylcyclohexyl-$OCCH_3$, with $\overset{O}{\|}$) + Br^-

10.11 Biological Methylating Reagents

If an organic chemist wanted to put a methyl group on a nucleophile (Nu^-), methyl iodide would most likely be the methylating agent used. Of the methyl halides, methyl iodide has the most easily displaced leaving group because I^- is the weakest base of the halide ions. The reaction would be a simple S_N2 reaction.

$$\overset{..}{Nu}^- + CH_3-I \longrightarrow CH_3-Nu + I^-$$

In a living cell, however, methyl iodide is not available. It is only slightly soluble in water, so it is not found in the predominantly aqueous environments of biological systems. Instead, biological systems use *S*-adenosylmethionine (SAM) and N^5-methyltetrahydrofolate as methylating agents; both of these compounds are soluble in water. Although they look much more complicated than methyl iodide, they perform the same function—they transfer a methyl group to a nucleophile. Notice that the methyl group in each of these methylating agents is attached to a positively charged atom. The positively charged atom readily accepts the electrons when the leaving group is displaced. In other words, the methyl groups are attached to very good leaving groups, allowing biological methylation to take place at a reasonable rate.

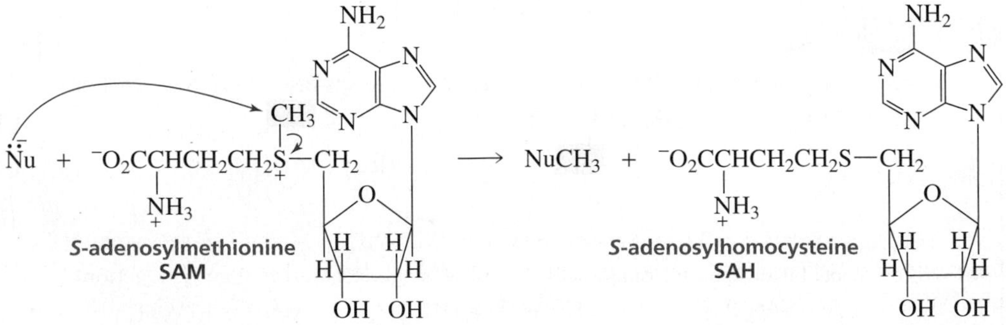

$$\overset{..}{Nu}^- + {}^-O_2CCHCH_2CH_2\overset{+}{S}-CH_2\cdots \longrightarrow NuCH_3 + {}^-O_2CCHCH_2CH_2S-CH_2\cdots$$

S-adenosylmethionine
SAM

S-adenosylhomocysteine
SAH

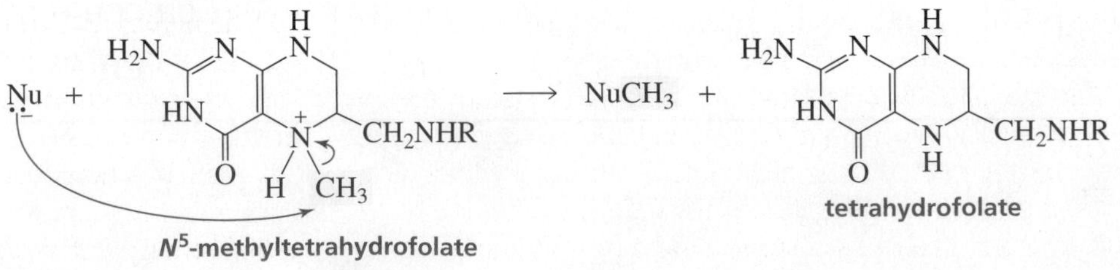

$Nu\!:^- + \;N^5\text{-methyltetrahydrofolate} \longrightarrow NuCH_3 + \;\text{tetrahydrofolate}$

N^5-**methyltetrahydrofolate**

tetrahydrofolate

ERADICATING TERMITES

Alkyl halides can be very toxic to biological organisms. For example, methyl bromide is used to kill termites and other pests. Methyl bromide works by methylating the NH$_2$ and SH groups of enzymes, thereby destroying the enzyme's ability to catalyze necessary biological reactions. Unfortunately, methyl bromide has been found to deplete the ozone layer (Section 9.7), so its production will be banned in developed countries in 2005 and in developing countries in 2015.

An example of a methylation reaction that takes place in biological systems is the conversion of noradrenaline (norepinephrine) to adrenaline (epinephrine). The reaction uses SAM to provide the methyl group. Noradrenaline and adrenaline are hormones that control glycogen metabolism; they are released into the bloodstream in response to stress. Adrenaline is more potent than noradrenaline.

S-ADENOSYLMETHIONINE: A NATURAL ANTIDEPRESSANT

S-Adenosylmethionine is sold in many health food and drug stores as a treatment for depression and arthritis. It is marketed under the name SAMe (pronounced Sammy). Although SAMe has been used clinically in Europe for more than two decades, it has not been rigorously evaluated in the United States and is not approved by the FDA. It can be sold, however, because the FDA does not prohibit the sale of most naturally occurring substances, as long as the marketer does not make therapeutic claims. SAMe has also been found to be effective in the treatment of liver diseases—diseases caused by alcohol and the hepatitis C virus. The attenuation of liver injuries is found to be accompanied by increased levels of glutathione in the liver. S-adenosylmethionine is required for the synthesis of glutathione—an important biological antioxidant (Section 17.6).

Summary

Alkyl halides undergo two kinds of **nucleophilic substitution reactions**: S$_N$2 and S$_N$1. In both reactions, a nucleophile substitutes for a halogen, which is called a **leaving group**. An S$_N$2 reaction is bimolecular—two molecules are involved in the rate-limiting step; an S$_N$1 reaction is unimolecular—one molecule is involved in the rate-limiting step.

The rate of an **S$_N$2 reaction** depends on the concentration of both the alkyl halide and the nucleophile. An S$_N$2 reaction is a one-step reaction: the nucleophile attacks the back side of the carbon that is attached to the halogen. The rate of an S$_N$2 reaction depends on steric hindrance: The bulkier the groups at the back side of the carbon undergoing attack, the slower is the reaction. Tertiary carbocations, therefore, cannot undergo S$_N$2 reactions. An S$_N$2 reaction takes place with **inversion of configuration**.

The rate of an **S$_N$1 reaction** depends only on the concentration of the alkyl halide. The halogen departs in the first step, forming a carbocation that is attacked by a nucleophile in the second step. The rate of an S$_N$1 reaction depends on the ease of carbocation formation. Tertiary alkyl halides, therefore, are more reactive than secondary alkyl halides because tertiary carbocations are more stable than secondary carbocations. Primary carbocations are so unstable that primary alkyl halides cannot undergo S_N1 reactions. An S_N1 reaction forms both inverted and noninverted products.

The rates of both S_N2 and S_N1 reactions are influenced by the nature of the leaving group. Weak bases are the best leaving groups because weak bases form the weakest bonds. Thus, the weaker the basicity of the leaving group, the faster the reaction will occur. Therefore, the relative reactivities of alkyl halides that differ only in the halogen atom are $RI > RBr > RCl > RF$ in both S_N2 and S_N1 reactions.

Basicity is a measure of how well a compound shares its lone pair with a proton. **Nucleophilicity** is a measure of how readily a compound is able to attack an electron-deficient atom. In general, the stronger base is the better nucleophile.

In addition to undergoing nucleophilic substitution reactions, alkyl halides also undergo elimination reactions: The halogen is removed from one carbon and a hydrogen is removed from an adjacent carbon. A double bond is formed between the two carbons from which the atoms are eliminated. Therefore, the product of an elimination reaction is an alkene. There are two important elimination reactions: E1 and E2.

An **E2 reaction** is a one-step reaction; the hydrogen and the halide ion are removed in the same step, so no intermediate is formed. In an **E1 reaction**, the alkyl halide dissociates, forming a carbocation intermediate. In a second step, a base removes a proton from a carbon that is adjacent to the positively charged carbon.

The major product of an elimination reaction is the more stable alkene—the alkene formed when a hydrogen is removed from the β-carbon that is bonded to the fewest hydrogens. If both E and Z isomers are possible for the product, the one with the bulkiest groups on opposite sides of the double bond is more stable; therefore, it will be formed in greater yield.

Predicting which products are formed when an alkyl halide undergoes a reaction begins with determining whether the conditions favor $S_N2/E2$ or $S_N1/E1$ reactions. $S_N2/E2$ reactions are favored by a high concentration of a good nucleophile/strong base, whereas $S_N1/E1$ reactions are favored by a poor nucleophile/weak base.

When $S_N2/E2$ reactions are favored, primary alkyl halides form primarily substitution products, secondary alkyl halides form both substitution and elimination products, and tertiary alkyl halides form only elimination products. When $S_N1/E1$ conditions are favored, secondary and tertiary alkyl halides form both substitution and elimination products; primary alkyl halides do not undergo $S_N1/E1$ reactions.

Summary of Reactions

1. S_N2 reaction: a one-step mechanism

Relative reactivities of alkyl halides: $CH_3X > 1° > 2° > 3°$

Only the inverted product is formed.

2. S_N1 reaction: a two-step mechanism with a carbocation intermediate

Relative reactivities of alkyl halides: $3° > 2° > 1° > CH_3X$

Both the inverted and noninverted products are formed.

3. E2 reaction: a one-step mechanism

Relative reactivities of alkyl halides: $3° > 2° > 1°$

Both E and Z stereoisomers are formed. The isomer with the bulkiest groups on opposite sides of the double bond will be formed in greater yield.

4. E1 reaction: a two-step mechanism with a carbocation intermediate

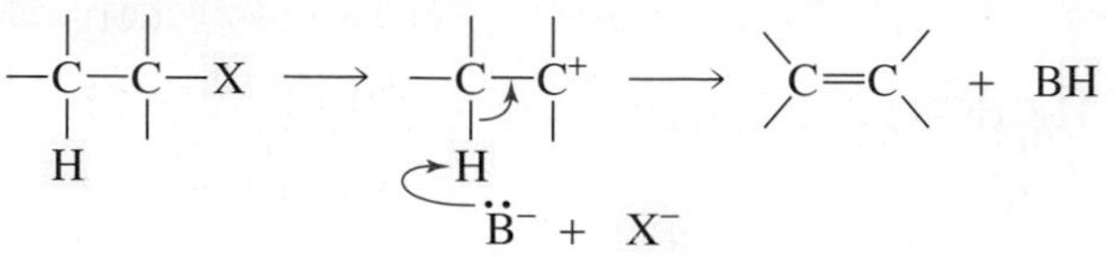

$$-\overset{\underset{\displaystyle |}{|}}{\underset{H}{C}}-\overset{\underset{\displaystyle |}{|}}{\underset{|}{C}}-X \longrightarrow -\overset{|}{\underset{|}{C}}\overset{+}{\underset{H}{C}} \longrightarrow \overset{}{\diagup}C=C\overset{}{\diagdown} + BH$$
$$\ddot{B}^- + X^-$$

Relative reactivities of alkyl halides: $3° > 2° > 1°$

Both *E* and *Z* stereoisomers are formed. The isomer with the bulkiest groups on opposite sides of the double bond will be formed in greater yield.

Competing S_N2 and E2 Reactions
  Primary alkyl halides: primarily substitution
  Secondary alkyl halides: substitution and elimination
  Tertiary alkyl halides: only elimination

Competing S_N1 and E1 Reactions
  Primary alkyl halides: cannot undergo S_N1 or E1 reactions
  Secondary alkyl halides: substitution and elimination
  Tertiary alkyl halides: substitution and elimination

Problems

21. Which reaction in each of the following pairs will take place more rapidly?

  a. $CH_3Br + HO^- \longrightarrow CH_3OH + Br^-$

  $CH_3Br + H_2O \longrightarrow CH_3OH + HBr$

  b. $CH_3I + HO^- \longrightarrow CH_3OH + I^-$

  $CH_3Cl + HO^- \longrightarrow CH_3OH + Cl^-$

  c. $CH_3Br + NH_3 \longrightarrow CH_3\overset{+}{N}H_3 + Br^-$

  $CH_3Br + H_2O \longrightarrow CH_3OH + HBr$

22. Give the product of the reaction of methyl bromide with each of the following nucleophiles:
  a. HO^- **b.** $^-NH_2$ **c.** H_2S **d.** HS^- **e.** CH_3O^- **f.** CH_3NH_2

23. Which is a better nucleophile?

  a. H_2O or HO^-

  b. NH_3 or $^-NH_2$

  c. $CH_3\overset{\overset{\displaystyle O}{\|}}{C}O^-$ or $CH_3CH_2O^-$

  d. 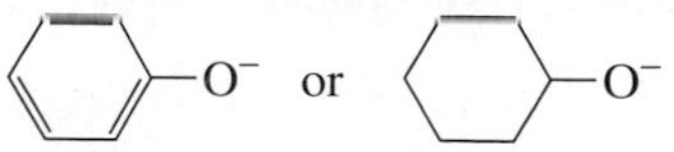 ⬡—O^- or ⬡—O^-

24. For each of the pairs in Problem 23, indicate which is a better leaving group.

25. What nucleophiles could be used to react with butyl bromide to prepare the following compounds?
  a. $CH_3CH_2CH_2CH_2OH$

  b. $CH_3CH_2CH_2CH_2OCH_3$

  c. $CH_3CH_2CH_2CH_2SCH_2CH_3$

  d. $CH_3CH_2CH_2CH_2C\equiv N$

  e. $CH_3CH_2CH_2CH_2O\overset{\overset{\displaystyle O}{\|}}{C}CH_3$

  f. $CH_3CH_2CH_2CH_2C\equiv CCH_3$

26. Which alkyl halide would you expect to be more reactive in an S_N2 reaction with a given nucleophile?

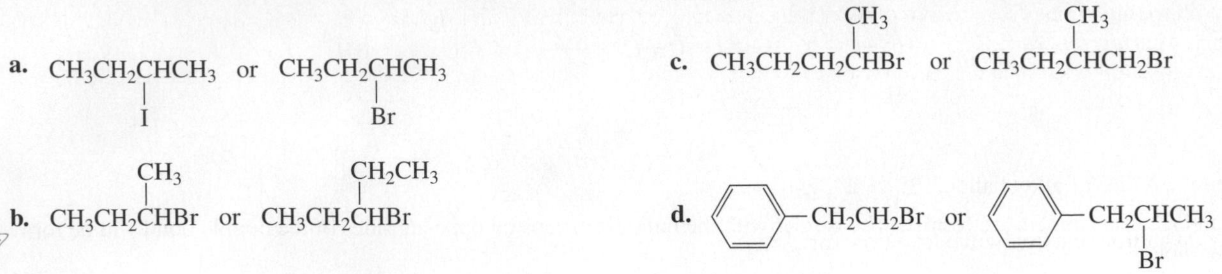

a. $CH_3CH_2CHCH_3$ or $CH_3CH_2CHCH_3$
       | |
       I   Br

b. CH_3CH_2CHBr or CH_3CH_2CHBr
              | |
              CH_3 CH_2CH_3

c. $CH_3CH_2CH_2CHBr$ or $CH_3CH_2CHCH_2Br$
                 | |
                 CH_3 CH_3

d. ⬡—CH_2CH_2Br or ⬡—CH_2CHCH_3
                                      |
                                      Br

27. For each of the pairs in Problem 26, which compound would be more reactive in an S_N1 reaction?

28. For each of the following reactions, give the substitution products; if the products can exist as stereoisomers, show what stereoisomers are obtained:

a. (*R*)-2-bromopentane + high concentration of CH_3O^-

b. (*R*)-2-bromopentane + CH_3OH

c. *trans*-1-bromo-4-methylcyclohexane + high concentration of CH_3O^-

d. *trans*-1-bromo-4-methylcyclohexane + CH_3OH

e. 3-bromo-3-methylpentane + CH_3OH

29. Give the major product obtained when each of the following alkyl halides undergoes an E2 reaction:

a. [cyclohexane with Cl]   b. [cyclohexane with CH_2Cl]   c. [cyclohexane with CH_3 and Cl]

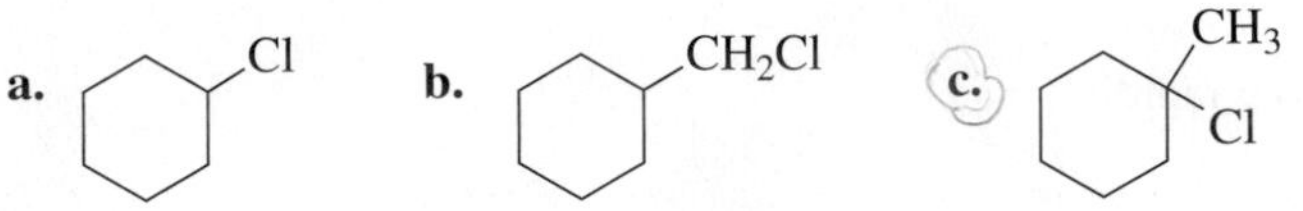

30. Give the stereoisomer that would be obtained in greater yield when each of the following alkyl halides undergoes an E2 reaction:

a. $CH_3CHCH_2CH_3$ b. $CH_3CHCH_2CH_3$ c. $CH_3CHCH_2CH_2CH_3$
         | | |
         Br   Cl   Cl

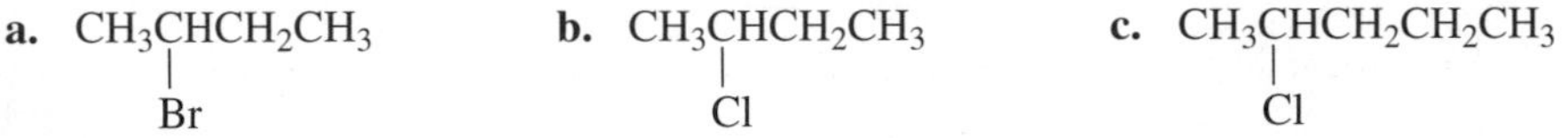

31. Which reactant will undergo an elimination reaction more rapidly?

a. $(CH_3)_3CCl$ $\xrightarrow[H_2O]{HO^-}$ b. $(CH_3)_3CBr$ $\xrightarrow[H_2O]{HO^-}$

$(CH_3)_3CI$ $\xrightarrow[H_2O]{HO^-}$ $(CH_3)_2CHBr$ $\xrightarrow[H_2O]{HO^-}$

32. a. Identify the three products that are formed when 2-bromo-2-methylpropane is dissolved in a mixture of 80% ethanol and 20% water.

b. Explain why the same products are obtained when 2-chloro-2-methylpropane is dissolved in a mixture of 80% ethanol and 20% water.

33. Starting with cyclohexane, how could the following compounds be prepared?

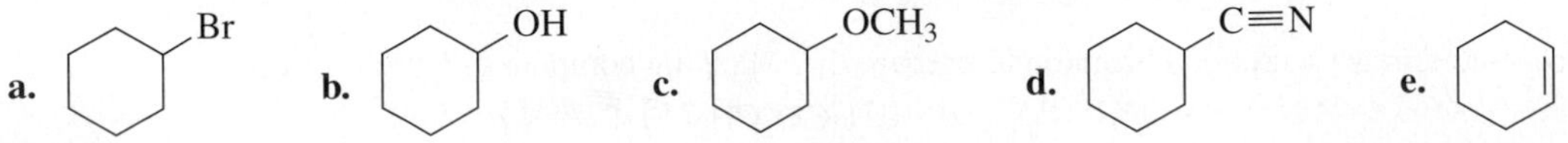

a. [cyclohexane Br]   b. [cyclohexane OH]   c. [cyclohexane OCH_3]   d. [cyclohexane C≡N]   e. [cyclohexene]

34. For each of the following reactions, give the major elimination product; if the product can exist as stereoisomers, indicate which stereoisomer is obtained in greater yield:

a. (*R*)-2-bromohexane + high concentration of HO^-

b. (*R*)-2-bromohexane + H_2O

c. 3-bromo-3-methylpentane + high concentration of HO^-

d. 3-bromo-3-methylpentane + H_2O

35. The rate of reaction of methyl iodide with quinuclidine was measured in nitrobenzene, and then the rate of reaction of methyl iodide with triethylamine was measured in the same solvent. The concentration of the reagents was the same in both experiments.
  a. Which reaction was faster?
  b. Which reaction had the larger rate constant?

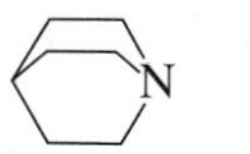

quinuclidine

$$CH_3CH_2NCH_2CH_3$$
with CH_2CH_3

triethylamine

36. Which substitution reaction will occur more rapidly?

a.

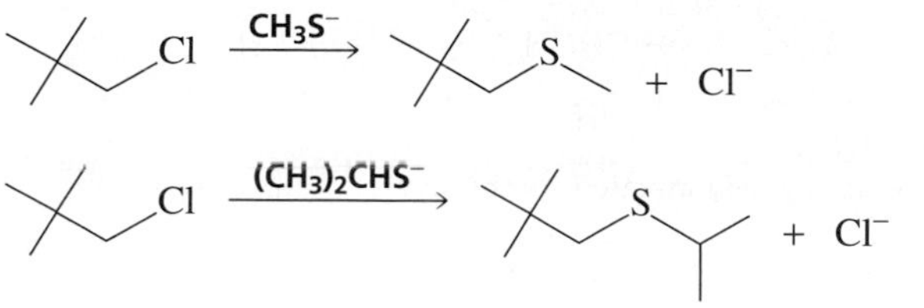

b.

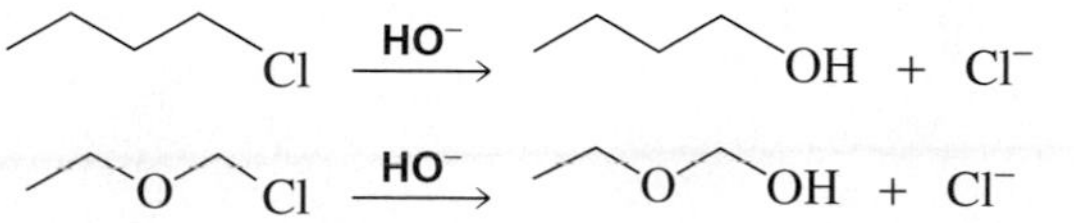

37. Which of the following is more reactive in an E2 reaction?

a.

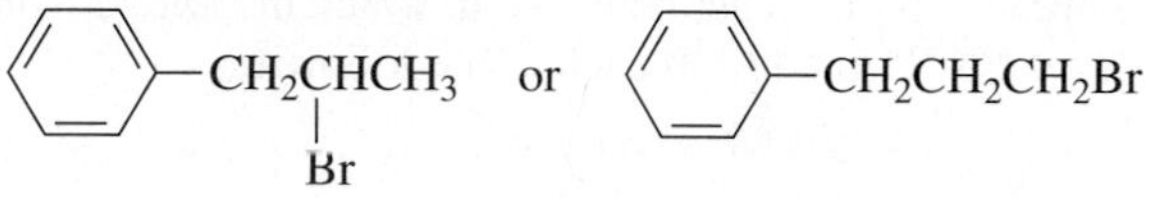

b. $CH_3CH_2CHCH_3$ (with Br) or $CH_2{=}CHCH_2CHCH_3$ (with Br)

38. a. Explain why 1-bromo-2,2-dimethylpropane has difficulty undergoing either S_N2 or S_N1 reactions.
  b. Can it undergo E2 and E1 reactions?

39. Which stereoisomer would be obtained in greatest yield from an E2 reaction of each of the following alkyl halides?

a.
$$CH_3C{-}CCH_2CH_3$$
with CH_3, CH_3 on top and CH_3, Br on bottom

b. $CH_3CH_2CH_2CHCHCH_2CH_3$ with CH_3 and I

40. An ether can be prepared by an S_N2 reaction of an alkyl halide with an alkoxide ion (RO^-). Which set of alkyl halide and alkoxide ion would give you a better yield of cyclopentyl methyl ether?

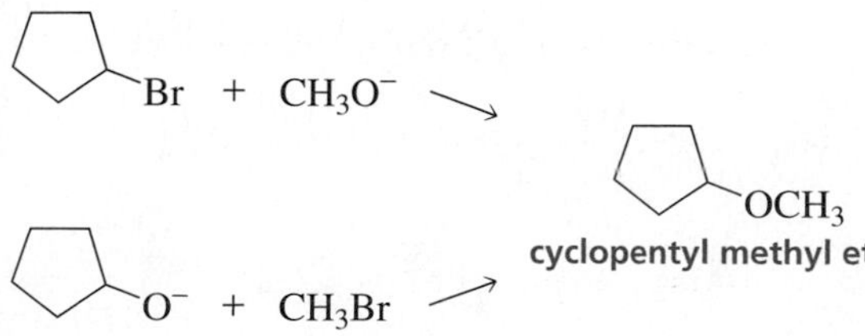

cyclopentyl methyl ether

41. Dr. Don T. Doit wanted to synthesize the anesthetic 2-ethoxy-2-methylpropane. He used ethoxide ion and 2-chloro-2-methylpropane for his synthesis and ended up with no ether. What was the product of his synthesis? What reagents should he have used?

$$CH_3CCH_3$$
with CH_3 on top and OCH_2CH_3 on bottom

2-ethoxy-2-methylpropane

42. Which compound undergoes an E1 reaction more rapidly, 3-bromocyclohexene or bromocyclohexane?

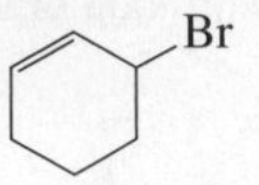

3-bromocyclohexene 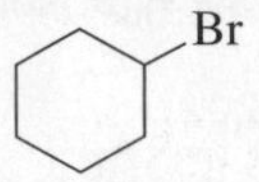   **bromocyclohexane**

43. In Section 10.11, we saw that *S*-adenosylmethionine (SAM) methylates the nitrogen atom of noradrenaline to form adrenaline, a more potent hormone. If SAM methylates an OH group on the benzene ring instead, it completely destroys noradrenaline's activity. Give the mechanism for the methylation of the OH group by SAM.

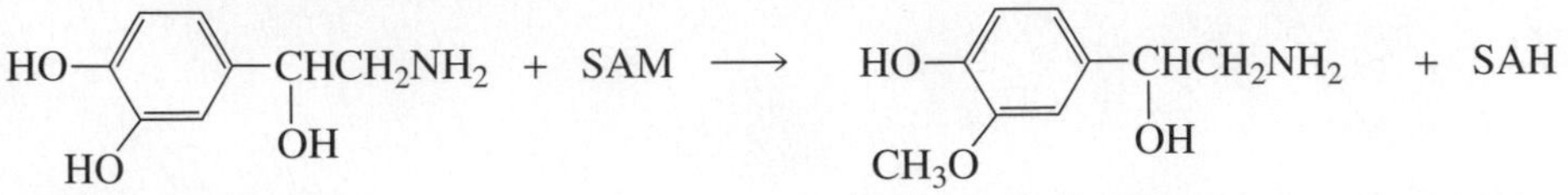

HO—⟨ ⟩—CHCH$_2$NH$_2$ + SAM ⟶ HO—⟨ ⟩—CHCH$_2$NH$_2$ + SAH
HO   OH   CH$_3$O   OH

noradrenaline   **a biologically inactive compound**

44. Show how the following compounds could be synthesized using the given starting materials:

a. $CH_3CH_2CH_2CH_2Br$ ⟶ $CH_3CH_2CH_2CH_2NH_2$ c.

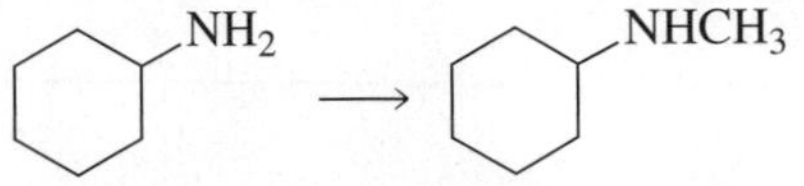

b.

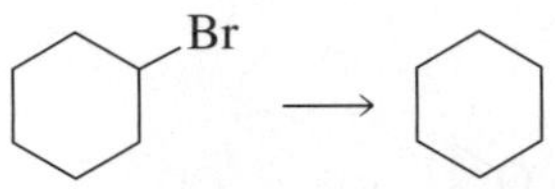

45. A cyclic compound can be formed by an intramolecular reaction. An intramolecular reaction is one in which the two reacting groups are in the same molecule. Give the structure of the ether that would be formed from each of the following intramolecular reactions.

a. $BrCH_2CH_2CH_2CH_2O^-$ ⟶ ether   b. $ClCH_2CH_2CH_2CH_2CH_2O^-$ ⟶ ether

11 Reactions of Alcohols, Amines, Ethers, and Epoxides

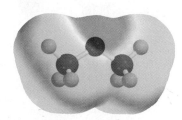

CH₃OH   **CH₃OCH₃**

I n Chapter 10, we saw that alkyl halides undergo substitution and elimination reactions because of their electron-withdrawing halogen atoms. Compounds with other electron-withdrawing groups also undergo substitution and elimination reactions. For example, an alcohol (ROH) has an electron withdrawing OH group. An OH group, however, is much more basic than a halogen, so we will see that it is much harder to displace.

11.1   Nomenclature of Alcohols

Before we look at the reactions of alcohols, we need to learn how to name them. **Alcohols** are compounds in which a hydrogen of an alkane has been replaced by an OH group. We have seen that alcohols are classified as **primary**, **secondary**, or **tertiary**, depending on whether the OH group is bonded to a primary, secondary, or tertiary carbon—the same way alkyl halides are classified (Section 3.5).

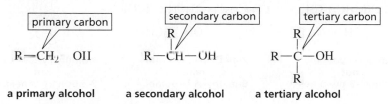

The common name of an alcohol consists of the name of the alkyl group to which the OH group is attached, followed by the word "alcohol."

CH₃CH₂OH   CH₃CH₂CH₂OH   CH₃CHOH
ethyl alcohol   **propyl alcohol** |
                                          CH₃
                                   isopropyl alcohol

methyl alcohol

ethyl alcohol

propyl alcohol

The OH group is the **functional group**—the center of reactivity of an alcohol. The IUPAC system uses the suffix "ol" to denote the OH group. Therefore, the systematic name of an alcohol is obtained by replacing the "e" at the end of the name of the parent hydrocarbon with the suffix "ol." This should remind you of the use of the suffix "ene" to denote the functional group of an alkene (Section 4.2).

$$CH_3OH \qquad CH_3CH_2OH$$
methanol \qquad **ethanol**

When necessary, the position of the functional group is indicated by a number.

$$CH_3CH_2CHCH_2CH_3$$
$$|$$
$$OH$$
3-pentanol

Let's review the rules used to name a compound that has a functional group suffix:

1. The parent hydrocarbon is the longest chain containing the functional group. The parent chain is numbered in the direction that gives the *functional group suffix the lowest possible number.*

$$\overset{1}{C}H_3\overset{2}{C}H\overset{3}{C}H_2\overset{4}{C}H_3 \qquad \overset{5}{C}H_3\overset{4}{C}H_2\overset{3}{C}H_2\overset{2}{C}H\overset{1}{C}H_2OH$$
$$| \qquad\qquad\qquad |$$
$$OH \qquad\qquad\qquad CH_2CH_3$$
2-butanol \qquad **2-ethyl-1-pentanol**

> The longest continuous chain has six carbons, but the longest continuous chain containing the OH functional group has five carbons so the compound is named as a pentanol.

2. If there is a functional group suffix and a substituent, the functional group suffix gets the lowest possible number.

$$\overset{1}{H}OCH_2\overset{2}{C}H_2\overset{3}{C}H_2Br \qquad \overset{4}{Cl}CH_2\overset{3}{C}H_2\overset{2}{C}H\overset{1}{C}H_3 \qquad \begin{array}{c} CH_3 \\ | \\ \overset{5}{C}H_3\overset{4}{C}\overset{3}{C}H_2\overset{2}{C}H\overset{1}{C}H_3 \end{array}$$
$$\qquad\qquad\qquad\qquad | \qquad\qquad\qquad | \quad\; |$$
$$\qquad\qquad\qquad\qquad OH \qquad\qquad\quad CH_3 \; OH$$

3-bromo-1-propanol \qquad **4-chloro-2-butanol** \qquad **4,4-dimethyl-2-pentanol**

3. If the same number for the functional group suffix is obtained in both directions, the chain is numbered in the direction that gives a substituent the lowest possible number. Notice that a number is not needed to designate the position of a functional group suffix in a cyclic compound, because it is assumed to be at the 1-position.

$$CH_3CHCHCH_2CH_3 \qquad CH_3CH_2CH_2CHCH_2CHCH_3$$
$$| \;\; | \qquad\qquad\qquad\qquad | \qquad\quad |$$
$$Cl \; OH \qquad\qquad\qquad\quad OH \qquad CH_3$$

2-chloro-3-pentanol \qquad **2-methyl-4-heptanol** \qquad **3-methylcyclohexanol**
not \qquad\qquad\qquad **not** \qquad\qquad\qquad **not**
4-chloro-3-pentanol \qquad **6-methyl-4-heptanol** \qquad **5-methylcyclohexanol**

4. If there is more than one substituent, the substituents are cited in alphabetical order.

$$\begin{array}{c} CH_2CH_3 \\ | \\ CH_3CHCH_2CHCH_2CHCH_3 \end{array}$$
$$| \qqu\qquad |$$
$$Br \qqu\qquad OH$$

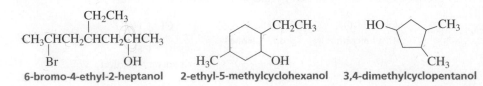

6-bromo-4-ethyl-2-heptanol \qquad **2-ethyl-5-methylcyclohexanol** \qquad **3,4-dimethylcyclopentanol**

Tutorial:
Nomenclature of alcohols

PROBLEM 1

Draw the structures of straight-chain alcohols that have from one to six carbons, and then give each of them a common name and a systematic name.

PROBLEM 2◆

Give each of the following compounds a systematic name, and indicate whether each is a primary, secondary, or tertiary alcohol:

Tutorial:
Summary of systematic nomenclature

a. $CH_3CH_2CH_2CH_2CH_2OH$

b.

CH₃ / HO— (cyclohexane with CH₃ and OH substituents)

c. $CH_3\overset{\underset{|}{CH_3}}{\underset{\underset{|}{OH}}{C}}CH_2CH_2CH_2Cl$

d. $CH_3CH_2CH_2\underset{\underset{|}{CH_2OH}}{CH}CH_2CH_3$

e. $CH_3\underset{\underset{|}{CH_3}}{CH}CH_2\underset{\underset{|}{OH}}{CH}CH_2CH_3$

f. $CH_3\underset{\underset{|}{CH_3}}{CH}CH_2\underset{\underset{|}{OH}}{CH}CH_2\underset{\underset{|}{CH_3}}{CH}CH_2CH_3$

PROBLEM 3◆

Write the structures of all the tertiary alcohols with molecular formula $C_6H_{14}O$, and give each a systematic name.

11.2 Substitution Reactions of Alcohols

An **alcohol** cannot undergo a nucleophilic substitution reaction because it has a strongly basic leaving group (HO^-) that cannot be displaced by a nucleophile.

$$CH_3—OH \ + \ Br^- \ \xrightarrow{\ \ \times\ \ } \ CH_3—Br \ + \ \underset{\text{strong base}}{HO^-}$$

An alcohol, however, can undergo a nucleophilic substitution reaction if its OH group is converted into a group that is a weaker base (i.e., a better leaving group). One way to convert an OH group into a weaker base is to protonate it. Protonation changes the leaving group from HO^- to H_2O, which is a weak enough base to be displaced by a nucleophile. The substitution reaction is slow and requires heat to take place in a reasonable period of time.

The weaker the base, the more easily it can be displaced.

$$CH_3—OH \ + \ HBr \ \rightleftharpoons \ CH_3—\underset{\underset{Br^-}{+}}{\overset{H}{O}}H \ \xrightarrow{\Delta} \ CH_3—Br \ + \ \underset{\text{weak base}}{H_2O}$$

Because the OH group of the alcohol has to be protonated before it can be displaced by a nucleophile, only weakly basic nucleophiles (I^-, Br^-, Cl^-) can be used in the substitution reaction. Moderately and strongly basic nucleophiles (NH_3, RNH_2, CH_3O^-) cannot be used because they would also be protonated in the acidic solution and, once protonated, would no longer be nucleophiles ($^+NH_4$, $RNH_3{}^+$) or would be poor nucleophiles (CH_3OH).

PROBLEM 4◆

Why are NH_3 and CH_3NH_2 no longer nucleophiles when they are protonated?

Primary, secondary, and tertiary alcohols all undergo nucleophilic substitution reactions with HI, HBr, and HCl to form alkyl halides.

$$CH_3CH_2CH_2OH + HI \xrightarrow{\Delta} CH_3CH_2CH_2I + H_2O$$

1-propanol
a primary alcohol

1-iodopropane

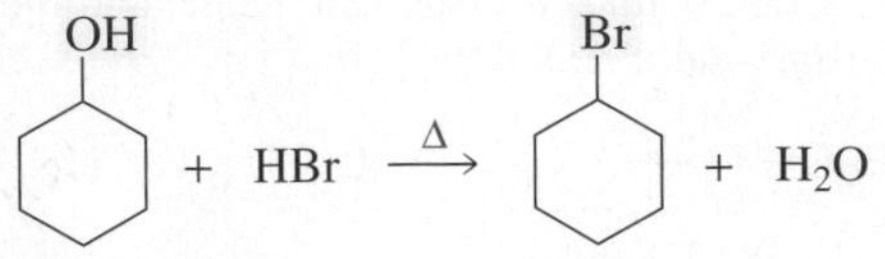

cyclohexanol
a secondary alcohol

bromocyclohexane

2-methyl-2-butanol
a tertiary alcohol

2-chloro-2-methylbutane

The mechanism of the substitution reaction depends on the structure of the alcohol. Secondary and tertiary alcohols undergo S$_N$1 reactions. The carbocation intermediate formed in the S$_N$1 reaction has two possible fates: It can combine with a nucleophile and form a substitution product, or it can lose a proton and form an elimination product. However, only the substitution product is actually obtained, because any alkene formed in an elimination reaction will undergo a subsequent addition reaction with HBr (Section 5.1) to form more of the substitution product.

Secondary and tertiary alcohols undergo S$_N$1 reactions with hydrogen halides.

mechanism of the S$_N$1 reaction

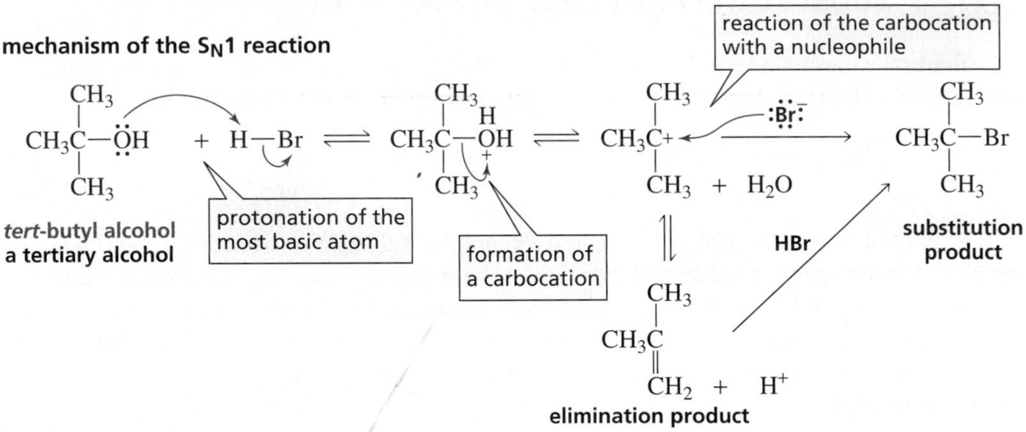

reaction of the carbocation with a nucleophile

tert-butyl alcohol
a tertiary alcohol

protonation of the most basic atom

formation of a carbocation

substitution product

HBr

elimination product

Carbocation stability: 3° > 2° > 1°.

Tertiary alcohols undergo substitution reactions with hydrogen halides faster than do secondary alcohols because tertiary carbocations are easier to form than secondary carbocations (Section 10.4). Thus, the reaction of a tertiary alcohol with a hydrogen halide proceeds readily at room temperature, whereas the reaction of a secondary alcohol with a hydrogen halide has to be heated to have the reaction occur at the same rate.

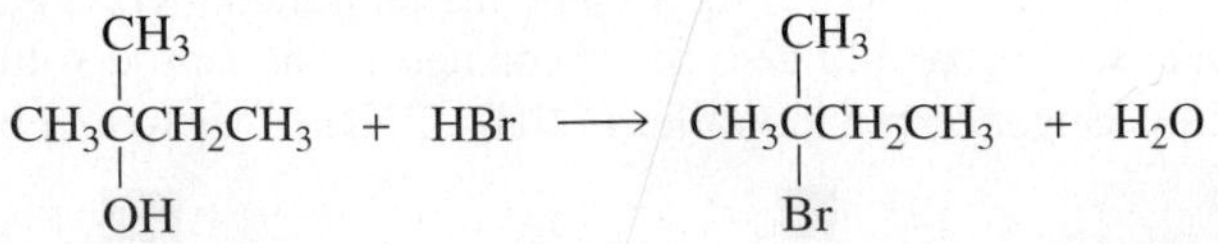

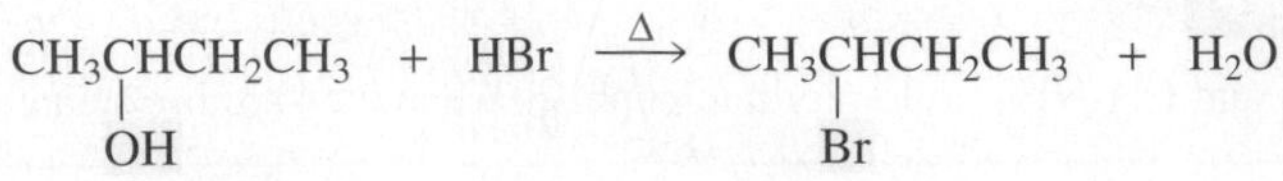

Primary alcohols cannot undergo S_N1 reactions because primary carbocations are too unstable to be formed (Section 10.10). Therefore, when a primary alcohol reacts with a hydrogen halide, it must do so in an S_N2 reaction—the nucleophile hits the back side of the carbon and displaces the leaving group.

Primary alcohols undergo S_N2 reactions with hydrogen halides.

mechanism of the S_N2 reaction

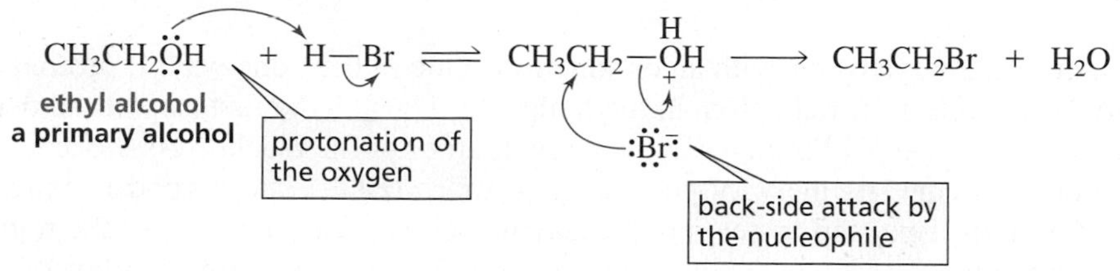

$$CH_3CH_2\ddot{O}H \; + \; H{-}Br \; \rightleftharpoons \; CH_3CH_2{-}\overset{H}{\underset{+}{O}}H \; \longrightarrow \; CH_3CH_2Br \; + \; H_2O$$

ethyl alcohol
a primary alcohol

protonation of the oxygen

$:\ddot{\underset{\cdot\cdot}{Br}}:^-$

back-side attack by the nucleophile

PROBLEM 5 SOLVED

Using the pK_a values of the conjugate acids of the leaving groups (the pK_a of HBr is -9; the pK_a of H_2O is 15.7; the pK_a of H_3O^+ is -1.7), explain the difference in reactivity of

a. CH_3Br and CH_3OH

b. $CH_3\overset{+}{O}H_2$ and CH_3OH

SOLUTION TO 5a The conjugate acid of the leaving group of CH_3Br is HBr; its pK_a is $= -9$; the conjugate acid of the leaving group of CH_3OH is H_2O; its pK_a is $= 15.5$. Because HBr is a much stronger acid than H_2O, Br^- is a much weaker base than HO^-. (Recall the stronger the acid, the weaker is its conjugate base.) Therefore, Br^- is a much better leaving group than HO^- causing CH_3Br to be much more reactive than CH_3OH.

PROBLEM 6◆

Give the major product of each of the following reactions:

a. $CH_3CH_2CHCH_3 \; + \; HBr \; \xrightarrow{\Delta}$
$|$
OH

b. 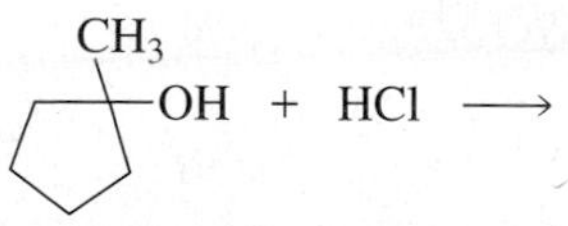 (cyclopentane with CH_3 and $-OH$ on same carbon) $+ \; HCl \; \longrightarrow$

PROBLEM 7 SOLVED

Show how 1-butanol can be converted into the following compounds:

a. $CH_3CH_2CH_2CH_2OCH_3$

d. $CH_3CH_2CH_2CH_2NHCH_2CH_3$

b. $CH_3CH_2CH_2CH_2O\overset{\overset{\displaystyle O}{\|}}{C}CH_2CH_3$

e. $CH_3CH_2CH_2CH_2C{\equiv}N$

SOLUTION TO 7a Because the OH group of 1-butanol is too basic to be substituted, the alcohol must first be converted into an alkyl halide. The alkyl halide has a leaving group that can be substituted by CH_3O^-, the nucleophile required to obtain the desired product.

$$CH_3CH_2CH_2CH_2OH \; \xrightarrow[\Delta]{HBr} \; CH_3CH_2CH_2CH_2Br \; \xrightarrow[\Delta]{CH_3O^-} \; CH_3CH_2CH_2CH_2OCH_3$$

11.3   Elimination Reactions of Alcohols: Dehydration

An alcohol can undergo an elimination reaction, forming an alkene by losing an OH from one carbon and an H from an adjacent carbon. Overall, this amounts to the elimination of a molecule of water. Loss of water from a molecule is called **dehydration**.

Dehydration of an alcohol requires an acid catalyst and heat. Sulfuric acid (H_2SO_4) is a commonly used acid catalyst.

$$CH_3CH_2CHCH_3 \underset{\Delta}{\overset{H_2SO_4}{\rightleftharpoons}} CH_3CH=CHCH_3 + H_2O$$
$$\underset{OH}{|}$$

An acid protonates the most basic atom in a molecule.

Secondary and tertiary alcohols undergo dehydration by an E1 pathway.

An acid always reacts with an organic molecule in the same way: It protonates the most basic (electron-rich) atom in the molecule. Thus, in the first step of the dehydration reaction, the acid protonates the oxygen atom of the alcohol. As we saw earlier, protonation converts the very poor leaving group (HO^-) into a good leaving group (H_2O). In the next step, water departs, leaving behind a carbocation. A base removes a proton from a β-carbon (a carbon adjacent to the positively charged carbon), forming an alkene and regenerating the acid catalyst. Notice that the dehydration reaction is an E1 reaction of a protonated alcohol.

mechanism of dehydration (E1)

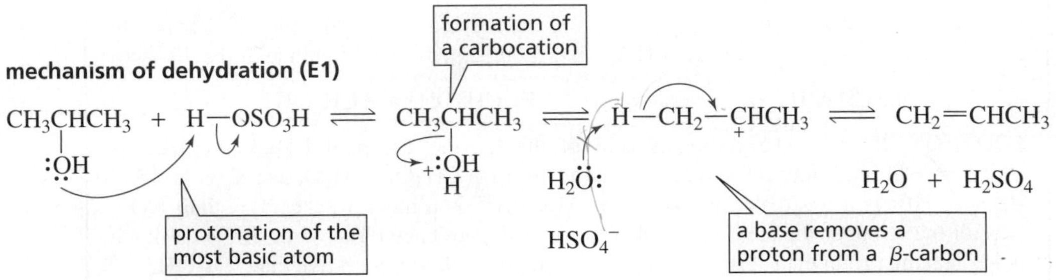

formation of a carbocation

$$CH_3CHCH_3 + H-OSO_3H \rightleftharpoons CH_3CHCH_3 \rightleftharpoons H-CH_2-\overset{+}{C}HCH_3 \rightleftharpoons CH_2=CHCH_3$$
$$\underset{:\ddot{O}H}{|} \qquad\qquad \underset{\overset{+}{:}\ddot{O}H}{|} \quad H_2\ddot{O}: \qquad\qquad H_2O + H_2SO_4$$
$$\qquad\qquad\qquad\qquad H \qquad HSO_4^-$$

protonation of the most basic atom

a base removes a proton from a β-carbon

As in the elimination reactions we have seen previously (Section 10.8), when more than one elimination product can be formed, the major product is the more stable alkene—the one obtained by removing a proton from the β-carbon that is bonded to the fewest hydrogens (Figure 11.1).

Movie: Dehydration

$$\underset{\underset{OH}{|}}{\overset{\overset{CH_3}{|}}{CH_3CCH_2CH_3}} \underset{\Delta}{\overset{H_3PO_4}{\rightleftharpoons}} \overset{\overset{CH_3}{|}}{CH_3C=CHCH_3} + \overset{\overset{CH_3}{|}}{CH_2=CCH_2CH_3} + H_2O$$
$$\qquad\qquad\qquad\qquad\qquad 84\% \qquad\qquad 16\%$$

GRAIN ALCOHOL AND WOOD ALCOHOL

When ethanol is ingested, it acts on the central nervous system. Ingesting moderate amounts affects one's judgment and lowers inhibitions. Higher amounts interfere with motor coordination and cause slurred speech and amnesia. Still higher amounts cause nausea and loss of consciousness. Ingesting very large amounts of ethanol interferes with spontaneous respiration and can be fatal.

The ethanol in alcoholic beverages is produced by the fermentation of glucose, obtained from grapes and from grains such as corn, rye, and wheat, which is why ethanol is also known as grain alcohol. Grains are cooked in the presence of malt (sprouted barley) to convert much of their starch into glucose. Yeast is added to convert glucose into ethanol and carbon dioxide (Section 19.5).

$$\underset{glucose}{C_6H_{12}O_6} \overset{yeast\ enzymes}{\longrightarrow} \underset{ethanol}{2\ CH_3CH_2OH} + 2\ CO_2$$

The kind of beverage produced (white or red wine, beer, scotch, bourbon, champagne) depends on the plant species being fermented, whether the CO_2 that is formed is allowed to escape, whether other substances are added, and how the beverage is purified (by sedimentation for wines, by distillation for scotch and bourbon).

The tax on liquor would make ethanol a prohibitively expensive laboratory reagent. Laboratory alcohol, therefore, is not taxed because ethanol is needed in a wide variety of commercial processes. It is, however, carefully regulated by the federal government to make certain that it is not used for the preparation of alcoholic beverages. Denatured alcohol—ethanol that has been made undrinkable by adding a denaturant such as benzene or methanol—is not taxed, but the added impurities make it unfit for many laboratory uses.

Methanol, also known as wood alcohol because at one time it was obtained by heating wood in the absence of oxygen, is highly toxic. Ingesting even very small amounts can cause blindness. Ingesting as little as an ounce can be fatal.

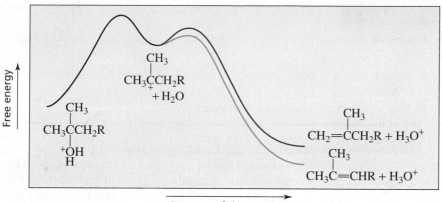

Because the rate-determining step in the dehydration of a secondary or a tertiary alcohol is formation of a carbocation intermediate, the rate of dehydration parallels the ease with which the carbocation is formed. Tertiary alcohols are the easiest to dehydrate because tertiary carbocations are more stable and are, therefore, easier to form than secondary and primary carbocations (Section 10.4).

relative ease of dehydration

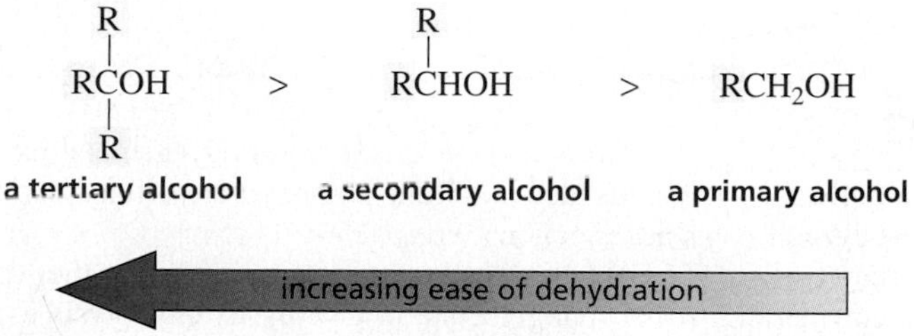

$$
\underset{\text{a tertiary alcohol}}{\overset{\overset{\displaystyle R}{|}}{\underset{\underset{\displaystyle R}{|}}{RCOH}}}
\quad > \quad
\underset{\text{a secondary alcohol}}{\overset{\overset{\displaystyle R}{|}}{RCHOH}}
\quad > \quad
\underset{\text{a primary alcohol}}{RCH_2OH}
$$

increasing ease of dehydration

PROBLEM 8◆

List the following alcohols in order of decreasing rate of dehydration in the presence of acid:

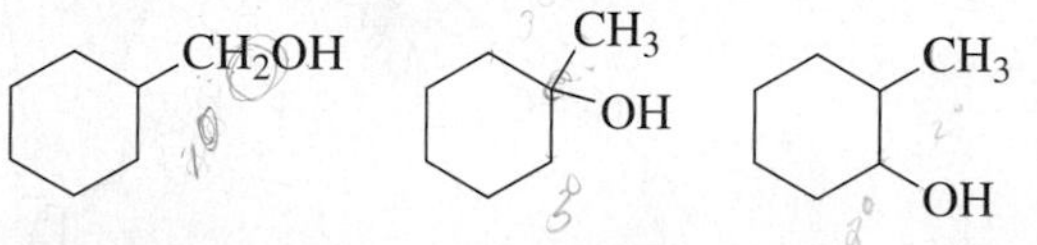

CH_2OH $\overset{CH_3}{\underset{OH}{}}$ $\overset{CH_3}{\underset{OH}{}}$

While the dehydration of a tertiary or a secondary alcohol is an E1 reaction, the dehydration of a primary alcohol is an E2 reaction because of the difficulty encountered in forming primary carbocations. A base removes the proton in the elimination reaction. An ether is obtained as the product of a competing S_N2 reaction, since primary alkyl halides are the ones most likely to form substitution products in S_N2/E2 reactions (Section 10.10).

mechanism of dehydration (E2) and competing substitution (S_N2)

Primary alcohols undergo dehydration by an E2 pathway.

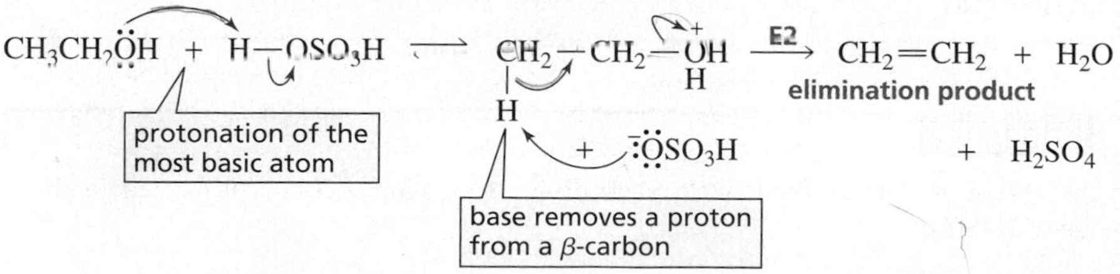

$$CH_3CH_2\ddot{O}H \; + \; H{-}OSO_3H \;\rightleftharpoons\; CH_2{-}CH_2{-}\overset{+}{\underset{\underset{H}{|}}{O}}H \xrightarrow{\text{E2}} CH_2{=}CH_2 \; + \; H_2O$$

elimination product

protonation of the most basic atom

$+ \; \ddot{\ddot{O}}SO_3H$ $+ \; H_2SO_4$

base removes a proton from a β-carbon

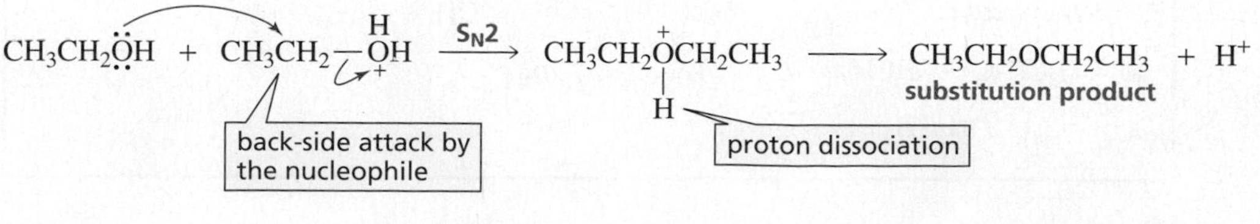

$$CH_3CH_2\ddot{O}H \; + \; CH_3CH_2{-}\overset{H}{\underset{\overset{+}{}}{O}}H \xrightarrow{\text{S}_N2} CH_3CH_2\overset{+}{\underset{\underset{H}{|}}{O}}CH_2CH_3 \longrightarrow CH_3CH_2OCH_2CH_3 \; + \; H^+$$

substitution product

back-side attack by the nucleophile

proton dissociation

PROBLEM 9

Heating an alcohol with H_2SO_4 is a good way to prepare a symmetrical ether such as diethyl ether.

a. Explain why it is not a good way to prepare an unsymmetrical ether such as ethyl propyl ether.

b. How would you synthesize ethyl propyl ether?

The products obtained from the acid-catalyzed elimination (dehydration) of an alcohol are identical to those obtained from elimination of an alkyl halide. That is, both the E and Z stereoisomers are obtained as products. More of the stereoisomer with the most bulkiest groups on opposite sides of the double bond is produced because, being more stable, the transition state leading to its formation is more stable so it is formed more rapidly (Section 10.8).

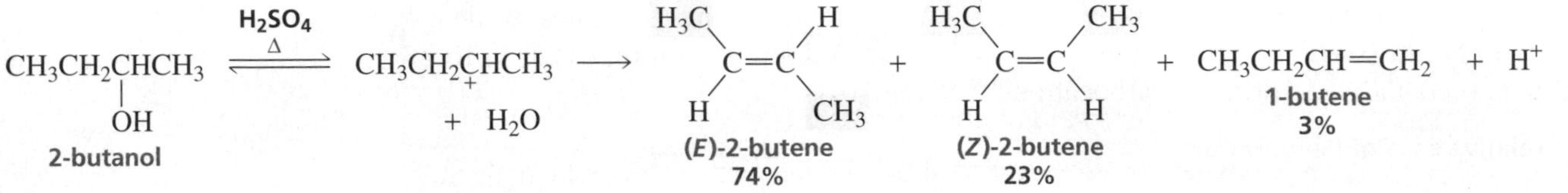

$$CH_3CH_2CHCH_3 \overset{H_2SO_4}{\underset{\Delta}{\rightleftharpoons}} CH_3CH_2\overset{+}{C}HCH_3 \longrightarrow$$

2-butanol, OH

$+ H_2O$

(E)-2-butene 74%

(Z)-2-butene 23%

$+ CH_3CH_2CH{=}CH_2$ 1-butene 3%

$+ H^+$

BIOLOGICAL DEHYDRATIONS

Dehydration reactions occur in many important biological processes. Instead of being catalyzed by strong acids, which would not be available to a cell, they are catalyzed by enzymes. Fumarase, for example, is the enzyme that catalyzes the dehydration of malate in the citric acid cycle.

The citric acid cycle is a series of reactions that oxidize compounds derived from carbohydrates, fatty acids, and amino acids (Section 19.6).

Enolase, another enzyme, catalyzes the dehydration of α-phosphoglycerate in glycolysis. Glycolysis is a series of reactions that prepares glucose for entry into the citric acid cycle.

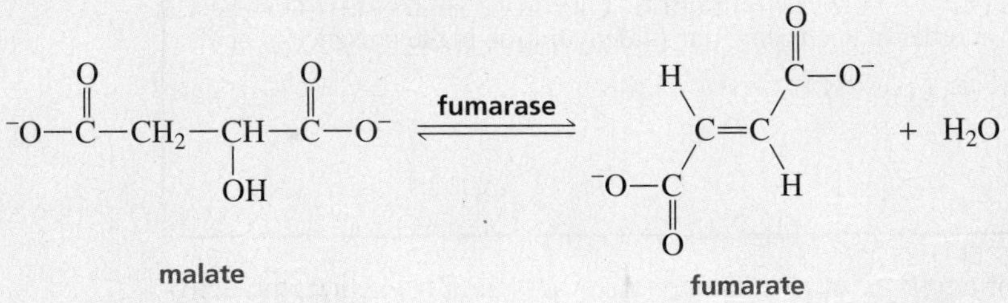

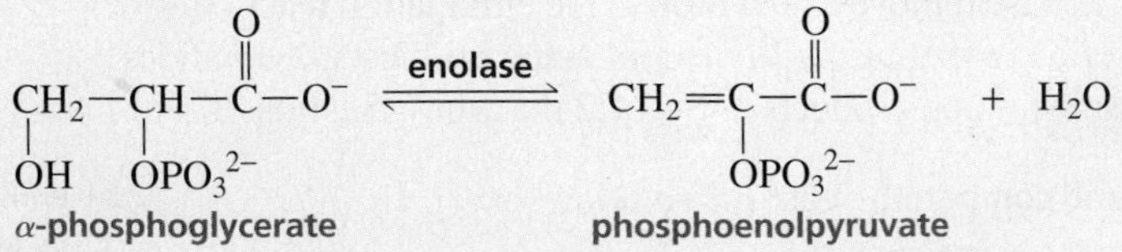

malate $\overset{fumarase}{\rightleftharpoons}$ fumarate $+ H_2O$

α-phosphoglycerate $\overset{enolase}{\rightleftharpoons}$ phosphoenolpyruvate $+ H_2O$

PROBLEM 10◆

Give the major product formed when each of the following alcohols is heated in the presence of H_2SO_4:

a. 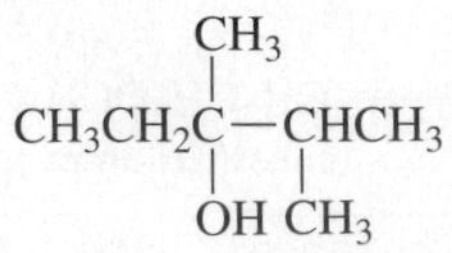 $CH_3CH_2\underset{\underset{OH}{|}}{\overset{\overset{CH_3}{|}}{C}}{-}\underset{\underset{CH_3}{|}}{CH}CH_3$

b.

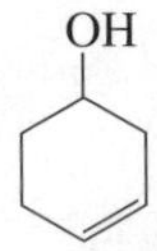

PROBLEM 11◆

The following compound is heated in the presence of H_2SO_4:

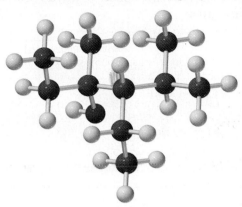

a. What constitutional isomer is produced in greatest yield?

b. What stereoisomer is produced in greatest yield?

We can summarize what we have learned about the mechanisms by which alcohols undergo substitution and elimination reactions: They react by S_N1 and E1 pathways, unless they cannot. In other words, 3° and 2° alcohols react by S_N1 and E1 pathways; but, since 1° alcohols cannot form primary carbocations, they have to react by S_N2 and E2 pathways.

> Alcohols and ethers undergo $S_N1/E1$ reactions unless they would have to form a primary carbocation, in which case they undergo $S_N2/E2$ reactions.

11.4   Oxidation of Alcohols

We have seen that a **reduction reaction** *increases* the number of C—H bonds (Section 5.12). Oxidation is the reverse of reduction. Therefore, an **oxidation reaction** *decreases* the number of C—H bonds (or increases the number of C—O bonds).

Secondary alcohols are oxidized to *ketones*. Chromic acid (H_2CrO_4) is the reagent commonly used to oxidize alcohols.

$$CH_3CH_2\overset{\overset{\displaystyle OH}{|}}{C}HCH_3 \xrightarrow{\textbf{H}_2\textbf{CrO}_4} CH_3CH_2\overset{\overset{\displaystyle O}{\|}}{C}CH_3$$

> Secondary alcohols are oxidized to ketones.

secondary alcohols   ketones

A *primary alcohol* is initially oxidized to an *aldehyde*. The reaction, however, does not stop at the aldehyde. Instead, the aldehyde is further oxidized to a *carboxylic acid*.

$$CH_3CH_2CH_2CH_2OH \xrightarrow{\textbf{H}_2\textbf{CrO}_4} \left[CH_3CH_2CH_2\overset{\overset{\displaystyle O}{\|}}{C}H\right] \xrightarrow[\text{oxidation}]{\text{further}} CH_3CH_2CH_2\overset{\overset{\displaystyle O}{\|}}{C}OH$$

a primary alcohol   an aldehyde   a carboxylic acid

> Primary alcohols are oxidized to aldehydes and carboxylic acids.

The oxidation of a primary alcohol can be stopped at the aldehyde if pyridinium chlorochromate (PCC) is used as the oxidizing agent in an anhydrous solvent such as dichloromethane (CH_2Cl_2).

$$CH_3CH_2CH_2CH_2OH \xrightarrow[\textbf{CH}_2\textbf{Cl}_2]{\textbf{PCC}} CH_3CH_2CH_2\overset{\overset{\displaystyle O}{\|}}{C}H$$

a primary alcohol   an aldehyde

Notice that the oxidation of either a primary or a secondary alcohol involves removal of a hydrogen from the carbon to which the OH is attached. The carbon bearing the OH group in a tertiary alcohol is not bonded to a hydrogen, so its OH group cannot be oxidized to a carbonyl group.

$$CH_3 - \overset{\overset{\displaystyle CH_3}{|}}{\underset{\underset{\displaystyle CH_3}{|}}{C}} - OH$$

cannot be oxidized to a carbonyl group

a tertiary alcohol

PROBLEM 12◆

Give the product formed from the reaction of each of the following compounds with chromic acid:

a. 3-pentanol **b.** 1-pentanol **c.** cyclohexanol **d.** benzyl alcohol

---

BLOOD ALCOHOL CONTENT

As blood passes through the arteries in the lungs, an equilibrium is established between the alcohol in one's blood and the alcohol in one's breath. So if the concentration of one is known, the concentration of the other can be estimated. The test that law enforcement agencies use to approximate a person's blood alcohol level is based on the oxidation of breath ethanol. The test employs a sealed glass tube that contains the oxidizing agent (sodium dichromate dissolved in sulfuric acid) impregnated onto an inert material. The ends of the tube are broken off, and one end of the tube is attached to a mouthpiece and the other to a balloon-type bag. The person undergoing the test blows into the mouthpiece until the bag is filled with air.

Any ethanol in the breath is oxidized as it passes through the column. When ethanol is oxidized, the red-orange dichromate ion ($Cr_2O_7^{2-}$) is reduced to green chromic ion (Cr^{3+}).

The greater the concentration of alcohol in the breath, the farther the green color spreads through the tube.

$$CH_3CH_2OH + Cr_2O_7^{2-} \xrightarrow{H^+} CH_3\overset{\overset{\displaystyle O}{\|}}{C}OH + Cr^{3+}$$

red orange   **green**

If the person fails this test—determined by the extent to which the green color spreads through the tube—a more accurate Breathalyzer test is administered. The Breathalyzer test also depends on the oxidation of breath ethanol, but it provides more accurate results because it is quantitative. In the test, a known volume of breath is bubbled through an acidic solution of sodium dichromate, and the concentration of the green chromic ion is measured precisely with a spectrophotometer (Section 6.12).

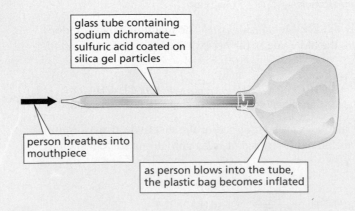

glass tube containing sodium dichromate–sulfuric acid coated on silica gel particles

person breathes into mouthpiece

as person blows into the tube, the plastic bag becomes inflated

11.5   Amines Do Not Undergo Substitution or Elimination Reactions

We have just seen that alcohols are much less reactive than alkyl halides in substitution and elimination reactions. Amines are even *less reactive* than alcohols. The relative reactivities of an alkyl fluoride (the least reactive of the alkyl halides because it has the poorest leaving group), an alcohol, and an amine can be appreciated by comparing the pK_a values of the conjugate acids of their leaving groups. (Recall that the weaker the acid, the stronger is its conjugate base and the poorer it is as a leaving group.) The leaving group of an amine ($^-NH_2$) is such a strong base that amines cannot undergo substitution or elimination reactions.

relative reactivities

$$RCH_2F \quad > \quad RCH_2OH \quad > \quad RCH_2NH_2$$

$$\begin{array}{ccc} HF & H_2O & NH_3 \\ pK_a = 3.2 & pK_a = 15.7 & pK_a = 36 \end{array}$$

The stronger the base, the poorer it is as a leaving group.

Protonation of the amino group makes it a better leaving group, but not nearly as good a leaving group as a protonated alcohol, which is ~13 pK_a units more acidic than a protonated amine. Therefore, unlike the leaving group of a protonated alcohol, the leaving group of a protonated amine cannot be replaced by a halide ion or dissociate to form a carbocation.

$$\begin{array}{cc} CH_3CH_2\overset{+}{O}H_2 & > \quad CH_3CH_2\overset{+}{N}H_3 \\ pK_a = -2.4 & pK_a = 11.2 \end{array}$$

Although amines cannot undergo substitution or elimination reactions, they are extremely important organic compounds. The lone pair on the nitrogen atom allows it to act as both a base and as a nucleophile.

Amines are the most common organic bases. We have seen that protonated amines have pK_a values of about 11 (Section 2.2) and that protonated anilines have pK_a values of about 5 (Sections 6.9 and 7.15). Neutral amines have very high pK_a values. For example, the pK_a of methylamine is 40.

$$CH_3CH_2CH_2\overset{+}{N}H_3 \qquad CH_3\overset{+}{N}H_2\!\!-\!\!CH_3 \qquad CH_3CH_2\overset{+}{N}H(CH_2CH_3)CH_2CH_3 \qquad C_6H_5\overset{+}{N}H_3 \qquad CH_3\!\!-\!\!C_6H_4\!\!-\!\!\overset{+}{N}H_3 \qquad CH_3NH_2$$

$$pK_a = 10.8 \qquad pK_a = 10.9 \qquad pK_a = 11.1 \qquad pK_a = 4.58 \qquad pK_a = 5.07 \qquad pK_a = 40$$

Amines react as nucleophiles in a wide variety of reactions. For example, they react as nucleophiles with alkyl halides in S_N2 reactions.

$$CH_3CH_2Br + CH_3NH_2 \longrightarrow CH_3CH_2\overset{+}{N}H_2CH_3 + Br^-$$

methylamine

We will see that they also react as nucleophiles with a wide variety of carbonyl compounds (Sections 12.7, 12.8, and 13.6).

PROBLEM 13

Why can protonated amino groups not be displaced by strongly basic nucleophiles such as HO^-?

ALKALOIDS

Alkaloids are amines found in the leaves, bark, roots, or seeds of plants. Examples include caffeine (found in tea leaves, coffee beans, and cola nuts), nicotine (found in tobacco leaves), and cocaine (obtained from the coca bush in the rainforest areas of Columbia, Peru, and Bolivia). Morphine is an alkaloid obtained from opium, the juice derived from a species of poppy (Section 22.3). Ephedrine, a bronchodilator, is obtained from *Ephedra sinica*, a plant found in China.

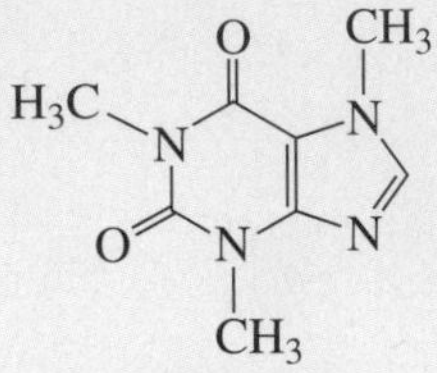

caffeine 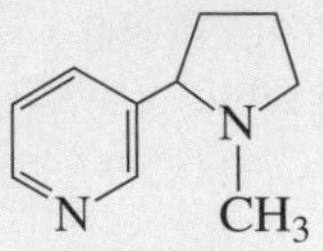   nicotine 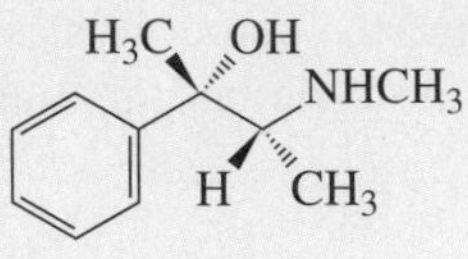   ephedrine 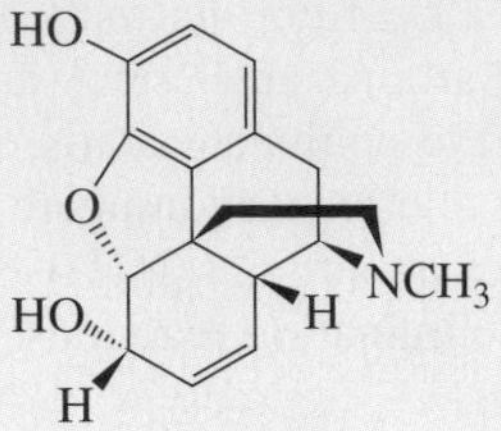   morphine

11.6 Nomenclature of Ethers

Ethers are compounds in which an oxygen is bonded to two alkyl substituents. The common name of an ether consists of the names of the two alkyl substituents (in alphabetical order), followed by the word "ether." The smallest ethers are almost always named by their common names.

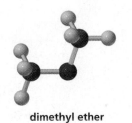

dimethyl ether

$CH_3OCH_2CH_3$
ethyl methyl ether

$CH_3CH_2OCH_2CH_3$
diethyl ether

tert-butil

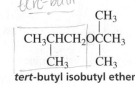

$CH_3CHCH_2OCCH_3$
tert-butyl isobutyl ether

The IUPAC system names an ether as an alkane with an RO substituent. The substituents are named by replacing the "yl" ending in the name of the alkyl substituent with "oxy."

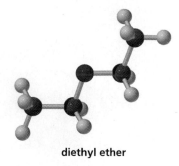

diethyl ether

CH_3O-
methoxy

CH_3CH_2O-
ethoxy

CH_3CHO-
CH_3
isopropoxy

CH_3CH_2CHO-
CH_3
sec-butoxy

CH_3CO-
CH_3
CH_3
tert-butoxy

$CH_3CHCH_2CH_3$
OCH_3
2-methoxybutane

$CH_3CH_2CHCH_2CH_2OCH_2CH_3$
CH_3
1-ethoxy-3-methylpentane

Tutorial:
Nomenclature of ethers

PROBLEM 14◆

a. Give the systematic name for each of the following ethers:

1. $CH_3OCH_2CH_3$

2. $CH_3CH_2OCH_2CH_3$

3. $CH_3CH_2CH_2CH_2CHCH_2CH_2CH_3$
   OCH_3

4. $CH_3CH_2CH_2OCH_2CH_2CH_2CH_3$

b. Do all of these ethers have common names?

c. What are their common names?

ANESTHETICS

Because diethyl ether (commonly known simply as ether) is a short-lived muscle relaxant, it has been widely used as an inhalation anesthetic. However, because it takes effect slowly and has a slow and unpleasant recovery period, other compounds, such as enflurane, isoflurane, and halothane, have

$$CH_3CH_2OCH_2CH_3 \qquad CF_3CHClOCHF_2$$
"ether"   **isoflurane**

Sodium pentothal (also called thiopental sodium) is commonly used as an intravenous anesthetic. The onset of anesthesia and the loss of consciousness occur within seconds of its administration. Care must be taken when administering sodium pentothal because the dose for effective anesthesia is 75% of the lethal dose. Because of its toxicity, it cannot be used as the sole anesthetic. It is generally used to induce anesthesia before an inhalation anesthetic is administered. Propofol is an anesthetic that has all the properties of the "perfect anesthetic": It can be used as the sole anesthetic by intravenous drip, it

replaced ether as an anesthetic. Diethyl ether is still used where there is a lack of trained anesthesiologists, because it is the safest anesthetic to administer by untrained hands. Anesthetics interact with the nonpolar molecules of cell membranes, causing the membranes to swell, which interferes with their permeability.

$$CHClFCF_2OCHF_2 \qquad CF_3CHClBr$$
enflurane   **halothane**

has a rapid and pleasant induction period and a wide margin of safety, and recovery from the drug also is rapid and pleasant.

sodium pentothal   **propofol**

Amputation of a leg without anesthetic in 1528.

Painting of the first use of anesthesia in 1846.

11.7 Substitution Reactions of Ethers

The OR group of an **ether** and the OH group of an alcohol have nearly the same basicity, because their conjugate acids have similar pK_a values. (The pK_a of CH_3OH is 15.5 and the pK_a of H_2O is 15.7.) Both groups are strong bases, so both are very poor leaving groups. Consequently, ethers and alcohols are equally unreactive toward nucleophilic substitution.

$$R-\ddot{O}-H \qquad R-\ddot{O}-R$$
an alcohol   **an ether**

Like alcohols, ethers can be activated by protonation. Ethers, therefore, can undergo nucleophilic substitution reactions with HBr or HI. As with alcohols, the reaction of ethers with hydrogen halides is slow, and the reaction mixture must be heated in order for the reaction to occur at a reasonable rate.

$$R-O-R' + HI \;\rightleftharpoons\; R-\overset{H}{\underset{+}{O}}-R' \;\xrightarrow{\Delta}\; R-I + R'-OH$$

poor leaving group I^-
good leaving group

The first step in the cleavage of an ether by HI or HBr is protonation of the ether's oxygen atom. This converts the very basic RO^- leaving group into the less basic ROH leaving group. What happens next in the mechanism depends on the structure of the ether. If departure of the leaving group creates a relatively stable carbocation (e.g., a tertiary carbocation), an S_N1 reaction occurs—the leaving group departs, and the halide ion combines with the carbocation.

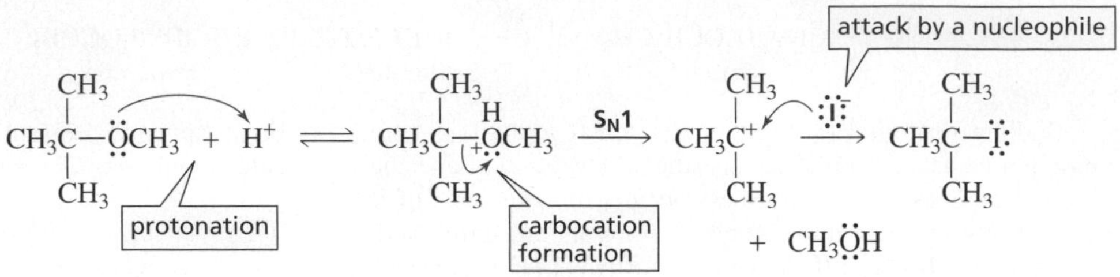

However, if departure of the leaving group would create an unstable carbocation (e.g., a methyl or a primary carbocation), the leaving group cannot depart. It has to be displaced by the halide ion. In other words, an S_N2 reaction occurs. In the S_N2 reaction, the halide ion preferentially attacks the less sterically hindered of the two alkyl groups.

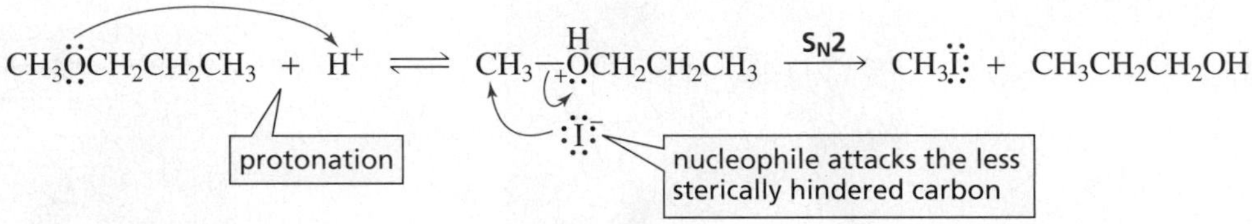

3-D Molecules:
Diethyl ether;
Tetrahydrofuran

Because the only reagents with which ethers react are hydrogen halides, ethers are frequently used as solvents. Some common ether solvents are shown in Table 11.1.

Table 11.1   Some Ethers That Are Used as Solvents

$CH_3CH_2OCH_2CH_3$		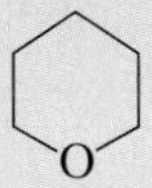	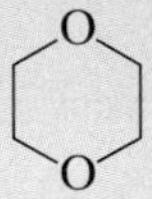	$CH_3OCH_2CH_2OCH_3$	$CH_3OC(CH_3)_3$
diethyl ether "ether"	tetrahydrofuran THF	tetrahydropyran	1,4-dioxane	1,2-dimethoxyethane DME	*tert*-butyl methyl ether MTBE

PROBLEM 15◆

Explain why methyl propyl ether forms both methyl iodide and propyl iodide when it is heated with excess HI.

PROBLEM 16◆   SOLVED

Give the major products that would be obtained from heating each of the following ethers with HI:

a. $CH_3CHCH_2OCH_2CH_3$
   |
   CH_3

b.

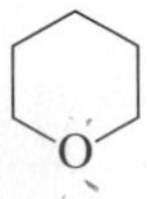

c. $CH_3COCH_2CH_3$

d.

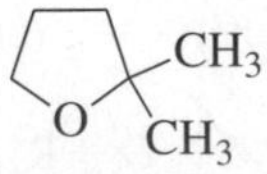

SOLUTION TO 16A   The reaction takes place by an S_N2 pathway because an S_N1 pathway would require the formation of either a primary carbocation or a methyl carbocation, both of which are too unstable to form. Iodide ion attacks the carbon of the methyl group because it is less sterically hindered than the carbon of the propyl group. Because there is excess HI, the alcohol product of the first reaction can react with HI in another S_N2 reaction. Thus, the major products of the overall reaction are methyl iodide and propyl iodide.

3-D Molecule:
[15]-Crown-5;
[12]-Crown-4

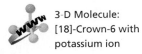

3-D Molecule:
[18]-Crown-6 with
potassium ion

$$CH_3CH_2CH_2\overset{..}{\underset{..}{O}}CH_3 \xrightarrow{\text{HI}} CH_3CH_2CH_2OH \xrightarrow{\text{HI}} CH_3CH_2CH_2I + H_2O$$
$$+ \quad CH_3I$$

AN UNUSUAL ANTIBIOTIC

Crown ethers are cyclic compounds that have several ether linkages. A crown ether specifically binds certain metal ions or organic molecules, depending on the size of its cavity. The crown ether is called the "host" and the species it binds is called the "guest." Because the ether linkages are chemically inert, the crown ether can bind the guest without reacting with it.

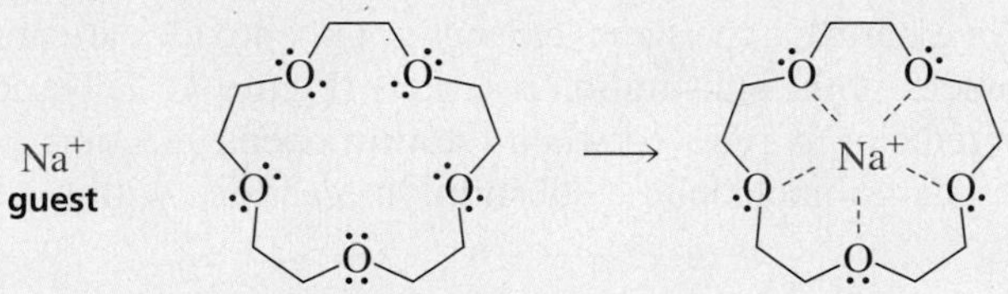

An antibiotic is a compound that interferes with the growth of microorganisms. Nonactin is a naturally occurring antibiotic that owes its biological activity to its ability to disrupt the carefully maintained electrolyte balance between the inside and outside of a cell. To achieve the gradient between potassium and sodium ions inside and outside the cell that is required for normal cell function, potassium ions are pumped in and sodium ions are pumped out. Nonactin disrupts this gradient by acting like a crown ether. Nonactin's diameter is such that it specifically binds potassium ions. The eight oxygens that point into the cavity and interact with K^+ are highlighted in the structure shown here. The outside of nonactin is nonpolar, so it can easily transport K^+ ions out of the cell through the nonpolar cell membrane. The decreased concentration of K^+ within the cell causes the bacterium to die.

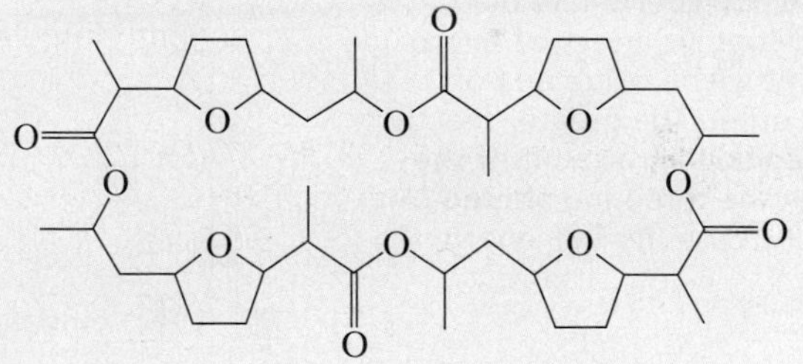

Na$^+$
guest

host
[15]-crown-5
cavity diameter = 1.7–2.2 Å

Na$^+$

nonactin

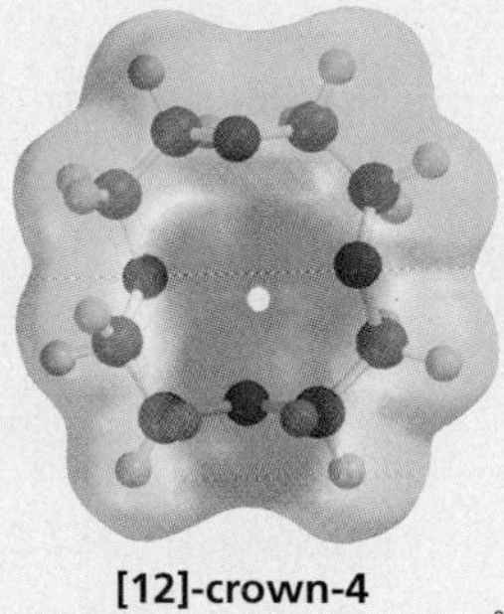

[12]-crown-4
cavity diameter = 1.2–1.5 Å

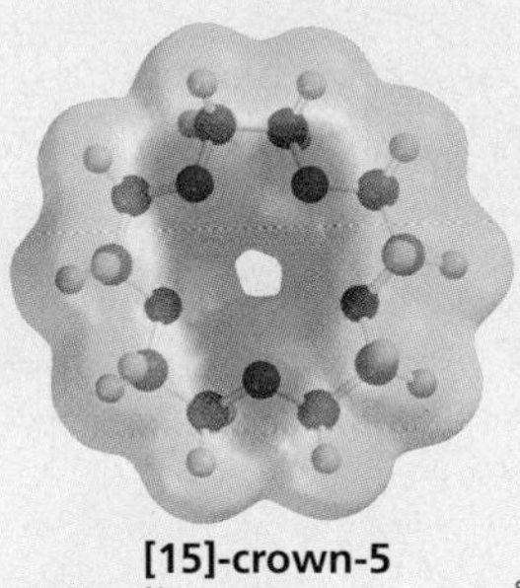

[15]-crown-5
cavity diameter = 1.7–2.2 Å

11.8   Reactions of Epoxides

Ethers in which the oxygen atom is incorporated into a three-membered ring are called **epoxides**. The common name of an epoxide is the common name of the alkene followed by "oxide," assuming that the oxygen atom is where the π bond of an alkene would be. The simplest epoxide is ethylene oxide.

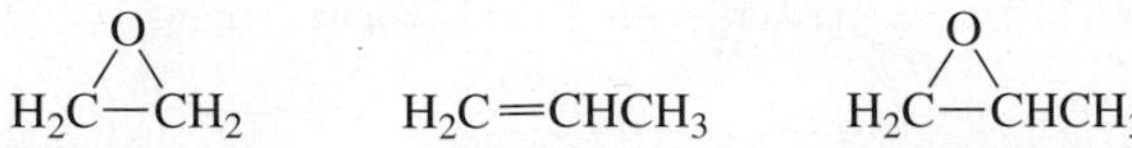

$$H_2C{=}CH_2$$
ethylene

$$H_2C\overset{\displaystyle O}{\overset{\diagup\diagdown}{-}}CH_2$$
ethylene oxide

$$H_2C{=}CHCH_3$$
propylene

$$H_2C\overset{\displaystyle O}{\overset{\diagup\diagdown}{-}}CHCH_3$$
propylene oxide

Alternatively, an epoxide can be named as an alkane, with an "epoxy" prefix that identifies the carbons to which the oxygen is attached.

H_2C—$CHCH_2CH_3$
1,2-epoxybutane

CH_3CH—$CHCH_3$
2,3-epoxybutane

H_2C—C—CH_3 / CH_3
1,2-epoxy-2-methylpropane

PROBLEM 17◆

Draw the structure of the following compounds:

a. cyclohexene oxide

b. 2,3-epoxy-2-methylpentane

An epoxide is formed from the reaction of an alkene with a *peroxyacid*. A **peroxyacid** has one more oxygen atom than a carboxylic acid. It is this oxygen atom that is inserted into the alkene in order to from the epoxide. Notice that the reaction increases the number of C—O bonds in the reactant. It is, therefore, an oxidation reaction (Section 11.4).

RCH=CH_2 + $RCOOH$ $\longrightarrow$ RCH—CH_2 + $RCOH$
an alkene   **a peroxyacid**   **an epoxide**   **a carboxylic acid**

Although an epoxide and an ether have the same leaving group, epoxides are much more reactive than ethers in nucleophilic substitution reactions (Figure 11.2) because the strain in the three-membered ring is relieved when the ring opens (Section 3.9). Epoxides, therefore, readily undergo nucleophilic substitution reactions with a wide variety of nucleophiles.

Figure 11.2 ▶
The reaction coordinate diagrams for nucleophilic attack of hydroxide ion on ethylene oxide and on diethyl ether. The greater reactivity of the epoxide is a result of the strain in the three-membered ring, which increases its free energy.

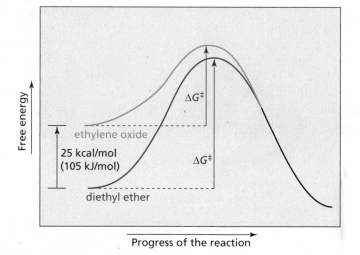

Epoxides, like other ethers, react with hydrogen halides. In the first step of the reaction, the oxygen atom is protonated by the acid. The protonated epoxide is then attacked by the halide ion. Because epoxides are so much more reactive than ethers, the reaction takes place readily at room temperature, unlike the reaction of an ether with a hydrogen halide that requires heat.

H_2C—CH_2 + H—Br $\rightleftharpoons$ H_2C—CH_2 + Br^- $\longrightarrow$ $HOCH_2CH_2Br$

protonation of the epoxide oxygen atom

back-side attack by the nucleophile

Protonated epoxides are so reactive that they can be opened by poor nucleophiles, such as H_2O and alcohols.

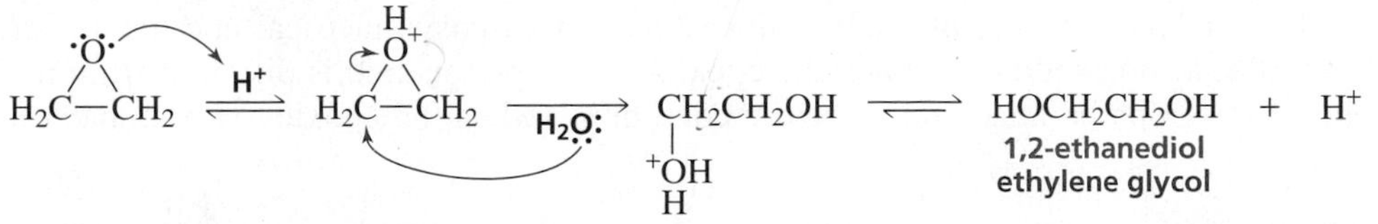

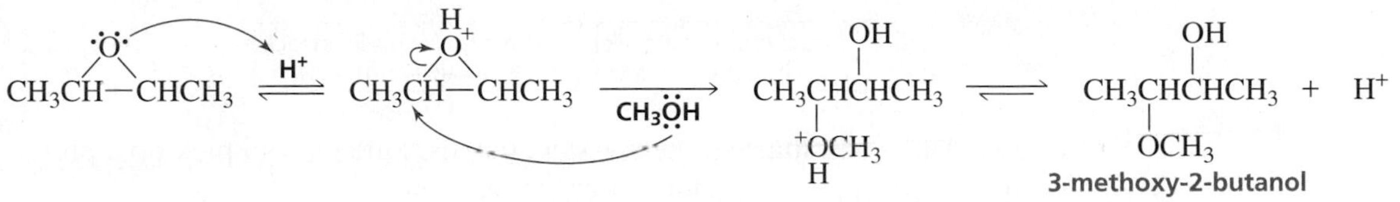

If different substituents are attached to the two carbons of the protonated epoxide (and the nucleophile is something other than H_2O), the product obtained from nucleophilic attack on the 2-position of the ring will be different from that obtained from attack on the 1-position. The major product is the one resulting from nucleophilic attack on the *more substituted* carbon.

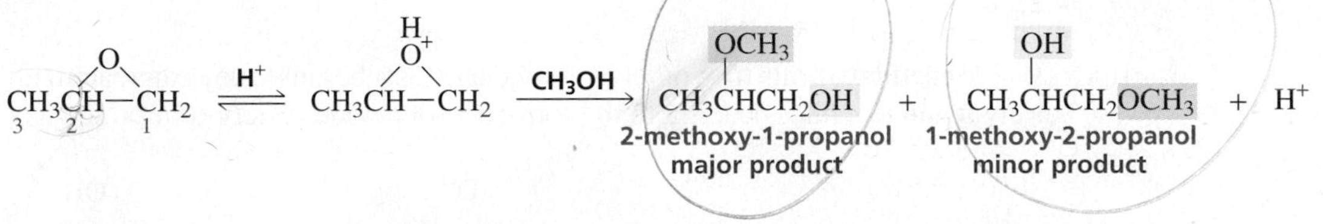

3-D Molecule:
Propylene oxide

The more substituted carbon is more likely to be attacked because, after the epoxide is protonated, it is so reactive that one of the C—O bonds begins to break before the nucleophile has an opportunity to attack. As the C—O bond starts to break, a partial positive charge develops on the carbon that is losing its share of the oxygen's electrons. The protonated epoxide breaks preferentially in the direction that puts the partial positive charge on the more substituted carbon, because a more substituted carbocation is more stable. (Recall that tertiary carbocations are more stable than secondary carbocations, which are more stable than primary carbocations.)

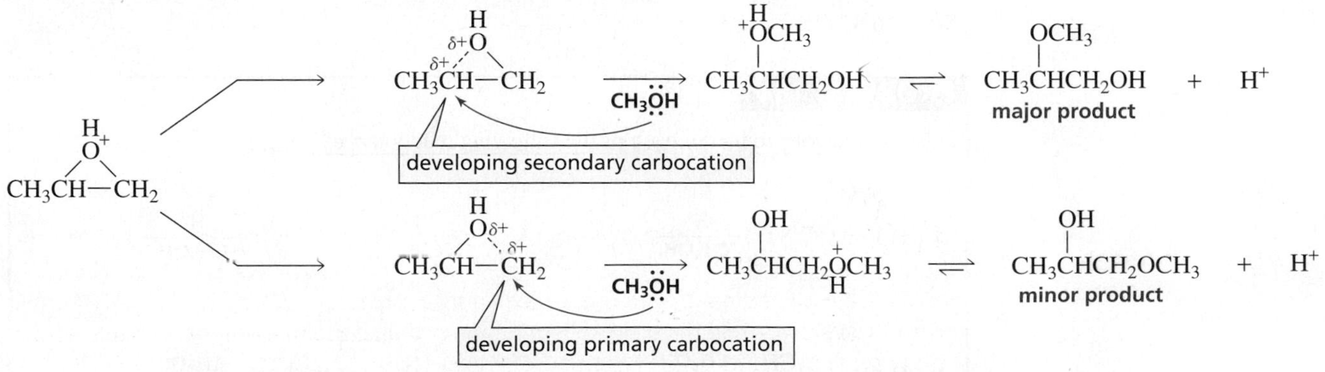

The best way to describe the reaction is to say that it occurs by a pathway that is partially S_N1 and partially S_N2. It is not a pure S_N1 reaction because a carbocation intermediate is not fully formed; it is not a pure S_N2 reaction because the leaving group begins to depart before the compound is attacked by the nucleophile.

Because of the strain in the three-membered ring, epoxides, unlike ethers, can undergo nucleophilic substitution reactions without first being protonated. When a nucleophile attacks an unprotonated epoxide, the reaction is a pure S_N2 reaction. That is, the C—O bond does not begin to break until the carbon is attacked by the nucleophile. In this case, the nucleophile is more likely to attack the *less substituted* carbon because the less substituted carbon is more accessible to attack. (It is less sterically hindered.) Thus, the site of nucleophilic attack on an unsymmetrical epoxide under neutral or basic conditions (when the epoxide *is not* protonated) is different from the site of nucleophilic attack under acidic conditions (when the epoxide *is* protonated).

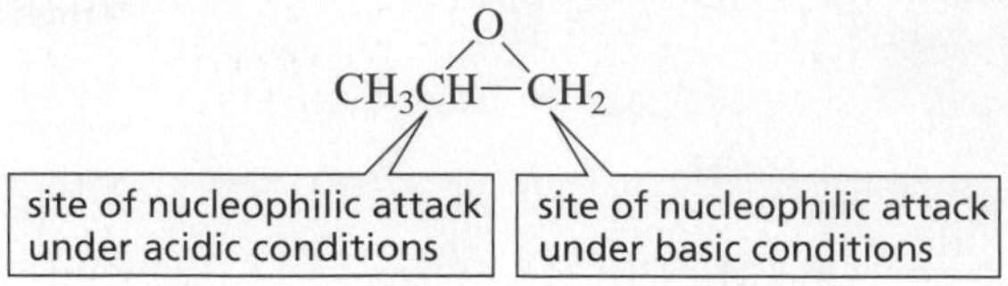

$$\overset{\text{O}}{\overset{\diagup \ \diagdown}{\text{CH}_3\text{CH—CH}_2}}$$

| site of nucleophilic attack under acidic conditions | site of nucleophilic attack under basic conditions |

After the nucleophile has attacked the epoxide, the alkoxide ion can pick up a proton from the solvent or from an acid added after the reaction is over.

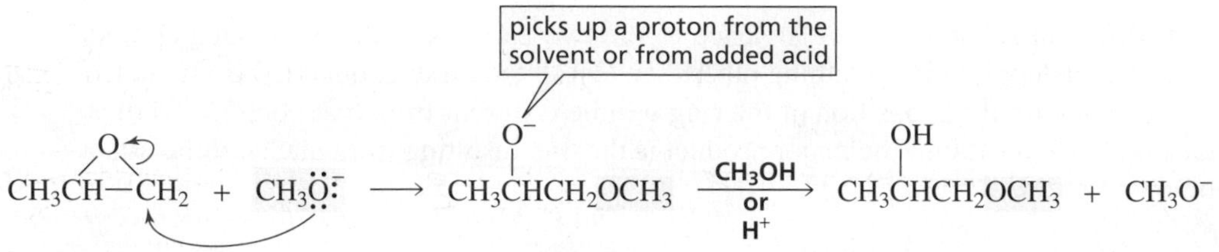

picks up a proton from the solvent or from added acid

$$\overset{\text{O}}{\overset{\diagup \ \diagdown}{\text{CH}_3\text{CH—CH}_2}} + \text{CH}_3\overset{..}{\underset{..}{\text{O}}}{:}^- \longrightarrow \text{CH}_3\overset{\text{O}^-}{\underset{|}{\text{CH}}}\text{CH}_2\text{OCH}_3 \xrightarrow[\text{H}^+]{\text{CH}_3\text{OH} \atop \text{or}} \text{CH}_3\overset{\text{OH}}{\underset{|}{\text{CH}}}\text{CH}_2\text{OCH}_3 + \text{CH}_3\text{O}^-$$

Epoxides are useful reagents to synthetic organic chemists because they can react with a wide variety of nucleophiles, leading to the formation of a wide variety of products.

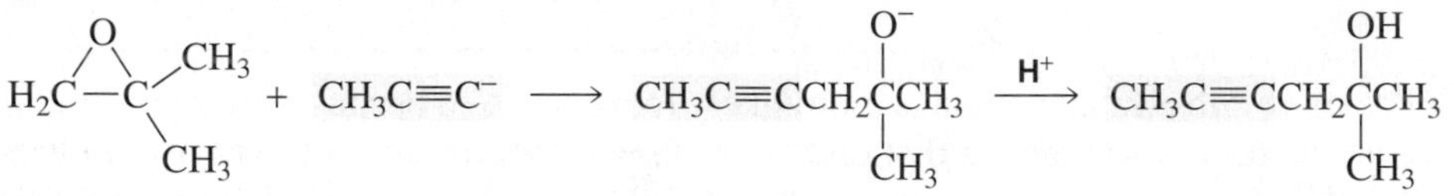

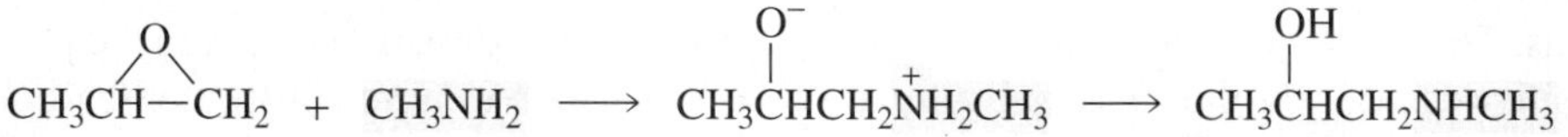

$$\overset{\text{O}}{\underset{\text{CH}_3}{\overset{\diagup \ \diagdown}{\text{H}_2\text{C—C}}}}\overset{\text{CH}_3}{} + \text{CH}_3\text{C}{\equiv}\text{C}^- \longrightarrow \text{CH}_3\text{C}{\equiv}\text{CCH}_2\overset{\text{O}^-}{\underset{\text{CH}_3}{\overset{|}{\text{C}}}}\text{CH}_3 \xrightarrow{\text{H}^+} \text{CH}_3\text{C}{\equiv}\text{CCH}_2\overset{\text{OH}}{\underset{\text{CH}_3}{\overset{|}{\text{C}}}}\text{CH}_3$$

$$\overset{\text{O}}{\overset{\diagup \ \diagdown}{\text{CH}_3\text{CH—CH}_2}} + \text{CH}_3\text{NH}_2 \longrightarrow \text{CH}_3\overset{\text{O}^-}{\underset{|}{\text{CH}}}\text{CH}_2\overset{+}{\text{N}}\text{H}_2\text{CH}_3 \longrightarrow \text{CH}_3\overset{\text{OH}}{\underset{|}{\text{CH}}}\text{CH}_2\text{NHCH}_3$$

Epoxides also are important in biological processes because they are reactive enough to be attacked by nucleophiles under the conditions found in living systems (Section 11.9).

PROBLEM 18◆

Give the major product of each of the following reactions:

a. 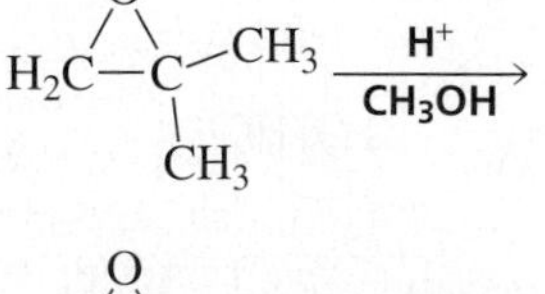 $\overset{\text{O}}{\underset{\text{CH}_3}{\overset{\diagup \ \diagdown}{\text{H}_2\text{C—C}}}}\text{—CH}_3 \xrightarrow{\text{H}^+ \atop \text{CH}_3\text{OH}}$

c. $\overset{\text{O}}{\underset{\text{H}_3\text{C} \quad \text{CH}_3}{\overset{\diagup \ \diagdown}{\text{H—C—C}}}}\text{—CH}_3 \xrightarrow{\text{H}^+ \atop \text{CH}_3\text{OH}}$

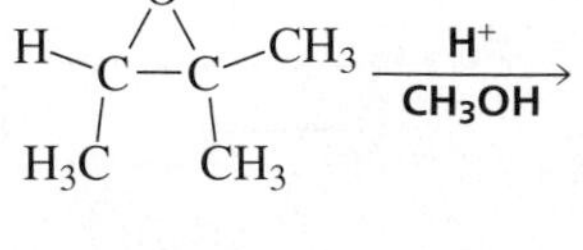

b. 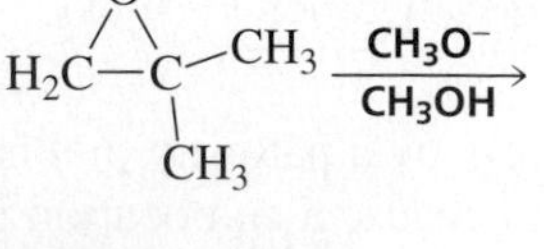 $\overset{\text{O}}{\underset{\text{CH}_3}{\overset{\diagup \ \diagdown}{\text{H}_2\text{C—C}}}}\text{—CH}_3 \xrightarrow{\text{CH}_3\text{O}^- \atop \text{CH}_3\text{OH}}$

d. 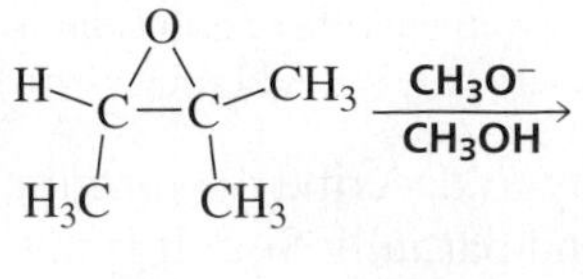 $\overset{\text{O}}{\underset{\text{H}_3\text{C} \quad \text{CH}_3}{\overset{\diagup \ \diagdown}{\text{H—C—C}}}}\text{—CH}_3 \xrightarrow{\text{CH}_3\text{O}^- \atop \text{CH}_3\text{OH}}$

PROBLEM 19◆

Would you expect the reactivity of a five-membered ring ether such as tetrahydrofuran (Table 11.1) to be more similar to an epoxide or to a noncyclic ether?

MUSTARD—A CHEMICAL WARFARE AGENT

Chemical warfare occurred for the first time in 1915, when Germany released chlorine gas against French and British forces in the battle of Ypres. For the remainder of World War I, both sides used a variety of chemical agents. One of the more common chemical warfare agents was mustard, a reagent that produces blisters over the surface of the body. Mustard is a very reactive compound because sulfur is a good nucleophile, so it easily displaces a chloride ion by an intramolecular S_N2 reaction, forming a cyclic sulfonium salt that reacts rapidly with a nucleophile. The sulfonium salt is particularly reactive because of the strained three-membered ring and the excellent (positively charged) leaving group.

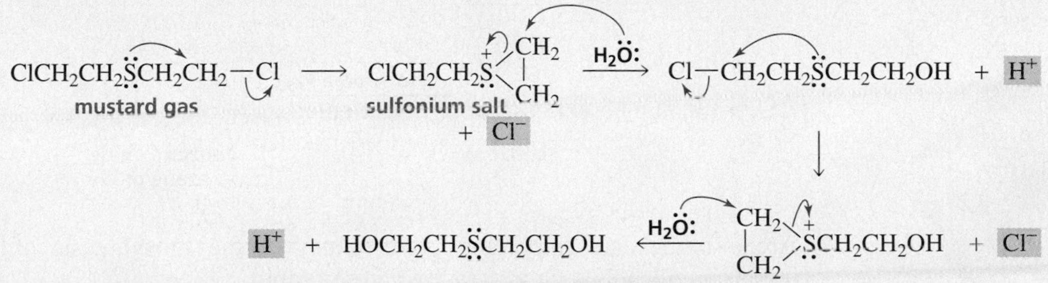

$$ClCH_2CH_2\ddot{S}CH_2CH_2-Cl \longrightarrow ClCH_2CH_2\overset{+}{S}\overset{CH_2}{\underset{CH_2}{<}} \xrightarrow{H_2\ddot{O}:} Cl-CH_2CH_2\ddot{S}CH_2CH_2OH + H^+$$

mustard gas — sulfonium salt + Cl⁻

$$H^+ + HOCH_2CH_2\ddot{S}CH_2CH_2OH \xleftarrow{H_2\ddot{O}:} \overset{CH_2}{\underset{CH_2}{>}}\overset{+}{S}CH_2CH_2OH + Cl^-$$

The blistering caused by mustard is due to the high local concentrations of HCl that are produced when water—or any other nucleophile—reacts with the chemical agent when it comes into contact with the skin or lungs. Autopsies of soldiers killed by mustard in World War I—estimated to be about 400,000— showed that inhaling vapors from the evaporating agent had caused blistering of the mucous membranes in their respiratory tracts. Rupturing of the blisters caused massive internal secondary infections. An international treaty in the 1980s banned the use of mustard and required that all stockpiles be destroyed.

PROBLEM 20◆

Finding that mustard interfered with bone marrow development caused chemists to look for less reactive mustards that might be used clinically. The following three compounds were studied:

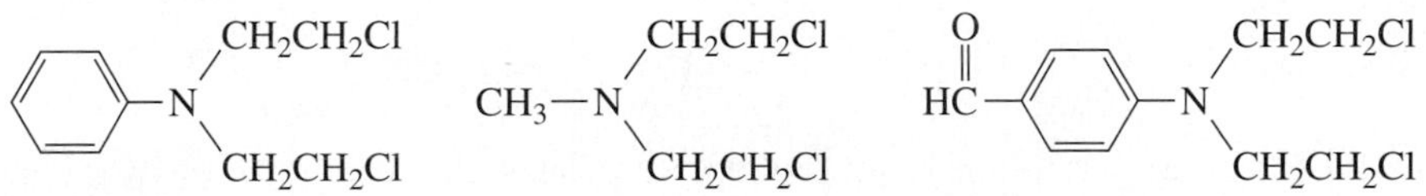

(structures: N,N-bis(2-chloroethyl)aniline with CH₂CH₂Cl groups; CH₃—N with two CH₂CH₂Cl groups; a para-substituted benzaldehyde (HC=O) with N(CH₂CH₂Cl)₂)

One was found to be too reactive, one was found to be too unreactive, and one was found to be too insoluble in water to be injected intravenously. Which is which? (*Hint:* Draw resonance contributors.)

ANTIDOTE TO A WAR GAS

Lewisite is a chemical warfare agent developed in 1917 by W. Lee Lewis, an American scientist. It rapidly penetrates clothing and skin and is poisonous because it contains arsenic, which combines with SH groups on enzymes, thereby inactivating them (Section 17.6). During World War II, the Allies were concerned that the Germans would use lewisite, so British scientists developed an antidote that the Allies called "British anti-lewisite" (BAL). BAL contains two SH groups that react with lewisite, thereby preventing it from reacting with the SH groups of enzymes.

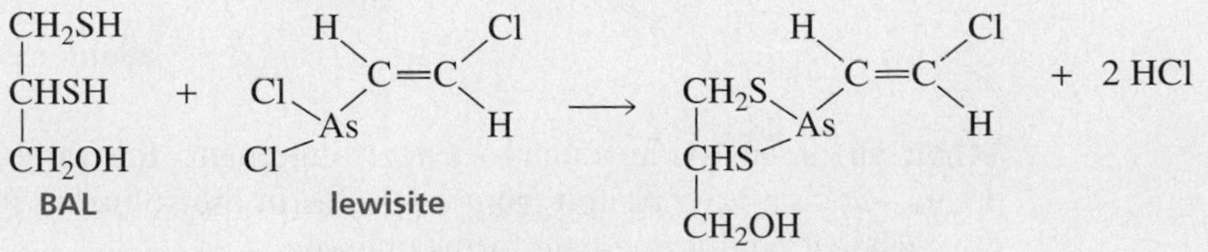

$$\begin{array}{c} CH_2SH \\ | \\ CHSH \\ | \\ CH_2OH \end{array} + \begin{array}{c} Cl \\ \quad \\ Cl \end{array}As-C=C\begin{array}{c} Cl \\ \quad \\ H \end{array} \quad (H) \longrightarrow \begin{array}{c} CH_2S \\ | \\ CHS \\ | \\ CH_2OH \end{array}As-C=C\begin{array}{c} Cl \\ \quad \\ H \end{array} (H) + 2\ HCl$$

BAL — lewisite

benzene

benzene oxide

11.9 Arene Oxides

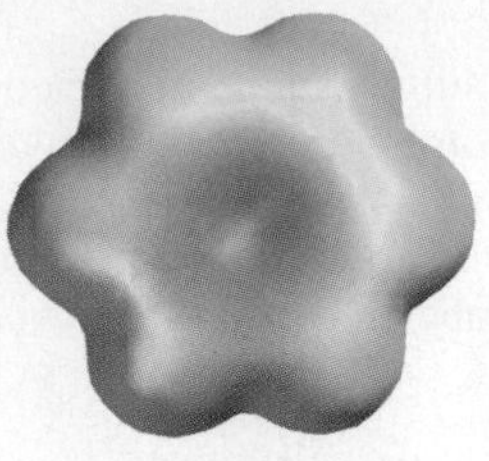

When an aromatic hydrocarbon such as benzene is ingested or inhaled, it is enzymatically converted into an *arene oxide* by an enzyme called cytochrome P_{450}. An **arene oxide** is a compound in which one of the "double bonds" of the aromatic ring has been converted into an epoxide. Formation of an arene oxide is the first step in changing an aromatic compound that enters the body as a foreign substance (e.g., cigarette smoke, drugs, automobile exhaust) into a more water-soluble compound that can eventually be eliminated.

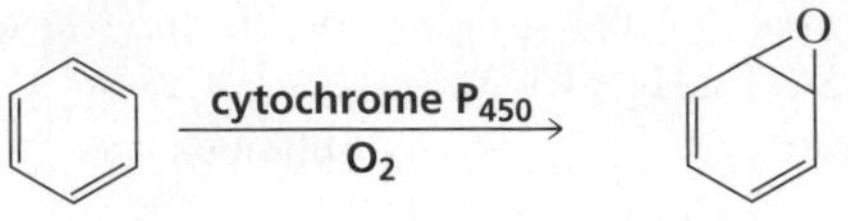

benzene $\xrightarrow[\text{O}_2]{\text{cytochrome P}_{450}}$ benzene oxide
an arene oxide

Arene oxides are important intermediates in the biosynthesis of biochemically important phenols such as tyrosine and serotonin.

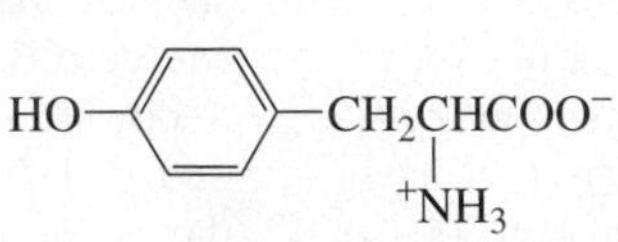

tyrosine
an amino acid

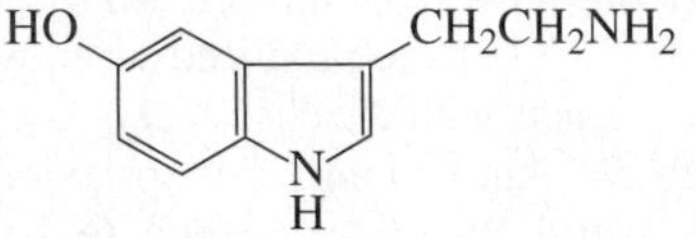

serotonin
a vasoconstrictor

An arene oxide can react in two different ways. It can react as a typical epoxide, undergoing attack by a nucleophile to form an addition product (Section 11.8). Alternatively, it can rearrange to form a phenol, something that epoxides such as ethylene oxide cannot do.

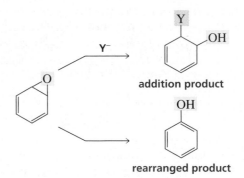

addition product

rearranged product

3-D Molecule:
Benzene oxide

When an arene oxide reacts with a nucleophile, the nucleophile attacks the three-member ring and forms an addition product.

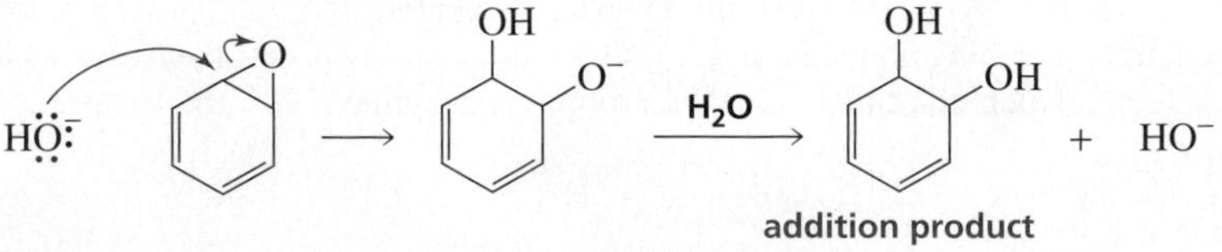

addition product

When an arene oxide undergoes rearrangement, the three-membered epoxide ring opens, picking up a proton from a species in the solution. Removal of a proton from the carbocation intermediate forms phenol.

benzene oxide   a carbocation   phenol

Because formation of the carbocation is the rate-determining step, the rate of formation of phenol depends on the stability of the carbocation. The more stable the carbocation, the easier it is to open the epoxide ring and form phenol.

Some aromatic hydrocarbons are carcinogens—compounds that cause cancer. Investigation has revealed, however, that it is not the hydrocarbon itself that is carcinogenic, but rather the arene oxide into which the hydrocarbon is converted. How do arene oxides cause cancer? We have seen that nucleophiles react with epoxides to form addition products. $2'$-Deoxyguanosine, a component of DNA, has a nucleophilic NH_2 group that is known to react with certain arene oxides. Once $2'$-deoxyguanosine becomes covalently attached to an arene oxide, $2'$-deoxyguanosine can no longer fit into the DNA double helix. Thus, the genetic code cannot be properly transcribed (Section 21.6), which can lead to mutations that cause cancer. Cancer results when cells lose their ability to control their growth and reproduction.

A segment of DNA

an arene oxide

covalently attached
to the arene oxide

$2'$-deoxyguanosine

Not all arene oxides are carcinogenic. Whether a particular arene oxide is carcinogenic depends on the relative rates of its two reaction pathways—rearrangement or reaction with a nucleophile. Arene oxide rearrangement leads to phenols that are not carcinogenic, whereas formation of addition products from nucleophilic attack by DNA can lead to cancer-causing products. Thus, if the rate of arene oxide rearrangement is greater than the rate of nucleophilic attack by DNA, the arene oxide will be harmless. If, however, the rate of nucleophilic attack is greater than the rate of rearrangement, the arene oxide will likely be a carcinogen.

Because the rate of arene oxide rearrangement depends on the stability of the carbocation formed in the first step of the rearrangement, *an arene oxide's cancer-causing potential depends on the stability of this carbocation.* If the carbocation is relatively stable, rearrangement will be fast and the arene oxide will most likely not be carcinogenic. On the other hand, if the carbocation is relatively unstable, rearrangement will be slow and the arene oxide, therefore, will exist long enough to undergo nucleophilic attack and be carcinogenic. This means that the more reactive the arene oxide (the more easily it opens to form a carbocation), the less likely it is to be carcinogenic.

The more stable the carbocation that is formed when the epoxide ring of an arene oxide opens, the less likely it is that the arene oxide is carcinogenic.

PROBLEM-SOLVING STRATEGY

Which compound is more likely to be carcinogenic?

or

OCH_3 NO_2

To determine the compound most likely to be carcinogenic, we must compare the stabilities of the carbocations formed when the epoxide rings open. The compound with the least stable carbocation is the one that is most apt to be carcinogenic.

The methoxy group stabilizes the carbocation by donating electrons into the ring by resonance. In contrast, the nitro group destabilizes the carbocation by withdrawing electrons from the ring by resonance. Thus, the nitro-substituted compound, with a less stable (harder-to-form) carbocation, will be less likely to undergo rearrangement to a harmless product. In addition, the electron-withdrawing nitro group increases the arene oxide's susceptibility to nucleophilic attack, the cancer-causing pathway.

Now continue on to Problem 21.

PROBLEM 21◆

Which compound is more likely to be carcinogenic? (*Hint:* Read the box on benzo[*a*]pyrene to see why the 4,5-epoxide is harmful.)

BENZO[*a*]PYRENE AND CANCER

Benzo[*a*]pyrene is one of the most carcinogenic of the aromatic hydrocarbons. It is formed whenever an organic compound is not completely burned. For example, benzo[*a*]pyrene is found in cigarette smoke, automobile exhaust, and charcoal-broiled meat. Several arene oxides can be formed from benzo[*a*]pyrene. The two most harmful are the 4,5-oxide and the 7,8-oxide. It has been suggested that people who develop lung cancer as a result of smoking may have a higher than normal concentration of cytochrome P_{450} in their lung tissue.

benzo[*a*]pyrene

4,5-benzo[*a*]pyrene oxide + **7,8-benzo[*a*]pyrene oxide**

a diol epoxide

The 4,5-oxide is harmful because it forms a carbocation that cannot be stabilized by electron delocalization without destroying the aromaticity of an adjacent benzene ring. Thus, the carbocation is relatively unstable, so the epoxide will tend not to open until it is attacked by a nucleophile. The 7,8-oxide is harmful because it reacts with water to form a diol, which then forms a diol epoxide. The diol epoxide does not readily undergo rearrangement (the harmless pathway), because it opens to a carbocation that is destabilized by the electron-withdrawing OH groups. Therefore, the diol epoxide can exist long enough to be attacked by nucleophiles (the carcinogenic pathway).

CHIMNEY SWEEPS AND CANCER

In 1775, a British physician named Percival Potts was the first to recognize that environmental factors can cause cancer, when he became aware that chimney sweeps had a higher incidence of scrotum cancer than the male population as a whole. He theorized that something in the chimney soot was causing cancer. We now know that it was benzo[*a*]pyrene.

Titch Cox, the chimney sweep responsible for cleaning the 800 chimneys at Buckingham Palace.

Summary

Alcohols and **ethers** have leaving groups that are stronger bases than halide ions, so alcohols and ethers are less reactive than alkyl halides and have to be protonated before they can undergo a substitution or an elimination reaction. **Epoxides** do not have to be activated by protonation, because ring strain increases their reactivity. The NH_2 group of an amine is such a strong base that amines cannot undergo substitution or elimination reactions.

Primary, secondary, and tertiary alcohols undergo nucleophilic substitution reactions with HI, HBr, and HCl to form alkyl halides. These are S_N2 reactions in the case of primary alcohols and S_N1 reactions in the case of secondary and tertiary alcohols.

An alcohol can be dehydrated if heated with an acid catalyst; **dehydration** is an E2 reaction in the case of primary alcohols and an E1 reaction in the case of secondary and tertiary alcohols. Tertiary alcohols are the easiest to dehydrate and primary alcohols are the hardest. The major product is the more substituted alkene. When the alkene can exist as stereoisomers, both the *E* and *Z* isomers are obtained, but the isomer with the bulkiest groups on opposite sides of the double bond predominates.

Secondary alcohols are oxidized to ketones. Primary alcohols are oxidized to carboxylic acids by chromic acid and to aldehydes by PCC.

Ethers can undergo nucleophilic substitution reactions with HBr or HI; if departure of the leaving group creates a relatively stable carbocation, an S_N1 reaction occurs; otherwise an S_N2 reaction occurs.

Epoxides undergo ring-opening reactions. Under basic conditions, the least sterically hindered carbon is attacked; under acidic conditions, the most substituted carbon is attacked. **Arene oxides** undergo rearrangement to form phenols or nucleophilic attack to form addition products. An arene oxide's cancer-causing potential depends on the stability of the carbocation formed during rearrangement.

Summary of Reactions

1. Conversion of an *alcohol* to an *alkyl halide* (Section 11.2).

$$ROH + HBr \xrightarrow{\Delta} RBr$$
$$ROH + HI \xrightarrow{\Delta} RI$$
$$ROH + HCl \xrightarrow{\Delta} RCl$$

relative rate: **tertiary > secondary > primary**

2. Dehydration of *alcohols* (Section 11.3).

$$-\underset{H}{\overset{|}{C}}-\underset{OH}{\overset{|}{C}}- \xrightarrow[\Delta]{H_2SO_4} C=C + H_2O$$

relative rate: **tertiary > secondary > primary**

3. Oxidation of *alcohols* (Section 11.4).

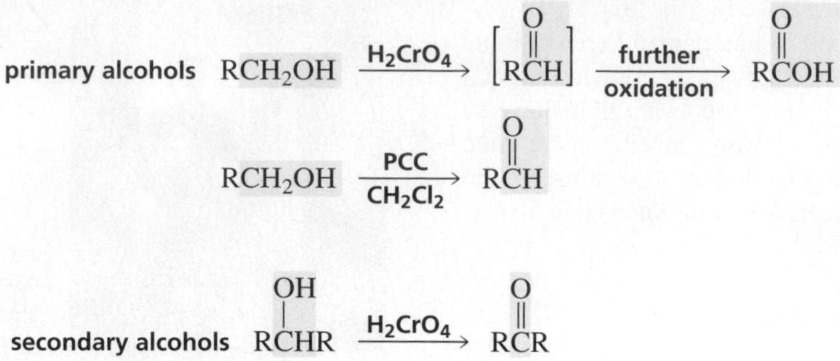

$$\text{primary alcohols} \quad RCH_2OH \xrightarrow{H_2CrO_4} \left[R\overset{O}{\overset{\|}{C}}H \right] \xrightarrow[\text{oxidation}]{\text{further}} R\overset{O}{\overset{\|}{C}}OH$$

$$RCH_2OH \xrightarrow[CH_2Cl_2]{PCC} R\overset{O}{\overset{\|}{C}}H$$

$$\text{secondary alcohols} \quad R\overset{OH}{\overset{|}{C}}HR \xrightarrow{H_2CrO_4} R\overset{O}{\overset{\|}{C}}R$$

4. Cleavage of *ethers* (Section 11.7).

$$ROR' + HX \xrightarrow{\Delta} ROH + R'X$$

$$HX = HBr \text{ or } HI$$

5. Ring-opening reactions of *epoxides* (Section 11.8)

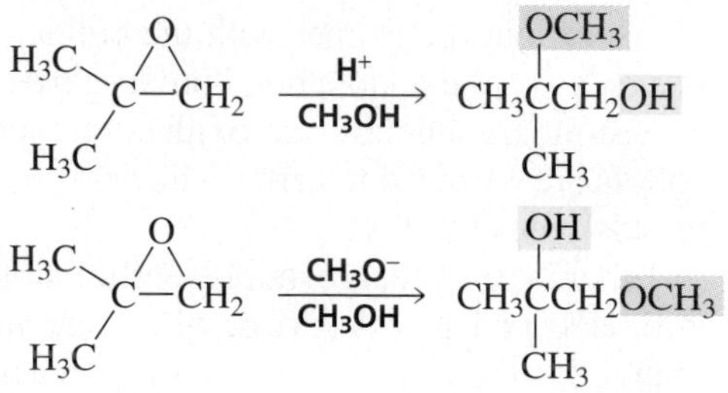

under acidic conditions, the nucleophile attacks the more substituted ring-carbon

under basic conditions, the nucleophile attacks the less sterically hindered ring-carbon

6. Reactions of *arene oxides*: ring opening and rearrangement (Section 11.9).

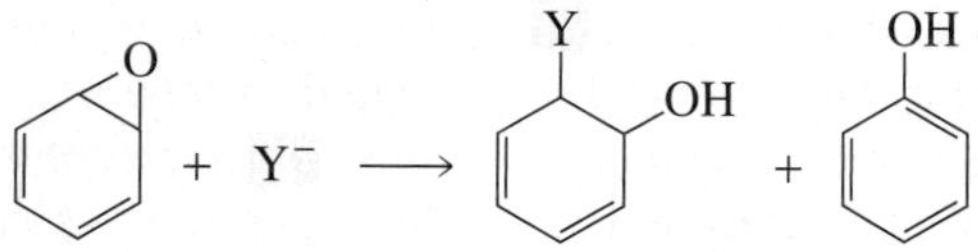

Problems

22. Give the product of each of the following reactions:

a. 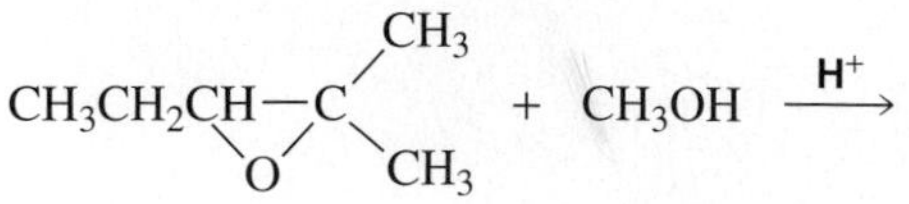 $CH_3CH_2CH\!-\!\underset{\underset{CH_3}{|}}{\overset{\overset{CH_3}{|}}{C}}$ $+ CH_3OH \xrightarrow{H^+}$ (with epoxide O)

b. $CH_3\underset{\underset{CH_3}{|}}{C}HCH_2OCH_3 + HI \xrightarrow{\Delta}$

c. cyclohexyl–$CH_2CH_2OH \xrightarrow{H_2CrO_4}$

d. $CH_3CH_2CH\!-\!\underset{\underset{CH_3}{|}}{\overset{\overset{CH_3}{|}}{C}}$ $+ CH_3OH \xrightarrow{CH_3O^-}$ (with epoxide O)

e. $CH_3\underset{\underset{CH_3}{|}}{C}H\!-\!\underset{\underset{OH}{|}}{\overset{\overset{CH_3}{|}}{C}}CH_3 \xrightarrow[\Delta]{H_2SO_4}$

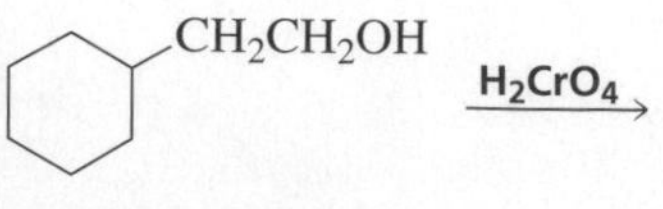

f. 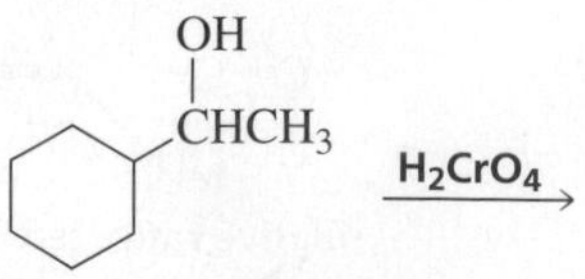 cyclohexyl–$\underset{\underset{OH}{|}}{C}HCH_3 \xrightarrow{H_2CrO_4}$

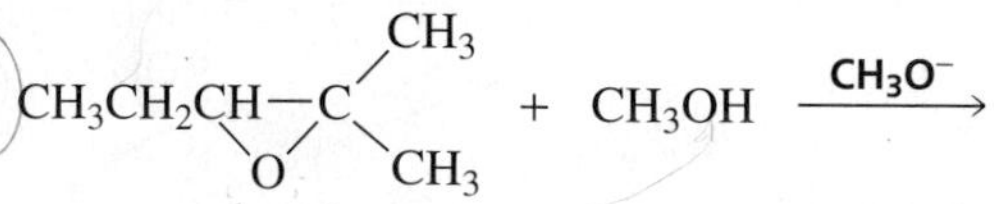

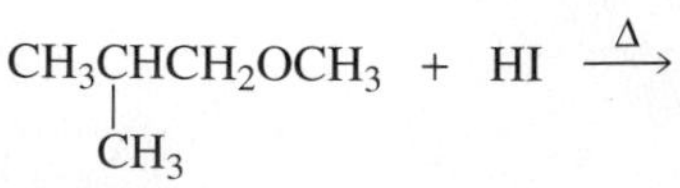

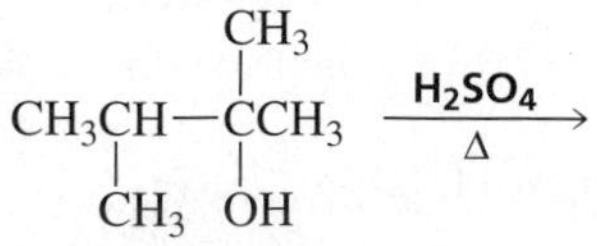

23. Give common and systematic names for each of the following ethers:

    a. $CH_3CHOCH_2CH_2CH_3$
            |
           CH_3
    b. $CH_3CH_2CH_2CH_2OCH_2CH_3$
    c. $CH_3CH_2CHOCH_3$
                  CH_3
    d. $CH_3CHOCHCH_3$
                 CH_3 CH_3

24. Indicate which alcohol will undergo dehydration more rapidly when heated with H_2SO_4.

a. 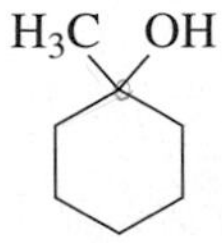 (cyclohexane with H_3C and OH) or (cyclohexane with CH_3 and OH)

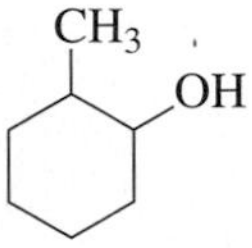

c. $CH_3CH_2CHCH_3$ or $CH_3CCH_2CH_3$
            | CH_3
            OH OH

b. 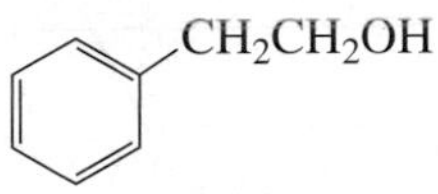 (benzene with CH_2CH_2OH) or (benzene with OH on $CHCH_3$)

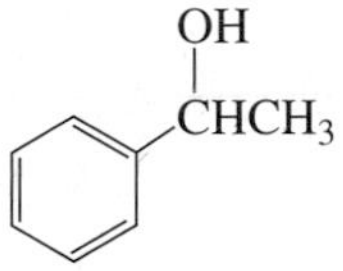

d. 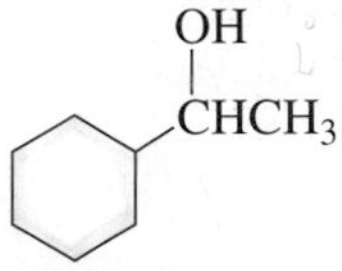 (cyclohexane with OH on $CHCH_3$) or (benzene with OH on $CHCH_3$)

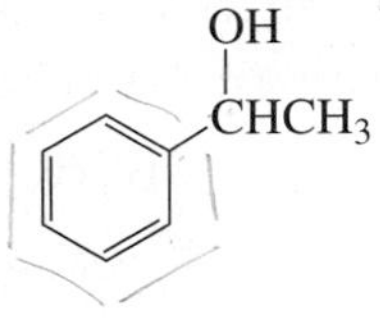

25. Name each of the following compounds:

    a. $CH_3CH_2CHOCH_2CH_3$
                $CH_2CH_2CH_2CH_3$

    b. 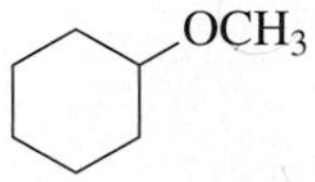 (cyclohexane with OCH_3)

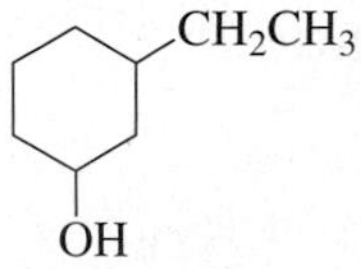

    c. $CH_3CHCH_2CH_2CH_2OH$
           |
           CH_3

    d. $CH_3CHOCH_2CH_2CHCH_3$
            | |
           CH_3 CH_3

    e. (cyclohexane with CH_2CH_3 and OH)

    f. $CH_3CHOCHCH_2CH_2CH_3$
                 |
              CH_3 (with CH_3 above)

26. Using the given starting material, any necessary inorganic reagents, and any carbon-containing compounds with no more than two carbon atoms, indicate how the following syntheses could be carried out:

    a. 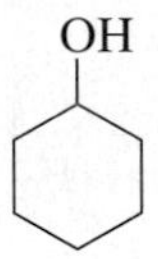 (cyclohexane with OH) → (cyclohexane)

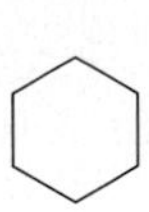

    b. $CH_3CH_2C\equiv CH$ → $CH_3CH_2C\equiv CCH_2CH_2OH$

27. Draw structures for the following:

    a. diisopropyl ether
    b. allyl vinyl ether
    c. *sec*-butyl isobutyl ether
    d. benzyl phenyl ether

28. If any of the ethers in Problem 27 can exist as stereoisomers, draw the stereoisomers.

29. Give the product of each of the following reactions:

    a. 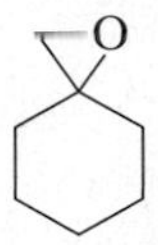 (spiro epoxide cyclohexane) $\xrightarrow[\text{CH}_3\text{OH}]{\text{CH}_3\text{O}^-}$

    d. 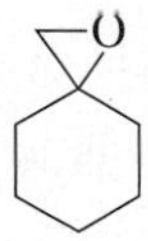 (spiro epoxide cyclohexane) $\xrightarrow[\text{CH}_3\text{OH}]{\text{H}^+}$

    b. $CH_3COCH_2CH_3 + HBr \xrightarrow{\Delta}$
       (with CH_3 above and CH_3 below)

    e. $CH_3CH_2CHOCCH_3 + HI \xrightarrow{\Delta}$
       (with CH_3 above and CH_3 CH_3 below)

    c. $CH_3CHCH_2CHCH_3 \xrightarrow{\text{H}_2\text{CrO}_4}$
          | |
        OH OH

    f. $HOCH_2CH_2CH_2CH_2OH \xrightarrow{\text{H}_2\text{CrO}_4}$

30. Draw structures for the following:
  a. *trans*-4-methylcyclohexanol

  b. 3-ethoxy-1-propanol

31. Give the product formed from the reaction of each of the following compounds with chromic acid:
  a. 3-methyl-2-pentanol   **b.** butanol   **c.** 2-methylcyclohexanol

32. Indicate which alcohol will undergo dehydration more rapidly when heated with H_2SO_4.

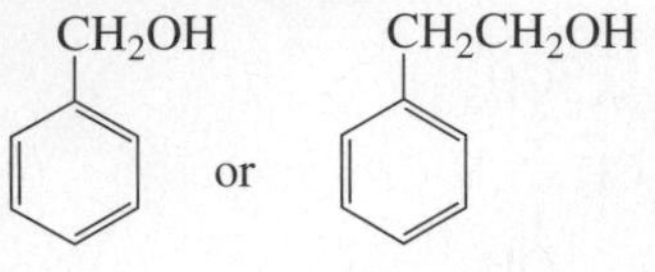

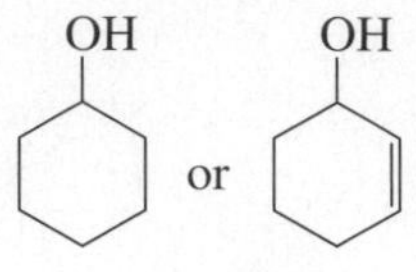

33. Propose a mechanism for the following reaction:

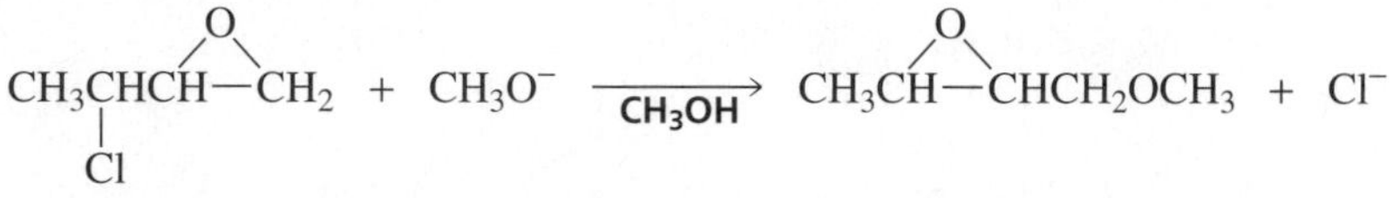

34. The observed relative reactivities of primary, secondary, and tertiary alcohols with a hydrogen halide are $3° > 2° > 1°$. If secondary alcohols underwent an S_N2 reaction rather than an S_N1 reaction with a hydrogen halide, what would be the relative reactivities of the three classes of alcohols?

35. Give the major product expected from the reaction of 1,2-epoxybutane with each of the following reagents:
  a. 0.1 M HCl   **b.** CH_3OH/H^+ **c.** CH_3OH/CH_3O^- **d.** 0.1 M NaOH

36. Name each of the following compounds:

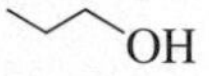

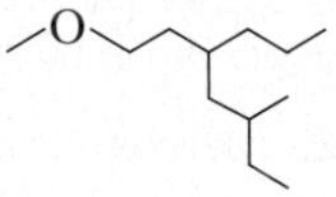

37. When ethyl ether is heated with excess HI for several hours, the only organic product obtained is ethyl iodide. Explain why ethyl alcohol is not obtained as a product.

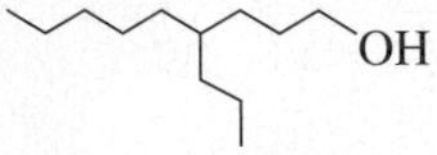

38. Ethylene oxide reacts readily with HO^- because of the strain in the three-membered ring. Cyclopropane has approximately the same amount of strain, but does not react with HO^-. Explain.

39. Propose a mechanism for each of the following reactions:

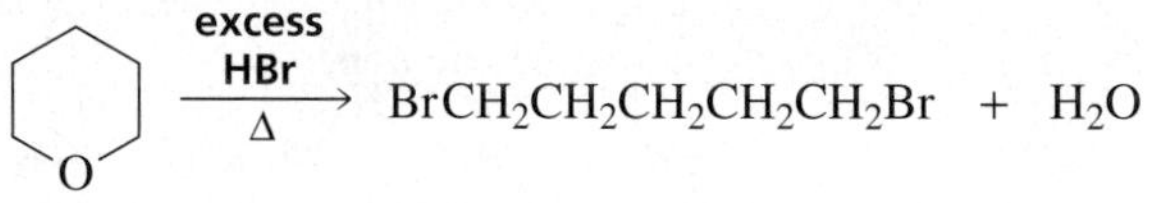

40. Triethylene glycol is one of the products obtained from the reaction of ethylene oxide and hydroxide ion. Propose a mechanism for its formation.

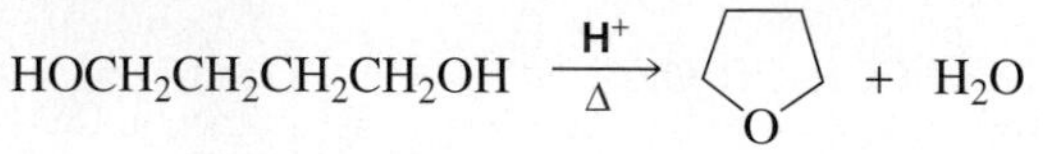

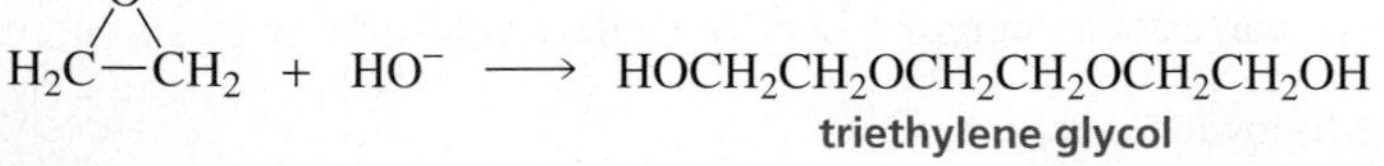

41. Explain why the major product obtained from the acid-catalyzed dehydration of 1-butanol is 2-butene.

42. What alkenes would you expect to be obtained from the acid-catalyzed dehydration of 1-hexanol?

43. Propose a mechanism for the following reaction:

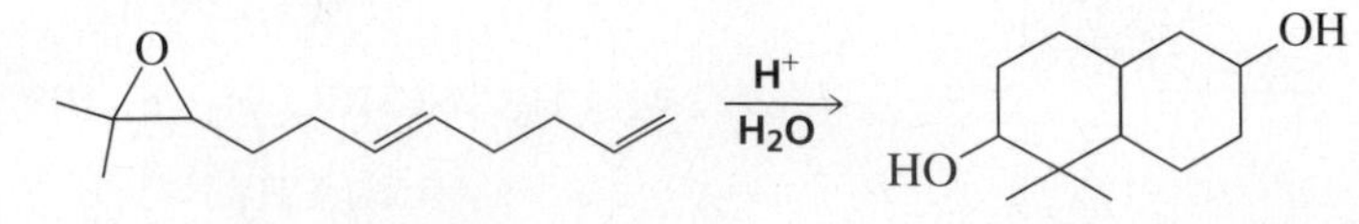

44. If the following ethers were synthesized directly from alcohols, which would be obtained in greatest yield?

    a. $CH_3OCH_2CH_2CH_3$ **b.** $CH_3CH_2OCH_2CH_2CH_3$ **c.** $CH_3CH_2OCH_2CH_3$ **d.** CH_3OCCH_3 (with CH_3 groups above and below)

45. Explain why (*S*)-2-butanol forms a racemic mixture when it is heated in sulfur acid.

46. Explain why more 1-naphthol, than 2-naphthol, is obtained from the rearrangement of naphthalene oxide.

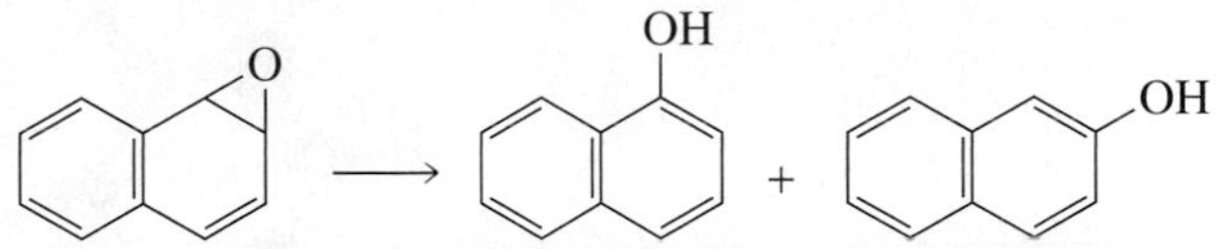

    naphthalene oxide   **1-naphthol**   **2-naphthol**
                             90%   **10%**

47. Three arene oxides can be obtained from phenanthrene.

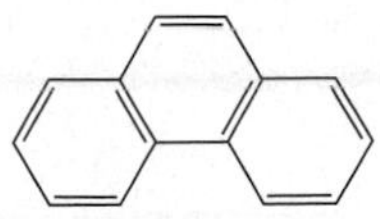

    phenanthrene

    a. Give the structures of the three phenanthrene oxides.
    b. What phenols can be obtained from each phenanthrene oxide?
    c. If a phenanthrene oxide can lead to the formation of more than one phenol, which phenol is obtained in greater yield?
    d. Which of the three phenanthrene oxides is most likely to be carcinogenic?

48. a. Propose a mechanism for the following reaction:

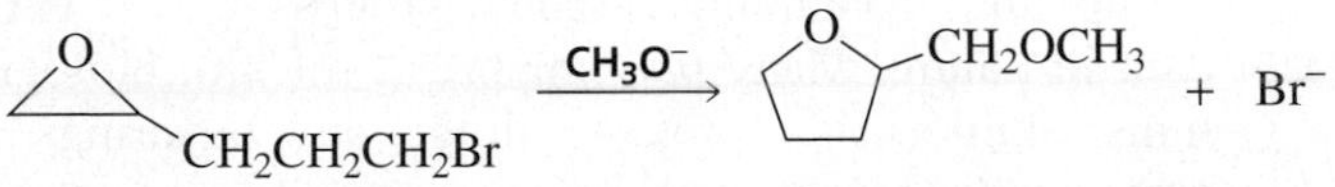

    b. A small amount of a product containing a six-membered ring is also formed. Give the structure of that product.
    c. Why is so little six-membered ring product formed?

49. Explain why the acid-catalyzed dehydration of an alcohol is a reversible reaction, whereas the base-promoted dehydrohalogenation of an alkyl halide is an irreversible reaction.

50. Two stereoisomers are obtained from the reaction of cyclopentene oxide and dimethylamine. The *R,R*-isomer is used in the manufacture of eclanamine, an antidepressant. What other isomer is obtained?

12 | Carbonyl Compounds I

Nucleophilic Acyl Substitution

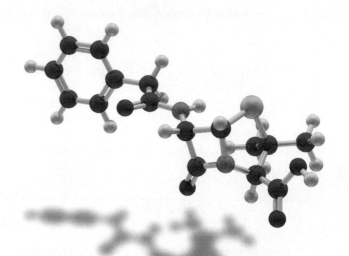

Penicillin G

Tutorial:
Functional groups

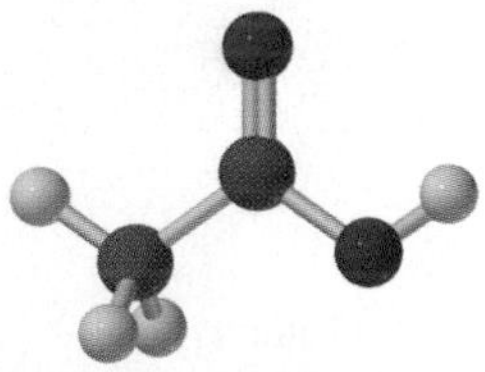

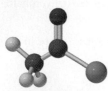

a carboxylic acid

an acyl chloride

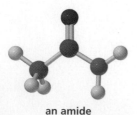

an ester

an amide

The **carbonyl group**—a carbon double bonded to an oxygen—is probably the most important functional group. Compounds containing carbonyl groups—called **carbonyl compounds**—are abundant in nature. Many play important roles in biological processes. Hormones, vitamins, amino acids, proteins, drugs, and flavorings are just a few of the carbonyl compounds that affect us daily.

An **acyl group** consists of a carbonyl group attached to an alkyl or an aryl group.

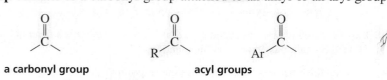

a carbonyl group   acyl groups

The substituent attached to the acyl group strongly affects the reactivity of the carbonyl compound. Carbonyl compounds can be divided into two classes: Class I carbonyl compounds are those in which the acyl group is attached to an atom or a group that *can* be replaced by another group. Carboxylic acids, acyl chlorides, esters, and amides belong to this class. All of these compounds contain a group (OH, Cl, OR, NH_2, NHR, NR_2) that can be replaced by a nucleophile. Acyl chlorides, esters, and amides are all called **carboxylic acid derivatives** because they differ from a carboxylic acid only in the nature of the group that has replaced the OH group of the carboxylic acid.

compounds with groups that can be replaced by a nucleophile

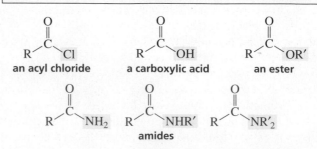

an acyl chloride   a carboxylic acid   an ester

amides

288

Class II carbonyl compounds are those in which the acyl group is attached to a group that *cannot* be readily replaced by another group. Aldehydes and ketones belong to this class. The H bonded to the acyl group of an aldehyde and the R bonded to the acyl group of a ketone cannot be readily replaced by a nucleophile.

an aldehyde   a ketone

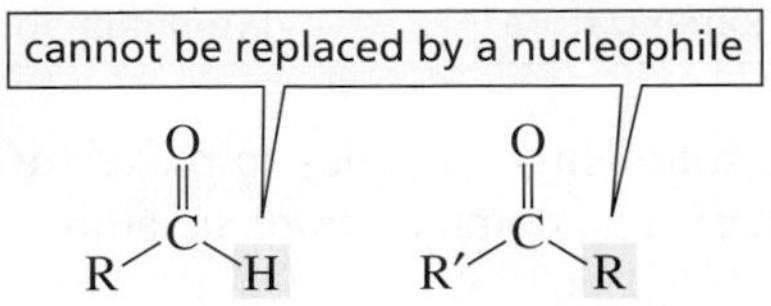

This chapter discusses the reactions of Class I carbonyl compounds. We will see that these compounds undergo substitution reactions, because they have an acyl group attached to a group that can be replaced by a nucleophile. The reactions of Class II carbonyl compounds—aldehydes and ketones—will be considered in Chapter 13, where we will see that these compounds *do not* undergo substitution reactions because their acyl group is attached to a group that *cannot* be replaced by a nucleophile.

12.1 Nomenclature

Carboxylic Acids

In systematic (IUPAC) nomenclature, a **carboxylic acid** is named by replacing the terminal "e" of the alkane name with "oic acid." For example, the one-carbon alkane is methane, so the one-carbon carboxylic acid is methan*oic acid*. Carboxylic acids in which a carboxyl group is attached to a ring are named by adding "carboxylic acid" to the name of the cyclic compound.

systematic name:	methanoic acid	ethanoic acid	propanoic acid	butanoic acid
common name:	formic acid	acetic acid	propionic acid	butyric acid

pentanoic acid
valeric acid

hexanoic acid
caproic acid

benzenecarboxylic acid
benzoic acid

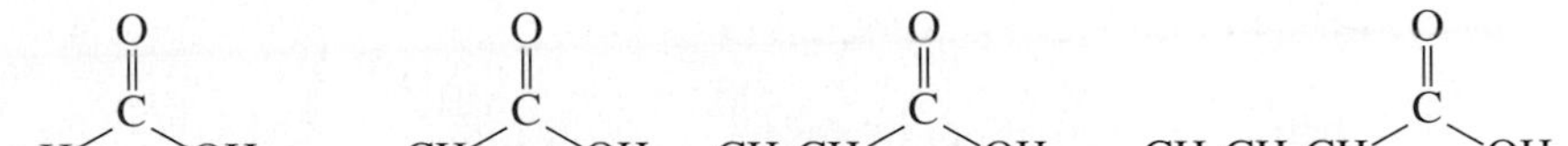

Carboxylic acids containing six or fewer carbons are frequently called by their common names. These names were chosen to describe some feature of the compound, usually its origin. For example, formic acid is found in ants, bees, and other stinging insects; its name comes from *formica*, which is Latin for "ant." Acetic acid—contained in vinegar—got its name from *acetum*, the Latin word for "vinegar." Propionic acid is the smallest acid that shows some of the characteristics of the larger fatty acids (Section 19.1); its name comes from the Greek words *pro* ("the first") and *pion* ("fat"). Butyric acid is found in rancid butter; the Latin word for "butter" is *butyrum*. Caproic acid is found in goat's milk; if you have the occasion to smell both a goat and caproic acid, you will find that they have similar odors. *Caper* is the Latin word for "goat."

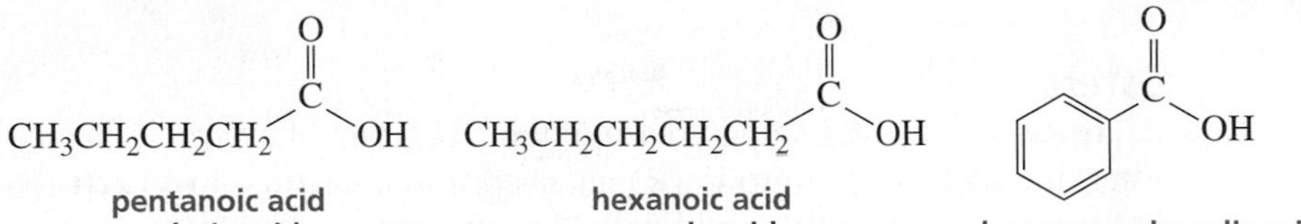

In systematic (IUPAC) nomenclature, the position of a substituent is designated by a number. The carbonyl carbon is always the C-1 carbon. In common nomenclature, the position of a substituent is designated by a lowercase Greek letter, and the carbonyl carbon is not given a designation. The carbon adjacent to the carbonyl carbon is the **α-carbon**, the carbon adjacent to the α-carbon is the β-carbon, and so on.

α = alpha
β = beta
γ = gamma
δ = delta
ε = epsilon

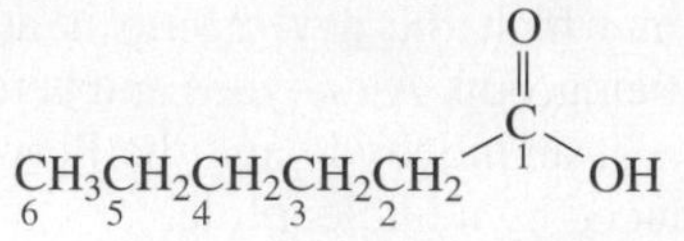

$$CH_3CH_2CH_2CH_2CH_2 \overset{\overset{O}{\|}}{\underset{}{C}} OH$$
6 5 4 3 2 1

systematic nomenclature

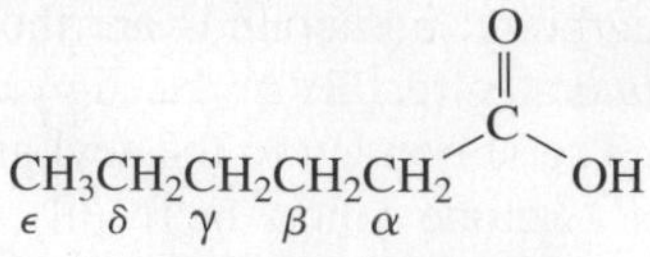

$$CH_3CH_2CH_2CH_2CH_2 \overset{\overset{O}{\|}}{\underset{}{C}} OH$$
ε δ γ β α

common nomenclature

Take a careful look at the following examples to make sure that you understand the difference between systematic and common nomenclature:

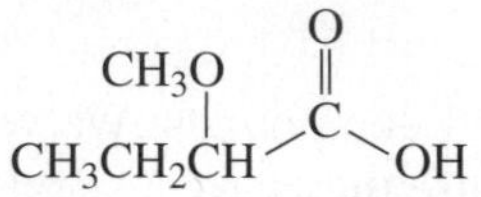

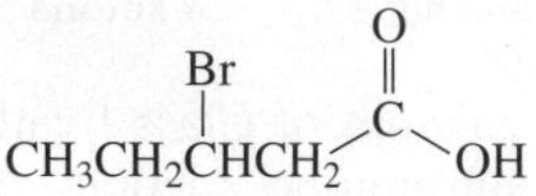

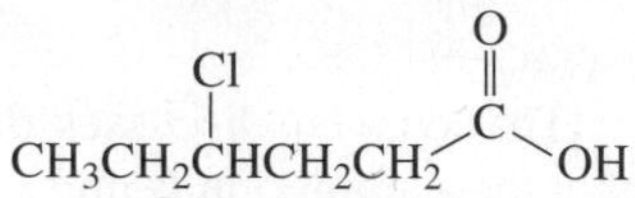

$$CH_3CH_2\overset{\overset{CH_3O}{|}}{CH} \overset{\overset{O}{\|}}{C} OH$$

systematic name: 2-methoxybutanoic acid
common name: α-methoxybutyric acid

$$CH_3CH_2\overset{\overset{Br}{|}}{CH}CH_2 \overset{\overset{O}{\|}}{C} OH$$

3-bromopentanoic acid
β-bromovaleric acid

$$CH_3CH_2\overset{\overset{Cl}{|}}{CH}CH_2CH_2 \overset{\overset{O}{\|}}{C} OH$$

4-chlorohexanoic acid
γ-chlorocaproic acid

The functional group of a carboxylic acid is called a **carboxyl group**.

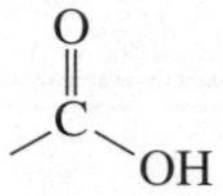

$$\overset{\overset{O}{\|}}{C} OH$$

a carboxyl group

—COOH   —CO₂H

carboxyl groups are frequently shown in abbreviated forms

Acyl Chlorides

An **acyl chloride** has a Cl in place of the OH group of a carboxylic acid. Acyl chlorides are named by using the acid name and replacing "ic acid" with "yl chloride."

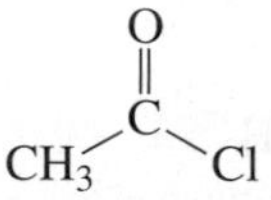

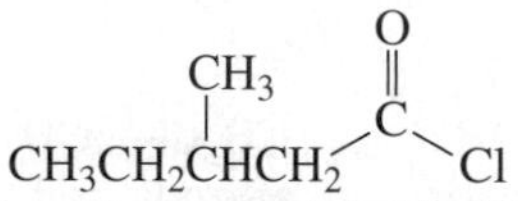

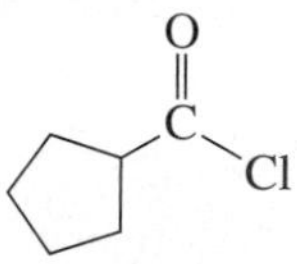

$$CH_3 \overset{\overset{O}{\|}}{C} Cl$$

systematic name: ethanoyl chloride
common name: acetyl chloride

$$CH_3CH_2\overset{\overset{CH_3}{|}}{CH}CH_2 \overset{\overset{O}{\|}}{C} Cl$$

3-methylpentanoyl chloride
β-methylvaleryl chloride

cyclopentanecarbonyl chloride

Esters

An **ester** has an OR group in place of the OH group of a carboxylic acid. In naming an ester, the name of the group (R) attached to the **carboxyl oxygen** is stated first, followed by the name of the acid with "ic acid" replaced by "ate."

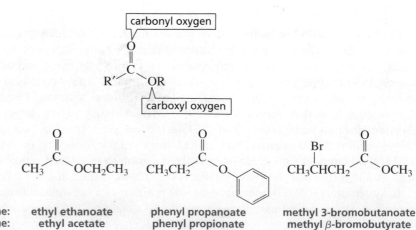

carbonyl oxygen

$$R' \overset{\overset{O}{\|}}{C} OR$$

carboxyl oxygen

$$CH_3 \overset{\overset{O}{\|}}{C} OCH_2CH_3$$

systematic name: ethyl ethanoate
common name: ethyl acetate

$$CH_3CH_2 \overset{\overset{O}{\|}}{C} O$$

phenyl propanoate
phenyl propionate

$$CH_3\overset{\overset{Br}{|}}{CH}CH_2 \overset{\overset{O}{\|}}{C} OCH_3$$

methyl 3-bromobutanoate
methyl β-bromobutyrate

Salts of carboxylic acids are named in the same way. The cation is named first, followed by the name of the acid with "ic acid" replaced by "ate."

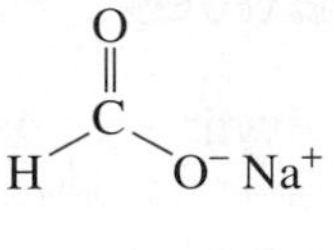

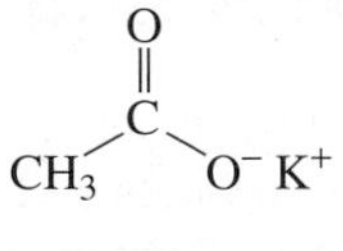

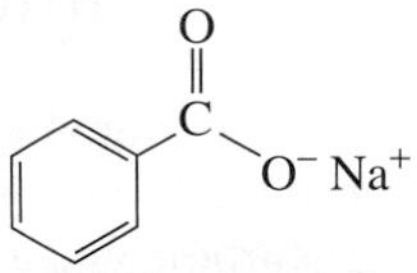

systematic name: sodium methanoate   potassium ethanoate   sodium benzenecarboxylate
common name: sodium formate   potassium acetate   sodium benzoate

Amides

An **amide** has an NH_2, NHR, or NR_2 group in place of the OH group of a carboxylic acid. Amides are named by using the acid name, replacing "oic acid," "ic acid," or "ylic acid" with "amide."

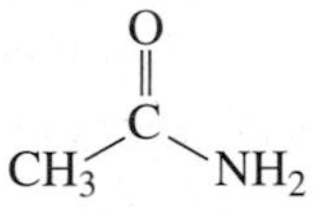

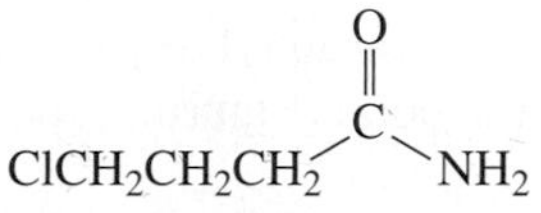

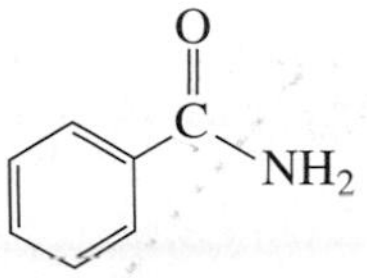

systematic name: ethanamide   4-chlorobutanamide   benzenecarboxamide
common name: acetamide γ-chlorobutyramide   benzamide

If a substituent is bonded to the nitrogen, the name of the substituent is stated first (if there is more than one substituent bonded to the nitrogen, they are stated alphabetically), followed by the name of the amide. The name of each substituent is preceded by a capital *N* to indicate that the substituent is bonded to a nitrogen.

3-D Molecule:
N-Methylbenzamide

***N*-cyclohexylpropanamide** 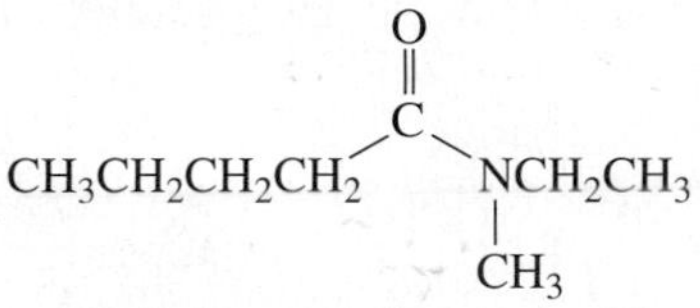   ***N*-ethyl-*N*-methylpentanamide**   ***N,N*-diethylbutanamide**

PROBLEM 1◆

Name the following compounds:

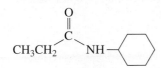

a. $CH_3CH_2CNH_2$

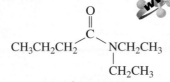

d. $CH_3CH_2CH_2CH_2CCl$

b. $CH_3CH_2CH_2COCH_2CHCH_3$ (with O double bond and CH_3 substituent)

e. $CH_3CH_2CH_2CH_2CH_2CN(CH_3)_2$

c. $CH_3CH_2CH_2CO^- K^+$

f. cyclopentane—COOH

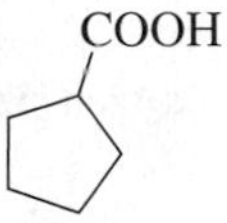

Tutorial:
Nomenclature of carboxylic acids and their derivatives

PROBLEM 2◆

Write the structure of each of the following compounds:

a. phenyl acetate
b. sodium acetate
c. *N*-benzylethanamide

d. ethyl 2-chloropentanoate
e. β-bromobutyramide
f. cyclohexanecarbonyl chloride

3-D Molecules:
Acetyl chloride;
Methyl acetate;
Acetic acid; Acetamide

12.2 Structures of Carboxylic Acids and Carboxylic Acid Derivatives

The **carbonyl carbon** in carboxylic acids and carboxylic acid derivatives is sp^2 hybridized. It uses its three sp^2 orbitals to form σ bonds to the carbonyl oxygen, the α-carbon, and a substituent (Y). The three atoms attached to the carbonyl carbon are in the same plane, and the bond angles are each approximately 120°.

The **carbonyl oxygen** is also sp^2 hybridized. One of its sp^2 orbitals forms a σ bond with the carbonyl carbon, and each of the other two sp^2 orbitals contains a lone pair. The remaining p orbital of the carbonyl oxygen overlaps the remaining p orbital of the carbonyl carbon to form a π bond (Figure 12.1).

Figure 12.1 ▶
Bonding in a carbonyl group. The π bond is formed by side-to-side overlap of a p orbital of carbon with a p orbital of oxygen.

Esters, carboxylic acids, and amides each have two major resonance contributors.

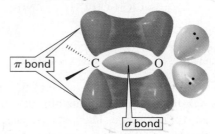

The resonance contributor on the right is more important for the amide than it is for the ester or the carboxylic acid. It is even less important for the acyl chloride, because the more electronegative the atom, the less it is able to accommodate a positive charge.

PROBLEM 3◆

Which is longer, the carbon–oxygen single bond in a carboxylic acid or the carbon–oxygen bond in an alcohol? Why?

PROBLEM 4◆

What are the relative lengths of the three carbon–oxygen bonds in methyl acetate?

12.3   Physical Properties of Carbonyl Compounds

The acid properties of carboxylic acids have been discussed previously (Sections 2.2 and 6.9). Recall that carboxylic acids have pK_a values of approximately 5 (Appendix II). The boiling points and other physical properties of carbonyl compounds are listed in Appendix I. Carbonyl compounds have the following relative boiling points:

relative Boiling Points

alcohol

amide > carboxylic acid ≫ ester ~ acyl chloride ~ aldehyde ~ ketone *> ether*

The boiling points of an ester, acyl chloride, ketone, and aldehyde with comparable molecular weights are similar, and they all are lower than the boiling point of an alcohol with a comparable molecular weight because only the alcohol molecules can form hydrogen bonds with each other. The boiling points of the carbonyl compounds are higher than the boiling point of the ether because of the polarity of the carbonyl group.

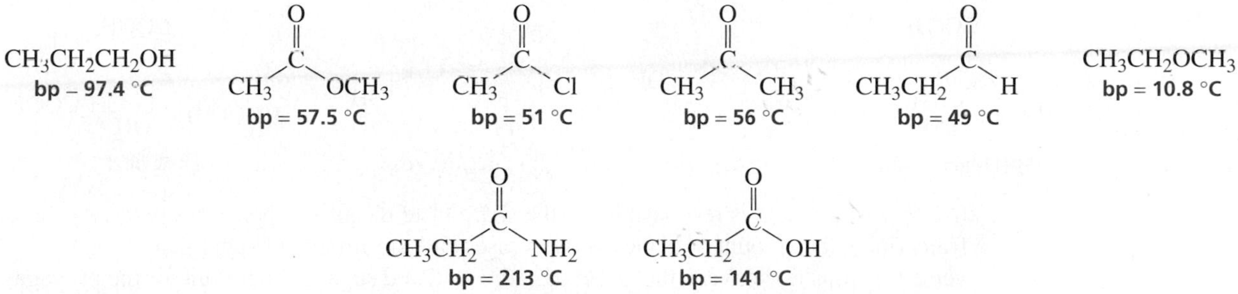

$CH_3CH_2CH_2OH$
bp – 97.4 °C

$CH_3CH_2OCH_3$
bp = 10.8 °C

Carboxylic acids have relatively high boiling points because they form intermolecular hydrogen bonds (hydrogen bonds between molecules), giving them larger effective molecular weights.

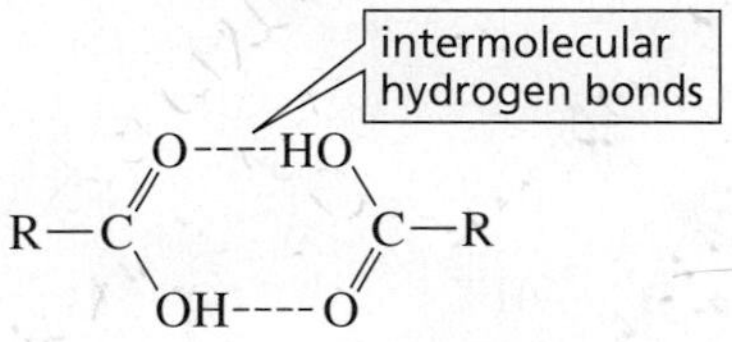

Amides have the highest boiling points, because they have strong dipole–dipole interactions because the resonance contributor with separated charges contributes significantly to the overall structure of the compound (Section 12.2). If the nitrogen of an amide is bonded to a hydrogen, hydrogen bonds will form between the molecules.

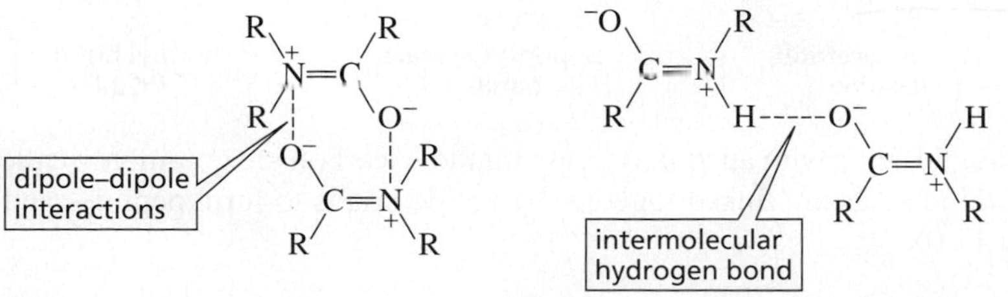

Carboxylic acid derivatives are soluble in solvents such as ethers, chlorinated alkanes, and aromatic hydrocarbons. Like alcohols and ethers, carbonyl compounds with fewer than four carbons are soluble in water (Section 3.7).

Esters and *N,N*-disubstituted amides are often used as solvents because they are polar, but they do not have reactive OH or NH_2 groups.

12.4 Naturally Occurring Carboxylic Acids and Carboxylic Acid Derivatives

Acyl halides are much more reactive than carboxylic acids and esters, which, in turn, are more reactive than amides. We will see the reason for this difference in reactivity in Section 12.5.

Because of their high reactivity, acyl halides are not found in nature. Carboxylic acids, on the other hand, being less reactive, are found widely in nature. For example, glucose is metabolized to pyruvic acid. (*S*)-(+)-Lactic acid is the compound responsible for the burning sensation felt in muscles during anaerobic exercise, and it is also found in sour milk. Spinach and other leafy green vegetables are rich in oxalic acid. Succinic acid and citric acid are important intermediates in the citric acid cycle, a series of reactions in biological systems that oxidize carbohydrates, fatty acids, and amino acids to CO_2 (Section 19.6). Citrus fruits are rich in citric acid; the concentration is greatest in lemons, less in grapefruit, and still less in oranges.

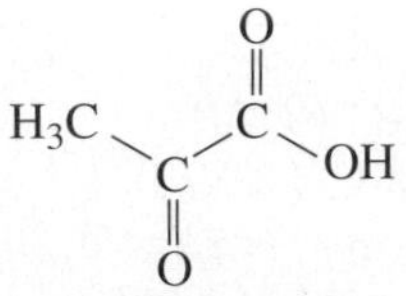

pyruvic acid

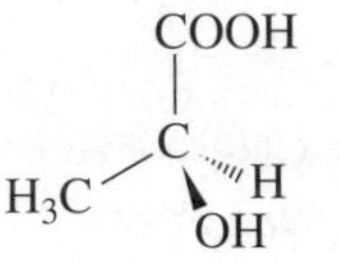

(*S*)-(+)-lactic acid

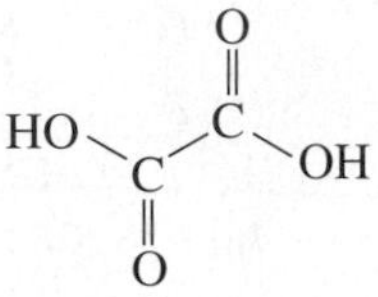

oxalic acid

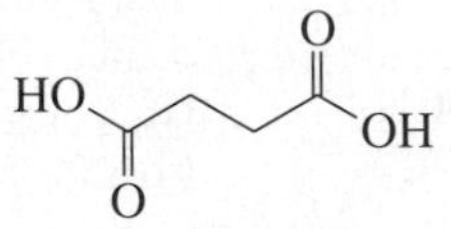

succinic acid

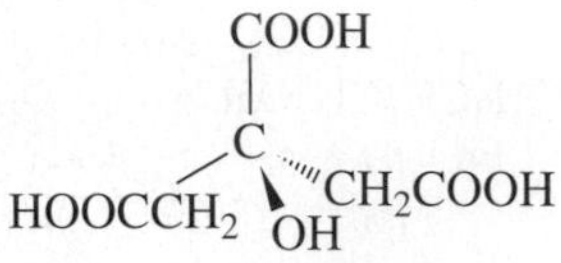

citric acid

(*S*)-(−)-Malic acid is responsible for the sharp taste of unripe apples and pears. As these fruits ripen, the amount of malic acid decreases and the amount of sugar increases. The inverse relationship between the levels of malic acid and sugar is important for the propagation of the plant: Animals will not eat the fruit until it becomes ripe—at which time its seeds are mature enough to germinate when they are scattered about. Prostaglandins are locally acting hormones that have several different physiological functions (Section 12.9), such as stimulating inflammation, causing hypertension, and producing pain and swelling.

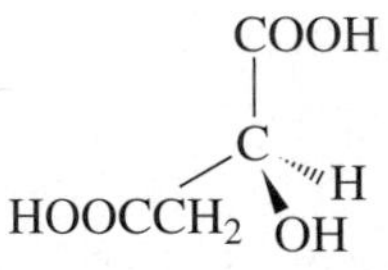

(*S*)-(−)-malic acid

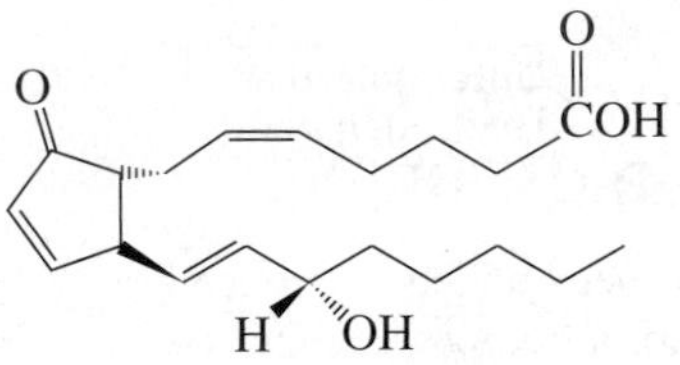

prostaglandin A$_2$

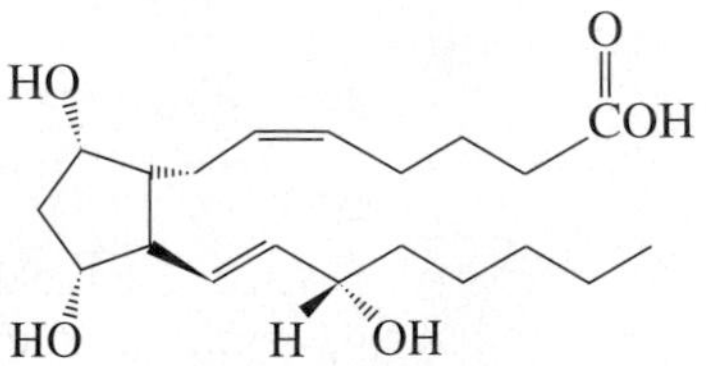

prostaglandin F$_{2\alpha}$

Esters are also commonly found in nature. Many of the fragrances of flowers and fruits are due to esters. (See Problem 29.)

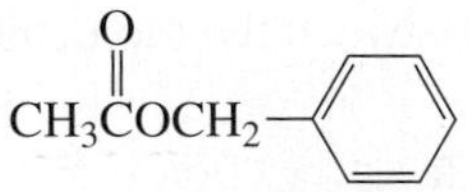

benzyl acetate
jasmine

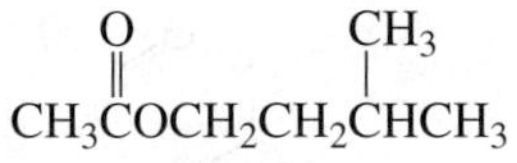

isopentyl acetate
banana

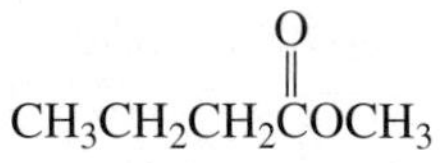

methyl butyrate
apple

Carboxylic acids with an amino group on the α-carbon are commonly called **amino acids**. Amino acids are linked together by amide bonds to form peptides and proteins (Section 17.0).

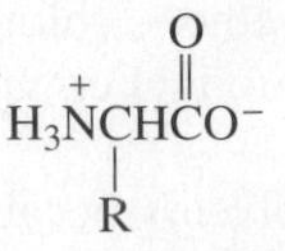

an amino acid

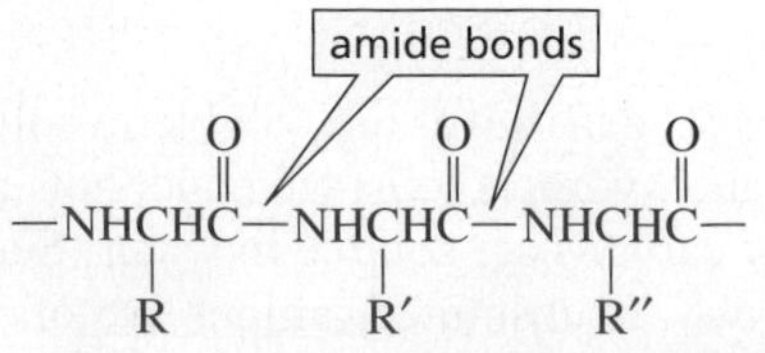

general structure for a peptide or a protein

THE DISCOVERY OF PENICILLIN

Sir Alexander Fleming (1881–1955) was born in Scotland. He was a professor of bacteriology at University College, London. The story is told that one day Fleming was about to throw away a culture of staphylococcal bacteria that had been contaminated by a rare strain of the mold *Penicillium notatum*. He noticed that the bacteria had not grown wherever there was a particle of mold. This suggested to him that the mold must have produced an antibacterial substance. Ten years later, Howard Florey and Ernest Chain isolated the active substance—penicillin G (Section 12.12)—but this delay allowed sulfa drugs to be the first antibiotics (Section 22.4). After penicillin G was found to cure bacterial infections in mice, it was used successfully in 1941 on nine cases of human bacterial infections. By 1943, penicillin G was being produced for the military and was first used for war casualties in Sicily and Tunisia. The drug became available to the civilian population in 1944. The pressure of the war made the determination of penicillin G's structure a priority because large quantities of the drug could not be synthesized until its structure was determined.

Fleming, Florey, and Chain shared the 1945 Nobel Prize in physiology or medicine. Chain also discovered penicillinase, the enzyme that destroys penicillin (Section 12.12). Although Fleming is generally given credit for the discovery of penicillin, there is clear evidence that the germicidal activity of the mold was recognized in the nineteenth century by Lord Joseph Lister (1827–1912), the English physician renowned for the introduction of aseptic surgery.

Sir Alexander Fleming (1881–1955) *was born in Scotland, the seventh of eight children of a farmer. In 1902, he received a legacy from an uncle that, together with a scholarship, allowed him to study medicine at the University of London. He subsequently became a professor of bacteriology there in 1928. He was knighted in 1944.*

Sir Howard W. Florey (1898–1968) *was born in Australia and received a medical degree from the University of Adelaide. He went to England as a Rhodes Scholar and studied at both Oxford and Cambridge Universities. He became a professor of pathology at the University of Sheffield in 1931 and then at Oxford in 1935. Knighted in 1944, he was given a peerage in 1965 that made him Baron Florey of Adelaide.*

Ernest B. Chain (1906–1979) *was born in Germany and received a Ph.D. from Friedrich-Wilhelm University in Berlin. In 1933, he left Germany for England because Hitler had come to power. He studied at Cambridge, and in 1935 Florey invited him to Oxford. In 1948, he became the director of an institute in Rome, but he returned to England in 1961 to become a professor at the University of London.*

DALMATIANS: DON'T TRY TO FOOL MOTHER NATURE

When amino acids are metabolized, the excess nitrogen is concentrated into uric acid, a compound with five amide bonds. A series of enzyme-catalyzed reactions degrades uric acid to ammonium ion. The extent to which uric acid is degraded in animals depends on the species. Birds, reptiles, and insects excrete excess nitrogen as uric acid. Mammals excrete excess nitrogen as allantoin. Excess nitrogen in aquatic animals is excreted as allantoic acid, urea, or ammonium salts.

uric acid
excreted by:
birds, reptiles, insects

allantoin

mammals

allantoic acid

marine vertebrates

urea

cartilaginous fish, amphibia

$^{+}NH_4X^{-}$
ammonium salt
marine invertebrates

Dalmatians, unlike other mammals, excrete high levels of uric acid. The reason for this is that breeders of Dalmatians have selected dogs that have no white hairs in their black spots, and the gene that causes the white hairs is linked to the gene that causes uric acid to be converted to allantoin. Dalmatians, therefore, are susceptible to gout (painful deposits of uric acid in joints).

Caffeine, another naturally occurring amide, is found in cocoa and in coffee beans. Penicillin G has two amide bonds; the four-membered ring amide is the reactive part of the molecule (Section 12.12).

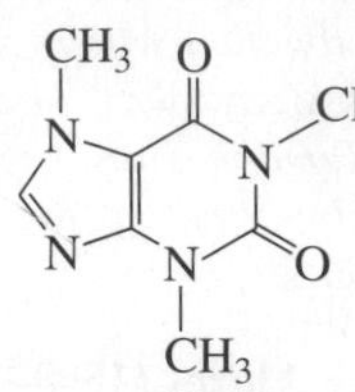

caffeine

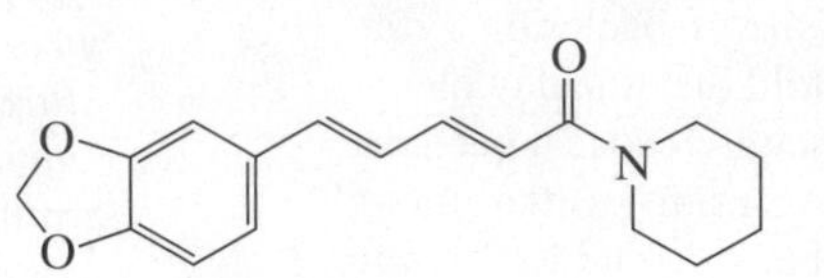

piperine
the major component of black pepper

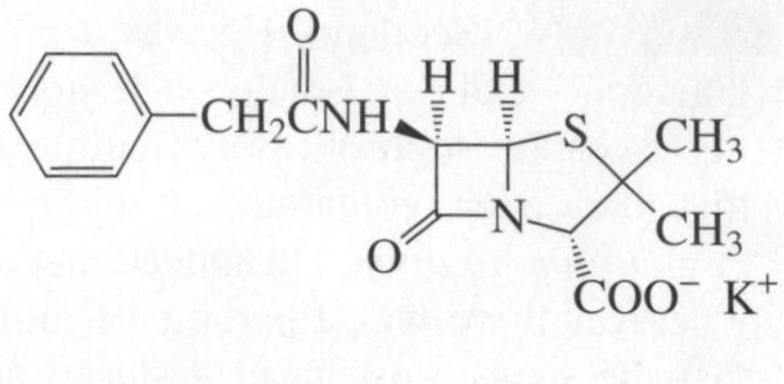

penicillin G

12.5 How Class I Carbonyl Compounds React

The reactivity of carbonyl compounds results from the polarity of the carbonyl group—oxygen is more electronegative than carbon. The carbonyl carbon, therefore, is electron deficient, so we can safely predict that it will be attacked by nucleophiles.

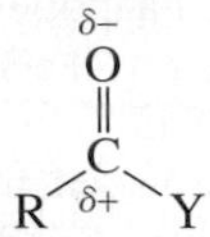

When a nucleophile attacks the carbonyl carbon of a carboxylic acid derivative, the carbon–oxygen π bond breaks and an intermediate is formed. The intermediate is called a **tetrahedral intermediate** because the sp^2 carbon in the reactant has become an sp^3 carbon in the intermediate. Generally, *a compound that has an sp^3 carbon bonded to an oxygen atom will be unstable if the sp^3 carbon is bonded to another electronegative atom.* The tetrahedral intermediate, therefore, is unstable because Y and Z are both electronegative atoms. A lone pair on the oxygen reforms the π bond, and either Y$^-$ or Z$^-$ is expelled with its bonding electrons.

> A compound that has an sp^3 carbon bonded to an oxygen atom generally will be unstable if the sp^3 carbon is bonded to a second electronegative atom.

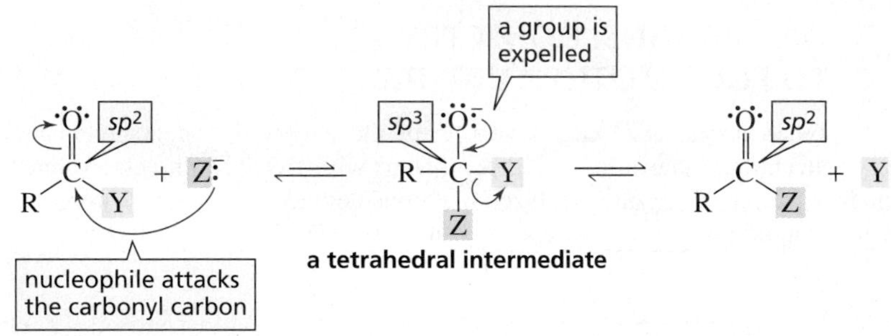

a group is expelled

a tetrahedral intermediate

nucleophile attacks the carbonyl carbon

Whether Y$^-$ or Z$^-$ is expelled depends on their relative basicities. The weaker base is expelled preferentially, making this another example of the principle we first saw in Section 10.3: *The weaker the base, the better it is as a leaving group.* Because a weak base does not share its electrons as well as a strong base does, a weaker base forms a weaker bond—one that is easier to break. If Z$^-$ is a much weaker base than Y$^-$, Z$^-$ will be expelled and the reaction can be written as follows:

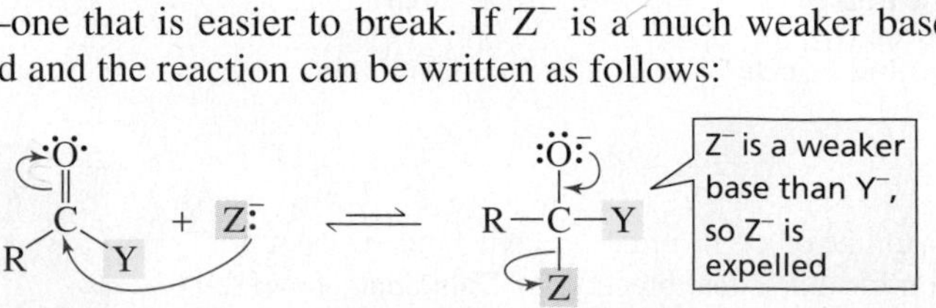

Z$^-$ is a weaker base than Y$^-$, so Z$^-$ is expelled

In this case, no new product is formed. The nucleophile attacks the carbonyl carbon, but the tetrahedral intermediate expels the attacking nucleophile and reforms the reactants.

On the other hand, if Y$^-$ is a much weaker base than Z$^-$, Y$^-$ will be expelled and a new product will be formed.

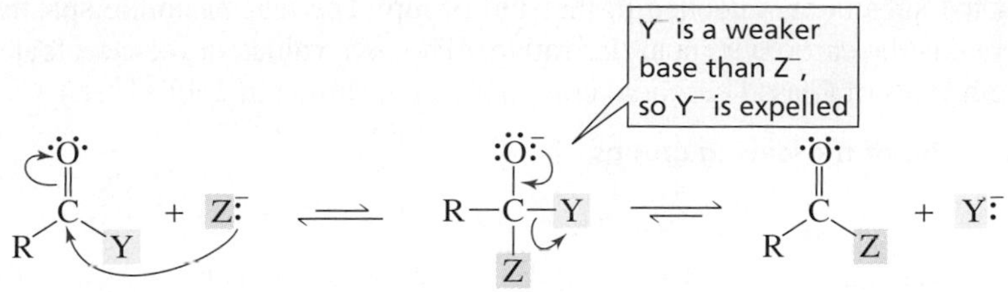

Y⁻ is a weaker base than Z⁻, so Y⁻ is expelled

This reaction is called a **nucleophilic acyl substitution reaction** because a nucleophile (Z⁻) has replaced the substituent (Y⁻) that was attached to the acyl group in the reactant.

If the basicities of Y⁻ and Z⁻ are similar, some molecules of the tetrahedral intermediate will expel Y⁻ and others will expel Z⁻. When the reaction is over, both reactant and product will be present.

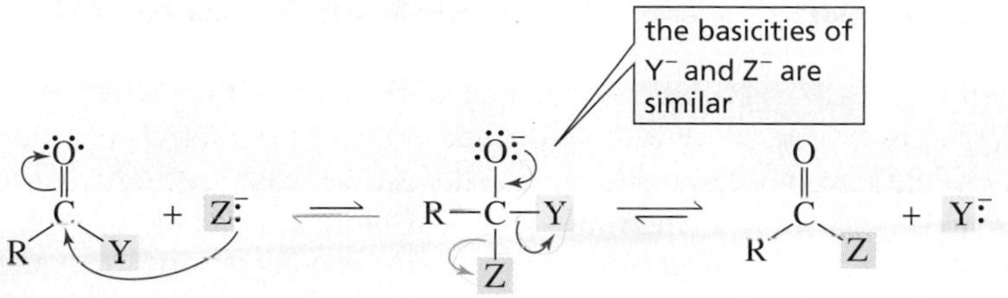

the basicities of Y⁻ and Z⁻ are similar

PROBLEM-SOLVING STRATEGY

The pK_a of HCl is -7; the pK_a of CH_3OH is 15.5. What is the product of the reaction of acetyl chloride with CH_3O^-?

In order to determine what the product of the reaction will be, we need to compare the basicities of the two groups that will be in the tetrahedral intermediate in order to see which one will be eliminated. Because HCl is a stronger acid than CH_3OH, Cl^- is a weaker base than CH_3O^-. Cl^-, therefore, will be eliminated from the tetrahedral intermediate, so the product of the reaction will be methyl acetate.

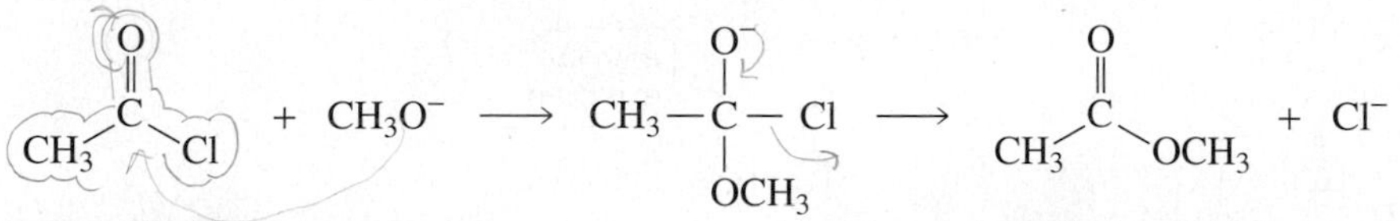

acetyl chloride   **methyl acetate**

Now continue on to Problem 5.

PROBLEM 5

a. The pK_a of HCl is -7; the pK_a of H_2O is 15.7. What is the product of the reaction of acetyl chloride with HO^-?

b. The pK_a of NH_3 is 36; the pK_a of H_2O is 15.7. What is the product of the reaction of acetamide with HO^-?

12.6   Relative Reactivities of Carboxylic Acids and Carboxylic Acid Derivatives

We have just seen that there are two steps in a nucleophilic acyl substitution reaction: formation of a tetrahedral intermediate and collapse of the tetrahedral intermediate. The weaker the base attached to the acyl group, the easier it is for *both steps* of the reaction to take place. In other words, the reactivity of a carboxylic acid derivative depends on the basicity of the substituent attached to the acyl group: The less basic the substituent, the more reactive is the carboxylic acid derivative. (The pK_a values of the conjugate acids of the leaving groups of Class I carbonyl compounds are shown in Table 12.1.)

relative basicities of the leaving groups

$$\boxed{\text{weakest base}} \quad Cl^- \; < \; {}^-OR \; \sim \; {}^-OH \; < \; {}^-NH_2 \quad \boxed{\text{strongest base}}$$

The weaker the base, the better it is as a leaving group.

relative reactivities of carboxylic acid derivatives

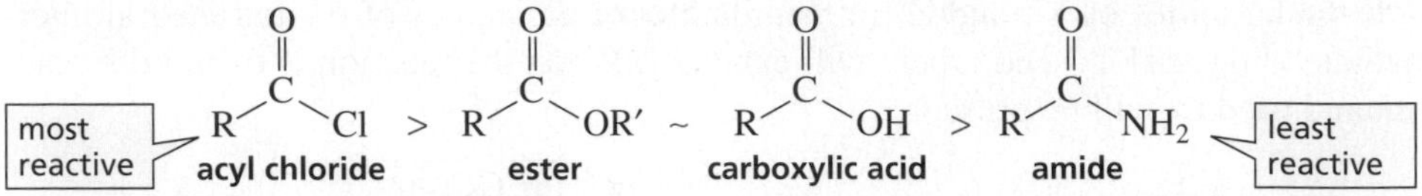

$$\boxed{\begin{array}{c}\text{most}\\\text{reactive}\end{array}} \quad \underset{\text{acyl chloride}}{R-\overset{O}{\overset{\|}{C}}-Cl} \; > \; \underset{\text{ester}}{R-\overset{O}{\overset{\|}{C}}-OR'} \; \sim \; \underset{\text{carboxylic acid}}{R-\overset{O}{\overset{\|}{C}}-OH} \; > \; \underset{\text{amide}}{R-\overset{O}{\overset{\|}{C}}-NH_2} \quad \boxed{\begin{array}{c}\text{least}\\\text{reactive}\end{array}}$$

relative reactivity:
acyl chloride > ester ~
carboxylic acid > amide

Table 12.1	The pK_a Values of the Conjugate Acids of the Leaving Groups of Carbonyl Compounds		
Carbonyl compound	**Leaving group**	**Conjugate acid of the leaving group**	**pK_a**
Class I			
$R-\overset{O}{\overset{\|}{C}}-Cl$	Cl^-	HCl	-7
$R-\overset{O}{\overset{\|}{C}}-OR'$	${}^-OR'$	$R'OH$	$\sim 15-16$
$R-\overset{O}{\overset{\|}{C}}-OH$	${}^-OH$	H_2O	15.7
$R-\overset{O}{\overset{\|}{C}}-NH_2$	${}^-NH_2$	NH_3	36
Class II			
$R-\overset{O}{\overset{\|}{C}}-H$	H^-	H_2	~ 40
$R-\overset{O}{\overset{\|}{C}}-R$	R^-	RH	~ 60

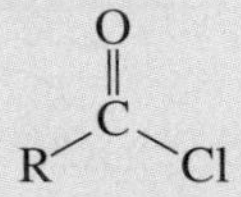

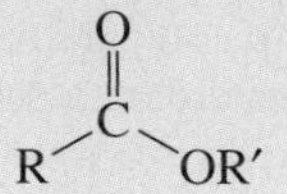

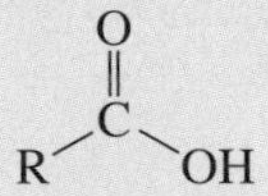

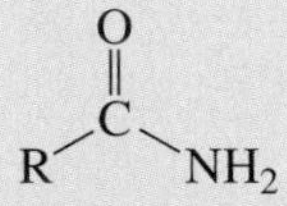

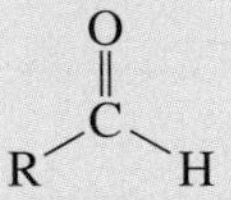

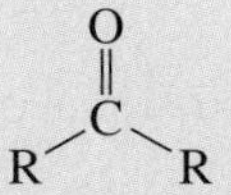

How does having a weak base attached to the acyl group make the *first* step of the nucleophilic acyl substitution reaction easier? A weaker base is a more electronegative base (Section 2.2). Therefore, it is better at withdrawing electrons inductively from the carbonyl carbon, which increases the carbonyl carbon's susceptibility to nucleophilic attack.

$$\underset{R \;\;\; \overset{\delta+}{} \;\;\; Y}{\overset{\overset{\delta-\!\cdot\ddot{O}\cdot}{\|}{C}}{}} \quad \boxed{\begin{array}{l}\text{inductive electron withdrawal by Y increases}\\\text{the electrophilicity of the carbonyl carbon}\end{array}}$$

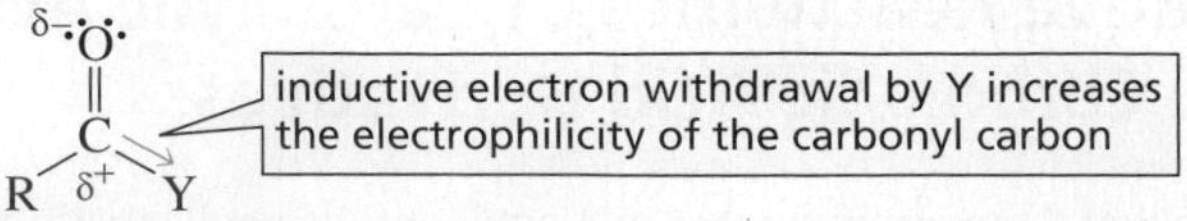

A weak base attached to the acyl group will also make the *second* step of the nucleophilic acyl substitution reaction easier because weak bases—which form weak bonds—are easier to eliminate when the tetrahedral intermediate collapses.

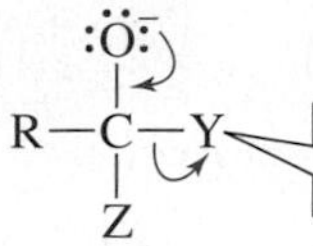

$$R-\overset{\overset{\displaystyle :\ddot{O}:}{|}}{\underset{\underset{\displaystyle Z}{|}}{C}}-Y \quad \boxed{\text{the weaker the base, the} \\ \text{easier it is to eliminate}}$$

In Section 12.5, we saw that in a nucleophilic acyl substitution reaction, the nucleophile that forms the tetrahedral intermediate must be a stronger base than the base that is already there. This means that *a carboxylic acid derivative can be converted into a less reactive carboxylic acid derivative, but not into one that is more reactive.* For example, an acyl chloride can be converted into an ester, because an alkoxide ion, such as methoxide ion, is a stronger base than a chloride ion.

$$R-\overset{\overset{\displaystyle O}{\|}}{C}-Cl \quad + \quad CH_3O^- \quad \longrightarrow \quad R-\overset{\overset{\displaystyle O}{\|}}{C}-OCH_3 \quad + \quad Cl^-$$

An ester, however, cannot be converted into an acyl chloride because a chloride ion is a weaker base than an alkoxide ion.

$$R-\overset{\overset{\displaystyle O}{\|}}{C}-OCH_3 \quad + \quad Cl^- \quad \longrightarrow \quad \text{no reaction}$$

For a carboxylic acid derivative to undergo a nucleophilic acyl substitution reaction, the incoming nucleophile must not be a much weaker base than the group that is to be replaced.

We can make the following general statement about the reactions of carboxylic acid derivatives: *A carboxylic acid derivative will undergo a nucleophilic acyl substitution reaction, provided that the newly added group in the tetrahedral intermediate is as strong or stronger a base than the group that was attached to the acyl group in the reactant.*

PROBLEM 6◆

What will be the product of a nucleophilic acyl substitution reaction—a new carboxylic acid derivative, a mixture of two carboxylic acid derivatives, or no reaction—if the new group in the tetrahedral intermediate is the following?

a. a stronger base than the group that was already there
b. a weaker base than the group that was already there
c. similar in basicity to the group that was already there

PROBLEM 7◆

Using the pK_a values in Table 12.1, predict the products of the following reactions:

a. $CH_3-\overset{\overset{\displaystyle O}{\|}}{C}-Cl \quad + \quad HO^- \quad \longrightarrow$

b. $CH_3-\overset{\overset{\displaystyle O}{\|}}{C}-OCH_3 \quad + \quad Cl^- \quad \longrightarrow$

Movie:
Nucleophilic acyl substitution

12.7 Reactions of Acyl Chlorides

Acyl chlorides react with alcohols to form esters, with water to form carboxylic acids, and with amines to form amides because in each case, the incoming nucleophile is a stronger base than the departing chloride ion.

$$\overset{\overset{\displaystyle O}{\|}}{C_6H_5-C-Cl} \quad + \quad CH_3OH \quad \longrightarrow \quad \overset{\overset{\displaystyle O}{\|}}{C_6H_5-C-OCH_3} \quad + \quad H^+ \quad + \quad Cl^-$$

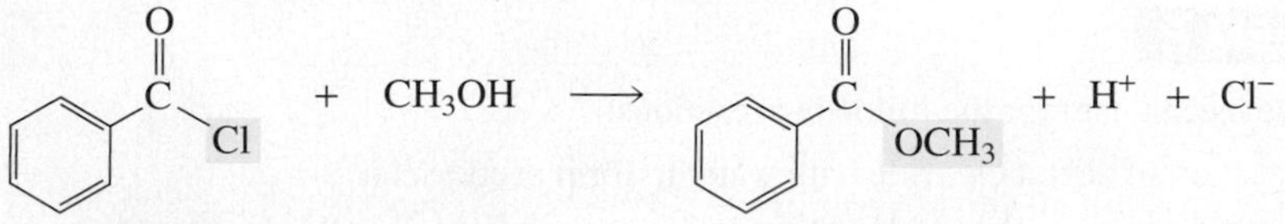

benzoyl chloride   **methyl benzoate**

acetyl chloride

3-D Molecule:
Benzoyl chloride

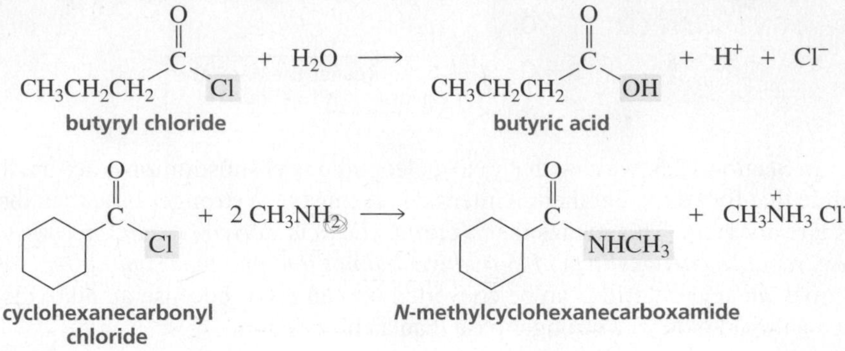

$$\underset{\textbf{butyryl chloride}}{CH_3CH_2CH_2 \overset{\displaystyle O}{\overset{\|}{C}} Cl} + H_2O \longrightarrow \underset{\textbf{butyric acid}}{CH_3CH_2CH_2 \overset{\displaystyle O}{\overset{\|}{C}} OH} + H^+ + Cl^-$$

$$\underset{\substack{\textbf{cyclohexanecarbonyl}\\\textbf{chloride}}}{\overset{\displaystyle O}{\overset{\|}{C}} Cl} + 2\ CH_3NH_2 \longrightarrow \underset{\textbf{N-methylcyclohexanecarboxamide}}{\overset{\displaystyle O}{\overset{\|}{C}} NHCH_3} + CH_3\overset{+}{N}H_3\ Cl^-$$

All carboxylic acid derivatives react by the same mechanism.

The tetrahedral intermediate eliminates the weakest base.

All carboxylic acid derivatives undergo nucleophilic acyl substitution reactions by the same mechanism. The nucleophile attacks the carbonyl carbon, forming a tetrahedral intermediate. If the nucleophile is a neutral molecule, there is a proton dissociation step before the tetrahedral intermediate collapses. (If the nucleophile is negatively charged, there is no need for the proton dissociation step; Section 12.5). When the tetrahedral intermediate collapses, it preferentially expels the weaker base.

mechanism for the conversion of an acyl chloride into an ester

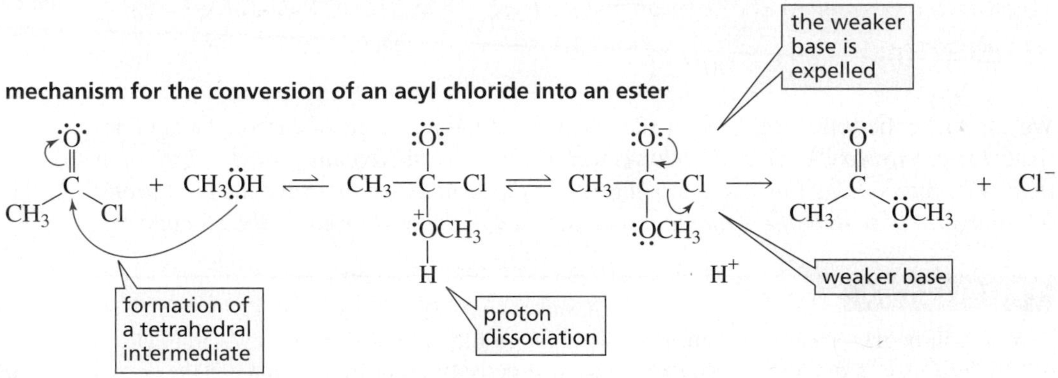

Notice that the reaction of an acyl chloride with an amine to form an amide is carried out with twice as much amine as acyl chloride, because the HCl generated in the reaction will protonate any amine that has yet to react. Protonated amines are not nucleophiles, so they cannot react with the acyl chloride. Therefore, there must be enough amine present to react with all the acyl chloride.

PROBLEM 8 **SOLVED**

Two amides are obtained from the reaction of acetyl chloride with a mixture of ethylamine and propylamine. Identify the amides.

SOLUTION   Either of the amines can react with acetyl chloride, so both *N*-ethylacetamide and *N*-propylacetamide are formed.

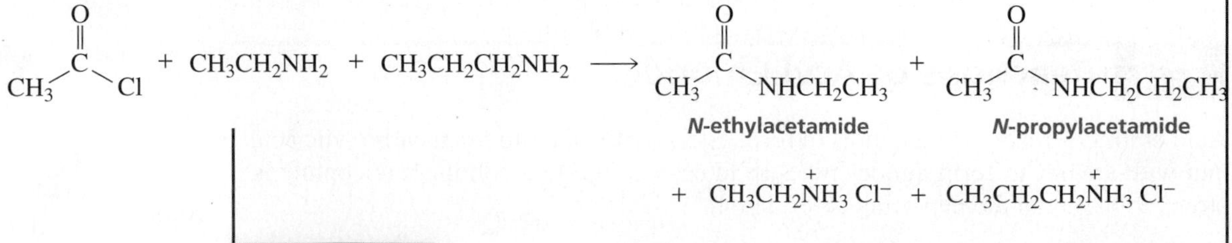

$$\underset{}{CH_3 \overset{\displaystyle O}{\overset{\|}{C}} Cl} + CH_3CH_2NH_2 + CH_3CH_2CH_2NH_2 \longrightarrow \underset{\textbf{N-ethylacetamide}}{CH_3 \overset{\displaystyle O}{\overset{\|}{C}} NHCH_2CH_3} + \underset{\textbf{N-propylacetamide}}{CH_3 \overset{\displaystyle O}{\overset{\|}{C}} NHCH_2CH_2CH_3}$$

$$+\ CH_3CH_2\overset{+}{N}H_3\ Cl^- +\ CH_3CH_2CH_2\overset{+}{N}H_3\ Cl^-$$

PROBLEM 9

Write the mechanism for the following reactions:

a. the reaction of acetyl chloride with water to form acetic acid

b. the reaction of acetyl chloride with methylamine to form *N*-methylacetamide

12.8   Reactions of Esters

Esters do not react with Cl^-, because Cl^- is a much weaker base than the RO^- leaving group of the ester (Table 12.1).

An ester reacts with water to form a carboxylic acid and an alcohol. This is an example of a *hydrolysis* reaction. A **hydrolysis reaction** is a reaction with water that converts one compound into two compounds *(lysis* is Greek for "breaking down").

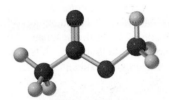

methyl acetate

a hydrolysis reaction

$$\underset{\substack{\text{methyl acetate}}}{CH_3\text{—}\overset{\overset{\displaystyle O}{\|}}{C}\text{—}OCH_3} + H_2O \underset{}{\overset{HCl}{\rightleftharpoons}} \underset{\substack{\text{acetic acid}}}{CH_3\text{—}\overset{\overset{\displaystyle O}{\|}}{C}\text{—}OH} + CH_3OH$$

An ester reacts with an alcohol to form a new ester and a new alcohol. This is an example of an **alcoholysis** reaction—a reaction with an alcohol that converts one compound into two compounds. This particular alcoholysis reaction is also called a **transesterification reaction** because one ester is converted to another ester.

a transesterification reaction

$$\underset{\substack{\text{methyl benzoate}}}{\text{⟨benzene ring⟩}\text{—}\overset{\overset{\displaystyle O}{\|}}{C}\text{—}OCH_3} + CH_3CH_2OH \overset{HCl}{\rightleftharpoons} \underset{\substack{\text{ethyl benzoate}}}{\text{⟨benzene ring⟩}\text{—}\overset{\overset{\displaystyle O}{\|}}{C}\text{—}OCH_2CH_3} + CH_3OH$$

Both the hydrolysis and the alcoholysis of an ester are very slow reactions because water and alcohols are poor nucleophiles and esters have very basic (poor) leaving groups. These reactions, therefore, are always catalyzed when carried out in the laboratory. Both hydrolysis and alcoholysis of an ester can be catalyzed by an acid (Section 12.9).

Esters also react with amines to form amides. A reaction with an amine that converts one compound into two compounds is called **aminolysis**.

an aminolysis reaction

$$\underset{\substack{\text{ethyl propionate}}}{CH_3CH_2\text{—}\overset{\overset{\displaystyle O}{\|}}{C}\text{—}OCH_2CH_3} + CH_3NH_2 \longrightarrow \underset{\substack{\textbf{\textit{N}-methylpropionamide}}}{CH_3CH_2\text{—}\overset{\overset{\displaystyle O}{\|}}{C}\text{—}NHCH_3} + CH_3CH_2OH$$

The reaction of an ester with an amine is not as slow as the reaction of an ester with water or an alcohol, because an amine is a better nucleophile. This is fortunate, because the reaction of an ester with an amine cannot be catalyzed by an acid. The acid will protonate the amine, and since a protonated amine is not a nucleophile, it cannot react with the ester.

NERVE IMPULSES, PARALYSIS, AND INSECTICIDES

After a nerve impulse is transmitted between cells, an ester called acetylcholine must be rapidly hydrolyzed to enable the recipient cell to receive another impulse.

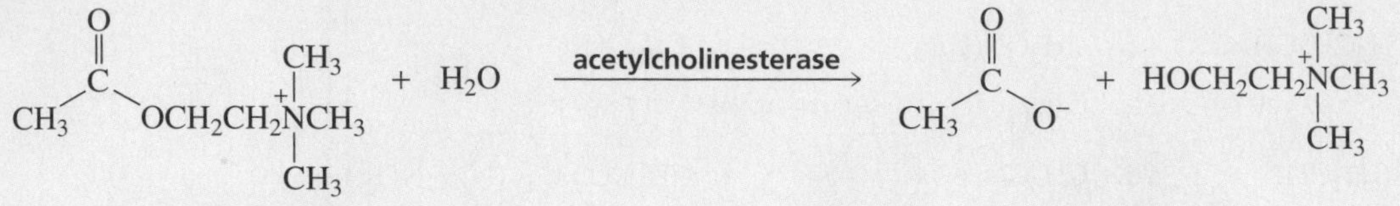

Acetylcholinesterase, the enzyme that catalyzes this hydrolysis, has a CH_2OH group that is necessary for its catalytic activity. Diisopropyl fluorophosphate (DFP), a military nerve gas used during World War II, inhibits acetylcholinesterase by undergoing a nucleophilic substitution reaction with the CH_2OH group.

When the enzyme is inhibited, the nerve impulses cannot be transmitted properly; thus, paralysis occurs. DFP is extremely toxic: Its LD_{50} (the lethal dose for 50% of the test animals) is only 0.5 mg/kg of body weight.

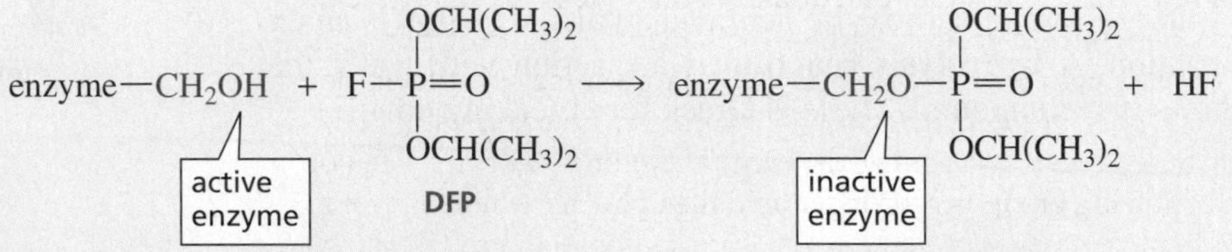

Malathion and parathion, compounds related to DFP, are used as insecticides. The LD_{50} of malathion is 2800 mg/kg. Parathion is more toxic, with an LD_{50} of 2 mg/kg.

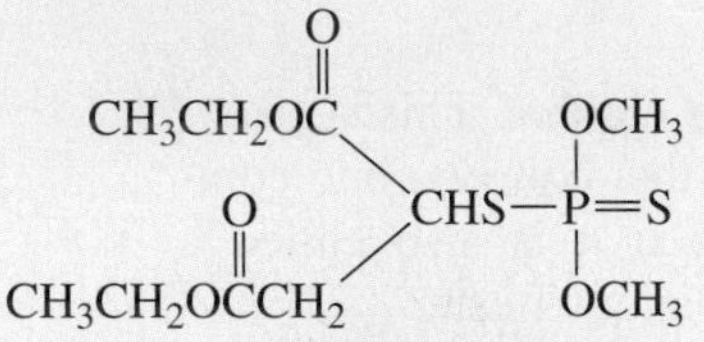

BIODEGRADABLE POLYMERS

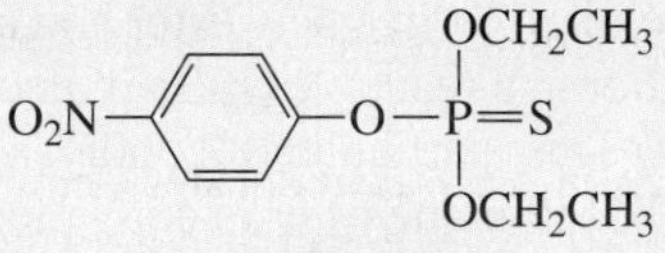

Biodegradable polymers are polymers that can be broken into small segments by enzyme-catalyzed reactions. The carbon–carbon bonds of chain-growth polymers are inert to enzyme-catalyzed reactions, so they are nonbiodegradable unless bonds that can be broken by an enzyme-catalyzed reaction are inserted into the polymer. Then when the polymer is buried as waste, enzymes that are produced by microorganisms present in the ground can degrade the polymer. One method used to make a biodegradable polymer involves inserting ester groups into it. For example, if the acetal shown below is added to an alkene that is undergoing radical polymerization, ester groups will be inserted into the polymer, forming "weak links" that are susceptible to enzyme-catalyzed hydrolysis.

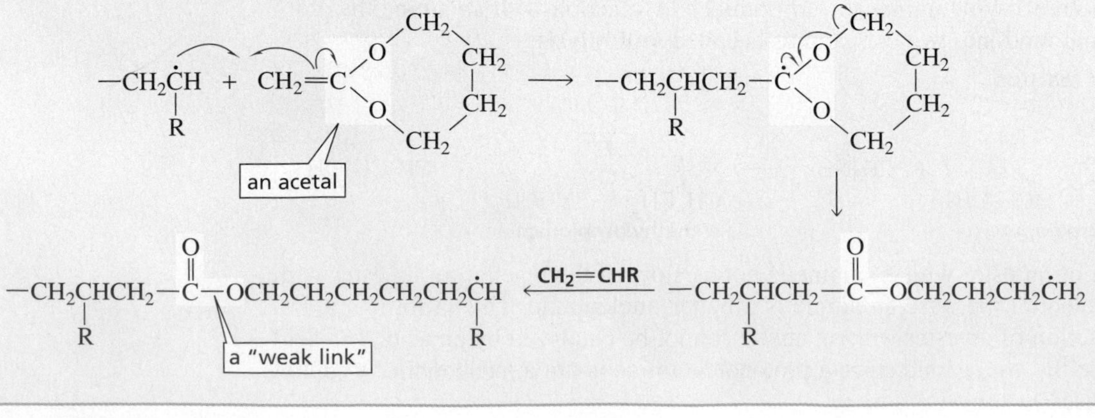

PROBLEM 11

Write a mechanism for the following reactions:

a. the noncatalyzed hydrolysis of methyl propionate
b. the aminolysis of phenyl formate, using methylamine

PROBLEM 12   SOLVED

a. List the following esters in order of decreasing reactivity toward hydrolysis:

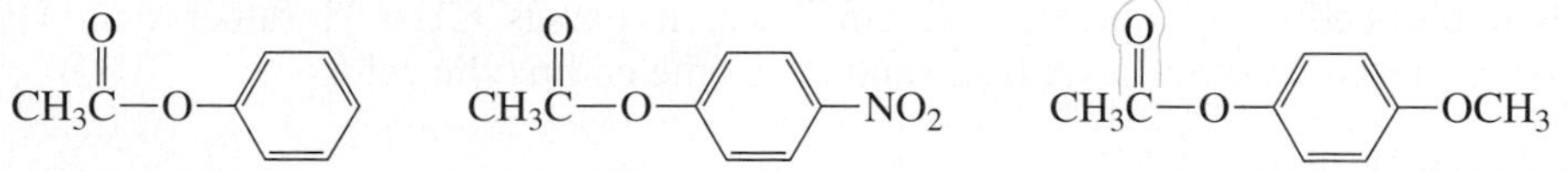

$$CH_3\overset{\displaystyle O}{\overset{\|}{C}}-O-\text{C}_6\text{H}_5 \qquad CH_3\overset{\displaystyle O}{\overset{\|}{C}}-O-\text{C}_6\text{H}_4-NO_2 \qquad CH_3\overset{\displaystyle O}{\overset{\|}{C}}-O-\text{C}_6\text{H}_4-OCH_3$$

b. How would the rate of hydrolysis of the *para*-methylphenyl ester compare with the rates of hydrolysis of these three esters?

SOLUTION TO 12a   Both formation of the tetrahedral intermediate and collapse of the tetrahedral intermediate are fastest for the ester with the electron-withdrawing nitro substituent and slowest for the ester with the electron-donating methoxy substituent. *Formation of the tetrahedral intermediate*: An electron-withdrawing substituent increases the ester's susceptibility to nucleophilic attack, and an electron-donating substituent decreases its susceptibility. *Collapse of the tetrahedral intermediate*: Electron withdrawal increases acidity and electron donation decreases acidity. Therefore, *para*-nitrophenol with a strong electron-withdrawing group is a stronger acid than phenol, which in turn is a stronger acid than *para*-methoxyphenol with a strong electron-donating group. Therefore, the *para*-nitrophenoxide ion is the weakest base and the best leaving group of the three, whereas the *para*-methoxyphenoxide ion is the strongest base and the worst leaving group. Thus,

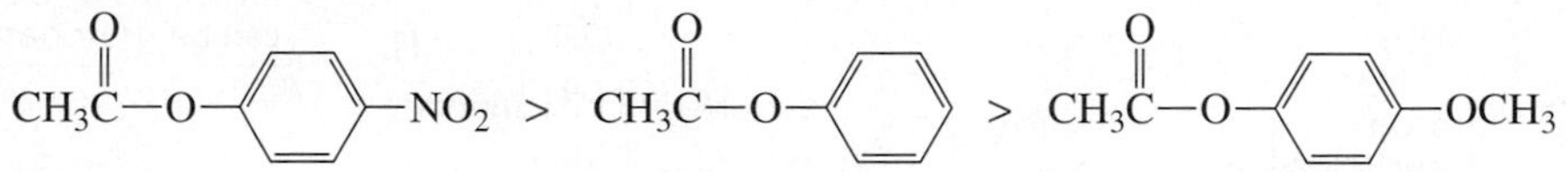

$$CH_3\overset{\displaystyle O}{\overset{\|}{C}}-O-\text{C}_6\text{H}_4-NO_2 \; > \; CH_3\overset{\displaystyle O}{\overset{\|}{C}}-O-\text{C}_6\text{H}_5 \; > \; CH_3\overset{\displaystyle O}{\overset{\|}{C}}-O-\text{C}_6\text{H}_4-OCH_3$$

SOLUTION TO 12b   The methyl substituent donates electrons inductively to the benzene ring, but donates electrons to a lesser extent than does the methoxy substituent that donates electrons by resonance. Therefore, the rate of hydrolysis of the methyl-substituted ester is slower than the rate of hydrolysis of the unsubstituted ester, but faster than the rate of hydrolysis of the methoxy-substituted ester.

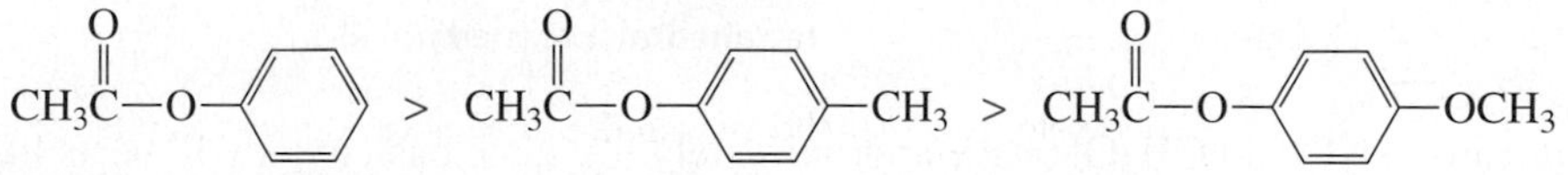

$$CH_3\overset{\displaystyle O}{\overset{\|}{C}}-O-\text{C}_6\text{H}_5 \; > \; CH_3\overset{\displaystyle O}{\overset{\|}{C}}-O-\text{C}_6\text{H}_4-CH_3 \; > \; CH_3\overset{\displaystyle O}{\overset{\|}{C}}-O-\text{C}_6\text{H}_4-OCH_3$$

12.9   Acid-Catalyzed Ester Hydrolysis

We have seen that esters hydrolyze slowly because water is a poor nucleophile and esters have very basic leaving groups. The rate of hydrolysis can be increased by acid.

When an acid is added to a reaction, the first thing that happens is the acid protonates the atom in the reactant that has the greatest electron density. The resonance contributors of the ester show that the atom with the greatest electron density is the carbonyl oxygen.

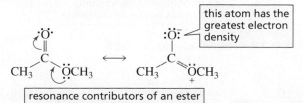

resonance contributors of an ester

this atom has the greatest electron density

Thus, in the first step in the mechanism for acid-catalyzed ester hydrolysis, the acid protonates the carbonyl oxygen.

In the second step of the mechanism, the nucleophile (H_2O) attacks the carbonyl carbon of the protonated carbonyl group. The protonated tetrahedral intermediate (tetrahedral intermediate I) that is formed is in equilibrium with its nonprotonated form (tetrahedral intermediate II). Either the OH or the OCH_3 can be protonated. Because the OH and OCH_3 groups have approximately the same basicity, both tetrahedral intermediate I (OH is protonated) and tetrahedral intermediate III (OCH_3 is protonated) are formed. When tetrahedral intermediate I collapses, it expels H_2O in preference to CH_3O^- (because H_2O is a weaker base) and reforms the ester. When tetrahedral intermediate III collapses, it expels CH_3OH rather than HO^- (because CH_3OH is a weaker base) and forms the carboxylic acid.

mechanism for acid-catalyzed ester hydrolysis

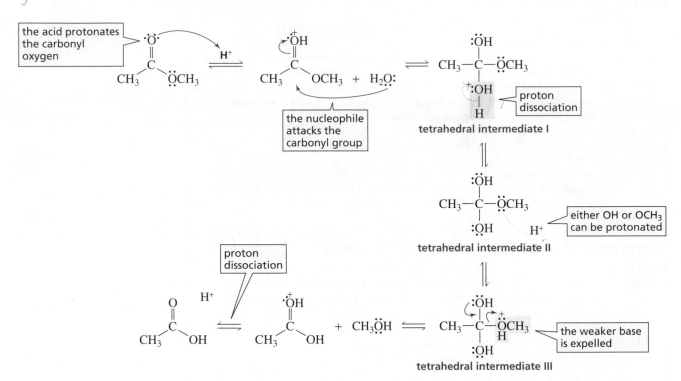

Because H_2O and CH_3OH have approximately the same basicity, it will be as likely for tetrahedral intermediate I to collapse to reform the ester as it will for tetrahedral intermediate III to collapse to form the carboxylic acid. Consequently, when the reaction has reached equilibrium, both ester and carboxylic acid will be present in approximately equal amounts.

both ester and carboxylic acid will be present in approximately equal amounts when the reaction has reached equilibrium

Excess water will force the equilibrium to the right.

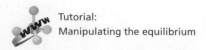

Tutorial:
Manipulating the equilibrium

The mechanism for the acid-catalyzed reaction of a carboxylic acid and an alcohol to form an ester and water is the exact reverse of the mechanism for the acid-catalyzed hydrolysis of an ester to form a carboxylic acid and an alcohol. If the ester is the desired product, the reaction should be carried out with excess alcohol in order to drive the equilibrium to the left.

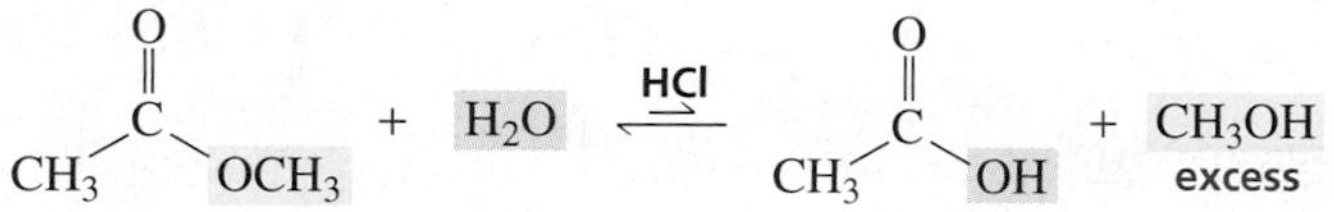

PROBLEM 13

Referring to the mechanism for the acid-catalyzed hydrolysis of methyl acetate, write the mechanism—showing all the curved arrows—for the acid-catalyzed reaction of acetic acid and methanol to form methyl acetate.

Now let's see how the acid increases the rate of ester hydrolysis. The acid is a catalyst. Recall that **catalyst** is a substance that increases the rate of a reaction without being consumed or changed in the overall reaction (Section 5.4). For a catalyst to increase the rate of a reaction, it must increase the rate of the slow step of the reaction. Changing the rate of a fast step will not affect the rate of the overall reaction. There are two relatively slow steps in the mechanism: formation of a tetrahedral intermediate and collapse of a tetrahedral intermediate. The other steps are fast steps. (Proton transfer to or from an electronegative atom such as oxygen or nitrogen is always a fast step.) The acid increases the rates of both slow steps.

The acid increases the rate of formation of the tetrahedral intermediate by protonating the carbonyl oxygen. Protonated carbonyl groups are more susceptible than nonprotonated carbonyl groups to nucleophilic attack because a positively charged oxygen is more electron withdrawing than a neutral oxygen. Increased electron withdrawal by the oxygen makes the carbonyl carbon more electron deficient, which increases its attractiveness to nucleophiles.

protonation of the carbonyl oxygen increases the susceptibility of the carbonyl carbon to nucleophilic attack

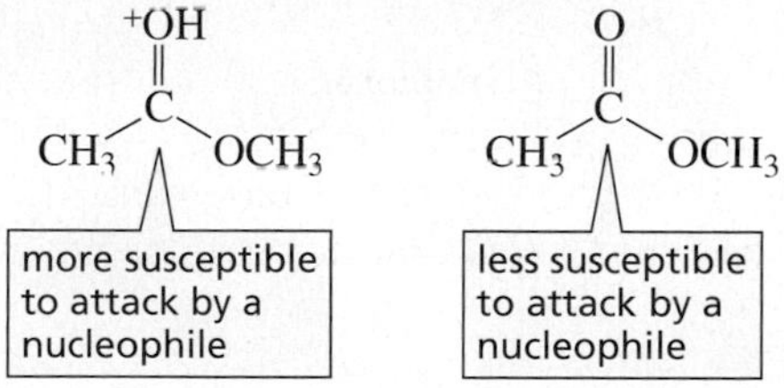

An acid catalyst increases the reactivity of a carbonyl group.

The acid increases the rate of collapse of the tetrahedral intermediate by decreasing the basicity of the leaving group, which makes it easier to eliminate. In the acid-catalyzed hydrolysis of an ester, the leaving group is ROH, a weaker base than the leaving group (RO⁻) in the uncatalyzed reaction.

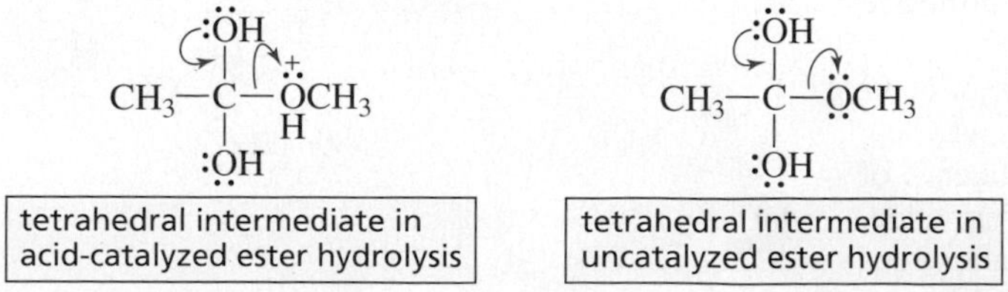

An acid catalyst can make a group a better leaving group.

ASPIRIN

Prostaglandins regulate a variety of physiological responses such as inflammation, blood pressure, blood clotting, fever, and pain. The enzyme prostaglandin synthase catalyzes the conversion of arachidonic acid into PGH_2, the precursor of all prostaglandins. A transesterification reaction that blocks prostaglandin synthesis is responsible for aspirin's activity as an anti-inflammatory agent.

$$\text{arachidonic acid} \xrightarrow{\text{prostaglandin synthase}} PGH_2 \begin{array}{l} \nearrow \text{prostaglandins} \\ \searrow \text{thromboxanes} \end{array}$$

Prostaglandin synthase is composed of two enzymes. One of the enzymes—cyclooxygenase—has a CH_2OH group that is necessary for enzymatic activity. The CH_2OH group reacts with aspirin (acetylsalicylic acid) in a transesterification reaction. This inactivates the enzyme. Prostaglandin therefore cannot be synthesized, and inflammation is suppressed.

Other anti-inflammatory drugs, such as ibuprofen (the active ingredient in Advil, Motrin, and Nuprin) and naproxen (the active ingredient in Aleve), also inhibit the synthesis of prostaglandins.

Both aspirin and these other nonsteroidal anti-inflammatory drugs (NSAIDs) inhibit the synthesis of all prostaglandins—those produced under normal physiological conditions and those produced in response to inflammation. The production of acid in the stomach is regulated by a prostaglandin. When prostaglandin synthesis stops, therefore, the acidity of the stomach can rise above normal levels. Celebrex, a relatively new drug, inhibits only the enzyme (cyclooxygenase-2) that produces prostaglandin in response to stress. Thus, inflammatory conditions now can be treated without some of the harmful side effects. This drug is known as a COX-2 inhibitor.

Aspirin also inhibits the synthesis of thromboxanes, compounds involved in blood clotting. Presumably, this is why low levels of aspirin have been reported to reduce the incidence of strokes and heart attacks that result from blood clot formation. Aspirin's activity as an anticoagulant is why doctors caution patients not to take aspirin for several days before surgery.

3-D Molecule:
Aspirin

PROBLEM 14◆

What products would be formed from the acid-catalyzed hydrolysis of the following esters?

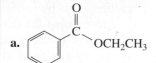

a.

b. 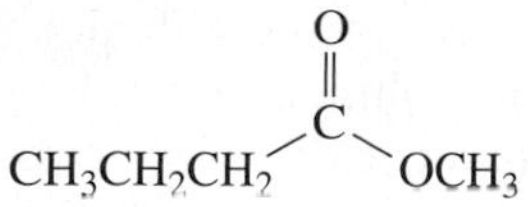 $CH_3CH_2CH_2$ — C(=O) — OCH_3

PROBLEM 15◆

What product would be formed from the acid-catalyzed hydrolysis of the following cyclic ester?

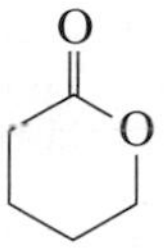

Transesterification

Transesterification—the reaction of an ester with an alcohol—is also catalyzed by acid. The mechanism for transesterification is identical to the mechanism for ester hydrolysis, except that the nucleophile is ROH rather than H_2O. As in hydrolysis, the leaving groups in the tetrahedral intermediate formed in transesterification have approximately the same basicity. Consequently, an excess of the reactant alcohol must be used to produce more of the desired product.

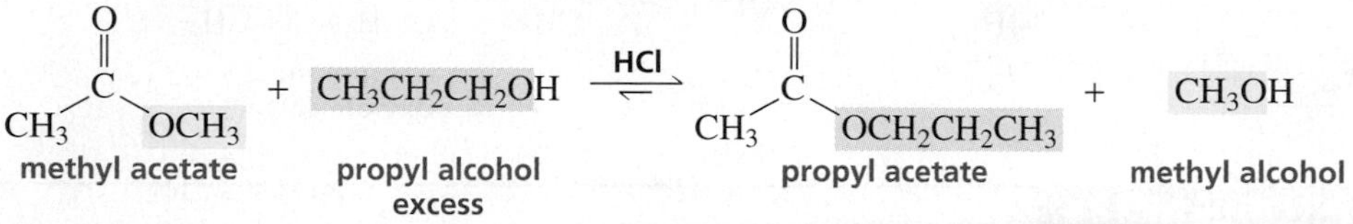

$$CH_3 - C(=O) - OCH_3 \ + \ CH_3CH_2CH_2OH \ \overset{HCl}{\rightleftharpoons} \ CH_3 - C(=O) - OCH_2CH_2CH_3 \ + \ CH_3OH$$

methyl acetate   propyl alcohol   propyl acetate   methyl alcohol
                    excess

PROBLEM 16◆

Give the products of the following reaction:

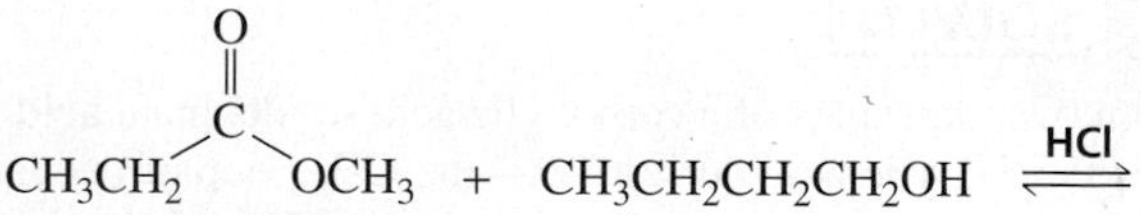

$$CH_3CH_2 - C(=O) - OCH_3 \ + \ CH_3CH_2CH_2CH_2OH \ \overset{HCl}{\rightleftharpoons}$$

PROBLEM 17

Write the mechanism for the acid-catalyzed transesterification reaction of methyl acetate with ethanol.

12.10   Reactions of Carboxylic Acids

Carboxylic acids can undergo nucleophilic acyl substitution reactions only when they are in their acidic forms. The basic form of a carboxylic acid does not undergo nucleophilic acyl substitution reactions because the negatively charged carboxylate ion is resistant to nucleophilic attack. Thus, carboxylate ions are even less reactive toward nucleophilic acyl substitution reactions than are amides.

acetic acid

relative reactivities toward nucleophilic acyl substitution

most reactive $R - C(=O) - OH$ > $R - C(=O) - NH_2$ > $R - C(=O) - O^-$ least reactive

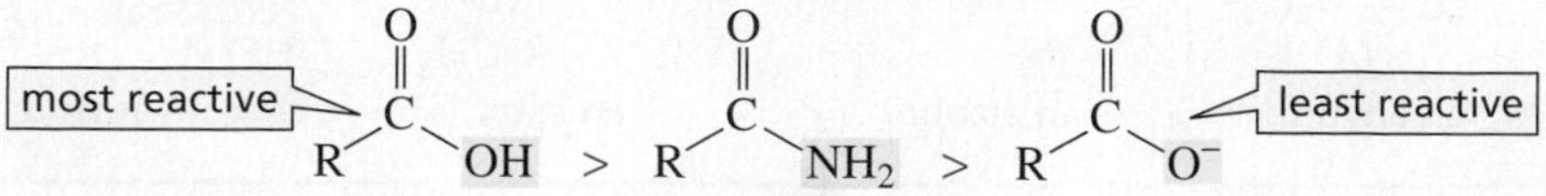

Carboxylic acids have approximately the same reactivity as esters—the HO^- leaving group of a carboxylic acid has approximately the same basicity as the RO^- leaving group of an ester. Therefore, like esters, carboxylic acids do not react with chloride ions.

Carboxylic acids react with alcohols to form esters. The reaction must be carried out in an acidic solution, not only to catalyze the reaction but also to keep the carboxylic acid in its acidic form so that it will react with the nucleophile. Because the tetrahedral intermediate formed in this reaction has two potential leaving groups of approximately the same basicity, the reaction must be carried out with excess alcohol to drive it toward products. Emil Fischer (Section 16.7) was the first to discover that an ester could be prepared by treating a carboxylic acid with excess alcohol in the presence of an acid catalyst, so the reaction is called a **Fischer esterification**.

$$
\underset{\substack{\text{acetic acid}}}{CH_3-\overset{\overset{\textstyle O}{\|}}{C}-OH} \;+\; \underset{\substack{\text{methyl alcohol}\\\text{excess}}}{CH_3OH} \;\underset{}{\overset{HCl}{\rightleftharpoons}}\; \underset{\substack{\text{methyl acetate}}}{CH_3-\overset{\overset{\textstyle O}{\|}}{C}-OCH_3} \;+\; H_2O
$$

Carboxylic acids do not undergo nucleophilic acyl substitution reactions with amines. Because a carboxylic acid is an acid and an amine is a base, the carboxylic acid immediately donates a proton to the amine when the two compounds are mixed. The ammonium carboxylate salt is the final product of the reaction; the carboxylate ion is unreactive and the protonated amine is not a nucleophile.

$$
\underset{}{CH_3-\overset{\overset{\textstyle O}{\|}}{C}-OH} \;+\; CH_3CH_2NH_2 \;\longrightarrow\; \underset{\substack{\text{an ammonium}\\\text{carboxylate salt}}}{CH_3-\overset{\overset{\textstyle O}{\|}}{C}-O^-\;\overset{+}{H_3}NCH_2CH_3}
$$

PROBLEM 18◆

Using an acyl chloride and an alcohol, show how the following esters could be synthesized:

a. methyl butyrate (odor of apples) **b.** octyl acetate (odor of oranges)

PROBLEM 19 SOLVED

Loss of water from two molecules of a carboxylic acid results in an **acid anhydride**. Acid anhydrides are also carboxylic acid derivatives—the OH group of the carboxylic acid has been replaced with a carboxylate group. Thus, the carboxylate group is the leaving group of an acid anhydride.

3-D Molecules:
Succinic acid;
Succinic anhydride

$$
\underset{}{R-\overset{\overset{\textstyle O}{\|}}{C}-OH \;\; HO-\overset{\overset{\textstyle O}{\|}}{C}-R} \;\longrightarrow\; \underset{\substack{\text{an acid anhydride}}}{R-\overset{\overset{\textstyle O}{\|}}{C}-O-\overset{\overset{\textstyle O}{\|}}{C}-R} \;+\; H_2O
$$

a carboxylate group

What is the product of the reaction of acid anhydride with

a. an alcohol? **b.** an amine?

SOLUTION TO 19a When an acid anhydride reacts with an alcohol, the two potential leaving groups in the tetrahedral intermediate will be a carboxylate ion and an alkoxide ion. The carboxylate ion is the weaker base, so the product of the reaction will be an ester and a carboxylic acid.

$$
\underset{\substack{\text{an ahydride}}}{R-\overset{\overset{\textstyle O}{\|}}{C}-O-\overset{\overset{\textstyle O}{\|}}{C}-R} \;+\; \underset{\substack{\text{an alcohol}}}{CH_3OH} \;\longrightarrow\; \underset{\substack{\text{an ester}}}{R-\overset{\overset{\textstyle O}{\|}}{C}-OCH_3} \;+\; \underset{\substack{\text{a carboxylic acid}}}{HO-\overset{\overset{\textstyle O}{\|}}{C}-R}
$$

12.11 Reactions of Amides

Amides are very unreactive compounds, which is comforting, since proteins—which impart strength to biological structures—are composed of amino acids linked together by amide bonds (Section 17.0). Amides do not react with chloride ions, alcohols, or water because, in each case, the incoming nucleophile is a weaker base than the leaving group of the amide (Table 12.1).

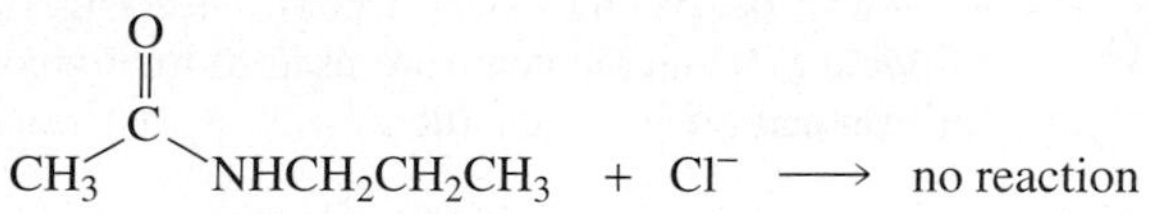

acetamide

$$CH_3\overset{\displaystyle O}{\underset{\displaystyle\|}{C}}NHCH_2CH_2CH_3 \;+\; Cl^- \longrightarrow \text{no reaction}$$

N-propylacetamide

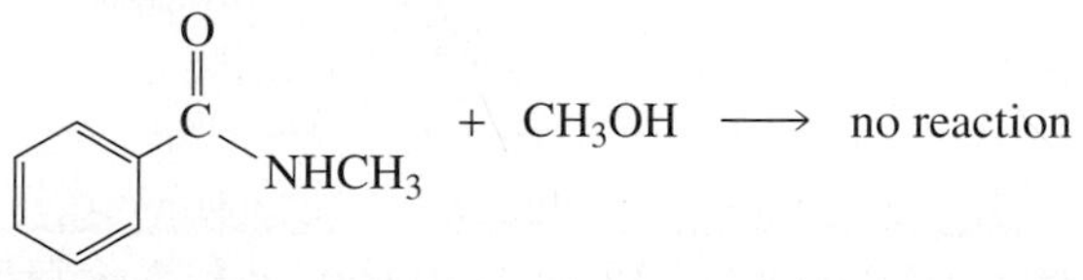

$$C_6H_5\overset{\displaystyle O}{\underset{\displaystyle\|}{C}}NHCH_3 \;+\; CH_3OH \longrightarrow \text{no reaction}$$

N-methylbenzamide

$$CH_3CH_2\overset{\displaystyle O}{\underset{\displaystyle\|}{C}}NHCH_2CH_3 \;+\; H_2O \longrightarrow \text{no reaction}$$

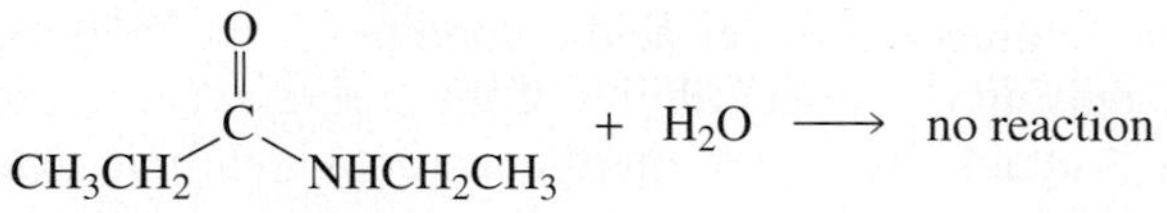

N-ethylpropanamide

Amides do, however, react with water and alcohols if the reaction mixture is heated in the presence of an acid. The reason for this will be explained in Section 12.12.

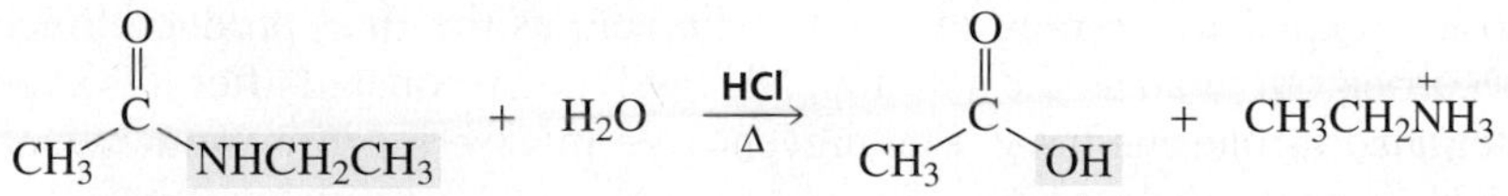

$$CH_3\overset{\displaystyle O}{\underset{\displaystyle\|}{C}}NHCH_2CH_3 \;+\; H_2O \;\xrightarrow[\Delta]{HCl}\; CH_3\overset{\displaystyle O}{\underset{\displaystyle\|}{C}}OH \;+\; CH_3CH_2\overset{+}{N}H_3$$

N-ethylacetamide

$$C_6H_5\overset{\displaystyle O}{\underset{\displaystyle\|}{C}}NHCH_3 \;+\; CH_3CH_2OH \;\xrightarrow[\Delta]{HCl}\; C_6H_5\overset{\displaystyle O}{\underset{\displaystyle\|}{C}}OCH_2CH_3 \;+\; CH_3\overset{+}{N}H_3$$

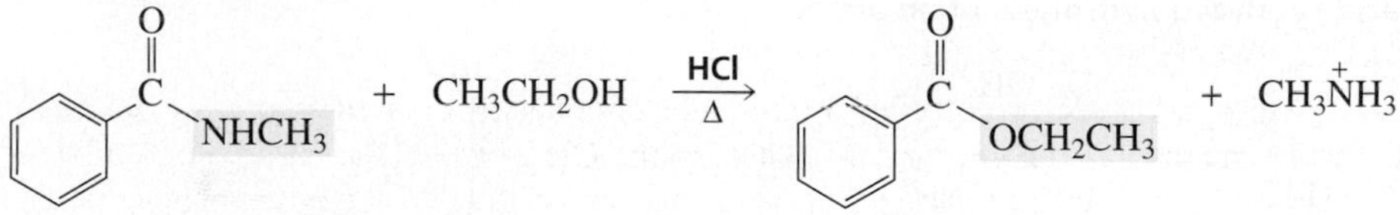

N-methylbenzamide

PROBLEM 20◆

What acyl chloride and what amine would be required to synthesize the following amines?

a. N-ethylbutanamide

b. N,N-dimethylbenzamide

PROBLEM 21◆

Which of the following reactions would lead to the formation of an amide?

1. 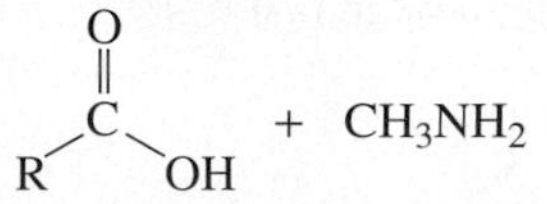 $R\overset{\displaystyle O}{\underset{\displaystyle\|}{C}}OH \;+\; CH_3NH_2$

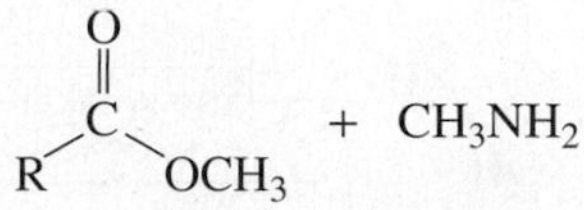

3. 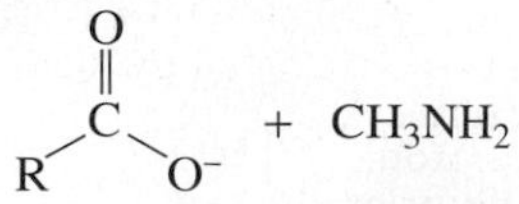 $R\overset{\displaystyle O}{\underset{\displaystyle\|}{C}}O^- \;+\; CH_3NH_2$

2. $R\overset{\displaystyle O}{\underset{\displaystyle\|}{C}}OCH_3 \;+\; CH_3NH_2$

4. 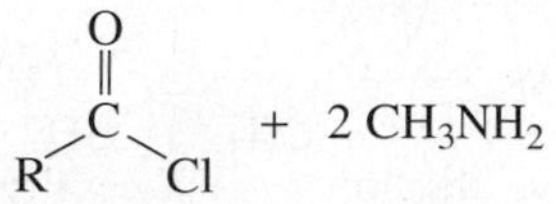 $R\overset{\displaystyle O}{\underset{\displaystyle\|}{C}}Cl \;+\; 2\, CH_3NH_2$

NATURE'S SLEEPING PILL

Melatonin, a naturally occurring amide, is a hormone that is synthesized by the pineal gland from the amino acid tryptophan. Melatonin regulates the dark–light clock that governs such things as the sleep–wake cycle, body temperature, and hormone production.

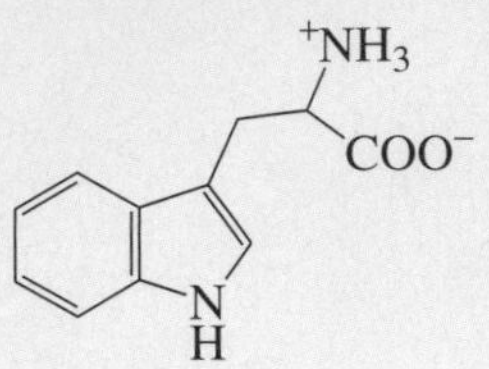

Melatonin levels increase from evening to night and then decrease as morning approaches. People with high levels of melatonin sleep longer and more soundly than those with low levels. The concentration of the hormone in the blood varies with age—6-year-olds have more than five times the concentration that 80-year-olds have—which is one of the reasons why young people have less trouble sleeping than older people. Melatonin supplements are used to treat insomnia, jet lag, and seasonal affective disorder.

tryptophan
an amino acid

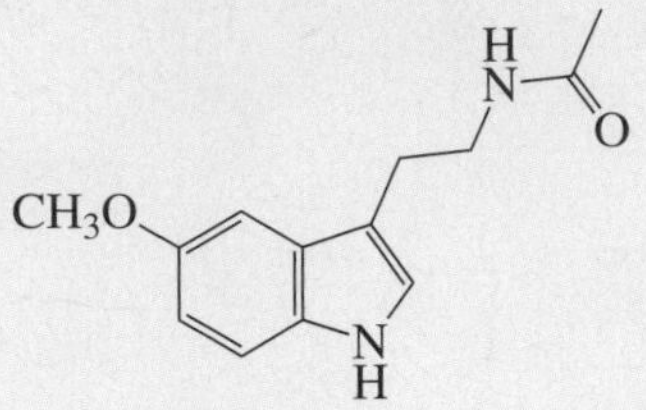

melatonin

12.12 Acid-Catalyzed Hydrolysis of Amides

When an amide is hydrolyzed under acidic conditions, the acid protonates the carbonyl oxygen, increasing the susceptibility of the carbonyl carbon to nucleophilic attack. Nucleophilic attack by water on the carbonyl carbon leads to tetrahedral intermediate I, which is in equilibrium with its nonprotonated form, tetrahedral intermediate II. Reprotonation can occur either on oxygen to reform tetrahedral intermediate I or on nitrogen to form tetrahedral intermediate III. Protonation on nitrogen is favored because the NH_2 group is a stronger base than the OH group. Of the two possible leaving groups in tetrahedral intermediate III (HO^- and NH_3), NH_3 is the weaker base, so it is expelled, forming the carboxylic acid as the final product. Since the reaction is carried out in an acidic solution, NH_3 will be protonated after it is expelled from the tetrahedral intermediate. This prevents the reverse reaction from occurring, because $^+NH_4$ is not a nucleophile.

mechanism for acid-catalyzed hydrolysis of an amide

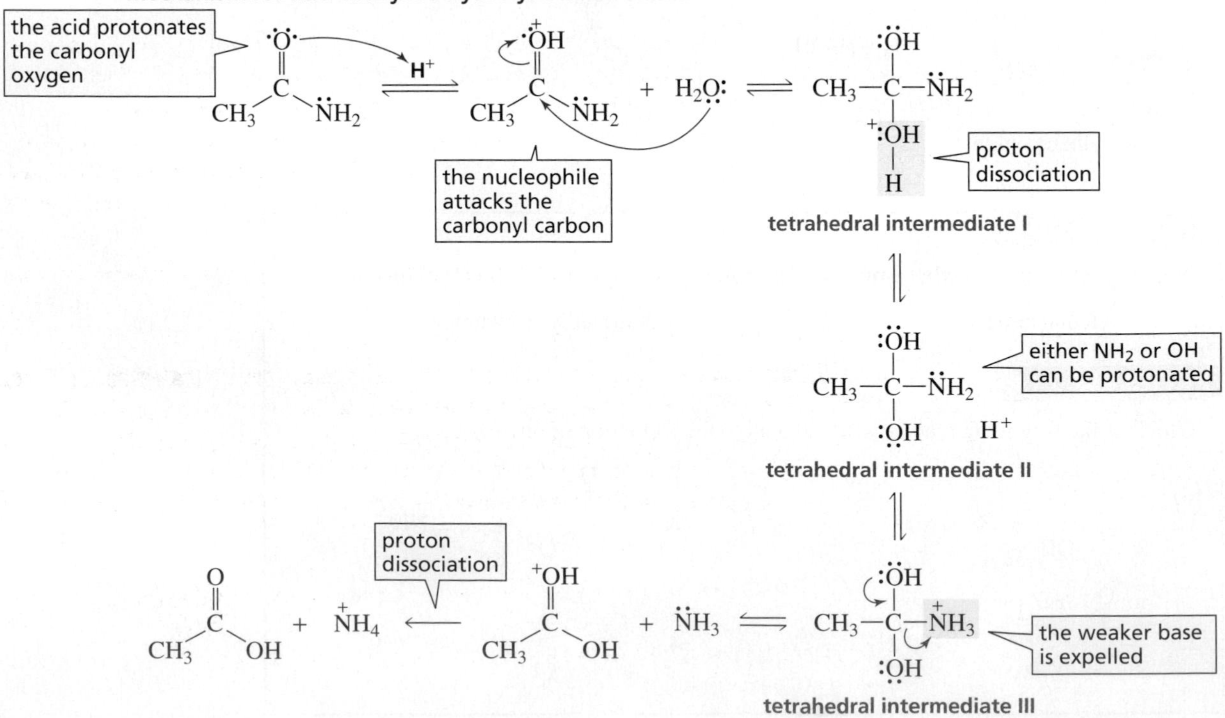

the acid protonates the carbonyl oxygen

the nucleophile attacks the carbonyl carbon

proton dissociation

tetrahedral intermediate I

either NH_2 or OH can be protonated

tetrahedral intermediate II

proton dissociation

the weaker base is expelled

tetrahedral intermediate III

PENICILLIN AND DRUG RESISTANCE

Penicillin contains an amide in a four-membered ring. The strain in the ring increases the amide's reactivity. It is thought that the antibiotic activity of penicillin results from its ability to put an acyl group on a CH_2OH group of an enzyme that is involved in the synthesis of bacterial cell walls. This inactivates the enzyme, and actively growing bacteria die because they are unable to synthesize functional cell walls. Penicillin has no effect on mammalian cells because mammalian cells are not enclosed by cell walls. To minimize hydrolysis of the four-membered ring during storage, penicillins are refrigerated.

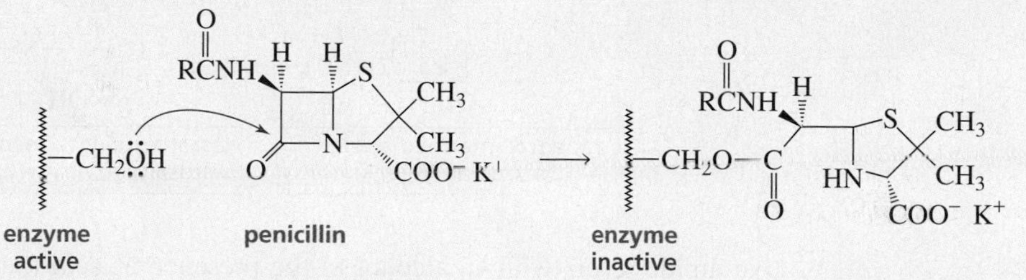

enzyme active   penicillin

enzyme inactive

Bacteria that are resistant to penicillin secrete penicillinase, an enzyme that catalyzes the hydrolysis of penicillin's four-membered ring. The ring-opened product has no antibacterial activity.

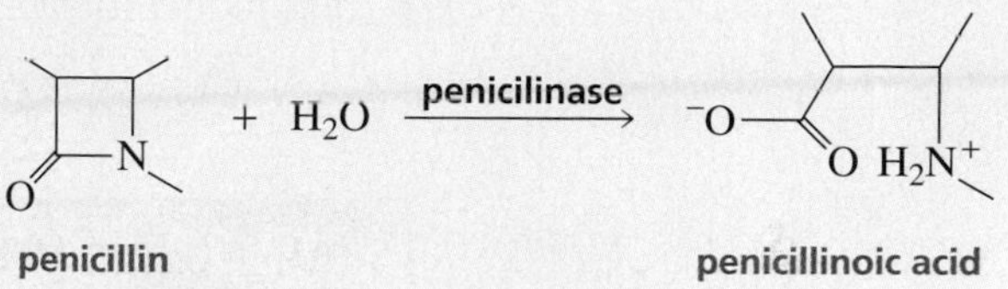

penicillin $+ H_2O$ $\xrightarrow{\text{penicilinase}}$ penicillinoic acid

PENICILLINS IN CLINICAL USE

More than 10 different penicillins are currently in clinical use. They differ only in the group (R) attached to the carbonyl group. Some of these penicillins are shown here. In addition to their structural differences, the penicillins differ in the organisms against which they are most effective. They also differ in their resistance to penicillinase. For example, ampicillin, a *synthetic* penicillin, is clinically effective against bacteria that are resistant to penicillin G, a *naturally occurring* penicillin. Almost 19% of humans are allergic to penicillin G.

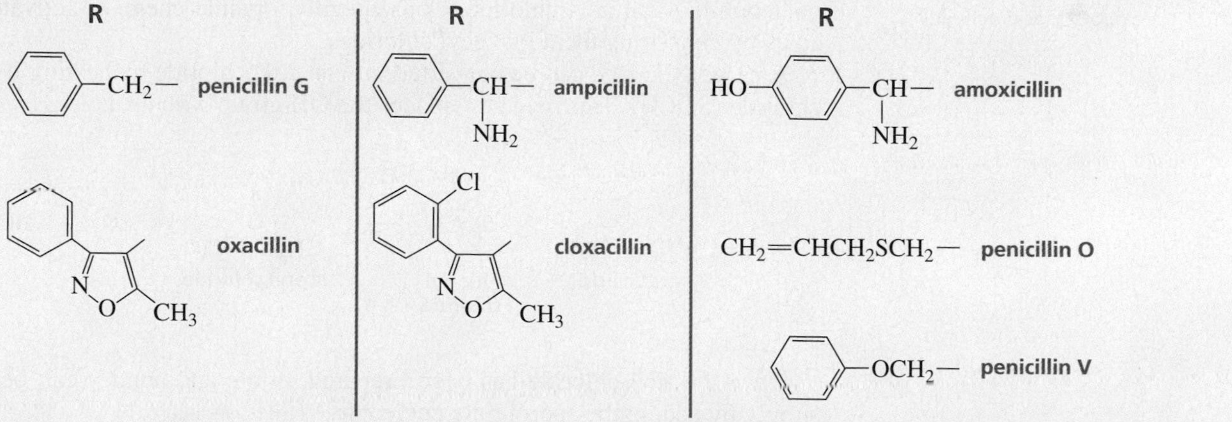

penicillin G

ampicillin

amoxicillin

oxacillin

cloxacillin

penicillin O

penicillin V

Penicillin V is a *semisynthetic* penicillin that is in clinical use. It is not a naturally occurring penicillin; nor is it a true synthetic penicillin because chemists don't synthesize it. The *Penicillium* mold synthesizes it after the mold is fed 2-phenoxyethanol, the compound it needs for the side chain.

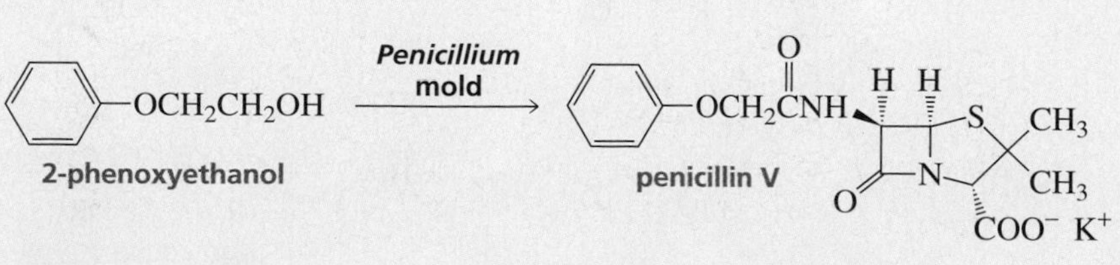

2-phenoxyethanol $\xrightarrow{\text{Penicillium mold}}$ penicillin V

Let's take a minute to see why an amide cannot be hydrolyzed without a catalyst. In the uncatalyzed reaction, the amide is not protonated. Therefore, water, a very poor nucleophile, must attack a neutral amide that is much less susceptible to nucleophilic attack than a protonated amide would be. In addition, the NH_2 group of the tetrahedral intermediate is not protonated in the uncatalyzed reaction. Therefore, HO^- is the group expelled from the tetrahedral intermediate (because HO^- is a weaker base than $^-NH_2$) which reforms the amide.

$$CH_3-\overset{\overset{\displaystyle :\ddot{O}H}{|}}{\underset{\underset{\displaystyle :\ddot{O}H}{|}}{C}}-\overset{+}{N}H_3 \qquad CH_3-\overset{\overset{\displaystyle :\ddot{O}H}{|}}{\underset{\underset{\displaystyle :\ddot{O}H}{|}}{C}}-\ddot{N}H_2$$

tetrahedral intermediate in acid-catalyzed amide hydrolysis	tetrahedral intermediate in uncatalyzed amide hydrolysis

An amide reacts with an alcohol in the presence of acid (page 309) for the same reason that it reacts with water in the presence of acid.

12.13 Synthesis of Carboxylic Acid Derivatives

Of the various classes of carbonyl compounds discussed in this chapter—acyl chlorides, esters, carboxylic acids, and amides—carboxylic acids are the most commonly available. However, we have seen that carboxylic acids are relatively unreactive toward nucleophilic acyl substitution reactions because the OH group of a carboxylic acid is a strong base and therefore a poor leaving group. Therefore, chemists need a way to activate carboxylic acids so that they can readily undergo nucleophilic acyl substitution reactions.

Because acyl chlorides are the most reactive of the carboxylic acid derivatives, the easiest way to synthesize any other carboxylic acid derivative is to add the appropriate nucleophile to an acyl chloride. Consequently, organic chemists activate carboxylic acids by converting them into acyl chlorides.

A carboxylic acid can be converted into an acyl chloride by heating it with thionyl chloride ($SOCl_2$). This reagent replaces the OH group with a Cl.

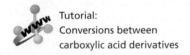

Tutorial:
Conversions between carboxylic acid derivatives

$$\underset{\substack{\text{acetic acid}}}{\underset{CH_3}{\overset{O}{\overset{||}{C}}}\diagdown OH} + \underset{\substack{\text{thionyl}\\\text{chloride}}}{SOCl_2} \xrightarrow{\Delta} \underset{\substack{\text{acetyl chloride}}}{\underset{CH_3}{\overset{O}{\overset{||}{C}}}\diagdown Cl} + SO_2 + HCl$$

Once the acyl chloride has been prepared, esters and amides can be synthesized simply by adding the appropriate nucleophile (Section 12.7).

$$\underset{R}{\overset{O}{\overset{||}{C}}}\diagdown Cl + ROH \longrightarrow \underset{\substack{\text{an ester}}}{\underset{R}{\overset{O}{\overset{||}{C}}}\diagdown OR} + HCl$$

$$\underset{R}{\overset{O}{\overset{||}{C}}}\diagdown Cl + 2\ RNH_2 \longrightarrow \underset{\substack{\text{an amide}}}{\underset{R}{\overset{O}{\overset{||}{C}}}\diagdown NHR} + \overset{+}{R}NH_3\ Cl^-$$

PROBLEM 22◆

How would you synthesize the following compounds starting with a carboxylic acid?

a.

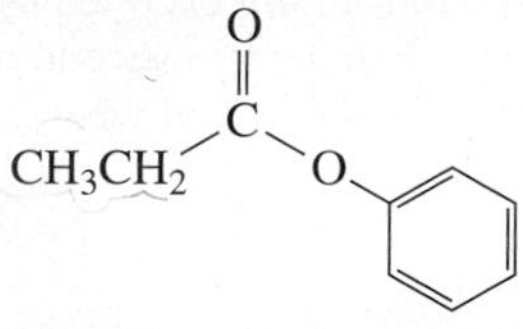

b.

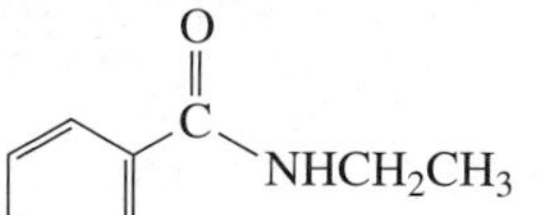

SYNTHETIC POLYMERS

Synthetic polymers play important roles in our daily lives. Polymers are compounds that are made by linking together many small molecules called monomers. We have seen that **chain-growth polymers** are made by adding monomers to the end of a growing chain (Section 5.16).

A second kind of synthetic polymer, called a **step-growth polymer**, is made with monomers that have reactive functional groups at each end. The functional groups form ester or amide bonds between the monomers. Nylon and Dacron are step-growth polymers. Nylon is a polyamide. Dacron is a polyester.

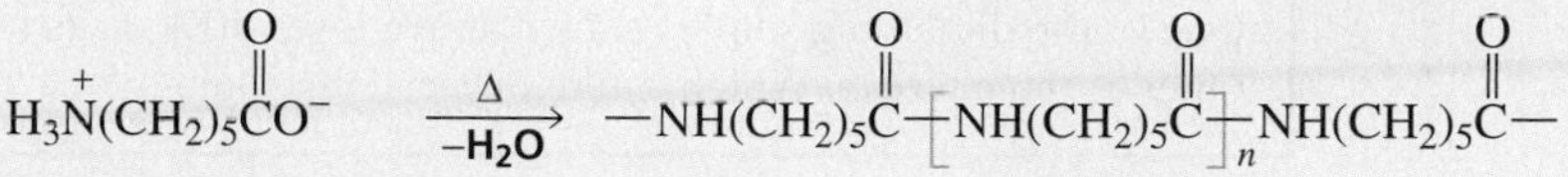

6-aminohexanoic acid

nylon 6
a polyamide

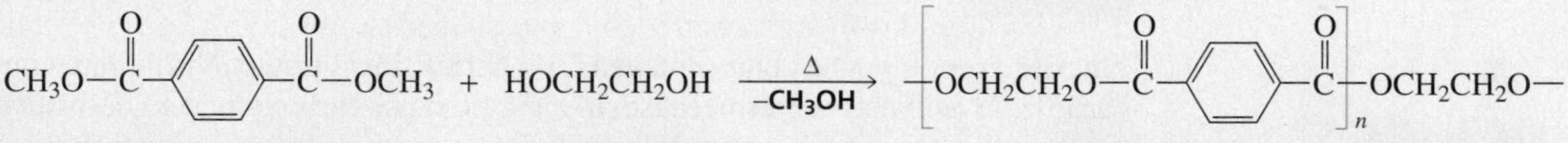

dimethyl terephthalate   1,2-ethanediol
ethylene glycol

poly(ethylene terephthalate)
Dacron®
a polyester

Synthetic polymers have taken the place of metals, fabrics, glass, ceramics, wood, and paper, allowing us to have a greater variety and larger quantities of materials than nature could have provided. New polymers are continually being designed to fit human needs. For example, Kevlar and Lexan are relatively new step-growth polymers. Kevlar has a tensile strength greater than steel. It is used for high-performance skis and bulletproof vests.

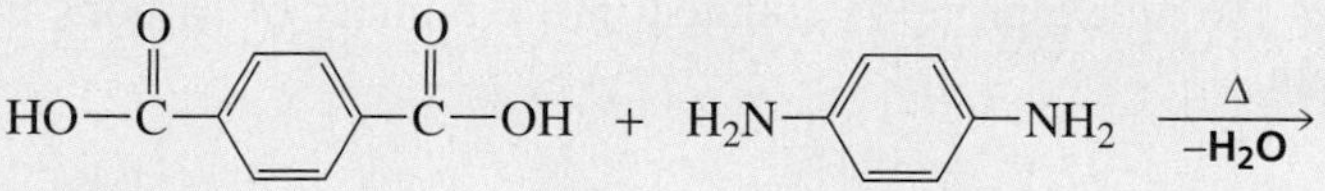

1,4-benzenedicarboxylic acid   1,4-diaminobenzene

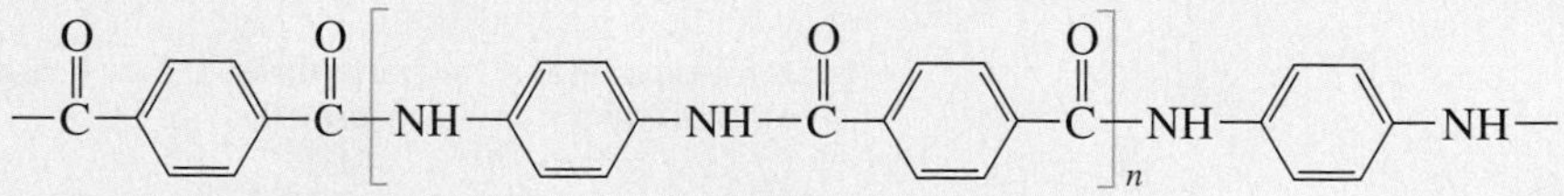

Kevlar®

Lexan is a strong and transparent polymer used for such things as traffic light lenses and compact disks.

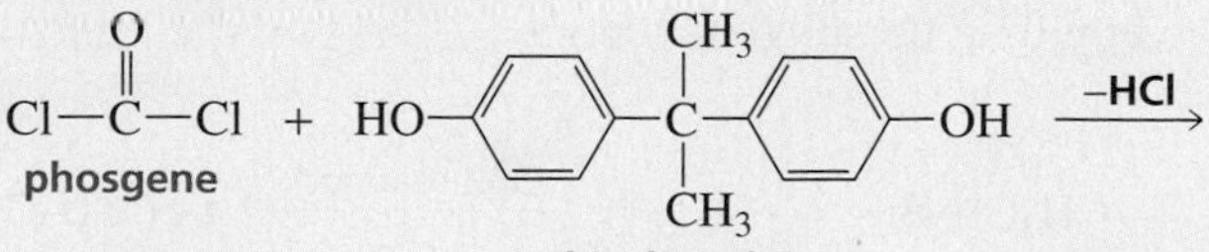

phosgene

bisphenol A

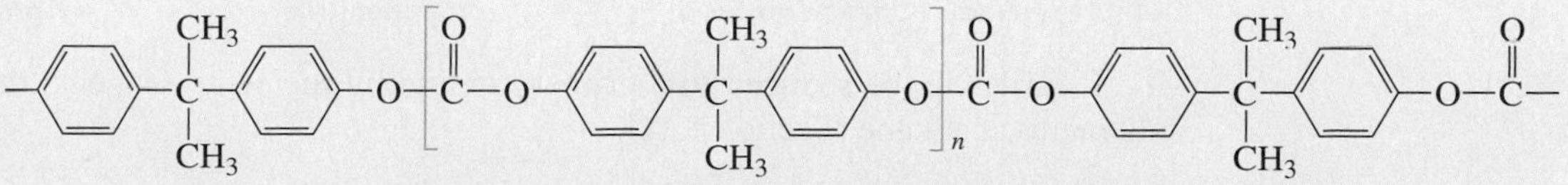

Lexan®

DISSOLVING SUTURES

Dissolving sutures, such as dexon and poly(diox-anone) (PDS), are synthetic polymers that now are routinely used in surgery. These sutures have many ester groups that are slowly hydrolyzed to small molecules that sub-

sequently are metabolized to compounds easily excreted by the body. Patients, therefore, do not have to undergo a second medical procedure to remove the sutures, as is required when traditional suture materials are used.

$$\left[OCH_2\overset{O}{\overset{\|}{C}}-OCH_2\overset{O}{\overset{\|}{C}} \right]_n$$

Dexon

$$\left[OCH_2CH_2OCH_2\overset{O}{\overset{\|}{C}}-OCH_2CH_2OCH_2\overset{O}{\overset{\|}{C}} \right]_n$$

PDS

Depending on their structures, these synthetic sutures lose 50% of their strength after two to three weeks and are completely absorbed within three to six months.

PROBLEM 23◆

One of the dissolving sutures shown above loses 50% of its strength in two weeks and the other in three weeks. Which suture material lasts longer?

12.14 Nitriles

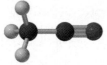

acetonitrile

Nitriles are compounds that contain a C≡N functional group. Nitriles are considered carboxylic acid derivatives because, like all the other carboxylic acid derivatives, they react with water to form carboxylic acids. They are even less reactive than amides, but hydrolyze when heated with water and an acid.

$$CH_3CH_2C\equiv N \ + \ H_2O \ \xrightarrow[\Delta]{HCl} \ \underset{CH_3CH_2}{\overset{O}{\overset{\|}{C}}}\!\!-OH \ + \ \overset{+}{N}H_4$$

Nitriles are named by adding "nitrile" to the parent alkane name. Notice that the triple-bonded carbon of the nitrile group is counted in the number of carbons in the longest continuous chain.

$$CH_3C\equiv N \qquad \underset{CH_3}{\overset{CH_3}{\overset{|}{CHCH_2CH_2CH_2C\equiv N}}} \qquad CH_2\!=\!CHC\equiv N$$

systematic name:	ethanenitrile	5-methylhexanenitrile	propenenitrile
common name:	acetonitrile		acrylonitrile

Tutorial:
Common terms pertaining to carboxylic acids and their derivatives

Nitriles can be prepared from an S_N2 reaction of alkyl halide with cyanide ion. Because a nitrile can be hydrolyzed to carboxylic acid, you now know how to convert an alkyl halide into a carboxylic acid. Notice that the carboxylic acid has one more carbon than the alkyl halide.

$$CH_3CH_2Br \ + \ {}^-C\equiv N \ \xrightarrow{S_N2 \ reaction} \ CH_3CH_2C\equiv N \ \xrightarrow[\Delta]{HCl, \ H_2O} \ \underset{CH_3CH_2}{\overset{O}{\overset{\|}{C}}}\!\!-OH$$

ethyl bromide   cyanide ion   **propanenitrile**   **propanoic acid**

A nitrile can be reduced to a primary amine by the same reagents that reduce an alkyne to an alkane (Section 5.12).

$$CH_3CH_2CH_2CH_2C\equiv N \ \xrightarrow{H_2}{Pt/C} \ CH_3CH_2CH_2CH_2CH_2NH_2$$

pentanenitrile   **pentylamine**

PROBLEM 24◆

Which alkyl halides form the following carboxylic acids after reacting with sodium cyanide and the product heated in an acidic aqueous solution?

a. butyric acid   **b.** 4-methylpentanoic acid

Summary

A **carbonyl group** is a carbon double-bonded to an oxygen; an **acyl group** is a carbonyl group attached to an alkyl or an aryl group. **Acyl chlorides**, **esters**, and **amides** are called **carboxylic acid derivatives** because they differ from a carboxylic acid only in the nature of the group that has replaced the OH group of the carboxylic acid.

Carbonyl compounds can be placed in one of two classes. Class I carbonyl compounds contain a group that can be replaced by another group; carboxylic acids and carboxylic acid derivatives belong to this class. Class II carbonyl compounds do not contain a group that can be replaced by another group; aldehydes and ketones belong to this class.

The reactivity of carbonyl compounds resides in the polarity of the carbonyl group; the carbonyl carbon has a partial positive charge that is attractive to nucleophiles. Class I carbonyl compounds undergo nucleophilic acyl substitution reactions: a nucleophile replaces the substituent that was attached to the acyl group in the reactant. All Class I carbonyl compounds react with nucleophiles in the same way: the nucleophile attacks the carbonyl carbon, forming an unstable tetrahedral intermediate, which reforms a carbonyl compound by eliminating the weakest base.

A carboxylic acid derivative will undergo a nucleophilic acyl substitution reaction provided that the newly added group in the tetrahedral intermediate is not a much weaker base than the group that was attached to the acyl group in the reactant. The weaker the base attached to the acyl group, the easier it is for both steps of the nucleophilic acyl substitution reaction to take place. The relative reactivities toward nucleophilic acyl substitution: acyl halides > esters and carboxylic acids > amides > carboxylate ions. Amides are unreactive compounds but do react with water and alcohols if the reaction mixture is heated in the presence of an acid. Nitriles are harder to hydrolyze than amides. Carboxylic acids are activated by being converted to acyl chlorides.

Hydrolysis, **alcoholysis**, and **aminolysis** are reactions in which water, alcohols, and amines, respectively, convert one compound into two compounds. A **transesterification reaction** converts one ester to another ester. Treating a carboxylic acid with excess alcohol and an acid catalyst is called a **Fischer esterification**.

The rate of hydrolysis or alcoholysis can be increased by acid. An acid increases the rate of formation of the tetrahedral intermediate by protonating the carbonyl oxygen, which increases the electrophilicity of the carbonyl group, and by decreasing the basicity of the leaving group, which makes it easier to eliminate.

Summary of Reactions

1. Reactions of acyl halides (Section 12.7)

$$ \underset{R}{\overset{O}{\|}}{C}{-}Cl \quad + \quad CH_3OH \quad \longrightarrow \quad \underset{R}{\overset{O}{\|}}{C}{-}OCH_3 \quad + \quad HCl $$

$$ \underset{R}{\overset{O}{\|}}{C}{-}Cl \quad + \quad H_2O \quad \longrightarrow \quad \underset{R}{\overset{O}{\|}}{C}{-}OH \quad + \quad HCl $$

$$ \underset{R}{\overset{O}{\|}}{C}{-}Cl \quad + \quad 2\,CH_3NH_2 \quad \longrightarrow \quad \underset{R}{\overset{O}{\|}}{C}{-}NHCH_3 \quad + \quad CH_3\overset{+}{N}H_3\;Cl^- $$

2. Reactions of esters (Sections 12.8–12.9)

$$\underset{R}{\overset{O}{\underset{OR}{\parallel}}}C + CH_3OH \;\rightleftharpoons^{HCl}\; \underset{R}{\overset{O}{\underset{OCH_3}{\parallel}}}C + ROH$$

$$\underset{R}{\overset{O}{\underset{OR}{\parallel}}}C + H_2O \;\rightleftharpoons^{HCl}\; \underset{R}{\overset{O}{\underset{OH}{\parallel}}}C + ROH$$

$$\underset{R}{\overset{O}{\underset{OR}{\parallel}}}C + CH_3NH_2 \;\longrightarrow\; \underset{R}{\overset{O}{\underset{NHCH_3}{\parallel}}}C + ROH$$

3. Reactions of carboxylic acids (Section 12.10)

$$\underset{R}{\overset{O}{\underset{OH}{\parallel}}}C + CH_3OH \;\rightleftharpoons^{HCl}\; \underset{R}{\overset{O}{\underset{OCH_3}{\parallel}}}C + H_2O$$

$$\underset{R}{\overset{O}{\underset{OH}{\parallel}}}C + CH_3NH_2 \;\longrightarrow\; \underset{R}{\overset{O}{\underset{O^-\ \overset{+}{H_3N}CH_3}{\parallel}}}C$$

4. Reactions of amides (Sections 12.11 and 12.12)

$$\underset{R}{\overset{O}{\underset{NH_2}{\parallel}}}C + H_2O \;\xrightarrow[\Delta]{HCl}\; \underset{R}{\overset{O}{\underset{OH}{\parallel}}}C + \overset{+}{NH_4}\,Cl^-$$

$$\underset{R}{\overset{O}{\underset{NH_2}{\parallel}}}C + CH_3OH \;\xrightarrow[\Delta]{HCl}\; \underset{R}{\overset{O}{\underset{OCH_3}{\parallel}}}C + \overset{++}{NH_4}\,Cl^-$$

5. Activation of carboxylic acids (Section 12.13)

$$\underset{R}{\overset{O}{\underset{OH}{\parallel}}}C + SOCl_2 \;\xrightarrow{\Delta}\; \underset{R}{\overset{O}{\underset{Cl}{\parallel}}}C + SO_2 + HCl$$

6. Hydrolysis of nitriles (Section 12.14)

$$RC{\equiv}N + H_2O \;\xrightarrow[\Delta]{HCl}\; \underset{R}{\overset{O}{\underset{OH}{\parallel}}}C + \overset{+}{NH_4}$$

Problems

25. Write a structure for each of the following compounds:

   a. *N,N*-dimethylhexanamide   **c.** cyclohexanecarbonyl chloride   **e.** sodium acetate

   b. 3,3-dimethylhexanamide   **d.** cycloheptanecarboxylic acid   **f.** propionyl chloride

26. Name the following compounds:

a. $\underset{\underset{CH_2CH_3}{|}}{CH_3CH_2CHCH_2CH_2CH_2}\overset{\overset{O}{\|}}{C}OH$

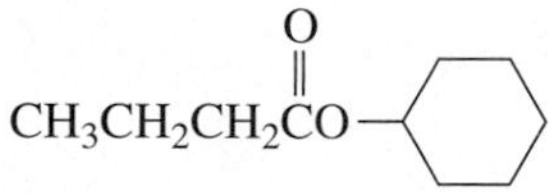

d. $CH_3CH_2CH_2\overset{\overset{O}{\|}}{C}N(CH_3)_2$

g. $CH_2{=}CHCH_2\overset{\overset{O}{\|}}{C}NHCH_3$

b. $CH_3CH_2\overset{\overset{O}{\|}}{C}OCH_2CH_2CH_3$

e. $CH_3CH_2CH_2CH_2\overset{\overset{O}{\|}}{C}Cl$

h. $\underset{H_3C}{\overset{\overset{CH_2CH_3}{|}}{\underset{}{C}}}\overset{}{\underset{CH_2COOH}{\cdots H}}$

c. $CH_3CH_2CH_2\overset{\overset{O}{\|}}{C}O{-}\bigcirc$

f. $CH_3CH_2\underset{\underset{CH_3}{|}}{CH}CH_2\overset{\overset{O}{\|}}{C}OCH_3$

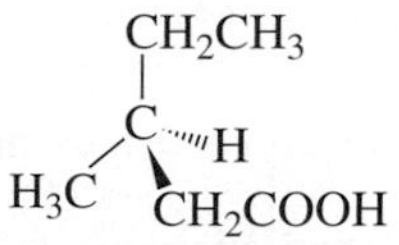

i. $CH_3CH_2CH_2CH_2C{\equiv}N$

27. What products would be formed from the reaction of acetyl chloride with the following reagents?
a. water
b. excess dimethylamine
c. excess aniline
d. cyclohexanol
e. 4-chlorophenol
f. isopropyl alcohol

28. a. List the following esters in order of decreasing reactivity in the first step of a nucleophilic acyl substitution reaction (formation of the tetrahedral intermediate):

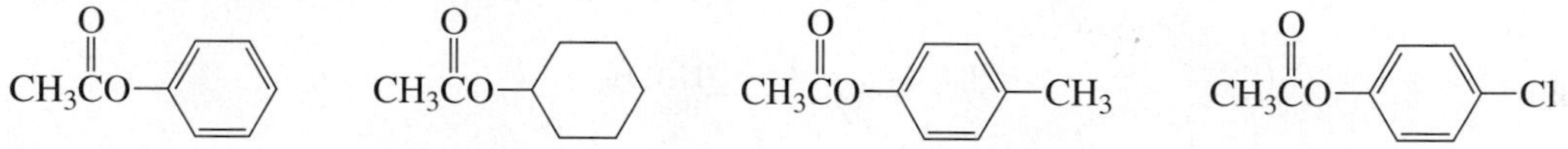

$CH_3\overset{\overset{O}{\|}}{C}O{-}\bigcirc$

A

$CH_3\overset{\overset{O}{\|}}{C}O{-}\bigcirc$

B

$CH_3\overset{\overset{O}{\|}}{C}O{-}\bigcirc{-}CH_3$

C

$CH_3\overset{\overset{O}{\|}}{C}O{-}\bigcirc{-}Cl$

D

b. List the same esters in order of decreasing reactivity in the last step of a nucleophilic acyl substitution reaction (collapse of the tetrahedral intermediate).

29. Using an alcohol for one method and an alkyl halide for the other, show two ways to make each of the following esters:
a. propyl acetate (odor of pears)
b. ethyl butyrate (odor of pineapple)
c. isopentyl acetate (odor of bananas)
d. methyl phenylethanoate (odor of honey)

30. Which compound would you expect to have a higher boiling point, the ester or the ketone?

$CH_3\overset{\overset{O}{\|}}{C}OCH_3 \qquad CH_3\overset{\overset{O}{\|}}{C}CH_2CH_3$

31. If propionyl chloride is added to one equivalent of methylamine, only a 50% yield of *N*-methylpropanamide is obtained. If, however, the acyl chloride is added to two equivalents of methylamine, the yield of *N*-methylpropanamide is almost 100%. Explain these observations.

32. What reagents would you use to convert methyl propanoate into the following compounds?
a. isopropyl propanoate
b. sodium propanoate
c. *N*-ethylpropanamide
d. propanoic acid

33. Aspartame, the sweetener used in the commercial products NutraSweet and Equal, is 160 times sweeter than sucrose. What products would be obtained if aspartame were hydrolyzed completely in an aqueous solution of HCl?

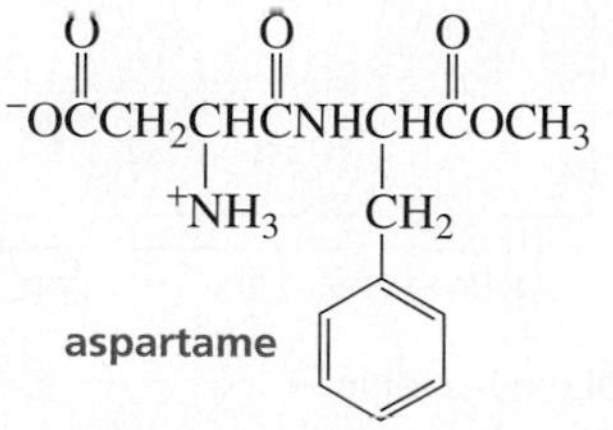

$\overset{-}{O}\overset{\overset{O}{\|}}{C}CH_2\underset{\underset{{}^{+}NH_3}{|}}{CH}\overset{\overset{O}{\|}}{C}NH\underset{\underset{CH_2}{|}}{CH}\overset{\overset{O}{\|}}{C}OCH_3$

aspartame $\bigcirc$

34. a. Which of the following reactions will not give the carbonyl product shown?

1. $CH_3\overset{\overset{O}{\|}}{C}NH_2 + Cl^- \longrightarrow CH_3\overset{\overset{O}{\|}}{C}Cl$

2. $CH_3\overset{\overset{O}{\|}}{C}OH + CH_3NH_2 \longrightarrow CH_3\overset{\overset{O}{\|}}{C}NHCH_3$

3. $CH_3\overset{\overset{\displaystyle O}{\|}}{C}OCH_3 + CH_3NH_2 \longrightarrow CH_3\overset{\overset{\displaystyle O}{\|}}{C}NHCH_3$

5. $CH_3\overset{\overset{\displaystyle O}{\|}}{C}Cl + H_2O \longrightarrow CH_3\overset{\overset{\displaystyle O}{\|}}{C}OH$

4. $CH_3\overset{\overset{\displaystyle O}{\|}}{C}OCH_3 + Cl^- \longrightarrow CH_3\overset{\overset{\displaystyle O}{\|}}{C}Cl$

6. $CH_3\overset{\overset{\displaystyle O}{\|}}{C}NHCH_3 + H_2O \longrightarrow CH_3\overset{\overset{\displaystyle O}{\|}}{C}OH$

b. Which of the reactions that do not occur can be made to occur if an acid catalyst is added to the reaction mixture?

35. Identify the major and minor products of the following reaction:

$$\text{(ring structure with CH}_3\text{, CHOH, CH}_2\text{CH}_3\text{ substituents and NH)} + CH_3\overset{\overset{\displaystyle O}{\|}}{C}Cl \longrightarrow$$

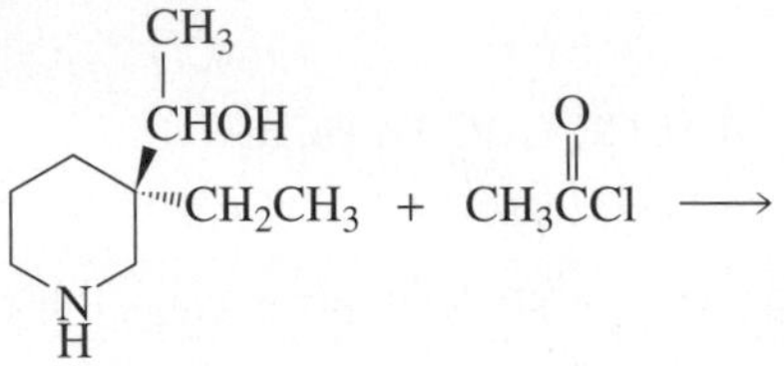

36. D. N. Kursanov, a Russian chemist, studied the hydrolysis of the following ester in a 1.0 M solution of sodium hydroxide, and he was able to prove that the bond that is broken in the reaction is the acyl C—O bond, rather than the alkyl C—O bond:

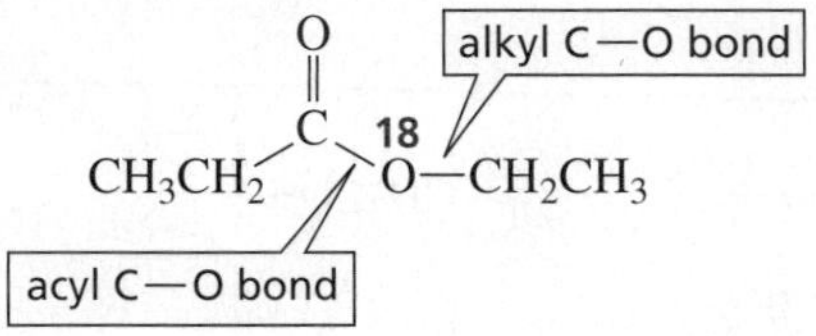

$$CH_3CH_2\overset{\overset{\displaystyle O}{\|}}{C}\overset{18}{O}-CH_2CH_3$$

alkyl C—O bond

acyl C—O bond

a. Which of the products contained the ^{18}O label?
b. What product would have contained the ^{18}O label if the alkyl C—O bond had broken?

37. Give the products of the following reactions:

a.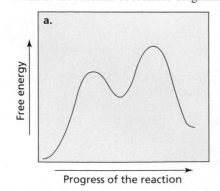
$\xrightarrow[\text{2. 2 CH}_3\text{NH}_2]{\text{1. SOCl}_2}$

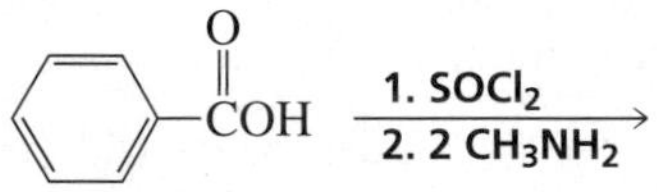

c.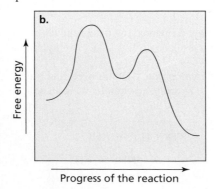
$\xrightarrow[\Delta]{\text{HCl}}$

b. $CH_3\overset{\overset{\displaystyle O}{\|}}{C}Cl + KF \longrightarrow$

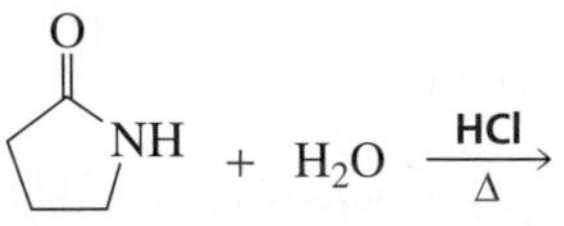

d.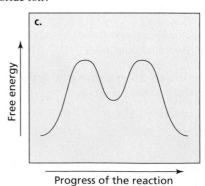
$\xrightarrow{\text{HCl}}$

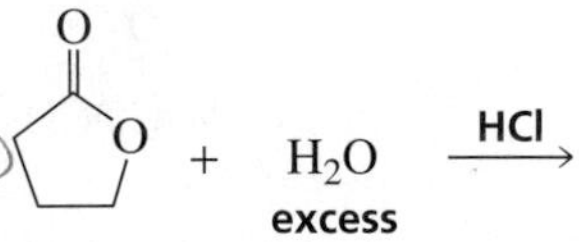

38. Which of the reaction coordinate diagrams represents the reaction of an ester with chloride ion?

a.	b.	c.
Free energy vs Progress of the reaction	Free energy vs Progress of the reaction	Free energy vs Progress of the reaction

39. Which ester is more reactive, methyl acetate or phenyl acetate?

40. List the following amides in order of decreasing reactivity toward acid-catalyzed hydrolysis:

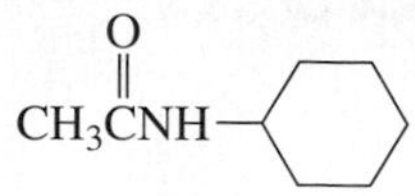

$$\underset{\textbf{A}}{CH_3\overset{\displaystyle O}{\overset{\|}{C}}NH-\!\!\bigcirc} \qquad \underset{\textbf{B}}{CH_3\overset{\displaystyle O}{\overset{\|}{C}}NH-\!\!\overset{NO_2}{\bigcirc}} \qquad \underset{\textbf{C}}{CH_3\overset{\displaystyle O}{\overset{\|}{C}}NH-\!\!\bigcirc-NO_2} \qquad \underset{\textbf{D}}{CH_3\overset{\displaystyle O}{\overset{\|}{C}}NH-\!\!\bigcirc}$$

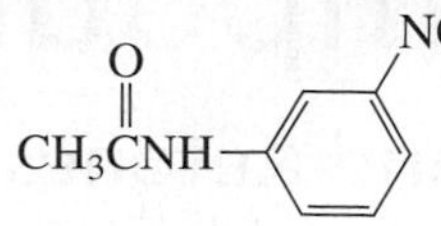

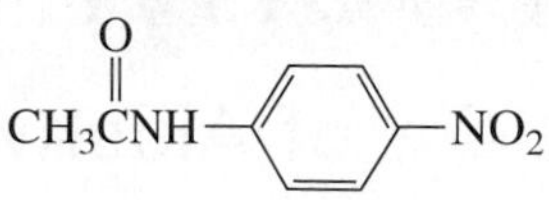

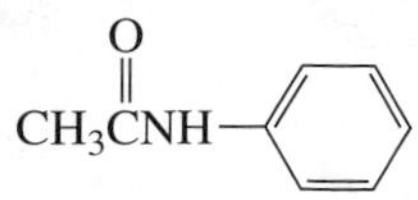

41. Give the products of the following reactions:

a. 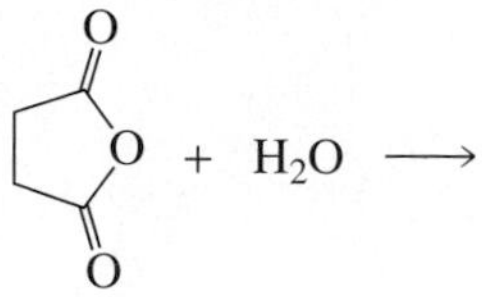 (cyclic anhydride) $+\ H_2O \longrightarrow$

b. 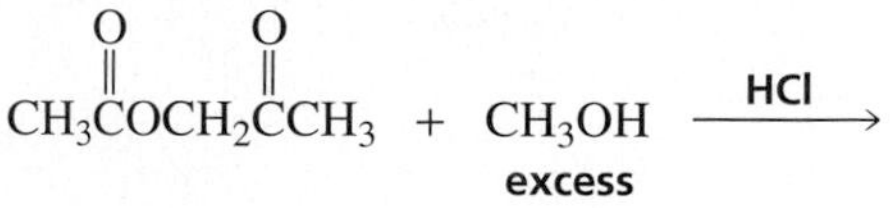 $CH_3\overset{\displaystyle O}{\overset{\|}{C}}OCH_2\overset{\displaystyle O}{\overset{\|}{C}}CH_3\ +\ \underset{\textbf{excess}}{CH_3OH}\ \xrightarrow{\ HCl\ }$

42. An aqueous solution of a primary or secondary amine reacts with an acyl chloride to form an amide as the major product. However, if the amine is tertiary, an amide is not formed. What product *is* formed? Explain.

43. Is the acid-catalyzed hydrolysis of acetamide a reversible or an irreversible reaction? Explain.

44. What product would you expect to obtain from each of the following reactions?

a.  $CH_3CH_2\overset{\displaystyle OH}{\overset{|}{C}}HCH_2CH_2CH_2\overset{\displaystyle O}{\overset{\|}{C}}OH \xrightarrow{\ HCl\ }$

b. 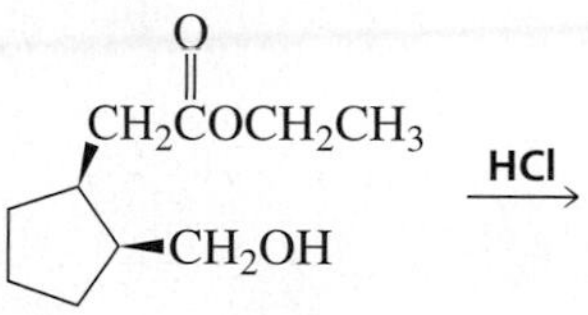 (cyclopentane ring with $CH_2\overset{\displaystyle O}{\overset{\|}{C}}OCH_2CH_3$ and $\text{-}CH_2OH$ substituents) $\xrightarrow{\ HCl\ }$

45. a. When a carboxylic acid is dissolved in isotopically labeled water (H_2O^{18}), the label is incorporated into both oxygens of the acid. Propose a mechanism to account for this.

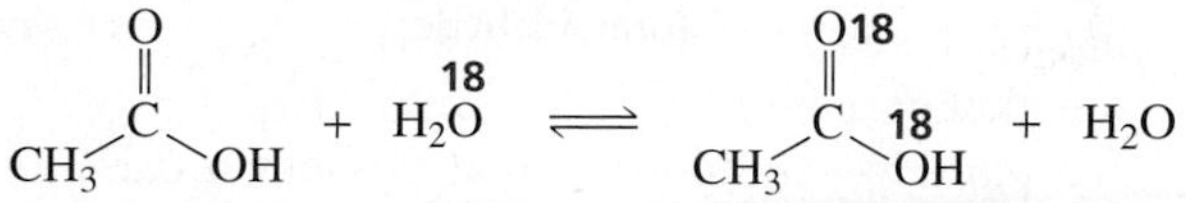

$$CH_3\overset{\displaystyle O}{\overset{\|}{C}}OH\ +\ H_2\overset{18}{O}\ \rightleftharpoons\ CH_3\overset{\displaystyle O18}{\overset{\|}{C}}\overset{18}{O}H\ +\ H_2O$$

b. If a carboxylic acid is dissolved in isotopically labeled methanol ($CH_3{}^{18}OH$) and an acid catalyst is added, where will the label reside in the product?

13 Carbonyl Compounds II

Reactions of Aldehydes and Ketones • More Reactions of Carboxylic Acid Derivatives

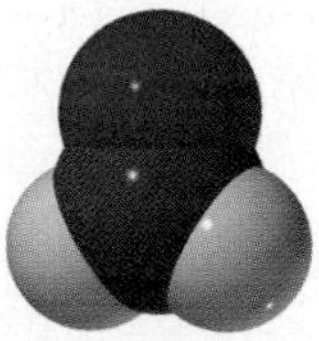

formaldehyde

acetaldehyde

acetone

formaldehyde

acetaldehyde

acetone

In Section 12.0, we saw that carbonyl compounds—compounds that possess a carbonyl group (C=O)—can be divided into two classes: Class I carbonyl compounds, which have a group that can be replaced by a nucleophile, and Class II carbonyl compounds, which do not have a group that can be replaced by a nucleophile. Class II carbonyl compounds comprise aldehydes and ketones.

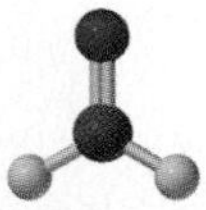

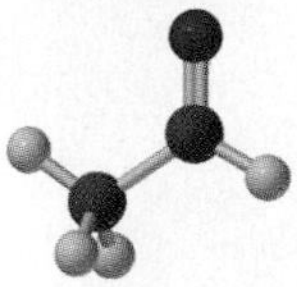

The carbonyl carbon of the simplest aldehyde, formaldehyde, is bonded to two hydrogens. The carbonyl carbon in all other **aldehydes** is bonded to a hydrogen and to an alkyl (or an aryl) group. The carbonyl carbon of a **ketone** is bonded to two alkyl (or aryl) groups. Aldehydes and ketones *do not have* a group that can be replaced by another group, because hydride ions (H⁻) and carbanions (R⁻) and are too basic to be displaced by nucleophiles under normal conditions.

formaldehyde

an aldehyde

$$\underset{\text{a ketone}}{\overset{\displaystyle O}{\underset{R}{\|}}\underset{R'}{C}}$$

a ketone

Aldehydes and ketones with fewer than five carbons are soluble in water. Additional physical properties are discussed in Section 12.3 and listed in Appendix I.

Many aldehydes and ketones are found in nature. Aldehydes have pungent odors, whereas ketones tend to smell sweet. Vanillin and cinnamaldehyde are examples of naturally occurring aldehydes. A whiff of vanilla extract will allow you to appreciate the pungent odor of vanilla. The ketones carvone and camphor are responsible for the characteristic sweet odors of spearmint leaves, caraway seeds, and the leaves of the camphor tree.

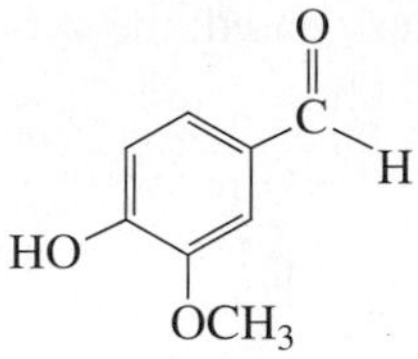

vanillin
vanilla flavoring

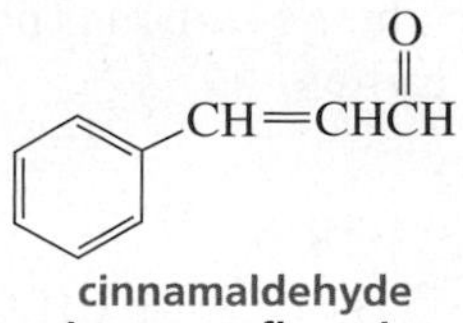

cinnamaldehyde
cinnamon flavoring

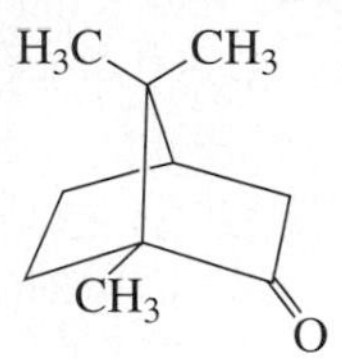

camphor

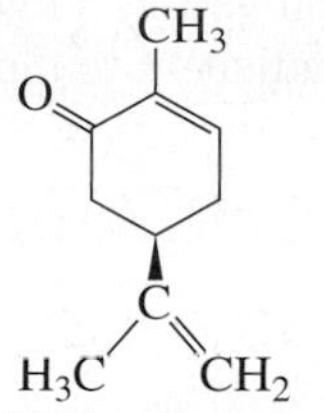

(R)-(–)-carvone
spearmint oil

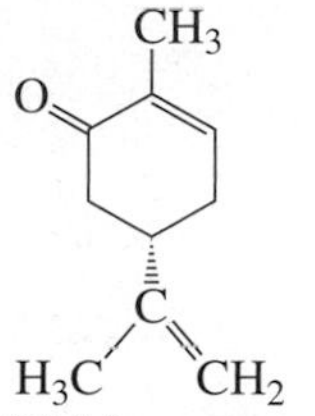

(S)-(+)-carvone
caraway seed oil

In ketosis, a pathological condition that can occur in people with diabetes, the body produces more acetoacetate than can be metabolized. The excess acetoacetate breaks down to acetone (a ketone) and CO_2. Ketosis can be recognized by the smell of acetone on a person's breath.

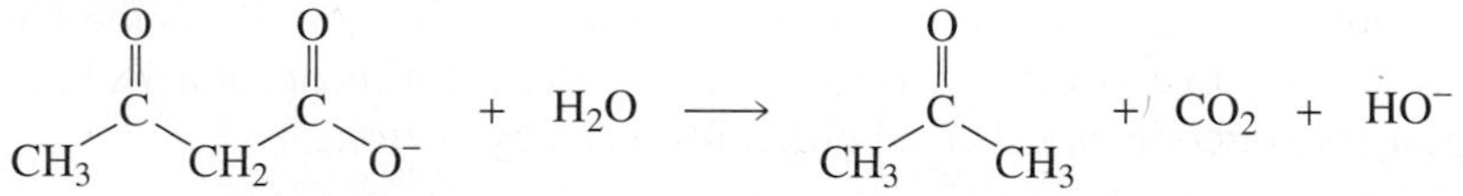

$$CH_3-\overset{O}{\underset{}{C}}-CH_2-\overset{O}{\underset{}{C}}-O^- \ + \ H_2O \ \longrightarrow \ CH_3-\overset{O}{\underset{}{C}}-CH_3 \ + \ CO_2 \ + \ HO^-$$

acetoacetate   **acetone**

3-D Molecules:
Formaldehyde;
Acetaldehyde;
Acetone

Two ketones that are of biological importance illustrate how a small difference in structure can be responsible for a large difference in biological activity: Progesterone is a female sex hormone synthesized primarily in the ovaries, whereas testosterone is a male sex hormone synthesized primarily in the testes.

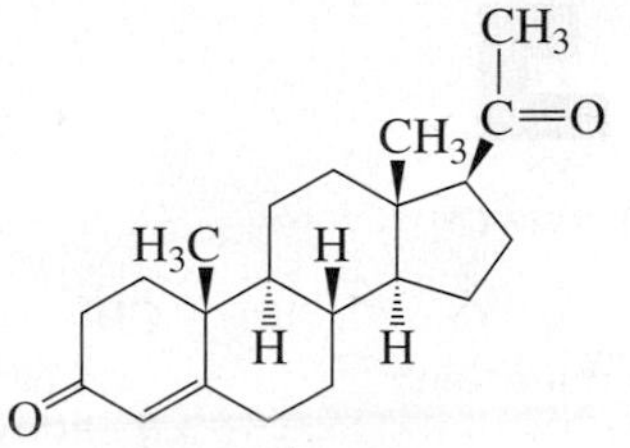

progesterone
a female sex hormone

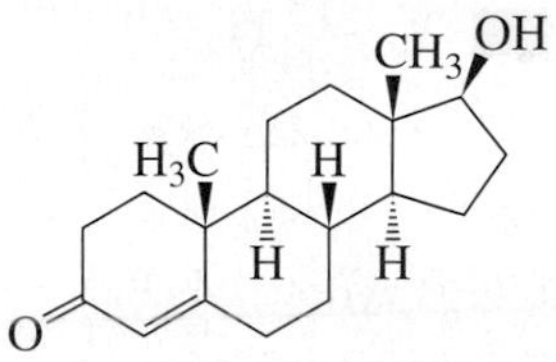

testosterone
a male sex hormone

13.1   Nomenclature

Aldehydes

The systematic name of an aldehyde is obtained by replacing the "e" on the end of the parent hydrocarbon with "al." For example, a one-carbon aldehyde is methan*al*; a two-carbon aldehyde is ethan*al*. The position of the carbonyl carbon does not have to be designated, because it is always at the end of the parent hydrocarbon and therefore always has the 1-position.

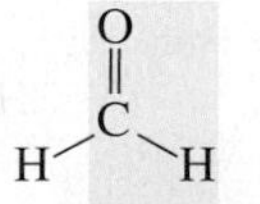

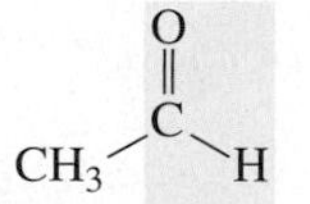

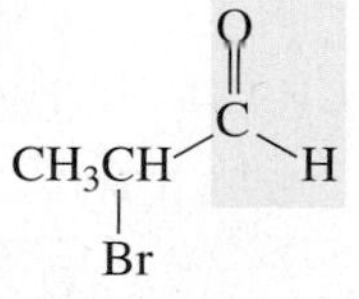

| systematic name: | **methanal** | **ethanal** | |
| common name: | **formaldehyde** | **acetaldehyde** | |

2-bromopropanal
α-bromopropionaldehyde

The common name of an aldehyde is the same as the common name of the corresponding carboxylic acid (Section 12.1), except that "ic acid" (or "oic acid") is replaced by "aldehyde." When common names are used, the position of a substituent is designated by a lowercase Greek letter. The carbonyl carbon is not designated; the carbon adjacent to the carbonyl carbon is the α-carbon.

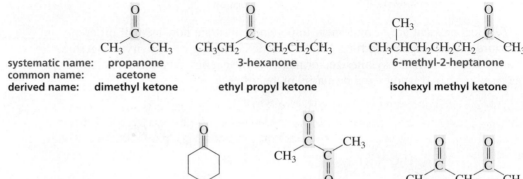

systematic name:	3-chlorobutanal	3-methylbutanal
common name:	β-chlorobutyraldehyde	isovaleraldehyde

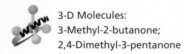
Ketones

The systematic name of a ketone is obtained by removing the "e" from the end of the name of the parent hydrocarbon and adding "one." The chain is numbered in the direction that gives the carbonyl carbon the smaller number. In the case of cyclic ketones, a number is not necessary because the carbonyl carbon is assumed to be at the 1-position. Frequently, derived names are used for ketones—the substituents attached to the carbonyl group are cited in alphabetical order, followed by "ketone."

systematic name:	propanone	3-hexanone	6-methyl-2-heptanone
common name:	acetone		
derived name:	dimethyl ketone	ethyl propyl ketone	isohexyl methyl ketone

systematic name:	cyclohexanone	butanedione	2,4-pentanedione
common name:			acetylacetone

Only a few ketones have common names. The smallest ketone, propanone, is usually referred to by its common name, acetone. Acetone is a common laboratory solvent.

PROBLEM 1◆

Why are numbers not used to designate the positions of the functional groups in propanone and butanedione?

PROBLEM 2◆

Name each of the following compounds:

a. $CH_3CH_2CHCH_2CH$ (with O double bond on terminal CH, CH_3 branch)

d. ⬡—$CH_2CH_2CH_2CH$ (with O double bond)

b. $CH_3CH_2CH_2CCH_2CH_2CH_3$ (with O double bond on central C)

e. $CH_3CH_2CHCH_2CH_2CH$ (with CH_2CH_3 branch, O double bond)

c. $CH_3CHCH_2CCH_2CH_2CH_3$ (with CH_3 branch, O double bond)

f. $CH_3CH_2CCH_2CH_2CHCH_3$ (with O double bond, CH_3 branch)

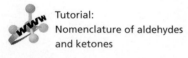

BUTANEDIONE: AN UNPLEASANT COMPOUND

Fresh perspiration is odorless. Bacteria that are always present on our skin produce lactic acid, thereby creating an acidic environment that allows other bacteria to break down the components of perspiration. This forms the compounds with the unappealing odors we associate with armpits and sweaty feet. One such compound is butanedione.

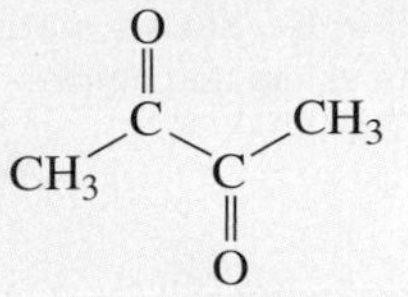

butanedione

13.2 Relative Reactivities of Carbonyl Compounds

We have seen that the carbonyl group is polar because oxygen, being more electronegative than carbon, has a greater share of the double bond's electrons (Section 12.5). The partial positive charge on the carbonyl carbon causes it to be attacked by nucleophiles. The electron deficiency of the carbonyl carbon is indicated by the blue area in the electrostatic potential maps.

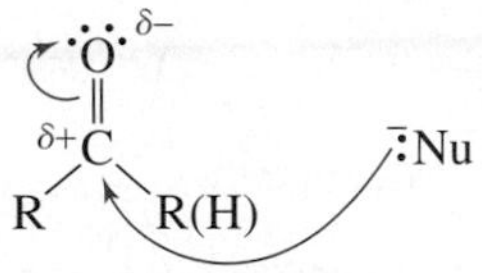

formaldehyde

The partial positive charge on the carbonyl carbon of a ketone is smaller than that on an aldehyde's carbonyl group because, relative to a hydrogen, an alkyl group is electron donating (Section 5.2). An aldehyde, therefore, is more reactive toward nucleophilic attack.

relative reactivities

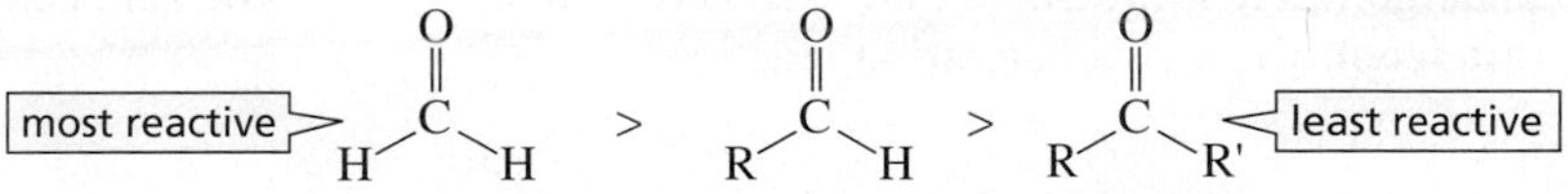

most reactive ▷   formaldehyde > an aldehyde > a ketone   ◁ least reactive

acetaldehyde

Steric factors also contribute to the greater reactivity of an aldehyde. The carbonyl carbon of an aldehyde is more accessible to the nucleophile than is the carbonyl carbon of a ketone, because the hydrogen attached to the carbonyl carbon of an aldehyde is smaller than the second alkyl group attached to the carbonyl carbon of a ketone.

For the same reason, ketones with small alkyl groups bonded to the carbonyl carbon are more reactive than ketones with large alkyl groups.

acetone

relative reactivities

most reactive ▷ H_3C—CO—CH_3 > H_3C—CO—$CHCH_3$ (CH_3) > CH_3CH(CH_3)—CO—$CHCH_3$ (CH_3)   ◁ least reactive

---

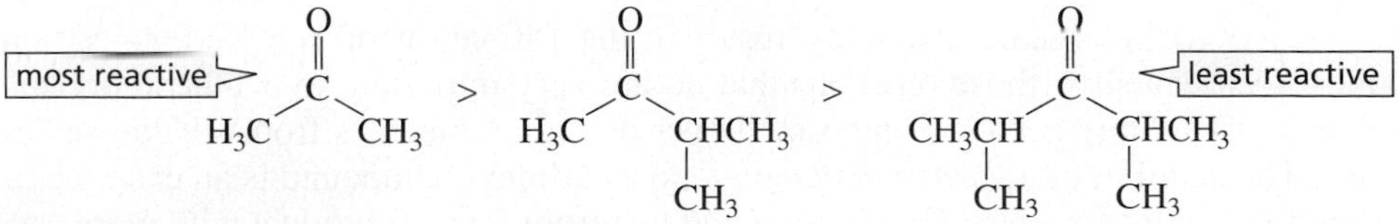

PROBLEM 3◆

Which ketone is more reactive

a. 2-heptanone or 4-heptanone?

b. 4-methyl-3-hexanone or 5-methyl-3-hexanone?

Aldehydes and ketones are less reactive than acyl chlorides, but are more reactive than esters and amides.

How does the reactivity of an aldehyde or a ketone toward nucleophiles compare with the reactivity of the carbonyl compounds whose reactions we looked at in Chapter 12? Aldehydes and ketones are in the middle—they are less reactive than acyl chlorides, but more reactive than esters, carboxylic acids, and amides.

relative reactivities of carbonyl compounds

acyl chloride > aldehyde > ketone > ester ~ carboxylic acid > amide > carboxylate ion

most reactive   least reactive

13.3 How Aldehydes and Ketones React

In Section 12.5, we saw that the carbonyl group of a carboxylic acid or a carboxylic acid derivative is attached to a group that can be replaced by another group. Therefore, these compounds react with nucleophiles to form substitution products.

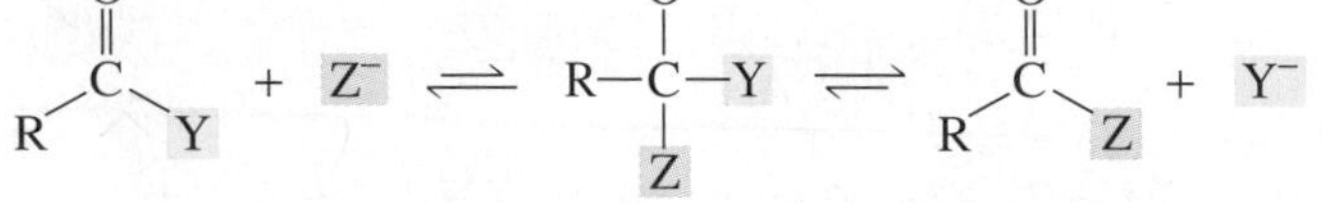

product of
nucleophilic acyl
substitution

Aldehydes and ketones undergo nucleophilic addition reactions.

Carboxylic acid derivatives undergo nucleophilic acyl substitution reactions.

In contrast, the carbonyl group of an aldehyde or a ketone is attached to a group that is too strong a base (H^- or R^-) to be eliminated under normal conditions, so it cannot be replaced by another group. Consequently, aldehydes and ketones react with nucleophiles to form addition products, not substitution products. Thus, aldehydes and ketones undergo **nucleophilic addition** reactions, whereas carboxylic acid derivatives undergo **nucleophilic acyl substitution** reactions.

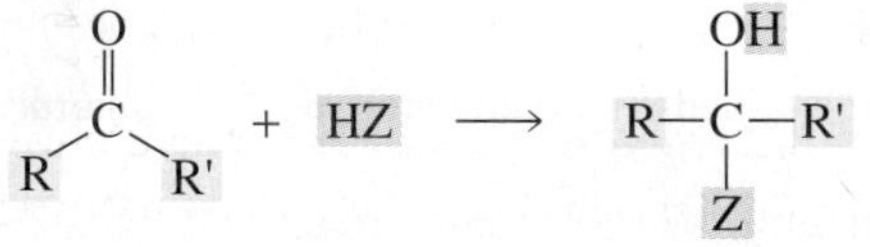

product of
nucleophilic
addition

13.4 Reactions of Carbonyl Compounds with Grignard Reagents

Few reactions in organic chemistry result in the formation of new carbon–carbon bonds. Consequently, those reactions that do are very important to synthetic organic chemists when they need to synthesize larger organic molecules from smaller molecules. The addition of a *carbon nucleophile* to a carbonyl compound is an example of a reaction that forms a new C—C bond and therefore forms a product with more carbon atoms than the starting material.

Grignard reagents are the most widely used carbon nucleophiles. They are prepared by adding an alkyl halide to magnesium shavings being stirred in ether (e.g., diethyl ether). The magnesium is inserted between the carbon and the halogen. Grignard reagents react as if they were carbanions. Recall that a carbanion is a species containing a negatively charged carbon (Section 1.4).

$$CH_3CH_2Br \xrightarrow[\text{Et}_2\text{O}]{\text{Mg}} CH_3CH_2MgBr$$

$$CH_3CH_2MgBr \quad \text{reacts as if it were} \quad CH_3\overset{..}{\overset{-}{C}}H_2 \quad \overset{+}{M}gBr$$

Grignard regents have a nucleophilic carbon because the carbon is bonded to an atom that is *less* electronegative than carbon. (See Table 1.3 on page 8.) The nucleophilic carbon of a Grignard reagent reacts with electrophiles.

less electronegative than carbon

$$\overset{\delta-}{CH_3CH_2}\overset{\delta+}{-M} + E^+ \longrightarrow CH_3CH_2-E + M^+$$

nucleophile   electrophile

Compare this with the carbon in alkyl halides, which is electrophilic because it is attached to an atom that is *more* electronegative than carbon. We have seen that the electrophilic carbon of an alkyl halide reacts with nucleophiles (Section 10.1).

more electronegative than carbon

$$\overset{\delta+}{CH_3CH_2}\overset{\delta-}{-Z} + Y^- \longrightarrow CH_3CH_2-Y + Z^-$$

electrophile   nucleophile

Grignard reagents are such strong bases that they will react immediately with any acid that is present in the reaction mixture—even with trace amounts of very weak acids (any compound with a hydrogen attached to an oxygen or to a nitrogen). When this happens, the organometallic compound is converted into an alkane.

$$\underset{\underset{Br}{|}}{CH_3CH_2CHCH_3} \xrightarrow[\text{Et}_2\text{O}]{\text{Mg}} \underset{\underset{MgBr}{|}}{CH_3CH_2CHCH_3} \xrightarrow{\text{H}_2\text{O}} CH_3CH_2CH_2CH_3$$

Therefore, it is important that only anhydrous reagents are used when organometallic compounds are being synthesized and when they react with other reagents.

Francis August Victor Grignard (1871–1935) *was born in France, the son of a sailmaker. He received a Ph.D. from the University of Lyons in 1901. His synthesis of the first Grignard reagent was announced in 1900. During the next five years, some 200 papers were published about Grignard reagents. He was a professor of chemistry at the University of Nancy and later at the University of Lyons. During World War I, Grignard was drafted into the French army, where he developed a method to detect chemical warfare agents.*

PROBLEM 4◆

Give the products of the following reactions:

a. $CH_3CH_2MgBr + H_2O \longrightarrow$

b. $CH_3CH_2MgBr + CH_3OH \longrightarrow$

c. $CH_3CH_2MgBr + CH_3NH_3 \longrightarrow$

Reactions of Aldehydes and Ketones with Grignard Reagents

Nucleophilic attack of a Grignard reagent on a carbonyl carbon forms an alkoxide ion that is complexed with magnesium ion. If the carbonyl compound is an aldehyde or a ketone, the alkoxide ion is stable because it does not have a group that can be eliminated. (The tetrahedral compound is stable because the newly formed sp^3 carbon is attached to an oxygen and to *no other* electronegative atom.) The reaction forms an addition product—the nucleophile has added to the carbonyl carbon. Addition of water or dilute acid breaks up the complex.

When a Grignard reagent reacts with *formaldehyde*, the addition product is a *primary alcohol*.

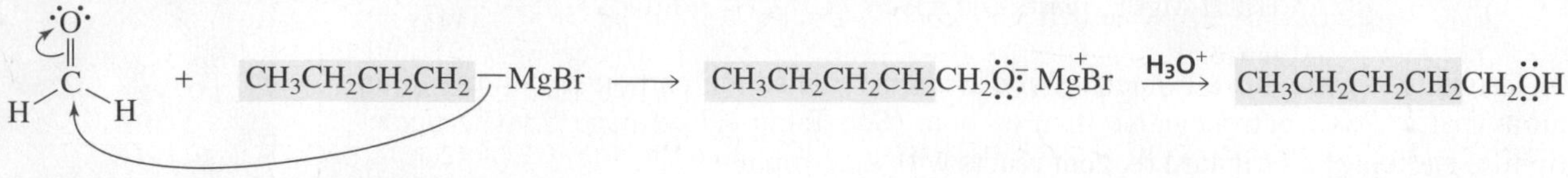

formaldehyde   +   butylmagnesium bromide   ⟶   an alkoxide ion $\xrightarrow{H_3O^+}$ 1–pentanol a primary alcohol

When a Grignard reagent reacts with an *aldehyde other than formaldehyde*, the addition product is a *secondary alcohol*.

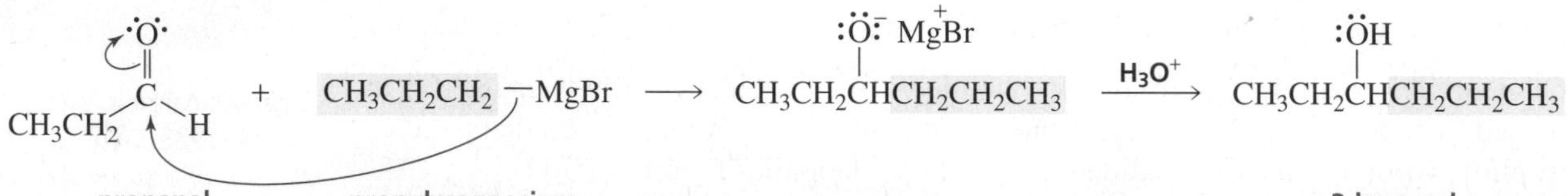

propanal   +   propylmagnesium bromide   ⟶ $\xrightarrow{H_3O^+}$ 3-hexanol a secondary alcohol

When a Grignard reagent reacts with a *ketone*, the addition product is a *tertiary alcohol*.

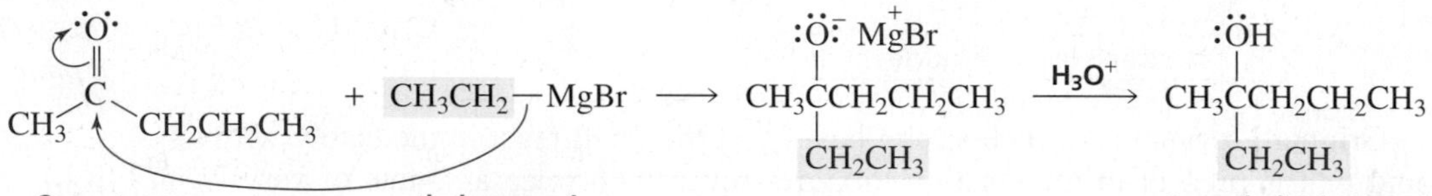

2-pentanone   +   ethylmagnesium bromide   ⟶ $\xrightarrow{H_3O^+}$ 3-methyl-3-hexanol a tertiary alcohol

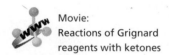

Movie:
Reactions of Grignard reagents with ketones

In the following reactions, numbers are used with the reagents to indicate that the acid is not added until after the Grignard reagent has reacted with the carbonyl compound:

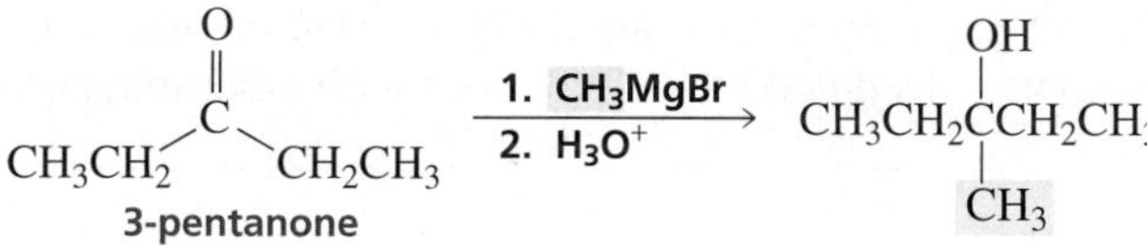

3-pentanone $\xrightarrow[\text{2. } H_3O^+]{\text{1. } CH_3MgBr}$ 3-methyl-3-pentanol

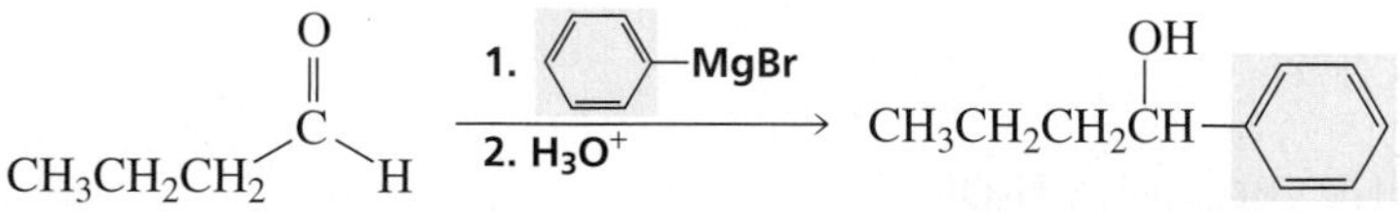

butanal $\xrightarrow[\text{2. } H_3O^+]{\text{1. } \text{—MgBr}}$ 1-phenyl-1-butanol

The reaction of a Grignard reagent with a carbonyl compound can produce compounds with a variety of structures because both the structure of the carbonyl compound and the structure of the Grignard reagent can be varied.

PROBLEM 5◆

What products would be formed when the following compounds react with CH_3MgBr followed by the addition of acid?

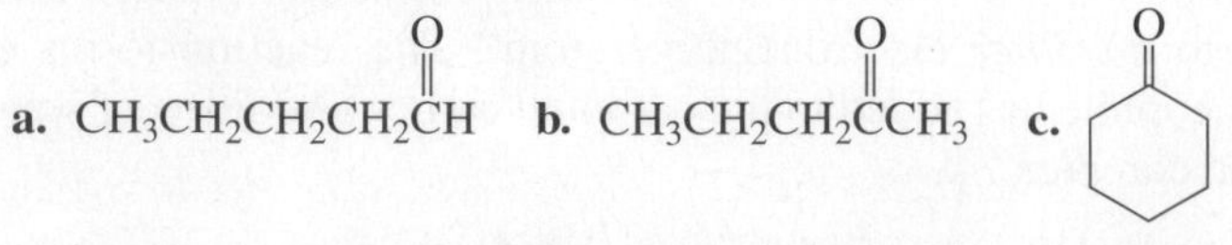

a. $CH_3CH_2CH_2CH_2\overset{O}{\overset{\|}{C}}H$ b. $CH_3CH_2CH_2\overset{O}{\overset{\|}{C}}CH_3$ c.

PROBLEM 6◆

We saw that 3-methyl-3-hexanol can be synthesized from the reaction of 2-pentanone with ethylmagnesium bromide. What two other combinations of ketone and Grignard reagent could be used to prepare the same tertiary alcohol?

Reactions of Esters and Acyl Chlorides with Grignard Reagents

Esters and acyl chlorides (Class I carbonyl compounds) undergo two successive reactions with a Grignard reagent. For example, when an ester reacts with a Grignard reagent, the first reaction is a *nucleophilic acyl substitution reaction* because an ester, unlike an aldehyde or a ketone, has a group that can be replaced by the Grignard reagent. The tetrahedral compound formed in the first step of the reaction is unstable because the newly formed sp^3 carbon is attached to an oxygen and to another electronegative atom; Section 12.5; thus, it is a tetrahedral intermediate. It expels a methoxide ion, forming a ketone. The reaction does not stop at the ketone, because we have just seen that a ketone undergoes a nucleophilic addition reaction with a Grignard reagent to form a tertiary alcohol, and a ketone is more reactive than an ester. Because the tertiary alcohol is formed as a result of two successive reactions with a Grignard reagent, the alcohol has two identical groups bonded to the tertiary carbon.

mechanism for the reaction of an ester with a Grignard reagent

$$CH_3CH_2 \overset{\ddot{O}:}{\underset{OCH_3}{\overset{\|}{C}}} + CH_3{-}MgBr \longrightarrow CH_3CH_2\overset{:\ddot{O}:^{-}\ \overset{+}{MgBr}}{\underset{CH_3}{\overset{|}{C}{-}OCH_3}} \longrightarrow CH_3CH_2 \overset{\ddot{O}:}{\underset{CH_3}{\overset{\|}{C}}} + CH_3O^{-}$$

an ester

a group is expelled from the tetrahedral intermediate

a ketone

product of nucleophilic acyl substitution

$CH_3{-}MgBr$

product of nucleophilic addition

$$CH_3CH_2\overset{:\ddot{O}H}{\underset{CH_3}{\overset{|}{C}CH_3}} \xleftarrow{H_3O^+} CH_3CH_2\overset{:\ddot{O}:^{-}\ \overset{+}{MgBr}}{\underset{CH_3}{\overset{|}{C}CH_3}}$$

a tertiary alcohol

Movie:
Reaction of a Grignard reagent with an ester

Tertiary alcohols are also formed from the reaction of two equivalents of a Grignard reagent with an acyl chloride—the first is a nucleophilic acyl substitution reaction with Cl being replaced by the Grignard reagent, and the second is a nucleophilic addition reaction.

$$CH_3CH_2CH_2 \overset{O}{\underset{Cl}{\overset{\|}{C}}} \xrightarrow[\textbf{2. } \mathbf{H_3O^+}]{\textbf{1. } \mathbf{2\ CH_3CH_2MgBr}} CH_3CH_2CH_2\overset{OH}{\underset{CH_2CH_3}{\overset{|}{C}CH_2CH_3}}$$

butyryl chloride

3-ethyl-3-hexanol

PROBLEM 7   SOLVED

a. Which of the following tertiary alcohols cannot be prepared from the reaction of an ester with excess Grignard reagent?

1. $CH_3\overset{OH}{\underset{CH_3}{\overset{|}{C}CH_3}}$

3. $CH_3CH_2\overset{OH}{\underset{CH_3}{\overset{|}{C}CH_2CH_2CH_3}}$

5. $CH_3\overset{OH}{\underset{CH_2CH_3}{\overset{|}{C}CH_2CH_2CH_2CH_3}}$

2. $CH_3\overset{OH}{\underset{CH_3}{\overset{|}{C}CH_2CH_3}}$

4. $CH_3CH_2\overset{OH}{\underset{CH_3}{\overset{|}{C}CH_2CH_3}}$

6.

b. For those alcohols that can be prepared by the reaction of an ester with excess Grignard reagent, what ester and what Grignard reagent should be used?

Tutorial:
Grignard reagents in synthesis

SOLUTION TO 7a A tertiary alcohol is obtained from the reaction of an ester with two equivalents of a Grignard reagent. Therefore, tertiary alcohols prepared in this way must have two identical substituents on the carbon to which the OH is bonded, because two substituents come from the Grignard reagent. Alcohols (3) and (5) cannot be prepared in this way because they do not have two identical substituents.

SOLUTION TO 7b(2) Methyl propanoate and excess methylmagnesium bromide.

PROBLEM 8◆

Which of the following secondary alcohols can be prepared from the reaction of methyl formate with excess Grignard reagent?

$$CH_3CH_2CHCH_3 \quad CH_3CHCH_3 \quad CH_3CHCH_2CH_2CH_3 \quad CH_3CH_2CHCH_2CH_3$$
$$\underset{OH}{|} \qquad\qquad \underset{OH}{|} \qquad\qquad \underset{OH}{|} \qquad\qquad \underset{OH}{|}$$

PROBLEM-SOLVING STRATEGY

Why does a Grignard reagent not add to the carbonyl carbon of a carboxylic acid?

Because we know that Grignard reagents add to carbonyl carbons, finding that a Grignard reagent does not add to the carbonyl carbon of a carboxylic acid means that it must react more rapidly with another part of the molecule. A carboxylic acid has an acidic proton that reacts rapidly with the Grignard reagent, converting it to an alkane.

Now continue on to Problem 9.

PROBLEM 9◆

Which of the following compounds will not undergo a nucleophilic addition reaction with a Grignard reagent?

SYNTHESIZING ORGANIC COMPOUNDS

Organic chemists synthesize compounds for many reasons: to study their properties, to answer a variety of chemical questions, or because they have useful properties. One reason chemists synthesize natural products is to provide us with greater supplies of these compounds than nature can produce. For example, Taxol—a compound that has been successful in treating ovarian and breast cancer—is extracted from the bark of *Taxus*, the yew tree found in the Pacific Northwest. The supply of natural Taxol is limited because yew trees are uncommon and grow very slowly and stripping the bark kills the tree. Moreover, the bark of one tree provides only one dose of the drug. In addition, *Taxus* forests serve as habitats for the spotted owl, an endangered species, so harvesting the trees would accelerate the owl's demise. Once chemists were successful in determining the structure of Taxol, efforts were undertaken to synthesize it in order to make it more widely available. Several syntheses have been successful.

Taxol

Once a compound has been synthesized, chemists can study its properties to learn how it works; then they can design and synthesize safer or more potent analogs. For example, chemists have found that the anticancer activity of Taxol is substantially reduced if its four ester groups are hydrolyzed. This gives one small clue as to how the molecule functions.

SEMISYNTHETIC DRUGS

Taxol is a difficult molecule to synthesize because of its complicated structure. Chemists have made the synthesis a lot easier by allowing the yew tree to carry out the first part of the synthesis. Chemists extract a precursor of the drug from the needles of the tree, and the precursor is converted to Taxol in the laboratory. Thus, the precursor is isolated from a renewable resource, whereas the drug itself could be obtained only by killing the tree. This is an example of how chemists have learned to synthesize compounds jointly with nature.

13.5   Reactions of Carbonyl Compounds with Hydride Ion

Reactions of Aldehydes and Ketones with Hydride Ion

Addition of a hydride ion to an aldehyde or ketone forms an alkoxide ion. Subsequent protonation by an acid produces an alcohol. The overall reaction adds H_2 to the carbonyl group. Recall that the addition of hydrogen to an organic compound is a **reduction reaction** (Section 5.12).

$$\underset{\underset{R}{}{}}{\overset{\displaystyle O}{\underset{R'}{C}}} + \ :H^- \longrightarrow R-\underset{\underset{H}{|}}{\overset{\overset{O^-}{|}}{C}}-R' \ \overset{H^+}{\rightleftharpoons} \ R-\underset{\underset{H}{|}}{\overset{\overset{OH}{|}}{C}}-R'$$

Aldehydes and ketones are generally reduced using sodium borohydride ($NaBH_4$) as the source of hydride ion. The reduction reaction involves a nucleophilic addition of hydride ion to the aldehyde or ketone. Aldehydes are reduced to primary alcohols, and ketones are reduced to secondary alcohols. Notice that the acid is not added to the reaction mixture until after the hydride ion has reacted with the carbonyl compound.

$$\underset{\substack{\text{butanal} \\ \text{an aldehyde}}}{CH_3CH_2CH_2\overset{\displaystyle O}{\overset{\|}{C}}H} \quad \xrightarrow[\text{2. H}_3\text{O}^+]{\text{1. NaBH}_4} \quad \underset{\substack{\text{1-butanol} \\ \text{a primary alcohol}}}{CH_3CH_2CH_2CH_2OH}$$

$$\underset{\substack{\text{2-pentanone} \\ \text{a ketone}}}{CH_3CH_2CH_2\overset{\displaystyle O}{\overset{\|}{C}}CH_3} \quad \xrightarrow[\text{2. H}_3\text{O}^+]{\text{1. NaBH}_4} \quad \underset{\substack{\text{2-pentanol} \\ \text{a secondary alcohol}}}{CH_3CH_2CH_2\overset{\overset{OH}{|}}{C}HCH_3}$$

PROBLEM 10◆

What alcohols are obtained from the reduction of the following compounds with sodium borohydride?

a. 2-methylpropanal   **b.** cyclohexanone   **c.** benzaldehyde   **d.** methyl phenyl ketone

---

Reactions of Carboxylic Acids and Carboxylic Acid Derivatives with Hydride Ion

Because Class I carbonyl compounds have a group that can be replaced by another group, they undergo two successive reactions with hydride ion, just like they undergo two successive reactions with a Grignard reagent (Section 13.4). For example, the reaction of an acyl chloride with sodium borohydride forms an alcohol.

$$\underset{\substack{\text{butanoyl chloride}}}{CH_3CH_2CH_2\overset{\displaystyle O}{\overset{\|}{C}}Cl} \quad \xrightarrow[\text{2. H}_3\text{O}^+]{\text{1. NaBH}_4} \quad \underset{\substack{\text{1-butanol}}}{CH_3CH_2CH_2CH_2OH}$$

First, the acyl chloride undergoes a nucleophilic acyl substitution reaction because the tetrahedral compound that is formed in the first step is unstable; it expels a chloride ion, forming an aldehyde. The aldehyde then undergoes a nucleophilic addition reaction with a second equivalent of hydride ion, forming an alkoxide ion, which gives a primary alcohol when it is protonated.

mechanism for the reaction of an acyl chloride with hydride ion

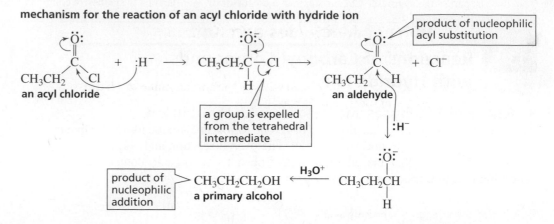

Sodium borohydride ($NaBH_4$) is not a sufficiently strong hydride donor to react with the less reactive (compared with aldehydes and ketones) esters, carboxylic acids, and amides, so these compounds must be reduced with lithium aluminum hydride ($LiAlH_4$), a more reactive hydride donor.

The reaction of an ester with $LiAlH_4$ produces two alcohols, one corresponding to the acyl portion of the ester and one corresponding to the alkyl portion.

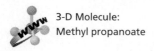
3-D Molecule:
Methyl propanoate

$$CH_3CH_2\overset{\overset{\displaystyle O}{\|}}{C}OCH_3 \xrightarrow[\text{2. } H_3O^+]{\text{1. } LiAlH_4} CH_3CH_2CH_2OH + CH_3OH$$

**methyl propanoate
an ester** **1-propanol** **methanol**

The reaction of a carboxylic acid with $LiAlH_4$ forms a single primary alcohol.

$$CH_3\overset{\overset{\displaystyle O}{\|}}{C}OH \xrightarrow[\text{2. } H_3O^+]{\text{1. } LiAlH_4} CH_3CH_2OH$$

acetic acid **ethanol**

Amides also undergo two successive additions of hydride ion when they react with $LiAlH_4$. Overall, the reaction converts a carbonyl group into a CH_2 group, so the product of the reaction is an amine. Primary, secondary, or tertiary amines can be formed, depending on the number of substituents bonded to the nitrogen of the amide. (Notice that H_2O rather than H_3O^+ is used in the second step of the reaction. If H_3O^+ is used, the product will be a protonated amine.)

$$\text{benzamide} \xrightarrow[\text{2. } H_2O]{\text{1. } LiAlH_4} \text{benzylamine}$$

benzamide **benzylamine
a primary amine**

$$CH_3\overset{\overset{\displaystyle O}{\|}}{C}NHCH_3 \xrightarrow[\text{2. } H_2O]{\text{1. } LiAlH_4} CH_3CH_2NHCH_3$$

***N*-methylacetamide** **ethylmethylamine
a secondary amine**

PROBLEM 11◆

What amides would you treat with $LiAlH_4$ in order to prepare the following amines?

a. benzylmethylamine   **b.** ethylamine   **c.** diethylamine   **d.** triethylamine

---

13.6 Reactions of Aldehydes and Ketones with Amines

An aldehyde or a ketone reacts with a *primary amine* to form an imine. An **imine** is a compound with a carbon–nitrogen double bond.

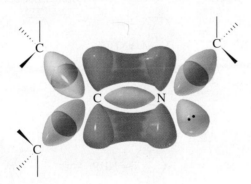

an aldehyde or   a primary amine   an imine
a ketone   a Schiff base

A $C=N$ group (Figure 13.1) is similar to a $C=O$ group (Figure 12.1 on page 292). The imine nitrogen is sp^2 hybridized. One of its sp^2 orbitals forms a σ bond with the imine carbon, one forms a σ bond with a substituent, and the third contains a lone pair. The p orbital of nitrogen and the p orbital of carbon overlap to form a π bond.

◀ **Figure 13.1**
Bonding in an imine.

An aldehyde or a ketone reacts with a *secondary amine* to form an enamine (pronounced "ENE-amine"). An **enamine** is a tertiary amine with a double bond in the α,β-position relative to the nitrogen atom. Notice that the double bond is in the part of the molecule that comes from the aldehyde or ketone. The name "enamine" comes from "ene" + "amine," with the "e" omitted in order to avoid two successive vowels.

an aldehyde or   a secondary amine   an enamine
a ketone

When you first look at the products of imine and enamine formation, they appear to be quite different. However, when you look at the mechanisms for the reactions, you will see that the mechanisms are exactly the same except for the site from which a proton is lost in the last step of the reaction.

Addition of a Primary Amine

Aldehydes and ketones react with primary amines to form imines.

Aldehyde and ketones react with primary amines to form imines. The reaction requires a trace amount of acid.

3-D Molecules:
The *N*-methylimine of *acetone*.

(reaction: benzaldehyde an aldehyde + CH₃CH₂NH₂ ethylamine a primary amine ⇌ [trace H⁺] an imine C=NCH₂CH₃ + H₂O)

(reaction: 3-pentanone a ketone + H₂NCH₂—⬡ benzylamine a primary amine ⇌ [trace H⁺] an imine + H₂O)

Tutorial:
Imine formation

In the first step of the mechanism, the amine attacks the carbonyl carbon. Gain of a proton by the alkoxide ion and loss of a proton by the ammonium ion forms a neutral tetrahedral intermediate. The neutral tetrahedral intermediate, called a *carbinolamine*, is in equilibrium with two protonated forms because either the nitrogen or the oxygen atom can be protonated. Water is expelled from the oxygen-protonated intermediate, forming a protonated imine that loses a proton to yield the imine.

mechanism for imine formation

nucleophile attacks the carbonyl carbon

N–protonated carbinolamine

proton dissociation

neutral tetrahedral intermediate a carbinolamine

proton dissociation

O–protonated carbinolamine

water is expelled

an imine   *a protonated imine*

Unlike the stable tetrahedral compounds that are formed when a Grignard reagent or a hydride ion adds to an aldehyde or a ketone, the tetrahedral compound formed when an amine adds to an aldehyde or ketone is unstable because the newly formed sp^3 carbon is bonded to an oxygen and to another electronegative atom (a nitrogen).

A compound with an *sp*³ carbon bonded to an oxygen and to another electronegative atom is unstable.

$$R-\underset{\underset{H}{|}}{\overset{\overset{OH}{|}}{C}}-R \quad R-\underset{\underset{CH_3}{|}}{\overset{\overset{OH}{|}}{C}}-R$$

stable tetrahedral compounds

$$R-\underset{\underset{NHCH_3}{|}}{\overset{\overset{OH}{|}}{C}}-R \quad R-\underset{\underset{OCH_3}{|}}{\overset{\overset{OH}{|}}{C}}-R$$

unstable tetrahedral compounds

Imine formation is reversible: In an acidic aqueous solution, an imine is hydrolyzed back to the carbonyl compound and amine. Notice that the amine is protonated because the solution is acidic.

An imine undergoes acid-catalyzed hydrolysis to form a carbonyl compound and a primary amine.

(reaction: ⬡—CH=NCH₂CH₃ + H₂O →[HCl] ⬡—CH=O + CH₃CH₂N⁺H₃)

Imine formation and hydrolysis are important reactions in biological systems. For example, we will see that imine hydrolysis is the reason DNA contains A, G, C, and T nucleotides, whereas RNA contains A, G, C, and U nucleotides (Section 21.9).

Addition of a Secondary Amine

Aldehydes and ketones react with secondary amines to form enamines. Like imine formation, the reaction requires a trace amount of an acid catalyst.

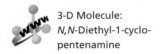

The mechanism for enamine formation is exactly the same as that for imine formation, until the last step of the reaction. When a primary amine reacts with an aldehyde or a ketone, the protonated imine loses a proton from nitrogen in the last step of the reaction, forming a neutral imine. However, when the amine is secondary, the positively charged nitrogen is not bonded to a hydrogen. A stable neutral molecule is obtained by removing a proton from the α-carbon of the compound derived from the carbonyl compound. An enamine is the result.

3-D Molecule:
N,N-Diethyl-1-cyclopentenamine

mechanism for enamine formation

In an aqueous acidic solution, an enamine is hydrolyzed back to the carbonyl compound and secondary amine, a reaction that is similar to the acid-catalyzed hydrolysis of an imine back to the carbonyl compound and primary amine. Again, the amine is protonated because the solution is acidic.

> **PROBLEM 12◆**
>
> Give the products of the following reactions. (A trace amount of acid is present in each reaction.)
>
> **a.** cyclopentanone + ethylamine   **c.** 3-pentanone + hexylamine
> **b.** cyclopentanone + diethylamine   **d.** 3-pentanone + cyclohexylamine

---

13.7   Reactions of Aldehydes and Ketones with Oxygen Nucleophiles

Addition of Water

Most hydrates are too unstable to be isolated.

The addition of water to an aldehyde or a ketone forms a *hydrate*. A **hydrate** is a molecule with two OH groups on the same carbon. Hydrates of aldehydes or ketones are generally too unstable to be isolated because the tetrahedral carbon is attached to two electron withdrawing (oxygen) atoms.

3-D Molecules:
Acetone; Acetone hydrate

$$
\underset{\substack{\text{an aldehyde or} \\ \text{a ketone}}}{\overset{\displaystyle O}{\underset{\displaystyle R}{\overset{\|}{C}}{}_{R\,(H)}}} \; + \; H_2O \; \rightleftharpoons \; \underset{\text{a hydrate}}{R-\overset{\displaystyle OH}{\underset{\displaystyle OH}{\overset{|}{C}}}-R\,(H)}
$$

Water is a poor nucleophile and, therefore, adds relatively slowly to a carbonyl group. The rate of the reaction can be increased by an acid catalyst because the acid makes the carbonyl carbon more susceptible to nucleophilic attack (Figure 13.2). Keep in mind that a catalyst affects the *rate* at which an aldehyde or a ketone is converted to a hydrate; it has no effect on the *amount* of aldehyde or ketone converted to hydrate.

mechanism for acid-catalyzed hydrate formation

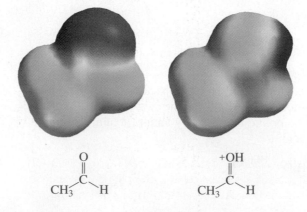

    the acid protonates the carbonyl oxygen   the nucleophile attacks the carbonyl carbon   proton dissociation

The extent to which an aldehyde or a ketone is hydrated in an aqueous solution depends on the substituents attached to the carbonyl compound. For example, only 0.2% of acetone is hydrated at equilibrium, but 99.9% of formaldehyde is hydrated. Bulky substituents and electron-donating substituents (i.e., the methyl groups of acetone) *decrease* the percentage of hydrate present at equilibrium, whereas small substituents and electron-withdrawing substituents (the hydrogens of formaldehyde) *increase* it.

Figure 13.2 ▶
The electrostatic potential maps show that the carbonyl carbon of the protonated aldehyde is more susceptible to nucleophilic attack (the blue is more intense) than the carbonyl carbon of the unprotonated aldehyde.

PRESERVING BIOLOGICAL SPECIMENS

A 37% solution of formaldehyde in water is known as *formalin*—commonly used in the past to preserve biological specimens. Because formaldehyde is an eye and skin irritant, it has been replaced in most biology laboratories by other preservatives. One preservative frequently used is a solution of 2 to 5% phenol in ethanol with added antimicrobial agents.

PROBLEM 13◆

When trichloroacetaldehyde is dissolved in water, almost all of it is converted to the hydrate. Chloral hydrate is a sedative that can be lethal. A cocktail laced with it is commonly known—in detective novels, at least—as a "Mickey Finn." Explain why an aqueous solution of trichloroacetaldehyde is almost all hydrate.

$$Cl_3C-\overset{\displaystyle O}{\overset{\|}{C}}-H \ + \ H_2O \ \longrightarrow \ Cl_3C-\overset{\displaystyle OH}{\underset{\displaystyle OH}{\overset{|}{\underset{|}{C}}}}-H$$

trichloroacetaldehyde　　　　　　　　**chloral hydrate**

PROBLEM 14◆

Which of the following ketones forms the most hydrate in an aqueous solution?

$$CH_3O-\!\!\!\!\bigcirc\!\!\!\!-\overset{\displaystyle O}{\overset{\|}{C}}-\!\!\!\!\bigcirc\!\!\!\!-OCH_3 \qquad \bigcirc\!\!\!\!-\overset{\displaystyle O}{\overset{\|}{C}}-\!\!\!\!\bigcirc \qquad O_2N-\!\!\!\!\bigcirc\!\!\!\!-\overset{\displaystyle O}{\overset{\|}{C}}-\!\!\!\!\bigcirc\!\!\!\!-NO_2$$

Addition of Alcohol

The product formed when one equivalent of an alcohol adds to an *aldehyde* is called a **hemiacetal**. The product formed when a second equivalent of alcohol is added is called an **acetal**. Like water, an alcohol is a poor nucleophile, so an acid catalyst is required for the reaction to take place at a reasonable rate.

$$\underset{\textbf{an aldehyde}}{\overset{\displaystyle O}{\overset{\|}{\underset{CH_3}{C}}\!\!\diagdown_{H}}} \ + \ CH_3OH \ \overset{\textbf{HCl}}{\rightleftharpoons} \ \underset{\textbf{a hemiacetal}}{CH_3-\overset{\displaystyle OH}{\underset{\displaystyle OCH_3}{\overset{|}{\underset{|}{C}}}}-H} \ \overset{\textbf{CH}_3\textbf{OH, HCl}}{\rightleftharpoons} \ \underset{\textbf{an acetal}}{CH_3-\overset{\displaystyle OCH_3}{\underset{\displaystyle OCH_3}{\overset{|}{\underset{|}{C}}}}-H} \ + \ H_2O$$

When the carbonyl compound is a *ketone* instead of an aldehyde, the addition products are called a **hemiketal** and a **ketal**, respectively.

$$\underset{\textbf{a ketone}}{\overset{\displaystyle O}{\overset{\|}{\underset{CH_3}{C}}\!\!\diagdown_{CH_3}}} \ + \ CH_3OH \ \overset{\textbf{HCl}}{\rightleftharpoons} \ \underset{\textbf{a hemiketal}}{CH_3-\overset{\displaystyle OH}{\underset{\displaystyle OCH_3}{\overset{|}{\underset{|}{C}}}}-CH_3} \ \overset{\textbf{CH}_3\textbf{OH, HCl}}{\rightleftharpoons} \ \underset{\textbf{a ketal}}{CH_3-\overset{\displaystyle OCH_3}{\underset{\displaystyle OCH_3}{\overset{|}{\underset{|}{C}}}}-CH_3} \ + \ H_2O$$

Hemi is the Greek word for "half." When one equivalent of alcohol has added to an aldehyde or a ketone, the compound is halfway to the final acetal or ketal, which contains groups from two equivalents of alcohol.

In the first step of acetal (or ketal) formation, the acid protonates the carbonyl oxygen, making the carbonyl carbon more susceptible to nucleophilic attack (Figure 13.2). Loss of a proton from the protonated tetrahedral intermediate gives the hemiacetal (or hemiketal). Because the reaction is carried out in an acidic solution, the hemiacetal

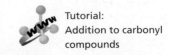

Tutorial:
Addition to carbonyl compounds

(or hemiketal) is in equilibrium with its protonated form. The two oxygen atoms of the hemiacetal (or hemiketal) are equally basic, so either one can be protonated. Loss of water from the tetrahedral intermediate with a protonated OH group forms an O-alkylated intermediate that is very reactive because of its positively charged oxygen atom. Nucleophilic attack on this compound by a second molecule of alcohol, followed by loss of a proton, forms the acetal (or ketal).

mechanism for acid-catalyzed acetal or ketal formation

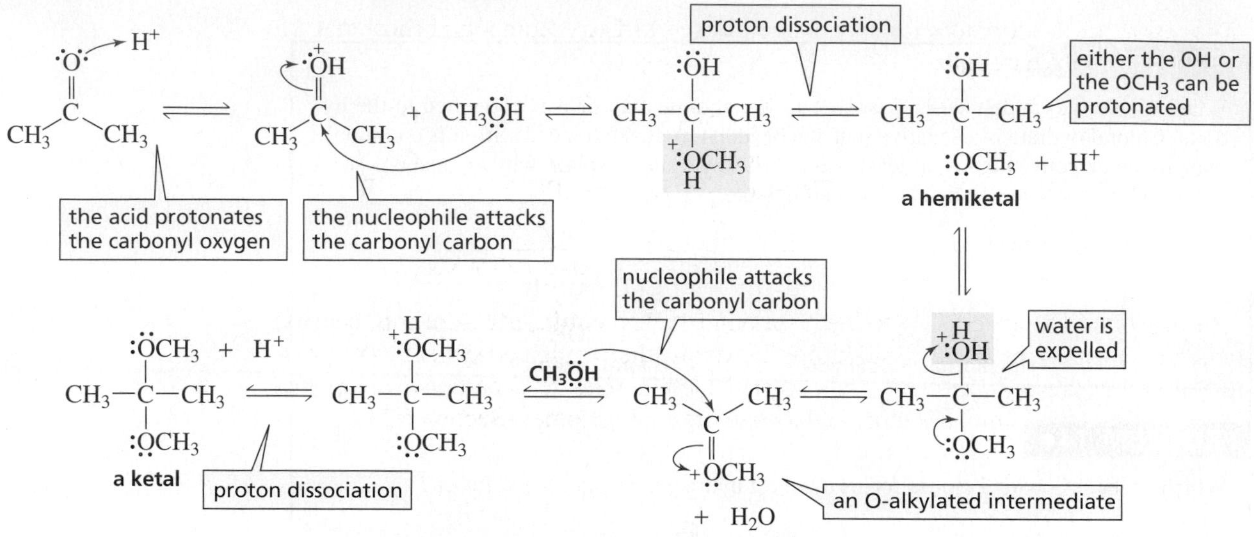

Although the tetrahedral carbon of an acetal or ketal is bonded to two oxygen atoms, causing us to predict that the acetal or ketal is not stable, the acetal or ketal can be isolated if the water eliminated from the hemiacetal (or hemiketal) is removed from the reaction mixture. This is because, if water is not available, the only compound the acetal or ketal can form is the O-alkylated species, which is less stable than the acetal or ketal.

The acetal or ketal can be hydrolyzed back to the aldehyde or ketone in an acidic aqueous solution.

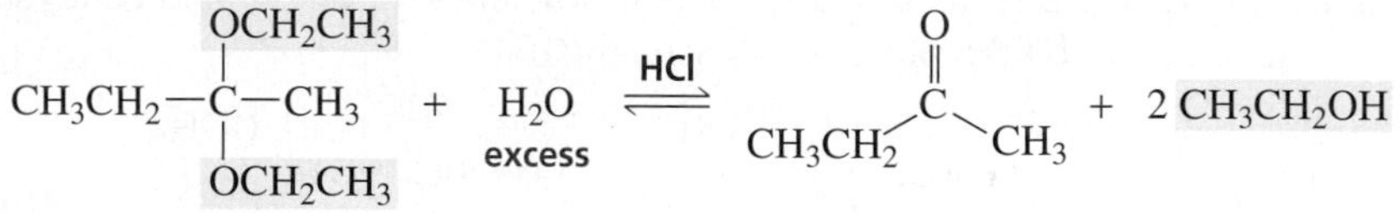

Tutorial:
Common terms, the addition of nucleophiles to carbonyl compounds

PROBLEM 15◆

Which of the following are

a. hemiacetals   **c.** hemiketals   **e.** hydrates

b. acetals   **d.** ketals

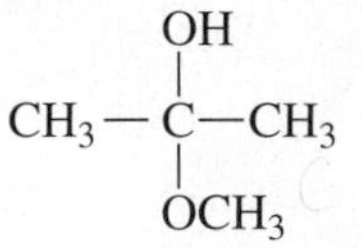

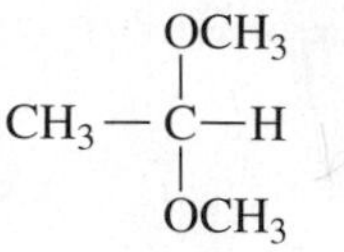

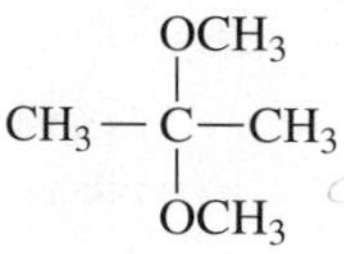

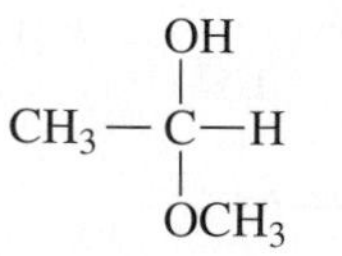

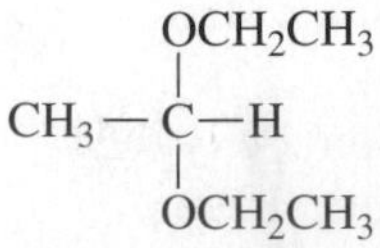

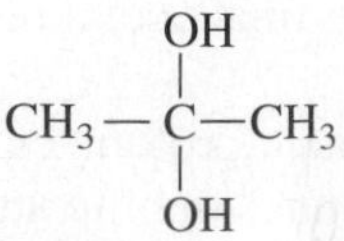

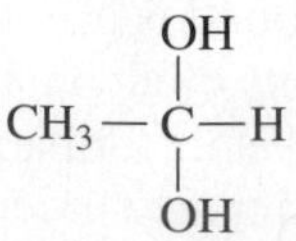

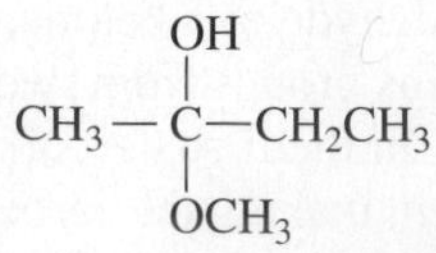

PROBLEM 16

Show the mechanism for the acid-catalyzed hydrolysis of an acetal.

13.8   Nucleophilic Addition to α,β-Unsaturated Carbonyl Compounds

The resonance contributors for an α,β-unsaturated carbonyl compound show that the β-carbon is an electrophile.

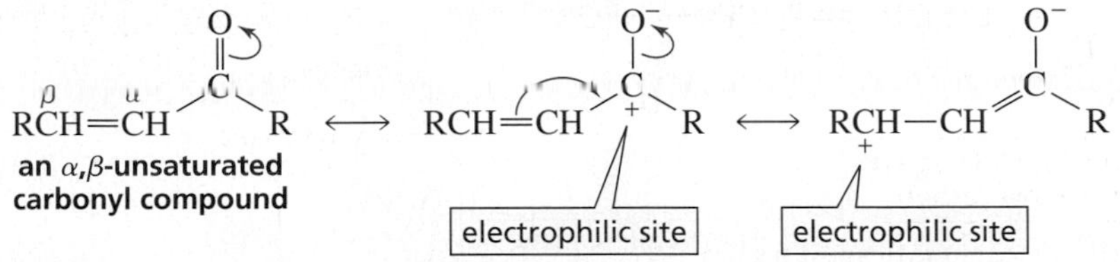

The addition of a nucleophile to the β-carbon is called **conjugate addition**, because addition occurs at the 1- and 4-positions (i.e., across the conjugated system). The product of 1,4-addition is an enol, and the enol tautomerizes to a carbonyl compound because the keto tautomer is more stable than the enol tautomer (Section 5.11).

conjugate addition

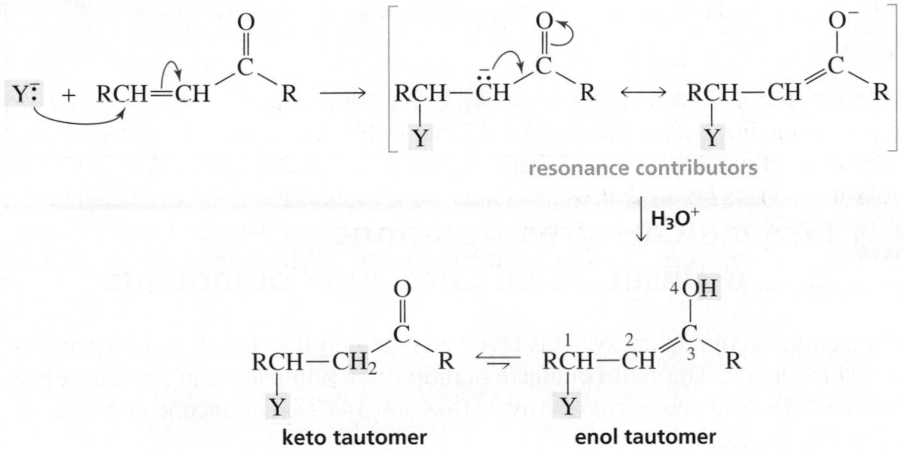

resonance contributors

keto tautomer   enol tautomer

Thus, the overall reaction is addition of the nucleophile to the β-carbon of the double bond and addition of a proton from the reaction mixture to the α-carbon.

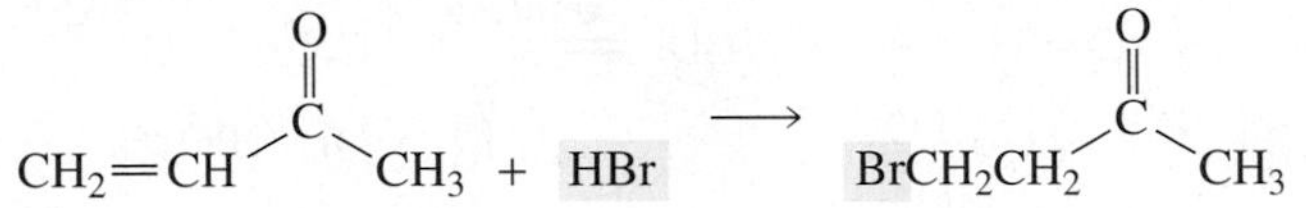

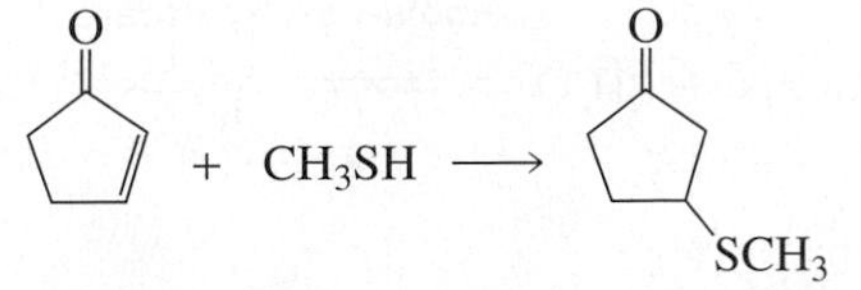

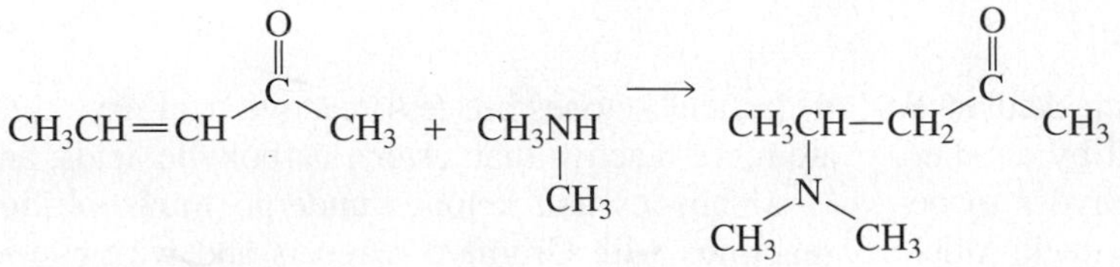

ANTICANCER DRUGS

Two compounds—vernolepin and helenalin—owe their effectiveness as anticancer drugs to conjugate addition reactions.

Cancer cells are cells that have lost their ability to control their growth; therefore, they proliferate rapidly. DNA polymerase is an enzyme that a cell needs to make a copy of its DNA for a new cell. Each of these anticancer drugs has two α,β-unsaturated carbonyl groups. When an SH group of DNA polymerase reacts with one of the α,β-unsaturated carbonyl groups, the enzyme is inactivated.

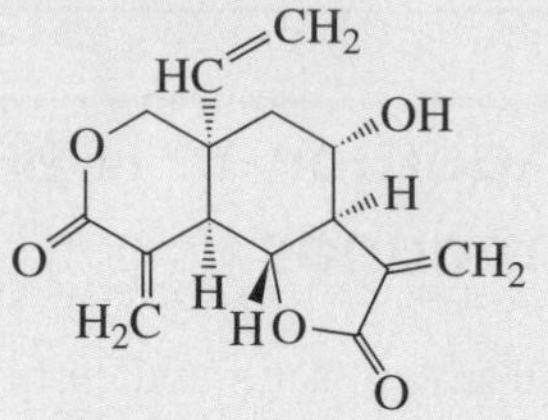

vernolepin

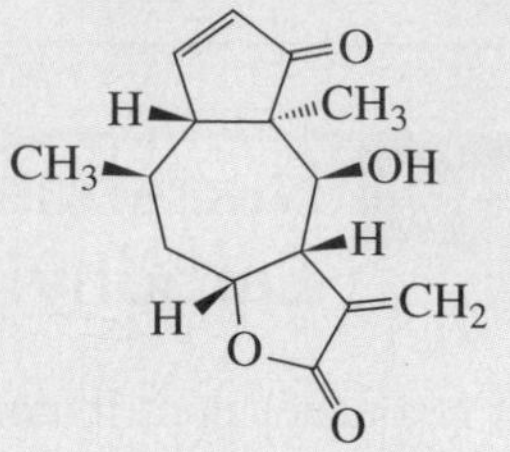

helenalin

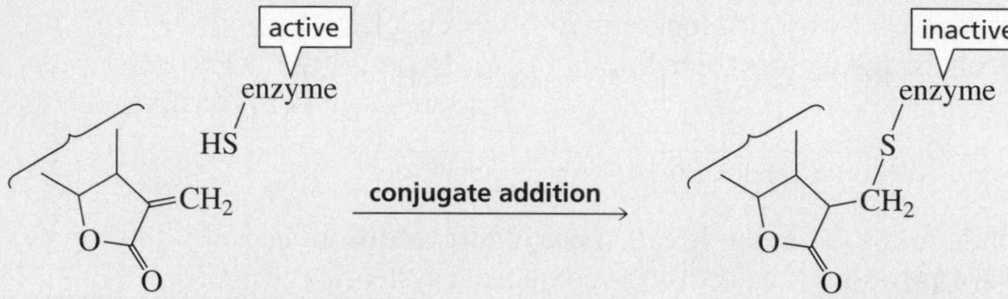

PROBLEM 17◆

Give the major product of each of the following reactions:

a. 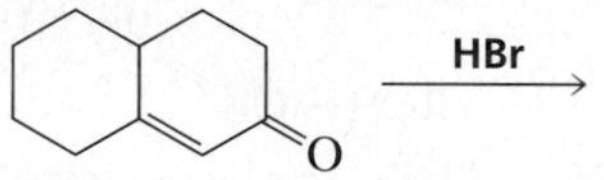 ⟶ HBr

b. 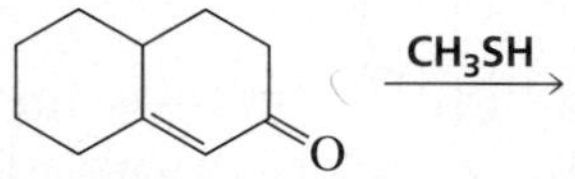 ⟶ CH$_3$SH

13.9 Enzyme-Catalyzed Additions to α,β-Unsaturated Carbonyl Compounds

Several reactions in biological systems involve conjugate addition to α,β-unsaturated carbonyl compounds. The first conjugate addition reaction occurs in gluconeogenesis—the synthesis of glucose from pyruvate (Section 14.11). The second occurs in the oxidation of fatty acids (Section 19.3).

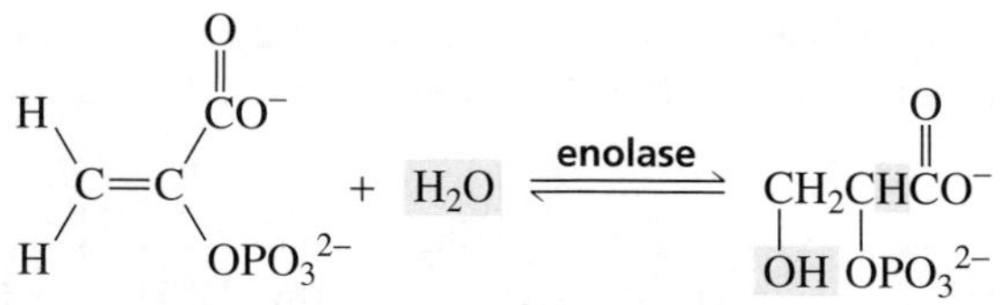

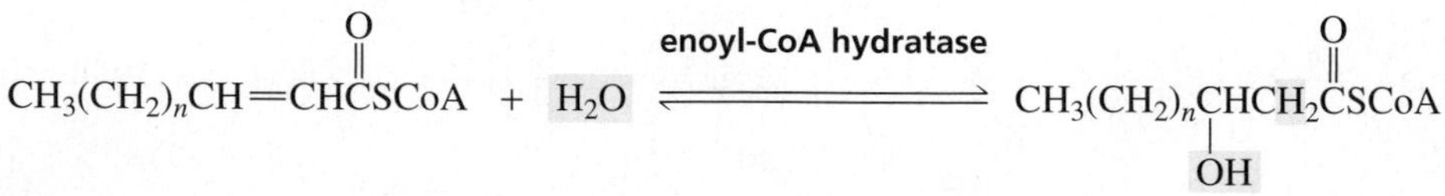

Summary

Aldehydes and **ketones** have an acyl group attached to a group (H or R) that cannot be readily replaced by another group. Steric and electronic factors cause aldehydes to be more reactive than ketones toward nucleophilic attack. Alde-

hydes and ketones are less reactive than acyl chlorides and are more reactive than esters, carboxylic acids, and amides.

Aldehydes and ketones undergo **nucleophilic addition reactions** with Grignard reagents and with hydride ion. In contrast, esters and acyl chlorides undergo **nucleophilic acyl substitution** reactions with Grignard reagents and with hydride ion that form an aldehyde or a ketone, which then undergoes a **nucleophilic addition** reaction with a second equivalent of the nucleophile. Notice that the tetrahedral intermediate formed by attack of a nucleophile on a carbonyl compound is stable if the newly formed tetrahedral carbon is not bonded to a second electronegative atom or group and is generally unstable if it is.

Grignard reagents react with aldehydes to form secondary alcohols, and with ketones, esters, and acyl chlorides to form tertiary alcohols. Aldehydes, acyl chlorides, and carboxylic acids are reduced to primary alcohols by hydride ion, ketones are reduced to secondary alcohols, and amides are reduced to amines.

Aldehydes and ketones react with primary amines to form **imines** and with secondary amines to form **enamines**. The mechanisms are the same, except for the site from which a proton is lost in the last step of the reaction. Imine and enamine formation are reversible; imines and enamines are hydrolyzed under acidic conditions back to the carbonyl compound and amine.

Aldehydes and ketones undergo acid-catalyzed addition of water to form hydrates. Most hydrates are too unstable to be isolated. Acid-catalyzed addition of alcohol to aldehydes forms **hemiacetals** and **acetals**, and to ketones forms **hemiketals** and **ketals**. Acetal and ketal formation are reversible.

Nucleophilic addition to the β-carbon of an α,β-unsaturated carbonyl compound is called **conjugate addition**.

Summary of Reactions

1. Reaction of *carbonyl compounds* with a Grignard reagent (Section 13.4).
   a. Reaction of *formaldehyde* with a Grignard reagent forms a primary alcohol:

$$
\underset{H}{\overset{O}{\underset{\Vert}{H-C}}} \xrightarrow[\text{2. H}_3\text{O}^+]{\text{1. CH}_3\text{MgBr}} CH_3CH_2OH
$$

   b. Reaction of an *aldehyde* (other than formaldehyde) with a Grignard reagent forms a secondary alcohol:

$$
\underset{R}{\overset{O}{\underset{\Vert}{C}}}\!\!{}_H \xrightarrow[\text{2. H}_3\text{O}^+]{\text{1. CH}_3\text{MgBr}} R-\overset{\text{OH}}{\underset{\text{CH}_3}{C}}-H
$$

   c. Reaction of a *ketone* with a Grignard reagent forms a tertiary alcohol:

$$
\underset{R}{\overset{O}{\underset{\Vert}{C}}}\!\!{}_{R'} \xrightarrow[\text{2. H}_3\text{O}^+]{\text{1. CH}_3\text{MgBr}} R-\overset{\text{OH}}{\underset{\text{CH}_3}{C}}-R'
$$

   d. Reaction of an *ester* with a Grignard reagent forms a tertiary alcohol with two identical substituents:

$$
\underset{R}{\overset{O}{\underset{\Vert}{C}}}\!\!{}_{OR'} \xrightarrow[\text{2. H}_3\text{O}^+]{\text{1. 2 CH}_3\text{MgBr}} R-\overset{\text{OH}}{\underset{\text{CH}_3}{C}}-CH_3
$$

   e. Reaction of an *acyl chloride* with a Grignard reagent forms a tertiary alcohol with two identical substituents:

$$
\underset{R}{\overset{O}{\underset{\Vert}{C}}}\!\!{}_{Cl} \xrightarrow[\text{2. H}_3\text{O}^+]{\text{1. 2 CH}_3\text{MgBr}} R-\overset{\text{OH}}{\underset{\text{CH}_3}{C}}-CH_3
$$

2. Reactions of *carbonyl compounds* with hydride ion donors (Section 13.5).
   a. Reaction of an *aldehyde* with sodium borohydride forms a primary alcohol:

$$
\underset{R}{\overset{O}{\underset{\Vert}{C}}}\!\!{}_H \xrightarrow[\text{2. H}_3\text{O}^+]{\text{1. NaBH}_4} RCH_2OH
$$

b. Reaction of a *ketone* with sodium borohydride forms a secondary alcohol:

$$\underset{R}{\overset{O}{\underset{\parallel}{C}}}\underset{R}{} \xrightarrow[\text{2. H}_3\text{O}^+]{\text{1. NaBH}_4} R-\overset{OH}{\underset{\mid}{CH}}-R$$

c. Reaction of an *acyl chloride* with sodium borohydride forms a primary alcohol:

$$\underset{R}{\overset{O}{\underset{\parallel}{C}}}\underset{Cl}{} \xrightarrow[\text{2. H}_3\text{O}^+]{\text{1. NaBH}_4} R-CH_2-OH$$

d. Reaction of an *ester* with lithium aluminum hydride forms two alcohols:

$$\underset{R}{\overset{O}{\underset{\parallel}{C}}}\underset{OR'}{} \xrightarrow[\text{2. H}_3\text{O}^+]{\text{1. LiAlH}_4} RCH_2OH + R'OH$$

e. Reaction of a *carboxylic acid* with lithium aluminum hydride forms a primary alcohol:

$$\underset{R}{\overset{O}{\underset{\parallel}{C}}}\underset{OH}{} \xrightarrow[\text{2. H}_3\text{O}^+]{\text{1. LiAlH}_4} R-CH_2-OH$$

f. Reaction of an *amide* with lithium aluminum hydride forms an amine:

$$\underset{R}{\overset{O}{\underset{\parallel}{C}}}\underset{NH_2}{} \xrightarrow[\text{2. H}_2\text{O}]{\text{1. LiAlH}_4} R-CH_2-NH_2$$

$$\underset{R}{\overset{O}{\underset{\parallel}{C}}}\underset{NHR'}{} \xrightarrow[\text{2. H}_2\text{O}]{\text{1. LiAlH}_4} R-CH_2-NHR'$$

$$\underset{R}{\overset{O}{\underset{\parallel}{C}}}\underset{\underset{\underset{R''}{\mid}}{NR'}}{} \xrightarrow[\text{2. H}_2\text{O}]{\text{1. LiAlH}_4} R-CH_2-\underset{\underset{R''}{\mid}}{N}-R'$$

3. Reactions of *aldehydes* and *ketones* with amines (Section 13.6).
   a. Reaction with a *primary amine* forms an imine:

$$\underset{R}{\overset{R}{}}C=O + H_2NR \underset{}{\overset{\text{trace}}{\underset{H^+}{\rightleftharpoons}}} \underset{R}{\overset{R}{}}C=NR + H_2O$$

   b. Reaction with a *secondary amine* forms an enamine:

$$\underset{-CH}{\overset{R}{}}C=O + RNHR \overset{\text{trace}}{\underset{H^+}{\rightleftharpoons}} \underset{-C}{\overset{R}{}}C-\underset{R}{\overset{R}{}}N + H_2O$$

4. Reactions of an *aldehyde* or a *ketone* with oxygen nucleophiles.
   a. Reaction of an *aldehyde* or a *ketone* with water forms a hydrate (Section 13.7):

$$\underset{R}{\overset{O}{\underset{\parallel}{C}}}\underset{R'}{} + H_2O \overset{\text{HCl}}{\rightleftharpoons} R-\overset{OH}{\underset{\underset{OH}{\mid}}{\overset{\mid}{C}}}-R'$$

   b. Reaction of an *aldehyde* or a *ketone* with excess alcohol forms an acetal or a ketal (Section 13.7):

$$\underset{R}{\overset{O}{\underset{\parallel}{C}}}\underset{R'}{} + 2\,R''OH \overset{\text{HCl}}{\rightleftharpoons} R-\overset{OH}{\underset{\underset{OR''}{\mid}}{\overset{\mid}{C}}}-R' \rightleftharpoons R-\overset{OR''}{\underset{\underset{OR''}{\mid}}{\overset{\mid}{C}}}-R' + H_2O$$

5. Reactions of α,β-unsaturated carbonyl compounds with nucleophiles (Section 13.8):

$$RCH{=}CHCR' + \underline{NuH} \longrightarrow RCHCH_2CR'$$
$$\underset{\text{conjugate addition}}{\overset{|}{Nu}}$$

Problems

18. Draw the structure for each of the following compounds:
  a. isobutyraldehyde
  b. 4-octanone
  c. 4-bromohexanal
  d. 4-bromo-3-heptanone
  e. 3-methylcyclohexanone
  f. 2,4-pentanedione

19. Give the products of each of the following reactions:

a. $CH_3CH_2CH + CH_3CH_2OH \xrightarrow[\text{excess}]{HCl}$

b. $CH_3CH_2CCH_3 \xrightarrow[\text{2. H}_3O^+]{\text{1. NaBH}_4}$

c. $CH_3CH_2CH_2COCH_2CH_3 \xrightarrow[\text{2. H}_3O^+]{\text{1. LiAlH}_4}$

d. 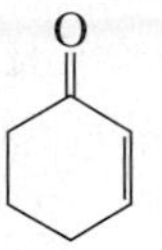 (cyclohexenone) $+$ HBr $\longrightarrow$

20. Give an example of each of the following:
  a. a hemiacetal   **b.** an imine   **c.** a ketal   **d.** an enamine

21. List the following compounds in order of decreasing reactivity toward nucleophilic attack:

$$\underset{CH_3}{CH_3CH_2CHCCH_2CH_3} \qquad CH_3CH_2CH \qquad \underset{CH_3\ CH_3}{CH_3CH_2CHCCCH_2CH_3(OCH_3)}$$

$$CH_3CH_2CCH_2CH_3 \qquad \underset{CH_3\ CH_3}{CH_3CH_2CHCCHCH_2CH_3} \qquad \underset{CH_3}{CH_3CHCH_2CCH_2CH_3}$$

22. Show the reagents required to form the primary alcohol.

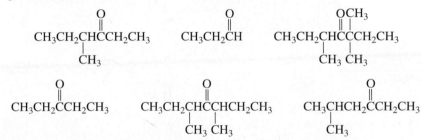

23. Fill in the boxes.

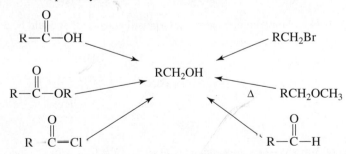

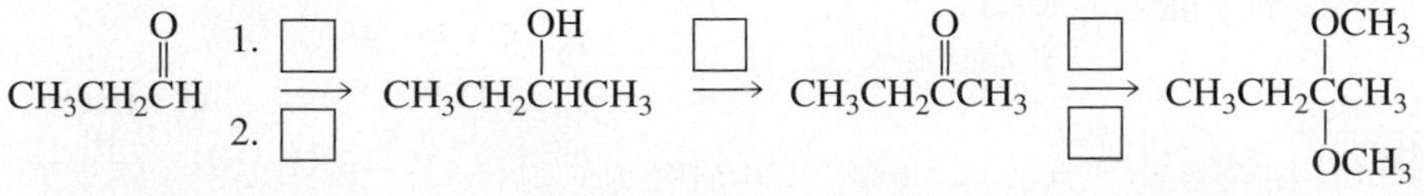

$$CH_3CH_2CH \underset{2.}{\overset{1.\ \square}{\rightleftharpoons}} \underset{box}{} CH_3CH_2CHCH_3(OH) \xrightarrow{\square} CH_3CH_2CCH_3(O) \underset{\square}{\overset{\square}{\rightleftharpoons}} CH_3CH_2CCH_3(OCH_3)(OCH_3)$$

24. Using cyclohexanone as the starting material, describe how each of the following compounds could be synthesized:
  a. 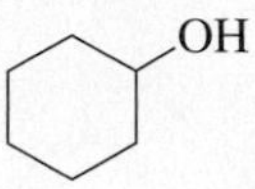 (cyclohexane-OH) **b.**  (cyclohexene) **c.** 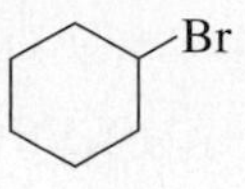 (cyclohexane-Br) **d.** 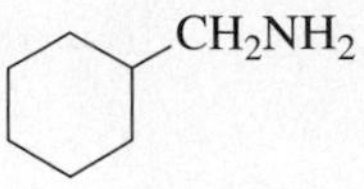 (cyclohexane-CH_2NH_2) **e.** 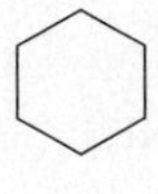 (cyclohexane) **f.** 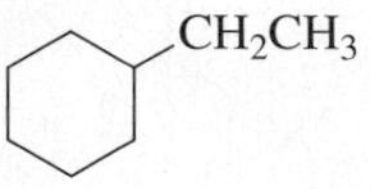 (cyclohexane-CH_2CH_3)

25. a. How many isomers are obtained from the reaction of 2-pentanone with ethylmagnesium bromide followed by treatment with aqueous acid?

  b. How many isomers are obtained from the reaction of 2-pentanone with methylmagnesium bromide followed by treatment with aqueous acid?

26. How would you convert *N*-methylbenzamide into the following compounds?

  a. *N*-methylbenzylamine   **b.** benzoic acid   **c.** methylbenzoate   **d.** benzyl alcohol

27. Propose a mechanism for the following reaction:

$$HOCH_2CH_2CH_2CH_2\overset{\overset{O}{\|}}{C}H \xrightarrow[CH_3OH]{HCl} \text{(pyran ring)}OCH_3$$

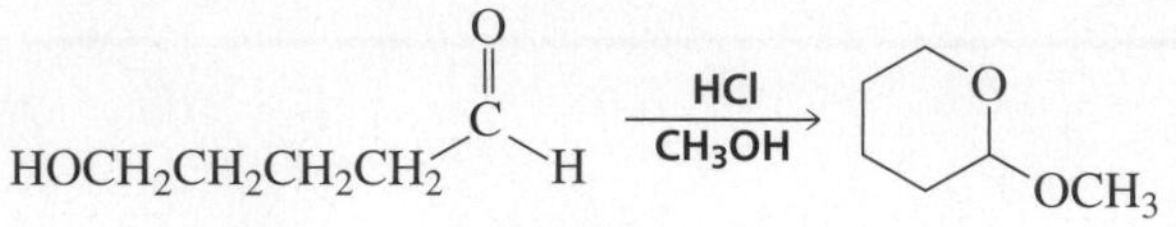

28. List the following compounds in order of decreasing amount of hydrate in an acidic aqueous solution:

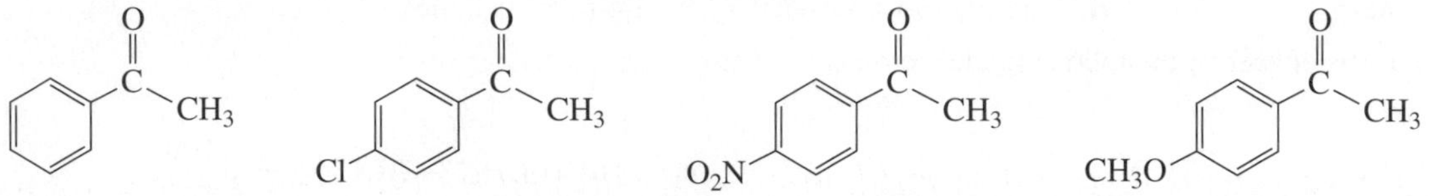

29. Fill in the boxes.

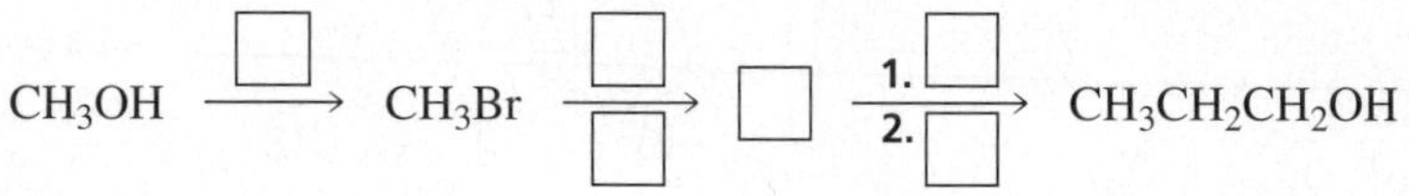

$$CH_3OH \xrightarrow{\square} CH_3Br \quad \frac{\square}{\square} \quad \square \quad \xrightarrow[2.]{1. \square} CH_3CH_2CH_2OH$$

30. Give the products of each of the following reactions:

  a. 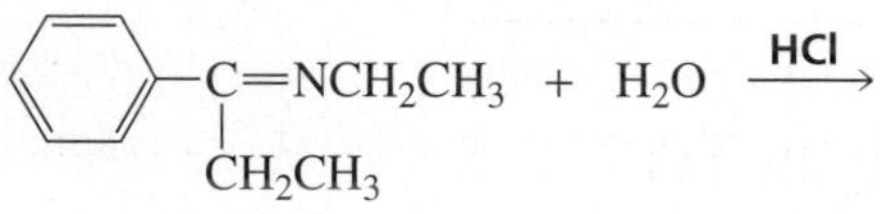 $\text{(phenyl)}-\overset{}{\underset{CH_2CH_3}{C}}=NCH_2CH_3 + H_2O \xrightarrow{HCl}$

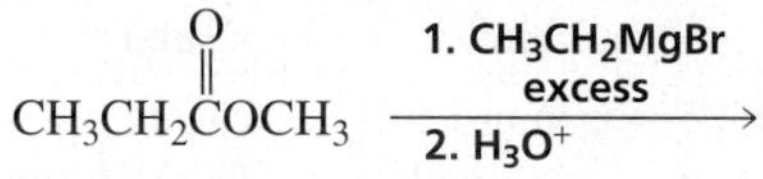

  b. 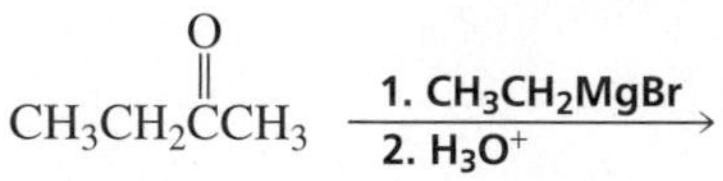 $CH_3CH_2\overset{\overset{O}{\|}}{C}CH_3 \xrightarrow[2. H_3O^+]{1.\ CH_3CH_2MgBr}$

  c. $CH_3CH_2\overset{\overset{O}{\|}}{C}OCH_3 \xrightarrow[2.\ H_3O^+]{\substack{1.\ CH_3CH_2MgBr \\ excess}}$

  d. 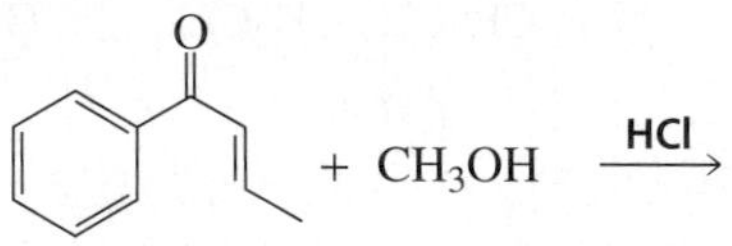 $\text{(phenyl ketone, enone)} + CH_3OH \xrightarrow{HCl}$

31. Propose a mechanism for the following reaction:

$$\text{(dihydropyran)} + CH_3CH_2OH \xrightarrow{HCl} \text{(pyran)}OCH_2CH_3$$

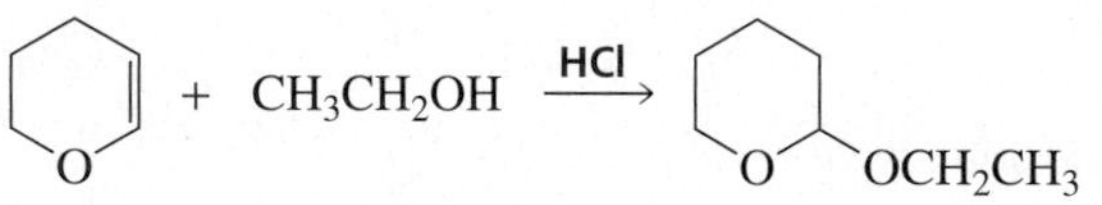

32. List three different sets of reagents (a carbonyl compound and a Grignard reagent) that could be used to prepare each of the following tertiary alcohols:

  a. 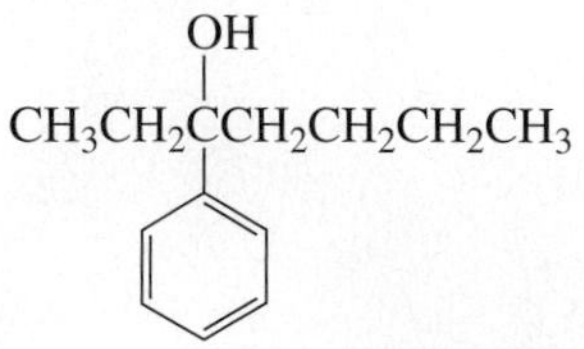 $CH_3CH_2\overset{\overset{OH}{|}}{\underset{\text{(phenyl)}}{C}}CH_2CH_2CH_2CH_3$

  b. 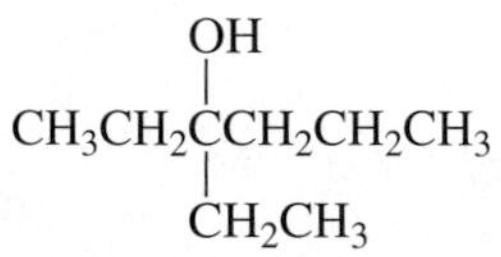 $CH_3CH_2\overset{\overset{OH}{|}}{\underset{CH_2CH_3}{C}}CH_2CH_2CH_3$

33. Give the product of the reaction of 3-methyl-2-cyclohexenone with each of the following reagents:

  a. CH_3CH_2SH **b.** HBr   **c.** H_2, Pd/C

34. Give the product of each of the following reactions:

  a. 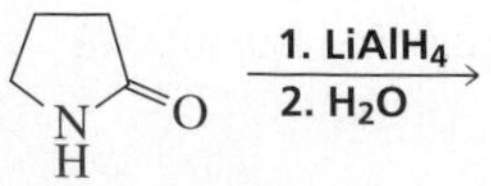 $\text{(2-pyrrolidinone)} \xrightarrow[2.\ H_2O]{1.\ LiAlH_4}$

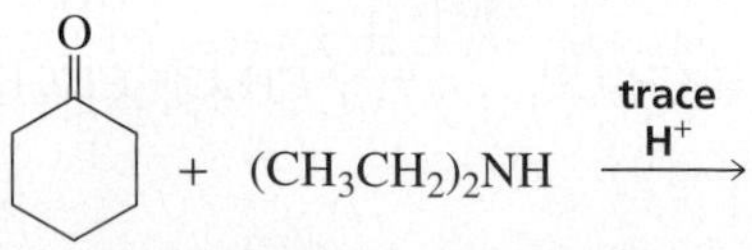

  b. 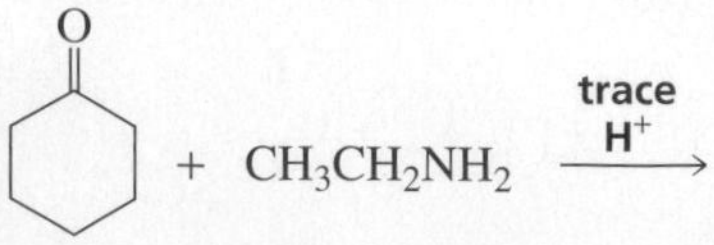 $\text{(cyclohexanone)} + CH_3CH_2NH_2 \xrightarrow{\substack{trace \\ H^+}}$

  c. $\text{(cyclohexanone)} + (CH_3CH_2)_2NH \xrightarrow{\substack{trace \\ H^+}}$

  d. 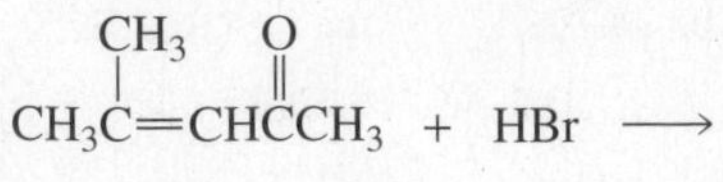 $CH_3\overset{\overset{CH_3}{|}}{C}=CH\overset{\overset{O}{\|}}{C}CH_3 + HBr \longrightarrow$

35. Indicate how the following compounds could be prepared from the given starting materials:

a. 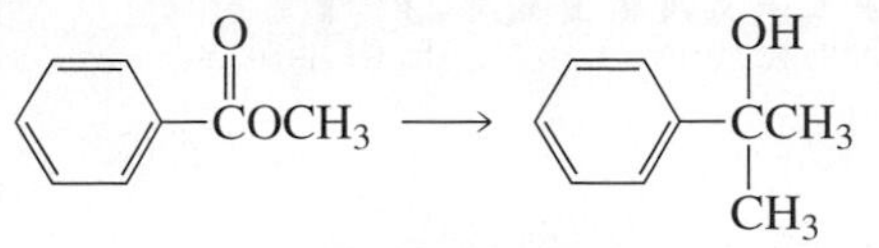 [benzene ring]–C(=O)OCH$_3$ $\longrightarrow$ [benzene ring]–C(OH)(CH$_3$)CH$_3$

b. 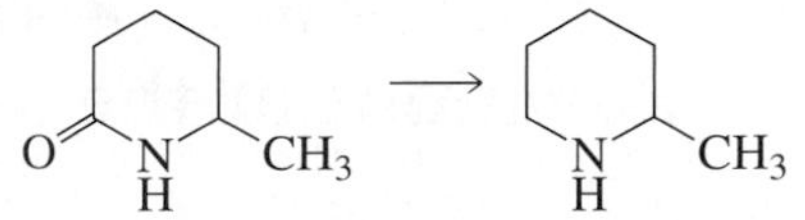 [2-methyl piperidinone] $\longrightarrow$ [2-methyl piperidine]

36. Give the products of the following reactions. Show all stereoisomers that are formed.

a. CH$_3$CH$_2$CCH$_2$CH$_2$CH$_2$CH$_3$ (with C=O) $\xrightarrow[\text{2. H}_3\text{O}^+]{\text{1. NaBH}_4}$

b. 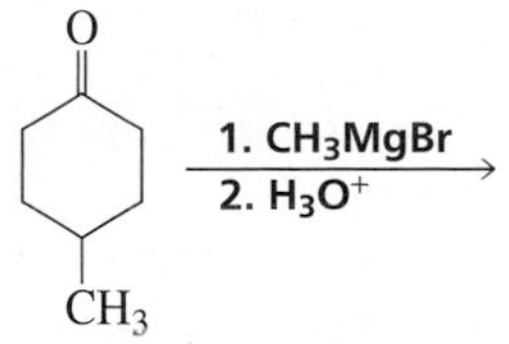 [4-methylcyclohexanone] $\xrightarrow[\text{2. H}_3\text{O}^+]{\text{1. CH}_3\text{MgBr}}$

37. Indicate how the following compounds could be prepared from the given starting materials:

a. CH$_3$CH$_2$CH$_2$CH$_2$Br $\longrightarrow$ CH$_3$CH$_2$CH$_2$CH$_2$COH (with C=O)

b. CH$_3$CH$_2$CH$_2$CH$_2$Br $\longrightarrow$ CH$_3$CH$_2$CH$_2$CH$_2$CH$_2$NH$_2$

38. Starting with N-benzylbenzamide, how would you make the following compounds?
a. dibenzylamine   **b.** benzoic acid   **c.** benzyl alcohol

39. Put the appropriate compound in each box.

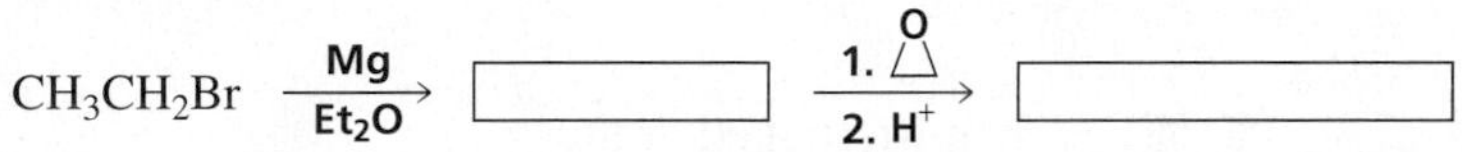

CH$_3$CH$_2$Br $\xrightarrow[\text{Et}_2\text{O}]{\text{Mg}}$ [ ] $\xrightarrow[\text{2. H}^+]{\text{1. } \triangle\text{O}}$ [ ]

40. What alcohol would be formed from the reaction of the following Grignard reagent with ethylene oxide followed by the addition of acid?

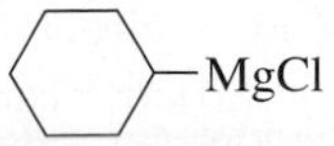

[cyclohexyl]–MgCl

41. a. Write the mechanism for the following reactions:
   1. the acid-catalyzed hydrolysis of an imine to a carbonyl compound and a primary amine.
   2. the acid-catalyzed hydrolysis of an enamine to a carbonyl compound and a secondary amine.
b. How do these mechanisms differ?

42. Which of the following alkyl halides could be successfully used to form a Grignard reagent?

a. HOCH$_2$CH$_2$CH$_2$CH$_2$Br   **b.** BrCH$_2$CH$_2$CH$_2$COH (with C=O)   **c.** CH$_3$NCH$_2$CH$_2$CH$_2$Br, with CH$_3$ below N

43. The pK_a values of oxaloacetic acid are 2.22 and 3.98.

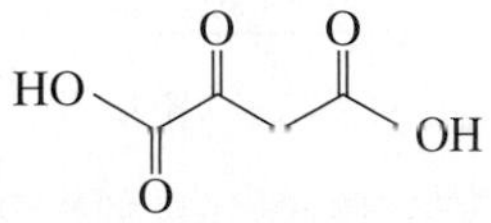

HO–C(=O)–CH$_2$(with C=O)–C(=O)–OH

oxaloacetic acid

a. Which carboxyl group is more acidic?
b. The amount of hydrate present in an aqueous solution of oxaloacetic acid depends on the pH of the solution: 95% at pH = 0, 81% at pH = 1.3, 35% at pH = 3.1, 13% at pH = 4.7, 6% at pH = 6.7, and 6% at pH = 12.7. Explain this pH dependence.

44. Propose a reasonable mechanism for the following reaction:

CH$_3$CCH$_2$CH$_2$COCH$_2$CH$_3$ $\xrightarrow[\text{2. H}_3\text{O}^+]{\text{1. CH}_3\text{MgBr}}$ [lactone with 2 CH$_3$ groups] + CH$_3$CH$_2$OH

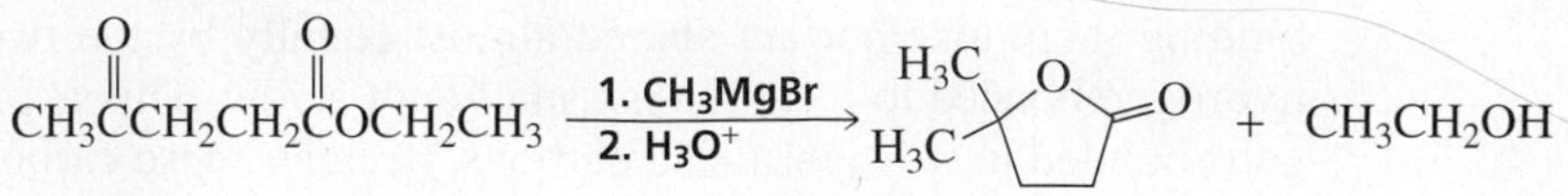

Carbonyl Compounds III

Reactions at the α-Carbon

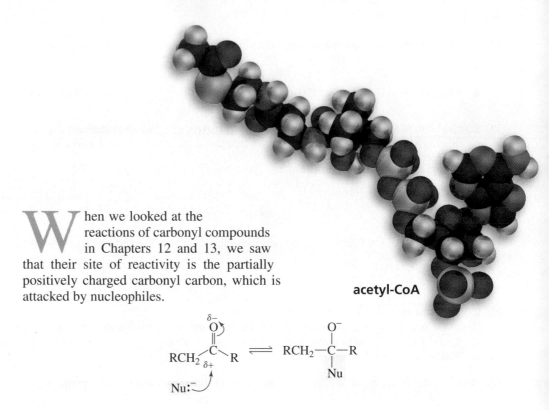

acetyl-CoA

When we looked at the reactions of carbonyl compounds in Chapters 12 and 13, we saw that their site of reactivity is the partially positively charged carbonyl carbon, which is attacked by nucleophiles.

Aldehydes, ketones, and esters have a second site of reactivity. A hydrogen bonded to a carbon *adjacent* to a carbonyl carbon is sufficiently acidic to be removed by a strong base. The carbon adjacent to a carbonyl carbon is called an **α-carbon.** A hydrogen bonded to an α-carbon is called an **α-hydrogen.**

14.1 Acidity of α-Hydrogens

Hydrogen and carbon have similar electronegativities, which means that the electrons binding them together are shared almost equally by the two atoms. Consequently, a hydrogen bonded to a carbon is usually not acidic. This is particularly true for hydrogens bonded to sp^3 hybridized carbons, because these carbons are the most similar to hydrogen in electronegativity (Section 5.13). The high pK_a of ethane is evidence of the low acidity of a hydrogen bonded to an sp^3 hybridized carbon.

$$CH_3CH_3$$

$$\boxed{pK_a \sim 60}$$

A hydrogen bonded to an sp^3 hybridized carbon that is adjacent to a carbonyl carbon is much more acidic than hydrogens bonded to other sp^3 hybridized carbons. For example, the pK_a value for dissociation of a proton from an α-carbon of an aldehyde or a ketone ranges from 16 to 20, and the pK_a value for dissociation of a proton from or α-carbon of an ester is about 25 (Table 14.1). Notice that, although an α-hydrogen is more acidic than most other carbon-bound hydrogens, it is less acidic than a hydrogen of water ($pK_a = 15.7$).

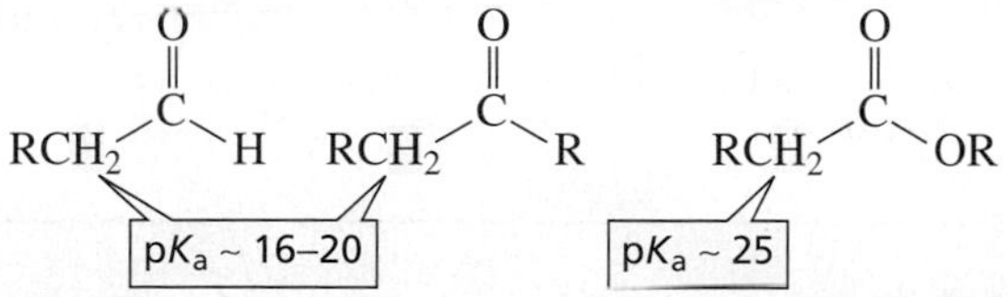

Table 14.1   The pK_a Values of Some Carbon Acids

	pK_a		pK_a
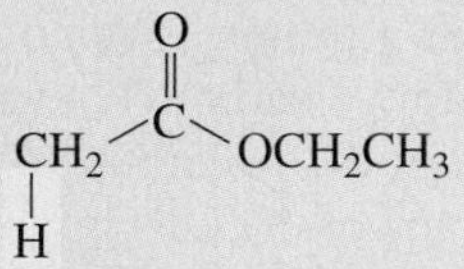	25	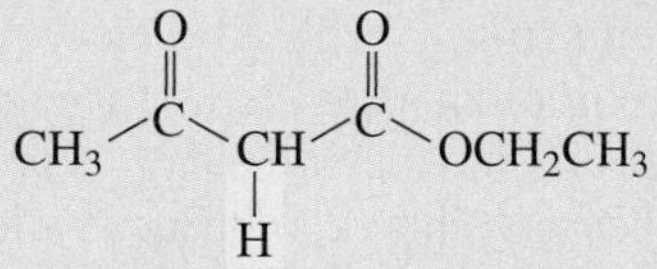	10.7
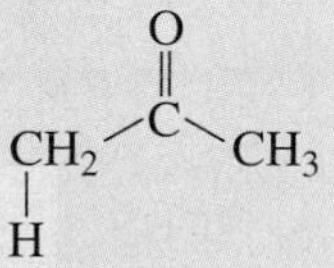	20	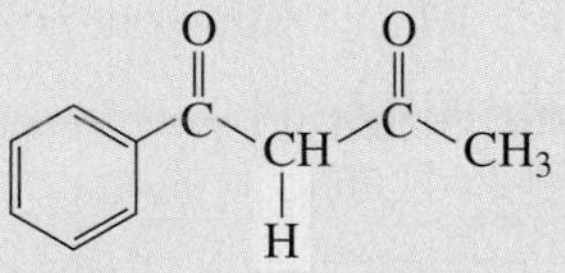	9.4
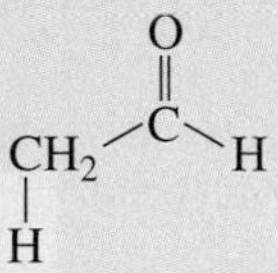	17	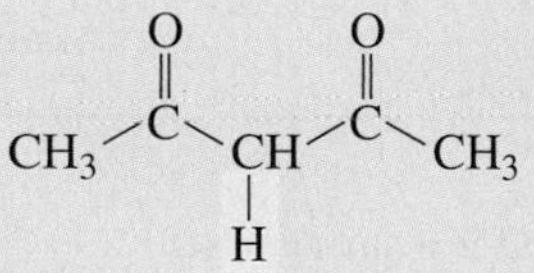	8.9
			5.9
	13.3		

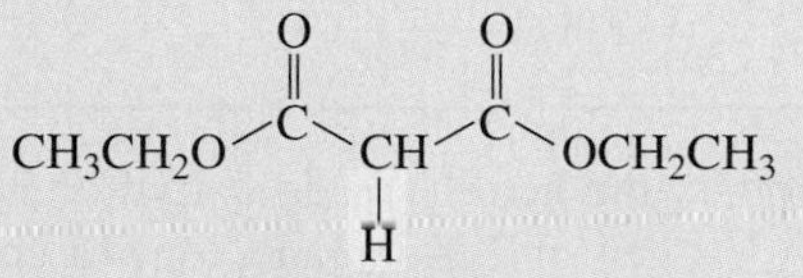

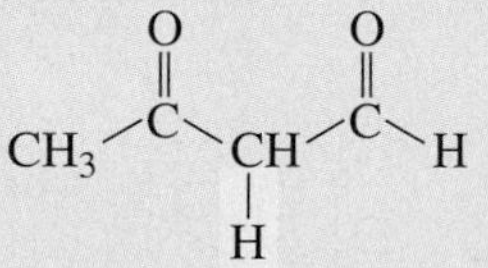

Why is a hydrogen that is bonded to an sp^3 hybridized carbon adjacent to a carbonyl carbon so much more acidic than hydrogens bonded to other sp^3 hybridized carbons? An α-hydrogen is more acidic because the base formed when a proton is removed from an α-carbon is more stable than a base formed when a proton is removed from other sp^3 hybridized carbons, and we have seen that the more stable the base, the stronger is its conjugate acid (Section 2.3).

Why is the base more stable? When a proton is removed from ethane, the electrons left behind reside solely on a carbon atom. Because carbon is not very electronegative, a carbanion is unstable. As a result, the pK_a of its conjugate acid is very high.

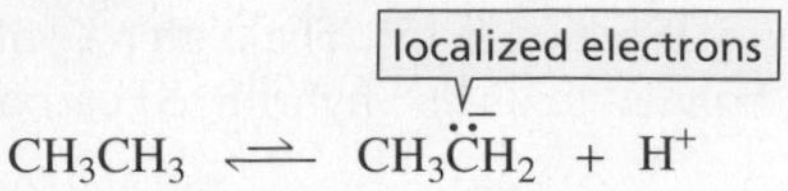

$$\boxed{\text{localized electrons}}$$

$$CH_3CH_3 \; \rightleftharpoons \; CH_3\ddot{C}H_2 \; + \; H^+$$

When a proton is removed from a carbon adjacent to a carbonyl carbon, the electrons left behind are delocalized onto the oxygen. The oxygen atom is better able to accommodate the electrons because it is more electronegative than carbon. Electron delocalization also increases the stability of the base (Section 6.7).

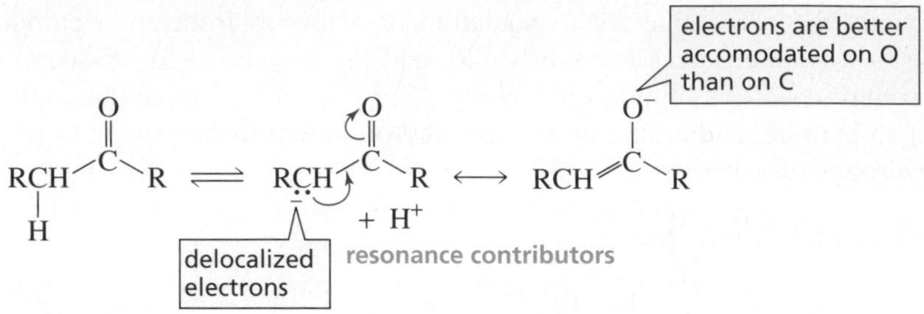

When a proton is removed from a carbon adjacent to a carbonyl carbon, the electrons left behind are delocalized onto the oxygen.

electrons are better accomodated on O than on C

delocalized electrons   resonance contributors

PROBLEM 1

The pK_a of a hydrogen on the sp^3 carbon of propene is 42, which is greater than that of any of the compounds listed in Table 14.1, but less than the pK_a of an alkane. Explain.

Now we can understand why aldehydes and ketones (p$K_a = 16 - 20$) are more acidic than esters (p$K_a = 25$). The electrons left behind when a proton is removed from the α-carbon of an ester are not as readily delocalized onto the carbonyl oxygen as are the electrons left behind when a proton is removed from the α-carbon of an aldehyde or a ketone. This is so because a lone pair on the oxygen of the OR group of the ester can also be delocalized onto the carbonyl oxygen. Thus, the two pairs of electrons compete for delocalization onto oxygen.

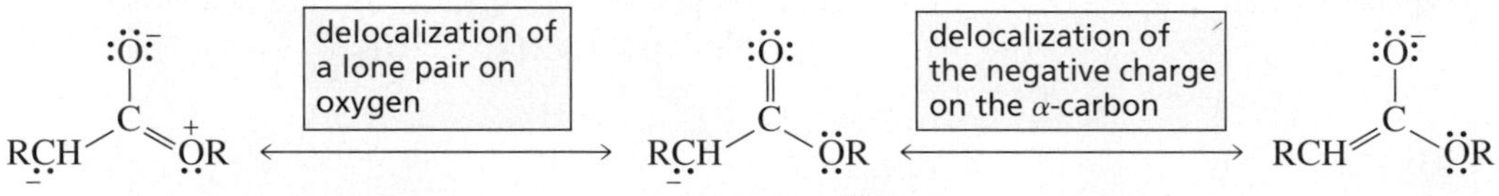

The α-hydrogen of a ketone or an aldehyde is more acidic than the α-hydrogen of an ester.

delocalization of a lone pair on oxygen

delocalization of the negative charge on the α-carbon

resonance contributors

PROBLEM 2◆

a. Which compound is a stronger acid?

b. Which compound has a greater pK_a value?

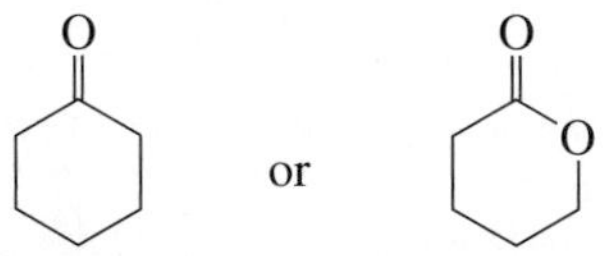

or

If the α-carbon is *between* two carbonyl groups, the acidity of an α-hydrogen is even greater (Table 14.1). For example, an α-hydrogen of 2,4-pentanedione, a compound with an α-hydrogen between two ketone carbonyl groups, has a pK_a value of 8.9. An α-hydrogen of ethyl 3-oxobutyrate, a compound with an α-carbon between a ketone carbonyl group and an ester carbonyl group, has a pK_a value of 10.7.

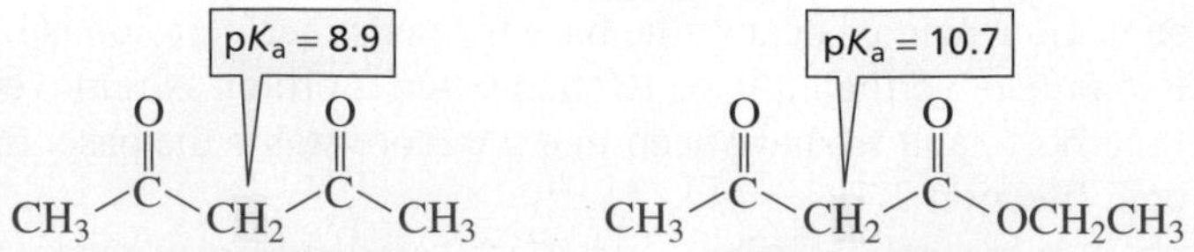

$\boxed{pK_a = 8.9}$

$\boxed{pK_a = 10.7}$

2,4-pentanedione
acetylacetone
a β-diketone

ethyl 3-oxobutyrate
ethyl acetoacetate
a β-keto ester

The acidity of α-hydrogens bonded to carbons flanked by two carbonyl groups increases because the electrons left behind when the proton is removed can be delocalized onto *two* oxygen atoms.

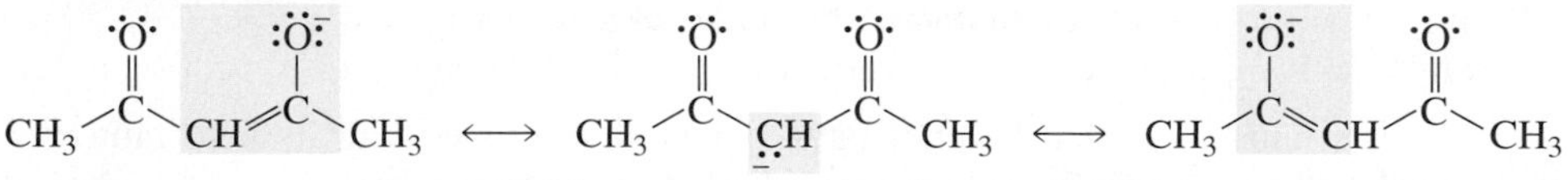

resonance contributors for the 2,4-pentanedione anion

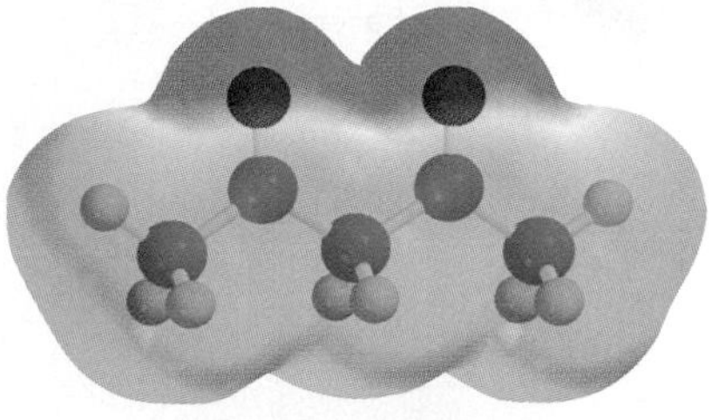

2,4-pentanedione

PROBLEM 3◆

Why is 2,4-pentanedione a stronger acid than ethyl 3-oxobutyrate?

PROBLEM-SOLVING STRATEGY

Explain why HO⁻ cannot remove a proton from the α-carbon of a carboxylic acid.

Finding that HO⁻ cannot remove a proton from the α-carbon of a carboxylic acid suggests that HO⁻ reacts with another portion of the molecule more rapidly. Because the proton on the carboxyl group is more acidic than the proton on the α-carbon, HO⁻ removes a proton from the carboxyl group rather than from the α-carbon.

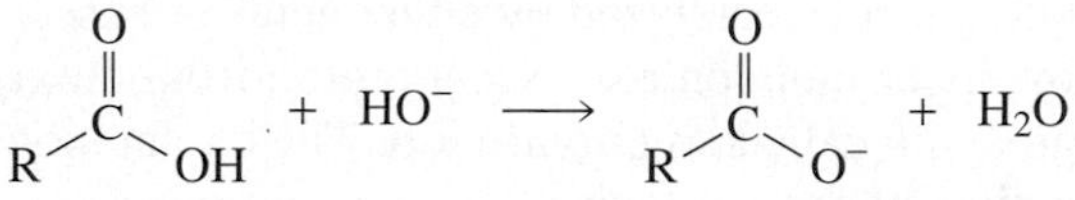

Now continue on to Problem 4.

PROBLEM 4◆

Explain why a proton can be removed from the α-carbon of *N,N*-dimethylethanamide but not from the α-carbon of either *N*-methylethanamide or ethanamide.

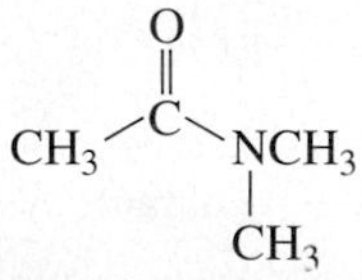

N,N-dimethylethanamide 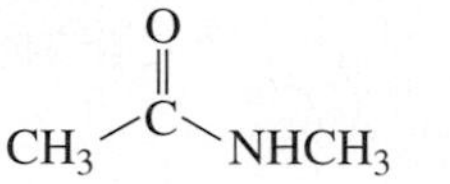   *N*-methylethanamide 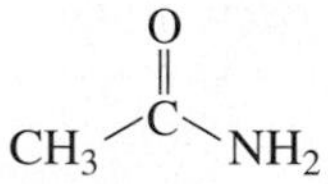   ethanamide

PROBLEM 5◆

Explain why HO⁻ cannot remove a proton from the α-carbon of an acyl chloride.

14.2 Keto–Enol Tautomers

A ketone exists in equilibrium with its enol tautomer. Recall that **tautomers** are isomers that are in rapid equilibrium (Section 5.11). Keto–enol tautomers differ in the location of a double bond and a hydrogen.

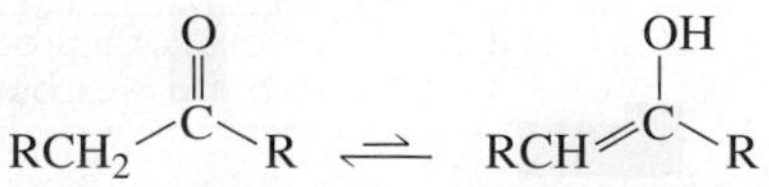

keto tautomer   **enol tautomer**

For most ketones, the enol tautomer is much less stable than the keto tautomer. For example, an aqueous solution of acetone exists as an equilibrium mixture of more than 99.9% keto tautomer and less than 0.1% enol tautomer.

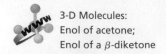

$$\underset{\substack{>99.9\% \\ \textbf{keto tautomer}}}{CH_3-\overset{\overset{\textstyle O}{\|}}{C}-CH_3} \;\rightleftharpoons\; \underset{\substack{<0.1\% \\ \textbf{enol tautomer}}}{CH_2=\overset{\overset{\textstyle OH}{|}}{C}-CH_3}$$

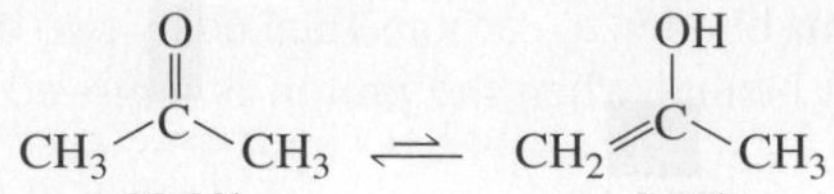

Phenol is unusual in that its enol tautomer is *more* stable than its keto tautomer because the enol tautomer is aromatic, but the keto tautomer is not.

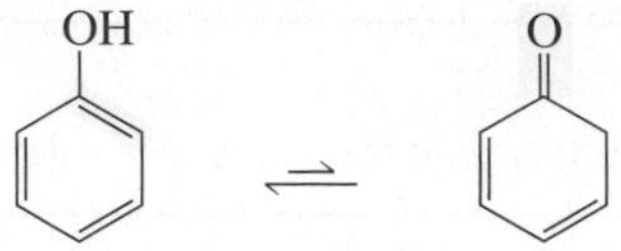

enol tautomer
aromatic   ⇌   **keto tautomer**
not aromatic

14.3 Enolization

The interconversion of keto and enol tautomers is called **tautomerization** or **enolization**. Enolization can be catalyzed by either acids or bases.

In a basic solution, hydroxide ion removes a proton from the α-carbon of the keto tautomer, forming an anion called an **enolate ion**. The enolate ion has two resonance contributors. Protonation on oxygen forms the enol tautomer.

base-catalyzed keto–enol interconversion

removal of a proton from the α-carbon

protonation of oxygen

keto tautomer   **enolate ion**   **enol tautomer**

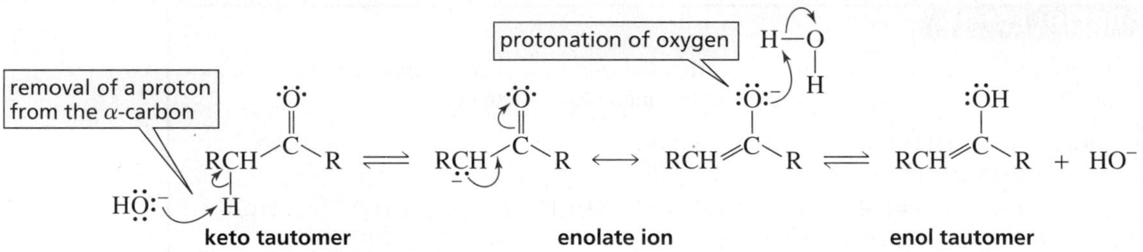

In an acidic solution, the carbonyl oxygen of the keto tautomer is protonated and water removes a proton from the α-carbon, forming the enol.

acid-catalyzed keto–enol interconversion

protonation of oxygen

keto tautomer   **enol tautomer**

removal of a proton from the α-carbon

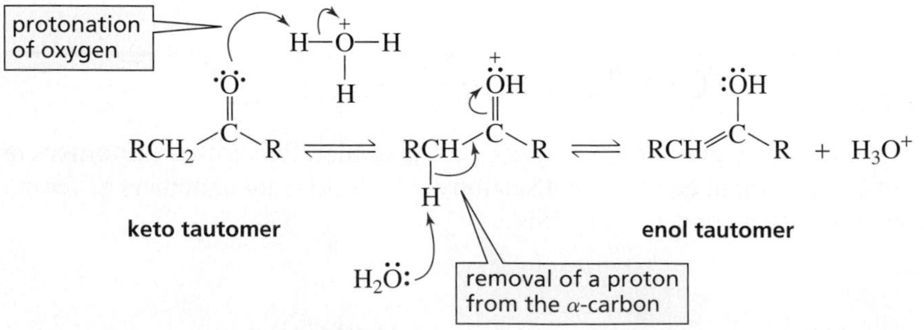

Notice that the steps are reversed in the base- and acid-catalyzed reactions. In the base-catalyzed reaction, the base removes the α-proton in the first step and the oxygen is protonated in the second step. In the acid-catalyzed reaction, the oxygen is protonated in the first step and the α-proton is removed in the second step.

PROBLEM 6◆

Draw the enol tautomers for each of the following compounds:

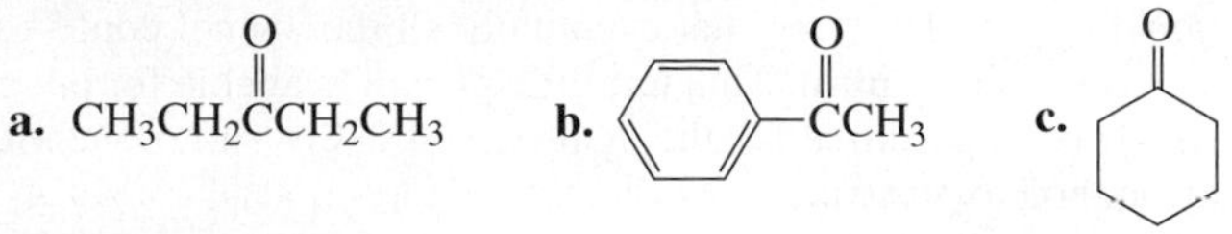

a. $CH_3CH_2\overset{O}{\overset{\|}{C}}CH_2CH_3$ b. (image: phenyl ring)$\overset{O}{\overset{\|}{C}}CH_3$ c. (image: cyclohexanone)

PROBLEM 7◆

Draw the two enol tautomers for the following compound. Which one is more stable?

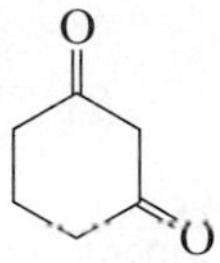

(image: cyclohexane-1,3-dione structure)

14.4   Alkylation of Enolate Ions

The resonance contributors of the enolate ion show that it has two electron-rich sites: the α-carbon and the oxygen.

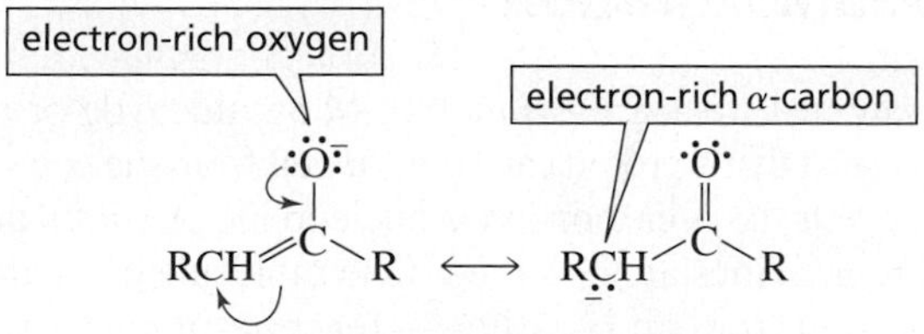

electron-rich oxygen

electron-rich α-carbon

$$RCH\overset{:\ddot{O}:^-}{\underset{}{=}}\overset{}{C}\diagdown R \longleftrightarrow R\ddot{C}H\overset{\cdot\ddot{O}\cdot}{\underset{}{-}}\overset{}{C}\diagdown R$$

resonance contributors of an enolate ion

www   3-D Molecule:
Enolate ion of acetone

Which nucleophilic site (C or O) reacts with an electrophile depends on the electrophile. Protonation occurs preferentially on oxygen because of the greater concentration of negative charge on the more electronegative oxygen atom. However, when the electrophile is something other than a proton, carbon is more likely to be the nucleophile because carbon is a better nucleophile than oxygen.

   Alkylation of the α-carbon of a carbonyl compound is an important reaction because it gives us another way to form a carbon–carbon bond. Alkylation is carried out by first removing a proton from the α-carbon with a strong base and then adding the appropriate alkyl halide. Because the alkylation is an S_N2 reaction, it works best with primary alkyl halides and methyl halides (Section 10.2).

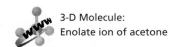

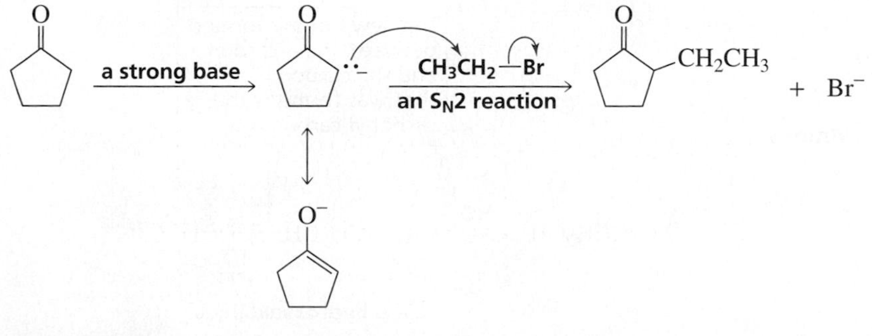

$$\text{(cyclopentanone)} \xrightarrow{\text{a strong base}} \text{(enolate)} \xrightarrow[\text{an } S_N2 \text{ reaction}]{CH_3CH_2\!-\!Br} \text{(2-ethylcyclopentanone)}\;+\;Br^-$$

(enolate O⁻ resonance form)

Enolate ions can be alkylated on the α-carbon.

PROBLEM 8

Draw the contributing resonance structures for the enolate ion of

a. 3-pentanone   b. cyclohexanone

THE SYNTHESIS OF ASPIRIN

In the first step in the industrial synthesis of aspirin, a phenolate ion reacts with carbon dioxide under pressure to form *o*-hydroxybenzoic acid (also known as salicylic acid). Salicylic acid reacts with acetic anhydride to form acetylsalicylic acid (aspirin).

During World War I, an American subsidiary of the Bayer Company bought as much phenol as it could from the international market, knowing that eventually all the phenol could be converted into aspirin. This left little phenol available for other countries to purchase for the synthesis of 2,4,6-trinitrophenol, a common explosive.

salicylic acid
o-hydroxybenzoic acid

acetylsalicylic acid
aspirin

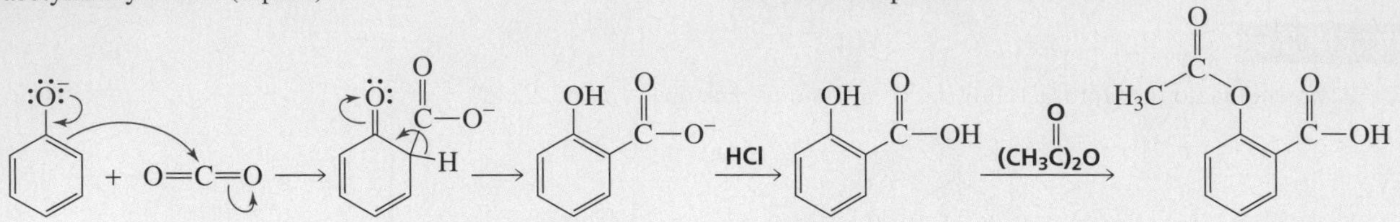

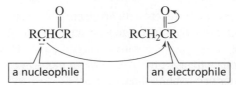

PROBLEM 9◆

Give the product that would be formed if the enolate ion of each compound in Problem 8 were treated with ethyl bromide.

14.5 The Aldol Addition

In Chapter 13, we saw that the carbonyl carbon of an aldehyde or a ketone is an electrophile. We have just seen that a proton can be removed from the α-carbon of an aldehyde or a ketone, which converts the α-carbon into a nucleophile. An **aldol addition** is a reaction in which *both* of these activities are observed: One molecule of a carbonyl compound—after a proton is removed from an α-carbon—reacts as a *nucleophile* and attacks the *electrophilic* carbonyl carbon of a second molecule of the carbonyl compound.

Thus, an aldol addition is a reaction between two molecules of an *aldehyde* or two molecules of a *ketone*. When the reactant is an aldehyde, the addition product is a β-hydroxyaldehyde, which is why the reaction is called an aldol addition ("ald" for aldehyde, "ol" for alcohol). Notice that the reaction forms a new C—C bond between the α-carbon of one molecule and the carbon that formerly was the carbonyl carbon of the other molecule.

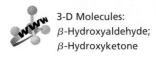

3-D Molecules: β-Hydroxyaldehyde; β-Hydroxyketone

aldol additions

a β-hydroxyaldehyde

a β-hydroxyketone

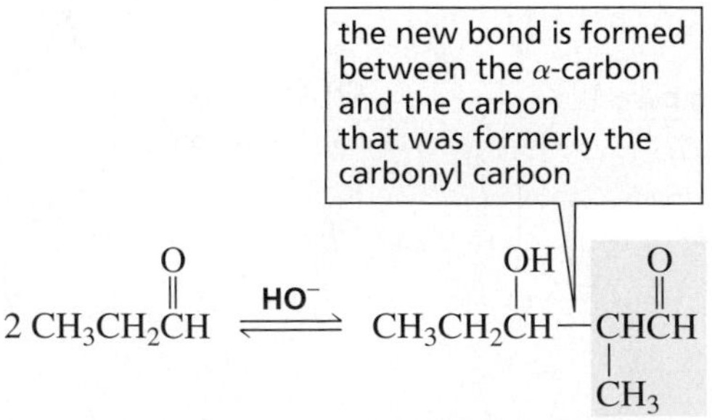

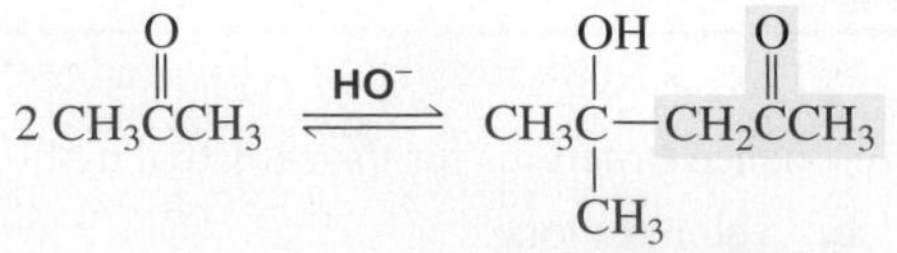

In the first step of an aldol addition, a base removes a proton from an α-carbon, creating an enolate ion (a nucleophile). The enolate ion adds to the carbonyl carbon of a second molecule of the carbonyl compound, and the resulting negatively charged oxygen is protonated by the solvent.

mechanism for the aldol addition

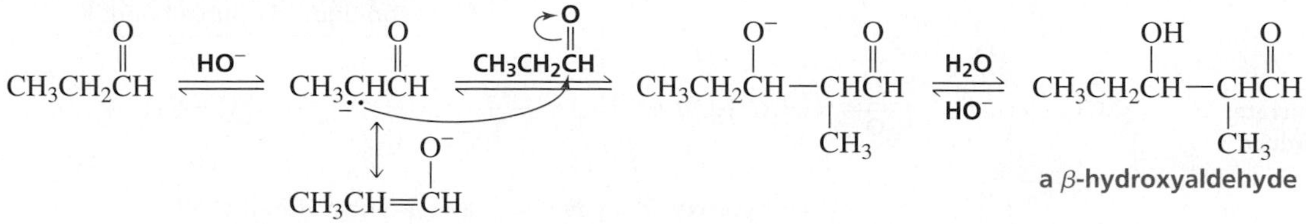

a β-hydroxyaldehyde

When the reactant in an aldol addition is a ketone, the addition product is a β-hydroxyketone.

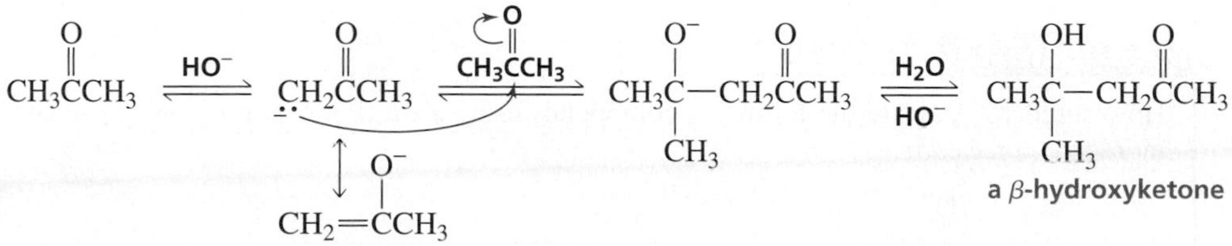

a β-hydroxyketone

Because an aldol addition reaction occurs between two molecules of the same carbonyl compound, the product has twice as many carbons as the reacting aldehyde or ketone.

> The new C—C bond formed in an aldol addition is between the α-carbon of one molecule and the carbon that formerly was the carbonyl carbon of the other molecule.

PROBLEM 10

Show the aldol addition product that would be formed from each of the following compounds:

a. $CH_3CH_2CH_2CH_2CH$ (with C=O)
$\quad$ O

c. $CH_3CH_2CCH_2CH_3$ (with C=O)
$\quad$ O

b. $CH_3CHCH_2CH_2CH$ (with C=O, and CH_3 branch)
$\quad\quad CH_3$

d. cyclohexanone structure

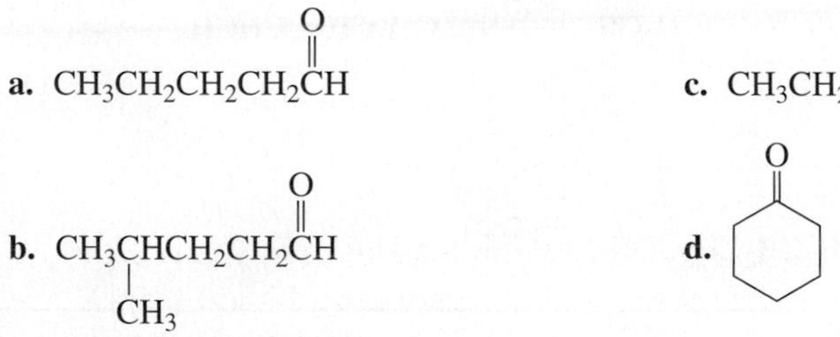

PROBLEM 11◆

For each of the following compounds, indicate the aldehyde or ketone from which it would be formed by an aldol addition:

a. 2-ethyl-3-hydroxyhexanal
b. 4-hydroxy-4-methyl-2-pentanone
c. 2,4-dicyclohexyl-3-hydroxybutanal
d. 5-ethyl-5-hydroxy-4-methyl-3-heptanone

14.6 Dehydration of Aldol Addition Products

We have seen that alcohols are dehydrated when they are heated with acid (Section 11.3). The β-hydroxyaldehyde and β-hydroxyketone products of aldol addition reactions are easier to dehydrate than many other alcohols because the double bond formed as the result of dehydration is conjugated with a carbonyl group. Conjugation increases the stability of the product (Section 6.7) and therefore makes it easier to form. If the product of an aldol addition is dehydrated, the overall reaction is called an **aldol condensation**. A **condensation reaction** is a reaction that combines two molecules while removing a small molecule (usually water or an alcohol). Notice that an aldol condensation forms an α,β-unsaturated aldehyde or an α,β-unsaturated ketone.

An aldol addition product loses water to form an aldol condensation product.

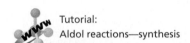

conjugated double bonds

$$2\ CH_3CH_2CH \xrightleftharpoons[H_2O]{HO^-} CH_3CH_2CH-CHCH \xrightarrow[\Delta]{H_3O^+} CH_3CH_2CH{=}C{-}CH\ +\ H_2O$$

a β-hydroxyaldehyde   an α,β-unsaturated aldehyde

PROBLEM 12◆

Give the product obtained from the aldol condensation of cyclohexanone.

PROBLEM 13 | **SOLVED**

How could you prepare the following compounds using a starting material containing no more than three carbons?

a. $CH_3CH_2CH_2COH$ (with =O)

b. $CH_3CH_3CH{=}CCH$ (with =O and CH_3)

SOLUTION TO 13a   A compound with the correct four-carbon skeleton can be obtained if a two-carbon aldehyde undergoes an aldol addition. Dehydration of the addition product forms an α,β-unsaturated aldehyde. Catalytic hydrogenation forms an aldehyde that is oxidized to the target compound by chromic acid (Section 11.4).

$$CH_3CH \xrightarrow{HO^-} CH_3CHCH_2CH \xrightarrow[\Delta]{H_3O^+} CH_3CH{=}CHCH$$

$$\downarrow H_2 \mid Pt/C$$

$$CH_3CH_2CH_2COH \xleftarrow{H_2CrO_4} CH_3CH_2CH_2CH$$

Tutorial:
Aldol reactions—synthesis

14.7 The Claisen Condensation

When two molecules of an *ester* undergo a condensation reaction, the reaction is called a **Claisen condensation**. The product of a Claisen condensation is a β-keto ester.

Ludwig Claisen (1851–1930) *was born in Germany and received a Ph.D. from the University of Bonn, studying under Kekulé. He was a professor of chemistry at the University of Bonn, Owens College (Manchester, England), the University of Munich, the University of Aachen, the University of Kiel, and the University of Berlin.*

the new bond is formed between the α-carbon and the carbon that formerly was the carbonyl carbon

$$2\ CH_3CH_2COCH_2CH_3 \xrightarrow[\text{2. HCl}]{\text{1. } CH_3CH_2O^-} CH_3CH_2C{-}CHCOCH_2CH_3\ +\ CH_3CH_2OH$$

$$\underset{CH_3}{|}$$

a β-keto ester

3-D Molecule:
β-Keto ester

As in an aldol addition, in a Claisen condensation, one molecule of carbonyl compound acts as a nucleophile, after an α-hydrogen is removed by a strong base and the nucleophile attacks the carbonyl carbon of a second molecule of ester.

The new bond is formed between the α-carbon of one molecule and the carbon that formerly was the carbonyl carbon of the other molecule.

mechanism for the Claisen condensation

$$CH_3\underset{\underset{H}{|}}{CH}COCH_3 \;\xrightarrow{CH_3\ddot{O}^-}\; CH_3\underset{\cdot\cdot}{CH}COCH_3 \;\xrightarrow{CH_3CH_2COCH_3}\; CH_3CH_2\underset{\underset{CH_3O}{|}}{C}{-}\underset{\underset{CH_3}{|}}{CH}COCH_3$$

$$+ \; CH_3OH$$

$$CH_3CH_2\underset{\underset{CH_3}{|}}{\overset{O}{C}}{-}\underset{|}{\overset{O}{CH}}COCH_3 \; + \; CH_3OH \;\xleftarrow{HCl}\; CH_3CH_2\underset{\underset{CH_3}{|}}{\overset{O}{C}}{-}\underset{|}{\overset{O}{CH}}COCH_3 \; + \; CH_3O^-$$

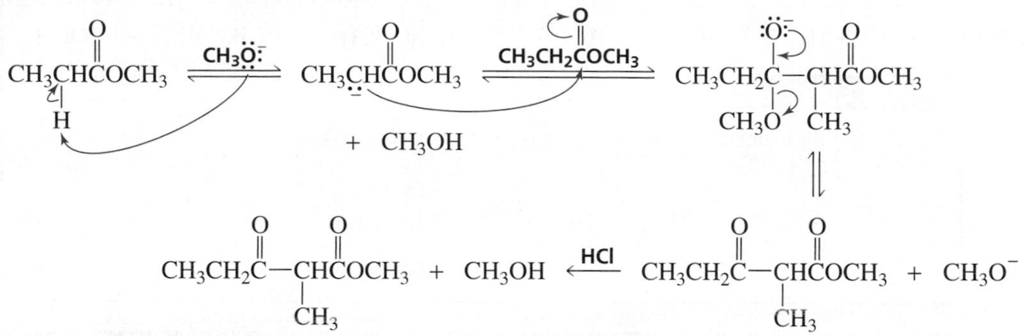

The base employed is the same as the leaving group of the ester. Therefore, the reactant is not changed if the base acts as a nucleophile and attacks the carbonyl group.

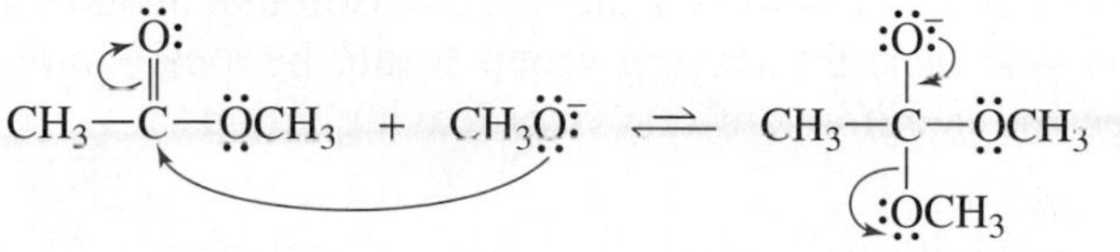

$$CH_3{-}\overset{\ddot{O}:}{\underset{|}{C}}{-}\ddot{O}CH_3 \; + \; CH_3\ddot{O}^- \;\rightleftharpoons\; CH_3{-}\underset{\underset{:\ddot{O}CH_3}{|}}{C}{-}\ddot{O}CH_3$$

After nucleophilic attack, the Claisen condensation and the aldol addition differ. In the Claisen condensation, the negatively charged oxygen reforms the carbon–oxygen π bond and expels the $^-$OR group. In the aldol addition, the negatively charged oxygen obtains a proton from the solvent.

Claisen condensation: **aldol addition**

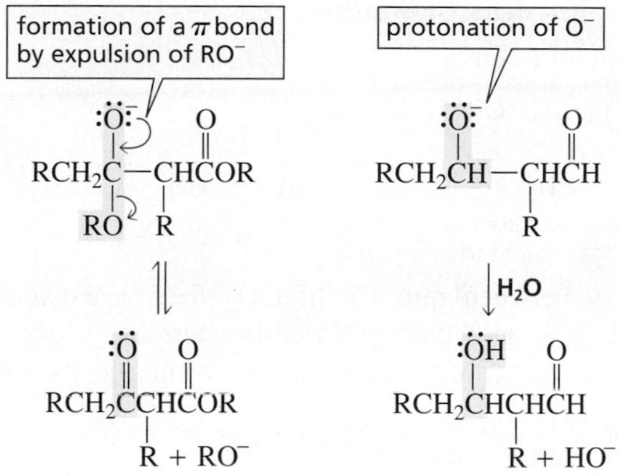

| formation of a π bond by expulsion of RO$^-$ | protonation of O$^-$ |

$$RCH_2\underset{\underset{RO}{|}}{\overset{:\ddot{O}:^-}{C}}{-}\underset{\underset{R}{|}}{\overset{O}{CH}}COR \qquad RCH_2\underset{\underset{R}{|}}{\overset{:\ddot{O}:^-}{CH}}{-}\overset{O}{CH}CH$$

$$\Big\updownarrow \qquad\qquad \Big\downarrow H_2O$$

$$RCH_2\overset{:\ddot{O}}{\underset{\underset{R + RO^-}{|}}{C}}\overset{O}{CH}COR \qquad RCH_2\overset{:\ddot{O}H}{\underset{\underset{R + HO^-}{|}}{CH}}\overset{O}{CH}CH$$

The difference between the last step of the Claisen condensation and the last step of the aldol addition arises from the difference between esters and aldehydes or ketones. The carbon to which the negatively charged oxygen is bonded in an ester is also bonded to a group that can be expelled. The carbon to which the negatively charged oxygen is bonded in an aldehyde or a ketone is not bonded to a group that can be expelled. Thus, the Claisen condensation is a nucleophilic substitution reaction, whereas the aldol addition is a nucleophilic addition reaction.

PROBLEM 14◆

Give the products of the following reactions:

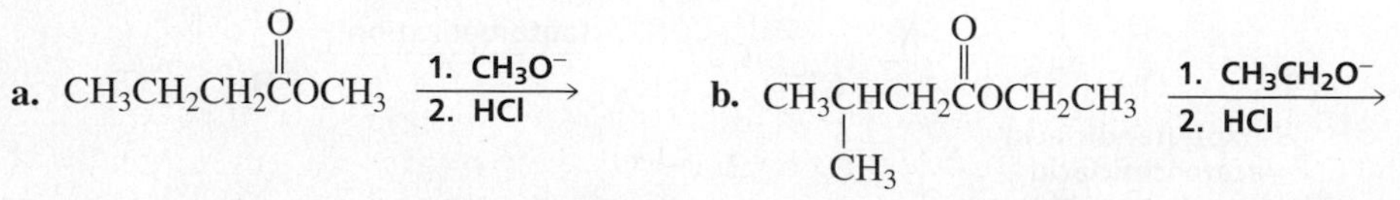

a. $CH_3CH_2CH_2\overset{O}{\overset{\|}{C}}OCH_3 \quad \xrightarrow[\text{2. HCl}]{\text{1. } CH_3O^-}$

b. $CH_3\underset{\underset{CH_3}{|}}{CH}CH_2\overset{O}{\overset{\|}{C}}OCH_2CH_3 \quad \xrightarrow[\text{2. HCl}]{\text{1. } CH_3CH_2O^-}$

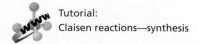

Tutorial:
Claisen reactions—synthesis

PROBLEM 15◆

Which of the following esters cannot undergo a Claisen condensation.

a. $CH_3CH=CHCOCH_3$ **b.** $HCOCH_3$ **c.** CH_3COCH_3 **d.** (phenyl)$-COCH_3$

PROBLEM 16◆

What starting materials were used to make the following β-keto ester?

$$CH_3CH_2CH_2CH_2CCHCOCH_3$$
$$| $$
$$CH_2CH_2CH_3$$

14.8 Decarboxylation of 3-Oxocarboxylic Acids

Carboxylate ions do not lose CO_2, for the same reason that alkanes such as ethane do not lose a proton—because the leaving group would be a carbanion. Carbanions are very strong bases and therefore are very poor leaving groups.

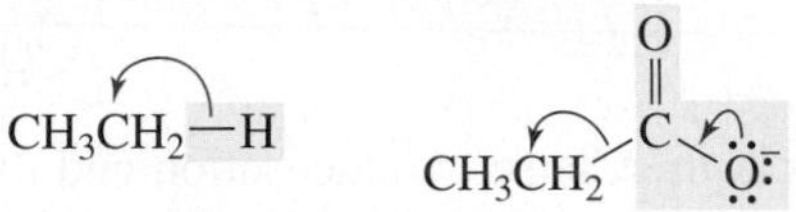

$$CH_3CH_2-H \qquad CH_3CH_2-C-O^-$$

If, however, the CO_2 group is bonded to a carbon that is adjacent to a carbonyl carbon, the CO_2 group can be removed because the electrons left behind can be delocalized onto the carbonyl oxygen. Consequently, 3-oxocarboxylate ions (carboxylate ions with a carbonyl group at the 3-position) lose CO_2 when they are heated. Loss of CO_2 from a molecule is called **decarboxylation**.

removing CO$_2$ from an α-carbon

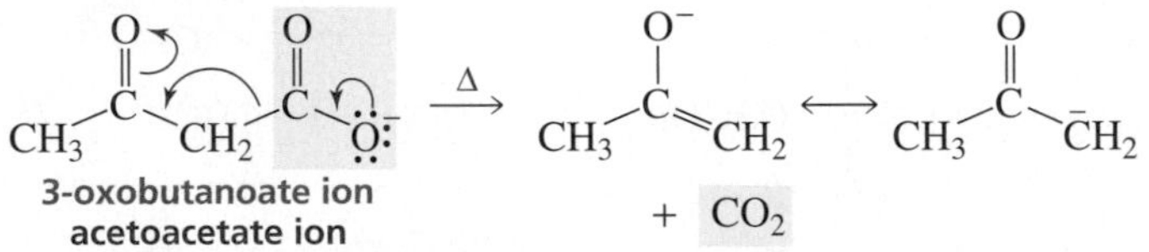

$$CH_3-C-CH_2-C-O^- \xrightarrow{\Delta} CH_3-C=CH_2 \longleftrightarrow CH_3-C-CH_2^-$$

3-oxobutanoate ion
acetoacetate ion

$+ CO_2$

Notice the similarity between removal of CO_2 from a 3-oxocarboxylate ion and removal of a proton from an α-carbon. In both reactions, a substituent—CO_2 in one case, H^+ in the other—is removed from an α-carbon and its bonding electrons are delocalized onto an oxygen.

removing a proton from an α-carbon

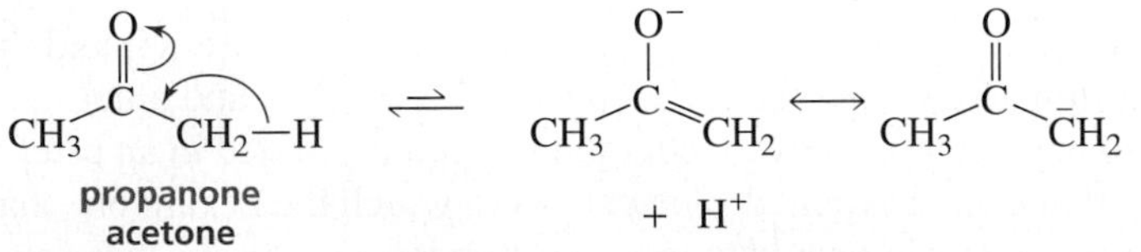

$$CH_3-C-CH_2-H \rightleftharpoons CH_3-C=CH_2 \longleftrightarrow CH_3-C-CH_2^-$$

propanone
acetone

$+ H^+$

Decarboxylation is even easier if the reaction is carried out under acidic conditions, because the reaction is catalyzed by an intramolecular transfer of a proton from the carboxyl group to the carbonyl oxygen. The enol that is formed immediately tautomerizes to a ketone.

3-Oxocarboxylic acids decarboxylate when heated.

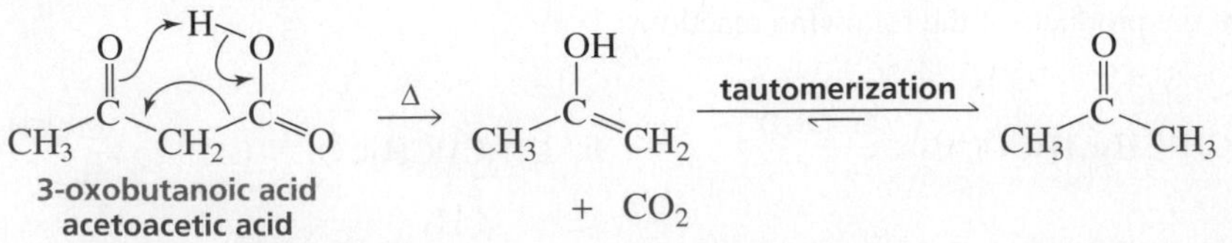

$$CH_3-C-CH_2-C=O \xrightarrow{\Delta} CH_3-C=CH_2 \xrightarrow{\text{tautomerization}} CH_3-C-CH_3$$

3-oxobutanoic acid
acetoacetic acid
a β-keto acid

$+ CO_2$

In summary, carboxylic acids with a carbonyl group at the 3-position (both β-ketocarboxylic acids and β-dicarboxylic acids) lose CO_2 when they are heated.

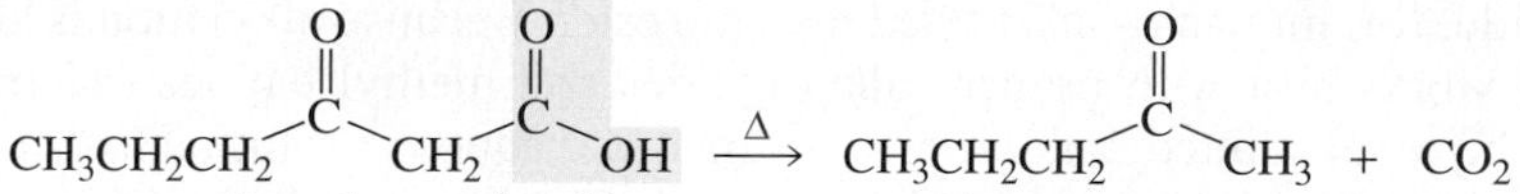

$$CH_3CH_2CH_2 \overset{O}{\underset{}{C}} CH_2 \overset{O}{\underset{}{C}} OH \xrightarrow{\Delta} CH_3CH_2CH_2 \overset{O}{\underset{}{C}} CH_3 + CO_2$$

3-oxohexanoic acid   2-pentanone

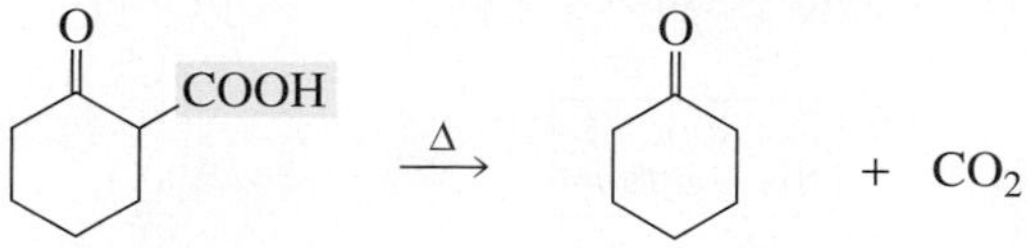

2-oxocyclohexane-   cyclohexanone
carboxylic acid

$$HO \overset{O}{\underset{}{C}} CH \overset{O}{\underset{}{C}} OH \xrightarrow{\Delta} CH_3CH_2 \overset{O}{\underset{}{C}} OH + CO_2$$
                    |
                    CH_3

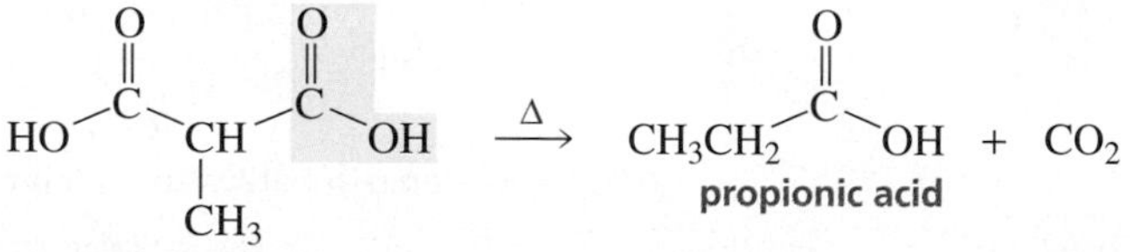

α-methylmalonic acid   propionic acid

PROBLEM 17◆

Which of the following compounds would be expected to lose CO_2 when heated?

a.

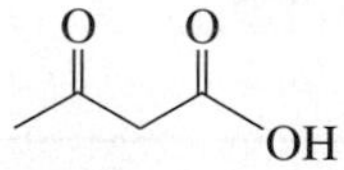

c. 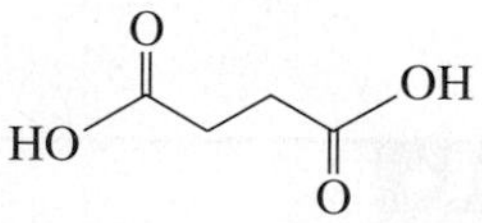 HO

b.

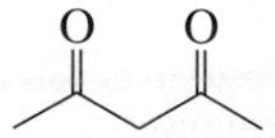

d. 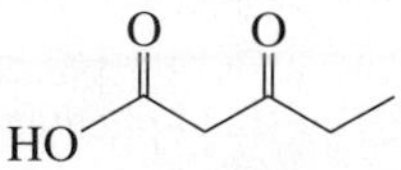 HO

14.9 The Malonic Ester Synthesis: Synthesis of Carboxylic Acids

A combination of two of the reactions discussed in this chapter—alkylation of an α-carbon and decarboxylation of a β-dicarboxylic acid—can be used to prepare carboxylic acids of any desired chain length. The procedure is called the **malonic ester synthesis** because the starting material for the synthesis is the diethyl ester of malonic acid. The first two carbons of the carboxylic acid being synthesized come from malonic ester, and the rest of the carboxylic acid comes from the alkyl halide used in the second step of the reaction.

malonic ester synthesis

$$C_2H_5O \overset{O}{\underset{}{C}} CH_2 \overset{O}{\underset{}{C}} OC_2H_5 \xrightarrow[\text{2. RBr}]{\text{1. }CH_3CH_2O^-} R-CH_2 \overset{O}{\underset{}{C}} OH$$
                                         3. HCl, H_2O, Δ

diethyl malonate
malonic ester

from malonic ester

from the alkyl halide

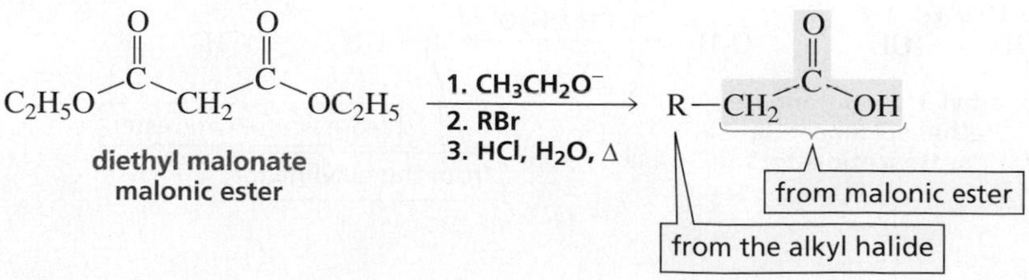

In the first part of the malonic ester synthesis, the α-carbon of the diester is alkylated (Section 14.4). A proton is easily removed from the α-carbon because it is flanked by two ester groups ($pK_a = 13$). The resulting α-carbanion reacts with an alkyl halide, forming an α-substituted malonic ester. Because alkylation is an S_N2 reaction, it works best with primary alkyl halides and methyl halides (Section 10.2). Heating the α-substituted malonic ester in an acidic aqueous solution hydrolyzes it to an α-substituted malonic acid, which, upon further heating, loses CO_2, *forming a carboxylic acid with two more carbons than the alkyl halide.*

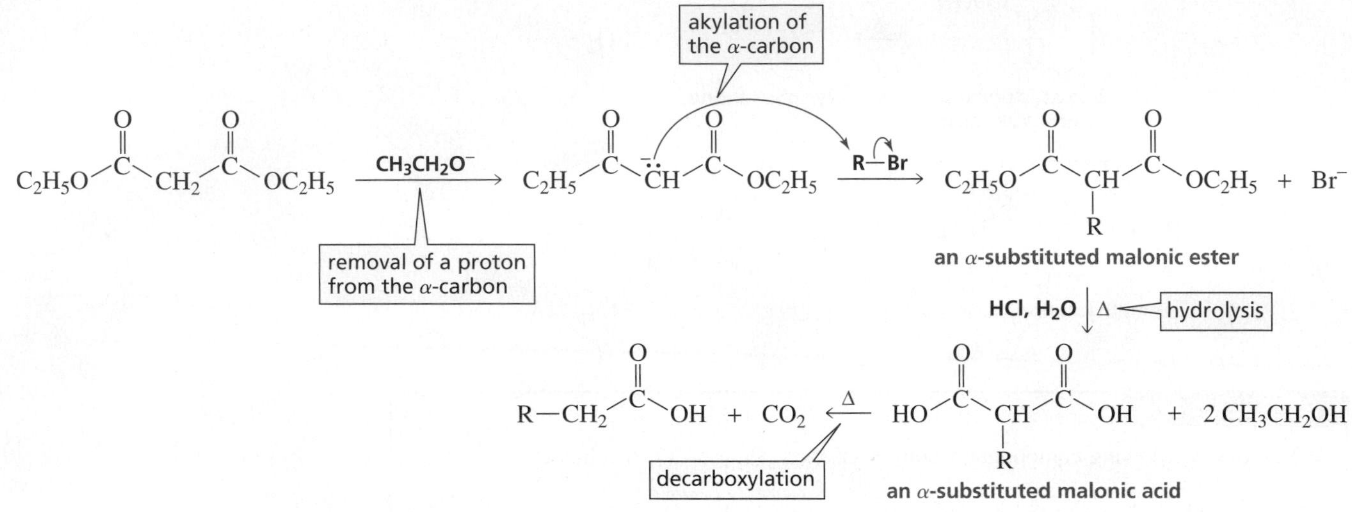

PROBLEM 18◆

What alkyl bromide should be used in the malonic ester synthesis of each of the following carboxylic acids?

a. propanoic acid **b.** 3-phenylpropanoic acid **c.** 4-methylpentanoic acid

---

14.10 The Acetoacetic Ester Synthesis: Synthesis of Methyl Ketones

The only difference between the acetoacetic ester synthesis and the malonic ester synthesis is the use of acetoacetic ester rather than malonic ester as the starting material. The difference in starting material causes the product of the **acetoacetic ester synthesis** to be a *methyl ketone* rather than a *carboxylic acid*. The carbonyl group of the methyl ketone and the carbon atoms on either side of it come from acetoacetic ester; the rest of the ketone comes from the alkyl halide used in the second step of the reaction.

acetoacetic ester synthesis

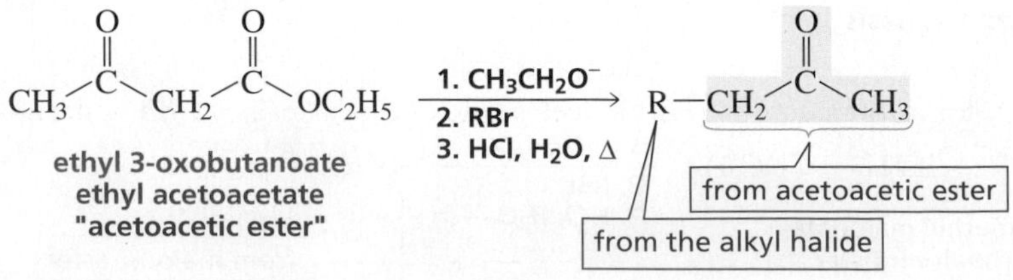

The mechanisms for the acetoacetic ester synthesis and the malonic ester synthesis are similar. The last step in the acetoacetic ester synthesis is decarboxylation of a substituted acetoacetic acid; the last step in the malonic ester synthesis is decarboxylation of a substituted malonic acid.

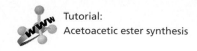

Tutorial:
Acetoacetic ester synthesis

alkylation of
the α-carbon

$$CH_3\overset{O}{\underset{}{C}}CH_2\overset{O}{\underset{}{C}}OC_2H_5 \xrightarrow{CH_3CH_2O^-} CH_3\overset{O}{\underset{}{C}}\overset{..}{C}H\overset{O}{\underset{}{C}}OC_2H_5 \xrightarrow{R-Br} CH_3\overset{O}{\underset{}{C}}\underset{\underset{R}{|}}{CH}\overset{O}{\underset{}{C}}OC_2H_5 + Br^-$$

removal of a proton
from the α-carbon

HCl, H_2O | Δ hydrolysis

$$CH_3\overset{O}{\underset{}{C}}CH_2-R + CO_2 \xleftarrow{\Delta} CH_3\overset{O}{\underset{}{C}}\underset{\underset{R}{|}}{CH}\overset{O}{\underset{}{C}}OH + CH_3CH_2OH$$

decarboxylation

PROBLEM 19 **SOLVED**

Starting with methyl propanoate, how could you prepare 4-methyl-3-heptanone?

$$\underset{\textbf{methyl propanoate}}{CH_3CH_2\overset{O}{\underset{}{C}}OCH_3} \xrightarrow{?} \underset{\underset{\textbf{4-methyl-3-heptanone}}{}}{CH_3CH_2\overset{O}{\underset{}{C}}\underset{\underset{CH_3}{|}}{CH}CH_2CH_2CH_3}$$

Tutorial:
Common terms: reactions at
the α-carbon

SOLUTION   Because the target molecule has four more carbon atoms than the starting material, a Claisen condensation appears to be a good way to start this synthesis. The Claisen condensation forms a β-keto ester that can easily be alkylated at the desired carbon because it is flanked by two carbonyl groups. Acid-catalyzed hydrolysis will form a 3-oxocarboxylic acid that will decarboxylate when it is heated.

$$CH_3CH_2\overset{O}{\underset{}{C}}OCH_3 \xrightarrow[\textbf{2. H}_3\textbf{O}^+]{\textbf{1. CH}_3\textbf{O}^-} CH_3CH_2\overset{O}{\underset{}{C}}-\underset{\underset{CH_3}{|}}{CH}-\overset{O}{\underset{}{C}}OCH_3 \xrightarrow[\textbf{2. CH}_3\textbf{CH}_2\textbf{CH}_2\textbf{Br}]{\textbf{1. CH}_3\textbf{O}^-} CH_3CH_2\overset{O}{\underset{}{C}}-\underset{\underset{CH_2CH_2CH_3}{|}}{\overset{\overset{CH_3}{|}}{C}}-\overset{O}{\underset{}{C}}OCH_3$$

Δ | HCl, H_2O

$$CH_3CH_2\overset{O}{\underset{}{C}}\underset{\underset{CH_3}{|}}{CH}CH_2CH_2CH_3$$

PROBLEM 20◆

What alkyl bromide should be used in the acetoacetic ester synthesis of each of the following methyl ketones?

a. 2-pentanone   **b.** 2-octanone   **c.** 4-phenyl-2-butanone

Many reactions that occur in biological systems involve reactions at the α-carbon. We will now look at a few examples.

A Biological Aldol Addition

Glucose, the most abundant sugar found in nature, is synthesized in biological systems from two molecules of pyruvate. The series of reactions that convert two molecules of pyruvate into glucose is called **gluconeogenesis**. The reverse process—the breakdown of glucose into two molecules of pyruvate—is called **glycolysis** (Section 19.4).

$$
2 \text{ CH}_3\overset{\overset{\displaystyle O}{\|}}{\text{C}}-\overset{\overset{\displaystyle O}{\|}}{\text{C}}\text{O}^-
\xrightleftharpoons[\;\;glycolysis\;\;]{gluconeogenesis}
\quad several\ steps \quad
\begin{array}{c}
\text{HC}{=}\text{O} \\
\text{H}-\!\!-\text{OH} \\
\text{HO}-\!\!-\text{H} \\
\text{H}-\!\!-\text{OH} \\
\text{H}-\!\!-\text{OH} \\
\text{CH}_2\text{OH}
\end{array}
$$

pyruvate   glucose

Because glucose has twice as many carbons as pyruvate, it should not be surprising that one of the steps in the biosynthesis of glucose is an aldol addition. An enzyme called aldolase catalyzes an aldol addition between dihydroxyacetone phosphate and glyceraldehyde-3-phosphate (Section 18.3). The product is fructose-1,6-diphosphate, which is subsequently converted to glucose.

3-D Molecule: Aldolase

$$
\begin{array}{c}
\text{CH}_2\text{OPO}_3{}^{2-} \\
| \\
\text{C}{=}\text{O} \\
| \\
\text{CH}_2\text{OH}
\end{array}
\qquad
\xrightleftharpoons[]{aldolase}
\qquad
\begin{array}{c}
\text{CH}_2\text{OPO}_3{}^{2-} \\
| \\
\text{C}{=}\text{O} \\
\text{HO}-\!\!-\text{H} \\
\text{H}-\!\!-\text{OH} \\
\text{H}-\!\!-\text{OH} \\
\text{CH}_2\text{OPO}_3{}^{2-}
\end{array}
$$

dihydroxyacetone phosphate   **fructose-1,6-diphosphate**

$$
\begin{array}{c}
\text{H}-\text{C}{=}\text{O} \\
\text{H}-\!\!-\text{OH} \\
\text{CH}_2\text{OPO}_3{}^{2-}
\end{array}
$$

glyceraldehyde-3-phosphate

PROBLEM 21

Propose a mechanism for the formation of fructose-1,6-diphosphate from dihydroxyacetone phosphate and glyceraldehyde-3-phosphate, using HO⁻ as the catalyst.

A Biological Aldol Condensation

Collagen is the most abundant protein in mammals, amounting to about one-fourth of the total protein. It is the major fibrous component of bone, teeth, skin, cartilage, and tendons. Individual collagen molecules—called tropocollagen—can be isolated only from tissues of young animals. As animals age, the individual molecules become cross-linked, which is why meat from older animals is tougher than meat from younger ones. Collagen cross-linking is an example of an aldol condensation.

Before collagen molecules can cross-link, the primary amino groups of the lysine residues of collagen must be converted to aldehyde groups. The enzyme that catalyzes this reaction is called lysyl oxidase. An aldol condensation between two aldehyde residues results in a cross-linked protein.

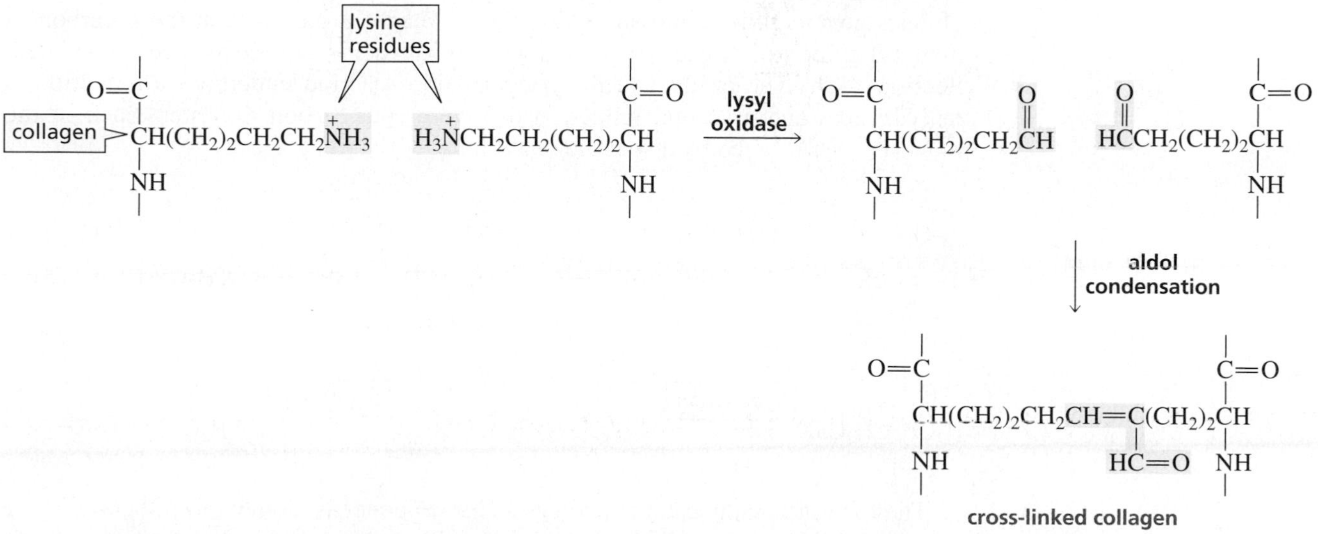

cross-linked collagen

A Biological Claisen Condensation

Fatty acids are long-chain, unbranched carboxylic acids (Section 20.1). Most naturally occurring fatty acids contain an even number of carbons because they are synthesized from acetic acid, which has two carbons.

Because the biological reaction occurs at physiological pH (7.3), the reactant is acetate—acetic acid without its proton. We have seen that carboxylate ions are very unreactive toward nucleophilic attack (Section 12.10). Biological systems activate carboxylate ions by converting them into *thioesters*. A **thioester** is an ester with sulfur in place of the carboxylate oxygen. This reaction requires ATP. ATP puts a leaving group on the carboxylate ion that can be replaced by the thiol (Section 19.2). A **thiol** is an alcohol with sulfur in place of the oxygen. The *thiol* used to make a thioester is called coenzyme A.

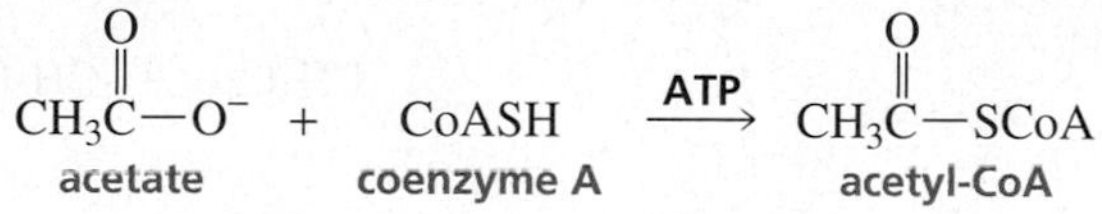

One of the necessary reactants for fatty acid synthesis is malonyl-CoA, which is obtained by carboxylation of acetyl-CoA.

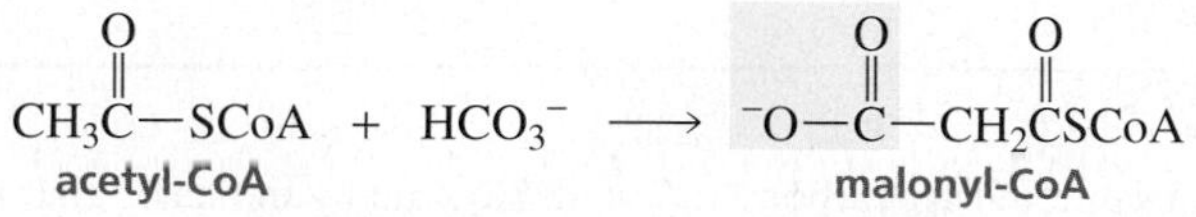

Before fatty acid synthesis can occur, the acyl groups of acetyl-CoA and malonyl-CoA are transferred to another thiol by means of a transesterification reaction (Section 12.8).

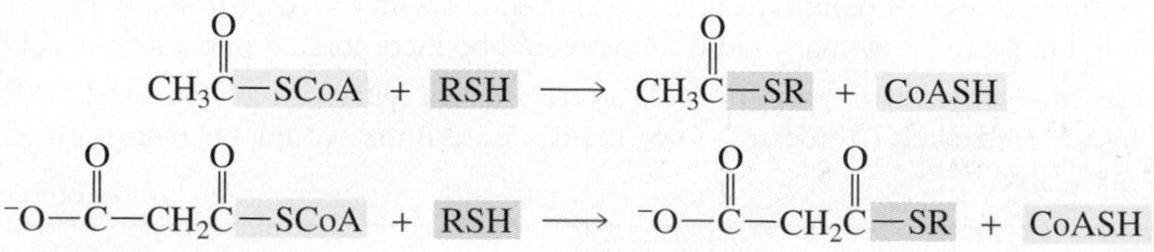

The first step in the biosynthesis of a fatty acid is a Claisen condensation between a molecule of acetyl thioester and a molecule of malonyl thioester. We have seen that the nucleophile needed for a Claisen condensation is obtained by using a strong base to remove an α-hydrogen. Strong bases are not available for biological reactions because they take place at neutral pH. So the required nucleophile is generated by removing CO_2—rather than a proton—from the α-carbon of malonyl thioester. (Recall that 3-oxocarboxylic acids are easily decarboxylated; Section 14.8.) The product of the condensation reaction undergoes a reduction, a dehydration, and a second reduction to give a four-carbon thioester. Each of the reactions is catalyzed by a specific enzyme.

$$CH_3\overset{O}{\overset{\|}{C}}-SR \quad \overset{..}{\text{:O:}}-\overset{O}{\overset{\|}{C}}-CH_2\overset{O}{\overset{\|}{C}}-SR \longrightarrow CH_3\overset{O}{\overset{\|}{C}}-CH_2\overset{O}{\overset{\|}{C}}-SR \longrightarrow CH_3\overset{:O:^-}{\overset{\|}{C}}-CH_2\overset{O}{\overset{\|}{C}}-SR$$

$$SR + CO_2 \qquad \boxed{\text{reduction}}$$

$$CH_3CH_2CH_2\overset{O}{\overset{\|}{C}}SR \overset{\boxed{\text{reduction}}}{\longleftarrow} CH_3CH=CH\overset{O}{\overset{\|}{C}}-SR \overset{\boxed{\text{dehydration}}}{\longleftarrow} CH_3\overset{OH}{\overset{|}{C}}H-CH_2\overset{O}{\overset{\|}{C}}-SR$$

The four-carbon thioester undergoes a Claisen condensation with another molecule of malonyl thioester. Again, the product of the condensation reaction undergoes a reduction, a dehydration, and a second reduction, this time to form a six-carbon thioester. The sequence of reactions is repeated, and each time two more carbons are added to the chain. This mechanism explains why naturally occurring fatty acids are unbranched and generally contain an even number of carbons.

$$CH_3CH_2CH_2\overset{O}{\overset{\|}{C}}-SR + {}^-O-\overset{O}{\overset{\|}{C}}-CH_2\overset{O}{\overset{\|}{C}}-SR \xrightarrow{\boxed{\begin{array}{c}\text{Claisen}\\\text{condensation}\end{array}}} CH_3CH_2CH_2\overset{O}{\overset{\|}{C}}-CH_2\overset{O}{\overset{\|}{C}}-SR + CO_2$$

$$\boxed{\begin{array}{l}\text{1. reduction}\\\text{2. dehydration}\\\text{3. reduction}\end{array}}$$

$$CH_3CH_2CH_2CH_2CH_2\overset{O}{\overset{\|}{C}}-SR$$

Once a thioester with the appropriate number of carbons is obtained, it undergoes a transesterification reaction with glycerol in order to form fats, oils, and phospholipids (Sections 20.3 and 20.5).

PROBLEM 22◆

Palmitic acid is a saturated 16-carbon fatty acid. How many moles of malonyl-CoA are required for the synthesis of one mole of palmitic acid?

PROBLEM 23◆

a. If the biosynthesis of palmitic acid were carried out with CD_3COSR and nondeuterated malonyl thioester, how many deuteriums would be incorporated into palmitic acid?

b. If the biosynthesis of palmitic acid were carried out with ${}^-OOCCD_2COSR$ and nondeuterated acetyl thioester, how many deuteriums would be incorporated into palmitic acid?

Summary

A hydrogen bonded to an **α-carbon** of an aldehyde, ketone, or ester is sufficiently acidic to be removed by a strong base, because the base formed when the proton is removed is stabilized by delocalization of its negative change onto an oxygen. Aldehydes and ketones ($pK_a = 16-20$) are more acidic than esters ($pK_a \sim 25$). **β-Diketones** ($pK_a \sim 9$) and **β-keto esters** ($pK_a \sim 11$) are even more acidic. The interconversion of **keto** and **enol tautomers** is called **tautomerization** or **enolization**; it can be catalyzed by acids or by bases. Generally, the **keto tautomer** is more stable. **Enolate ions** can be alkylated on the α-carbon.

In an **aldol addition**, the enolate ion of an aldehyde or a ketone attacks the carbonyl carbon of a second molecule of aldehyde or ketone, forming a β-hydroxyaldehyde or a β-hydroxyketone. The new C—C bond forms between the α-carbon of one molecule and the carbon that formerly was the carbonyl carbon of the other molecule. The product of an aldol addition can be dehydrated to give an **aldol condensation** product. In a **Claisen condensation**, the enolate ion of an ester attacks the carbonyl carbon of a second molecule of ester, eliminating an ⁻OR group to form a β-keto ester.

Carboxylic acids with a carbonyl group at the 3-position **decarboxylate** when they are heated. Carboxylic acids can be prepared by a **malonic ester synthesis**; the α-carbon of the diester is alkylated and the α-substituted malonic ester undergoes acid-catalyzed hydrolysis and decarboxylation; the resulting carboxylic acid has two more carbons than the alkyl halide. Similarly, methyl ketones can be prepared by an **acetoacetic ester synthesis**; the carbonyl group and the carbons on either side of it come from acetoacetic ester, and the rest of the methyl ketone comes from the alkyl halide.

Summary of Reactions

1. Base-catalyzed enolization (Section 14.3).

2. Acid-catalyzed enolization (Section 14.3).

3. Alkylation of an enolate ion (Section 14.4).

4. Aldol addition (Section 14.5).

5. Dehydration of the product of an aldol addition (Section 14.6).

6. Claisen condensation (Sections 14.7).

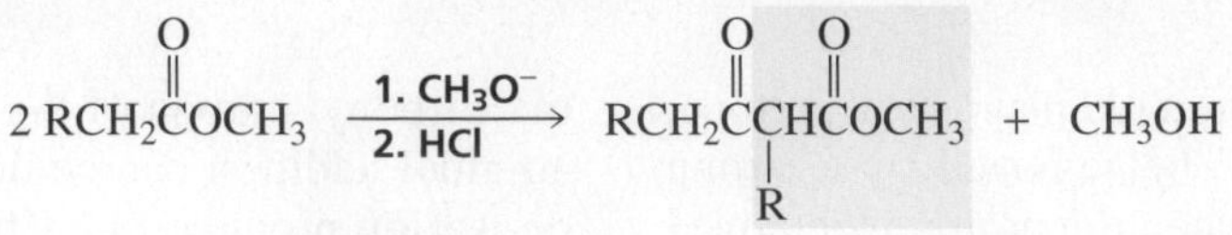

$$2 \ RCH_2\overset{O}{\underset{\|}{C}}OCH_3 \xrightarrow[\textbf{2. HCl}]{\textbf{1. CH}_3\textbf{O}^-} RCH_2\overset{O}{\underset{\|}{C}}\overset{O}{\underset{R}{\underset{|}{C}H}}\overset{O}{\underset{\|}{C}}OCH_3 \ + \ CH_3OH$$

7. Decarboxylation of a 3-oxocarboxylic acid (Section 14.8).

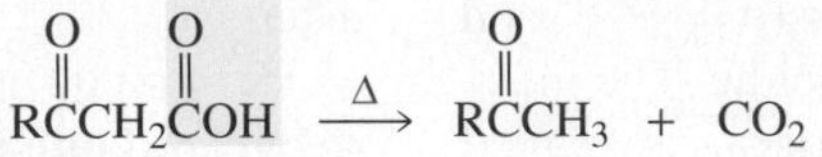

$$R\overset{O}{\underset{\|}{C}}CH_2\overset{O}{\underset{\|}{C}}OH \xrightarrow{\Delta} R\overset{O}{\underset{\|}{C}}CH_3 \ + \ CO_2$$

8. Malonic ester synthesis: preparation of carboxylic acids (Section 14.9).

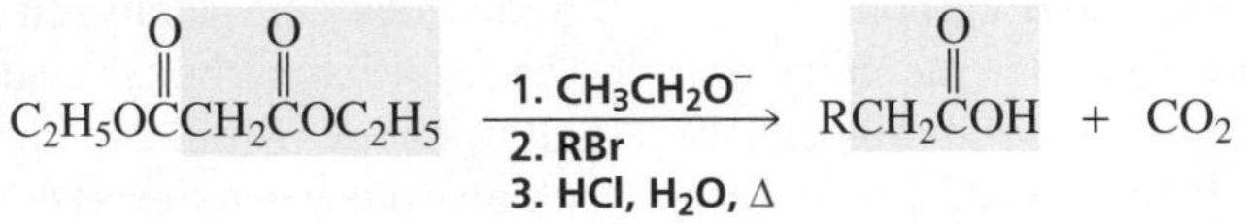

$$C_2H_5O\overset{O}{\underset{\|}{C}}CH_2\overset{O}{\underset{\|}{C}}OC_2H_5 \xrightarrow[\substack{\textbf{2. RBr} \\ \textbf{3. HCl, H}_2\textbf{O, }\Delta}]{\textbf{1. CH}_3\textbf{CH}_2\textbf{O}^-} RCH_2\overset{O}{\underset{\|}{C}}OH \ + \ CO_2$$

9. Acetoacetic ester synthesis: preparation of methyl ketones (Section 14.10).

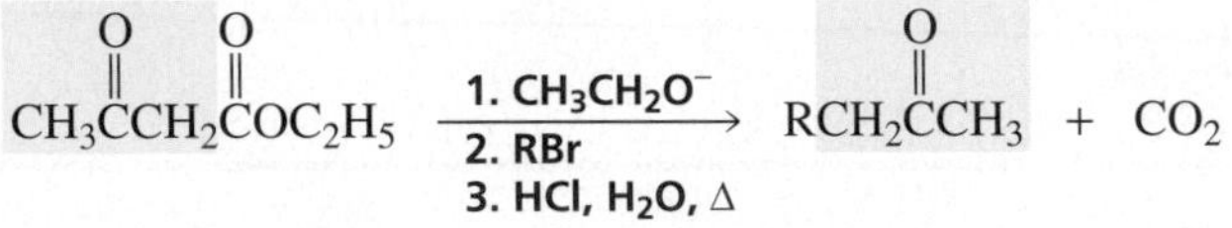

$$CH_3\overset{O}{\underset{\|}{C}}CH_2\overset{O}{\underset{\|}{C}}OC_2H_5 \xrightarrow[\substack{\textbf{2. RBr} \\ \textbf{3. HCl, H}_2\textbf{O, }\Delta}]{\textbf{1. CH}_3\textbf{CH}_2\textbf{O}^-} RCH_2\overset{O}{\underset{\|}{C}}CH_3 \ + \ CO_2$$

Problems

24. Write a structure for each of the following:
   a. a β-keto ester
   b. the enol tautomer of cyclopentanone
   c. the carboxylic acid obtained from the malonic ester synthesis when the alkyl halide is propyl bromide

25. Draw the enol tautomers for each of the following compounds. If the compound has more than one enol tautomer, indicate which one is more stable.

a. 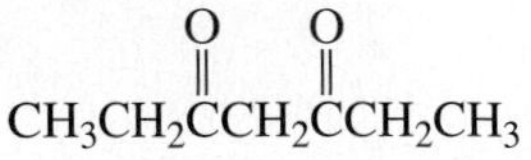 $CH_3CH_2\overset{O}{\underset{\|}{C}}CH_2\overset{O}{\underset{\|}{C}}CH_2CH_3$
   b. (phenyl)$-CH_2\overset{O}{\underset{\|}{C}}CH_3$
   c. (cyclohexanone with CH₃ substituent)

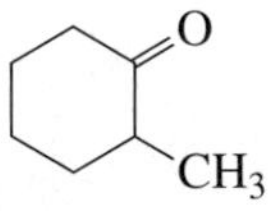

26. List the compounds in each of the following groups in order of decreasing acidity:

a. 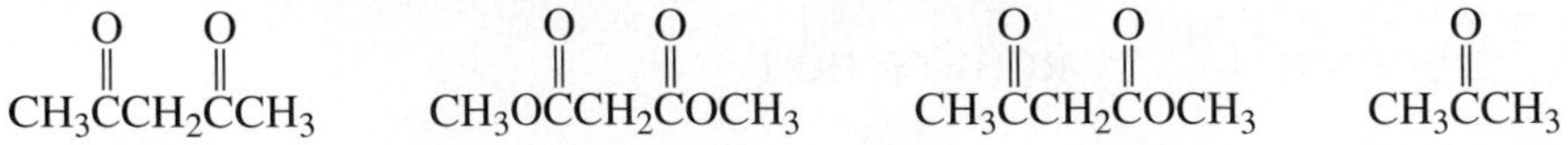 $CH_3\overset{O}{\underset{\|}{C}}CH_2\overset{O}{\underset{\|}{C}}CH_3$ $\quad$ $CH_3O\overset{O}{\underset{\|}{C}}CH_2\overset{O}{\underset{\|}{C}}OCH_3$ $\quad$ $CH_3\overset{O}{\underset{\|}{C}}CH_2\overset{O}{\underset{\|}{C}}OCH_3$ $\quad$ $CH_3\overset{O}{\underset{\|}{C}}CH_3$

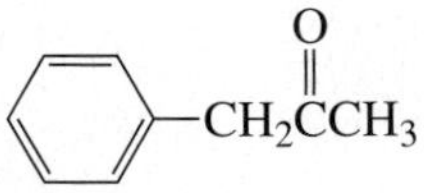

b. (δ-valerolactam with NCH₃) (δ-valerolactone) (cyclohexanone)

c. $CH_2\!=\!CH_2$ $\quad$ CH_3CH_3 $\quad$ $CH_3\overset{O}{\underset{\|}{C}}H$ $\quad$ $HC\!\equiv\!CH$

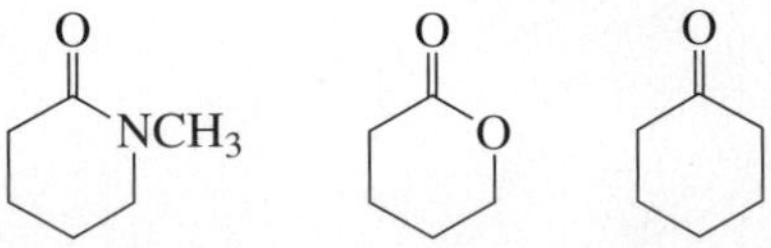

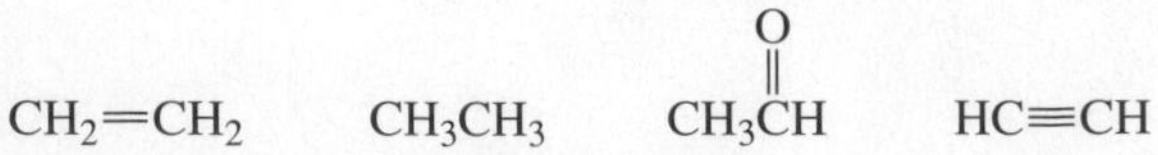

27. Show how hexanoic acid can be prepared from a malonic ester synthesis.

28. A β,γ-unsaturated carbonyl compound rearranges to a more stable conjugated α,β-unsaturated compound in the presence of either an acid or a base.
    a. Propose a mechanism for the base-catalyzed rearrangement.
    b. Propose a mechanism for the acid-catalyzed rearrangement.

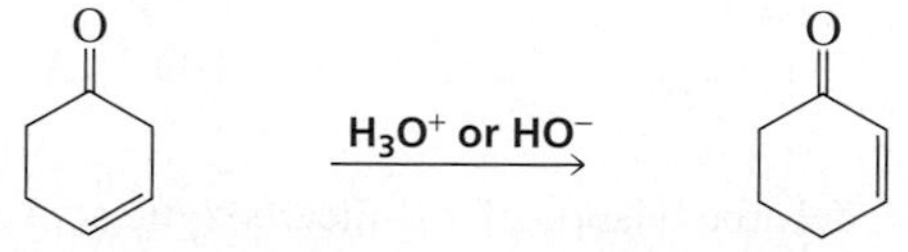

a β,γ-unsaturated
carbonyl compound

an α,β-unsaturated
carbonyl compound

29. Arachidic acid is a saturated 20-carbon fatty acid. How many moles of malonyl-CoA are required for the synthesis of one mole of arachidic acid?

30. **a.** If the biosynthesis of arachidic acid were carried out with CD_3COSR and nondeuterated malonyl thioester, how many deuteriums would be incorporated into arachidic acid?
    b. If the biosynthesis of arachidic acid were carried out with $^-OOCCD_2COSR$ and nondeuterated acetyl thioester, how many deuteriums would be incorporated into arachidic acid?

31. Give the structures of the four β-keto esters that would be obtained from a mixture of methyl acetate and methyl propanoate in a solution of $NaOCH_3$ in methanol.

32. Show how the following compounds could be synthesized from the given starting materials:

    a. 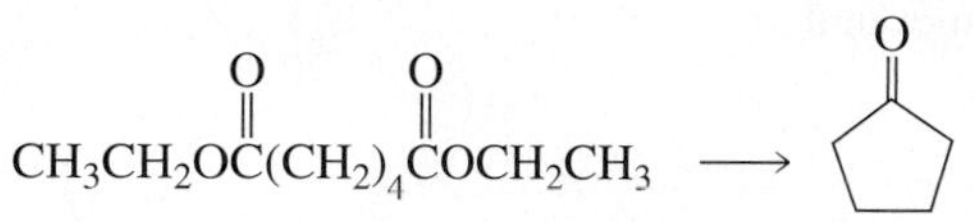 $CH_3CH_2OC(CH_2)_4COCH_2CH_3 \longrightarrow$

    b. 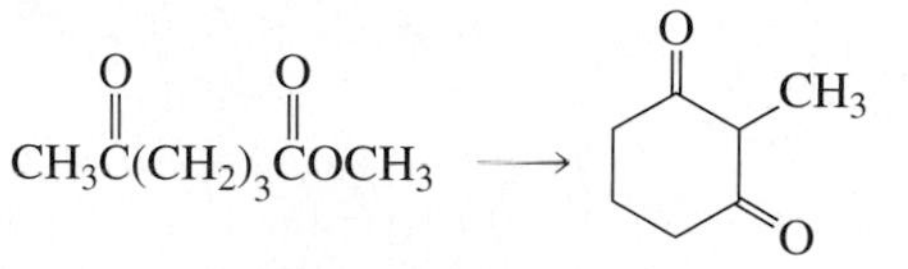 $CH_3C(CH_2)_3COCH_3 \longrightarrow$

33. Explain why the following carboxylic acid cannot be prepared by the malonic ester synthesis:

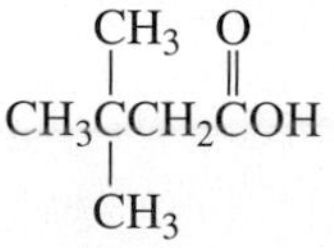

34. Both 2,6-heptanedione and 2,8-nonanedione form a product with a six-membered ring when treated with sodium hydroxide. Give the structure of the six-membered ring products.

35. **a.** Explain why a racemic mixture of 2-methyl-1-phenyl-1-butanone is formed when (R)-2-methyl-1-phenyl-1-butanone is dissolved in a basic aqueous solution.
    b. Give an example of another ketone that would form a racemic mixture in a basic aqueous solution.

36. An intramolecular Claisen condensation is called a Dieckmann condensation. Give the mechanism for the following Dieckmann condensation.

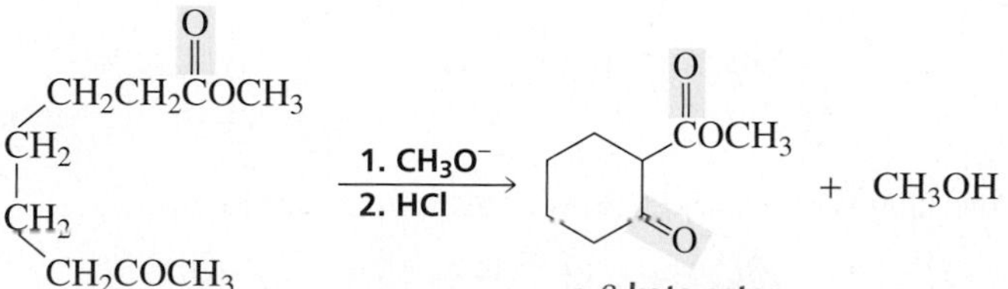

a 1,7-diester

a β-keto ester

$+ \ CH_3OH$

37. What product is formed when a 1,6-diester undergoes a Dieckmann condensation?

38. Give the major product of the following reaction:

39. a. Draw the enol tautomer of 2,4-pentanedione.
  b. Most ketones form less than 1% enol in an aqueous solution. Explain why the enol tautomer of 2,4-pentanedione is much more prevalent (15%).

40. Give the products of the following reactions:
  a. diethyl heptanedioate: (1) sodium ethoxide; (2) HCl
  b. diethyl 2-ethylhexanedioate: (1) sodium ethoxide; (2) HCl
  c. diethyl malonate: (1) sodium ethoxide; (2) isobutyl bromide; (3) HCl, H_2O + Δ
  d. 2,7-octanedione + sodium hydroxide

41. Which would require a higher temperature, decarboxylation of a β-dicarboxylic acid or decarboxylation of a β-keto acid?

42. What compound is formed when a dilute solution of cyclohexanone is shaken with NaOD in D_2O for several hours?

43. Give the structures of the four β-hydroxyaldehydes that would be obtained from a mixture of butanal and pental in a basic aqueous solution.

44. Ninhydrin reacts with an amino acid to form a purple-colored compound. Propose a mechanism to account for the formation of the colored compound.

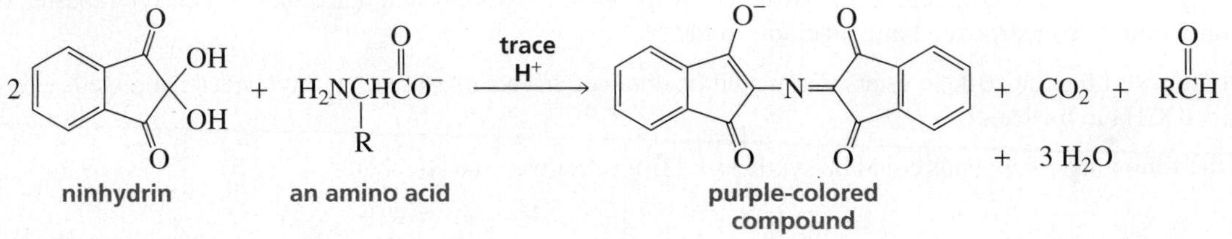

15 Determining the Structures of Organic Compounds

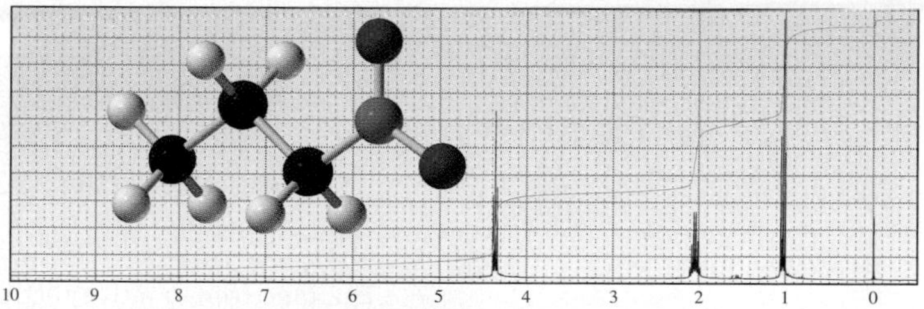

1-Nitropropane

Determining the structures of organic compounds is an important part of organic chemistry. Whenever a chemist synthesizes a compound, its structure must be confirmed. For example, we saw that a ketone is formed when an alkyne undergoes the acid-catalyzed addition of water (Section 5.11). But how was it determined that the product of that reaction is actually a ketone?

Scientists search the world for new compounds with physiological activity. If a promising compound is found, its structure needs to be determined. Without knowing its structure, chemists cannot design ways to synthesize the compound, nor can they undertake studies to provide insights into its biological behavior.

At one time, determining the structure of an organic compound was a daunting task and required a relatively large amount of the compound—a real problem for the analysis of compounds that were difficult to obtain.

Today, a number of different instrumental techniques are used to determine the structures of organic compounds. These techniques can be performed quickly on small amounts of a compound. We have already discussed one such technique: ultraviolet/visible (UV/Vis) spectroscopy, which provides information about organic compounds with conjugated double bonds (Section 6.10). In this chapter, we will look at two more instrumental techniques:

- infrared (IR) spectroscopy, which allows us to determine the *kinds of functional groups* in a compound, and
- nuclear magnetic resonance (NMR) spectroscopy, which provides information about the carbon–hydrogen framework of a compound.

We will be referring to different classes of organic compounds as we discuss the various instrumental techniques; these classes are listed inside the back cover of the book for easy reference.

15.1 Spectroscopy and the Electromagnetic Spectrum

Spectroscopy is the study of the interaction of matter and *electromagnetic radiation*. A continuum of different types of electromagnetic radiation—each type associated with a particular energy range—constitutes the electromagnetic spectrum (Figure 15.1).

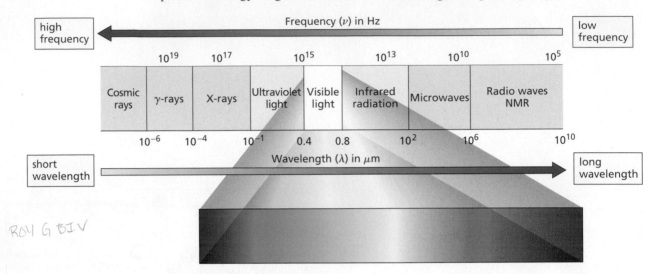

▲ **Figure 15.1**
The electromagnetic spectrum.

The electromagnetic spectrum is made up of the following components:

- *Cosmic rays*, which consist of radiation discharged by the sun, have the highest energy.
- *γ-Rays* (gamma rays) are emitted from the nuclei of certain radioactive elements and, because of their high energy, can severely damage biological organisms.
- *X-Rays,* somewhat lower in energy than γ-rays, are less harmful, except in high doses. Low-dose X-rays are used to examine the internal structure of organisms. The denser the tissue, the more it blocks X-rays.
- *Ultraviolet (UV) light* is responsible for sunburns, and repeated exposure can cause skin cancer by damaging DNA molecules in skin cells.
- *Visible light* is the electromagnetic radiation we see.
- We feel *infrared radiation* as heat.
- We cook with *microwaves* and use them in radar.
- *Radio waves* have the lowest energy. They are used for radio and television communication, digital imaging, remote controls, and wireless linkages for laptop computers. Radio waves are also used in NMR spectroscopy and in magnetic resonance imaging (MRI).

Electromagnetic radiation can be characterized by either its frequency (ν) or its wavelength (λ). **Frequency** is defined as the number of wave crests that pass by a given point in one second. **Wavelength** is the distance from any point on one wave to the corresponding point on the next wave.

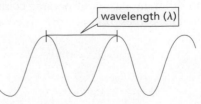

The relationship between the energy (E) and the frequency (or the wavelength) of the electromagnetic radiation is described by the equation

$$E = h\nu = \frac{hc}{\lambda}$$

where h is the proportionality constant called *Planck's constant*, named after the German physicist who discovered the relationship. The frequency of electromagnetic radiation, therefore, is equal to the speed of light (c) divided by the radiation's wavelength:

$$\nu = \frac{c}{\lambda} \qquad c = 3 \times 10^{10}\ \text{cm/s}$$

Wavenumber ($\widetilde{\nu}$) is another way to describe the *frequency* of electromagnetic radiation, and the one most often used in infrared spectroscopy. It is the number of waves in one centimeter. Therefore, it has units of reciprocal centimeters. The relationship between wavenumber (in cm^{-1}) and wavelength (in μm) is given by the equation

$$\widetilde{\nu}(\text{cm}^{-1}) = \frac{10^4}{\lambda(\mu m)} \qquad (\text{because } 1\ \mu m = 10^{-4}\ \text{cm})$$

So *high frequencies*, *large wavenumbers*, and *short wavelengths* are associated with *high energy*.

> High frequencies, large wavenumbers, and short wavelengths are associated with high energy.

PROBLEM 1◆

One of the following depicts the waves associated with infrared radiation, and one depicts the waves associated with ultraviolet light. Which is which?

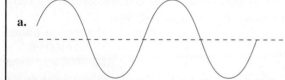 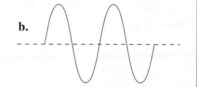

a.
b.

PROBLEM 2◆

a. Which is higher in energy, electromagnetic radiation with wavenumber $100\ \text{cm}^{-1}$ or with wavenumber $2000\ \text{cm}^{-1}$?

b. Which is higher in energy, electromagnetic radiation with wavelength $9\ \mu m$ or with wavelength $8\ \mu m$?

c. Which is higher in energy, electromagnetic radiation with wavenumber $3000\ \text{cm}^{-1}$ or with wavelength $2\ \mu m$?

15.2 Infrared Spectroscopy

When we say that a bond between two atoms has a certain length, we are stating an average length because a bond behaves as if it were a vibrating spring connecting two atoms. A bond vibrates with both stretching and bending motions. A *stretch* is a vibration occurring along the line of the bond that changes the bond length. A *bend* is a vibration that does *not* occur along the line of the bond that changes the bond angle.

Tutorial:
IR stretching and bending

H————Cl

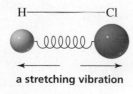

a stretching vibration

Each stretching and bending vibration of a bond in a molecule occurs with a characteristic frequency. **Infrared radiation** has just the right energy to correspond to the energy of the stretching and bending vibrations of organic molecules. When a compound is bombarded with radiation of a frequency that exactly matches the frequency of one of its vibrations, the molecule absorbs energy. By experimentally determining the wavenumbers of the energy absorbed by a particular compound, we can ascertain what kinds of bonds it has. For example, the stretching vibration of a $C=O$ bond absorbs energy with wavenumber ~ 1700 cm^{-1}, whereas the stretching vibration of an $O-H$ bond absorbs energy with wavenumber ~ 3450 cm^{-1} (Figure 15.2).

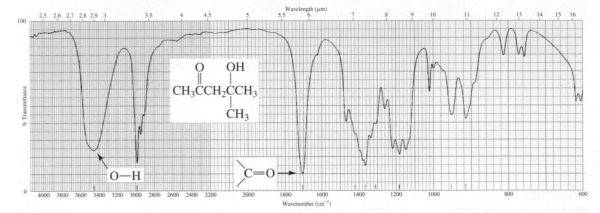

Figure 15.2 ▶
The infrared spectrum of 4-hydroxy-4-methyl-2-pentanone.

An infrared (IR) spectrum can be divided into two areas. The left-hand two-thirds of the spectrum (4000–1400 cm^{-1}) is where most of the functional groups show absorption bands. This is called the **functional group region**. The right-hand third (1400–600 cm^{-1}) of the spectrum is called the **fingerprint region** because it is characteristic of the compound as a whole, just as a fingerprint is characteristic of an individual.

Because it takes more energy to stretch a bond than to bend it, **absorption bands** for stretching vibrations are found in the functional group region (4000–1400 cm^{-1}), whereas absorption bands for bending vibrations are typically found in the fingerprint region (1400–600 cm^{-1}). Stretching vibrations, therefore, are the most useful vibrations in determining what kinds of bonds a molecule has. The IR **stretching frequencies** associated with different types of bonds are shown in Table 15.1.

It takes more energy to stretch a bond than to bend it.

Table 15.1	Important IR Stretching Frequencies	
Type of bond	**Wavenumber (cm^{-1})**	**Intensity**
$C\equiv N$	2260–2220	medium
$C\equiv C$	2260–2100	medium to weak
$C=C$	1680–1600	medium
$C=N$	1650–1550	medium
⬡	~ 1600 and ~ 1500–1430	strong to weak
$C=O$	1780–1650	strong
$C-O$	1250–1050	strong
$C-N$	1230–1020	medium
$O-H$ (alcohol)	3650–3200	strong, broad
$O-H$ (carboxylic acid)	3300–2500	strong, very broad
$N-H$	3500–3300	medium, broad
$C-H$	3300–2700	medium

15.3   Characteristic Infrared Absorption Bands

IR spectra can be quite complex because the stretching and bending vibrations of each bond in a molecule can give rise to an absorption band. Therefore, organic chemists generally do not try to identify all the absorption bands in an IR spectrum. In this chapter, we will look at some characteristic absorption bands so you will be able to tell something about the structure of a compound that gives a particular IR spectrum.

Stronger bonds show absorption bands at larger wavenumbers.

Effect of Bond Order

A C≡C bond is stronger than a C=C bond, so a C≡C bond stretches at a higher frequency (~2100 cm⁻¹) than does a C=C bond (~1650 cm⁻¹). C—C bonds show stretching vibrations in the region from 1200 to 800 cm⁻¹, but these vibrations are weak and very common, so they are of little value in identifying compounds. Similarly, a C=O bond stretches at a higher frequency (~1700 cm⁻¹) than does a C—O bond (~1100 cm⁻¹), and a C≡N bond stretches at a higher frequency (~2200 cm⁻¹) than does a C=N bond (~1600 cm⁻¹), which in turn stretches at a higher frequency than does a C—N bond (~1100 cm⁻¹) (Table 15.1).

PROBLEM 3◆

Which will occur at a larger wavenumber

a. a C≡C stretch or a C=C stretch?   **c.** a C=N stretch or a C≡N stretch?

b. a C—H stretch or a C—H bend?   **d.** a C=O stretch or a C—O stretch?

Resonance Effects

Table 15.1 shows a range of wavenumbers for each stretch because the exact position of the absorption band depends on other structural features of the molecule. For example, the IR spectrum in Figure 15.3 shows that the carbonyl group (C=O) of 2-pentanone absorbs at 1720 cm⁻¹, whereas the IR spectrum in Figure 15.4 shows that the carbonyl group of 2-cyclohexenone absorbs at a lower frequency (1680 cm⁻¹). 2-Cyclohexenone's carbonyl group absorbs at a lower frequency because it has less double-bond character due to electron delocalization. Because a single bond is weaker than a double bond, a carbonyl group with

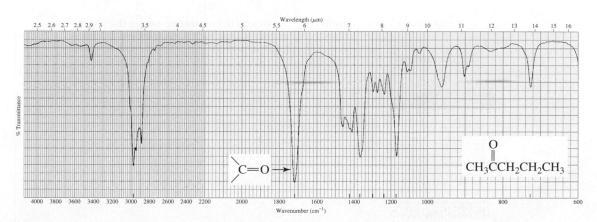

▲ **Figure 15.3**
The IR spectrum of 2-pentanone. The intense absorption band at ~1720 cm⁻¹ indicates a C=O bond.

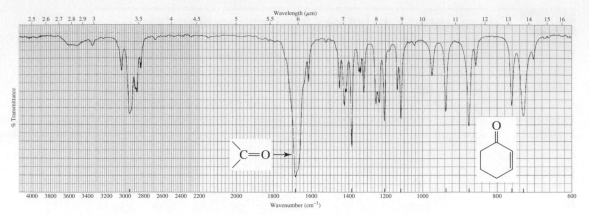

▲ **Figure 15.4**
The IR spectrum of 2-cyclohexenone. Electron delocalization gives its carbonyl group less double-bond character, so it absorbs at a lower frequency (~1680 cm^{-1}) than does a carbonyl group with localized electrons (~1720 cm^{-1}).

significant single-bond character will stretch at a lower frequency than will one with little or no single-bond character.

CH$_3$CCH$_2$CH$_2$CH$_3$

2-pentanone

\C=O at 1720 cm^{-1}

2-cyclohexenone

\C=O at 1680 cm^{-1}

PROBLEM-SOLVING STRATEGY

Which will occur at a larger wavenumber, the C—N stretch of an amine or the C—N stretch of an amide?

To answer this kind of question, we need to see if electron delocalization causes the bonds to be other than pure single bonds. When we do that, we see that electron delocalization causes the C—N bond of the amide to have partial double-bond character. The C—N stretch of an amide, therefore, will occur at a larger wavenumber.

R—NH$_2$

no electron delocalization

electron delocalization causes the C — N bond to have partial double-bond character

Now continue on to Problem 4.

PROBLEM 4◆

Which will occur at a larger wavenumber

a. the C—O stretch of phenol or the C—O stretch of cyclohexanol?

b. the C=O stretch of a ketone or the C=O stretch of an amide?

c. the stretch or the bend of the C—O bond in ethanol?

PROBLEM 5◆

Which would show an absorption band at a larger wavenumber: a carbonyl group bonded to an sp^3 hybridized carbon or a carbonyl group bonded to an sp^2 hybridized carbon?

15.4 | The Intensity and Shape of Absorption Bands

The **intensity of an absorption band** depends on the polarity of the bond: The more polar the bond, the greater is the intensity of the absorption. An O—H bond will show a more intense absorption band than an N—H bond, because the O—H bond is more polar. Likewise, an N—H bond will show more intense absorption than a C—H bond, because the N—H bond is more polar.

The more polar the bond, the more intense is the absorption.

relative bond polarities
relative intensities of IR absorption

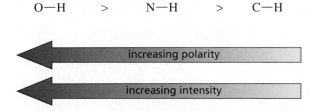

The **shape of an absorption band** can be helpful in identifying the compound responsible for an IR spectrum. For example, both O—H and N—H bonds stretch at wavenumbers above 3100 cm^{-1}, but the shapes of their stretches are distinctive. Notice the difference in the shape of these absorption bands in the IR spectra of 1-hexanol (Figure 15.5), pentanoic acid (Figure 15.6), and isopentylamine (Figure 15.7).

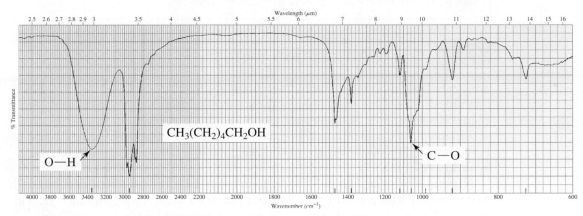

▲ **Figure 15.5**
The IR spectrum of 1-hexanol.

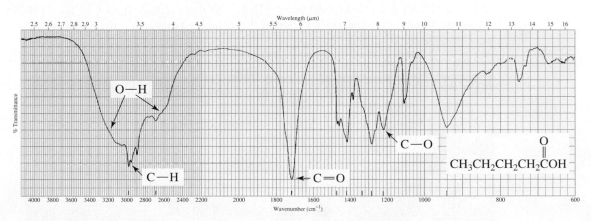

▲ **Figure 15.6**
The IR spectrum of pentanoic acid.

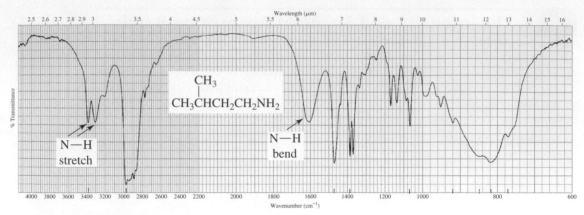

▲ **Figure 15.7**
The IR spectrum of isopentylamine. The double peak at ~3300 cm⁻¹ indicates the presence of two N—H bonds.

The position, intensity, and shape of an absorption band are helpful in identifying functional groups.

An N—H absorption band ($\sim$3300 cm^{-1}) is narrower and less intense than an O—H absorption band ($\sim$3300 cm^{-1}), and the O—H absorption band of a carboxylic acid ($\sim$3300 $-$ 2500 cm^{-1}) is broader than the O—H absorption band of an alcohol.

PROBLEM 6

Why is the C—O absorption band of 1-hexanol at a smaller wavenumber (1060 cm^{-1}) than the C—O absorption band of pentanoic acid (1220 cm^{-1})?

15.5 C—H Absorption Bands

A C—H bond is stronger when the carbon is sp hybridized than when it is sp^2 hybridized, which in turn is stronger than when the carbon is sp^3 hybridized. More energy is needed to stretch a stronger bond, and this is reflected in the C—H stretch absorption bands, which occur at $\sim$3300 cm^{-1} if the carbon is sp hybridized, at $\sim$3100 cm^{-1} if the carbon is sp^2 hybridized, and at $\sim$2900 cm^{-1} if the carbon is sp^3 hybridized (Table 15.2).

A useful step in the analysis of a spectrum entails looking at the absorption bands in the vicinity of 3000 cm^{-1}. Figures 15.8 and 15.9 show the IR spectra for methylcyclohexane and cyclohexene. The only absorption band in the vicinity of 3000 cm^{-1} in Figure 15.8 is slightly to the right of that value. This tells us that the compound has hydrogens bonded to sp^3 carbons, but none bonded to sp^2 or to sp carbons. The spectrum in Figure 15.9 shows absorption bands slightly to the left and slightly to the right of 3000 cm^{-1}, indicating that the compound that produced that spectrum contains hydrogens bonded to sp^2 and sp^3 carbons.

Tutorial:
IR spectra

Table 15.2	IR Absorptions of Carbon–Hydrogen Bonds
Carbon–Hydrogen Stretching Vibrations	**Wavenumber (cm⁻¹)**
C≡C—H	~3300
C=C—H	3100–3020
C—C—H	2960–2850
R—C(=O)—H	~2820 and ~2720

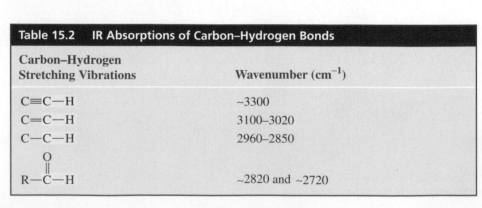

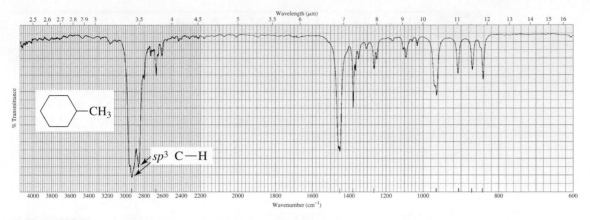

▲ Figure 15.8
The IR spectrum of methylcyclohexane.

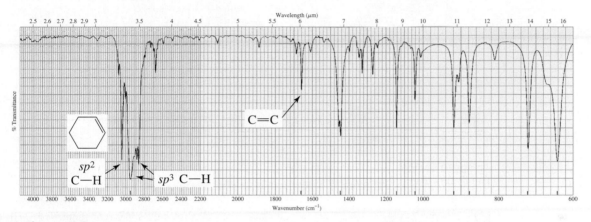

▲ Figure 15.9
The IR spectrum of cyclohexene.

The stretch of the C—H bond in an aldehyde group shows two absorption bands, one at ~2820 cm^{-1} and the other at ~2720 cm^{-1} (Figure 15.10). This makes aldehydes relatively easy to identify because essentially no other absorption occurs at these wavenumbers.

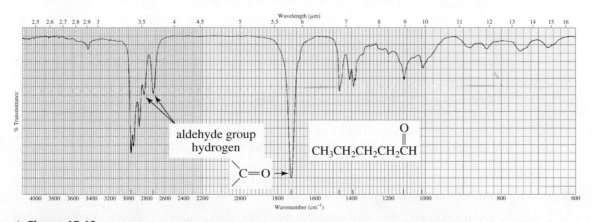

▲ Figure 15.10
The IR spectrum of pentanal. The absorptions at ~2820 and ~2720 cm^{-1} readily identify an aldehyde group. Note also the intense absorption band at ~1730 cm^{-1} indicating a C=O bond.

15.6 Absence of Absorption Bands

The absence of an absorption band can be as useful as the presence of a band in identifying a compound by IR spectroscopy. For example, the spectrum in Figure 15.11 shows a strong absorption at $\sim$1100 cm^{-1}, indicating the presence of a C—O bond (Table 15.1). Clearly, the compound is not an alcohol because there is no absorption above 3100 cm^{-1}. Nor is it an ester or any other kind of carbonyl compound because there is no absorption at $\sim$1700 cm^{-1}. The compound has no C≡C, C=C, C≡N, C=N, or C—N bonds. We may deduce, then, that the compound is an ether. Its C—H absorption bands show that it has hydrogens only on sp^3 hybridized carbons. The compound is actually diethyl ether.

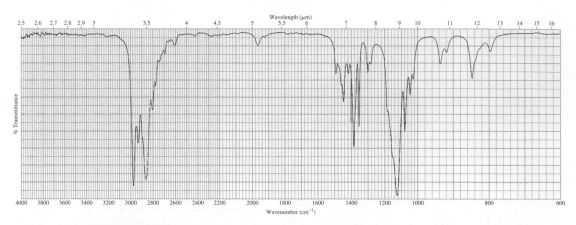

▲ **Figure 15.11**
The IR spectrum of diethyl ether.

PROBLEM 7◆

a. An oxygen-containing compound shows an absorption band at $\sim$1700 cm^{-1} and no absorption bands at $\sim$3300 cm^{-1}, $\sim$2700 cm^{-1}, or $\sim$1100 cm^{-1}. What class of compound is it?

b. A nitrogen-containing compound shows no absorption band at $\sim$3400 cm^{-1} and no absorption bands between 1700 cm^{-1} and 1600 cm^{-1}. What class of compound is it?

15.7 Identifying Infrared Spectra

We will now look at a few IR spectra and see what we can determine about the structure of the compounds that give rise to the spectra. We might not be able to identify the compound precisely, but when we are told what it is, its structure should fit with our observations.

Compound 1. The absorption in the 3000 cm^{-1} region in Figure 15.12 indicates that hydrogens are attached to sp^2 carbons (3050 cm^{-1}) but not to sp^3 carbons. The absorptions at 1600 cm^{-1} and 1460 cm^{-1} indicate that the compound has a benzene ring. The absorptions at 2810 cm^{-1} and 2730 cm^{-1} show that the compound is an aldehyde. The absorption band for the carbonyl group (C=O) is lower (1700 cm^{-1}) than normal (1720 cm^{-1}), so the carbonyl group has partial single-bond character. Thus, it must be attached directly to the benzene ring. The compound is benzaldehyde.

Compound 2. The absorptions in the 3000 cm^{-1} region in Figure 15.13 indicate that hydrogens are attached to sp^3 carbons (2950 cm^{-1}) but not to sp^2 carbons. The shape of the strong absorption band at 3300 cm^{-1} is characteristic of an O—H group of an alcohol. The absorption at $\sim$2100 cm^{-1} indicates that the compound has a triple bond.

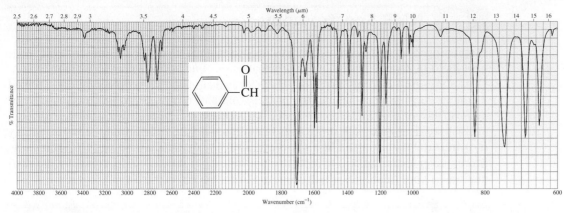

▲ **Figure 15.12**
The IR spectrum of Compound 1.

The sharp absorption band at 3300 cm^{-1} indicates that the compound has a hydrogen on an *sp* hybridized carbon, so we know the compound is a terminal alkyne. It is 2-propyn-1-ol.

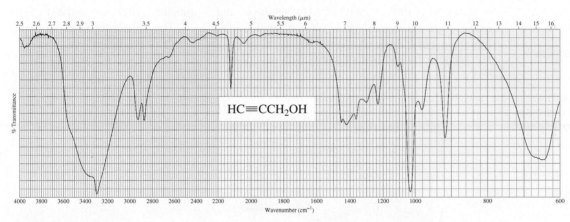

▲ **Figure 15.13**
The IR spectrum of Compound 2.

PROBLEM 8

How could IR spectroscopy distinguish between the following?

a. a ketone and an aldehyde

b. benzene and toluene

c. cyclohexene and cyclohexane

d. a primary amine and a tertiary amine

PROBLEM 9

For each of the following pairs of compounds, give one absorption band that could be used to distinguish between them:

a. $CH_3CH_2COCH_3$ and CH_3CH_2COH **c.** $CH_3CH_2C\equiv CCH_3$ and $CH_3CH_2C\equiv CH$

b. CH_3CH_2COH and $CH_3CH_2CH_2OH$ **d.** ⬡ and ⬡

15.8 NMR Spectroscopy

A spinning charged ^1H or ^{13}C nucleus generates a magnetic field, similar to the magnetic field of a small bar magnet. In the absence of an applied magnetic field, the nuclear spins are randomly oriented. However, when a sample is placed in an applied magnetic field (Figure 15.14), the nuclei twist and turn to align themselves *with* or *against* the field of the larger magnet. More energy is needed for a proton to align against the field than with it. Protons that align with the field are in the lower-energy **α-spin state**; protons that align against the field are in the higher-energy **β-spin state**. More nuclei are in the α-spin state than in the β-spin state. The difference in the populations is very small, but is sufficient to form the basis of NMR spectroscopy.

Figure 15.14 ▶
In the absence of an applied magnetic field, the spins of the nuclei are randomly oriented. In the presence of an applied magnetic field, the spins of the nuclei line up with or against the field.

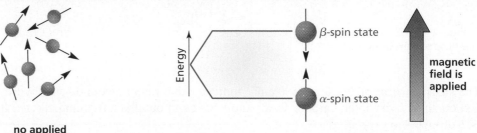

no applied
magnetic field

Earth's magnetic field is 5 × 10⁻⁵ T, measured at the equator. Its maximum surface magnetic field is 7 × 10⁻⁵ T, measured at the south magnetic pole.

The energy difference (ΔE) between the α- and β-spin states depends on the strength of the **applied magnetic field (B_0)**, measured in tesla (T). The greater the strength of the magnetic field to which the nuclei are exposed, the greater is the difference in energy between the α- and β-spin states (Figure 15.15).

Figure 15.15 ▶
The greater the strength of the applied magnetic field, the greater is the difference in energy between the α- and β-spin states.

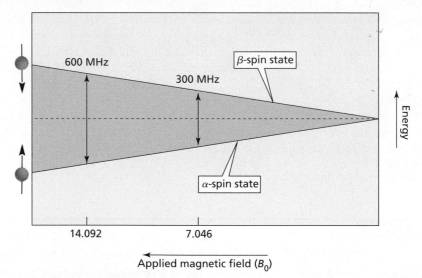

When a sample is subjected to radiation whose energy corresponds to the difference in energy (ΔE) between the α- and β-spin states, nuclei in the α-spin state are promoted to the β-spin state. When the nuclei return to their original state, they emit signals whose frequency depends on the difference in energy (ΔE) between the α- and β-spin states. The NMR spectrometer detects these signals and displays them as a plot of signal frequency versus intensity—an NMR spectrum. It is because the nuclei are *in resonance* with the radiation that the term "**nuclear magnetic resonance (NMR)**" came into being. In this context, "resonance" refers to the flipping back and forth of nuclei between the α- and β-spin states in response to the radiation.

Today's NMR spectrometers operate at frequencies between 60 and 900 MHz. The **operating frequency** of a particular spectrometer depends on the strength of the built-in magnet (Figure 15.15). The greater the operating frequency of the instrument—and the stronger the magnet—the better is the resolution of the NMR spectrum.

NIKOLA TESLA (1856–1943)

Nikola Tesla was born in Croatia, the son of a clergyman. He immigrated to the United States in 1884 and became a citizen in 1891. He was a proponent of alternating current and bitterly fought Edison, who promoted direct current. Although Tesla did not win his dispute with Marconi over which of them invented the radio, Tesla is given credit for developing neon and fluorescent lighting, the electron microscope, the refrigerator motor, and the Tesla coil, a type of transformer.

Nikola Tesla in his laboratory.

15.9   Shielding

The frequency at which a signal occurs depends on the strength of the applied magnetic field (Figure 15.15). Thus, if all the protons in an organic compound experienced the same applied magnetic field, they would all give signals with the same frequency in response to a given applied magnetic field. If this were the case, all NMR spectra would consist of one signal, which would tell us nothing about the structure of the compound, except that it contains hydrogens.

A nucleus, however, is embedded in a cloud of electrons that partly *shields* it from the applied magnetic field. Fortunately for chemists, the **shielding** varies for different protons within a molecule. In other words, all the protons do not experience the same applied magnetic field.

What causes shielding? In a magnetic field, the electrons circulate about the nuclei and induce a local magnetic field that acts in opposition to the applied magnetic field and subtracts from it. What the nuclei "sense" through the surrounding electronic environment—the **effective magnetic field**—is, therefore, somewhat smaller than the applied field:

$$B_{\text{effective}} = B_{\text{applied}} - B_{\text{local}}$$

This means that the greater the electron density of the environment in which the proton is located, the greater B_{local} is, so the more the proton is shielded from the applied magnetic field. Thus, protons in electron-dense environments sense a *smaller effective magnetic field* and, therefore, will require a *lower frequency* to come into resonance because ΔE is smaller (Figure 15.15). Protons in electron-poor environments sense a *larger effective magnetic field* and, therefore, will require a *higher frequency* to come into resonance, because ΔE is larger.

An NMR spectrum gives a signal for each proton in a different environment. Protons in electron-rich environments are more shielded and appear at lower frequencies—on the right-hand side of the spectrum (Figure 15.16). Protons in electron-poor environments are less shielded and appear at higher frequencies—on the left-hand side of the spectrum. For example, the signal for the methyl protons of CH_3F occurs at a higher frequency than the signal for the methyl protons of CH_3Br because fluorine is more electron withdrawing than bromine; thus, its protons are in a less electron-rich environment.

The terms "upfield" and "downfield" are used to describe the position of a signal: **Upfield** means farther to the right-hand side of the spectrum, and **downfield** means farther to the left-hand side of the spectrum.

> The larger the magnetic field sensed by the proton, the higher is the frequency of the signal.

Figure 15.16 ▶
Shielded nuclei come into
resonance at lower frequencies
than deshielded nuclei.

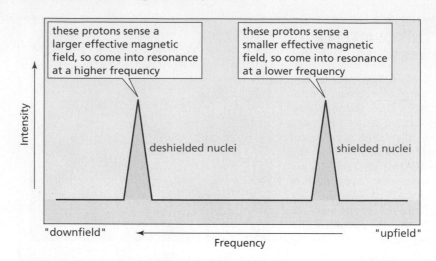

these protons sense a
larger effective magnetic
field, so come into resonance
at a higher frequency

these protons sense a
smaller effective magnetic
field, so come into resonance
at a lower frequency

deshielded nuclei

shielded nuclei

"downfield"   Frequency   "upfield"

deshielded = less shielded.

15.10 The Number of Signals in the ^{1}H NMR Spectrum

Protons in the same environment are called **chemically equivalent protons**. For example, 1-bromopropane has three different sets of chemically equivalent protons. The three methyl protons are chemically equivalent because of rotation about the C—C bond. The two methylene protons on the middle carbon are chemically equivalent, and the two methylene protons on the carbon bonded to the bromine atom make up the third set of chemically equivalent protons.

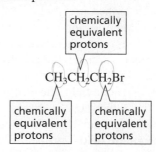

chemically
equivalent
protons

$CH_3CH_2CH_2Br$

chemically
equivalent
protons

chemically
equivalent
protons

Each set of chemically equivalent protons gives rise to a signal.

Each set of chemically equivalent protons in a compound gives rise to a signal in the ^{1}H NMR spectrum of that compound. Because 1-bromopropane has three sets of chemically equivalent protons, it has three signals in its ^{1}H NMR spectrum.

2-Bromopropane has two sets of chemically equivalent protons and, therefore, it has two signals in its ^{1}H NMR spectrum; the six methyl protons in 2-bromopropane are equivalent, so they give rise to only one signal. Ethyl methyl ether has three sets of chemically equivalent protons: the methyl protons on the carbon adjacent to the oxygen, the methylene (CH_2) protons on the carbon adjacent to the oxygen, and the methyl protons on the carbon that is one carbon removed from the oxygen. The chemically equivalent protons in the following compounds are designated by the same letter:

$\overset{a}{CH_3}\overset{b}{CH_2}\overset{c}{CH_2}Br$
three signals

$\overset{a}{CH_3}\overset{b}{CH}\overset{a}{CH_3}$
$|$
Br
two signals

$\overset{a}{CH_3}\overset{c}{CH_2}O\overset{b}{CH_3}$
three signals

$\overset{a}{CH_3}O\overset{a}{CH_3}$
one signal

$\overset{a}{CH_3}$
$|$
$\overset{a}{CH_3}\overset{}{C}O\overset{b}{CH_3}$
$|$
CH_3
a
two signals

You can tell how many sets of chemically equivalent protons a compound has from the number of signals in its ^{1}H NMR spectrum.

PROBLEM 10◆

How many signals would you expect to see in the ^{1}H NMR spectrum of each of the following compounds?

a. $CH_3CH_2CH_2CH_3$

c. $CH_3CH_2CH_2\overset{\overset{\textstyle O}{\|}}{C}CH_3$

e. $CH_3\underset{\underset{\textstyle CH_3}{|}}{C}HCH_2\underset{\underset{\textstyle CH_3}{|}}{C}HCH_3$

b. $BrCH_2CH_2Br$

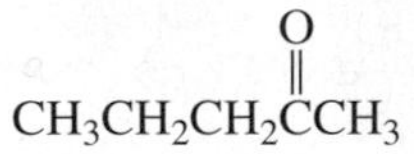

d. 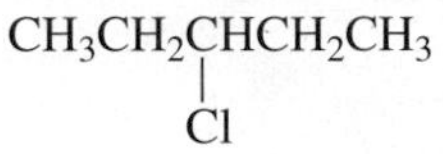 $CH_3CH_2\underset{\underset{\textstyle Cl}{|}}{C}HCH_2CH_3$

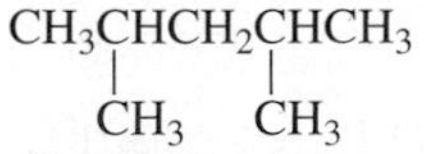

f. 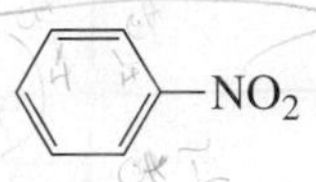  —NO_2

PROBLEM 11◆

How could you distinguish the ^{1}H NMR spectra of the following compounds?

a. $CH_3OCH_2OCH_3$

b. CH_3OCH_3

c. 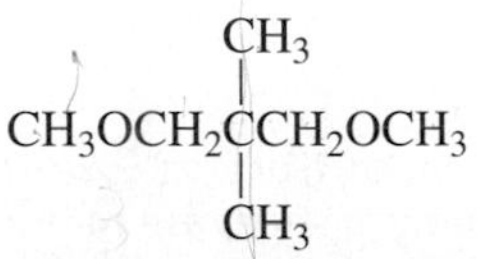 $CH_3OCH_2\underset{\underset{\textstyle CH_3}{|}}{\overset{\overset{\textstyle CH_3}{|}}{C}}CH_2OCH_3$

15.11 The Chemical Shift

A small amount of an inert **reference compound** is added to the sample tube containing the compound whose NMR spectrum is to be taken. The positions of the signals in an NMR spectrum are defined according to how far they are from the signal of the reference compound. The most commonly used reference compound is tetramethylsilane (TMS).

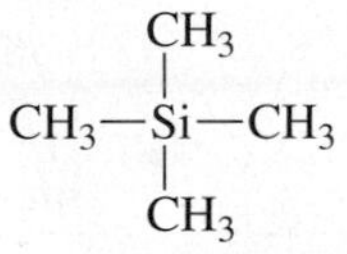

$$CH_3-\underset{\underset{\textstyle CH_3}{|}}{\overset{\overset{\textstyle CH_3}{|}}{Si}}-CH_3$$

tetramethylsilane
TMS

3-D Molecule:
Tetramethylsilane

The methyl protons of TMS are in a more electron-dense environment than are most protons in organic molecules because silicon is less electronegative than carbon (electronegativities of 1.8 and 2.5, respectively). Consequently, the signal for the methyl protons of TMS is at a lower frequency than most other signals (i.e., it appears to the right of the other signals).

The position at which a signal occurs in an NMR spectrum is called the *chemical shift*. The **chemical shift** (δ) is a measure of how far the signal is from the reference (TMS) signal. The chemical shift is determined by measuring the distance from the reference signal (in hertz) and dividing by the operating frequency of the instrument (in megahertz). Because the units are Hz/MHz, a chemical shift has units of parts per million (ppm) of the operating frequency:

$$\delta = \text{chemical shift (ppm)} = \frac{\text{distance downfield from TMS (Hz)}}{\text{operating frequency of the spectrometer (MHz)}}$$

Most proton chemical shifts fall in the range from 0 to 10 ppm.

The ^{1}H NMR spectrum for 1-bromo-2,2-dimethylpropane in Figure 15.17 shows that the chemical shift of the methyl protons is at 1.05 ppm and the chemical shift of the methylene protons is at 3.28 ppm. *Notice that low-frequency (upfield, shielded) signals have small δ (ppm) values, whereas high-frequency (downfield, deshielded) signals have large δ values.*

The greater the chemical shift (δ) the higher is the frequency.

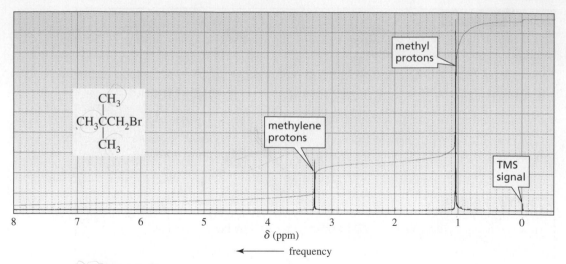

▲ Figure 15.17
^{1}H NMR spectrum of 1-bromo-2,2-dimethylpropane. The TMS signal is a reference signal from which chemical shifts are measured; it defines the zero position on the scale.

The chemical shift (δ) is independent of the operating frequency of the spectrometer.

The advantage of the δ scale is that the chemical shift of a given nucleus is *independent of the operating frequency of the NMR spectrometer*. Thus, the chemical shift of the methyl protons of 1-bromo-2,2-dimethylpropane is at 1.05 ppm in both a 60-MHz and a 360-MHz instrument. The following diagram will help you keep track of the terms associated with NMR spectroscopy:

protons in electron-poor environments	protons in electron-dense environments
deshielded protons	shielded protons
downfield	upfield
high frequency	low frequency
large δ values	small δ values

$\longleftarrow \delta$
$\longleftarrow$ frequency

PROBLEM 12◆

A signal has been reported to occur at 600 Hz downfield from the TMS signal in an NMR spectrometer with a 300-MHz operating frequency.

a. What is the chemical shift of the signal?

b. What would its chemical shift be in an instrument operating at 100 MHz?

PROBLEM 13◆

If two signals differ by 1.5 ppm in a 300-MHz spectrometer, by how much do they differ in a 100-MHz spectrometer?

PROBLEM 14◆

Where would you expect to find the ^{1}H NMR signal of $(CH_3)_2Mg$ relative to the TMS signal? (*Hint:* Magnesium is even less electronegative than silicon.)

15.12 The Relative Positions of ^{1}H NMR Signals

The ^{1}H NMR spectrum of 1-bromo-2,2-dimethylpropane in Figure 15.17 has two signals because the compound has two different kinds of protons. The methylene protons
are in a less electron-dense environment than are the methyl protons because the methylene protons are closer to the electron-withdrawing bromine. Because the methylene
protons are in a less electron-dense environment, they are less shielded from the
applied magnetic field. The signal for these protons, therefore, occurs at a higher
frequency than the signal for the more shielded methyl protons. *Remember that the
right-hand side of an NMR spectrum is the low-frequency side, where protons in
electron-dense environments (more shielded) show a signal. The left-hand side is the
high-frequency side, where protons in electron-poor environments (less shielded) show
a signal* (Figure 15.16).

We would expect the ^{1}H NMR spectrum of 1-nitropropane to have three signals
because the compound has three different kinds of protons. The closer the protons
are to the electron-withdrawing nitro group, the less they are shielded from the applied magnetic field, so the higher the frequency at which their signals will appear.
Thus, the protons closest to the nitro group show a signal at the highest frequency
(4.37 ppm), and the ones farthest from the nitro group show a signal at the lowest
frequency (1.04 ppm).

**Electron withdrawal causes NMR
signals to appear at higher frequencies
(at larger δ values).**

| 1.04 ppm | 2.07 ppm |

4.37 ppm

$$CH_3CH_2CH_2NO_2$$

Compare the chemical shifts of the methylene protons immediately adjacent to the
halogen in each of the following alkyl halides. The position of the signal depends on
the electronegativity of the halogen—the more electronegative the halogen, the higher
is the frequency of the signal. Thus, the signal for the methylene protons adjacent to
fluorine (the most electronegative of the halogens) occurs at the highest frequency,
whereas the signal for the methylene protons adjacent to iodine (the least electronegative of the halogens) occurs at the lowest frequency.

$$CH_3CH_2CH_2CH_2CH_2F \qquad CH_3CH_2CH_2CH_2CH_2Cl \qquad CH_3CH_2CH_2CH_2CH_2Br \qquad CH_3CH_2CH_2CH_2CH_2I$$

| 4.50 ppm | | 3.50 ppm | | 3.40 ppm | | 3.20 ppm |

PROBLEM 15

a. Which set of protons in each of the following compounds is the least shielded?

1. $CH_3CH_2CH_2Cl$ **2.** $CH_3CH_2\overset{\displaystyle O}{\overset{\displaystyle \|}{C}}OCH_3$ **3.** $CH_3CHCHBr$

$\qquad\qquad\qquad\qquad\qquad\qquad\qquad\qquad\qquad\qquad$ Br Br

b. Which set of protons in each compound is the most shielded?

PROBLEM 16◆

One of the spectra in Figure 15.18 is due to 1-chloropropane, and the other to 1-iodopropane.
Which is which?

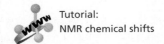

Tutorial:
NMR chemical shifts

Figure 15.18 ▶
¹H NMR spectra for Problem 16.

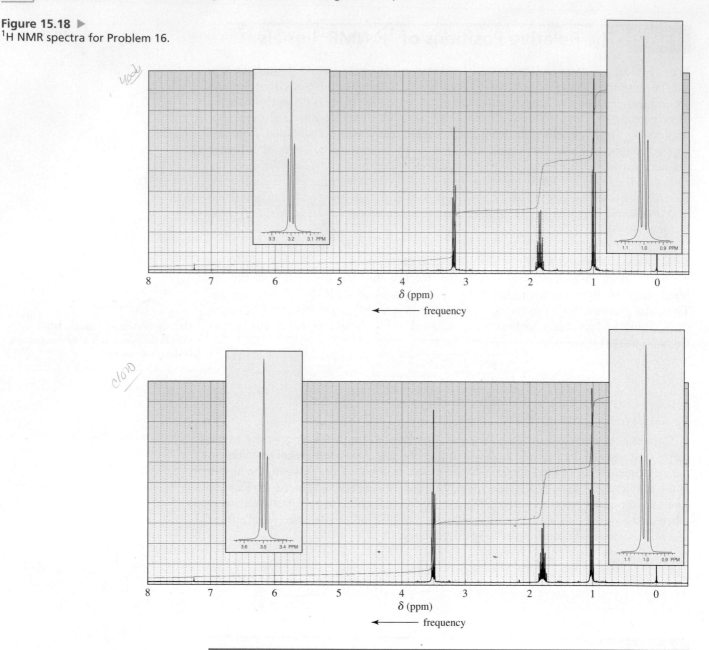

15.13 Characteristic Values of Chemical Shifts

Approximate values of chemical shifts for different kinds of protons are shown in Table 15.3. An ¹H NMR spectrum can be divided into six regions. Rather than memorizing chemical shift values, if you remember the kinds of protons that are in each region, you will be able to tell what kinds of protons a molecule has from a quick look at its NMR spectrum.

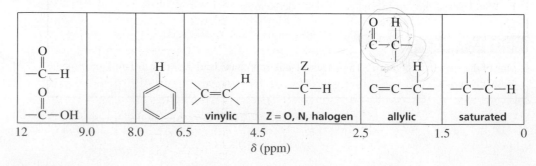

Table 15.3   Approximate Values of Chemical Shifts for ^{1}H NMR[a]

Type of proton	Approximate chemical shift (ppm)	Type of proton	Approximate chemical shift (ppm)
$(CH_3)_4Si$	0	 ⟨benzene ring⟩—H	6.5–8
—CH_3	0.9	$\overset{\displaystyle O}{\overset{\|}{—C}}$—H	9.0–10
—CH_2—	1.3	I—$\overset{\|}{\underset{\|}{C}}$—H	2.5–4
—$\overset{\|}{CH}$—	1.4		
—$\overset{\|}{C}$=$\overset{\|}{C}$—CH_3	1.7	Br—$\overset{\|}{\underset{\|}{C}}$—H	2.5–4
$\overset{\displaystyle O}{\overset{\|}{—C}}$—$CH_3$	2.1	Cl—$\overset{\|}{\underset{\|}{C}}$—H	3–4
⟨benzene ring⟩—CH_3	2.3	F—$\overset{\|}{\underset{\|}{C}}$—H	4–4.5
—C≡C—H	2.4	RNH_2	Variable, 1.5–4
R—O—CH_3	3.3	ROH	Variable, 2–5
R—$\overset{\|}{\underset{\displaystyle R}{C}}$=$CH_2$	4.7	ArOH	Variable, 4–7
R—$\overset{\|}{\underset{\displaystyle R}{C}}$=$\overset{\|}{\underset{\displaystyle R}{C}}$—H	5.3	$\overset{\displaystyle O}{\overset{\|}{—C}}$—OH	Variable, 10–12
		$\overset{\displaystyle O}{\overset{\|}{—C}}$—$NH_2$	Variable, 5–8

[a]The values are approximate because they are affected by neighboring substituents.

Table 15.3 shows that the chemical shift of methyl protons is at a lower frequency (0.9 ppm) than is the chemical shift of methylene protons (1.3 ppm) in a similar environment and that the chemical shift of methylene protons is at a lower frequency than is the chemical shift of a methine proton (1.4 ppm) in a similar environment. (When an sp^3 carbon is bonded to only one hydrogen, the hydrogen is called a **methine hydrogen**.) For example, the ^{1}H NMR spectrum of butanone shows three signals. The signal for the **a** protons of butanone is the signal at the lowest frequency because the protons are farthest from the electron-withdrawing carbonyl group. (In correlating an NMR spectrum with a structure, the set of protons responsible for the signal at the lowest frequency will be labeled **a**, the next set will be labeled **b**, the next set **c**, etc.) The **b** and **c** protons are the same distance from the carbonyl group, but the signal for the **b** protons is at a lower frequency

In a similar environment, the signal for methyl protons occurs at a lower frequency than the signal for methylene protons, which in turn occurs at a lower frequency than the signal for a methine proton.

because methyl protons appear at a lower frequency than do methylene protons in a similar environment.

$$
\underset{\substack{a \quad c \quad b}}{CH_3CH_2\overset{\displaystyle O}{\overset{\displaystyle \|}{C}}CH_3}
\qquad
\underset{\substack{\\ \\ \underset{a}{CH_3}}}{\overset{\substack{b \quad c \quad a}}{CH_3OCHCH_3}}
$$

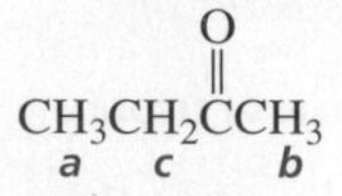

butanone

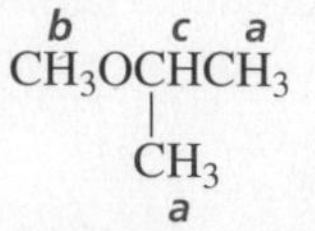

2-methoxypropane

The signal for the **a** protons of 2-methoxypropane is the signal at the lowest frequency in the ^{1}H NMR spectrum of this compound because these protons are farthest from the electron-withdrawing oxygen. The **b** and **c** protons are the same distance from the oxygen, but the signal for the **b** protons appears at a lower frequency because, in a similar environment, methyl protons appear at a lower frequency than does a methine proton.

PROBLEM 17◆

In each of the following compounds, which of the underlined protons has the greater chemical shift (i.e., the higher frequency signal)?

a. CH$_3$CHCHBr
  | |
  Br Br

b. CH$_3$CHOCH$_3$
        |
       CH$_3$

c. CH$_3$CH$_2$CHCH$_3$
              |
              Cl

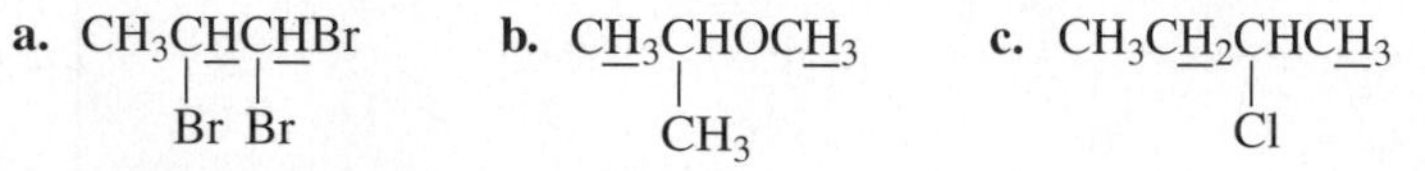

PROBLEM 18◆

In each of the following pairs of compounds, which of the underlined protons has the greater chemical shift (i.e., the higher frequency signal)?

a. CH$_3$CH$_2$CH$_2$Cl or CH$_3$CH$_2$CHCH$_3$
                              |
                              Cl

b. CH$_3$CH$_2$CH$_2$Cl or CH$_3$CH$_2$CH$_2$Br

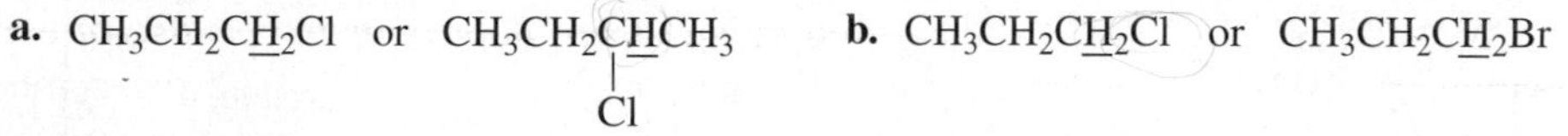

PROBLEM 19

Without referring to Table 15.3, label the proton that gives the signal at the lowest frequency *a*, the next lowest *b*, etc.

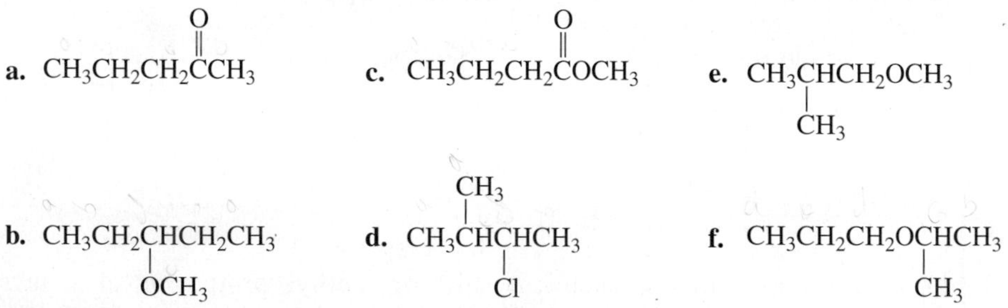

a. CH$_3$CH$_2$CH$_2$$\overset{\displaystyle O}{\overset{\displaystyle \|}{C}}CH_3$

c. CH$_3$CH$_2$CH$_2$$\overset{\displaystyle O}{\overset{\displaystyle \|}{C}}OCH_3$

e. CH$_3$CHCH$_2$OCH$_3$
          |
         CH$_3$

b. CH$_3$CH$_2$CHCH$_2$CH$_3$
              |
             OCH$_3$

d. CH$_3$CHCHCH$_3$
       (CH$_3$)
          |
          Cl

f. CH$_3$CH$_2$CH$_2$OCHCH$_3$
                      |
                     CH$_3$

15.14 Integration of NMR Signals

The two signals in the ^{1}H NMR spectrum of 1-bromo-2,2-dimethylpropane in Figure 15.17 are not the same size because *the area under each signal is proportional to the number of protons that gives rise to the signal*. (The spectrum is shown again in

Figure 15.19.) The area under the signal occurring at the lower frequency is larger because the signal is caused by *nine* methyl protons, while the smaller, higher-frequency signal results from *two* methylene protons.

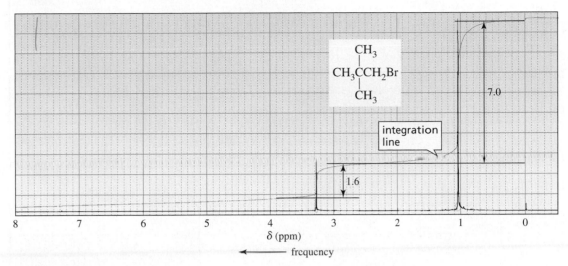

▲ **Figure 15.19**
Analysis of the integration line in the ^{1}H NMR spectrum of 1-bromo-2,2-dimethylpropane.

The area under each signal can be determined by integration. The integrals can be reported digitally or can be displayed by a line of integration superimposed on the original spectrum as in Figure 15.19. The height of each integration step is proportional to the number of protons that give rise to the signal. By measuring the heights of the integration steps, you can determine that the relative number of protons is approximately $1.6 : 7.0 = 1 : 4.4$. (The measured integrals are approximate because of experimental error.) The ratios are multiplied by a number that will cause all the numbers to be close to whole numbers—in this case, we multiply by 2—as there can be only whole numbers of protons. That means that the ratio of protons in the compound is $2 : 8.8$, which is rounded to $2 : 9$.

The **integration** tells us the *relative* number of protons that give rise to each signal, not the *absolute* number. For example, integration could not distinguish between 1,1-dichloroethane and 1,2-dichloro-2-methylpropane because both compounds would show an integral ratio of $1 : 3$.

CH$_3$—CH—Cl   CH$_3$—C—CH$_2$Cl
| |
Cl   Cl

CH$_3$ (above central C)

1,1-dichloroethane   **1,2-dichloro-2-methylpropane**
ratio of protons = 1:3   **ratio of protons 2:6 = 1:3**

PROBLEM 20

How would integration distinguish the ^{1}H NMR spectra of the following compounds?

CH$_3$ CH$_3$ CH$_2$Br
| | |
CH$_3$—C—CH$_2$Br   CH$_3$—C—CH$_2$Br   CH$_3$—C—CH$_2$Br
| | |
CH$_3$ Br   CH$_2$Br

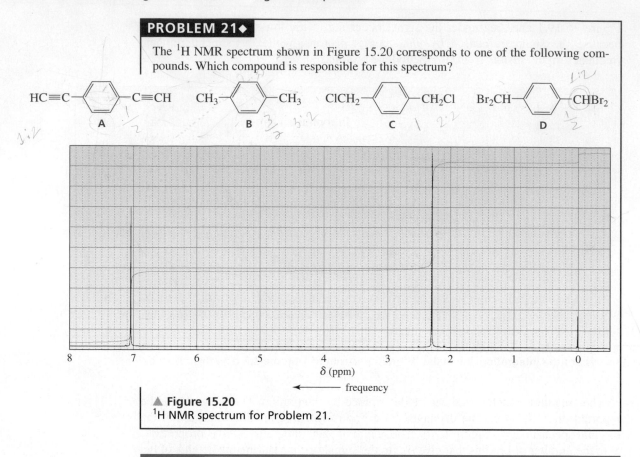

PROBLEM 21◆

The ^{1}H NMR spectrum shown in Figure 15.20 corresponds to one of the following compounds. Which compound is responsible for this spectrum?

▲ **Figure 15.20**
^{1}H NMR spectrum for Problem 21.

15.15 Splitting of the Signals

Notice that the shapes of the signals in the ^{1}H NMR spectrum of 1,1-dichloroethane (Figure 15.21) are different from the shapes of the signals in the ^{1}H NMR spectrum of 1-bromo-2,2-dimethylpropane (Figure 15.17). Both signals in Figure 15.17 are **singlets** (each has a single peak), whereas the signal for the methyl protons of 1,1-dichloroethane (the lower-frequency signal) is split into two peaks (a **doublet**), and the signal for the methine proton is split into four peaks (a **quartet**). (Magnifications of the doublet and quartet are shown as insets in Figure 15.21.)

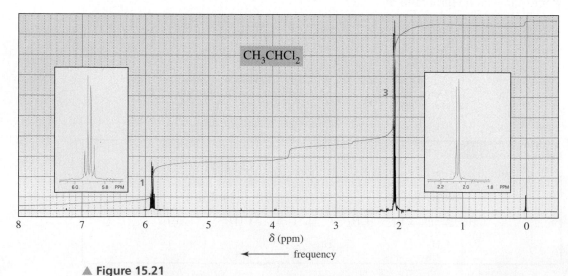

▲ **Figure 15.21**
^{1}H NMR spectrum of 1,1-dichloroethane. The higher-frequency signal is an example of a quartet; the lower-frequency signal is a doublet.

Splitting is caused by protons bonded to adjacent carbons. The splitting of a signal is described by the **N + 1 rule**, where N is the number of *equivalent* protons bonded to *adjacent* carbons. By "equivalent protons," we mean that the protons bonded to an adjacent carbon are equivalent to each other, but not equivalent to the proton giving rise to the signal. Both signals in Figure 15.17 are singlets because neither the carbon adjacent to the methyl groups nor that adjacent to the methylene group in 1-bromo-2,2-dimethylpropane is bonded to any protons ($N + 1 = 0 + 1 = 1$). In contrast, in Figure 15.21, the carbon adjacent to the methyl group in 1,1-dichloroethane is bonded to one proton, so the signal for the methyl protons is split into a doublet ($N + 1 = 1 + 1 = 2$). The carbon adjacent to the carbon bonded to the methine proton is bonded to three equivalent protons, so the signal for the methine proton is split into a quartet ($N + 1 = 3 + 1 = 4$). The number of peaks in a signal is called the **multiplicity** of the signal. Protons that split each other's signal are called **coupled protons**. Coupled protons are on adjacent carbons.

Keep in mind that it is not the number of protons giving rise to a signal that determines the multiplicity of the signal; rather, it is the number of protons bonded to the immediately adjacent carbons that determines the multiplicity. For example, the signal for the *a* protons in the following compound will be split into three peaks (a **triplet**) because the adjacent carbon is bonded to two hydrogens. The signal for the *b* protons will appear as a quartet because the adjacent carbon is bonded to three hydrogens, and the signal for the *c* protons will be a singlet.

$$\underset{a \quad\quad b \quad\quad\quad c}{CH_3CH_2\overset{\overset{\displaystyle O}{\|}}{C}OCH_3}$$

A signal for a proton is never split by *equivalent* protons. For example, the ^1H NMR spectrum of bromomethane shows one singlet. The three methyl protons are chemically equivalent, and chemically equivalent protons do not split each other's signal. The four protons in 1,2-dichloroethane are also chemically equivalent, so its ^1H NMR spectrum also shows one singlet.

$$CH_3Br \quad\quad\quad\quad ClCH_2CH_2Cl$$
bromomethane   1,2-dichloroethane

> each compound has an NMR spectrum that shows one singlet because equivalent protons do not split each other's signals

PROBLEM 22♦

The ^1H NMR spectra of two carboxylic acids with molecular formula $C_3H_5O_2Cl$ are shown in Figure 15.22. Identify the carboxylic acids. (The "offset" notation means that the signal has been moved to the right by the indicated amount).

a.

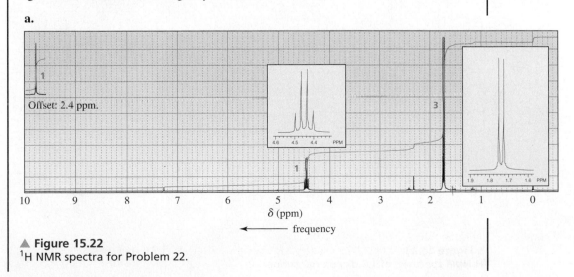

δ (ppm)

← frequency

▲ **Figure 15.22**
^1H NMR spectra for Problem 22.

Sidebar notes:

An ^1H NMR signal is split into N + 1 peaks, where N is the number of equivalent protons bonded to adjacent carbons.

Tutorial:
NMR signal splitting

Equivalent protons do not split each other's signal.

b.

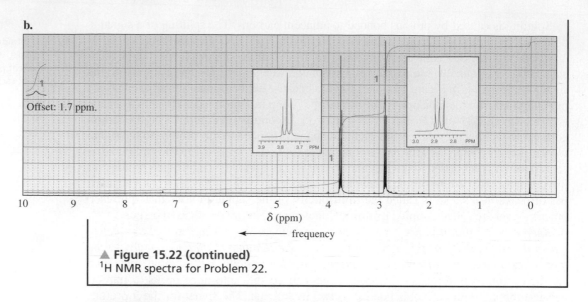

▲ **Figure 15.22 (continued)**
^{1}H NMR spectra for Problem 22.

15.16 More Examples of ^{1}H NMR Spectra

There are two signals in the ^{1}H NMR spectrum of 1,3-dibromopropane (Figure 15.23). The signal for the H_b protons is split into a triplet by the two hydrogens on the adjacent carbon. The H_a protons have two adjacent carbons that are bonded to protons. The protons on one adjacent carbon are equivalent to the protons on the other adjacent carbon. Because the two sets of protons are equivalent, the $N + 1$ rule is applied to both sets at the same time. In other words, N is equal to the sum of the equivalent protons on both carbons. So the signal for the H_a protons is split into a quintet ($4 + 1 = 5$).

The ^{1}H NMR spectrum of isopropyl butanoate shows five signals (Figure 15.24). The signal for the H_a protons is split into a triplet by the H_c protons. The signal for the H_b protons is split into a doublet by the H_e proton. The signal for the H_d protons is split into a triplet by the H_c protons, and the signal for the H_e proton is split into a septet by the H_b protons. The signal for the H_c protons is split by both the H_a and H_d protons. Because the H_a and H_d protons are not equivalent, the $N + 1$ rule has to be

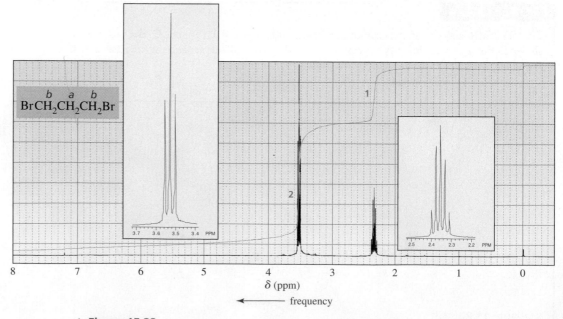

▲ **Figure 15.23**
^{1}H NMR spectrum of 1,3-dibromopropane.

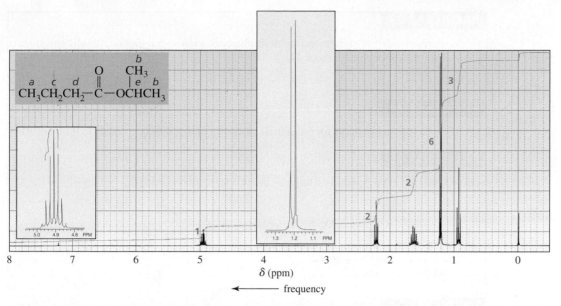

▲ **Figure 15.24**
^{1}H NMR spectrum of isopropyl butanoate.

applied separately to each set. Thus, the signal for the H_c protons will be split into a quartet by the H_a protons, and each of these four peaks will be split into a triplet by the H_d protons: $(N_a + 1)(N_d + 1) = (4)(3) = 12$. As a result, the signal for the H_c protons is a **multiplet** (a signal that is more complex than a triplet, quartet, quintet, etc.).

There are five sets of chemically equivalent protons in ethylbenzene (Figure 15.25). We see the expected triplet for the H_a protons and the quartet for the H_b protons. (This is a characteristic pattern for an ethyl group.) The five protons on the benzene ring are not all in the same environment, so we expect to see three signals for them: one for the H_c protons, one for the H_d protons, and one for the H_e protons. However, we do not see three distinct signals because their environments are not sufficiently different to allow them to appear as separate signals.

Notice that the signals for the benzene ring protons in Figure 15.25 occur in the 7.0–8.5 ppm region. Other kinds of protons usually do not resonate in this region, so signals in this region of an ^{1}H NMR spectrum indicate that the compound probably contains an aromatic ring.

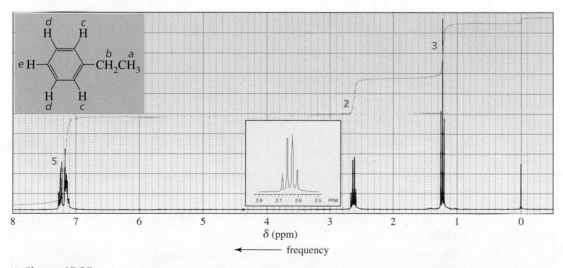

▲ **Figure 15.25**
^{1}H NMR spectrum of ethylbenzene. The signals for the H_c, H_d, and H_e protons overlap.

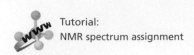

Tutorial:
NMR spectrum assignment

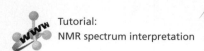

Tutorial:
NMR spectrum interpretation

PROBLEM 23

Indicate the number of signals and the multiplicity of each signal in the ^{1}H NMR spectrum of each of the following compounds:

a. $ICH_2CH_2CH_2Br$ **b.** $ClCH_2CH_2CH_2Cl$ **c.** $ICH_2CH_2CHBr_2$

PROBLEM 24

Predict the splitting patterns for the signals given by each of the compounds in Problem 10.

PROBLEM 25

How could ^{1}H NMR spectra distinguish the following compounds?

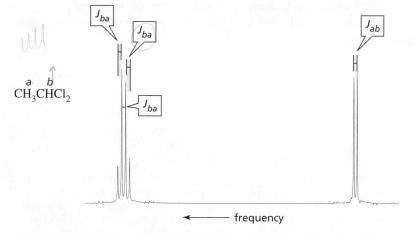

A   B   C

PROBLEM 26

How would the ^{1}H NMR spectra for the four compounds with molecular formula $C_3H_6Br_2$ differ?

15.17   Coupling Constants

The distance, in hertz, between two adjacent peaks of a split NMR signal is called the **coupling constant** (denoted by *J*). The coupling constant for H_a being split by H_b is denoted by J_{ab}. The signals of coupled protons (protons that split each other's signal) have the same coupling constant; in other words, $J_{ab} = J_{ba}$ (Figure 15.26). Coupling constants are useful in analyzing complex NMR spectra because protons on adjacent carbons can be identified by their identical coupling constants. Coupling constants range from 0 to 15 Hz.

Figure 15.26 ▶
The H_a and H_b protons of 1,1-dichloroethane are coupled protons, so their signals have the same coupling constant, $J_{ab} = J_{ba}$.

Let's now summarize the kind of information that can be obtained from an ^{1}H NMR spectrum:

1. The number of signals indicates the number of different kinds of protons that are in the compound.

2. The position of a signal indicates the kind of proton(s) responsible for the signal (methyl, methylene, methine, allylic, vinylic, aromatic, etc.) and the kinds of neighboring substituents.

3. The integration of the signal tells the relative number of protons responsible for the signal.

4. The multiplicity of the signal ($N + 1$) tells the number of protons (N) bonded to adjacent carbons.

5. The coupling constants identify coupled protons.

PROBLEM-SOLVING STRATEGY

Identify the compound with molecular formula $C_9H_{10}O$ that gives the IR and ^1H NMR spectra in Figure 15.27.

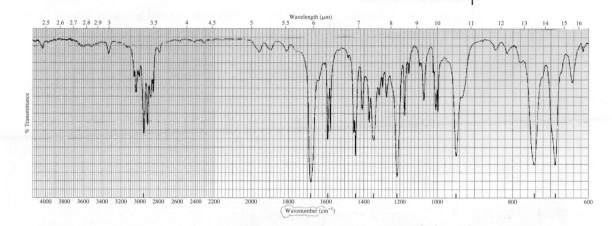

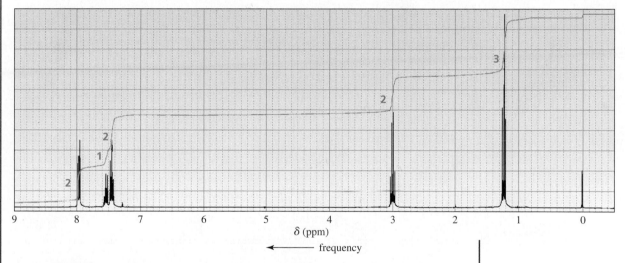

▲ **Figure 15.27**
IR and ^1H NMR spectra for this problem-solving strategy.

The best way to approach this kind of problem is to identify whatever structural features you can from the molecular formula and the IR spectrum and then use the information from the ^1H NMR spectrum to expand on that knowledge. From the molecular formula and IR spectrum, we learn that the compound is a ketone: It has a carbonyl group at ~1680 cm^{-1}, only one oxygen, and no absorption bands at ~2820 and ~2720 cm^{-1} that would indicate an aldehyde. The carbonyl group absorption is at a lower frequency than normal, which suggests that it has partial single-bond character as a result of electron delocalization, indicating that it is attached to an sp^2 carbon. The compound contains a benzene ring (>3000 cm^{-1}, ~1600 cm^{-1}, and 1440 cm^{-1}), and it has hydrogens bonded to sp^3 carbons (<3000 cm^{-1}). In the NMR spectrum, the triplet at ~1.2 ppm and the quartet at ~3.0 ppm indicate the presence of an ethyl group that is attached to an electron-withdrawing group.

The signals in the 7.4–8.0 ppm region confirm the presence of a benzene ring. From this information, we can conclude that the compound is the following ketone. The integration ratio (5 : 2 : 3) confirms this answer.

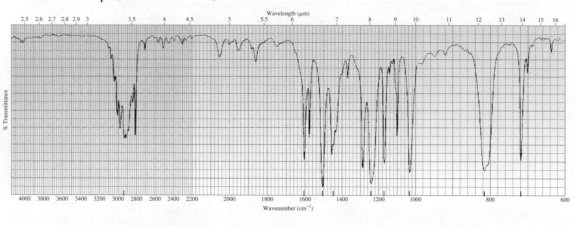

Now continue on to Problem 27.

PROBLEM 27◆

Identify the compound with molecular formula $C_8H_{10}O$ that gives the IR and ^1H NMR spectra shown in Figure 15.28.

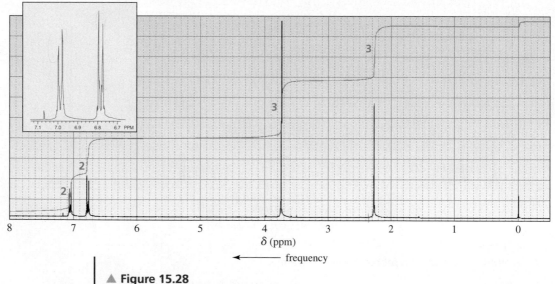

▲ **Figure 15.28**
IR and ^1H NMR spectra for Problem 27.

15.18 ^{13}C NMR Spectroscopy

The number of signals in a ^{13}C NMR spectrum tells you how many different kinds of carbons a compound has—just as the number of signals in an ^1H NMR spectrum tells you how many different kinds of hydrogens a compound has. The principles behind

^{1}H NMR and ^{13}C NMR spectroscopy are essentially the same. There are, however, some differences that make ^{13}C NMR easier to interpret.

One advantage to ^{13}C NMR spectroscopy is that the chemical shifts range over about 220 ppm (Table 15.4), compared with over about 12 ppm for ^{1}H NMR (Table 15.3). This means that signals are less likely to overlap. Notice that aldehyde (190–200 ppm) and ketone (205–220 ppm) carbonyl groups can be easily distinguished from other carbonyl groups.

The reference compound used in ^{13}C NMR is TMS, the same reference compound used in ^{1}H NMR. A disadvantage of ^{13}C NMR spectroscopy is that, unless special techniques are used, the area under a ^{13}C NMR signal is **not** proportional to the number of carbons giving rise to the signal. Thus, the number of carbons giving rise to a ^{13}C NMR signal cannot routinely be determined by integration.

Table 15.4	Approximate Values of Chemical Shifts for ^{13}C NMR		
Type of carbon	**Approximate chemical shift (ppm)**	**Type of carbon**	**Approximate chemical shift (ppm)**
$(CH_3)_4Si$	0	C—I	0–40
R—CH$_3$	8–35	C—Br	25–65
R—CH$_2$—R	15–50	C—Cl C—N C—O	35–80 40–60 50–80
R—CH—R (R)	20–60	R, —N\ C=O	165–175
R—C—R (R, R)	30–40	R, RO\ C=O	165–175
≡C	65–85	R, HO\ C=O	175–185
=C	100–150	R, H\ C=O	190–200
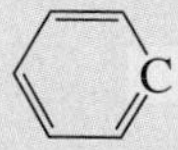 (benzene ring) C	110–170	R, R\ C=O	205–220

The ^{13}C NMR spectrum of 2-butanol is shown in Figure 15.29. 2-Butanol has carbons in four different environments, so there are four signals in the spectrum. The relative positions of the signals depend on the same factors that determine the relative positions of the proton signals in an ^{1}H NMR spectrum. Carbons in electron-dense environments produce low-frequency signals, and carbons close to electron-withdrawing groups produce high-frequency signals. This means that the signals for the carbons of 2-butanol are in the same relative order that we would expect for the signals of the protons bonded to those carbons in the ^{1}H NMR spectrum. Thus, the carbon of the methyl group farthest away from the electron-withdrawing OH group gives the lowest-frequency signal. As the frequency increases, the other methyl carbon comes next, followed by the methylene carbon; the carbon attached to the OH group gives the highest- frequency signal.

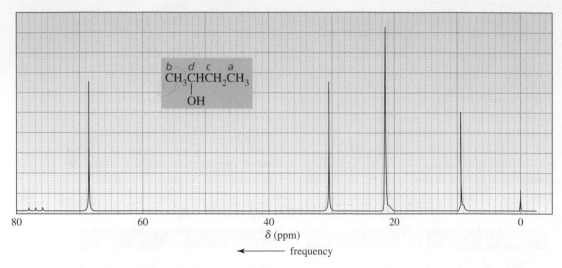

▲ Figure 15.29
Proton-decoupled ^{13}C NMR spectrum of 2-butanol.

The signals in a ^{13}C NMR spectrum are not split by neighboring carbons. If, however, the spectrometer is run in a *proton-coupled* mode, each signal will be split by the *hydrogens* bonded to the carbon that produces the signal. The multiplicity of the signal is determined by the $N + 1$ rule. The **proton-coupled ^{13}C NMR spectrum** of 2-butanol is shown in Figure 15.30. (The triplet at 78 ppm is produced by the solvent, $CDCl_3$.) The signals for the methyl carbons are each split into a quartet because each methyl carbon is bonded to three hydrogens ($3 + 1 = 4$). The signal for the methylene carbon is split into a triplet ($2 + 1 = 3$) and the signal for the carbon bonded to the OH group is split into a doublet ($1 + 1 = 2$).

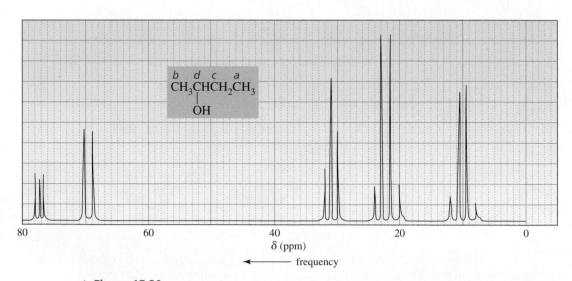

▲ Figure 15.30
Proton-coupled ^{13}C NMR spectrum of 2-butanol. If the spectrometer is run in a proton-coupled mode, splitting is observed in a ^{13}C NMR spectrum.

PROBLEM 28◆

Answer the following questions for each of the compounds:

a. How many signals are in its ^{13}C NMR spectrum?

b. Which signal is at the lowest frequency?

1. CH$_3$CH$_2$CH$_2$Br

3. CH$_3$CH$_2$$\overset{\overset{\displaystyle O}{\|}}{C}OCH_3$

5. CH$_3$$\overset{\displaystyle}{\underset{\displaystyle Br}{C}}HCH_3$

2. CH$_3$CH$_2$OCH$_3$

4. CH$_3$$\overset{\overset{\displaystyle CH_3}{|}}{\underset{\underset{\displaystyle CH_3}{|}}{C}}OCH_3$

6. CH$_3$$\overset{\overset{\displaystyle O}{\|}}{C}CH_2CH_2$$\overset{\overset{\displaystyle O}{\|}}{C}CH_3$

PROBLEM 29

Describe the proton-coupled ^{13}C NMR spectrum for compounds 1, 2, and 3 in Problem 28, showing relative values (not absolute values) of chemical shifts.

PROBLEM-SOLVING STRATEGY

Identify the compound with molecular formula C$_9$H$_{10}$O$_2$ that gives the following ^{13}C NMR spectrum:

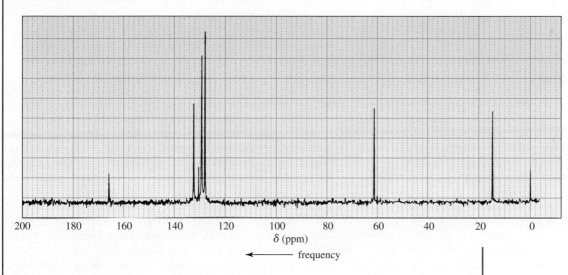

First, pick out the signals that can be absolutely identified. For example, the two oxygen atoms in the molecular formula and the carbonyl carbon signal at 166 ppm indicate that the compound is an ester. The four signals at about 130 ppm suggest that the compound has a benzene ring with a single substituent. (One signal is for the carbon to which the substituent is attached, one is for the ortho carbons, one is for the meta carbons, and one is for the para carbon.) Now subtract the molecular formula of those pieces from the molecular formula of the compound. Subtracting C$_6$H$_5$ and CO$_2$ from the molecular formula leaves C$_2$H$_5$, the molecular formula of an ethyl substituent. Therefore, we know that the compound is either ethyl benzoate or phenyl propanoate.

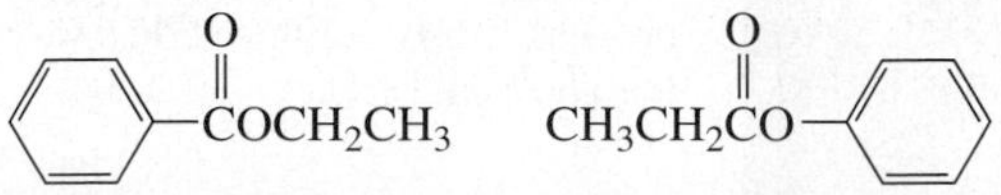

ethyl benzoate   **phenyl propanoate**

Seeing that the methylene group is at about 60 ppm indicates that it is adjacent to an oxygen. Thus, the compound is ethyl benzoate.

Now continue on to Problem 30.

PROBLEM 30◆

Identify the compound in Figure 15.31 from its molecular formula and its ^{13}C NMR spectrum.

a. $C_{11}H_{22}O$

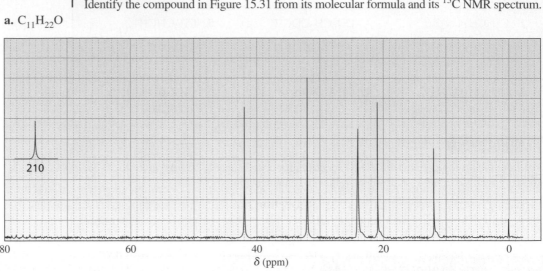

Figure 15.31 ▶
The ^{13}C NMR spectra for Problem 30.

MAGNETIC RESONANCE IMAGING

NMR has become an important tool in medical diagnosis because it allows physicians to probe internal organs and structures without invasive surgical methods or the harmful ionizing radiation of X-rays. When NMR was first introduced into clinical practice in 1981, the selection of an appropriate name was a matter of some debate. Because some members of the public associate nuclear processes with harmful radiation, the "N" was dropped from the medical application of NMR, which is known as **magnetic resonance imaging (MRI)**. The spectrometer is called an **MRI scanner**.

An MRI scanner consists of a magnet sufficiently large to accommodate an entire person, along with additional coils for exciting the nuclei, modifying the magnetic field, and receiving signals. Different tissues yield different signals, and the signals can be attributed to a specific site of origin in the patient, allowing a cross-sectional image of the patient's body to be constructed.

An MRI may be taken in any plane, essentially independently of the patient's position, allowing optimum visualization of the anatomical feature of interest. By contrast, the plane of a computed tomography (CT) scan, which uses X-rays, is defined by the position of the patient within the machine and is usually perpendicular to the long axis of the body. CT scans in other planes may be obtained only if the patient is a skilled contortionist.

Most of the signals in an MRI scan originate from the hydrogens of water molecules because these hydrogens are far more abundant in tissues than are the hydrogens of organic compounds. The difference in the way water is bound in different tissues is what produces much of the variation in signal among organs, as well as the variation between healthy and diseased tissue (Figure 15.32). MRI scans, therefore, can sometimes provide much more information than images obtained by other means. For example, MRI can provide detailed images of blood vessels. Flowing fluids, such as blood, respond differently to excitation in an MRI scanner than do stationary tissues, and they normally do not produce a signal. However, the data may be processed to eliminate signals from motionless structures, thereby showing signals only from the fluids. This technique is sometimes used instead of more invasive methods to examine the vascular tree. It is now possible to completely suppress the signal from certain types of tissue (usually fat). It is also possible to differentiate intracellular and extracellular edema, which is important in assessing patients suspected of having suffered strokes.

NMR spectroscopy using ^{31}P is not yet in routine clinical use, but it is being used widely in clinical research. It is of particular interest because ATP and ADP (Section 19.1) are involved in most metabolic processes, so it will provide a way to investigate cellular metabolism.

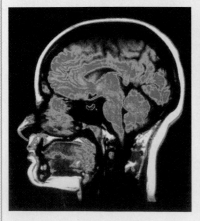
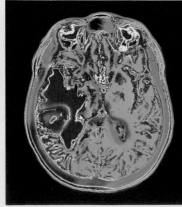

◀ **Figure 15.32**
(a) MRI of a normal brain. The pituitary is highlighted (pink). (b) MRI of an axial section through the brain showing a tumor (purple) surrounded by damaged, fluid-filled tissue (red).

Summary

Spectroscopy is the study of the interaction of matter and **electromagnetic radiation**. A continuum of different types of electromagnetic radiation constitutes the electromagnetic spectrum. High-energy radiation is associated with *high frequencies, large wavenumbers,* and *short wavelengths.*

Infrared spectroscopy identifies the kinds of functional groups in a compound. Each stretching and bending vibration of a bond occurs with a characteristic frequency. It takes more energy to stretch a bond than to bend it. When a compound is bombarded with radiation of a frequency that exactly matches the frequency of one of its vibrations, the molecule absorbs energy and exhibits an **absorption band**. The position, intensity, and shape of an absorption band help identify functional groups. Stronger bonds show absorption bands at larger wavenumbers. The intensity of an absorption band depends on the polarity of the bond.

NMR spectroscopy is used to identify the carbon–hydrogen framework of an organic compound. When a sample is placed in a magnetic field, protons aligning with the field are in the lower-energy α-spin state; those aligning against the field are in the higher-energy β-spin state. The energy difference between the spin states depends on the strength of the **applied magnetic field**. When subjected to radiation with energy corresponding to the energy difference between the spin states, nuclei in the α-spin state are promoted to the β-spin state. When they return to their original state, they emit signals whose frequency depends on the difference in energy between the spin states. An **NMR**

spectrometer detects and displays these signals as a plot of their frequency versus their intensity—an **NMR spectrum**.

The number of signals in an ^{1}H NMR spectrum indicates the number of different kinds of protons in a compound. The **chemical shift** is a measure of how far the signal is from the reference (TMS) signal. The larger the magnetic field sensed by the proton, the higher is the frequency of the signal. The electron density of the environment in which the proton is located **shields** the proton from the applied magnetic field. Therefore, a proton in an electron-dense environment shows a signal at a lower frequency than a proton near electron-withdrawing groups. Thus, the position of a signal indicates the kind of proton(s) responsible for the signal and the kinds of neighboring substituents. **Integration** tells us the relative number of protons that give rise to each signal.

The **multiplicity** of a signal indicates the number of protons bonded to adjacent carbons. Multiplicity is described by the **N + 1 rule**, where N is the number of equivalent protons bonded to adjacent carbons. The **coupling constant** (J) is the distance between two adjacent peaks of a split NMR signal. Coupled protons have the same coupling constant.

The number of signals in a ^{13}C NMR spectrum tells how many different kinds of carbons a compound has. Carbons in electron-dense environments produce low-frequency signals; carbons close to electron-withdrawing groups produce high-frequency signals. ^{13}C NMR signals are not normally split by neighboring carbons, unless the spectrometer is run in a proton-coupled mode.

Problems

31. For each of the following pairs of compounds, identify one IR absorption band that could be used to distinguish between them:

a. $CH_3CH_2\overset{\overset{\displaystyle O}{\|}}{C}H$ and $CH_3CH_2\overset{\overset{\displaystyle O}{\|}}{C}CH_3$

b. 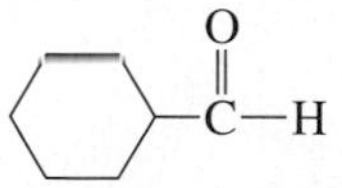 ⬡—$\overset{\overset{\displaystyle O}{\|}}{C}$—H and ⬡—$\overset{\overset{\displaystyle O}{\|}}{C}$—H

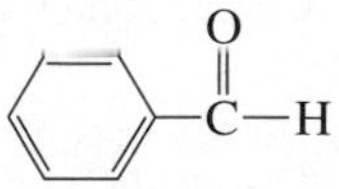

c. $CH_3CH_2\overset{\overset{\displaystyle O}{\|}}{C}NH_2$ and $CH_3CH_2\overset{\overset{\displaystyle O}{\|}}{C}OCH_3$

d. $CH_3CH_2CH_2OH$ and $CH_3CH_2OCH_3$

32. For each of the IR spectra in Figures 15.33 and 15.34, four compounds are shown. In each case, indicate which of the four compounds is responsible for the spectrum.

a. $CH_3CH_2CH_2C{\equiv}CCH_3$ $CH_3CH_2CH_2CH_2OH$ $CH_3CH_2CH_2CH_2C{\equiv}CH$ $CH_3CH_2CH_2\overset{\overset{\displaystyle O}{\|}}{C}OH$

b. $CH_3CH_2\overset{\overset{\displaystyle O}{\|}}{C}OH$ $CH_3CH_2\overset{\overset{\displaystyle O}{\|}}{C}OCH_2CH_3$ $CH_3CH_2\overset{\overset{\displaystyle O}{\|}}{C}H$ $CH_3CH_2\overset{\overset{\displaystyle O}{\|}}{C}CH_3$

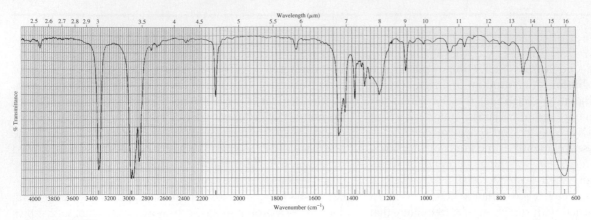

▲ **Figure 15.33**
The IR spectrum for Problem 32a.

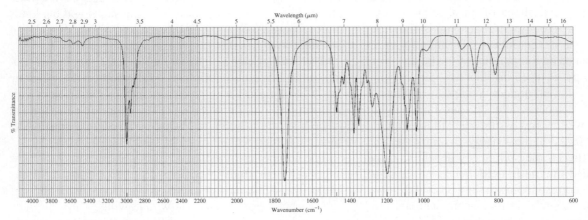

▲ **Figure 15.34**
The IR spectrum for Problem 32b.

33. Norlutin and Enovid are ketones that suppress ovulation. Consequently, they have been used clinically as contraceptives. For which of these compounds would you expect the infrared carbonyl absorption (C=O stretch) to be at a higher frequency? Explain.

Norlutin®   Enovid®

34. For each of the following pairs of compounds, identify one IR absorption band that could be used to distinguish between them:

a. $CH_3CH_2COCH_3$ and $CH_3CH_2CCH_3$

c. CH_3CH_2CH=$CHCH_3$ and CH_3CH_2C≡CCH_3

b. [structure] and [structure]

d. CH_3CH_2CH=CH_2 and $CH_3CH_2CH_2CH_3$

35. a. How could you use IR spectroscopy to determine that the following reaction had occurred?

$$\xrightarrow[\text{HO}^-, \Delta]{\text{NH}_2\text{NH}_2}$$

b. After purifying the product, how could you determine that all the NH_2NH_2 had been removed?

36. For each of the IR spectra in Figures 15.35, 15.36, and 15.37, indicate which of the given compounds is responsible for the spectrum:

a. $CH_3CH_2CH{=}CH_2$ $CH_3CH_2CH_2CH_2OH$ $CH_2{=}CHCH_2CH_2OH$ $CH_3CH_2CH_2OCH_3$ $CH_3CH_2CH_2\overset{O}{\overset{\|}{C}}OH$

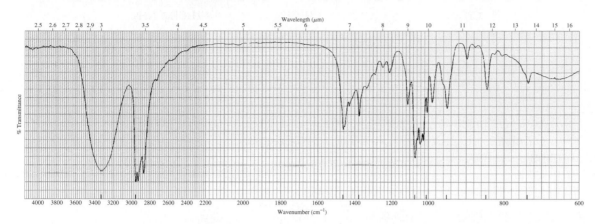

◀ **Figure 15.35**
The IR spectrum for Problem 36a.

b.

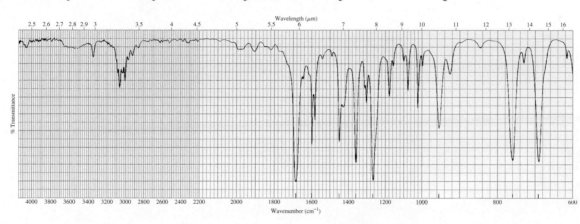

◀ **Figure 15.36**
The IR spectrum for Problem 36b.

c.

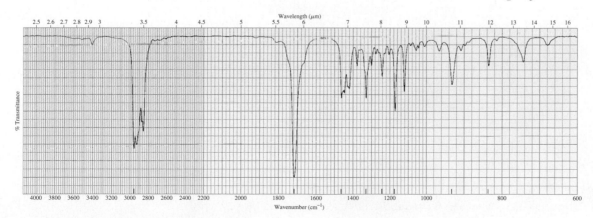

◀ **Figure 15.37**
The IR spectrum for Problem 36c.

37. Each of the IR spectra shown in Figure 15.38 is the spectrum of one of the following compounds. Identify the compound that is responsible for each spectrum.

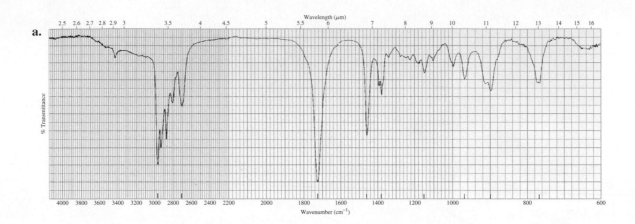

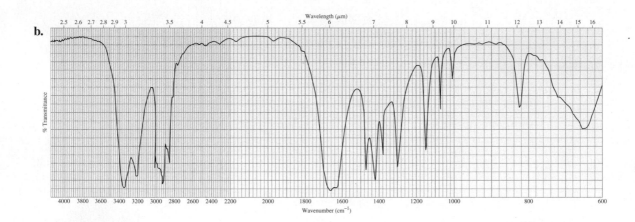

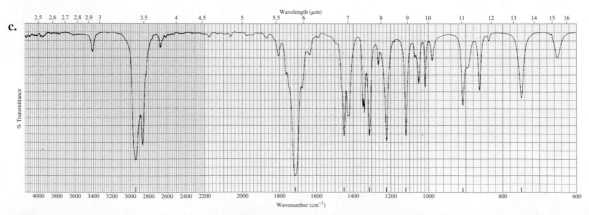

▲ **Figure 15.38**
The IR spectra for Problem 37.

38. The IR spectrum shown in Figure 15.39 is the spectrum of one of the following compounds. Identify the compound.

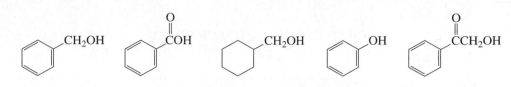

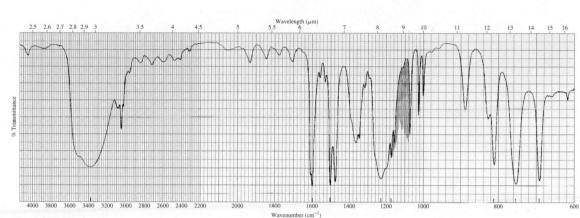

▲ **Figure 15.39**
The IR spectrum for Problem 38.

39. How could IR spectroscopy distinguish between 1,5-hexadiene and 2,4-hexadiene?

40. How does the frequency of the electromagnetic radiation used in NMR spectroscopy compare with that used in IR and UV/Vis spectroscopy?

41. How many signals are produced by each of the following compounds in its
   a. ^{1}H NMR spectrum?
   b. ^{13}C NMR spectrum?

42. Label each set of chemically equivalent protons, using *a* for the set that will be at the lowest frequency (farthest upfield) in the ^{1}H NMR spectrum, *b* for the next lowest, etc. Indicate the multiplicity of each signal.

  a. CH_3CHNO_2
          CH_3

  c. $CH_3CHCCH_2CH_2CH_3$
          CH_3

  e. $ClCH_2CCHCl_2$
          CH_3

  b. $CH_3CH_2CH_2OCH_3$

  d. $CH_3CH_2CH_2CCH_2Cl$

  f. $ClCH_2CH_2CH_2CH_2CH_2Cl$

43. Identify the following compounds. (Relative integrals are given from left to right across the spectrum.)
   a. The ^{1}H NMR spectrum of a compound with molecular formula $C_4H_{10}O_2$ has two singlets with an area ratio of 2 : 3.
   b. The ^{1}H NMR spectrum of a compound with molecular formula $C_8H_6O_2$ has two singlets with an area ratio of 1 : 2.

44. Match each of the ^{1}H NMR spectra with one of the following compounds:

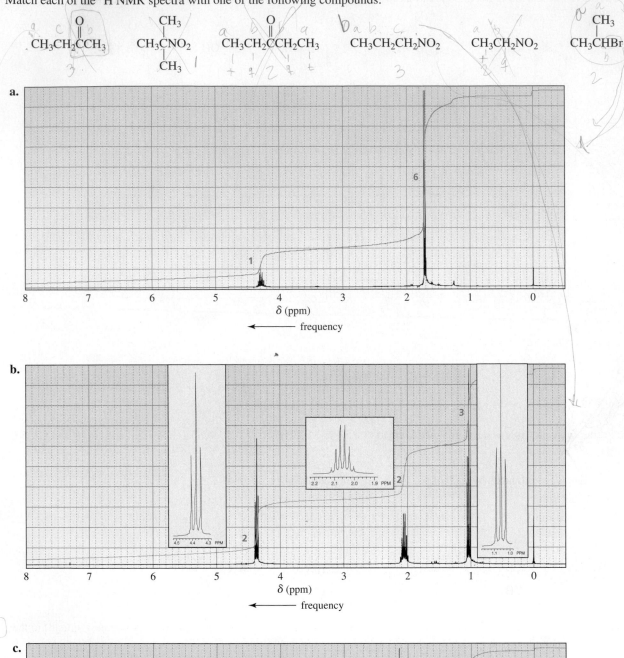

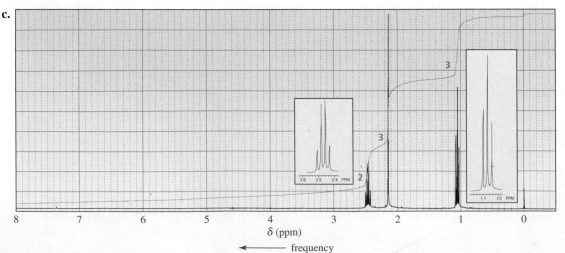

45. How can 1,2-, 1,3-, and 1,4-dinitrobenzene be distinguished by

    a. ^{1}H NMR spectroscopy?   **b.** ^{13}C NMR spectroscopy?

46. The ^{1}H NMR spectra of three isomers with molecular formula C_4H_9Br are shown here. Which isomer produces which spectrum?

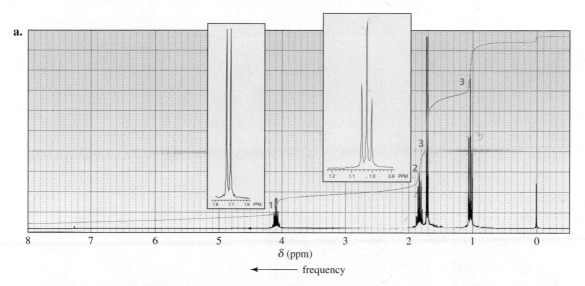

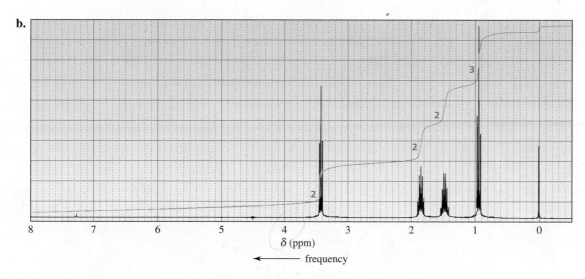

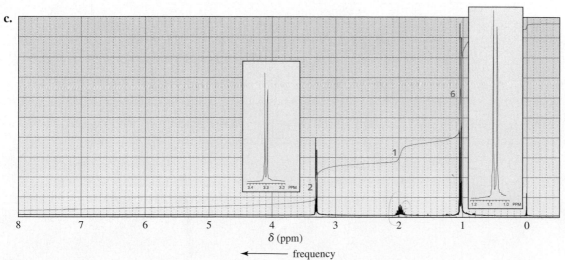

47. How could ^{1}H NMR distinguish between the compounds in each of the following pairs?

a. $CH_3CH_2CH_2OCH_3$ and $CH_3CH_2OCH_2CH_3$

c.
$$\underset{\underset{CH_3}{|}}{CH_3CH}-\underset{\underset{CH_3}{|}}{CHCH_3} \quad \text{and} \quad CH_3\underset{\underset{CH_3}{|}}{\overset{\overset{CH_3}{|}}{C}}CH_2CH_3$$

2 señales   13 señales

b. $BrCH_2CH_2CH_2Br$ and $BrCH_2CH_2CH_2NO_2$

d.
$$CH_3-\underset{\underset{CH_3}{|}}{\overset{\overset{CH_3}{|}}{C}}-\overset{\overset{O}{||}}{C}-OCH_3 \quad \text{and} \quad CH_3-\underset{\underset{OCH_3}{|}}{\overset{\overset{OCH_3}{|}}{C}}-CH_3$$

señales   2 señales

48. Identify each of the following compounds from the ^{1}H NMR data and molecular formula. The number of hydrogens responsible for each signal is shown in parentheses.

a. $C_4H_8Br_2$ 1.97 ppm (6) singlet
                  3.89 ppm (2) singlet

b. C_8H_9Br 2.01 ppm (3) doublet
                 5.14 ppm (1) quartet
                 7.35 ppm (5) broad singlet

c. $C_5H_{10}O_2$ 1.15 ppm (3) triplet
                 1.25 ppm (3) triplet
                 2.33 ppm (2) quartet
                 4.13 ppm (2) quartet

49 Compound A, with molecular formula C_4H_9Cl, shows two signals in its ^{13}C NMR spectrum. Compound B, an isomer of compound A, shows four signals, and in the proton-coupled mode, the signal farthest downfield is a doublet. Identify compounds A and B.

50. How could ^{1}H NMR spectroscopy distinguish between the following compounds?

a. ⬡ and ⬡

b. CH_3—⬡—CH_3 and CH_3—⬡—OCH_3

51. The ^{1}H NMR spectra of three isomers with molecular formula $C_7H_{14}O$ are shown here. Which isomer produces which spectrum?

a.

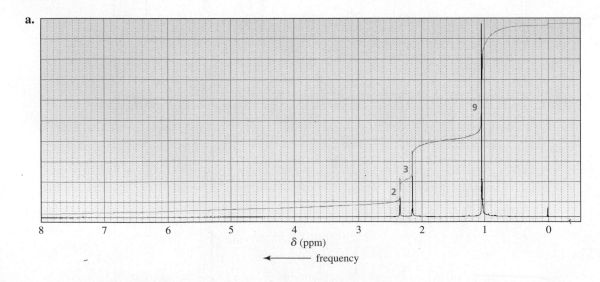

b.

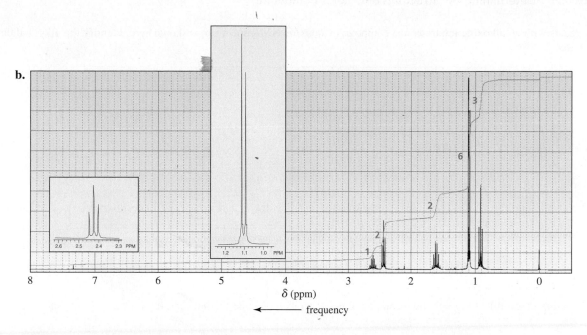

δ (ppm)

$\longleftarrow$ frequency

c.

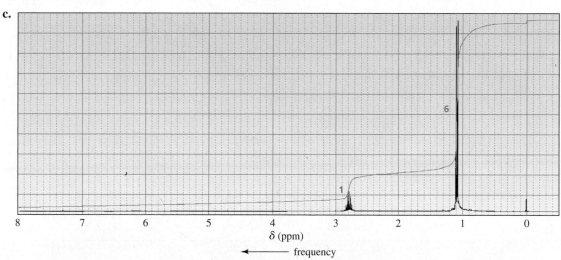

δ (ppm)

$\longleftarrow$ frequency

52. How could you use ^{1}H NMR spectroscopy to distinguish among the following esters?

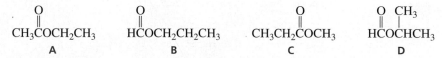

CH$_3$COCH$_2$CH$_3$ HCOCH$_2$CH$_2$CH$_3$ CH$_3$CH$_2$COCH$_3$ HCOCHCH$_3$
    A   B   C   D

53. Identify the compound with molecular formula C$_7$H$_{14}$O that gives the following proton-coupled ^{13}C NMR spectrum.

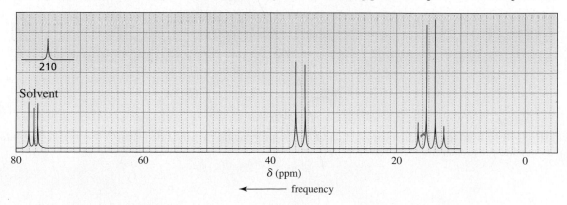

δ (ppm)

$\longleftarrow$ frequency

54. An alkyl halide reacts with an alkoxide ion to form a compound whose ^{1}H NMR spectrum is shown here. Identify the alkyl halide and the alkoxide ion.

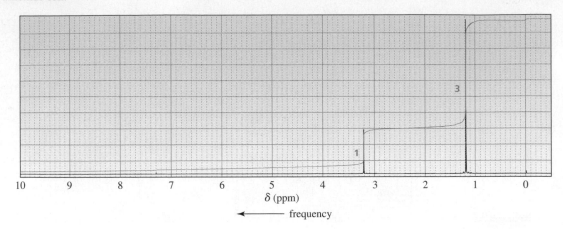

55. Determine the structure of the following unknown compound based on its molecular formula and its IR and ^{1}H NMR spectra.

$C_6H_{12}O_2$

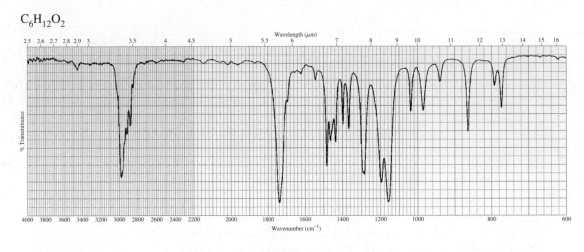

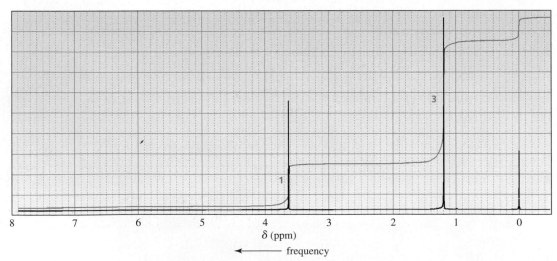

56. How could ^{1}H NMR be used to prove that the addition of HBr to propene follows the rule that says that the electrophile adds to the sp^2 carbon bonded to the greater number of hydrogens (Section 5.3).

16 | Carbohydrates

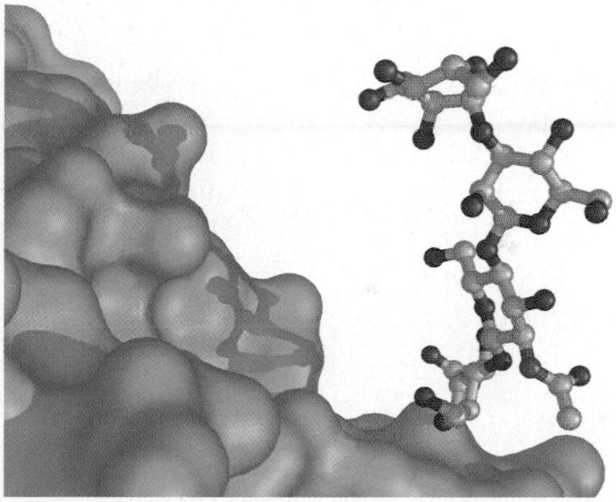

Bioorganic compounds are organic compounds found in biological systems. There is great similarity between the organic reactions chemists carry out in the laboratory and those performed by nature inside a living cell. In other words, bioorganic reactions can be thought of as organic reactions that take place in tiny flasks called cells.

Most bioorganic compounds have more complicated structures than those of the organic compounds you are used to seeing, but do not let the structures fool you into thinking that their chemistry is equally complicated. One reason the structures of bioorganic compounds are more complicated is that bioorganic compounds must be able to recognize each other, and much of their structure is for that purpose—a function called **molecular recognition**.

The first group of bioorganic compounds we will look at are *carbohydrates*—the most abundant class of compounds in the biological world, making up more than 50% of the dry weight of the Earth's biomass. Carbohydrates are important constituents of all living organisms and have a variety of different functions. Some are important structural components of cells; others act as recognition sites on cell surfaces. For example, the first event in all our lives was a sperm recognizing a carbohydrate on the surface of an egg's wall. Other carbohydrates serve as a major source of metabolic energy. For example, the leaves, fruits, seeds, stems, and roots of plants contain carbohydrates that plants use for their own metabolic needs and that then serve the metabolic needs of the animals that eat the plants.

Carbohydrates are polyhydroxy aldehydes such as D-glucose, polyhydroxy ketones such as D-fructose, and compounds formed by linking up polyhydroxy aldehydes or polyhydroxy ketones (Section 16.13). The chemical structures of carbohydrates are commonly represented by *wedge-and-dash structures* or by *Fischer projections*.

D-glucose

D-fructose

A **Fischer projection**, represents an asymmetric center as the point of intersection of two perpendicular lines; horizontal lines represent bonds that point toward the viewer and vertical lines represent bonds that extend back from the plane of the paper away from the viewer.

wedge-and-dash structure   Fischer projection

D-glucose
a polyhydroxy aldehyde

wedge-and-dash structure   Fischer projection

D-fructose
a polyhydroxy ketone

3-D Molecules:
D-Glucose;
D-Fructose

The most abundant carbohydrate in nature is glucose. Animals obtain glucose from food that contains glucose, such as plants. Plants produce glucose by *photosynthesis*. During photosynthesis, plants take up water through their roots and use carbon dioxide from the air to synthesize glucose and oxygen. Because photosynthesis is the reverse of the process used by organisms to obtain energy (the oxidation of glucose to carbon dioxide and water) plants require energy to carry out photosynthesis. Plants obtain the energy they need for photosynthesis from sunlight, captured by chlorophyll molecules in green plants. Photosynthesis uses the CO_2 that animals exhale as waste and generates the O_2 that animals inhale to sustain life. Nearly all the oxygen in the atmosphere has been released by photosynthetic processes.

$$\underset{\text{glucose}}{C_6H_{12}O_6} + 6\,O_2 \underset{\text{photosynthesis}}{\overset{\text{oxidation}}{\rightleftharpoons}} 6\,CO_2 + 6\,H_2O + \text{energy}$$

16.1   Classification of Carbohydrates

The terms *carbohydrate*, *saccharide*, and *sugar* are often used interchangeably. There are two classes of carbohydrates: *simple carbohydrates* and *complex carbohydrates*. **Simple carbohydrates** are **monosaccharides** (single sugars), whereas **complex carbohydrates** contain two or more monosaccharides linked together. **Disaccharides** have two monosaccharides linked together, **oligosaccharides** have three to 10 monosaccharides (*oligos* is Greek for "few") linked together, and **polysaccharides** have more than 10 monosaccharides linked together. Disaccharides, oligosaccharides, and polysaccharides can be broken down to monosaccharides by hydrolysis.

$$\underset{\text{polysaccharide}}{-M-M-M-M-M-M-M-M-M-} \overset{\text{hydrolysis}}{\longrightarrow} \underset{\text{monosaccharide}}{x\,M}$$

A *monosaccharide* can be a polyhydroxy aldehyde such as D-glucose or a polyhydroxy ketone such as D-fructose. Polyhydroxy aldehydes are called **aldoses** ("ald" is for aldehyde; "ose" is the suffix for a sugar), whereas polyhydroxy ketones are called **ketoses**. Monosaccharides are also classified according to the number of carbons they contain: Monosaccharides with three carbons are **trioses**, those with four carbons are **tetroses**, those with five carbons are **pentoses**, and those with six and seven carbons are **hexoses** and **heptoses**, respectively. Therefore, a six-carbon polyhydroxy aldehyde such as D-glucose is an aldohexose, whereas a six-carbon polyhydroxy ketone such as D-fructose is a ketohexose.

PROBLEM 1◆

Classify the following monosaccharides:

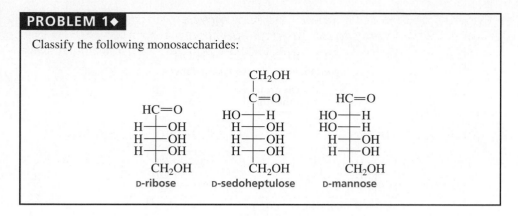

D-ribose   D-sedoheptulose   D-mannose

16.2   The D and L Notation

The smallest aldose, and the only one whose name does not end in "ose," is glyceraldehyde, an aldotriose.

asymmetric center

HOCH₂CHCHO
|
OH

glyceraldehyde

A carbon to which four different groups are attached is an asymmetric center.

Because glyceraldehyde has an asymmetric center, it can exist as a pair of enantiomers. We know that the isomer on the left has the *R* configuration because the arrow drawn from the highest priority substituent (OH) to the next highest priority substituent (CH=O) is clockwise (Section 8.6). To determine the configuration of a Fischer projection, after drawing an arrow from the highest priority to the next highest priority substituent, we need to look at the lowest priority substituent. Only if the lowest priority substituent is on a *vertical bond* does a clockwise arrow specify *R* and a counterclockwise arrow *S*. If the smallest priority substituent is on a *horizontal bond*, as it is here, then the answer you get from the direction of the arrow is opposite of the usual answer—clockwise is *S* and counterclockwise is *R*.

clockwise is *R*

CH=O

HO—C—ʹʹʹH
|
CH₂OH

(R)-(+)-glyceraldehyde

CH=O

H—C—OH
|
HOCH₂

(S)-(−)-glyceraldehyde

perspective formulas

H is on a horizontal bond so counterclockwise is *R*

HC=O
H——OH
CH₂OH

(R)-(+)-glyceraldehyde

HC=O
HO——H
CH₂OH

(S)-(−)-glyceraldehyde

Fischer projections

The notations D and L are used to describe the configurations of carbohydrates. In a Fischer projection of a monosaccharide, the carbonyl group is always placed on top (in the case of aldoses) or as close to the top as possible (in the case of ketoses). From its structure, you can see that galactose has four asymmetric centers (C-2, C-3, C-4, and C-5). *If the OH group attached to the bottom-most asymmetric center (the carbon that is second from the bottom) is on the right, the compound is a D-sugar. If that OH group is on the left, the compound is an L-sugar.* Almost all sugars found in nature are D-sugars. Notice that the mirror image of a D-sugar is an L-sugar.

$$\begin{array}{ccc}
\text{HC}=\text{O} & & \text{HC}=\text{O} \\
\text{H}\!-\!\!\!-\!\text{OH} & & \text{HO}\!-\!\!\!-\!\text{H} \\
\text{CH}_2\text{OH} & & \text{CH}_2\text{OH} \\
\text{D-glyceraldehyde} & & \text{L-glyceraldehyde}
\end{array}$$

the OH group is on the right

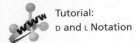

$$\begin{array}{ccc}
\text{HC}=\text{O} & & \text{HC}=\text{O} \\
\text{H}\!-\!\!\!-\!\text{OH} & & \text{HO}\!-\!\!\!-\!\text{H} \\
\text{HO}\!-\!\!\!-\!\text{H} & & \text{H}\!-\!\!\!-\!\text{OH} \\
\text{HO}\!-\!\!\!-\!\text{H} & & \text{H}\!-\!\!\!-\!\text{OH} \\
\text{H}\!-\!\!\!-\!\text{OH} & & \text{HO}\!-\!\!\!-\!\text{H} \\
\text{CH}_2\text{OH} & & \text{CH}_2\text{OH} \\
\text{D-galactose} & & \text{L-galactose}
\end{array}$$

the OH group is on the right

mirror image of D-galactose

Emil Fischer and his colleagues studied carbohydrates in the late nineteenth century, when techniques for determining the configurations of compounds were not available. Fischer arbitrarily assigned the *R*-configuration to the dextrorotatory isomer of glyceraldehyde that we call D-glyceraldehyde. He turned out to be correct: D-Glyceraldehyde is (*R*)-(+)-glyceraldehyde, and L-glyceraldehyde is (*S*)-(−)-glyceraldehyde.

Like *R* and *S*, D and L indicate the configuration of an asymmetric center, but they do not indicate whether the compound rotates polarized light to the right (+) or to the left (−) (Section 8.7). For example, D-glyceraldehyde is dextrorotatory, whereas D-lactic acid is levorotatory. In other words, optical rotation, like melting or boiling points, is a physical property of a compound, whereas "*R*, *S*, D, and L" are conventions humans use to indicate the configuration of a molecule.

$$\begin{array}{ccc}
\text{HC}=\text{O} & & \text{COOH} \\
\text{H}\!-\!\!\!-\!\text{OH} & & \text{H}\!-\!\!\!-\!\text{OH} \\
\text{CH}_2\text{OH} & & \text{CH}_3 \\
\text{D-(+)-glyceraldehyde} & & \text{D-(−)-lactic acid}
\end{array}$$

The common name of the monosaccharide, together with the D or L designation, completely defines its structure because the configurations of all the asymmetric centers are implicit in the common name.

PROBLEM 2

Draw Fischer projections of L-glucose and L-fructose.

16.3 Configurations of Aldoses

Aldotetroses have two asymmetric centers and therefore four stereoisomers. The aldotetroses are called erythrose and threose. Two of the stereoisomers are D-sugars and two are L-sugars.

$$\begin{array}{cccc}
\text{HC}=\text{O} & \text{HC}=\text{O} & \text{HC}=\text{O} & \text{HC}=\text{O} \\
\text{H}\!-\!\!\!-\!\text{OH} & \text{HO}\!-\!\!\!-\!\text{H} & \text{HO}\!-\!\!\!-\!\text{H} & \text{H}\!-\!\!\!-\!\text{OH} \\
\text{H}\!-\!\!\!-\!\text{OH} & \text{HO}\!-\!\!\!-\!\text{H} & \text{H}\!-\!\!\!-\!\text{OH} & \text{HO}\!-\!\!\!-\!\text{H} \\
\text{CH}_2\text{OH} & \text{CH}_2\text{OH} & \text{CH}_2\text{OH} & \text{CH}_2\text{OH} \\
\text{D-erythrose} & \text{L-erythrose} & \text{D-threose} & \text{L-threose}
\end{array}$$

Aldopentoses have three asymmetric centers and therefore eight stereoisomers (four pairs of enantiomers), whereas aldohexoses have four asymmetric centers and 16 stereoisomers (eight pairs of enantiomers). The four D-aldopentoses and the eight D-aldohexoses are shown in Table 16.1.

Table 16.1   Configurations of the D-Aldoses

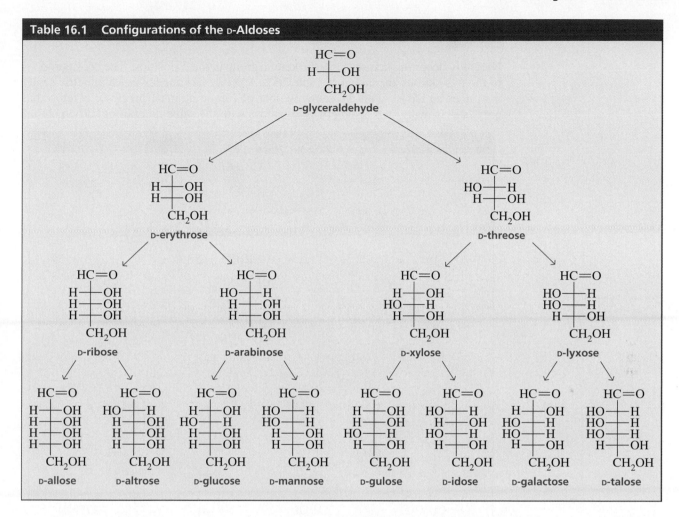

Diasteromers that differ in configuration at only one asymmetric center are called **epimers**. For example, D-ribose and D-arabinose are C-2 epimers (they differ in configuration only at C-2); D-idose and D-talose are C-3 epimers.

D-Glucose, D-mannose, and D-galactose are the most common aldohexoses in biological systems. An easy way to learn their structures is to memorize the structure of D-glucose and then remember that D-mannose is the C-2 epimer of D-glucose and D-galactose is the C-4 epimer of D-glucose.

D-Mannose is the C-2 epimer of D-glucose.

D-Galactose is the C-4 epimer of D-glucose.

PROBLEM 3◆

a. What sugar is the C-3 epimer of D-xylose?

b. What sugar is the C-5 epimer of D-allose?

16.4 Configurations of Ketoses

Naturally occurring ketoses have the ketone group in the 2-position. The configurations of the D-2-ketoses are shown in Table 16.2. A ketose has one fewer asymmetric center than does an aldose with the same number of carbon atoms. Therefore, a ketose has only half as many stereoisomers as an aldose with the same number of carbon atoms.

Table 16.2   Configurations of the D-Ketoses

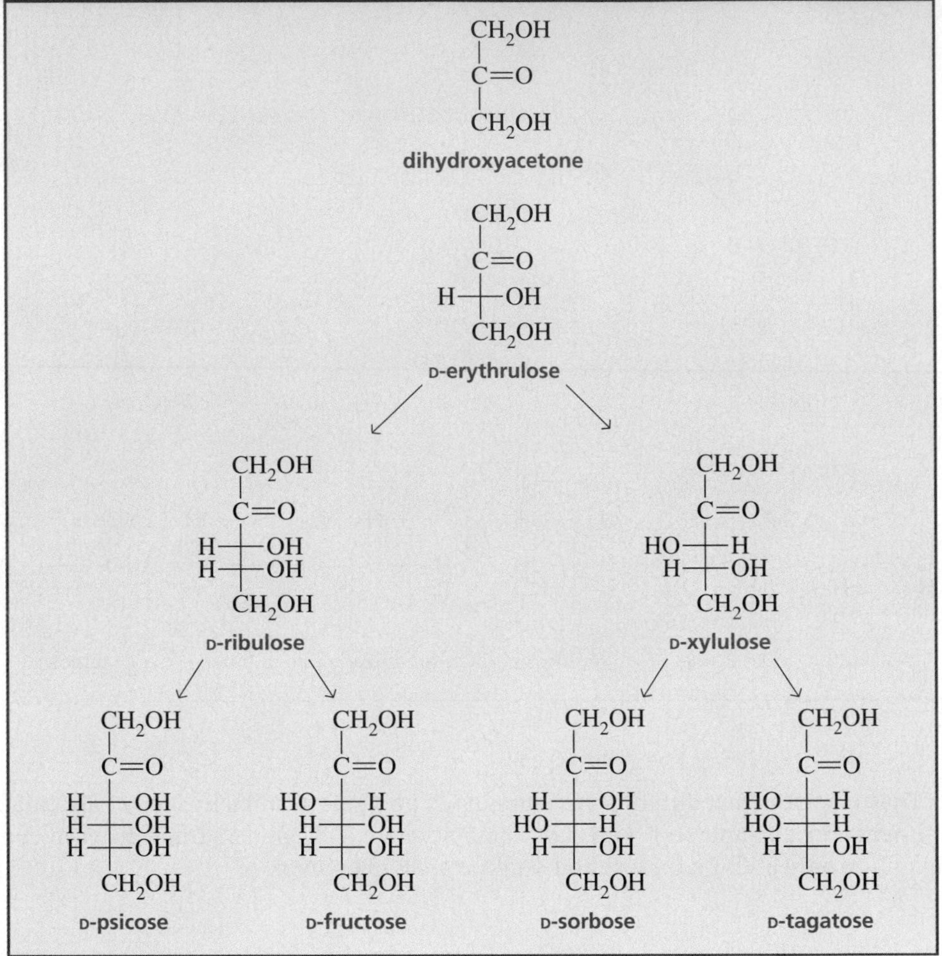

dihydroxyacetone

D-erythrulose

D-ribulose   D-xylulose

D-psicose   D-fructose   D-sorbose   D-tagatose

PROBLEM 4◆

What sugar is the C-3 epimer of D-fructose?

PROBLEM 5◆

How many stereoisomers are possible for
a. a 2-ketoheptose?   **b.** an aldoheptose?   **c.** a ketotriose?

16.5 Oxidation–Reduction Reactions of Monosaccharides

Because monosaccharides contain *alcohol* functional groups and *aldehyde* or *ketone* functional groups, the reactions of monosaccharides are an extension of what you have already learned about the reactions of alcohols, aldehydes, and ketones. For example, an aldehyde group in a monosaccharide can be oxidized or reduced and can react with nucleophiles to form imines, hemiacetals, and acetals. When you read the sections that deal with the reactions of monosaccharides, you will find cross-references to the sections in which the same reactivity for simple organic compounds is discussed. As you study, refer back to these sections; they will make learning about carbohydrates a lot easier and will give you a good review of some chemistry that you have already learned about.

Reduction

The carbonyl group of aldoses and ketoses can be reduced by sodium borohydride ($NaBH_4$; Section 13.5). The product of the reduction is a polyalcohol, known as an **alditol**. Reduction of an aldose forms one alditol. For example, the reduction of D-mannose forms D-mannitol, the alditol found in mushrooms, olives, and onions. Reduction of a ketose forms two alditols because the reaction creates a new asymmetric center in the product. For example, the reduction of D-fructose forms both D-mannitol and D-glucitol, the C-2 epimer of D-mannitol. D-Glucitol—also called sorbitol—is about 60% as sweet as sucrose. It is found in plums, pears, cherries, and berries and is used as a sugar substitute in the manufacture of candy.

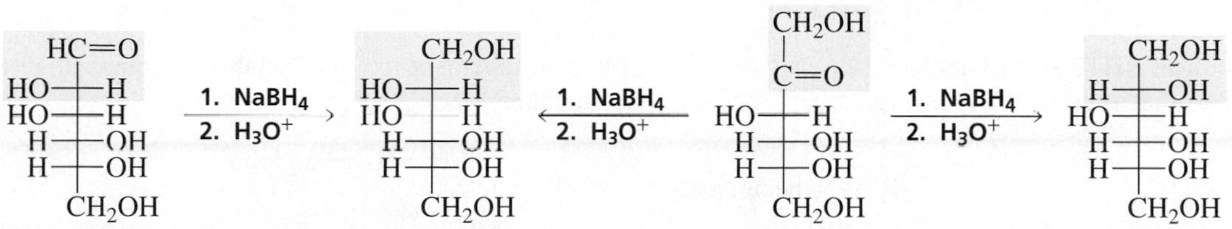

D-mannose → D-mannitol an alditol ← D-fructose → D-glucitol an alditol

D-Xylitol—obtained from the reduction of D-xylose—is used as a sweetening agent in cereals and in "sugarless" gum.

PROBLEM 6◆

What products are obtained from the reduction of

a. D-idose? **b.** D-sorbose?

PROBLEM 7◆

What other monosaccharide is reduced only to the alditol obtained from the reduction of D-galactose?

PROBLEM 8◆

What monosaccharide is reduced to two alditols, one of which is the alditol obtained from the reduction of D-talose?

Oxidation

Aldoses can be distinguished from ketoses by observing what happens to the color of an aqueous solution of bromine when it is added to the sugar. Br_2 is a mild oxidizing agent and easily oxidizes the aldehyde group, but it cannot oxidize ketones or alcohols. Consequently, if a small amount of an aqueous solution of Br_2 is added to an unknown monosaccharide, the reddish-brown color of Br_2 will disappear if the monosaccharide is an aldose because Br_2 will be reduced to Br^- which is colorless. If the red color persists, indicating no reaction with Br_2, the monosaccharide is a ketose. The product of the oxidation reaction is an **aldonic acid**.

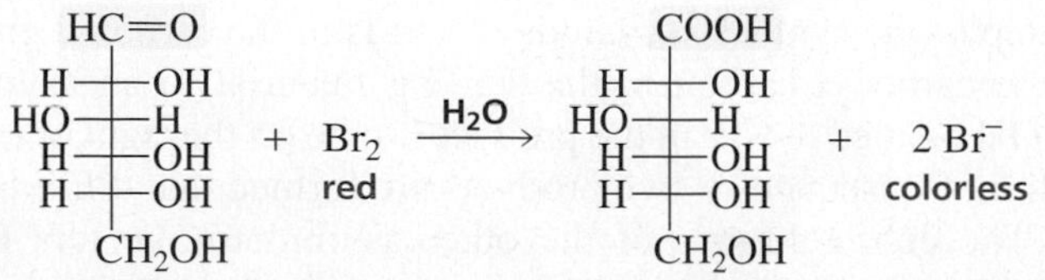

D-glucose + Br_2 (red) $\xrightarrow{H_2O}$ D-gluconic acid an aldonic acid + 2 Br^- (colorless)

If an oxidizing agent stronger than Br_2 is used (such as HNO_3), the primary alcohol will also be oxidized. The product that is obtained when both the aldehyde and the primary alcohol groups of an aldose are oxidized is called an **aldaric acid**. (In an ald**on**ic acid, **on**e end is oxidized. In an ald**ar**ic acid, both ends **ar**e oxidized.)

D-glucose → D-glucaric acid, an aldaric acid (HNO_3, Δ)

MEASURING THE BLOOD GLUCOSE LEVELS OF DIABETICS

Glucose reacts with an NH_2 group of hemoglobin to form an imine (Section 13.6) that subsequently undergoes an irreversible rearrangement to a more stable α-aminoketone known as hemoglobin-A_{IC}.

Diabetes results when the body does not produce sufficient insulin or when the insulin it produces does not properly stimulate its target cells. Because insulin is the hormone that maintains the proper level of glucose in the blood, diabetics have increased blood glucose levels. The amount of hemoglobin-A_{IC} formed is proportional to the concentration of glucose in the blood, so diabetics have a higher concentration of hemoglobin-A_{IC} than nondiabetics. Thus, measuring the hemoglobin-A_{IC} level is a way to determine whether the blood glucose level of a diabetic is being controlled.

Cataracts, a common complication in diabetics, are caused by the reaction of glucose with the NH_2 group of proteins in the lens of the eye. It is thought that the arterial rigidity common in old age may be attributable to a similar reaction of glucose with the NH_2 group of proteins.

PROBLEM 9◆

a. Name an aldohexose other than D-glucose that is oxidized to D-glucaric acid by nitric acid.

b. What is another name for D-glucaric acid?

c. Name another pair of aldohexoses that are oxidized to identical aldaric acids.

16.6 Chain Elongation: The Kiliani–Fischer Synthesis

Heinrich Kiliani (1855–1945) *was born in Germany. He received a Ph.D. from the University of Munich, studying under Professor Emil Erlenmeyer. Kiliani became a professor of chemistry at the University of Freiburg.*

The carbon chain of an aldose can be increased by one carbon in a **Kiliani–Fischer synthesis**. In other words, tetroses can be converted into pentoses, and pentoses can be converted into hexoses.

In the first step of the synthesis, cyanide ion adds to the carbonyl group. This reaction converts the carbonyl carbon in the starting material to an asymmetric center. Therefore, the OH bonded to C-2 in the product can be on the right or on the left in the Fischer projection. Consequently, two products are formed that differ only in configuration at C-2. The configurations of the other asymmetric centers do not change, because no bond to any of the asymmetric centers is broken during the course of the reaction. The $C\equiv N$ bond is reduced to an imine, using a partially deactivated palladium catalyst so that the imines are not further reduced to amines (Section 5.12). The imines can then be hydrolyzed to aldoses (Section 13.6).

the modified Kiliani–Fischer synthesis

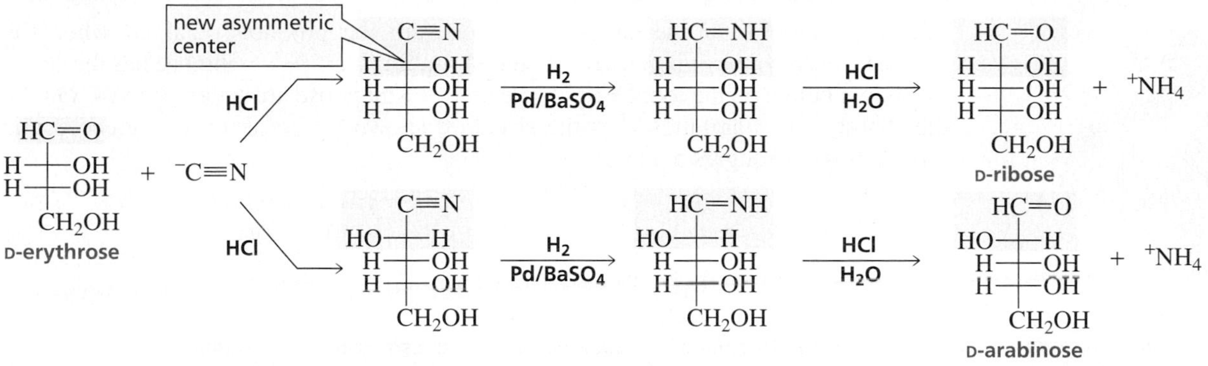

Notice that the synthesis leads to a pair of C-2 epimers because of the formation of the new asymmetric center.

PROBLEM 10◆

What monosaccharides would be formed in a Kiliani–Fischer synthesis starting with

a. D-xylose?   **b.** L-threose?

The Kiliani–Fischer synthesis leads to a pair of C-2 epimers.

16.7   Stereochemistry of Glucose: The Fischer Proof

In 1891, Emil Fischer determined the stereochemistry of glucose using one of the most brilliant examples of reasoning in the history of chemistry. He chose (+)-glucose for his study because it is the most common monosaccharide found in nature.

Fischer knew that (+)-glucose is an aldohexose, but 16 different structures can be written for an aldohexose. Which of them represents the structure of (+)-glucose? Fischer considered the eight stereoisomers that had the C-5 OH group on the right in the Fischer projection (the stereoisomers shown below that we now call the D-sugars). One of these is (+)-glucose, and its mirror image is (−)-glucose. It would not be possible to determine whether (+)-glucose was D-glucose or L-glucose until 1951. Fischer used the following information to determine glucose's stereochemistry—that is, to determine the configuration of each of its asymmetric centers.

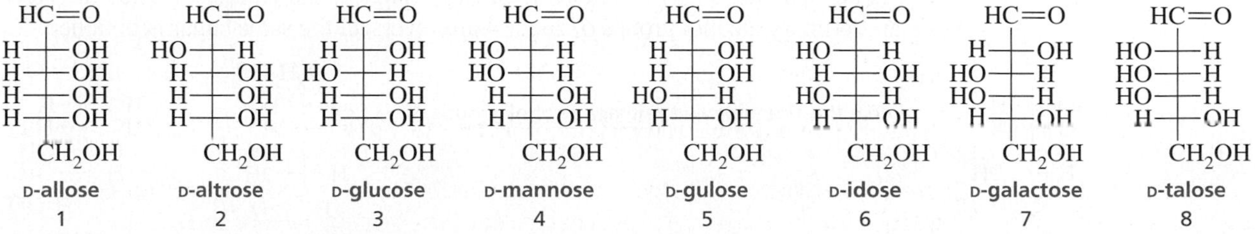

1. When the Kiliani–Fischer synthesis is performed on the sugar known as (−)-arabinose, the two sugars known as (+)-glucose and (+)-mannose are obtained. This means that (+)-glucose and (+)-mannose are C-2 epimers; in other words, they have the same configuration at C-3, C-4, and C-5. Consequently, (+)-glucose and (+)-mannose have to be one of the following pairs: sugars 1 and 2, 3 and 4, 5 and 6, or 7 and 8.

Emil Fischer (1852–1919) *was born in a village near Cologne, Germany. He became a chemist against the wishes of his father, a successful merchant, who wanted him to enter the family business. He was a professor of chemistry at the Universities of Erlangen, Würzburg, and Berlin. In 1902, he received the Nobel Prize in chemistry for his work on sugars. During World War I, he organized German chemical production. Two of his three sons died in that war.*

2. (+)-Glucose and (+)-mannose are both oxidized by nitric acid to optically active aldaric acids. The aldaric acids of sugars 1 and 7 would not be optically active because each has a plane of symmetry. (A compound containing a plane of symmetry is achiral—it has a superimposable mirror image; Section 8.9.) Excluding sugars 1 and 7 means that (+)-glucose and (+)-mannose must be sugars 3 and 4 or 5 and 6.

3. Because (+)-glucose and (+)-mannose are the products obtained when the Kiliani–Fischer synthesis is carried out on (−)-arabinose, if (−)-arabinose has the structure shown below on the left, (+)-glucose and (+)-mannose are sugars 3 and 4. On the other hand, if (−)-arabinose has the structure shown on the right, (+)-glucose and (+)-mannose are sugars 5 and 6.

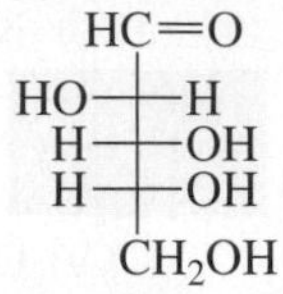

$$
\begin{array}{cc}
\text{HC=O} & \text{HC=O} \\
\text{HO}\!-\!\text{H} & \text{H}\!-\!\text{OH} \\
\text{H}\!-\!\text{OH} & \text{HO}\!-\!\text{H} \\
\text{H}\!-\!\text{OH} & \text{H}\!-\!\text{OH} \\
\text{CH}_2\text{OH} & \text{CH}_2\text{OH}
\end{array}
$$

the structure of (–)-arabinose if (+)-glucose and (+)-mannose are sugars 3 and 4   **the structure of (–)-arabinose if (+)-glucose and (+)-mannose are sugars 5 and 6**

When (−)-arabinose is oxidized with nitric acid, the aldaric acid that is obtained is optically active. This means that the aldaric acid does *not* have a plane of symmetry. Therefore, (−)-arabinose must have the structure shown on the left because the aldaric acid of the sugar on the right has a plane of symmetry. Thus, (+)-glucose and (+)-mannose are represented by sugars 3 and 4.

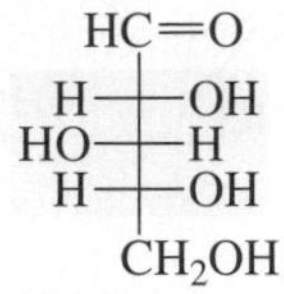

4. The last step in the Fischer proof was to determine whether (+)-glucose is sugar 3 or sugar 4. To answer this question, Fischer had to develop a chemical method that would interchange the aldehyde and primary alcohol groups of an aldohexose. When he chemically interchanged the aldehyde and primary alcohol groups of the sugar known as (+)-glucose, he obtained an aldohexose that was different from (+)-glucose. When he chemically interchanged the aldehyde and primary alcohol groups of (+)-mannose, he still had (+)-mannose. Therefore, he concluded that (+)-glucose is sugar 3 because reversing the aldehyde and alcohol groups of sugar 3 leads to a different sugar (L-gulose).

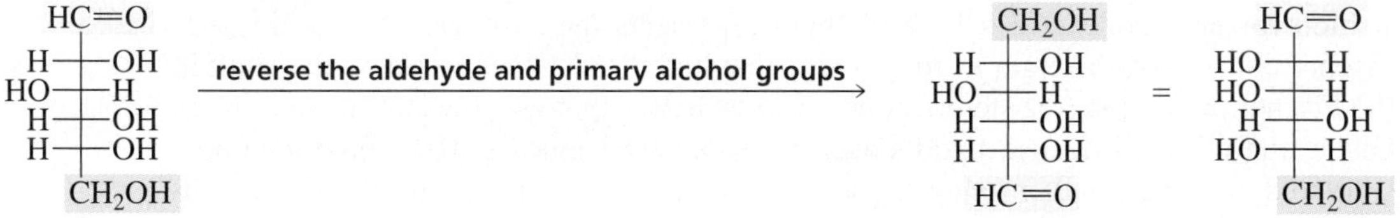

$$
\begin{array}{c}
\text{HC=O} \\
\text{H}\!-\!\text{OH} \\
\text{HO}\!-\!\text{H} \\
\text{H}\!-\!\text{OH} \\
\text{H}\!-\!\text{OH} \\
\text{CH}_2\text{OH}
\end{array}
\xrightarrow{\text{reverse the aldehyde and primary alcohol groups}}
\begin{array}{c}
\text{CH}_2\text{OH} \\
\text{H}\!-\!\text{OH} \\
\text{HO}\!-\!\text{H} \\
\text{H}\!-\!\text{OH} \\
\text{H}\!-\!\text{OH} \\
\text{HC=O}
\end{array}
=
\begin{array}{c}
\text{HC=O} \\
\text{HO}\!-\!\text{H} \\
\text{HO}\!-\!\text{H} \\
\text{H}\!-\!\text{OH} \\
\text{HO}\!-\!\text{H} \\
\text{CH}_2\text{OH}
\end{array}
$$

D-glucose   **L-gulose drawn upside down**   **L-gulose**

If (+)-glucose is sugar 3, (+)-mannose must be sugar 4. As predicted, when the aldehyde and primary alcohol groups of sugar 4 are reversed, the same sugar is obtained.

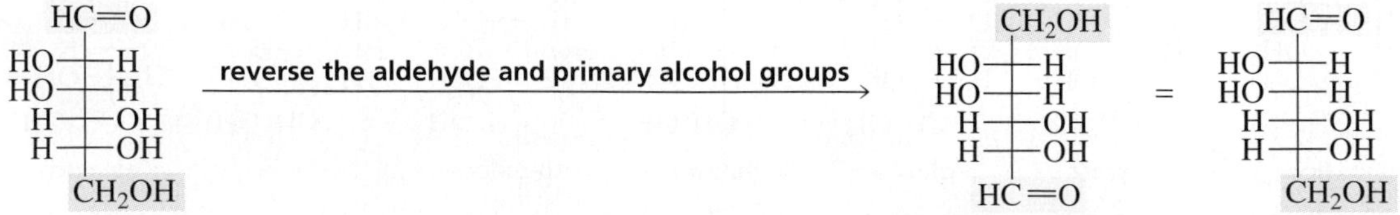

$$
\begin{array}{c}
\text{HC=O} \\
\text{HO}\!-\!\text{H} \\
\text{HO}\!-\!\text{H} \\
\text{H}\!-\!\text{OH} \\
\text{H}\!-\!\text{OH} \\
\text{CH}_2\text{OH}
\end{array}
\xrightarrow{\text{reverse the aldehyde and primary alcohol groups}}
\begin{array}{c}
\text{CH}_2\text{OH} \\
\text{HO}\!-\!\text{H} \\
\text{HO}\!-\!\text{H} \\
\text{H}\!-\!\text{OH} \\
\text{H}\!-\!\text{OH} \\
\text{HC=O}
\end{array}
=
\begin{array}{c}
\text{HC=O} \\
\text{HO}\!-\!\text{H} \\
\text{HO}\!-\!\text{H} \\
\text{H}\!-\!\text{OH} \\
\text{H}\!-\!\text{OH} \\
\text{CH}_2\text{OH}
\end{array}
$$

D-mannose   **D-mannose drawn upside down**   **D-mannose**

Using similar reasoning, Fischer went on to determine the stereochemistry of the other aldohexoses. He received the Nobel Prize in chemistry in 1902 for this achievement. His original guess that (+)-glucose is a D-sugar was later shown to be correct, so all of his structures are correct. If he had been wrong and (+)-glucose had been an L-sugar, his contribution to the stereochemistry of aldoses would still have had the same significance, but all his stereochemical assignments would have been reversed.

GLUCOSE/DEXTROSE

André Dumas first used the term "glucose" in 1838 to refer to the sweet compound that comes from honey and grapes. Later, Kekulé (Section 6.1) decided that it should be called dextrose because it was dextrorotatory. When Fischer studied the sugar, he called it glucose, and chemists have called it glucose ever since, although dextrose is often found on food labels.

Jean-Baptiste–André Dumas (1800–1884) *was born in France. Apprenticed to an apothecary, he left to study chemistry in Switzerland. He became a professor of chemistry at the University of Paris and at the Collège de France. He was the first French chemist to teach laboratory courses. In 1848, he left science for a political career. He became a senator, master of the French mint, and mayor of Paris.*

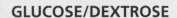

16.8 Cyclic Structure of Monosaccharides: Hemiacetal Formation

D-Glucose exists in three different forms: the open-chain form that we have been discussing (Table 16.1) and two cyclic forms—α-D-glucose and β-D-glucose. We know that the two cyclic forms are different, because they have different melting points and different specific rotations (Section 8.8).

How can D-glucose exist in a cyclic form? In Section 13.7, we saw that an aldehyde reacts with an equivalent of an alcohol to form a hemiacetal. A monosaccharide such as D-glucose has an aldehyde group and several alcohol groups. The alcohol group bonded to C-5 of D-glucose reacts with the aldehyde group. To see that C-5 is in the proper position to attack the aldehyde group, we need to convert the Fischer projection of D-glucose to a flat ring structure. To do this, draw the primary alcohol group *up* from the back left-hand corner. Groups on the *right* in a Fischer projection are *down* in the cyclic structure, whereas groups on the *left* in a Fischer projection are *up* in the cyclic structure. The reaction forms two cyclic (six-membered ring) hemiacetals.

Groups on the *right* in a Fischer projection are *down* in a planar cyclic structure

Groups on the *left* in a Fischer projection are *up* in a planar cyclic structure

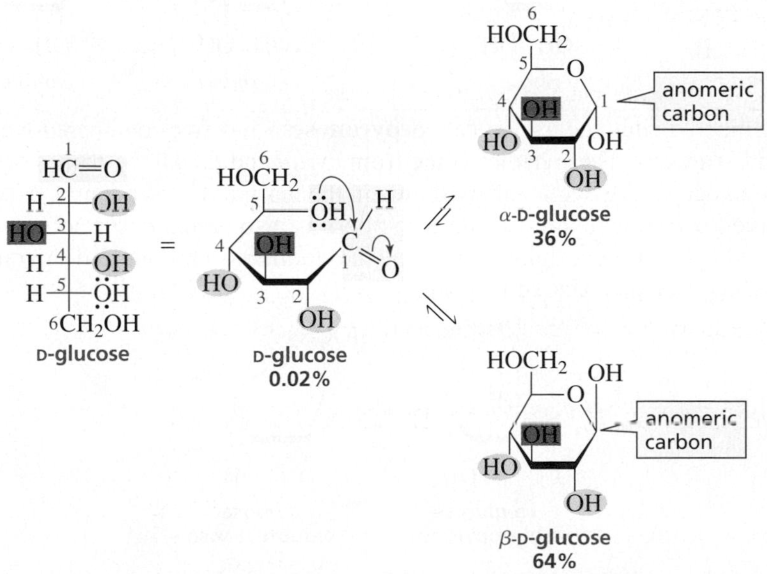

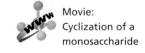

Movie: Cyclization of a monosaccharide

Why are there two different cyclic forms? Two different hemiacetals are formed because the carbonyl carbon of the open-chain aldehyde becomes a new asymmetric center in the cyclic hemiacetal. If the OH group bonded to the new asymmetric center is up, the hemiacetal is α-D-glucose; if the OH group is down, the hemiacetal is β-D-glucose. The mechanism for cyclic hemiacetal formation is the same as the mechanism for hemiacetal formation between individual aldehyde and alcohol molecules (Section 13.7).

α-D-Glucose and β-D-glucose are called anomers. **Anomers** are two sugars that differ in configuration only at the carbon that was the carbonyl carbon in the open-chain form. This carbon is called the **anomeric carbon**. The prefixes α- and β- denote the configuration about the anomeric carbon. Anomers, like epimers, differ in configuration at only one carbon atom. Notice that the anomeric carbon is the only carbon in the molecule that is bonded to two oxygens.

In an aqueous solution, the open-chain aldehyde is in equilibrium with the two cyclic hemiacetals. Formation of the cyclic hemiacetals proceeds nearly to completion, so very little glucose exists in the open-chain form (about 0.02%). The sugar still undergoes the reactions discussed in the previous sections because the reagents react with the small amount of open-chain aldehyde that is present. As the aldehyde reacts, the equilibrium shifts to offset the disturbance in the equilibrium. As a result, more open-chain aldehyde is formed which can then undergo reaction. Eventually, all the glucose molecules react by way of the open-chain aldehyde.

When crystals of pure α-D-glucose are dissolved in water, the specific rotation gradually changes from $+112.2°$ to $+52.7°$. When crystals of pure β-D-glucose are dissolved in water, the specific rotation gradually changes from $+18.7°$ to $+52.7°$. This change in rotation occurs because, in water, the hemiacetal opens to form the aldehyde and, when the aldehyde recyclizes, both α-D-glucose and β-D-glucose can be formed. Eventually, the three forms of glucose reach equilibrium concentrations. The specific rotation of the equilibrium mixture is $+52.7°$—this is why the same specific rotation will result, regardless of whether the crystals originally dissolved in water are α-D-glucose or β-D-glucose. A slow change in optical rotation to an equilibrium value is called **mutarotation**.

If an aldose can form a five- or a six-membered ring, it will exist predominantly as a cyclic hemiacetal in solution. D-Ribose forms five-membered ring hemiacetals: α-D-ribose and β-D-ribose.

pyran

furan

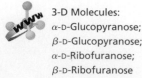

Six-membered-ring sugars are called **pyranoses**, and five-membered-ring sugars are called **furanoses**. These names come from *pyran* and *furan*, the names of the five- and six-membered-ring cyclic ethers shown in the margin. Consequently, α-D-glucose is also called α-D-glucopyranose and α-D-ribose is also called α-D-ribofuranose. The prefix α-indicates the configuration about the anomeric carbon, and "pyranose" or "furanose" indicates the size of the ring.

Like an aldose, if a ketose can form a five-or six-membered ring, it will exist predominantly in solution as a cylic pyranose or furanose. For example, D-fructose is a furanose in the disaccharide known as sucrose (Section 16.12).

Because five-membered rings are close to planar, furanoses are well represented by the planar structures. The planar representations, however, are structurally misleading for pyranoses because a six-membered ring is not flat, but exists preferentially in a chair conformation (Section 3.10).

PROBLEM 11 SOLVED

4-Hydroxy- and 5-hydroxyaldehydes exist primarily in the cyclic hemiacetal form. Give the structure of the cyclic hemiacetal formed by each of the following:

a. 4-hydroxybutanal   **b.** 5-hydroxypentanal

SOLUTION TO 11a   Draw the reactant with the alcohol and carbonyl groups on both sides of the molecule. Then look to see what size ring will be formed. Two cyclic products are obtained because the carbonyl carbon of the reactant has been converted into a new asymmetric center in the product.

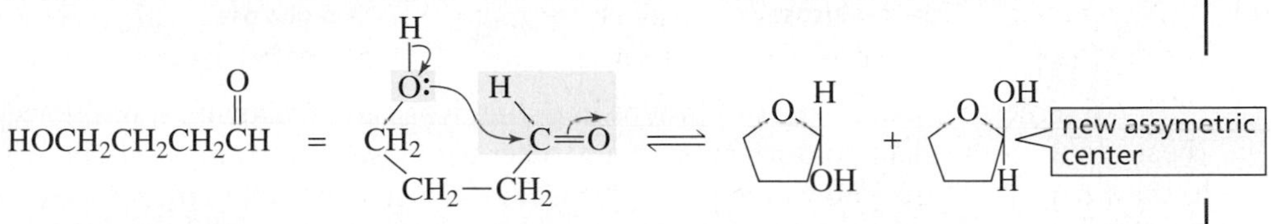

16.9 Stability of Glucose

Drawing D-glucose in its chair conformation shows why it is the most common aldohexose in nature. To convert the planar representation of D-glucose into a chair conformation, start by drawing the chair so that the backrest is on the left and the footrest is on the right. (It would be helpful to build a molecular model.) Then place the ring oxygen at the back right-hand corner and the primary alcohol group in the equatorial position. The primary alcohol group is the largest of all the substituents, and large substituents are more stable in the equatorial position because there is less steric strain in that position (Section 3.11). Because the OH group bonded to C-4 is trans to the primary alcohol group (this is easily seen in the flat six-membered ring representations), the C-4 OH group is also in the equatorial position. (Recall from Section 3.12 that 1,2-diequatorial substituents are trans to one another.) The C-3 OH group is trans to the C-4 OH group, so the C-3 OH group is also in the equatorial position. As you move around the ring, you will find that all the OH substituents in β-D-glucose are in equatorial positions. The axial positions are all occupied by hydrogens, which require little space and therefore experience little steric strain. No other aldohexose exists in such a strain-free conformation. This means that β-D-glucose is the most stable of all the aldohexoses, so it is not surprising that it is the most prevalent aldohexose in nature.

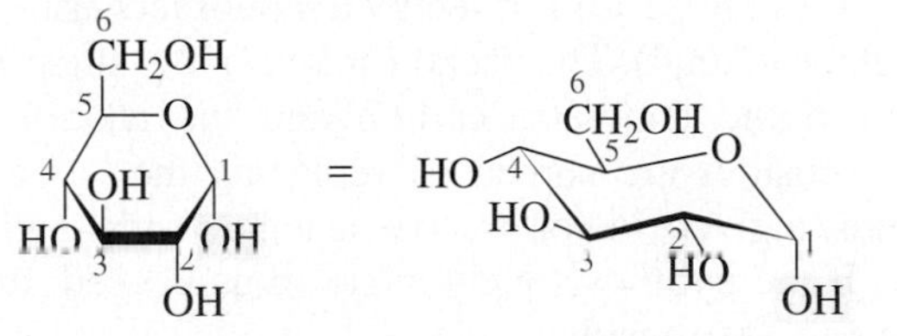

α-D-glucose
chair conformation

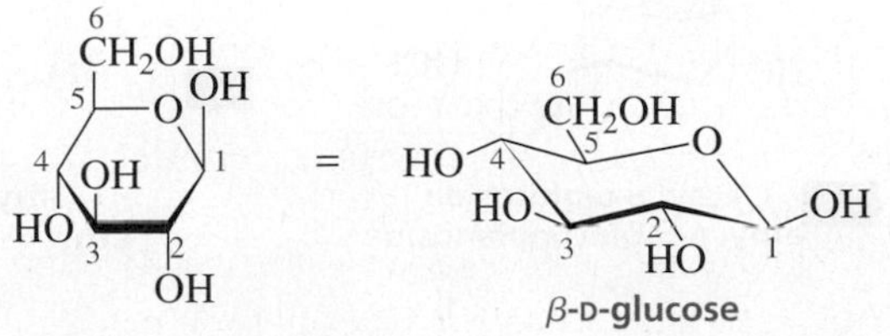

β-D-glucose
chair conformation

The *α*-position is to the right in a Fischer projection, down in a planar representation, and axial in a chair conformation.

The *β*-position is to the left in a Fischer projection, up in a planar representation, and equatorial in a chair conformation.

Why is there more β-D-glucose than α-D-glucose in an aqueous solution at equilibrium? The OH group bonded to the anomeric carbon is in the equatorial position in β-D-glucose, whereas it is in the axial position in α-D-glucose. Therefore, β-D-glucose is more stable than α-D-glucose, so β-D-glucose predominates at equilibrium in an aqueous solution.

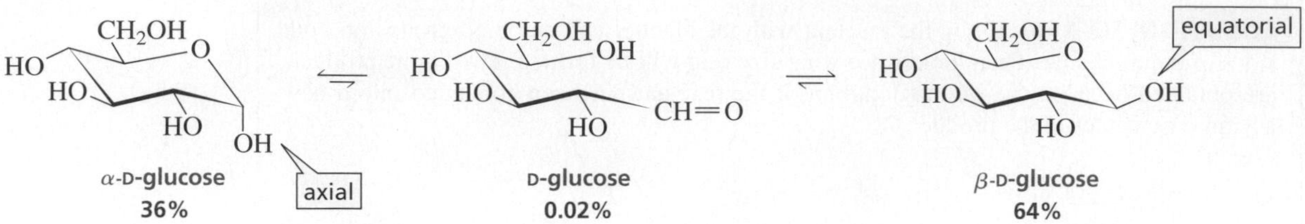

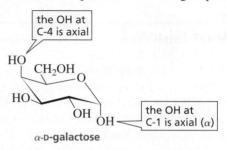

α-D-glucose
36%

D-glucose
0.02%

β-D-glucose
64%

If you remember that all the OH groups in β-D-glucose are in equatorial positions, it is easy to draw the chair conformation of any other pyranose. For example, if you want to draw α-D-galactose, you would put all the OH groups in equatorial positions, except the OH groups at C-4 (because galactose is a C-4 epimer of glucose) and at C-1 (because it is the α-anomer). You would put these two OH groups in axial positions.

the OH at
C-4 is axial

the OH at
C-1 is axial (α)

α-D-galactose

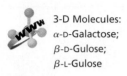

PROBLEM 12◆ | **SOLVED**

Which OH groups are in the axial position in

a. β-D-mannopyranose? **b.** β-D-idopyranose? **c.** α-D-allopyranose?

SOLUTION TO 12a All the OH groups in β-D-glucose are in equatorial positions. Because β-D-mannose is a C-2 epimer of β-D-glucose, only the C-2 OH group of β-D-mannose will be in the axial position.

16.10 Formation of Glycosides

Just like a hemiacetal (or hemiketal) reacts with an alcohol to form an acetal (or ketal), the cyclic hemiacetal (or hemiketal) formed by a monosaccharide can react with an alcohol to form an acetal (or ketal). The acetal (or ketal) of a sugar is called a **glycoside**, and the bond between the anomeric carbon and the alkoxy oxygen is called a **glycosidic bond**. Glycosides are named by replacing the "e" ending of the sugar's name with "ide." Thus, a glycoside of glucose is a glucoside, a glycoside of galactose is a galactoside, etc. If the pyranose or furanose name is used, the acetal (or ketal) is called a **pyranoside** or a **furanoside**.

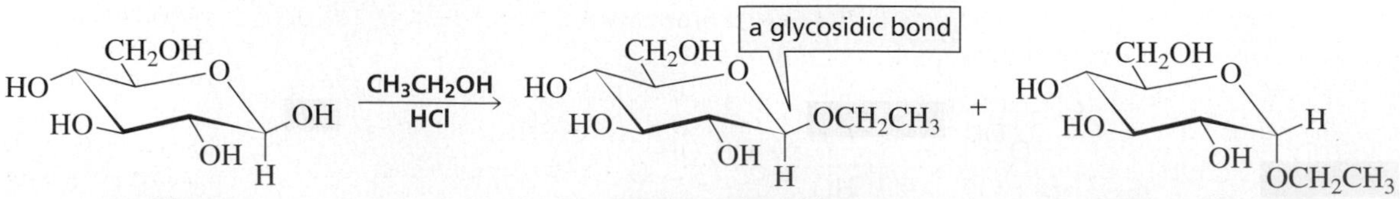

a glycosidic bond

β-D-glucose
β-D-glucopyranose

CH₃CH₂OH
HCl

ethyl β-D-glucoside
ethyl β-D-glucopyranoside

ethyl α-D-glucoside
ethyl α-D-glucopyranoside

Notice that the reaction of a single anomer with an alcohol leads to the formation of both the α- and β-glycosides. The mechanism of the reaction shows why both glycosides are formed. The OH group bonded to the anomeric carbon becomes protonated in the acidic solution, and a lone pair on the ring oxygen helps expel a molecule of water. The anomeric carbon is now sp^2 hybridized, so that part of the molecule is planar (Section 8.13). When the alcohol comes in from the top of the plane, the β-glycoside is formed; when the alcohol comes in from the bottom of the plane, the α-glycoside is formed. Notice that the mechanism is the same as that shown for acetal formation in Section 13.7.

mechanism of glycoside formation

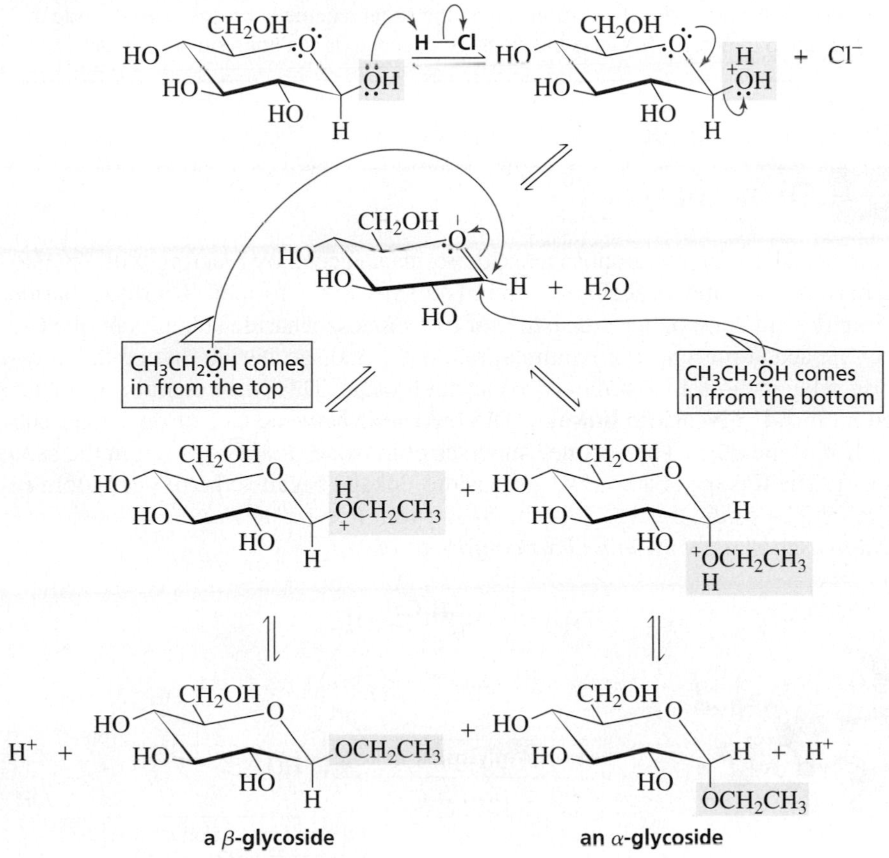

CH₃CH₂OH comes in from the top

CH₃CH₂OH comes in from the bottom

a β-glycoside

an α-glycoside

16.11 Reducing and Nonreducing Sugars

Glycosides are acetals (or ketals). Therefore, they are not in equilibrium with the open-chain aldehyde (or ketone) in aqueous solution. Without being in equilibrium with a compound with a carbonyl group, they cannot be oxidized by Br_2. Glycosides, therefore, are nonreducing sugars—they cannot reduce Br_2.

In contrast, hemiacetals (or hemiketals) are in equilibrium with the open-chain sugars in aqueous solution, so they can reduce Br_2. In summary, as long as a sugar has an aldehyde, a ketone, a hemiacetal, or a hemiketal group, it is able to reduce an oxidizing agent and therefore is classified as a **reducing sugar**. Without one of these groups, it is a **nonreducing sugar**.

A sugar with an aldehyde, a ketone, a hemiacetal, or a hemiketal group is a reducing sugar. A sugar without one of these groups is a nonreducing sugar.

PROBLEM 13◆ SOLVED

Name the following compounds, and indicate whether each is a reducing sugar or a nonreducing sugar:

a. [structure diagram with CH$_2$OH, HO, O, OCH$_2$CH$_2$CH$_3$, OH, HO]

b. [structure diagram with HO, CH$_2$OH, O, HO, OH, OCH$_3$]

SOLUTION TO 13a The only OH group in an axial position in (a) is the one at C-3. Therefore, this sugar is the C-3 epimer of D-glucose, which is D-allose. The substituent at the anomeric carbon is in the β-position. Thus, the sugar's name is propyl β-D-alloside or propyl β-D-allopyranoside. Because the sugar is an acetal, it is a nonreducing sugar.

16.12 Disaccharides

If the hemiacetal group of a monosaccharide forms an acetal by reacting with an alcohol group of another monosaccharide, the glycoside that is formed is a disaccharide. **Disaccharides** are compounds consisting of two monosaccharide subunits hooked together by an acetal linkage. For example, maltose is a disaccharide that contains two D-glucose subunits hooked together by an acetal linkage. This particular acetal linkage is called an **α-1,4′-glycosidic linkage**. The linkage is between C-1 of one sugar subunit and C-4 of the other. The "prime" superscript indicates that C-4 is not in the same ring as C-1. The linkage is an α-1,4′-glycosidic linkage because the oxygen atom involved in the glycosidic linkage is in the α-position. *Remember that the α-position is axial when a sugar is shown in a chair conformation.*

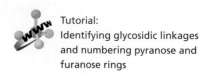

Tutorial:
Identifying glycosidic linkages and numbering pyranose and furanose rings

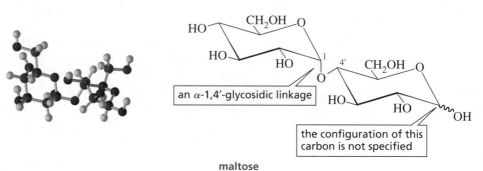

maltose

Note that the structure of maltose is shown without specifying the configuration of the anomeric carbon that is not an acetal (the anomeric carbon of the subunit on the right marked with a wavy line), because maltose can exist in both the α and β forms. In α-maltose, the OH group bonded to this anomeric carbon is in the axial position. In β-maltose, the OH group is in the equatorial position.

Cellobiose is another disaccharide that contains two D-glucose subunits. Cellobiose differs from maltose in that the two glucose subunits are hooked together by a **β-1,4′-glycosidic linkage**. Thus, the only difference in the structures of maltose and cellobiose is the configuration of the glycosidic linkage. Like maltose, cellobiose exists in both α and β forms because the OH group bonded to the anomeric carbon not involved in acetal formation can be in either the axial position (in α-cellobiose) or the equatorial position (in β-cellobiose).

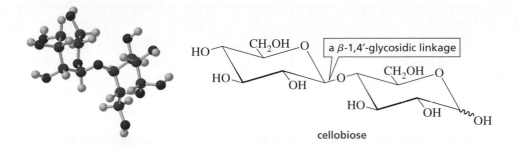

a β-1,4'-glycosidic linkage

cellobiose

Lactose is a disaccharide found in milk. Lactose constitutes 4.5% of cow's milk by weight and 6.5% of human milk. One of the subunits of lactose is D-galactose, and the other is D-glucose. The D-galactose subunit is an acetal, and the D-glucose subunit is a hemiacetal. The subunits are joined by a β-1,4'-glycosidic linkage.

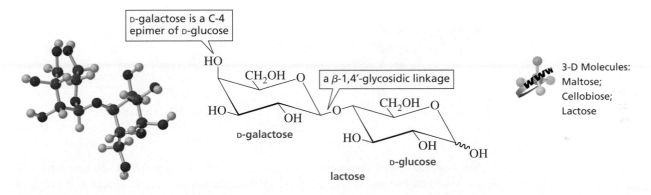

D-galactose is a C-4 epimer of D-glucose

a β-1,4'-glycosidic linkage

D-galactose

D-glucose

lactose

3-D Molecules:
Maltose;
Cellobiose;
Lactose

LACTOSE INTOLERANCE

Lactase is an enzyme that specifically breaks the β-1,4'-glycosidic linkage of lactose. Cats and dogs lose their intestinal lactase when they become adults; they are then no longer able to digest lactose. Consequently, when they are fed milk or milk products, the undegraded lactose causes digestive problems such as bloating, abdominal pain, and diarrhea. These problems occur because only monosaccharides can pass into the bloodstream, so lactose has to pass undigested into the large intestine. When humans have stomach flu or other intestinal disturbances, they can temporarily lose their lactase, thereby becoming lactose intolerant. Some humans lose their lactase permanently as they mature. Lactose intolerance is much more common in people whose ancestors came from non-dairy-producing countries. For example, only 3% of Danes, but 97% of Thais, are lactose intolerant.

The most common disaccharide is sucrose (table sugar). Sucrose is obtained from sugar beets and sugarcane. About 90 million tons of sucrose are produced in the world each year. Sucrose consists of a D-glucose subunit and a D-fructose subunit linked by a glycosidic bond between C-1 of glucose (in the α-position) and C-2 of fructose (in the β-position).

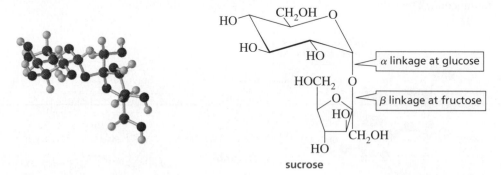

sucrose

3-D Molecule:
Sucrose

Sucrose has a specific rotation of +66.5°. When it is hydrolyzed, the resulting 1 : 1 mixture of glucose and fructose has a specific rotation of −22.0°. Because the sign of the rotation changes when sucrose is hydrolyzed, a 1 : 1 mixture of glucose and fructose is called *invert sugar*. Honeybees have the enzyme that catalyzes the hydrolysis of sucrose, so the honey they produce is a mixture of sucrose, glucose, and fructose. Because fructose is sweeter than sucrose, invert sugar is sweeter than sucrose. Some foods are advertised as containing fructose instead of sucrose, which means that they achieve the same level of sweetness with a lower sugar content.

16.13 Polysaccharides

Polysaccharides contain as few as 10 or as many as several thousand monosaccharide units joined together by glycosidic linkages. The most common polysaccharides are starch and cellulose.

Starch is the major component of flour, potatoes, rice, beans, corn, and peas. Starch is a mixture of two different polysaccharides: amylose (about 20%) and amylopectin (about 80%). Amylose is composed of unbranched chains of D-glucose units joined by α-1,4′-glycosidic linkages.

three subunits of amylose

Amylopectin is a branched polysaccharide. Like amylose, it is composed of chains of D-glucose units joined by α-1,4′-glycosidic linkages. Unlike amylose, however, amylopectin also contains **α-1,6′-glycosidic linkages**. These linkages create the branches in the polysaccharide (Figure 16.1).

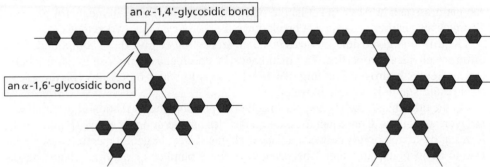

five subunits of amylopectin

an α-1,6′-glycosidic linkage

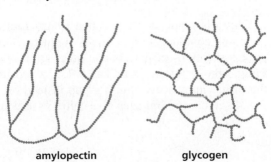

an α-1,4′-glycosidic bond

an α-1,6′-glycosidic bond

◀ **Figure 16.1**
Branching in amylopectin. The hexagons represent glucose units. They are joined by α-1,4′- and α-1,6′-glyclosidic bonds.

Living cells oxidize D-glucose in the first of a series of processes that provide them with energy (Section 19.4). When animals have more D-glucose than they need for energy, they convert excess D-glucose into a polymer called glycogen. When an animal needs energy, glycogen is broken down into individual D-glucose molecules. Plants convert excess D-glucose into starch.

Glycogen has a structure similar to that of amylopectin, but glycogen has more branches (Figure 16.2). The high degree of branching in glycogen has important physiological effects. When the body needs energy, many individual glucose units can be simultaneously removed from the ends of many branches.

◀ **Figure 16.2**
A comparison of the branching in amylopectin and glycogen.

amylopectin   glycogen

WHY THE DENTIST IS RIGHT

Bacteria found in the mouth have an enzyme that converts sucrose into a polysaccharide called dextran. Dextran is made up of glucose units joined mainly through α-1,3′- and α-1,6′-glycosidic linkages. About 10% of dental plaque is composed of dextran. This is the chemical basis of why your dentist cautions you not to eat candy.

Cellulose is the structural material of higher plants. Cotton, for example, is composed of about 90% cellulose, and wood is about 50% cellulose. Like amylose, cellulose is composed of unbranched chains of D-glucose units. Unlike amylose, however, the glucose units in cellulose are joined by β-1,4'-glycosidic linkages rather than by α-1,4'-glycosidic linkages.

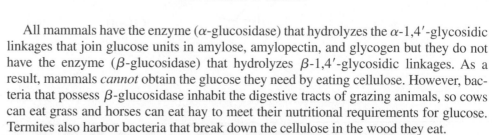

a β-1,4'-glycosidic linkage

three subunits of cellulose

All mammals have the enzyme (α-glucosidase) that hydrolyzes the α-1,4'-glycosidic linkages that join glucose units in amylose, amylopectin, and glycogen but they do not have the enzyme (β-glucosidase) that hydrolyzes β-1,4'-glycosidic linkages. As a result, mammals *cannot* obtain the glucose they need by eating cellulose. However, bacteria that possess β-glucosidase inhabit the digestive tracts of grazing animals, so cows can eat grass and horses can eat hay to meet their nutritional requirements for glucose. Termites also harbor bacteria that break down the cellulose in the wood they eat.

The different glycosidic linkages in starch and cellulose give these compounds very different physical properties. The α-linkages in starch cause amylose to form a helix that promotes hydrogen bonding of its OH groups to water molecules (Figure 16.3). As a result, starch is soluble in water.

On the other hand, the β-linkages in cellulose promote the formation of intramolecular hydrogen bonds. Consequently, these molecules line up in linear arrays (Figure 16.4), forming hydrogen bonds between adjacent chains. These large aggregates cause cellulose to be insoluble in water. The strength of these bundles of polymer chains makes cellulose an effective structural material. Processed cellulose is also used for the production of paper and cellophane.

▲ **Figure 16.3**
The α-1,4'-glycosidic linkages in amylose cause it to form a left-handed helix. Many of its OH groups form hydrogen bonds with water molecules.

Figure 16.4 ▶
The β-1,4'-glycosidic linkages in cellulose form intramolecular hydrogen bonds, which cause the molecules to line up in linear arrays.

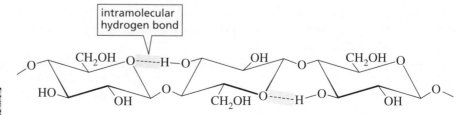

intramolecular hydrogen bond

Strands of cellulose in a plant fiber.

Chitin is a polysaccharide that is structurally similar to cellulose. It is the major structural component of the shells of crustaceans (e.g., lobsters, crabs, and shrimp) and the exoskeletons of insects and other arthropods. It is also the structural material of fungi. Like cellulose, chitin has β-1,4'-glycosidic linkages. It differs from cellulose in that it has an *N*-acetylamino group instead of an OH group at the C-2 position. The β-1,4'-glycosidic linkages give chitin its structural rigidity.

The shell of this bright orange crab from Australia is composed largely of chitin.

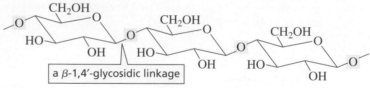

three subunits of chitin

CONTROLLING FLEAS

Several different drugs have been developed to help pet owners control fleas. One of these drugs is lufenuron, the active ingredient in Program. Lufenuron interferes with the production of chitin. Because the exoskeleton of a flea is composed primarily of chitin, a flea cannot live if it cannot make chitin.

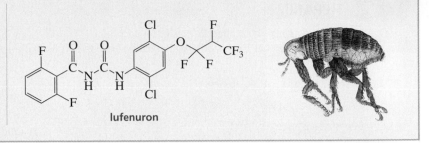

lufenuron

PROBLEM 14

What is the main structural difference between

a. amylose and cellulose?

b. amylose and amylopectin?

c. amylopectin and glycogen?

d. cellulose and chitin?

16.14   Some Naturally Occurring Products Derived from Carbohydrates

Deoxy sugars are sugars in which one of the OH groups is replaced by a hydrogen (*deoxy* means "without oxygen"). 2-Deoxyribose is an important example of a deoxy sugar; it is missing the oxygen at the C-2 position. D-Ribose is the sugar component of ribonucleic acid (RNA), whereas 2-deoxyribose is the sugar component of deoxyribonucleic acid (DNA).

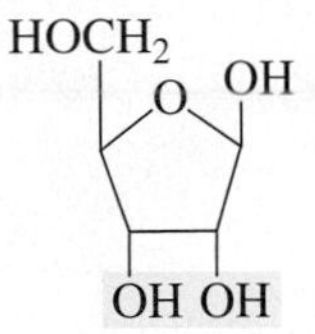

β-D-**ribose** β-D-**2-deoxyribose**

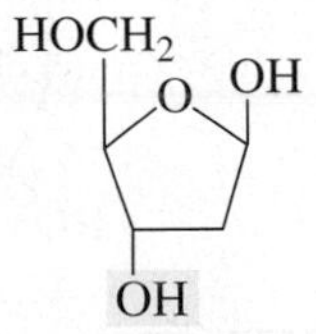

In **amino sugars**, one of the OH groups is replaced by an amino group. *N*-Acetylglucosamine, the subunit of chitin, is an example of an amino sugar (Section 16.13). Some important antibiotics contain amino sugars. For example, the three subunits of the antibiotic gentamicin are deoxyamino sugars. Notice that the middle subunit is missing the ring oxygen, so it really isn't a sugar.

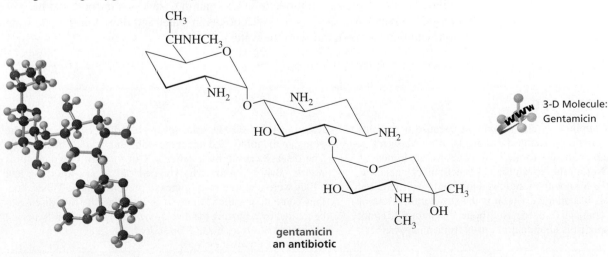

**gentamicin
an antibiotic**

3-D Molecule:
Gentamicin

HEPARIN

Heparin is a polysaccharide that is found principally in cells that line arterial walls. It is released when an injury occurs in order to prevent excessive blood clot formation. Heparin is widely used clinically as an anticoagulant.

heparin

Sir Walter Norman Haworth (1883–1950) *was the first to synthesize vitamin C and was the one who named it ascorbic acid. He was born in England and received a Ph.D. from the University of Göttingen in Germany and later was a professor of chemistry at the Universities of Durham and Birmingham in Great Britain. During World War II, he worked on the atomic bomb project. He received the Nobel Prize in chemistry in 1937 and was knighted in 1947.*

L-Ascorbic acid (vitamin C) is synthesized from D-glucose in plants and in the livers of most vertebrates. Humans, monkeys, and guinea pigs do not have the enzymes necessary for the biosynthesis of vitamin C, so they must obtain the vitamin in their diets. The L-configuration of ascorbic acid refers to the configuration at C-5, which was C-2 in D-glucose.

D-glucose → (enzymes) → **L-configuration** **L-ascorbic acid vitamin C** $pK_a = 4.17$ → (oxidation) → **L-dehydroascorbic acid**

Although L-ascorbic acid does not have a carboxylic acid group, it is an acidic compound because the pK_a of the C-3 OH group is 4.17. L-Ascorbic acid is readily oxidized—notice that it loses hydrogens—to L-dehydroascorbic acid, which is also physiologically active. If the five-membered ring is opened by hydrolysis, all vitamin C activity is lost. Therefore, not much intact vitamin C survives in food that has been thoroughly cooked. Worse, if the food is cooked in water and then drained, the water-soluble vitamin is thrown out with the water!

VITAMIN C

Vitamin C is an antioxidant because it prevents oxidation reactions by radicals. We have seen that vitamin C traps radicals formed in aqueous environments (Section 9.8). Not all the physiological functions of vitamin C are known. What is known, though, is that vitamin C is required for the synthesis of collagen, which is the structural protein of skin, tendons, connective tissue, and bone. If vitamin C is not present in the diet (it is abundant in citrus fruits and tomatoes), lesions appear on the skin, severe bleeding occurs about the gums, in the joints, and under the skin, and wounds heal slowly. The disease caused by a deficiency of vitamin C is known as *scurvy*. British sailors who shipped out to sea after the late 1700s were required to eat limes to prevent scurvy. This is how they came to be called "limeys." Scurvy was the first disease to be treated by adjusting the diet. *Scorbutus* is Latin for "scurvy"; *ascorbic*, therefore, means "no scurvy."

16.15   Carbohydrates on Cell Surfaces

The surfaces of many cells contain short oligosaccharide chains that allow the cells to interact with each other, as well as with invading viruses and bacteria. These oligosaccharides are linked to the surface of the cell by the reaction of an OH or an NH_2 group of a protein with the anomeric carbon of a cyclic sugar. Proteins bonded to oligosaccharides are called **glycoproteins**.

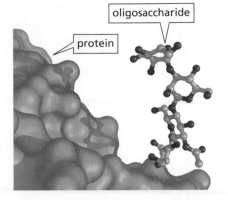

oligosaccharide

protein

glycoproteins

O-protein

NH-protein

Carbohydrates on the surfaces of cells provide a way for cells to recognize one another. The interaction between surface carbohydrates has been found to play a role in many diverse activities, such as infection, the prevention of infection, fertilization, inflammatory diseases like rheumatoid arthritis and septic shock, and blood clotting. For example, the goal of the HIV protease inhibitor drugs is to prevent HIV from recognizing and penetrating cells. The fact that several known antibiotics contain amino sugars (Section 16.14) suggests that they function by recognizing target cells. Carbohydrate interactions also are involved in the regulation of cell growth, so changes in membrane glycoproteins are thought to be correlated with malignant transformations.

Blood type (A, B, or O) is determined by the nature of the sugar bound to the protein on the surfaces of red blood cells. Each type of blood is associated with a different carbohydrate structure (Figure 16.5). Type AB blood has the carbohydrate structure of both type A and type B.

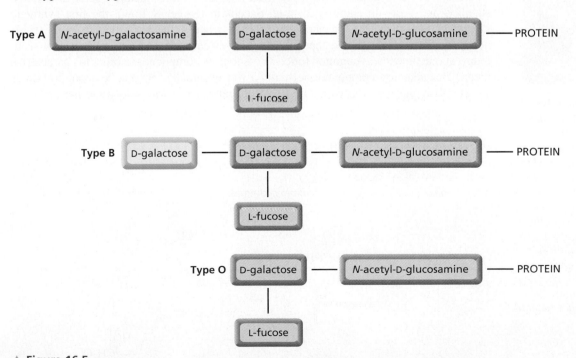

▲ **Figure 16.5**
Blood type is determined by the nature of the sugar on the surfaces of red blood cells.
Fucose is 6-deoxygalactose.

Antibodies are proteins that are synthesized by the body in response to a foreign substance, called an *antigen*. Interaction with the antibody either causes the antigen to precipitate or flags it for destruction by immune system cells. This is why, for example, blood cannot be transferred from one person to another unless the carbohydrate portions of the donor and acceptor are compatible. Otherwise the donated blood will be considered a foreign substance.

Looking at Figure 16.5, we can see why the immune system of type A people recognizes type B blood as foreign and vice versa. The immune system of people with type A, B, or AB blood does not, however, recognize type O blood as foreign, because the carbohydrate in type O blood is also a component of types A, B, and AB blood. Thus, anyone can accept type O blood, so people with type O blood are called universal donors. People with type AB blood can accept types AB, A, B, and O blood, so people with type AB blood are referred to as universal acceptors.

PROBLEM 15

From the nature of the carbohydrate bound to red blood cells, answer the following questions:

a. People with type O blood can donate blood to anyone, but they cannot receive blood from everyone. From whom can they *not* receive blood?

b. People with type AB blood can receive blood from anyone, but they cannot give blood to everyone. To whom can they *not* give blood?

16.16 Synthetic Sweeteners

For a molecule to taste sweet, it must bind to a receptor on a taste bud cell of the tongue. The binding of this molecule causes a nerve impulse to pass from the taste bud to the brain, where the molecule is interpreted as being sweet. Sugars differ in their degree of "sweetness." The relative sweetness of glucose is 1.00, that of sucrose is 1.45, and that of fructose, the sweetest of all sugars, is 1.65. Lactose, the sugar found in milk, is only one sixth as sweet as glucose.

Developers of synthetic sweeteners must consider several factors—such as toxicity, stability, and cost—in addition to taste. Saccharin (Sweet'N Low), the first synthetic sweetener, was discovered accidentally in 1878 when a chemist found that one of his newly synthesized compounds had an extremely sweet taste. (As strange as it may seem today, at one time it was common for chemists to taste compounds in order to characterize them.) He called this compound saccharin; it was eventually found to be about 300 times sweeter than glucose. Notice that, in spite of its name, saccharin is *not* a saccharide.

Saccharin was discovered in **Ira Remsen's (1846–1927)** *laboratory. Remsen was born in New York and, after receiving an M.D. from Columbia University, he decided to become a chemist. He earned a Ph.D. in Germany and then returned to the United States in 1872 to accept a faculty position at Williams College. In 1876, he became a professor of chemistry at the newly established Johns Hopkins University, where he initiated the first center for chemical research in the United States. He later became the second president of Johns Hopkins.*

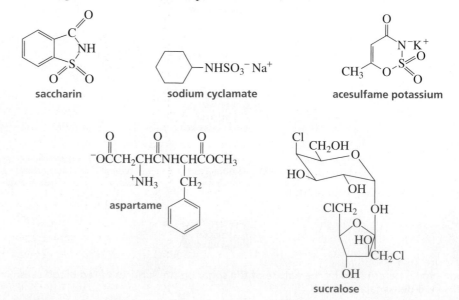

Because it has no caloric value, saccharin became an important substitute for sucrose when it became commercially available in 1885. The chief nutritional problem in the West was—and still is—the overconsumption of sugar and its consequences: obesity, heart disease, and dental decay. Saccharin is also important to diabetics, who must limit their consumption of sucrose and glucose. Although the toxicity of saccharin was not studied carefully when the compound first became available to the public (our current concern with toxicity is a fairly recent development), extensive studies done since then have shown saccharin to be a safe sugar substitute. In 1912, saccharin was temporarily banned in the United States, not because of any concern about its toxicity, but because of a concern that people would miss out on the nutritional benefits of sugar.

Sodium cyclamate became a widely used nonnutritive sweetener in the 1950s, but was banned in the United States some 20 years later in response to two studies that appeared to show that large amounts of sodium cyclamate cause liver cancer in mice.

Aspartame (NutraSweet, Equal) is about 200 times sweeter than sucrose; it was approved by the U.S. Food and Drug Administration (FDA) in 1981. Because aspartame contains a phenylalanine subunit, it should not be used by people with the genetic disease known as PKU (Section 19.5).

Acesulfame potassium (Sweet and Safe, Sunette, Sweet One) was approved in 1988. Also called ascesulfame-K, it, too, is about 200 times sweeter than glucose. It has less aftertaste than saccharine and is more stable than aspartame.

Sucralose (Splenda) is 600 times sweeter than glucose; it is the most recently approved (1991) synthetic sweeter. It maintains its sweetness at temperatures used in baking and in foods stored for long periods. Sucralose is made from sucrose by selectively replacing three of sucrose's OH groups with chlorines. During the chlorination process, the 4-position of the glucose ring becomes inverted, so sucralose is a galactoside, not a glucoside. The body does not recognize sucralose as a carbohydrate so it is not metabolized—it is eliminated from the body unchanged.

The fact that these synthetic sweeteners have such different structures shows that the sensation of sweetness is not induced by a single molecular shape.

Tutorial:
Carbohydrates
Common Terms

ACCEPTABLE DAILY INTAKE

The FDA sets an acceptable daily intake (ADI) for many of the food ingredients it clears for use. The ADI is the amount of the substance that people can consume safely each day of their lives. For example, the ADI for acesul-fame-K is 15 mg/kg/day. This means a 132 lb person can consume one-half pound of acesulfame-K every day—the amount that would be found in two gallons of an artificially sweetened beverage. The ADI for sucralose is also 15 mg/kg/day.

Summary

Carbohydrates are the most abundant class of compounds in the biological world. They are polyhydroxy aldehydes (**aldoses**) and polyhydroxy ketones (**ketoses**) or compounds formed by linking up aldoses and ketoses. D and L notations describe the configuration of the bottom-most asymmetric center of a **monosaccharide**; the configurations of the other carbons are implicit in the common name. Most naturally occurring sugars are D-sugars. Naturally occurring ketoses have the ketone group in the 2-position. **Epimers** differ in configuration at only one asymmetric center.

Reduction of an aldose forms one **alditol**; reduction of a ketose forms two alditols. Br_2 oxidizes aldoses, but not ketoses. Aldoses are oxidized to **aldonic acids** or to **aldaric acids**. The **Kiliani–Fischer synthesis** increases the carbon chain of an aldose by one carbon—it forms C-2 epimers.

The aldehyde or keto group of a **monosaccharide** reacts with one of its OH groups to form cyclic hemiacetals or hemiketals: Glucose forms α-D-glucose and β-D-glucose—called **anomers** because they differ in configuration only at the **anomeric carbon**—the carbon that was the carbonyl carbon in the open-chain form. Six-membered-ring sugars are **pyranoses**; five-membered-ring sugars are **furanoses**. The most abundant monosaccharide in nature is D-glucose. All the OH groups in β-D-glucose are in equatorial positions. A slow change in optical rotation to an equilibrium value is called **mutarotation**.

The cyclic hemiacetal (or hemiketal) can react with an alcohol to form an acetal (or ketal), called a **glycoside**. If the name "pyranose" or "furanose" is used, the acetal is called a **pyranoside** or a **furanoside**. The bond between the anomeric carbon and the alkoxy oxygen is called a **glycosidic bond**. **Disaccharides** consist of two monosaccharide subunits hooked together by an acetal linkage. Maltose has an **α-1,4′-glycosidic linkage**; cellobiose has a **β-1,4′-glycosidic linkage**. The most common disaccharide is sucrose; it consists of a D-glucose subunit and a D-fructose subunit linked by their anomeric carbons.

Polysaccharides contain as few as 10 or as many as several thousand monosaccharides joined together by glycosidic linkages. Starch is composed of amylose and amylopectin.

Amylose has unbranched chains of D-glucose units joined by α-1,4′-glycosidic linkages. Amylopectin, too, has chains of D-glucose units joined by α-1,4′-glycosidic linkages, but it also has **α-1,6′-glycosidic linkages** that create branches. Glycogen is similar to amylopectin, but it has more branches. Cellulose has unbranched chains of D-glucose units joined by β-1,4′-glycosidic linkages. The α-linkages cause amylose to form a helix; the β-linkages allow the molecules of cellulose to form intramolecular hydrogen bonds.

The surfaces of many cells contain short oligosaccharide chains that allow the cells to interact with each other. These oligosaccharides are linked to the cell surface by protein groups. Proteins bonded to oligosaccharides are called **glycoproteins**.

Summary of Reactions

1. Reduction (Section 16.5)

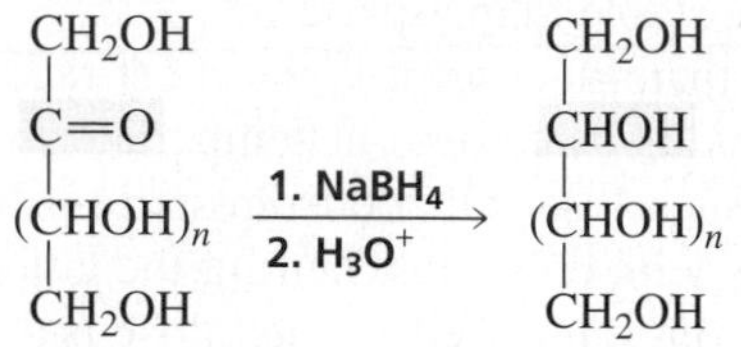

2. Oxidation (Section 16.5)

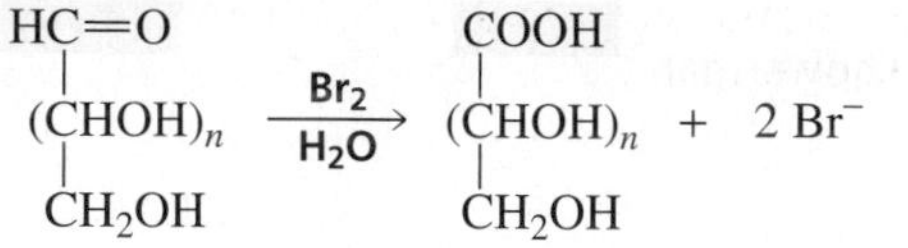

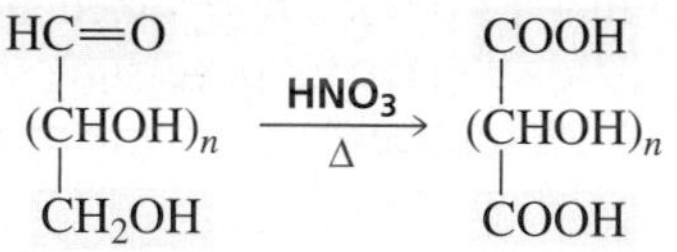

3. Chain elongation: the Kiliani–Fischer synthesis (Section 16.6)

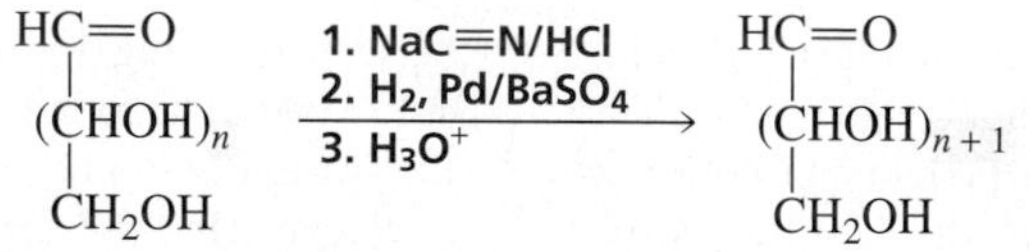

4. Glycoside formation (Section 16.10)

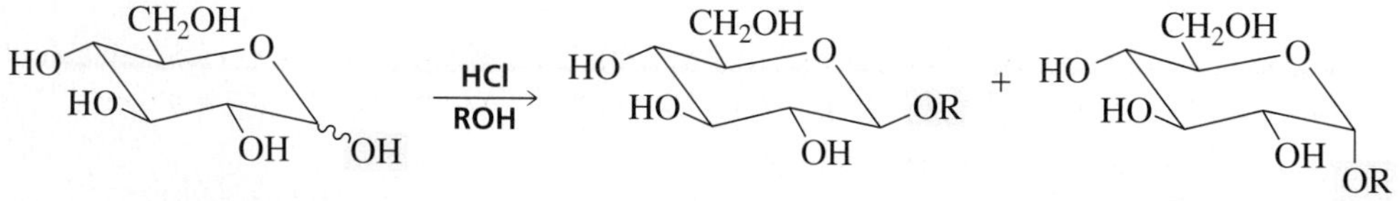

Problems

16. Give the product or products that are obtained when D-galactose reacts with the following:
    a. nitric acid
    b. NaBH$_4$ followed by H$_3$O$^+$
    c. Br$_2$ in water
    d. ethanol + HCl

17. What sugar is the C-4 epimer of L-gulose?

18. Identify the following sugars:
    a. An aldopentose that is not D-arabinose forms D-arabinitol when it is reduced with NaBH$_4$.
    b. A sugar that is not D-altrose forms D-altraric acid when it reacts with nitric acid.
    c. A ketose, when reduced with NaBH$_4$, forms D-altritol and D-allitol.

19. Answer the following questions about the eight aldopentoses:
    a. Which are enantiomers?   **b.** Which form an optically active compound when oxidized with nitric acid?

20. What other monosaccharide is reduced only to the alditol obtained from the reduction of D-talose?

21. What monosaccharide is reduced to two alditols, one of which is the alditol obtained from the reduction of D-mannose?

22. What monosaccharides would be formed in a Kiliani–Fischer synthesis starting with
    a. D-xylose?   **b.** L-threose?

23. Name the following compounds, and indicate whether each is a reducing sugar or a nonreducing sugar:

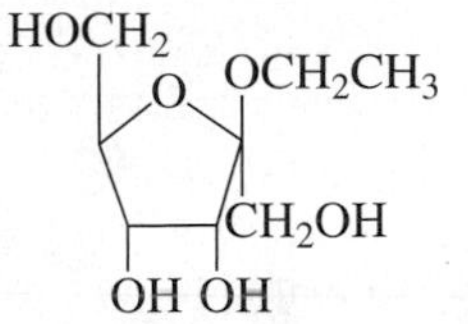

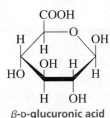

24. The reaction of D-ribose with one equivalent of methanol plus HCl forms four products. Give the structures of the products.

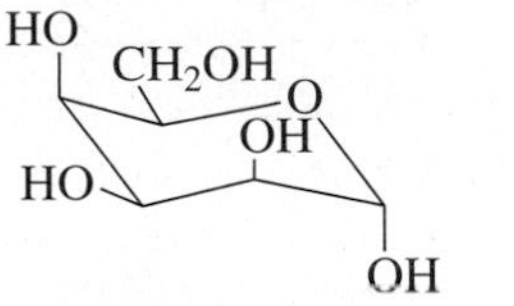

25. Dr. Isent T. Sweet isolated a monosaccharide and determined that it had a molecular weight of 150. Much to his surprise, he found that it was not optically active. What is the structure of the monosaccharide?

26. Indicate whether each of the following is D-glyceraldehyde or L-glyceraldehyde, assuming that the horizontal bonds point toward you and the vertical bonds point away from you:

    a. $HC=O$ $HOCH_2\!-\!\!\!\!-\!OH$ H
    b. H $HO\!-\!\!\!\!-\!CH_2OH$ $HC=O$
    c. CH_2OH $HO\!-\!\!\!\!-\!H$ $HC=O$

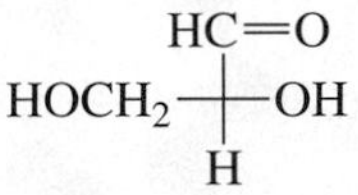

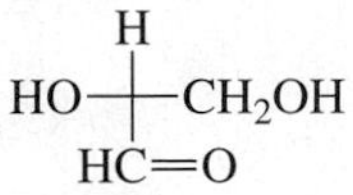

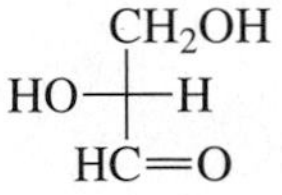

27. D-Glucuronic acid is found widely in plants and animals. One of its functions is to detoxify poisonous HO-containing compounds by reacting with them in the liver to form glucuronides. Glucuronides are water soluble and therefore readily excreted. After ingestion of a poison such as turpentine, morphine, or phenol, the glucuronides of these compounds are found in the urine. Give the structure of the glucuronide formed by the reaction of β-D-glucuronic acid and phenol.

$$COOH$$

β-D-**glucuronic acid**

28. In order to synthesize D-galactose, Professor Amy Losse went to the stockroom to get some D-lyxose to use as a starting material. She found that the labels had fallen off the bottles containing D-lyxose and D-xylose. How could she determine which bottle contained D-lyxose?

29. Hyaluronic acid, a component of connective tissue, is the fluid that lubricates the joints. It is an alternating polymer of N-acetyl-D-glucosamine and D-glucuronic acid joined by β-1,3′-glycosidic linkages. Draw a short segment of hyaluronic acid.

30. How many aldaric acids are obtained from the 16 aldohexoses?

31. Give the structure of the cyclic hemiacetal formed by each of the following:
    a. 4-hydroxypentanal   **b.** 4-hydroxyheptanal

32. Explain why the C-3 OH group of vitamin C is more acidic than the C-2 OH group.

33. Calculate the percentages of α-D-glucose and β-D-glucose present at equilibrium from the specific rotations of α-D-glucose, β-D-glucose, and the equilibrium mixture. Compare your values with those given in Section 16.8. (*Hint:* The specific rotation of the mixture equals the specific rotation of α-D-glucose times the fraction of glucose present in the α-form plus the specific rotation of β-D-glucose times the fraction of glucose present in the β-form.)

34. Predict whether D-altrose exists preferentially as a pyranose or a furanose. (*Hint:* The most stable arrangement for a five-membered ring is for all the adjacent substituents to be trans.)

17 | Amino Acids, Peptides, and Proteins

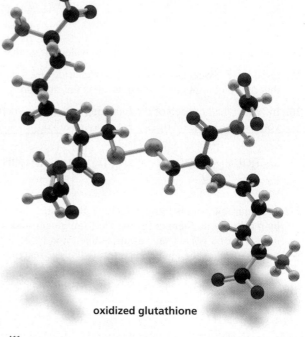

oxidized glutathione

The three kinds of polymers prevalent in nature are polysaccharides, proteins, and nucleic acids. We have already studied polysaccharides, which are naturally occurring polymers of sugar subunits (Section 16.1). We will now look at proteins and the structurally similar, but shorter, peptides.

Peptides and **proteins** are polymers of **amino acids** linked together by amide bonds. An amino acid is a carboxylic acid with an amino group on the α-carbon. The repeating units are called **amino acid residues**.

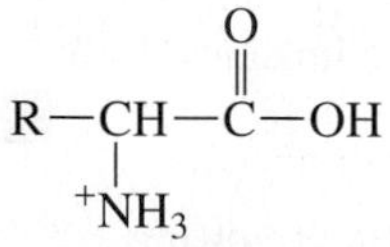

$$\underset{\substack{\text{an amino acid}}}{R-\underset{\underset{{}^+NH_3}{|}}{CH}-\overset{\overset{O}{\|}}{C}-OH}$$

amide bonds

$$-NHCH\overset{\overset{O}{\|}}{C}-\underset{R}{NHCH}\overset{\overset{O}{\|}}{C}-\underset{R'}{NHCH}\overset{\overset{O}{\|}}{C}-\underset{R''}{NHCH}\overset{\overset{O}{\|}}{C}-$$

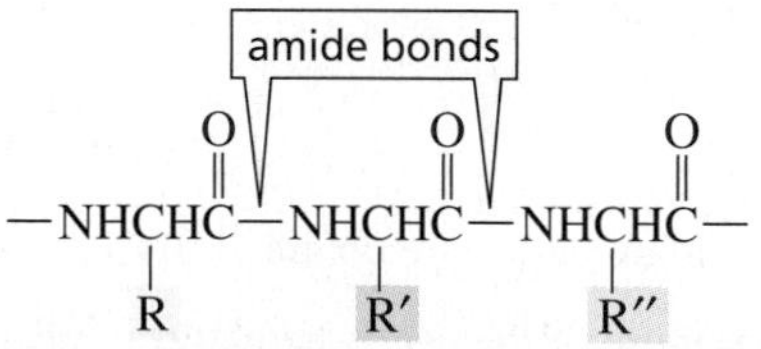

amino acids are linked together by amide bonds

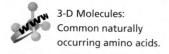

3-D Molecules: Common naturally occurring amino acids.

A **dipeptide** contains two amino acid residues, a **tripeptide** contains three, and a **polypeptide** contains many amino acid residues. Proteins are naturally occurring polypeptides that are made up of 40 to 4000 amino acid residues. Proteins and peptides serve many functions in biological systems (Table 17.1).

Proteins can be divided roughly into two classes. **Fibrous proteins** contain long chains of polypeptides that occur in bundles; these proteins are insoluble in water. All structural proteins are fibrous proteins. **Globular proteins** tend to have roughly spherical shapes and are soluble in water. Essentially all enzymes are globular proteins.

Table 17.1   Examples of the Many Functions of Proteins in Biological Systems

Structural proteins	These proteins impart strength to biological structures or protect organisms from their environment. For example, collagen is the major component of bones, muscles, and tendons; and keratin is the major component of hair, hooves, feathers, fur, and the outer layer of skin.
Protective proteins	Snake venoms and plant toxins protect their owners from disease. Blood-clotting proteins protect the vascular system when it is injured. Antibodies and peptide antibiotics protect us from disease.
Enzymes	Enzymes are proteins that catalyze the reactions that occur in living systems.
Hormones	Some of the hormones, such as insulin, that regulate the reactions that occur in living systems are proteins.
Proteins with physiological functions	These proteins are responsible for physiological functions such as the transport and storage of oxygen in the body, the storage of oxygen in the muscles, and the contraction of muscles.

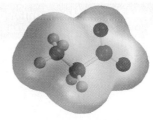

glycine

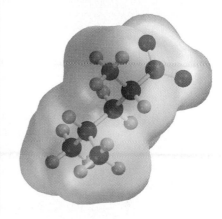

leucine

17.1   Classification and Nomenclature of Amino Acids

The structures of the 20 most common naturally occurring amino acids and the frequency with which each occurs in proteins are shown in Table 17.2. The amino acids differ only in the substituent (R) attached to the α-carbon. The wide variation in these substituents—called **side chains**—is what gives proteins their great structural diversity and, as a consequence, their great functional diversity (Table 17.1).

The amino acids are almost always called by their common names. Often, the name tells you something about the amino acid. For example, glycine got its name as a result of its sweet taste (*glykos* is Greek for "sweet"). Asparagine was first found in asparagus, and tyrosine was isolated from cheese (*tyros* is Greek for "cheese"). Table 17.2 shows that each of the amino acids has both a three-letter abbreviation (the first three letters of the name, in most cases) and a single-letter abbreviation.

Notice that, in spite of its name, isoleucine does *not* have an isobutyl substituent; it has a *sec*-butyl substituent. Leucine is the amino acid that has an isobutyl substituent. Proline is the only amino acid that is a secondary amine.

Ten amino acids are *essential amino acids*; these are denoted by red asterisks (*) in Table 17.2. We humans must obtain these 10 **essential amino acids** from our diets because we either cannot synthesize them at all or cannot synthesize them in adequate amounts to allow us to thrive. For example, we must have a dietary source of phenylalanine because we cannot synthesize benzene rings. However, we do not need tyrosine in our diets, because we can synthesize the necessary amounts from phenylalanine. Although humans can synthesize arginine, it is needed for growth in greater amounts than can be synthesized. So arginine is an essential amino acid for children, but a nonessential amino acid for adults.

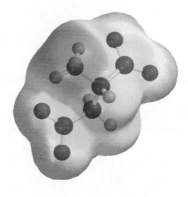

aspartate

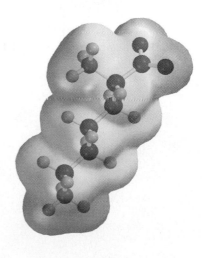

lysine

Table 17.2 **The Most Common Naturally Occurring Amino Acids**
The amino acids are shown in the form that predominates at physiological pH (7.3).

	Formula	Name	Abbreviations		Average relative abundance in proteins
Aliphatic side chain amino acids	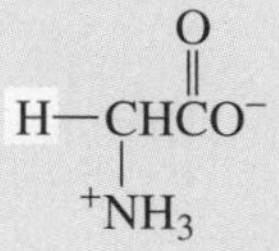 $H-CHCO^-$ with $^+NH_3$	Glycine	Gly	G	7.5%
	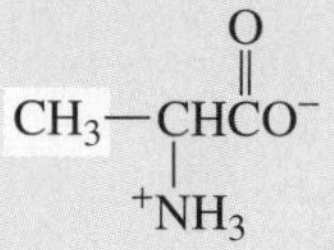 CH_3-CHCO^- with $^+NH_3$	Alanine	Ala	A	9.0%
	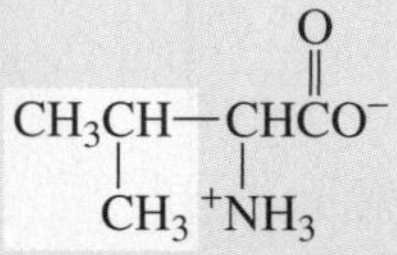 $CH_3CH-CHCO^-$ with CH_3 and $^+NH_3$	Valine*	Val	V	6.9%
	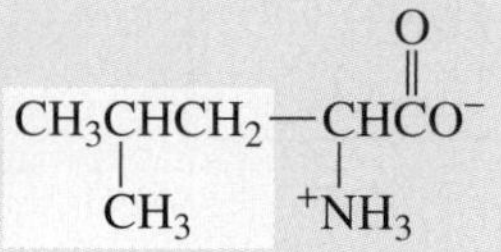 $CH_3CHCH_2-CHCO^-$ with CH_3 and $^+NH_3$	Leucine*	Leu	L	7.5%
	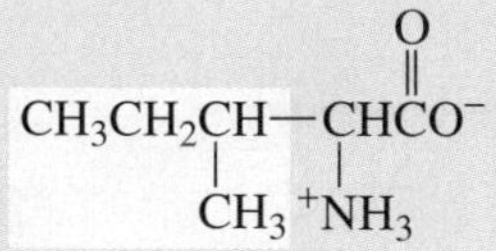 $CH_3CH_2CH-CHCO^-$ with CH_3 and $^+NH_3$	Isoleucine*	Ile	I	4.6%
Hydroxy-containing amino acids	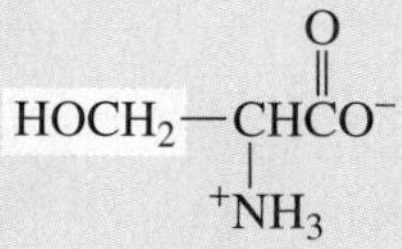 $HOCH_2-CHCO^-$ with $^+NH_3$	Serine	Ser	S	7.1%
	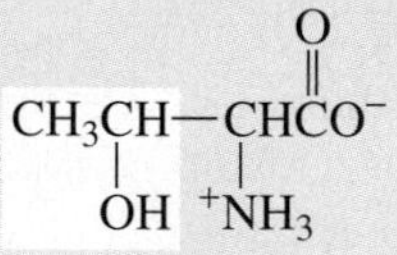 $CH_3CH-CHCO^-$ with OH and $^+NH_3$	Threonine*	Thr	T	6.0%
Sulfur-containing amino acids	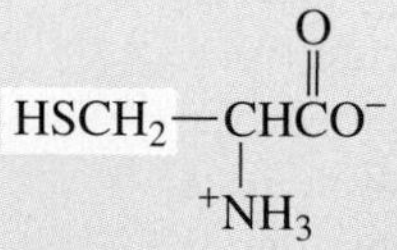 $HSCH_2-CHCO^-$ with $^+NH_3$	Cysteine	Cys	C	2.8%
	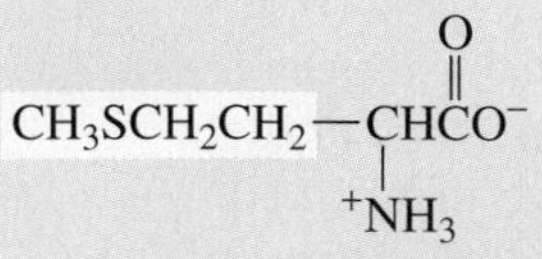 $CH_3SCH_2CH_2-CHCO^-$ with $^+NH_3$	Methionine*	Met	M	1.7%
Acidic amino acids	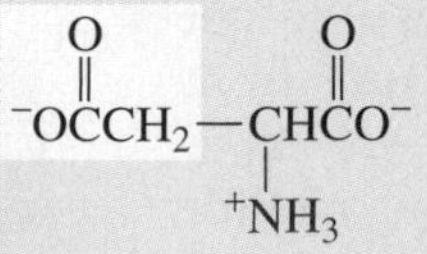 $^-OCCH_2-CHCO^-$ with $^+NH_3$	Aspartate (aspartic acid)	Asp	D	5.5%

*Essential amino acids

Table 17.2 (continued)

	Formula	Name	Abbreviations		Average relative abundance in proteins
Amides of acidic amino acids	$^-OCCH_2CH_2-CHCO^-$ (O, O, $^+NH_3$)	Glutamate (glutamic acid)	Glu	E	6.2%
	$H_2NCCH_2-CHCO^-$ (O, O, $^+NH_3$)	Asparagine	Asn	N	4.4%
	$H_2NCCH_2CH_2-CHCO^-$ (O, O, $^+NH_3$)	Glutamine	Gln	Q	3.9%
Basic amino acids	$H_3NCH_2CH_2CH_2CH_2-CHCO^-$ (O, $^+NH_3$)	Lysine*	Lys	K	7.0%
	$H_2NCNHCH_2CH_2CH_2-CHCO^-$ ($^+NH_2$, O, $^+NH_3$)	Arginine*	Arg	R	4.7%
Benzene-containing amino acids	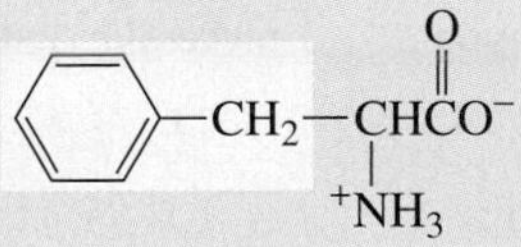 —CH_2-CHCO^- (O, $^+NH_3$)	Phenylalanine*	Phe	F	3.5%
	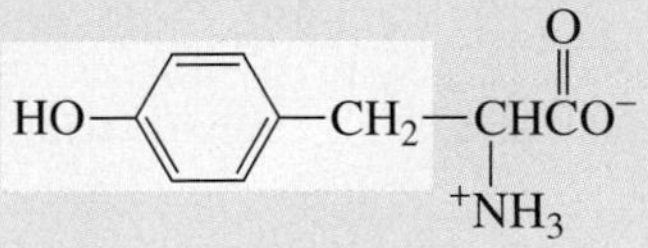 HO—CH_2-CHCO^- (O, $^+NH_3$)	Tyrosine	Tyr	Y	3.5%
Heterocylic amino acids	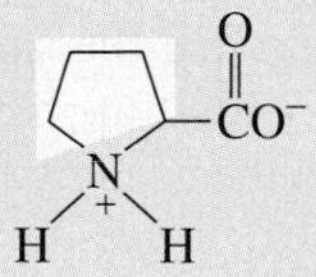 CO^- (proline ring, O, $\overset{+}{N}$, H H)	Proline	Pro	P	4.6%
	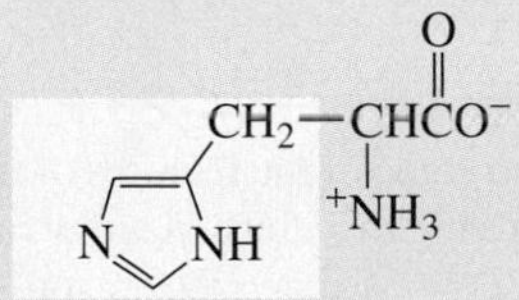 CH_2-CHCO^- (O, $^+NH_3$, imidazole N NH)	Histidine*	His	H	2.1%
	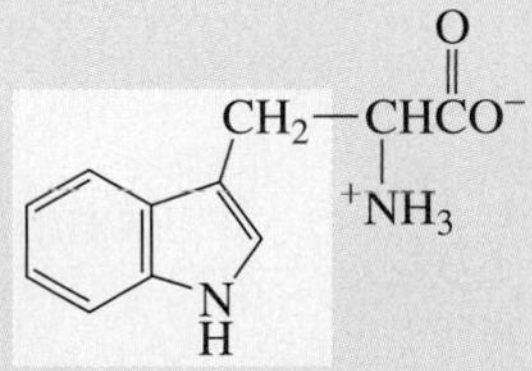 CH_2-CHCO^- (O, $^+NH_3$, indole N H)	Tryptophan*	Trp	W	1.1%

*Essential amino acids

PROTEINS AND NUTRITION

Proteins are an important component of our diets. Dietary protein is hydrolyzed in the body to individual amino acids. Some of these amino acids are used to synthesize proteins needed by the body, some are broken down further to supply energy to the body, and some are used as starting materials for the synthesis of nonprotein compounds that the body needs, such as thyroxine (Section 7.7), adrenaline, and melanin (Section 19.5).

Not all proteins contain the same amino acids. Most proteins from meat and dairy products contain all the amino acids needed by the body. However, most proteins from vegetable sources are *incomplete* proteins; they contain too little of one or more essential amino acids to support growth. For example, bean protein is deficient in methionine and wheat protein is deficient in lysine. Therefore, a balanced diet must contain proteins from different sources.

17.2 Configuration of Amino Acids

The *α-carbon* of all the naturally occurring amino acids, except glycine, is an asymmetric center. Therefore, 19 of the 20 amino acids listed in Table 17.2 can exist as enantiomers. The D and L notation used for monosaccharides (Section 16.2) is also used for amino acids. An amino acid drawn in a Fischer projection with the carboxyl group on the top and the R group on the bottom of the vertical axis is a **D-amino acid** if the amino group is on the right and an **L-amino acid** if the amino group is on the left. Unlike monosaccharides, where the D isomer is the one found in nature, most amino acids found in nature have the L configuration.

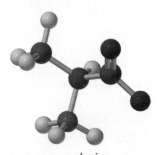

**L-alanine
an amino acid**

Naturally occurring amino acids have the L-configuration.

D-glyceraldehyde

L-glyceraldehyde

D-amino acid

L-amino acid

Why D-sugars and L-amino acids? While it makes no difference which isomer nature "selected" to be synthesized, it is important that the same isomer be synthesized by all organisms. For example, if mammals ended up having L-amino acids, then L-amino acids would need to be the isomers synthesized by the organisms upon which mammals depend for food.

AMINO ACIDS AND DISEASE

The Chamorro people of Guam have a high incidence of a syndrome that resembles amyotrophic lateral sclerosis (ALS or Lou Gehrig's disease) with elements of Parkinson's disease and dementia. This syndrome developed during World War II when, as a result of food shortages, the tribe ate large quantities of *Cycas circinalis* seeds. These seeds contain β-methylamino-L-alanine, an amino acid that binds to glutamate receptors. When monkeys are given β-methylamino-L-alanine,

they develop some of the features of this syndrome. There is hope that, by studying this unusual amino acid, we may gain an understanding of how ALS and Parkinson's disease arise.

L-alanine

β-methylamino-L-alanine

A PEPTIDE ANTIBIOTIC

Gramicidin S is an antibiotic produced by a strain of bacteria. It is a cyclic decapeptide. Notice that it contains ornithine, an amino not listed in Table 17.2 because it occurs rarely in nature. Ornithine resembles lysine, but has one less methylene group in its side chain. Notice also that the antibiotic contains two D-amino acids.

```
            ┌── L-Val ──┐
      L-Pro          L-Orn
   L-Phe                L-Leu
      L-Leu          D-Phe
      D-Orn          L-Pro
            └── L-Val ──┘
         gramicidin S
```

$$\overset{+}{H_3}NCH_2CH_2CH_2\overset{\overset{\displaystyle}{|}}{\underset{\underset{\displaystyle +NH_3}{|}}{C}}HCO^-$$

ornithine

PROBLEM 1◆

Which isomer—(R)-alanine or (S)-alanine—is L-alanine?

PROBLEM 2◆   SOLVED

Threonine has two asymmetric centers and therefore has four stereoisomers.

```
    COO⁻              COO⁻              COO⁻              COO⁻
H ──┼── ⁺NH₃     H₃N⁺──┼── H      H ──┼── ⁺NH₃     H₃N⁺──┼── H
H ──┼── OH       HO ──┼── H       HO ──┼── H       H ──┼── OH
    CH₃               CH₃               CH₃               CH₃
     1                 2                 3                 4
```

Naturally occurring L-threonine is (2S,3R)-threonine. Which of the stereoisomers is L-threonine?

SOLUTION   Stereoisomer #1 has the R configuration at both C-2 and C-3 because, in both cases, the arrow drawn from the highest to the next highest priority substituent is counterclockwise. In both cases, counterclockwise signifies R because the lowest priority substituent (H) is on a horizontal bond. Therefore, (2S,3R)-threonine has the opposite configuration at C-2 and the same configuration at C-3 compared with stereoisomer #1. Thus, L-threonine is stereoisomer #4. Notice that the ⁺NH₃ group is on the left, just as we would expect for the Fischer projection of an L-amino acid.

PROBLEM 3◆

Do any other amino acids in Table 17.2 have more than one asymmetric center?

17.3   Acid–Base Properties of Amino Acids

Every amino acid has a carboxyl group and an amino group, and each group can exist in an acidic form or a basic form, depending on the pH of the solution in which the amino acid is dissolved. Some amino acids, such as aspartate and glutamate, also have an ionizable side chain (Table 17.3).

We have seen that compounds exist primarily in their acidic forms (with their protons) in solutions that are more acidic than their pK_a values and primarily in their basic forms (without their protons) in solutions that are more basic than their pK_a values (Section 2.4). The carboxyl groups of the amino acids have pK_a values of approximately 2, and the protonated amino groups have pK_a values near 9. Both groups, therefore, will be in their acidic forms in a very acidic solution (pH ~ 0). At pH = 7,

The acidic form (with the proton) predominates if the pH of the solution is less than the pK_a of the compound, and the basic form (without the proton) predominates if the pH of the solution is greater than the pK_a of the compound.

the pH of the solution is greater than the pK_a of the carboxyl group, but less than the pK_a of the protonated amino group. The carboxyl group, therefore, will be in its basic form and the amino group will be in its acidic form. In a strongly basic solution (pH ~ 11), both groups will be in their basic forms.

$$\underset{\substack{\text{+}NH_3 \\ pH = 0}}{R-CH-\overset{\displaystyle O}{\overset{\|}{C}}-OH} \rightleftharpoons \underset{\substack{\text{+}NH_3 + H^+ \\ \text{a zwitterion} \\ pH = 7}}{R-CH-\overset{\displaystyle O}{\overset{\|}{C}}-O^-} \rightleftharpoons \underset{\substack{NH_2 + H^+ \\ pH = 11}}{R-CH-\overset{\displaystyle O}{\overset{\|}{C}}-O^-}$$

Tutorial:
Basic nitrogens in histidine and arginine

Table 17.3 The pKₐ Values of Amino Acids

Amino acid	pK_a α-COOH	pK_a α-NH$_3^+$	pK_a side chain
Alanine	2.34	9.69	—
Arginine	2.17	9.04	12.48
Asparagine	2.02	8.84	—
Aspartic acid	2.09	9.82	3.86
Cysteine	1.92	10.46	8.35
Glutamic acid	2.19	9.67	4.25
Glutamine	2.17	9.13	—
Glycine	2.34	9.60	—
Histidine	1.82	9.17	6.04
Isoleucine	2.36	9.68	—
Leucine	2.36	9.60	—
Lysine	2.18	8.95	10.79
Methionine	2.28	9.21	—
Phenylalanine	2.16	9.18	—
Proline	1.99	10.60	—
Serine	2.21	9.15	—
Threonine	2.63	9.10	—
Tryptophan	2.38	9.39	—
Tyrosine	2.20	9.11	10.07
Valine	2.32	9.62	—

Notice that an amino acid can never exist as an uncharged compound, regardless of the pH of the solution. To be uncharged, an amino acid would have to lose a proton from an $^+NH_3$ group with a pK_a of about 9 before it would lose a proton from a COOH group with a pK_a of about 2. This clearly is impossible: A weak acid ($pK_a = 9$) cannot be more acidic than a strong acid ($pK_a = 2$). Therefore, at physiological pH (7.3), an amino acid such as alanine exists as a dipolar ion, called a *zwitterion*. A **zwitterion** is a compound that has a negative charge on one atom and a positive charge on a nonadjacent atom. (The name comes from *zwitter*, German for "hermaphrodite" or "hybrid.")

PROBLEM 4

Why are the carboxylic acid groups of the amino acids so much more acidic ($pK_a \sim 2$) than a carboxylic acid such as acetic acid ($pK_a = 4.76$)?

PROBLEM 5

Unlike most amines and carboxylic acids, amino acids are insoluble in diethyl ether. Explain.

PROBLEM 6◆ SOLVED

Draw the form in which each of the following amino acids predominantly exists at physiological pH (7.3):

a. aspartic acid   **b.** glutamine   **c.** arginine

SOLUTION TO 6a   Both carboxyl groups are in their basic forms because the pH of the solution is greater than their pK_a values. The protonated amino group is in its acidic form because the pH of the solution is less than its pK_a value.

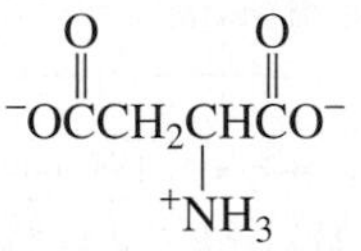

$$\overset{O}{\overset{\|}{}}\;\;\overset{O}{\overset{\|}{}}$$
$$^-OCCH_2CHCO^-$$
$$\underset{^+NH_3}{|}$$

PROBLEM 7

Draw the form in which glutamic acid predominantly exists in a solution with the following pH:

a. pH = 0   **b.** pH = 3   **c.** pH = 6   **d.** pH = 11

17.4   The Isoelectric Point

The **isoelectric point** (pI) of an amino acid is the pH at which it has no net charge. In other words, it is the pH at which the amount of positive charge on an amino acid exactly balances the amount of negative charge:

pI (isoelectric point) = pH at which there is no net charge

The pI of an amino acid that does *not* have an ionizable side chain—such as alanine—is midway between its two pK_a values.

An amino acid will be overall positively charged if the pH of the solution is less than the pI of the amino acid and will be overall negatively charged if the pH of the solution is greater than the pI of the amino acid.

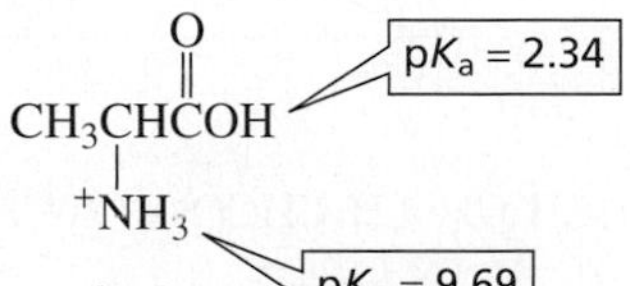

$$CH_3CHCOH \quad pK_a = 2.34$$

alanine   pK_a = 9.69

$$pI = \frac{2.34 + 9.69}{2} = \frac{12.03}{2} = 6.02$$

The pI of an amino acid that *has* an ionizable side chain is the average of the pK_a values of the similarly ionizing groups (e.g., positively charged groups ionizing to uncharged groups, or uncharged groups ionizing to negatively charged groups). For example, the pI of lysine is the average of the pK_a values of the two groups that are positively charged in their acidic form and uncharged in their basic form. The pI of glutamic acid, on the other hand, is the average of the pK_a values of the two groups that are uncharged in their acidic form and negatively charged in their basic form.

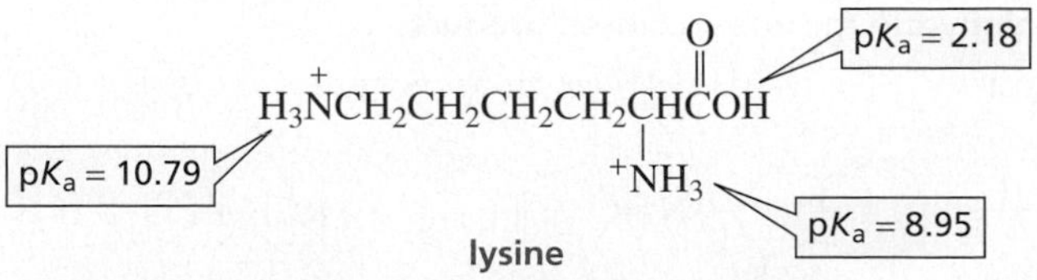

lysine   pK_a = 10.79, pK_a = 2.18, pK_a = 8.95

glutamic acid   pK_a = 4.25, pK_a = 2.19, pK_a = 9.67

$$pI = \frac{8.95 + 10.79}{2} = \frac{19.74}{2} = 9.87$$

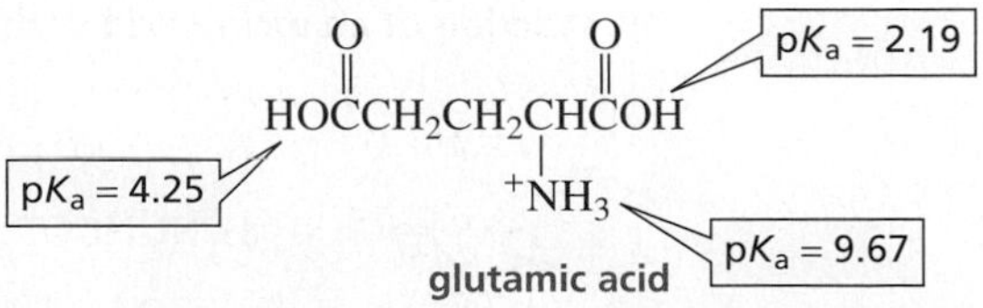

$$pI = \frac{2.19 + 4.25}{2} = \frac{6.44}{2} = 3.22$$

PROBLEM 8◆

Calculate the pI of the following amino acids:

a. asparagine   **c.** serine

b. arginine   **d.** aspartic acid

PROBLEM 9◆

a. Which amino acid has the lowest pI value?

b. Which amino acid has the highest pI value?

17.5   Separation of Amino Acids

Electrophoresis

Tutorial:
Electrophoresis and pI

A mixture of amino acids can be separated on the basis of their pI values by **electrophoresis**. In this technique, a few drops of a solution of an amino acid mixture are applied to the middle of a piece of filter paper or to a gel. When the paper or the gel is placed in a buffered solution between two electrodes and an electric field is applied (Figure 17.1), an amino acid with a pI greater than the pH of the solution will have an overall positive charge and will migrate toward the cathode (the negative electrode). The farther the amino acid's pI is from the pH of the buffer, the more positive the amino acid will be and the farther it will migrate toward the cathode. An amino acid with a pI less than the pH of the buffer will have an overall negative charge and will migrate toward the anode (the positive electrode). If two molecules have the same charge, the larger one will move more slowly during electrophoresis because the same charge has to move a greater mass.

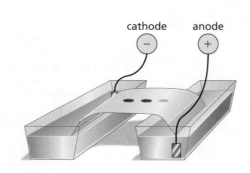
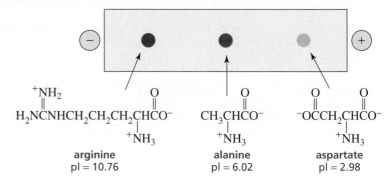

▲ **Figure 17.1**
Arginine, alanine, and aspartate separated by electrophoresis at pH = 5.

Since amino acids are colorless, how can we detect them after they have been separated? After the amino acids are separated, the filter paper is sprayed with ninhydrin and dried in a warm oven. Most amino acids form a purple product when they are heated with ninhydrin. The number of different kinds of amino acids in the mixture is determined by the number of colored spots on the filter paper (Figure 17.1). The individual amino acids are identified by their location on the paper compared with a standard.

the reaction of an amino acid with ninhydrin to form a colored product

---

PROBLEM 10◆

What aldehyde is formed when valine is treated with ninhydrin?

Ion-Exchange Chromatography

A technique called **ion-exchange chromatography** cannot only identify amino acids, but it can also determine the relative amounts of the amino acids present in the mixture. This technique employs a column packed with an insoluble resin. A solution of a mixture of amino acids is loaded onto the top of the column, and a series of aqueous solutions of increasing pH are poured through the column. The amino acids separate because they flow through the column at different rates, as explained below.

The structure of one commonly used resin is shown in Figure 17.2. If a mixture of lysine and glutamate in a solution with a pH of 6 were loaded onto the column, glutamate would travel down the column rapidly because its negatively charged side chain would be repelled by the negatively charged sulfonic acid groups of the resin. The positively charged side chain of lysine, on the other hand, would cause that amino acid to be retained on the column. This kind of resin is called a **cation-exchange resin** because it exchanges the Na^+ counterions of the SO_3^- groups for the positively charged species that are added to the column. In addition, the relatively nonpolar nature of the column causes it to retain nonpolar amino acids longer than polar amino acids.

Cations bind most strongly to cation-exchange resins.

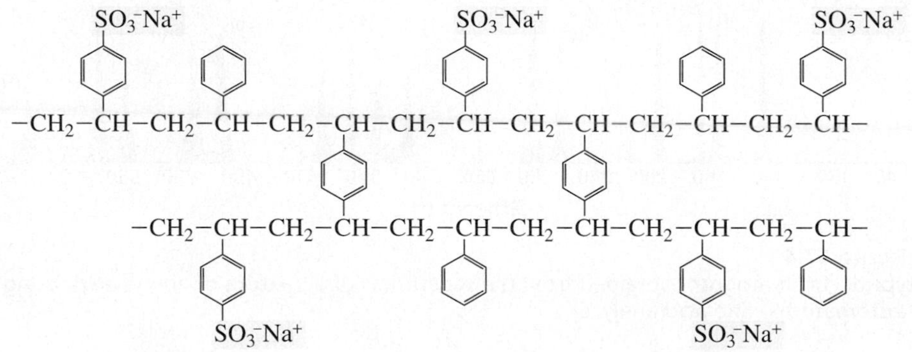

◀ **Figure 17.2**
A section of a cation-exchange resin.

An **amino acid analyzer** is an instrument that automates ion-exchange chromatography. When a solution of an amino acid mixture passes through the column of an amino acid analyzer containing a cation-exchange resin, the amino acids move through the column at different rates, depending on their overall charge. The solution leaving the column is collected in fractions, which are collected often enough that a different amino acid ends up in each fraction (Figure 17.3). If ninhydrin is added to each of the fractions, the concentration of the amino acid in each fraction can be determined by using visible spectroscopy (Section 6.10) to measure the intensity of the absorbance of the colored compound formed by the reaction of the amino acid with ninhydrin (Figure 17.4).

PROBLEM 11

Explain the order of elution (with an aqueous solution of pH 4) of each of the following pairs of amino acids on a column packed with the resin shown in Figure 17.2:

a. glutamate before threonine   **b.** alanine before leucine

Figure 17.3 ▶
Separation of amino acids by
ion-exchange chromatography.

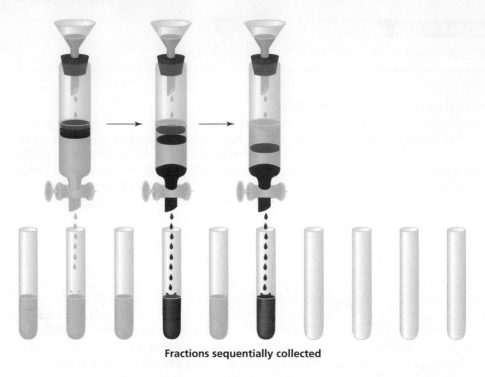

Fractions sequentially collected

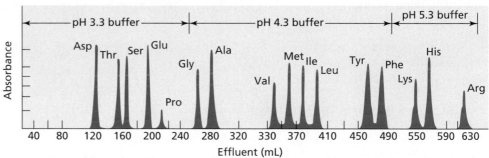

▲ **Figure 17.4**
A typical chromatogram obtained from the separation of a mixture of amino acids using
an automated amino acid analyzer.

WATER SOFTENERS: EXAMPLES OF CATION-EXCHANGE CHROMATOGRAPHY

Water softeners contain a column with a cation-exchange resin that has been flushed with concentrated sodium chloride. The presence of calcium and magnesium ions in water is what causes water to be "hard" (Section 20.3). When water passes through the column, the resin binds magnesium and calcium ions more tightly than it binds sodium ions. In this way, the water softener removes magnesium and calcium ions from water, replacing them with sodium ions. The resin must be recharged from time to time by flushing it with concentrated sodium chloride to replace the bound magnesium and calcium ions with sodium ions.

17.6 Peptide Bonds and Disulfide Bonds

Peptide bonds and disulfide bonds are the only covalent bonds that hold amino acid residues together in a peptide or a protein.

Peptide Bonds

The amide bonds that link amino acid residues are called **peptide bonds**. By convention, peptides and proteins are written with the free amino group (the **N-terminal amino acid**) on the left and the free carboxyl group (the **C-terminal amino acid**) on the right.

$$
\overset{+}{H_3}NCHCO^- \;+\; \overset{+}{H_3}NCHCO^- \;+\; \overset{+}{H_3}NCHCO^-
$$
$$
\qquad R \qquad\qquad\quad R' \qquad\qquad\quad R''
$$

(arrow down)

$$
\overset{+}{H_3}NCHC\!-\!NHCHC\!-\!NHCHCO^- \;+\; 2\,H_2O
$$
$$
\quad R \qquad\quad R' \qquad\quad R''
$$

| peptide bonds |

| the N-terminal amino acid | | the C-terminal amino acid |

a tripeptide

When the identities of the amino acids in a peptide are known but their sequence is not known, the amino acids are written separated by commas. When the sequence of amino acids is known, the amino acids are written separated by hyphens. In the following pentapeptide shown on the right, valine is the N-terminal amino acid and histidine is the C-terminal amino acid. The amino acids are numbered starting with the N-terminal end. The glutamate residue is referred to as Glu 4 because it is the fourth amino acid from the N-terminal end.

Glu, Cys, His, Val, Ala

| the pentapeptide contains the indicated amino acids, but their sequence is not known |

Val-Cys-Ala-Glu-His

| the amino acids in the pentapeptide have the indicated sequence |

A peptide bond has partial double-bond character because of electron delocalization (Section 12.2).

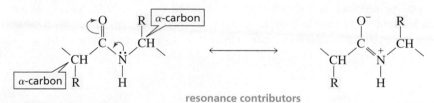

resonance contributors

The partial double-bond character prevents free rotation about the bond, so the carbon and nitrogen atoms of the peptide bond and the two atoms to which each is attached are held rigidly in a plane (Figure 17.5).

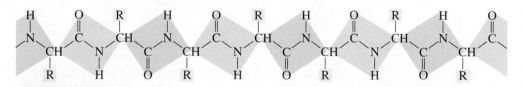

◀ **Figure 17.5**
A segment of a polypeptide chain. The plane defined by each peptide bond is indicated. Notice that the R groups bonded to the α-carbons are on alternate sides of the peptide backbone.

PROBLEM 12

Draw Gly-Val and Val-Gly.

PROBLEM 13

Draw the tetrapeptide Ala-Thr-Asp-Asn and indicate the peptide bonds.

PROBLEM 14◆

Using the three-letter abbreviations, write the six tripeptides that contain Ala, Gly, and Met.

PROBLEM 15

Which bonds in the backbone of a peptide can rotate freely?

ENKEPHALINS

Enkephalins are pentapeptides that are synthesized by the body to control pain. They
decrease the body's sensitivity to pain by binding to receptors
in certain brain cells. Part of the three-dimensional structures
of enkephalins must be similar to those of morphine and
related painkillers such as Demerol because they bind to the
same receptors (Section 22.3).

Tyr-Gly-Gly-Phe-Leu	Tyr-Gly-Gly-Phe-Met
leucine enkephalin	**methionine enkephalin**

Disulfide Bonds

When thiols are oxidized under mild conditions, they form disulfides. A **disulfide** is a
compound with an S—S bond.

$$2\ R—SH \xrightarrow{\text{mild oxidation}} RS—SR$$

a thiol   **a disulfide**

Because thiols can be oxidized to disulfides, disulfides can be reduced to thiols.

$$RS—SR \xrightarrow{\text{reduction}} 2\ R—SH$$

a disulfide   **a thiol**

Disulfides are reduced to thiols. Thiols are oxidized to disulfides.

Cysteine is an amino acid that contains a thiol group. Two cysteine molecules therefore can be oxidized to a disulfide. This disulfide is called cystine.

$$2\ \underset{\underset{\text{a thiol}}{\text{cysteine}}}{HSCH_2\overset{\overset{O}{\|}}{C}HCO^-} \xrightarrow{\text{mild oxidation}} \underset{\underset{\text{a disulfide}}{\text{cystine}}}{{}^-O\overset{\overset{O}{\|}}{C}CHCH_2S—SCH_2\overset{\overset{O}{\|}}{C}HCO^-}$$

$${}^+NH_3 \qquad\qquad\qquad {}^+NH_3 \qquad\qquad {}^+NH_3$$

Two cysteine residues in a protein can be oxidized to a disulfide. This is known as a
disulfide bridge. Disulfide bridges contribute to the overall shape of a protein by
holding the cysteine residues in close proximity, as shown in Figure 17.6.

Figure 17.6 ▶
Disulfide bridges cross-linking
portions of a peptide.

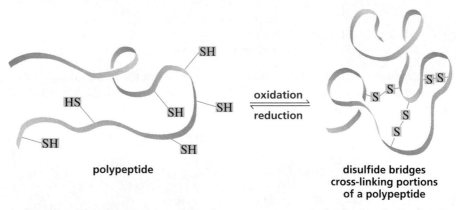

polypeptide   **disulfide bridges**
cross-linking portions
of a polypeptide

Insulin, a hormone secreted by the pancreas, controls the level of glucose in the
blood by regulating glucose metabolism. Insulin is a polypeptide with two peptide
chains. It has three disulfide brides, two of which hold the chains together.

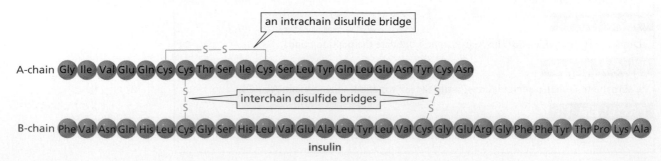

insulin

HAIR: STRAIGHT OR CURLY?

Hair is made up of a protein known as keratin. Keratin contains an unusually large number of cysteine residues, which give it many disulfide bridges to maintain its three-dimensional structure. People can alter the structure of their hair (if they think that it is either too straight or too curly) by changing the location of these disulfide bridges. This is accomplished by first applying a reducing agent to the hair to reduce all the disulfide bridges in the protein strands. Then the hair is given the desired shape (using curlers to curl it or combing it straight to uncurl it), and an oxidizing agent is applied that forms new disulfide bridges. The new disulfide bridges maintain the hair's new shape. When this treatment is applied to straight hair, it is called a "permanent." When it is applied to curly hair, it is called "hair straightening."

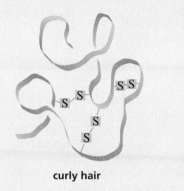

curly hair

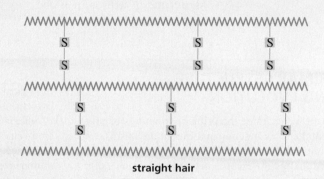

straight hair

PEPTIDE HORMONES

Bradykinin, vasopressin, and oxytocin are nonapeptides. Like insulin, they are peptide hormones. Bradykinin inhibits the inflammation of tissues. Vasopressin controls blood pressure by regulating the contraction of smooth muscle. It is also an antidiuretic. Oxytocin induces labor in pregnant women and stimulates milk production in nursing mothers. Vasopressin and oxytocin both have an intrachain disulfide bond, and their C-terminal amino acids contain amide rather than carboxyl groups. Notice that the C-terminal amide group is indicated by writing "NH_2" after the name of the C-terminal amino acid. In spite of their very different physiological effects, vasopressin and oxytocin differ only by two amino acids.

bradykinin   Arg-Pro-Pro-Gly-Phe-Ser-Pro-Phe-Arg

vasopressin   Cys-Tyr-Phe-Gln-Asn-Cys-Pro-Arg-Gly-NH_2
                    | |
                    S————————————S

oxytocin   Cys-Tyr-Ile-Gln-Asn-Cys-Pro-Leu-Gly-NH_2
                    | |
                    S————————————S

PROBLEM 16◆

Glutathione is a tripeptide that destroys harmful oxidizing agents in the body—compounds thought to be responsible for some of the effects of aging. Glutathione destroys the oxidizing agents by reducing them. Consequently, glutathione is oxidized.

$$\overset{COO^-}{\underset{H_3\overset{+}{N}CHCH_2CH_2}{|}}\overset{O}{\underset{}{C}}-NHCHC-NHCH_2CO^-$$

          CH_2

glutathione SH

a. What amino acids make up the tripeptide?
b. Draw the structure of oxidized glutathione (*Hint:* See the model on page 434.)
c. What is unusual about glutathione's structure? (If you can't answer this question, draw the structure you would expect for the tripeptide, and compare your structure with the actual structure of glutathione.)

3-D Molecules:
Glutathione;
Oxidized glutathione

17.7 Protein Structure

Protein molecules are described by several levels of structure. The **primary structure** of a protein is the sequence of amino acids in the chain and the location of all the disulfide bridges. The **secondary structure** describes the repetitive conformations assumed by segments of the backbone chain of the protein. The **tertiary structure** describes the three-dimensional structure of the entire polypeptide. If a protein has more than one polypeptide chain, it also has quaternary structure. The **quaternary structure** of a protein is the way the individual protein chains are arranged with respect to each other.

PRIMARY STRUCTURE AND EVOLUTION

When we examine the primary structures of proteins that carry out the same function in different organisms, we can relate the number of amino acid differences between the proteins to the taxonomic differences between the species. For example, cytochrome *c*, a protein that transfers electrons in biological oxidations, has about 100 amino acid residues. Yeast cytochrome *c* differs by 48 amino acids from horse cytochrome *c*, whereas duck cytochrome *c* differs by only two amino acids from chicken cytochrome *c*. Chickens and turkeys have cytochrome *c*'s with identical primary structures. Humans and chimpanzees also have identical cytochrome *c*'s, differing by one amino acid from the cytochrome *c* of the rhesus monkey.

17.8 Determining the Primary Structure of a Peptide or a Protein

One of the most widely used methods to identify the N-terminal amino acid of a peptide or a protein is to treat the protein with phenyl isothiocyanate (PITC), more commonly known as **Edman's reagent**. This reagent reacts with the N-terminal amino group, and the N-terminal amino acid is cleaved off as a PTH–amino acid, leaving behind a peptide with one fewer amino acid.

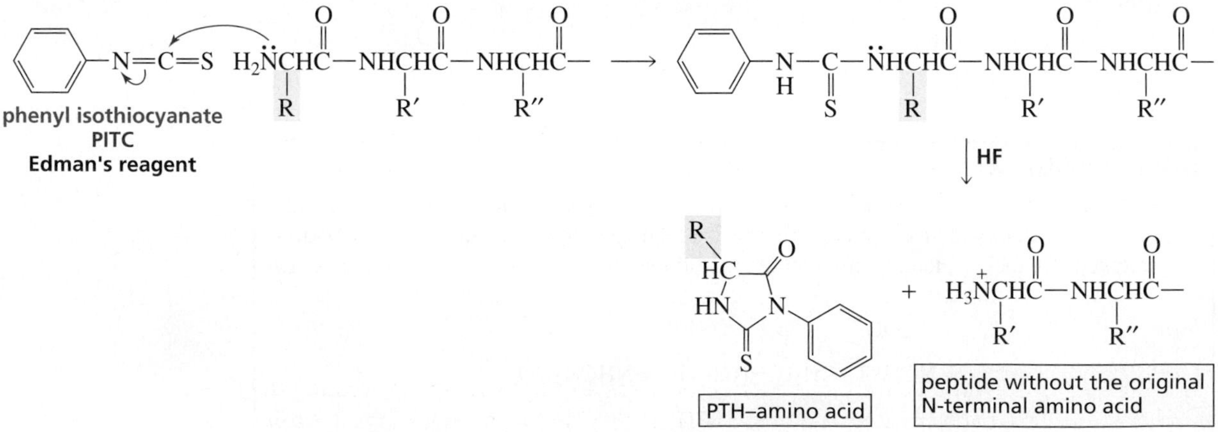

Because each amino acid has a different substituent (R), each amino acid forms a different PTH–amino acid. The particular PTH–amino acid can be identified by comparison with known PTH–amino acids. Several successive Edman degradations can be carried out on a protein, but the entire primary sequence cannot be determined in this way because side products accumulate that interfere with the results. An automated instrument known as a *sequenator* allows about 50 successive Edman degradations to be carried out on a protein.

PROBLEM 17

In determining the primary structure of insulin, what would lead you to conclude that it had more than one polypeptide chain?

Once the N-terminal amino acid has been identified, a sample of the protein is hydrolyzed with dilute acid. This treatment, called **partial hydrolysis**, hydrolyzes only some of the peptide bonds. The resulting fragments are separated, and the amino acid composition of each is determined using electrophoresis or an amino acid analyzer (Section 17.2). The sequence of the original protein can then be determined by lining up the peptides and looking for points of overlap.

PROBLEM-SOLVING STRATEGY

A nonapeptide undergoes partial hydrolysis to give peptides whose amino acid compositions are shown. Reaction of the intact nonapeptide with Edman's reagent releases PTH–Leu. What is the sequence of the nonapeptide?

a. Pro, Ser   **c.** Met, Ala, Leu   **e.** Glu, Ser, Val, Pro   **g.** Met, Leu

b. Gly, Glu   **d.** Gly, Ala   **f.** Glu, Pro, Gly   **h.** His, Val

Because we know that the N-terminal amino acid is Leu, we need to look for a fragment that contains Leu. Fragment (g) tells us that Met is next to Leu and fragment (c) tells us that Ala is next to Met. Now we look for a fragment that contains Ala. Fragment (d) contains Ala and tells us that Gly is next to Ala. From fragment (b), we know that Glu comes next. Glu is in both fragments (e) and (f). Fragment (e) has three amino acids we have yet to place in the growing peptide (Ser, Val, Pro), but fragment (f) has only one, so from fragment (f), we know that Pro is the next amino acid. Fragment (a) tells us that the next amino acid is Ser. Now we can use fragment (e). Fragment (e) tells us that the next amino acid is Val, and fragment (h) tells us that His is the last (C-terminal) amino acid. Thus, the amino acid sequence of the nonapeptide is

Leu-Met-Ala-Gly-Glu-Pro-Ser-Val-His

Now continue on to Problem 18.

Insulin was the first protein for which the primary sequence was determined. This was done in 1953 by **Frederick Sanger**, *who received the 1958 Nobel Prize in chemistry for his work. Sanger was born in England in 1918 and received a Ph.D. from Cambridge University, where he has worked for his entire career. He also received a share of the 1980 Nobel Prize in chemistry (Section 21.10) for being the first to sequence a DNA molecule (with 5375 nucleotide pairs).*

PROBLEM 18◆

A decapeptide undergoes partial hydrolysis to give peptides whose amino acid compositions are shown. Reaction of the intact decapeptide with Edman's reagent releases PTH-Gly. What is the sequence of the decapeptide?

a. Ala, Trp   **c.** Pro, Val   **e.** Trp, Ala, Arg   **g.** Glu, Ala, Leu

b. Val, Pro, Asp   **d.** Ala, Glu   **f.** Arg, Gly   **h.** Met, Pro, Leu, Glu

The C-terminal amino acid of a peptide or protein can be identified by treating the protein with carboxypeptidase A. A **peptidase** is an enzyme that catalyzes the hydrolysis of a peptide bond. Carboxypeptidase A catalyzes the hydrolysis the C-terminal peptide bond, cleaving off the C-terminal amino acid, as long as it is *not* arginine or lysine.

site where carboxypeptidase cleaves

$$\underset{R}{-\text{NHCHC}}\overset{O}{\underset{}{\|}}-\underset{R'}{\text{NHCHC}}\overset{O}{\underset{}{\|}}-\underset{R''}{\text{NHCHCO}^-}\overset{O}{\underset{}{\|}}$$

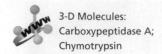

3-D Molecules:
Carboxypeptidase A;
Chymotrypsin

Trypsin and chymotrypsin are peptidases that catalyze the hydrolysis of specific peptide bonds (Table 17.4). Trypsin, for example, catalyzes the hydrolysis of the peptide bond on the C-side (right-hand side) of only arginine or lysine residues.

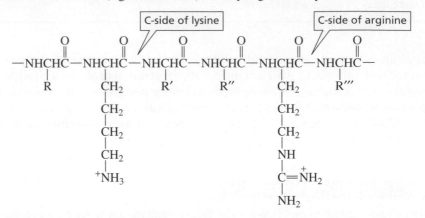

Table 17.4	Specificity of Peptide or Protein Cleavage
Reagent	**Specificity**
Chemical reagents	
Edman's reagent	removes the N-terminal amino acid
Cyanogen bromide	hydrolyzes on the C-side of Met
Peptidases*	
Carboxypeptidase A	removes the C-terminal amino acid (not Arg or Lys)
Trypsin	hydrolyzes on the C-side of Arg and Lys
Chymotrypsin	hydrolyzes on the C-side of amino acids that contain aromatic six-membered rings (Phe, Tyr, Trp)

*Cleavage will not occur if Pro is on either side of the bond to be hydrolyzed.

Thus, trypsin will catalyze the hydrolysis of three peptide bonds in the following peptide, creating a hexapeptide, a dipeptide, and two tripeptides.

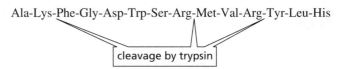

Chymotrypsin catalyzes the hydrolysis of the peptide bond on the C-side of amino acids that contain aromatic six-membered rings (Phe, Tyr, Trp).

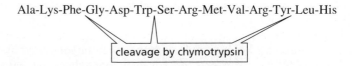

Peptidases will not catalyze the hydrolysis of a peptide bond if proline is at the hydrolysis site. Enzymes recognize the appropriate hydrolysis site by its shape and charge, and the cyclic structure of proline causes the hydrolysis site to have an unrecognizable three-dimensional shape (Section 18.1).

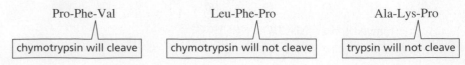

Cyanogen bromide (BrC≡N) causes the hydrolysis of the peptide bond on the C-side of a methionine residue. Because cyanogen bromide is not a protein and therefore does not recognize the substrate by its shape, cyanogen bromide will still cleave the peptide bond if proline is at the cleavage site.

Ala-Lys-Phe-Gly-Asp-Trp-Ser-Arg-Met-Val-Arg-Tyr-Leu-His

| cleavage by cyanogen bromide |

The first step in the mechanism for cleavage of a peptide bond by cyanogen bromide is attack by the highly nucleophilic sulfur of methionine on cyanogen bromide. Formation of a five-membered ring with departure of the weakly basic leaving group is followed by acid-catalyzed hydrolysis of the imine (Section 13.6), which cleaves the protein. Further hydrolysis can cause the cyclic ester to open to a carboxyl group and an alcohol group (Section 12.9).

mechanism for the cleavage of a peptide bond by cyanogen bromide

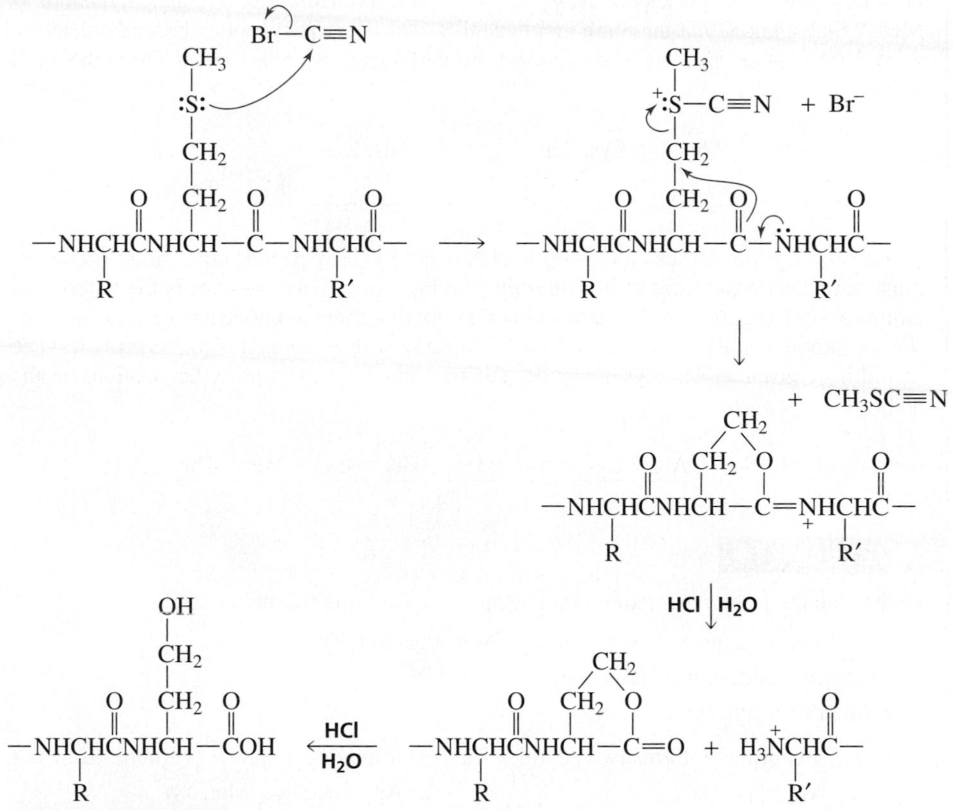

PROBLEM 19◆

Indicate the peptides that would result from cleavage by the indicated reagent:

a. His-Lys-Leu-Val-Glu-Pro-Arg-Ala-Gly-Ala by trypsin
b. Leu-Gly-Ser-Met-Phe-Pro-Tyr-Gly-Val by chymotrypsin

PROBLEM 20 | SOLVED

Determine the amino acid sequence of a polypeptide from the following results:

Acid hydrolysis gives Ala, Arg, His, 2 Lys, Leu, 2 Met, Ser, Thr, Val.
Carboxypeptidase A releases Val.
Edman's reagent releases PTH-Leu.

Cleavage with cyanogen bromide gives three peptides with the following amino acid compositions:

1. His, Lys, Met
2. Thr, Val
3. Ala, Arg, Leu, Lys, Met, Ser

Trypsin-catalyzed hydrolysis gives three peptides and a single amino acid:

1. Arg, Leu, Ser
2. Met, Thr, Val
3. Lys
4. Ala, His, Lys, Met

SOLUTION Acid hydrolysis shows that the polypeptide has 11 amino acids. The N-terminal amino acid is Leu (Edman's reagent), and the C-terminal amino acid is Val (carboxypeptidase A).

Leu __ __ __ __ __ __ __ __ __ Val

Because cyanogen bromide cleaves on the C-side of Met, any peptide containing Met must have Met as its C-terminal amino acid, and the peptide that does not contain Met must be the C-terminal peptide. We know that peptide 3 is the N-terminal peptide because it contains Leu. Since it is a hexapeptide, we know that the sixth amino acid in the polypeptide is Met. We also know that the ninth amino acid is Met because cyanogen bromide cleavage gave the dipeptide Thr, Val. The cyanogen bromide data also tells us that Thr is the tenth amino acid.

Ala, Arg, Lys, Ser   His, Lys

Leu __ __ __ __ Met __ __ Met Thr Val

Because trypsin cleaves on the C-side of Arg and Lys, any peptide containing Arg or Lys must have that amino acid as its C-terminal amino acid. Therefore, Arg is the C-terminal amino acid of peptide 1, so we now know that the first three amino acids are Leu-Ser-Arg. We also know that the next two are Lys-Ala because if they were Ala-Lys, trypsin cleavage would have given an Ala, Lys dipeptide. The trypsin data also identify the positions of His and Lys.

Leu Ser Arg Lys Ala Met His Lys Met Thr Val

PROBLEM 21◆

Determine the primary structure of an octapeptide from the following data:

Acid hydrolysis gives 2 Arg, Leu, Lys, Met, Phe, Ser, Tyr.

Carboxypeptidase A releases Ser.

Edman's reagent releases Leu.

Cyanogen bromide forms two peptides with the following amino acid compositions:

1. Arg, Phe, Ser
2. Arg, Leu, Lys, Met, Tyr

Trypsin forms the following two amino acids and two peptides:

1. Arg
2. Ser
3. Arg, Met, Phe
4. Leu, Lys, Tyr

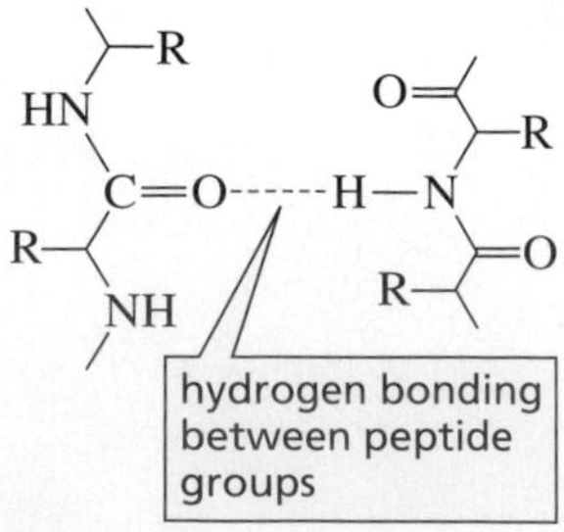

17.9 Secondary Structure of Proteins

Secondary structure describes the repetitive conformations assumed by segments of the backbone chain of a peptide or protein. In other words, the secondary structure describes how segments of the backbone fold. The conformations are stabilized by hydrogen bonding between peptide groups—between the hydrogen attached to a nitrogen of one amino acid residue and the carbonyl oxygen of another.

hydrogen bonding between peptide groups

α-Helix

One type of secondary structure is the **α-helix**. In an α-helix, the backbone of the polypeptide coils around the long axis of the protein molecule (Figure 17.7). The substituents on the α-carbons of the amino acids protrude outward from the helix, thereby minimizing steric hindrance. Each hydrogen attached to an amide nitrogen is hydrogen bonded to a carbonyl oxygen of an amino acid four amino acids away. Recall that a hydrogen bond can occur between a hydrogen bonded a nitrogen and a lone pair of an oxygen (Section 3.7).

3-D Molecule:
An α-helix

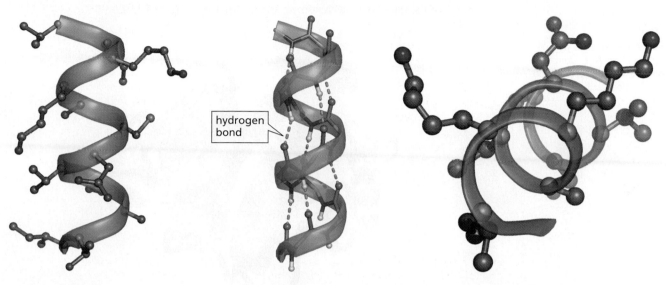

hydrogen bond

▲ **Figure 17.7**
(a) A segment of a protein in an α-helix. (b) The helix is stabilized by hydrogen bonding between peptide groups. (c) Looking up the longitudinal axis of an α-helix.

β-Pleated Sheet

The second type of secondary structure is the **β-pleated sheet**. In a β-pleated sheet, the polypeptide backbone is extended in a zigzag structure resembling a series of pleats. The hydrogen bonding in a β-pleated sheet occurs between neighboring peptide chains (Figure 17.8).

3-D Molecule:
β-pleated sheet

◀ **Figure 17.8**
Segment of a β-pleated sheet drawn to illustrate its pleated character.

N-terminal   C-terminal

R—CH   HC—R
  C—O····H—N
H—N   C=O
  HC—R   R—CH
O=C   N—H
  N—H····O=C
R—CH   HC—R
  C=O····H—N
H—N   C=O
  HC—R   R—CH

C-terminal   N-terminal

Because the substituents (R) on the α-carbons of the amino acids on adjacent chains are close to each other, the chains can get close enough together to form hydrogen bonds only if the substituents are small. Silk, for example, a protein with a large number of relatively small amino acids (glycine and alanine), has large segments of β-pleated sheets.

Wool and the fibrous protein of muscle are examples of proteins with secondary structures that are almost all α-helices. Consequently, these proteins can be stretched. In contrast, proteins with secondary structures that are predominantly β-pleated sheets, such as silk and spider webs, cannot be stretched because a β-pleated sheet is almost fully extended.

Generally, less than half of a protein's backbone is in an α-helix or a β-pleated sheet. The rest of the backbone has a nonrepetitive structure (Figure 17.9).

Figure 17.9 ▶
The backbone structure of carboxypeptidase A: α-helical segments are purple; β-pleated sheets are indicated by flat green arrows pointing in the N $\longrightarrow$ C direction.

17.10 | Tertiary Structure of Proteins

The **tertiary structure** of a protein is the three-dimensional arrangement of all the atoms in the protein (Figure 17.10). Proteins fold spontaneously in solution to maximize their stability. Every time there is a stabilizing interaction between two

Figure 17.10 ▶
The three-dimensional structure of carboxypeptidase A.

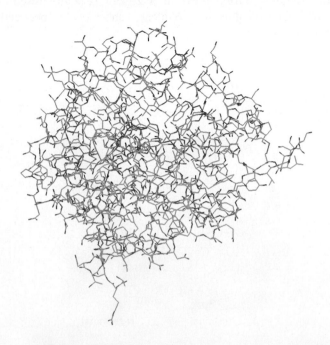

Max Ferdinand Perutz *and* **John Cowdery Kendrew** *were the first to determine the tertiary structure of a protein. Using X-ray diffraction, they determined the tertiary structure of myoglobin (1957) and hemoglobin (1959). For this work, they shared the 1962 Nobel Prize in chemistry.*

atoms, free energy is released. The more free energy released, the more stable is the protein. So a protein tends to fold in a way that maximizes the number of stabilizing interactions.

The stabilizing interactions include disulfide bonds, hydrogen bonds, electrostatic attractions (attractions between opposite charges), and hydrophobic interactions (attractions between nonpolar groups). Stabilizing interactions can occur between peptide groups (atoms in the backbone of the protein), between α-substituents, and between peptide groups and α-substituents (Figure 17.11). Because the α-substituents help determine how a protein folds, the tertiary structure of a protein is determined by its primary structure.

Max Perutz *was born in Austria in 1914. In 1936, because of the rise of Nazism, he moved to England. He received a Ph.D. from, and became a professor at, Cambridge University. He worked on the three-dimensional structure of hemoglobin and assigned the work on myoglobin (a smaller protein) to* **John Kendrew (1917–1997)**. *Kendrew was born in England and was educated at Cambridge University.*

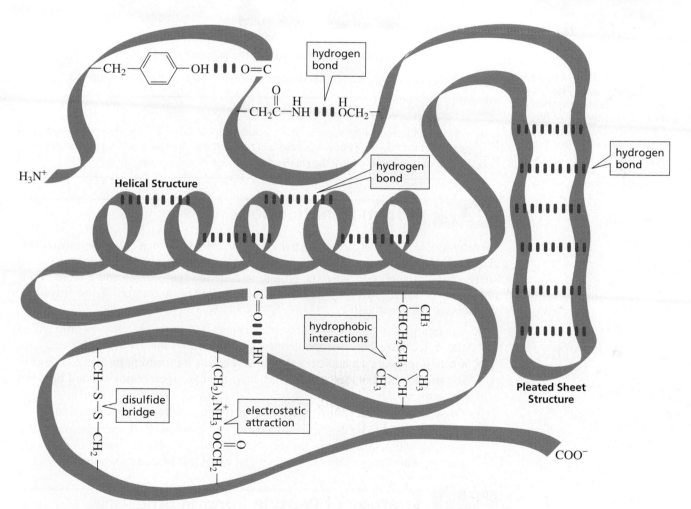

▲ **Figure 17.11**
Stabilizing interactions responsible for the tertiary structure of a protein.

Most proteins exist in aqueous environments. Therefore, they tend to fold in a way that exposes the maximum number of polar groups to the aqueous environment and that buries the nonpolar groups in the interior of the protein, away from water.

PROBLEM 22

How would a protein that resides in the interior of a membrane fold, compared with the water-soluble protein just discussed? (*Hint:* Membranes are nonpolar; see Section 19.5.)

17.11 Quaternary Structure of Proteins

Some proteins have more than one peptide chain. The individual chains are called **subunits**. The subunits are held together by the same kinds of interactions that hold the individual protein chains in a particular three-dimensional conformation: hydrophobic interactions, hydrogen bonding, and electrostatic attractions. The **quaternary structure** of a protein describes the way the subunits are arranged in space. Some of the possible arrangements of the six subunits of a hexamer are shown here:

possible quaternary structures for a hexamer

Hemoglobin has four subunits; its quaternary structure is shown in Figure 17.12.

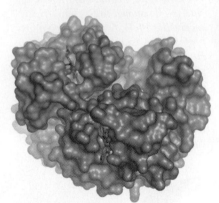

▲ **Figure 17.12**
A representation of the quaternary structure of hemoglobin. The orange and green represent the polypeptide chains; there are two identical orange subunits and two identical green subunits. Two of the porphyrin rings (gray beads) are visible (Section 7.3), ligated to iron (pink) and bonded to oxygen (red).

> **PROBLEM 23◆**
>
> **a.** Which of the following water-soluble proteins would have the greatest percentage of polar amino acids—a spherical protein, a cigar-shaped protein, or a subunit of a hexamer?
> **b.** Which of these would have the smallest percentage of polar amino acids?

17.12 Protein Denaturation

Destroying the highly organized tertiary structure of a protein is called **denaturation**. Anything that breaks the bonds responsible for maintaining the three-dimensional shape of the protein will cause the protein to denature (unfold). Because these bonds are weak, proteins are easily denatured. The following are some of the ways that proteins can be denatured:

- Changing the pH denatures proteins because it changes the charges on many of the side chains. This disrupts electrostatic attractions and hydrogen bonds.
- Certain reagents such as urea denature proteins by forming hydrogen bonds to the protein groups that are stronger than the hydrogen bonds formed between the groups.
- Organic solvents denature proteins by disrupting hydrophobic interactions.
- Proteins can also be denatured by heat or by agitation. Both increase molecular motion, which can disrupt the attractive forces. A well-known example is the change that occurs to the white of an egg when it is heated or whipped.

17.13 Strategy of Peptide Bond Synthesis: N-Protection and C-Activation

If the structure of a polypeptide is known, one can attempt to synthesize it. A problem arises when one tries to make a particular peptide bond because amino acids have two functional groups, For example, suppose you wanted to make the dipeptide Gly-Ala. That dipeptide is only one of four possible dipeptides that could be formed from alanine and glycine.

$$\overset{+}{H_3N}CH_2\overset{O}{\overset{\|}{C}}-NHCHC\overset{O}{\overset{\|}{C}}O^-$$
$$\underset{CH_3}{}$$
Gly-Ala

$$\overset{+}{H_3N}CHC\overset{O}{\overset{\|}{C}}-NHCHC\overset{O}{\overset{\|}{C}}O^-$$
$$\underset{CH_3}{} \quad \underset{CH_3}{}$$
Ala-Ala

$$\overset{+}{H_3N}CH_2\overset{O}{\overset{\|}{C}}-NHCH_2C\overset{O}{\overset{\|}{C}}O^-$$
Gly-Gly

$$\overset{+}{H_3N}CHC\overset{O}{\overset{\|}{C}}-NHCH_2C\overset{O}{\overset{\|}{C}}O^-$$
$$\underset{CH_3}{}$$
Ala-Gly

If the amino group of the amino acid that is to be on the N-terminal end (in this case, Gly) is protected, it will not be available to form a peptide bond. If the carboxyl group of this same amino acid is activated before the second amino acid is added, the amino group of the added amino acid (in this case, Ala) will react with the activated carboxyl group of glycine in preference to reacting with a nonactivated carboxyl group of another alanine molecule.

Oxytocin was the first small peptide to be synthesized. Its synthesis was achieved in 1953 by **Vincent du Vigneaud (1901–1978)**, *who later synthesized vasopressin. Du Vigneaud was born in Chicago and was a professor at George Washington University Medical School and later at Cornell University Medical College. For synthesizing these nonapeptides, he received the Nobel Prize in chemistry in 1955.*

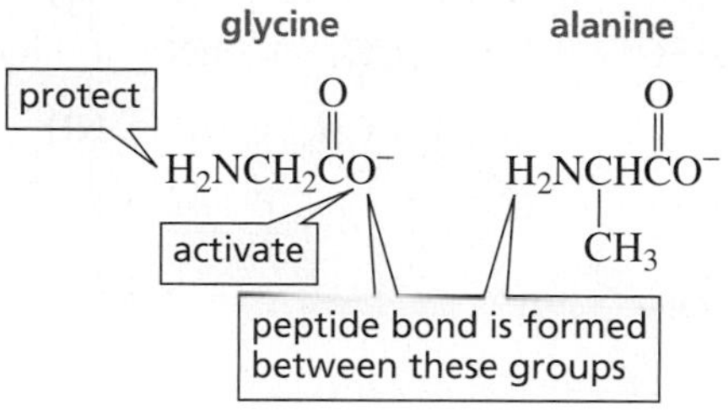

The reagent that is most often used to protect the amino group of an amino acid is di-*tert*-butyl dicarbonate. Its popularity is due to the ease with which the protecting group can be removed when the need for protection is over.

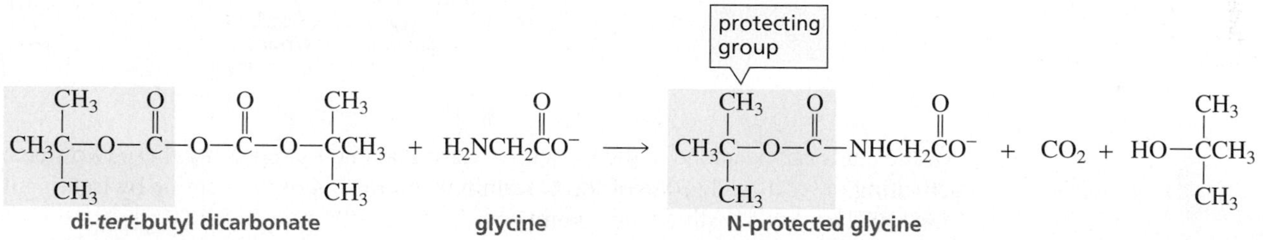

The reagent most often used to activate the carboxyl group is dicyclohexylcarbodiimide (DCC). DCC activates a carboxyl group by putting a good leaving group on the carbonyl carbon.

The N-terminal amino acid must have its amino group protected and its carboxyl group activated.

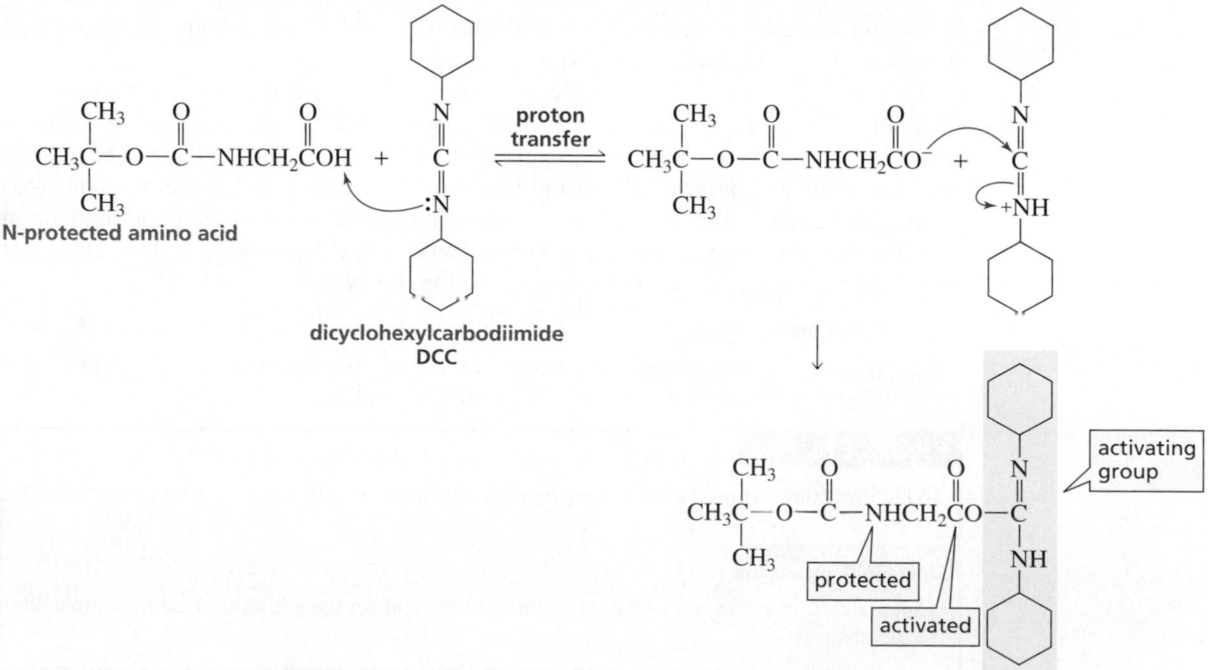

After the amino acid has its N-terminal group protected and its C-terminal group activated, the second amino acid is added. The unprotected amino group of the second amino acid attacks the activated carboxyl group and forms a new peptide bond.

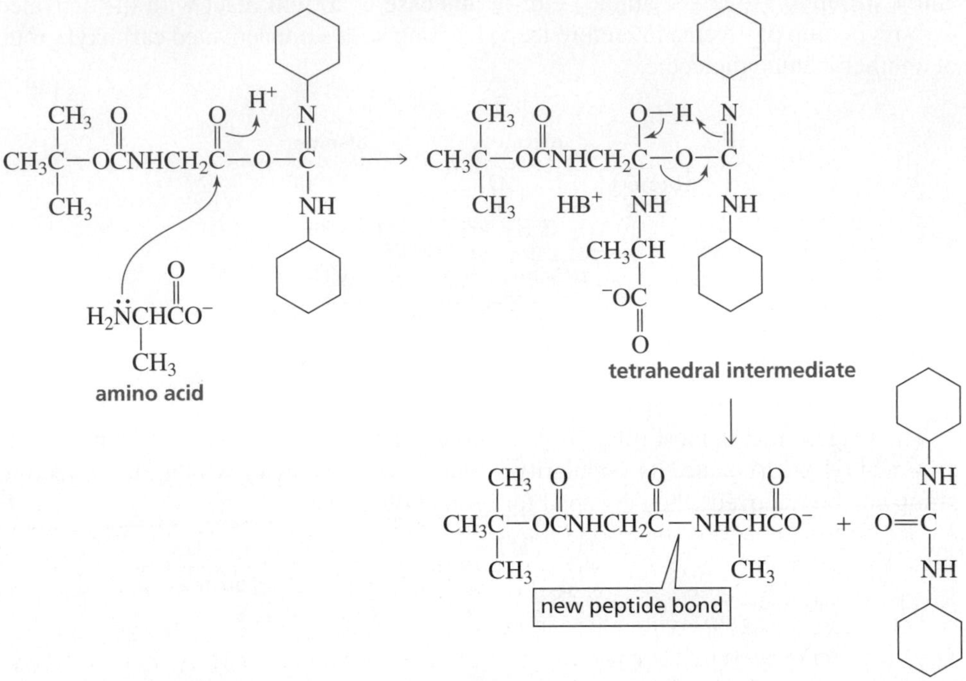

Amino acids can be added to the growing C-terminal end by repeating these two steps: activating the carboxyl group of the C-terminal amino acid of the peptide by treating it with DCC and then adding a new amino acid.

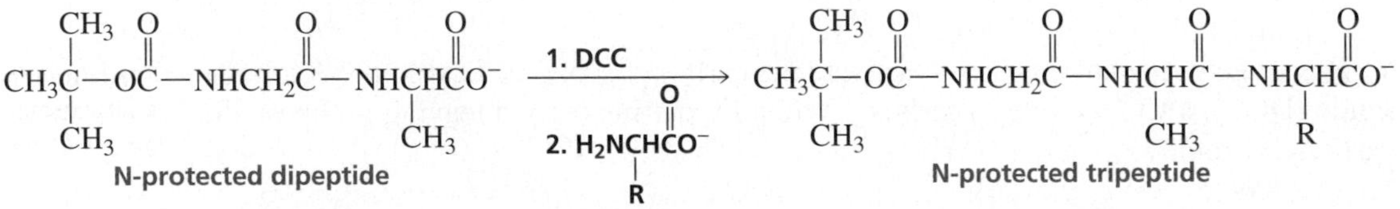

When the desired number of amino acids has been added to the chain, the protecting group on the N-terminal amino acid is removed.

Theoretically, one should be able to make as long a peptide as desired with this technique. Reactions do not produce 100% yields, however, and the yields are further decreased during the purification process. After each step of the synthesis, the peptide must be purified to prevent subsequent unwanted reactions with leftover reagents. Assuming that each amino acid can be added to the growing end of the peptide chain in an 80% yield, the overall yield of a nonapeptide would be only 17%. It is clear that large polypeptides could never be synthesized in this way.

Number of amino acids	2	3	4	5	6	7	8	9
Overall yield	80%	64%	51%	41%	33%	26%	21%	17%

PROBLEM 24◆

What dipeptides would be formed by heating a mixture of valine and N-protected leucine?

PROBLEM 25◆

Calculate the overall yield of a nonapeptide if the yield for the addition of each amino acid to the chain is 70%.

Since the early 1980s, it has been possible to synthesize proteins by genetic engineering techniques. Strands of DNA can be introduced into bacterial cells, causing the cells to produce large amounts of a desired protein (Section 21.12). For example, mass quantities of human insulin are produced from genetically modified *E. coli*.

NUTRASWEET

The synthetic sweetener aspartame, or NutraSweet (Section 16.16), is the methyl ester of a dipeptide of L-aspartate and L-phenylalanine. The ethyl ester of the same dipeptide is not sweet. If a D-amino acid is substituted for either of the L-amino acids of aspartame, the resulting dipeptide is bitter rather than sweet.

$$\overset{+}{H_3N}\overset{O}{\underset{CH_2}{\underset{|}{CH}C}}-NH\overset{O}{\underset{CH_2}{\underset{|}{CH}COCH_3}}$$

COO⁻

aspartame
NutraSweet

Summary

Peptides and **proteins** are polymers of **amino acids** linked together by **peptide** (amide) **bonds**. A **dipeptide** contains two amino acid residues and a **polypeptide** contains many amino acid residues. Proteins have 40 to 4000 amino acid residues. The **amino acids** differ only in the substituent attached to the α-carbon. Most amino acids found in nature have the L configuration.

The carboxyl groups of the amino acids have pK_a values of ~2 and the protonated amino groups have pK_a values of ~9. At physiological pH, an amino acid exists as a **zwitterion**. A few amino acids have side chains with ionizable hydrogens. The **isoelectric point** (pI) of an amino acid is the pH at which the amino acid has no net charge. A mixture of amino acids can be separated based on their pI's by **electrophoresis** or by **ion-exchange chromatography**. An **amino acid analyzer** is an instrument that automates ion-exchange chromatography.

The amide bonds that link amino acid residues are called **peptide bonds**. Two cysteine residues can be oxidized to a **disulfide bridge**. By convention, peptides and proteins are written with the free amino group (the **N-terminal amino acid**) on the left and the free carboxyl group (the **C-terminal amino acid**) on the right.

The **primary structure** of a protein is the sequence of its amino acids and the location of all its disulfide bridges.

The N-terminal amino acid of a peptide or protein can be determined with **Edman's reagent**. The C-terminal amino acid can be identified with carboxypeptidase A. **Partial hydrolysis** hydrolyzes only some of the peptide bonds. A **peptidase** catalyzes the hydrolysis of a peptide bond.

The **secondary structure** of a protein describes how local segments of the protein's backbone fold. An α-helix and a β-pleated sheet are secondary structures. The **tertiary structure** of a protein is the three-dimensional arrangement of all the atoms in the protein. A protein folds so as to maximize the number of stabilizing interactions: **disulfide bonds, hydrogen bonds, electrostatic attractions**, and **hydrophobic interactions**. The **quaternary structure** of a protein describes the way the **subunits**—individual protein chains—are arranged with respect to each other.

To synthesize a peptide bond, the amino group of the N-terminal amino acid must be protected (by *t*-BOC) and its carboxyl group activated (with DCC). The second amino acid is added to form a dipeptide. Amino acids can be added to the growing C-terminal end by activating the carboxyl group of the C-terminal amino acid with DCC and adding a new amino acid.

Problems

26. Draw the form in which each of the following amino acids predominantly exists at physiological pH (7.3):
   a. lysine   **b.** arginine   **c.** tyrosine

27. What is the pI of aspartame?

28. Indicate the peptides that would result from cleavage by the indicated reagent:
   a. Val-Arg-Gly-Met-Arg-Ala-Ser by carboxypeptidase A
   b. Ser-Phe-Lys-Met-Pro-Ser-Ala-Asp by cyanogen bromide
   c. Arg-Ser-Pro-Lys-Lys-Ser-Glu-Gly by trypsin

29. a. Which isomer—(*R*)-aspartate or (*S*)-aspartate—is L-aspartate?
   b. Can a general statement be made relating *R* and *S* to D and L?

30. Aspartame has a pI of 5.9. Draw its most prevalent form at physiological pH.

31. Draw the form of aspartic acid that predominates at
   a. pH = 1.0   **b.** pH = 2.6   **c.** pH = 6.0   **d.** pH = 11.0

32. a. Why is the pK_a of the glutamate side chain greater than the pK_a of the aspartate side chain?
   b. Why is the pK_a of the arginine side chain greater than the pK_a of the lysine side chain?

33. Explain the difference in the pK_a values of the carboxyl groups of alanine, serine, and cysteine.

34. Which would be a more effective buffer at physiological pH, a solution of 0.1 M glycylglycylglycylglycine or a solution of 0.2 M glycine?

35. a. At pH = 11, only one of arginine's nitrogen atoms is protonated. Which one is it?
   b. At pH = 5, one of the nitrogens in histidine's five-membered ring is protonated. Which one is it? (*Hint:* Localized electrons are more apt to be protonated than delocalized electrons.)

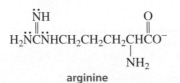

arginine 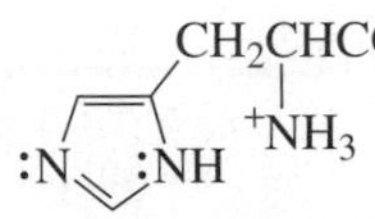   histidine

36. Identify the location and type of charge on the hexapeptide Lys-Ser-Asp-Cys-His-Tyr at
   a. pH = 7   **b.** pH = 5   **c.** pH = 9

37. a. Which amino acid has the greatest amount of negative charge at pH = 6.20?
   b. Which amino acid—glycine or methionine—has a greater negative charge at pH = 6.20?

38. Explain the order of elution (with a buffer of pH 4) of each of the following pairs of amino acids on a column packed with the cation-exchange resin shown in Figure 17.2:
   a. aspartate before serine   **c.** valine before leucine
   b. glycine before alanine   **d.** tyrosine before phenylalanine

39. Reaction of a polypeptide with carboxypeptidase A releases Met. The polypeptide undergoes partial hydrolysis to give the following peptides. What is the sequence of the polypeptide?
   a. Ser, Lys, Trp   **e.** Met, Ala, Gly   **i.** Lys, Ser
   b. Gly, His, Ala   **f.** Ser, Lys, Val   **j.** Glu, His, Val
   c. Glu, Val, Ser   **g.** Glu, His   **k.** Trp, Leu, Glu
   d. Leu, Glu, Ser   **h.** Leu, Lys, Trp   **l.** Ala, Met

40. Glycine has pK_a values of 2.3 and 9.6. Would you expect the pK_a values of glycylglycine to be higher or lower than these values?

41. The following polypeptide was hydrolyzed by trypsin:

Gly-Ser-Asp-Ala-Leu-Pro-Gly-Ile-Thr-Ser-Arg-Asp-Val-Ser-Lys-Val-Glu-Tyr-Phe-Glu-Ala-Gly-Arg-Ser-Glu-Phe-Lys-Glu-Pro-Arg-Leu-Tyr-Met-Lys-Val-Glu-Gly-Arg-Pro-Val-Ser-Ala-Gly-Leu-Trp

   a. How many fragments are obtained from the peptide?
   b. In what order would the fragments be eluted from an anion-exchange column using a buffer of pH = 5?

42. Why is proline never found in an α-helix?

43. Explain why the following are not found in an α-helix: two adjacent glutamates, two adjacent aspartates, or a glutamate adjacent to an aspartate.

44. After breaking the disufide bridges in a polypeptide, it is found to have two polypeptide chains with the following primary sequences:

Val-Met-Tyr-Ala-Cys-Ser-Phe-Ala-Glu-Ser

Ser-Cys-Phe-Lys-Cys-Trp-Lys-Tyr-Cys-Phe-Arg-Cys-Ser

Treatment of the original intact polypeptide with chymotrypsin yields the following peptides:
   a. Ala, Glu, Ser   **d.** Arg, Ser, Cys
   b. 2 Phe, 2 Cys, Ser   **e.** Ser, Phe, 2 Cys, Lys, Ala, Trp
   c. Tyr, Val, Met   **f.** Tyr, Lys

Determine the positions of the disulfide bridges in the original polypeptide.

45. Determine the amino acid sequence of a polypeptide from the following results:
  a. Complete hydrolysis of the peptide yields the following amino acids: Ala, Arg, Gly, 2 Lys, Met, Phe, Pro, 2 Ser, Tyr, Val.
  b. Treatment with Edman's reagent gives PTH-Val.
  c. Carboxypeptidase A releases Ala.
  d. Treatment with cyanogen bromide yields the following two peptides:
    1. Ala, 2 Lys, Phe, Pro, Ser, Tyr
    2. Arg, Gly, Met, Ser, Val
  e. Treatment with trypsin yields the following three peptides:
    1. Gly, Lys, Met, Tyr
    2. Ala, Lys, Phe, Pro, Ser
    3. Arg, Ser, Val
  f. Treatment with chymotrypsin yields the following three peptides:
    1. 2 Lys, Phe, Pro
    2. Arg, Gly, Met, Ser, Tyr, Val
    3. Ala, Ser

46. What products are obtained when aspartame (page 459) is hydrolyzed in the presence of an acid catalyst?

47. Dr. Kim S. Tree was preparing a manuscript for publication in which she reported that the pI of the tripeptide Lys-Lys-Lys was 10.6. One of her students pointed out that there must be an error in her calculations because the pK_a of the ε-amino group of lysine is 10.8 and the pI of the tripeptide has to be greater than any of its individual pK_a values. Was the student correct?

48. Show the steps in the synthesis of the tetrapeptide Leu-Phe-Lys-Val.

49. a. Calculate the overall yield of a nonapeptide if the yield for the addition of each amino acid to the chain is 70%.
  b. What would be the overall yield of a peptide containing 15 amino acid residues if the yield for the incorporation of each is 80%?

50. The C-terminal end of a protein extends into the aqueous environment surrounding the protein. The C-terminal amino acids are Gln, Asp, 2 Ser, and three nonpolar amino acids. Assuming that the $\Delta G°$ for the formation of a hydrogen bond is -3 kcal/mol and the $\Delta G°$ for removal of a hydrophobic group from water is -4 kcal/mol, calculate the $\Delta G°$ for folding the C-terminal end of the protein into the interior of the protein under the following conditions:
  a. Each of the polar groups form one intramolecular hydrogen bond.
  b. All but two of the polar groups form intramolecular hydrogen bonds.

18 Enzymes, Coenzymes, and Vitamins

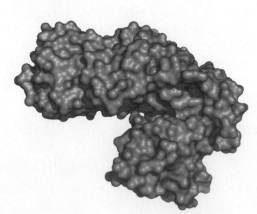

E ssentially all organic reactions that occur in bi-
ological systems require
a catalyst. Most biological catalysts are **enzymes**, which are
globular proteins (Section 17.0).
Enzymes are extraordinarily good
catalysts—they can increase the rate of a
reaction by as much as 10^{16}!

18.1 Enzyme-Catalyzed Reactions

The reactant of an enzyme-catalyzed reaction is called a **substrate**. The enzyme binds
the substrate in a pocket called an **active site** (Figure 18.1).

$$\text{substrate} \xrightarrow{\textbf{enzyme}} \text{product}$$

Figure 18.1 ▶
A substrate bound at the active
site of an enzyme.

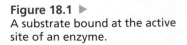

All the bond-making and bond-breaking steps of the reaction occur while the substrate is at this site. Enzymes are specific for the substrate whose reaction they catalyze (Section 8.14). All enzymes, however, do not have the same degree of specificity. Some are specific for a single compound and will not tolerate even the slightest variation in structure, whereas some enzymes catalyze the reaction of an entire family of compounds with related structures. The specificity of an enzyme for its substrate is an example of the phenomenon known as **molecular recognition**—the ability of one molecule to recognize another.

The specificity of an enzyme results from the particular **amino acid side chains** (α-substituents) that are at the active site (Section 17.1). For example, an amino acid with a negatively charged side chain can associate with a positively charged group on the substrate, an amino acid side chain with a hydrogen-bond donor can associate with a hydrogen-bond acceptor on the substrate, and a hydrophobic amino acid side chain can associate with a hydrophobic group on the substrate.

The following are some of the most important factors that contribute to the remarkable catalytic ability of enzymes:

- Reacting groups are brought together at the active site in the proper orientation for reaction.

- Some of the amino acid side chains of the enzyme serve as catalysts. These are positioned near the substrate precisely where they are needed for catalysis.

- Amino acid side chains can stabilize transition states and intermediates (Section 4.8), which makes them easier to form.

We will now look at the mechanisms of two enzyme-catalyzed reactions in order to understand how the amino acid side chains at the active site act as catalytic groups. In examining these reactions, notice that they are similar to reactions that you have seen organic compounds undergo. *If you refer back to sections referenced throughout this chapter, you will be able to see that much of the organic chemistry you have learned applies to the reactions of compounds found in the biological world.*

18.2   Mechanism for Glucose-6-Phosphate Isomerase

The names of most enzymes end in "ase" and the enzyme's name tells you the reaction that it catalyzes. For example, the enzyme glucose-6-phosphate isomerase catalyzes the isomerization of glucose-6-phosphate to fructose-6-phosphate. Recall that the open-chain form of glucose is an aldohexose, whereas the open-chain form of fructose is a ketohexose (Sections 16.3 and 16.4). Therefore, this enzyme converts an aldose to a ketose. The enzyme is known to have at least three catalytic groups at its active site, one functioning as an acid catalyst and two acting as bases catalysts (Figure 18.2). The reaction takes place as follows:

- Because the sugars exist predominantly in their cyclic forms in solution, the first step of the reaction is a ring-opening reaction. A base catalyst (a histidine side chain) removes a proton from the OH group on C-1, and an **acid catalyst** (a protonated lysine side chain) aids the departure of the leaving group by protonating it, thereby making it a weaker base and, therefore, a better leaving group (Section 12.9).

- In the second step of the reaction, a **base catalyst** (apparently a glutamate residue) removes a proton from the α-carbon of the aldehyde. Recall that α-hydrogens are relatively acidic (Section 14.1).

- In the next step, the enol is converted to a ketone (Section 14.3).

- In the final step of the reaction, the conjugate base of the acid catalyst employed in the first step catalyzes ring closure.

A proton is donated to the reactant in an acid-catalyzed reaction.

A proton is removed from the reactant in a base-catalyzed reaction.

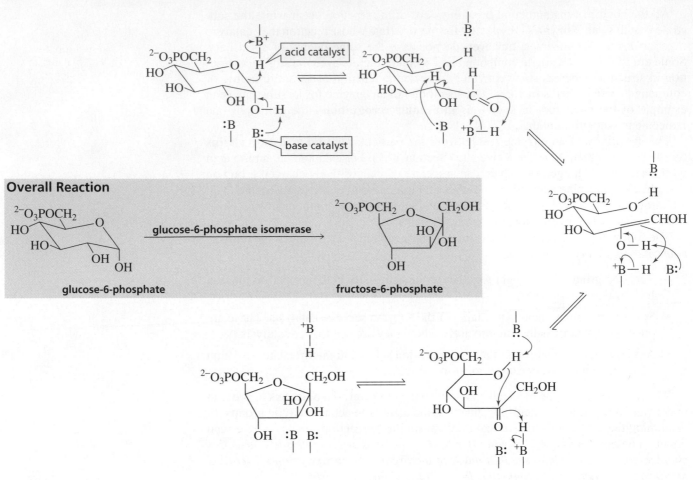

▲ **Figure 18.2**
Proposed mechanism for the isomerization of glucose-6-phosphate to fructose-6-diphosphate.

PROBLEM 1 SOLVED

Which of the following amino acid side chains can aid the departure of a leaving group by protonating it?

—CH$_2$CH$_2$SCH$_3$ —CH(CH$_3$)$_2$ —CH$_2$ (imidazolium)   —CH$_2$COH (with C=O)

    1   **2**   **3**   **4**

SOLUTION   Neither **1** nor **2** have an acidic proton, so these amino acid side chains cannot aid the departure of a leaving group. Both **3** and **4** have an acidic proton, so these amino acid side chains can aid the departure of a leaving group.

PROBLEM 2◆

Which of the following amino acid side chains can help remove a proton from the α-carbon of an aldehyde?

—CH$_2$CNH$_2$ (with C=O)   (phenol)—O$^-$ —CH$_2$ (imidazole)   —CH$_2$CO$^-$ (with C=O)

    1   **2**   **3**   **4**

18.3 Mechanism for Aldolase

Aldolase is the enzyme that converts fructose-1,6-diphosphate into two three-carbon compounds, glyceraldehyde-3-phosphate and dihydroxyacetone phosphate (Figure 18.3).

▼ Figure 18.3
Proposed mechanism for the aldolase-catalyzed cleavage of fructose-1,6-diphosphate to glyceraldehyde-3-phosphate and dihydroxyacetone phosphate.

Overall Reaction

fructose-1,6-diphosphate $\rightleftharpoons$ glyceraldehyde-3-phosphate + dihydroxyacetone phosphate

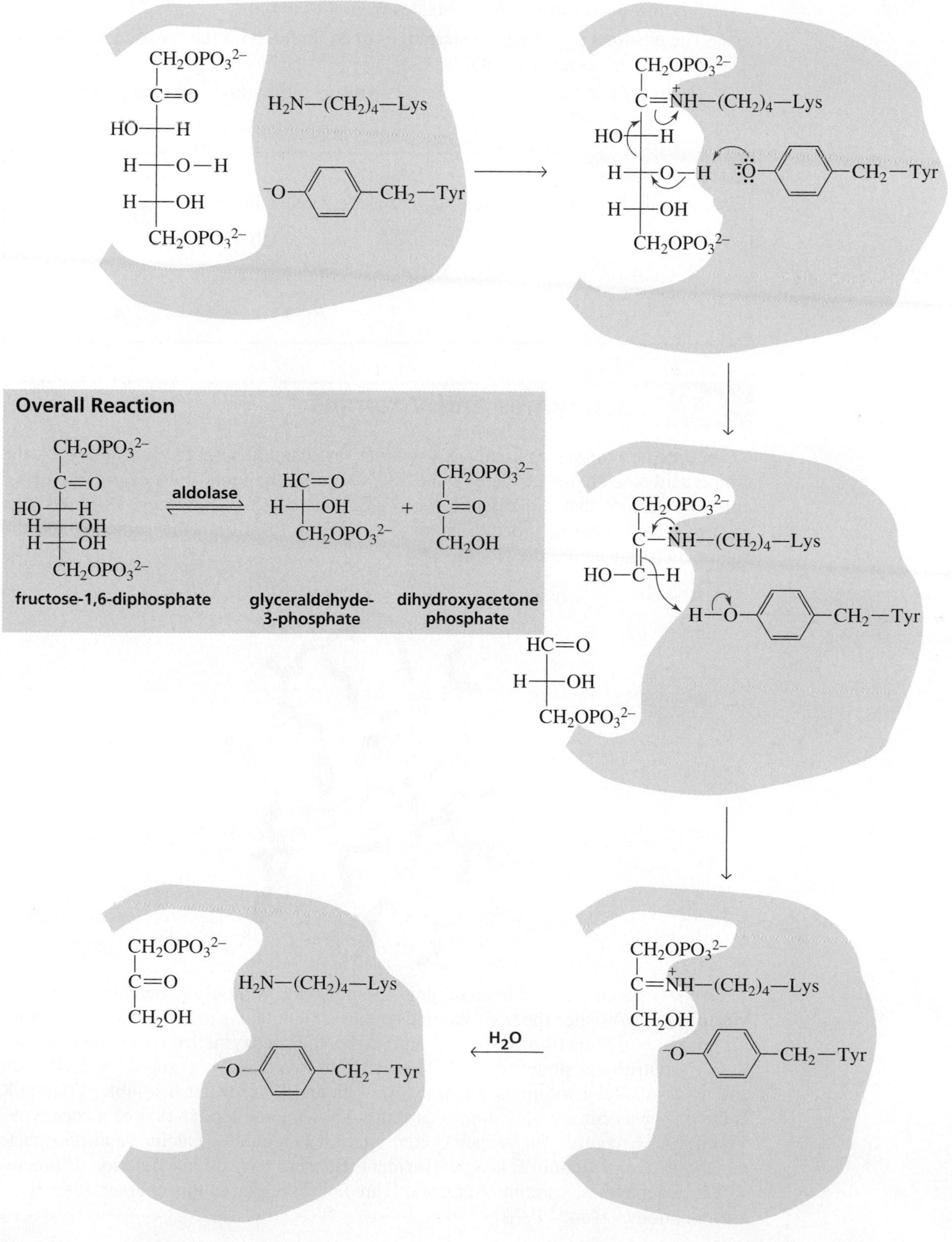

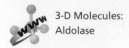

3-D Molecules:
Aldolase

The enzyme is called aldolase because the reverse reaction is an aldol addition reaction (Section 14.5). The reaction takes place as follows:

- In the first step of the reaction, fructose-1,6-diphosphate forms an imine with a lysine residue at the active site of the enzyme (Section 13.6).

- A basic catalyst (a side chain of tyrosine) removes a proton in a step that cleaves the bond between C-3 and C-4. One of the products (glyceraldehyde-3-phosphate) has now been formed. It dissociates from the enzyme.

- The enamine intermediate rearranges to an imine, with the tyrosine residue now functioning as an acid catalyst.

- Hydrolysis of the imine (Section 13.6) releases dihydroxyacetone phosphate, the other three-carbon product.

PROBLEM 3◆

Which of the following amino acid side chains can form an imine with the substrate?

$$-CH_2\overset{\overset{\displaystyle O}{\|}}{C}NH_2 \qquad -(CH_2)_4NH_2 \qquad -(CH_2)_3NH\overset{\overset{\displaystyle NH}{\|}}{C}NH_2 \qquad -CH_2OH$$

$$\mathbf{1} \qquad\qquad\qquad \mathbf{2} \qquad\qquad\qquad \mathbf{3} \qquad\qquad\qquad \mathbf{4}$$

18.4 Coenzymes and Vitamins

Cofactors help enzymes catalyze reactions that cannot be catalyzed solely by the amino acid side chains of the enzyme. Cofactors can be metal ions or organic molecules. Cofactors that are organic molecules are called **coenzymes**. Enzymes that require a coenzyme in order to catalyze a reaction bind both the substrate and the coenzyme at the active site (Figure 18.4).

Figure 18.4 ▶
A substrate and a coenzyme bound at the active site of an enzyme.

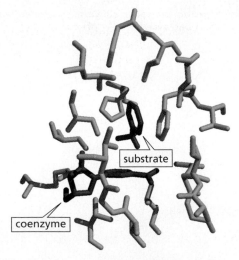

Coenzymes are derived from organic compounds commonly known as *vitamins*. A **vitamin** is a substance the body cannot synthesize, but that is needed in small amounts for normal body function. The body synthesizes the coenzyme from the vitamin.

Early nutritional studies divided vitamins into two classes: water-soluble vitamins and water-insoluble vitamins. Vitamins A, D, E, and K are water insoluble. Vitamin K is the only water-insoluble vitamin currently known to be a precursor of a coenzyme. Vitamin A is required for proper vision, vitamin D regulates calcium and phosphate metabolism, and vitamin E is an antioxidant. Because they do not function as precursors for coenzymes, vitamins A, D, and E are not discussed in this chapter. (See, however, Sections 9.6 and 20.7.)

All the water-soluble vitamins except vitamin C are precursors for coenzymes. In spite of its name, vitamin C is not actually a vitamin because it is required in fairly high amounts and most mammals are able to synthesize it (Section 16.14). Humans and guinea pigs cannot synthesize it, however, so it must be included in their diets. We have seen that vitamin C is a radical inhibitor (Section 9.6).

It is hard to overdose on water-soluble vitamins because the body can readily eliminate any reasonable excess. One can, however, overdose on water-insoluble vitamins because they are *not* easily eliminated by the body and can accumulate in cell membranes and other nonpolar components of the body.

VITAMIN B₁

Christiaan Eijkman (1858–1930) was a member of a medical team that was sent to the East Indies to study beriberi in 1886. At that time, all diseases were thought to be caused by microorganisms. When the microorganism that caused beriberi could not be found, the team left the East Indies. Eijkman stayed behind to become the director of a new bacteriological laboratory.

In 1896, Eijkman accidentally discovered the cause of beriberi when he noticed that chickens used in the laboratory had developed symptoms characteristic of the disease. He found that the symptoms had developed when a cook had started feeding the chickens rice meant for hospital patients. The symptoms disappeared when a new cook resumed feeding chicken feed to the chickens. Later it was recognized that thiamine (vitamin B₁) is present in rice hulls but not in polished rice. For this work, Eijkman shared the 1929 Nobel Prize in physiology or medicine with Frederick Hopkins.

Christiaan Eijkman

"VITAMINE"—AN AMINE REQUIRED FOR LIFE

Sir Frederick G. Hopkins (1861–1947) was born in England. He was the first to suggest that diseases, such as rickets and scurvy, might result from the absence of substances in the diet that are needed only in very small quantities. The first such compound recognized to be essential in the diet was an amine (thiamine), which led to the incorrect assumption that all such compounds were amines. They therefore, were called vitamines ("amines required for life"). The *e* was later dropped from the name. Hopkins' hypothesis later became known as the "vitamin concept," for which he received a share of the 1929 Nobel Prize in physiology or medicine. He also originated the concept of essential amino acids.

Sir Frederick Hopkins

18.5   Niacin: The Vitamin Needed for Many Oxidation–Reduction Reactions

An enzyme that catalyzes an oxidation or a reduction reaction requires a coenzyme because none of the amino acid side chains are oxidizing or reducing agents. The coenzyme functions as the oxidizing or reducing agent. The enzyme's role is to hold the substrate and coenzyme together so that the oxidation or reduction reaction can take place.

The coenzyme most commonly used by enzymes *to catalyze oxidation reactions* is **nicotinamide adenine dinucleotide (NAD^+)**; NAD^+ is composed of two nucleotides linked together through their phosphate groups. A **nucleotide** consists of a heterocycle attached to C-1 of a phosphorylated ribose (Section 21.1). A **heterocycle** is a cyclic compound in which one or more of the ring atoms is an atom other than carbon.

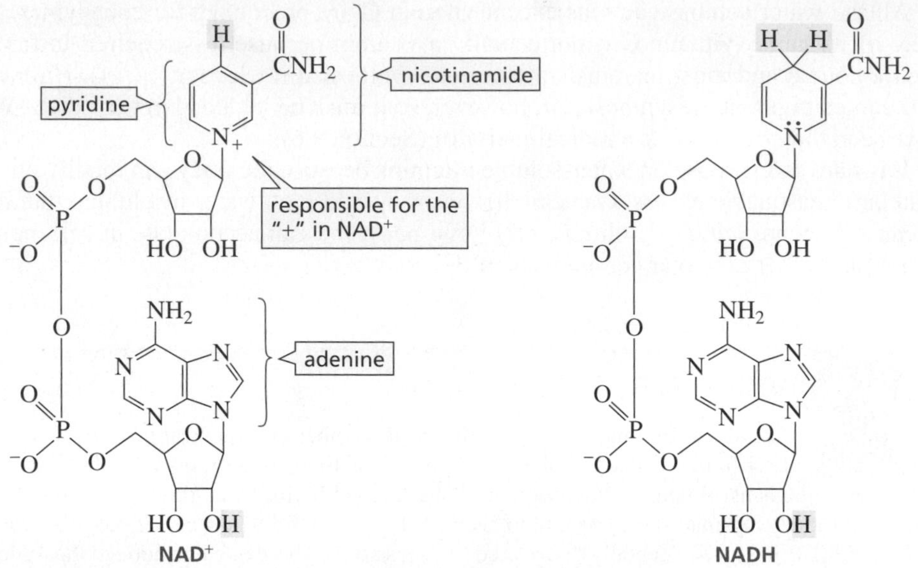

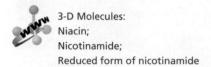

The structures labeled **NAD⁺** and **NADH** are shown at the top of the page, with labels for *pyridine*, *nicotinamide*, *adenine*, and the note "responsible for the '+' in NAD⁺".

3-D Molecules:
Niacin;
Nicotinamide;
Reduced form of nicotinamide

The heterocycle of one of the nucleotides of NAD⁺ is nicotinamide, and the heterocycle of the other is adenine. This accounts for the coenzyme's name (nicotinamide adenine dinucleotide). The positive charge in the name indicates the positively charged nitrogen of the pyridine ring. Niacin (nicotinic acid) is the portion of the coenzyme that the body cannot synthesize—it is a vitamin. Therefore, we must include niacin in our diets so the body can use it to synthesize NAD⁺.

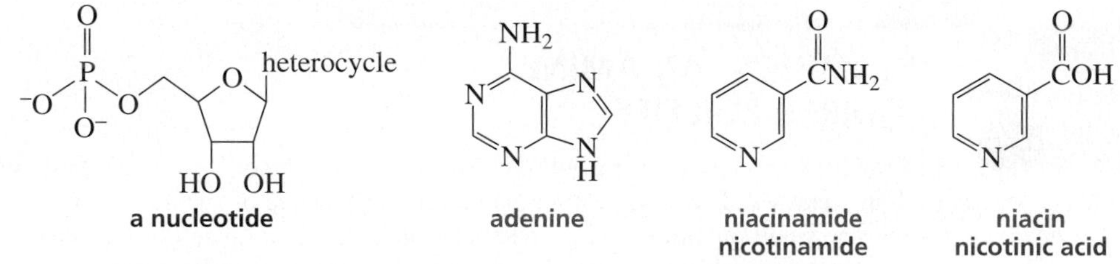

a nucleotide   **adenine**   **niacinamide**
                                     nicotinamide   **niacin**
                                                          nicotinic acid

The adenine nucleotide for the coenzyme is provided by ATP.

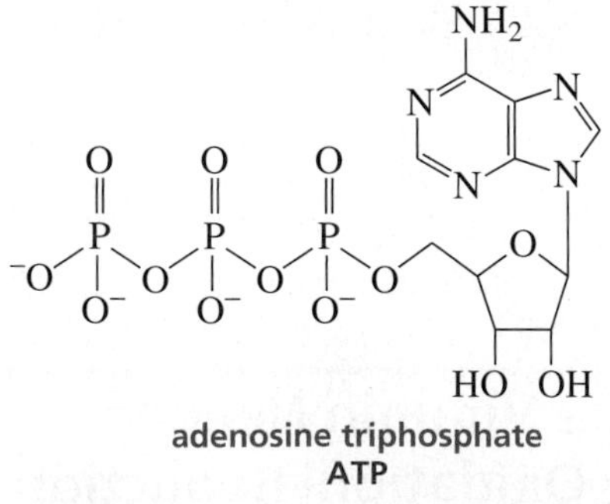

adenosine triphosphate
ATP

NIACIN

When bread companies started adding nicotinic acid to their bread, they insisted that its name be changed to niacin because nicotinic acid sounded too much like nicotine and they did not want their vitamin-enriched bread to be associated with a harmful substance.

NAD⁺ is an oxidizing agent.
NADH is a reducing agent.

When NAD⁺ oxidizes a substrate, the coenzyme is reduced to NADH. **NADH** is a reducing agent; it is used as a coenzyme by certain enzymes *that catalyze reduction reactions.*

$$\text{substrate}_{\text{reduced}} + \text{NAD}^+ \xrightleftharpoons{\text{enzyme}} \text{substrate}_{\text{oxidized}} + \text{NADH} + \text{H}^+$$

Malate dehydrogenase is an example of an enzyme that catalyzes an oxidation reaction; it catalyzes the oxidation of the *secondary alcohol group* of malate to a *ketone*. The enzyme requires the coenzyme NAD^+; NAD^+ is the oxidizing agent. Many enzymes that catalyze oxidation reactions are called **dehydrogenases**. Recall that the number of $C-H$ bonds decreases in an oxidation reaction (Section 11.4). Dehydrogenases oxidize substrates by removing hydrogen from them.

$$^-OCCH_2CH-CO^- + \boxed{NAD^+} \xrightleftharpoons[\text{dehydrogenase}]{\text{malate}} \ ^-OCCH_2C-CO^- + \boxed{NADH} + H^+$$

malate   oxaloacetate

How does this oxidation reaction take place? When a substrate is being *oxidized*, it donates a hydride ion (H^-) to the 4-position of the pyridine ring of NAD^+. The rest of the NAD^+ molecule is important for binding the coenzyme to the proper site on the enzyme. A basic amino acid side chain of the enzyme can help the reaction by removing a proton from the oxygen atom of the substrate.

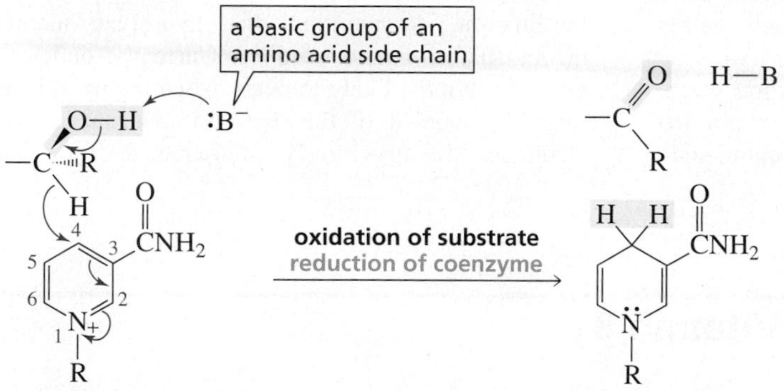

a basic group of an amino acid side chain

oxidation of substrate
reduction of coenzyme

PROBLEM 4◆

What is the product of the following reaction?

$$^-OCCHCHCH_2CO^- + NAD^+ \xrightarrow{\text{isocitrate dehydrogenase}}$$

OH

isocitrate

The mechanism for reduction by NADH is the reverse of the mechanism for oxidation by NAD^+. If a substrate is being *reduced*, the dihydropyridine ring of NADH donates a hydride ion from its 4-position to the substrate. An acidic amino acid side chain of the enzyme aids the reaction by donating a proton to the substrate.

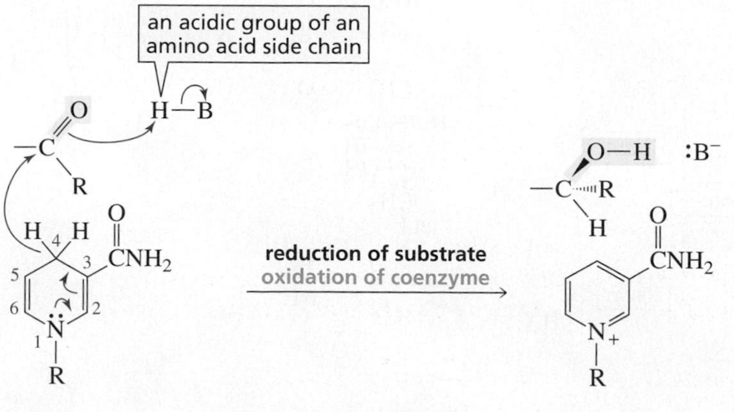

an acidic group of an amino acid side chain

reduction of substrate
oxidation of coenzyme

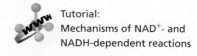

Tutorial:
Mechanisms of NAD^+- and NADH-dependent reactions

Because NADH reduces a compound by donating a hydride ion, it can be considered the biological equivalent of $NaBH_4$ or $LiAlH_4$—the hydride donors that we have seen used as reducing reagents in nonbiological reactions (Section 13.5).

PROBLEM 5◆

What is the product of the following reaction?

$$CH_3\overset{O}{\overset{\|}{C}}-\overset{O}{\overset{\|}{C}}O^- \xrightarrow[\text{NADH + H}^+]{\text{enzyme}}$$

NIACIN DEFICIENCY

A deficiency in niacin causes pellagra, a disease that begins with dermatitis and ultimately causes insanity and death. More than 120,000 cases of pellagra were reported in the United States in 1927, mainly in the rural south, where a corn-rich diet was prevalent. Although corn contains little niacin, it does contain nicotinamide; however, nicotinamide cannot be absorbed intestinally. The Mexican Indians soak corn in limewater—conditions that hydrolyze nicotinamide to niacin—before baking tortillas. Currently, deficiencies of the vitamin are most likely to occur when highly polished rice is a major component of the diet. Deficiencies are also seen in alcoholics who are severely malnourished.

18.6 Vitamin B₂

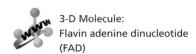

3-D Molecule:
Flavin adenine dinucleotide
(FAD)

Flavin adenine dinucleotide (FAD) is another coenzyme used *to oxidize substrates.* For example, FAD is the coenzyme used by succinate dehydrogenase to oxidize succinate to fumarate.

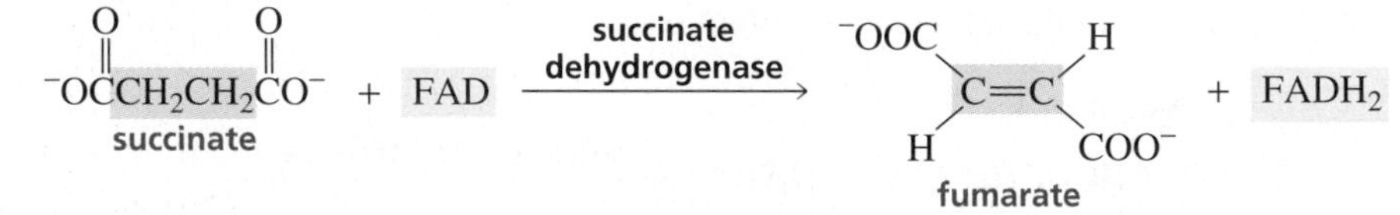

$$^-OCCH_2CH_2CO^- + \text{FAD} \xrightarrow{\substack{\text{succinate} \\ \text{dehydrogenase}}} \underset{\text{fumarate}}{{}^-OOC}\text{C=C}^{\text{H}}_{\text{COO}^-} + \text{FADH}_2$$

succinate

As its name indicates, FAD is a dinucleotide in which one of the heterocycles is flavin and the other is adenine. Notice that instead of ribose, the flavin nucleotide has a reduced ribose (ribitol). Flavin plus ribitol is the vitamin known as *riboflavin* or vitamin B₂. A vitamin B₂ deficiency causes inflammation of the skin.

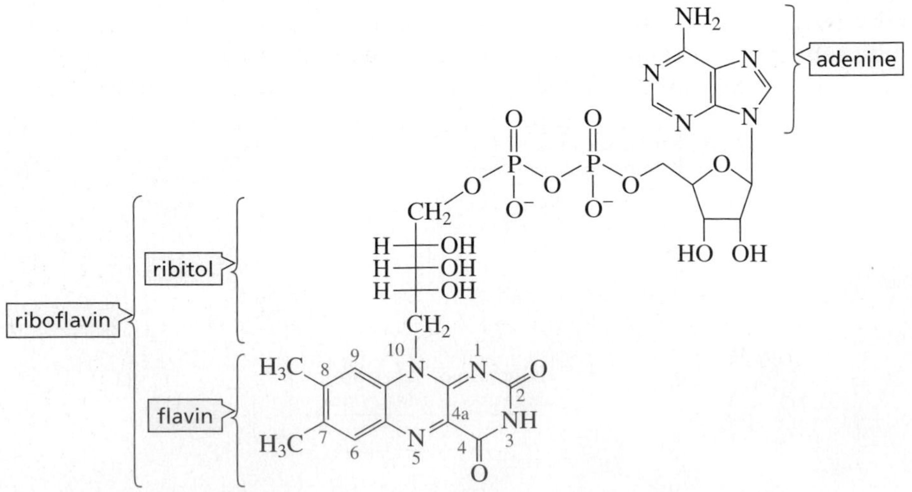

FAD

When FAD oxidizes a compound, FAD is reduced to FADH₂.

FAD is an oxidizing agent.

FADH₂ is a reducing agent.

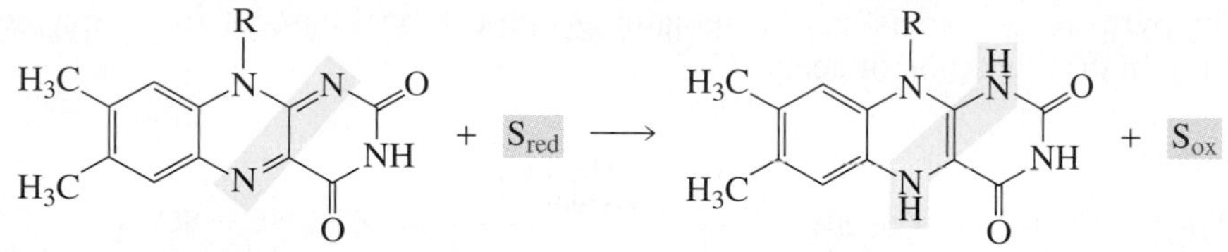

$$\text{FAD} + S_{red} \longrightarrow \text{FADH}_2 + S_{ox}$$

How can we tell which enzymes use FAD and which use NAD⁺ as the oxidizing coenzyme? A rough guideline is that NAD⁺ is the coenzyme used in enzyme-catalyzed oxidation reactions involving carbonyl compounds (alcohols being oxidized to ketones, aldehydes, or carboxylic acids), whereas FAD is the coenzyme used in other types of oxidations.

PROBLEM 6◆

How many conjugated double bonds are there in

a. FAD?   **b.** FADH₂?

PROBLEM 7◆

What is the product of the following reaction? (*Hint:* See Section 17.6).

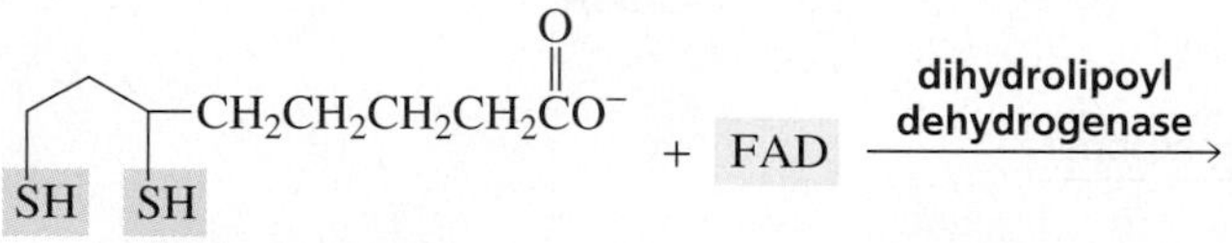

$$\text{dihydrolipoate} + \text{FAD} \xrightarrow{\text{dihydrolipoyl dehydrogenase}}$$

18.7 Vitamin B₁

Thiamine was the first of the B vitamins to be identified, so it became known as vitamin B₁ (Section 18.4). Vitamin B₁ is used to form the coenzyme **thiamine pyrophosphate (TPP)**. TPP is the coenzyme required by enzymes that catalyze *the transfer of a two-carbon fragment from one species to another.*

thiamine pyrophosphate
TPP

Thiamine pyrophosphate (TPP) is required by enzymes that catalyze the transfer of a two-carbon fragment from one species to another.

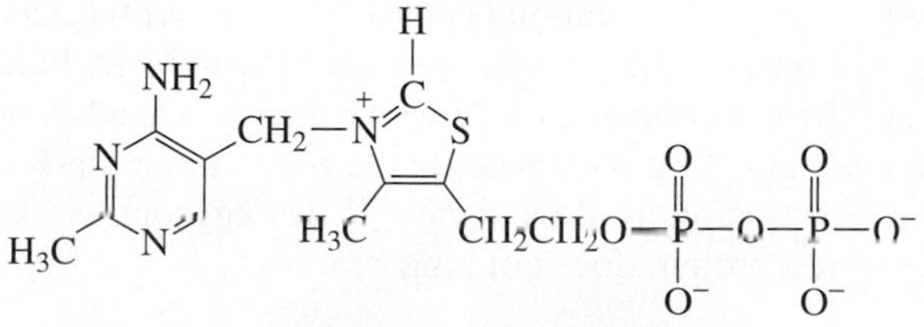

Pyruvate decarboxylase is an enzyme that requires thiamine pyrophosphate. Pyruvate decarboxylate catalyzes the decarboxylation of pyruvate and transfers the remaining two-carbon fragment to a proton, resulting in the formation of acetaldehyde.

$$\text{pyruvate} + \text{H}^+ \xrightarrow[\text{TPP}]{\text{pyruvate decarboxylase}} \text{acetaldehyde} + \text{CO}_2$$

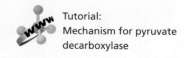

Tutorial:
Mechanism for pyruvate
decarboxylase

The *pyruvate dehydrogenase system* is a group of three enzymes and five coenzymes. One of the coenzymes is TPP. The overall reaction catalyzes the decarboxylation of pyruvate and transfers the remaining two-carbon fragment to coenzyme A, resulting in the formation of acetyl-CoA.

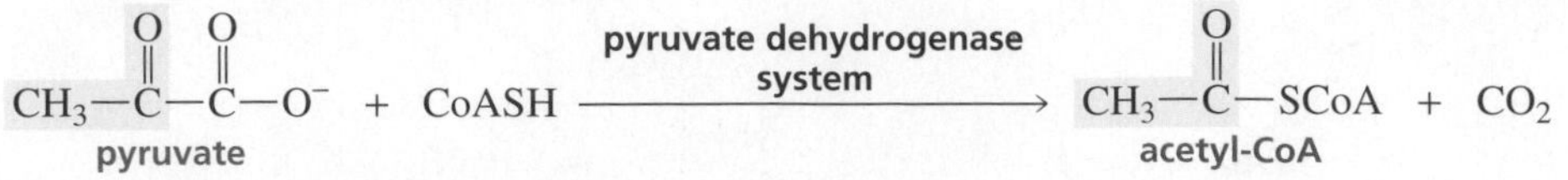

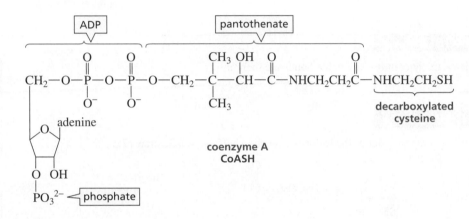

$$CH_3-\overset{O}{\underset{}{C}}-\overset{O}{\underset{}{C}}-O^- \ + \ CoASH \xrightarrow[\text{system}]{\text{pyruvate dehydrogenase}} CH_3-\overset{O}{\underset{}{C}}-SCoA \ + \ CO_2$$

pyruvate   acetyl-CoA

Coenzyme A (CoASH) is the coenzyme that activates a carboxylic acid by converting it into a thioester. Recall that carboxylate ions are very unreactive (Section 12.6). Thioesters are much more reactive because they have very good leaving groups. The vitamin needed to make CoASH is pantothenate.

Coenzyme A (CoASH) activates a carboxylic acid.

3-D Molecule:
Coenzyme A

ADP   pantothenate

$$CH_2-O-\overset{O}{\underset{O^-}{P}}-O-\overset{O}{\underset{O^-}{P}}-O-CH_2-\overset{CH_3}{\underset{CH_3}{C}}-\overset{OH}{\underset{}{CH}}-\overset{O}{\underset{}{C}}-NHCH_2CH_2\overset{O}{\underset{}{C}}-NHCH_2CH_2SH$$

adenine

O OH

$$PO_3{}^{2-}$$ phosphate

coenzyme A
CoASH

decarboxylated cysteine

PROBLEM 8♦

a. What two carbon fragment does pyruvate decarboxylase transfer to a proton?

b. What two carbon fragment does the pyruvate dehydrogenase system transfer to coenzyme A?

18.8 Vitamin H

3-D Molecule:
Biotin

Vitamin H (**biotin**) is an unusual vitamin because it can be synthesized by bacteria that live in the intestines. Consequently, biotin does not have to be included in our diet and deficiencies are rare. Biotin deficiencies, however, can be found in people who maintain a diet high in raw eggs. Egg whites contain a protein that binds biotin tightly and thereby prevents it from acting as a coenzyme. When eggs are cooked, the protein is denatured; the denatured protein does not bind biotin.

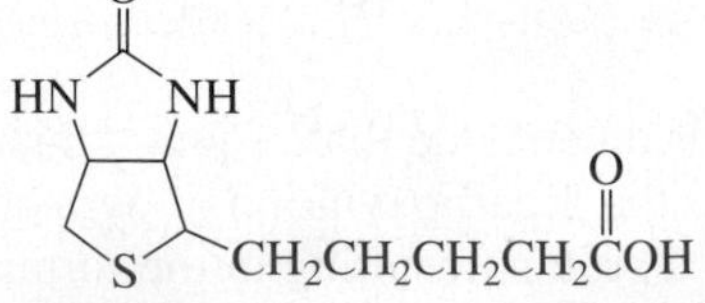

$$CH_2CH_2CH_2CH_2COH$$

biotin

Biotin is the coenzyme required by enzymes that catalyze *carboxylation of a carbon adjacent to a carbonyl group*. Therefore, the enzymes that require biotin as a coenzyme are called carboxylases. For example, acetyl-CoA carboxylase converts acetyl-CoA into malonyl-CoA. Biotin-requiring enzymes use bicarbonate (HCO_3^-) for the source of the carboxyl group that becomes attached to the substrate. They also require ATP and Mg^{2+}.

Biotin is required by enzymes that catalyze the carboxylation of a carbon adjacent to a carbonyl group.

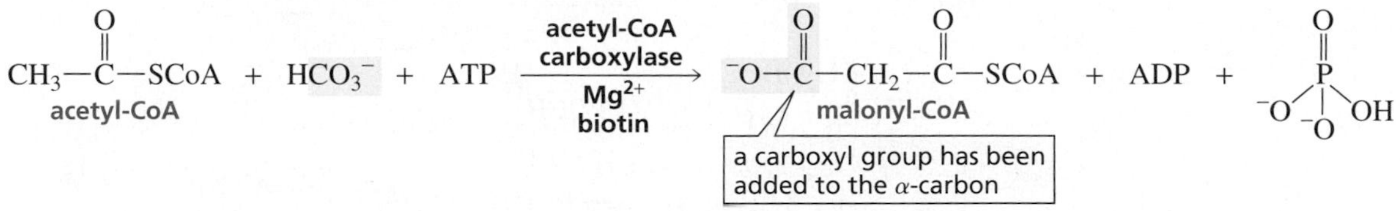

PROBLEM 9◆

Pyruvate carboxylase is a biotin-requiring enzyme. The enzyme's substrate is pyruvate. What is the product of the enzyme-catalyzed reaction?

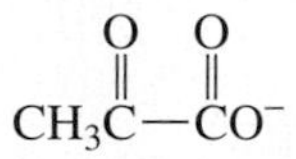

pyruvate

PROBLEM-SOLVING STRATEGY

How many moles of acetyl-CoA must be converted to malonyl-CoA in order to synthesize one mole of palmitic acid, a 16-carbon saturated fatty acid?

To answer this question, we need to recall how fatty acids are biosynthesized (Section 14.11). The biosynthesis starts with the reaction of a molecule of acetyl thioester with a molecule of malonyl thioester to form a four-carbon fatty acid. Each subsequent two-carbon unit is added by a molecule of malonyl thioester. Twelve more carbons are needed to form palmitic acid, so another six molecules of malonyl-CoA are required. Therefore, the synthesis of palmitic acid requires that 7 moles of acetyl-CoA be converted to malonyl-CoA.

Now continue on to Problem 10.

PROBLEM 10◆

How many moles of acetyl-CoA must be converted to malonyl-CoA in order to synthesize one mole of arachidic acid, a 20-carbon saturated fatty acid?

PROBLEM 11 SOLVED

How many moles of ATP are needed to make one mole of palmitic acid?

SOLUTION   One mole of a 16-carbon fatty acid is synthesized from one mole of acetyl-CoA and 7 moles of malonyl-CoA. Each mole of malonyl-CoA that is synthesized from acetyl-CoA requires one mole of ATP for the carboxylation reaction. Therefore, 7 moles of ATP are needed to make one mole of palmitic acid.

18.9 Vitamin B₆

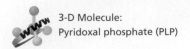

3-D Molecule:
Pyridoxal phosphate (PLP)

The coenzyme **pyridoxal phosphate (PLP)** is derived from the vitamin known as vitamin B_6 or pyridoxine. (Pyridoxal's "al" suffix indicates its aldehyde group.) A deficiency in vitamin B_6 causes anemia; severe deficiencies can cause seizures and death.

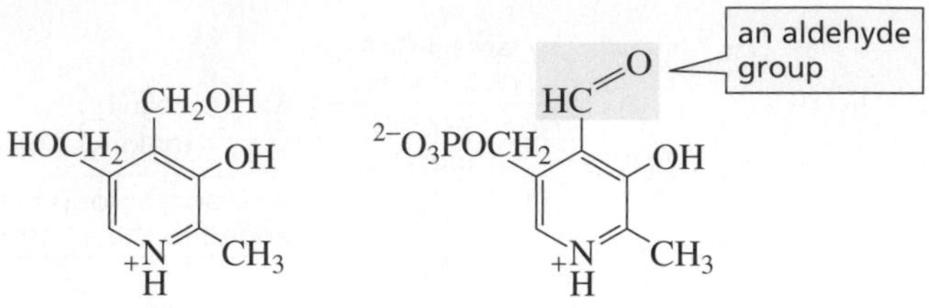

pyridoxine
vitamin B_6

an aldehyde group

pyridoxal phosphate
PLP

Pyridoxal phosphate (PLP) is required by enzymes that catalyze certain reactions of amino acids.

PLP is required by enzymes that catalyze certain reactions of amino acids. One of the most common of these reactions is transamination.

transamination

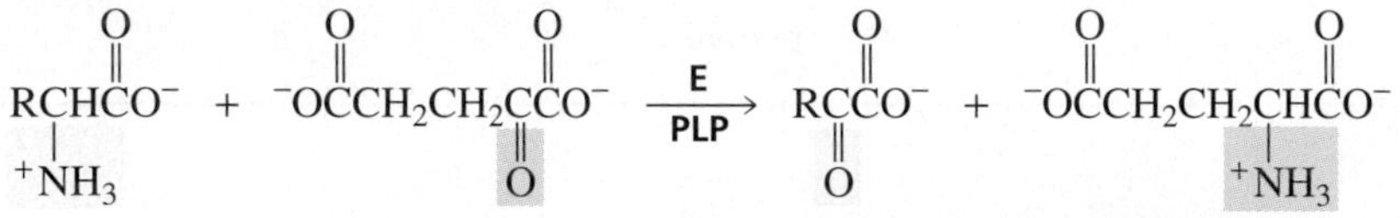

α-**ketoglutarate**   **glutamate**

The first reaction in the metabolism of most amino acids is the replacement of the amino group of the amino acid by a ketone group. This is called **transamination** because the amino group that is removed from the amino acid is not lost, but is *transferred* to the ketone group of α-ketoglutarate, thereby forming glutamate (an amino acid). The enzymes that catalyze transaminations are called *aminotransferases*. Transamination allows the amino groups of all the amino acids to be collected into a single amino acid (glutamate) so that excess nitrogen can be easily excreted.

PROBLEM 12◆

α-Keto acids other than α-ketoglutarate can be used to accept the amino group of amino acids in enzyme-catalyzed transaminations. What amino acids are formed from the following α-keto acids?

a. 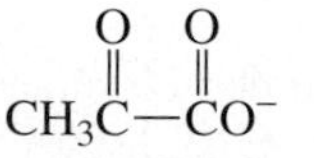 $CH_3\overset{O}{\underset{}{C}}{-}\overset{O}{\underset{}{C}}O^-$
pyruvate

b. 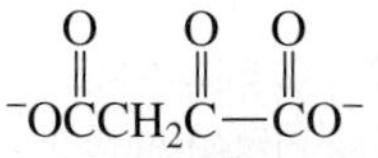 $^-O\overset{O}{\underset{}{C}}CH_2\overset{O}{\underset{}{C}}{-}\overset{O}{\underset{}{C}}O^-$
oxaloacetate

HEART ATTACKS: ASSESSING THE DAMAGE

After a heart attack, aminotransferases (and other enzymes) leak from the damaged cells of the heart into the bloodstream. The severity of the damage done to the heart can be determined from the concentrations of alanine aminotransferase and aspartate aminotransferase in the bloodstream.

18.10   Vitamin B$_{12}$

Enzymes that catalyze certain rearrangement reactions require **coenzyme B$_{12}$**, a coenzyme derived from vitamin B$_{12}$. The vitamin has a cyano group (or HO$^-$ or H$_2$O) coordinated with cobalt. In coenzyme B$_{12}$, this group is replaced by a 5′-deoxyadenosyl group.

3-D Molecules:
Coenzyme B$_{12}$,
Vitamin B$_{12}$

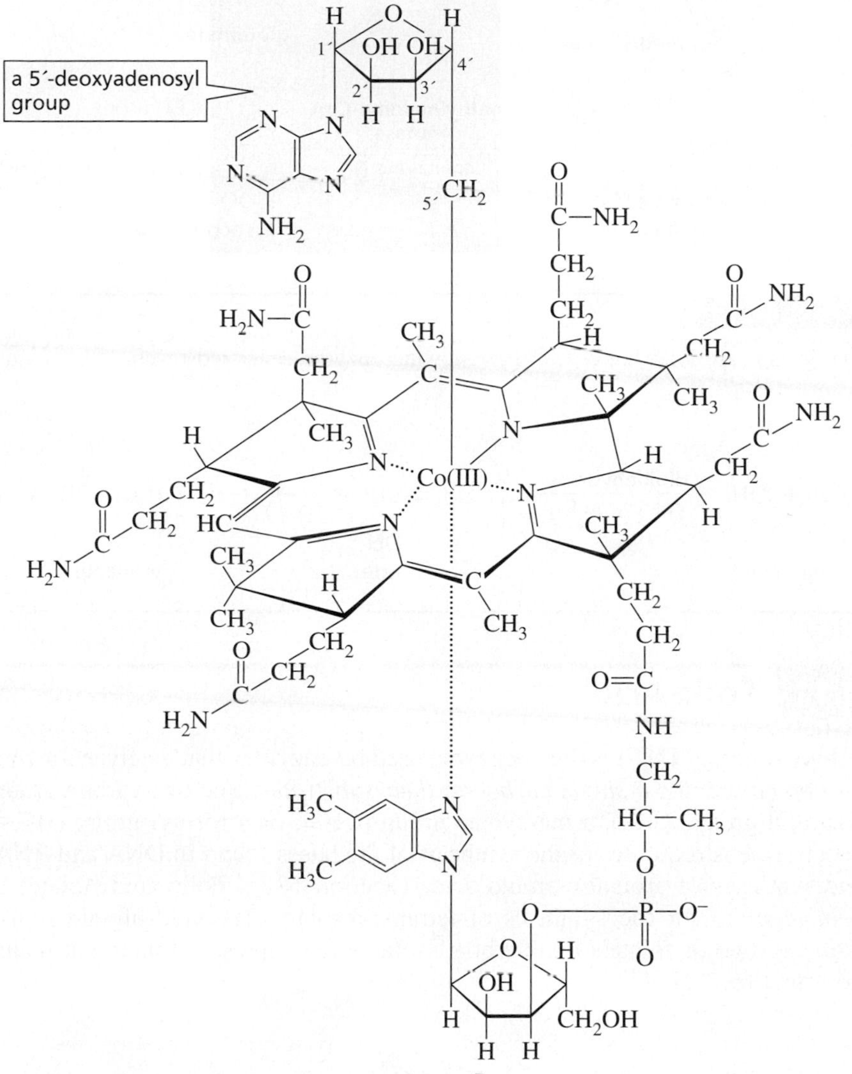

coenzyme B$_{12}$

Animals and plants cannot synthesize vitamin B$_{12}$. In fact, only a few microorganisms can synthesize it. Humans must obtain vitamin B$_{12}$ from their diet, particularly from meat. Because vitamin B$_{12}$ is needed in only very small amounts, deficiencies caused by insufficient consumption of the vitamin are rare, but they have been found in vegetarians who eat no animal products. Most deficiencies are caused by an inability to absorb the vitamin by the intestines. The deficiency causes pernicious anemia.

Coenzyme B$_{12}$ is required by enzymes that catalyze *reactions in which a group (Y) bonded to one carbon changes places with a hydrogen bonded to an adjacent carbon.*

Coenzyme B$_{12}$ is required by enzymes that catalyze the exchange of a hydrogen bonded to one carbon with a group bonded to an adjacent carbon.

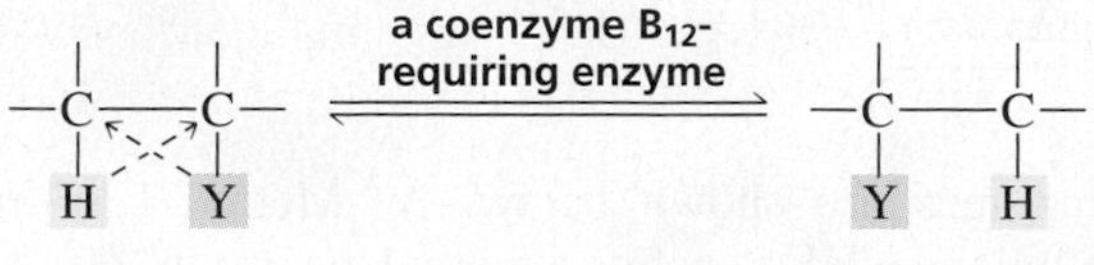

For example, in each of the following reactions, a COO^- group bonded to one carbon changes places with an H of an adjacent methyl group.

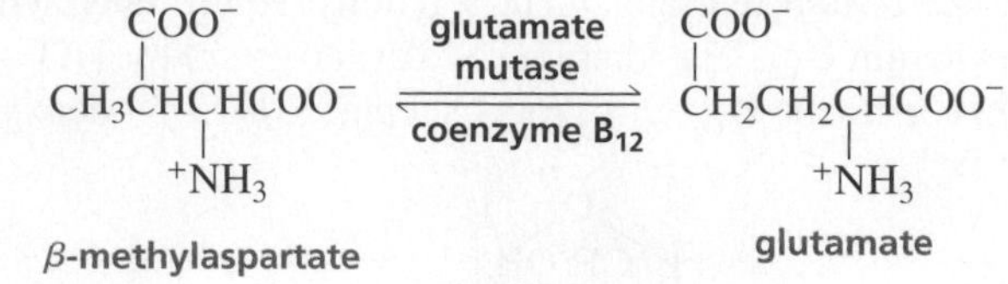

$$CH_3CHCHCOO^- \overset{\text{glutamate}}{\underset{\text{coenzyme B}_{12}}{\rightleftharpoons}} CH_2CH_2CHCOO^-$$

β-methylaspartate   **glutamate**

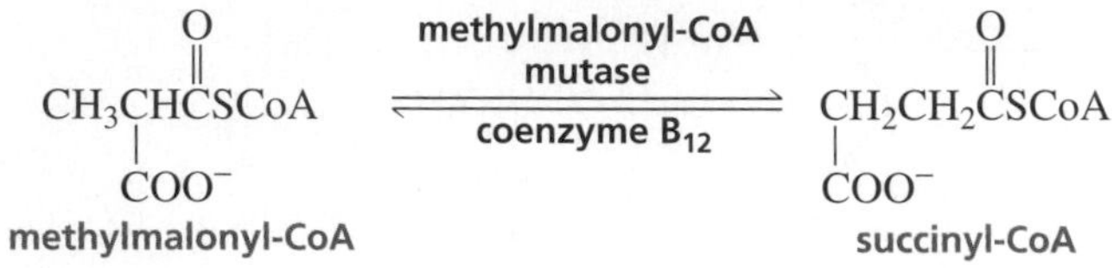

$$CH_3CHCSCoA \overset{\text{methylmalonyl-CoA}}{\underset{\text{coenzyme B}_{12}}{\overset{\text{mutase}}{\rightleftharpoons}}} CH_2CH_2CSCoA$$

methylmalonyl-CoA   **succinyl-CoA**

PROBLEM 13◆

What groups are interchanged in the following enzyme-catalyzed reaction that requires coenzyme B_{12}?

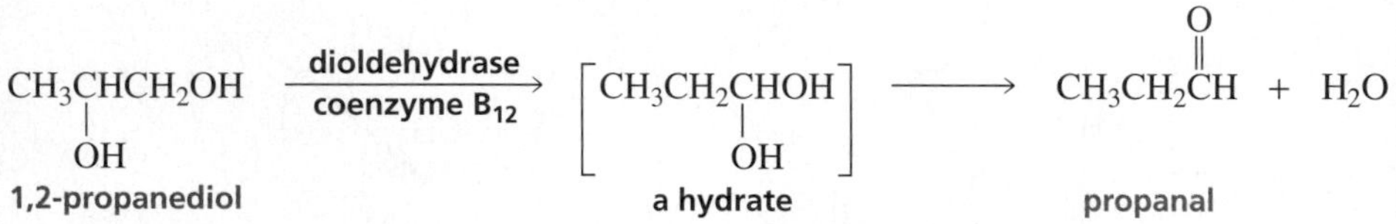

$$CH_3CHCH_2OH \overset{\text{dioldehydrase}}{\underset{\text{coenzyme B}_{12}}{\longrightarrow}} \left[CH_3CH_2CHOH \atop \quad\quad OH \right] \longrightarrow CH_3CH_2CH + H_2O$$

1,2-propanediol   **a hydrate**   **propanal**

18.11 Folic Acid

3-D Molecule:
Tetrahydrofolate (THF)

Tetrahydrofolate (THF) is the coenzyme required by enzymes that catalyze the transfer of a group containing one carbon to their substrates.

Tetrahydrofolate (THF) is the coenzyme used by enzymes that catalyze *the transfer of a group containing a single carbon to their substrates*. The one-carbon group can be a methyl group (CH_3), a methylene group (CH_2), or a formyl group ($HC=O$). The coenzyme is required for the synthesis of the bases found in DNA and RNA and for the synthesis of aromatic amino acids (Section 17.1). Folic acid (folate) is the vitamin required for the synthesis of tetrahydrofolate. Tetrahydrofolate is formed by reducing two of folate's double bonds. Bacteria synthesize folate, but mammals cannot (Section 22.4).

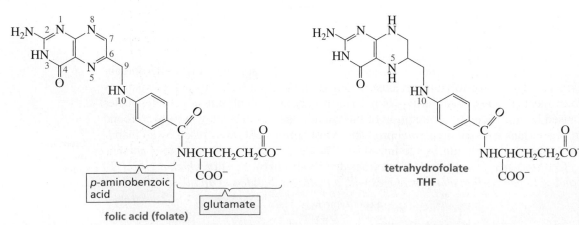

Three THF-coenzymes are shown below: N^5-Methyl-THF transfers a methyl group, N^5, N^{10}-methylene-THF transfers a methylene group, N^5, N^{10}-methenyl-THF transfers a formyl group.

N^5-methyl-THF   **N^5,N^{10}-methylene-THF**   **N^5,N^{10}-methenyl-THF**

PROBLEM 14◆

Why is the coenzyme called tetrahydrofolate?

The heterocyclic bases in RNA are adenine, guanine, cytosine, and uracil (A, G, C, and U); the heterocyclic bases in DNA are adenine, guanine, cytosine, and thymine (A, G, C, and T). In other words, the heterocyclic bases in RNA and DNA are the same, except that RNA contains U's, whereas DNA contains T's. Why DNA contains T's instead of U's is explained in Section 21.9.

Thymidylate synthase is the enzyme that catalyzes the synthesis of T's from U's. The enzyme requires N^5,N^{10}-methylene-THF as a coenzyme.

THF has been oxidized to DHF

dUMP   **N^5,N^{10}-methylene-THF**   thymidylate synthase   **dTMP**   **dihydrofolate DHF**

R′ = 2′-deoxyribose-5-phosphate

Even though the only structural difference between a T heterocyclic base and a U heterocyclic base is a *methyl* group, a T is synthesized by first transferring a *methylene* group to a U. The methylene group is then reduced to a methyl group. The coenzyme is the reducing agent. It, therefore, is oxidized (it loses hydrogens). The oxidized coenzyme is dihydrofolate.

Tutorial:
Mechanism for catalysis
by thymidylate synthase

CANCER CHEMOTHERAPY

Cancer is associated with rapidly growing and proliferating cells. Because cells cannot multiply if they cannot synthesize DNA, several cancer chemotherapeutic agents have been developed to inhibit thymidylate synthase. If a cell cannot make thymidine (T), it cannot synthesize DNA.

A common anticancer drug that inhibits thymidylate synthase is 5-fluorouracil. The enzyme reacts with 5-fluorouracil the same way it reacts with uracil. However, the fluorine substituent causes 5-fluoruracil to become permanently attached to the enzyme (because the base cannot remove a F^+ in an elimination reaction), blocking the active site of the enzyme so it can no longer bind uracil. Therefore, thymidine can no longer be synthesized, and without thymidine, DNA cannot be synthesized.

Unfortunately, most anticancer drugs cannot discriminate between diseased and normal cells. As a result, cancer chemotherapy is accompanied by terrible side effects. However, cancer

cells undergo uncontrolled cell division; thus, because they are dividing more rapidly than normal cells, they are harder hit by cancer-fighting chemotherapeutic agents.

5-fluorouracil 5-FU

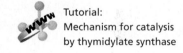

the enzyme has become irreversibly attached to the substrate

THE FIRST ANTIBIOTICS

Sulfonamides—commonly known as sulfa drugs—were introduced clinically in 1934 as the first effective antibiotics (Section 22.4). Notice that sulfanilamide is structurally similar to *p*-aminobenzoic acid.

Sulfanilamide acts by inhibiting the enzyme that incorporates *p*-aminobenzoic acid into folate. Because the enzyme cannot tell the difference between sulfanilamide and *p*-aminobenzoic acid, both compounds compete for the active site of the enzyme. Humans are not adversely affected by the drug because they do not synthesize folate—they get all their folate from their diets.

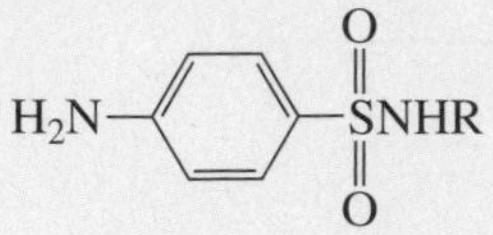

a sulfonamide 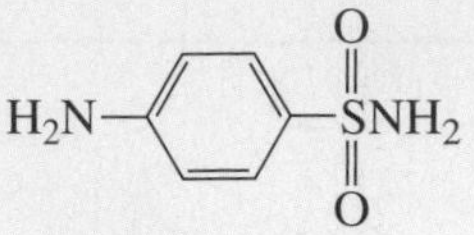   sulfanilamide 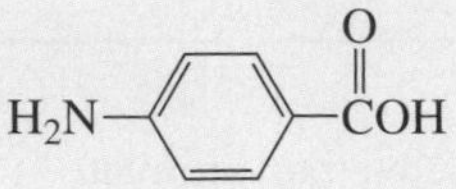   *p*-aminobenzoic acid

PROBLEM 15◆

What amino acid is formed by the following reaction?

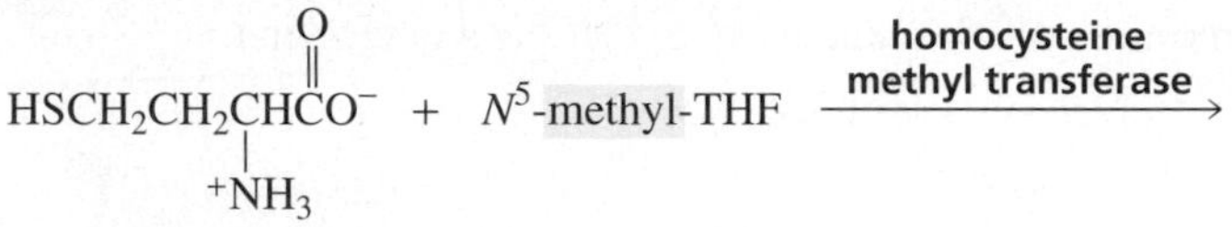

$$\text{HSCH}_2\text{CH}_2\overset{\overset{+}{\text{NH}_3}}{\underset{}{\text{CHCO}^-}} \;+\; N^5\text{-methyl-THF} \;\xrightarrow{\text{homocysteine methyl transferase}}$$

homocysteine

PROBLEM 16◆

What is the source of the methyl group in thymidine?

18.12 Vitamin K

3-D Molecule:
Vitamin K

Vitamin K is required for proper clotting of blood. The letter K comes from *koagulation*, which is German for "clotting." In order for blood to clot, blood-clotting proteins must bind Ca^{2+}. Vitamin K is required for proper Ca^{2+} binding. The vitamin is found in the leaves of green plants. Vitamin K deficiencies are rare because the vitamin is synthesized by intestinal bacteria. **Vitamin KH$_2$** is the coenzyme that is formed from vitamin K.

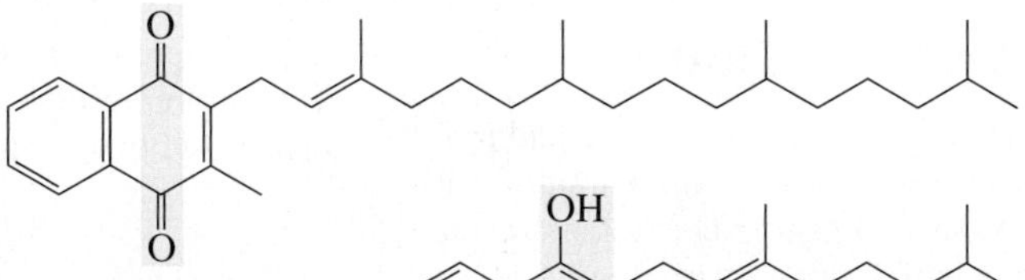

vitamin K

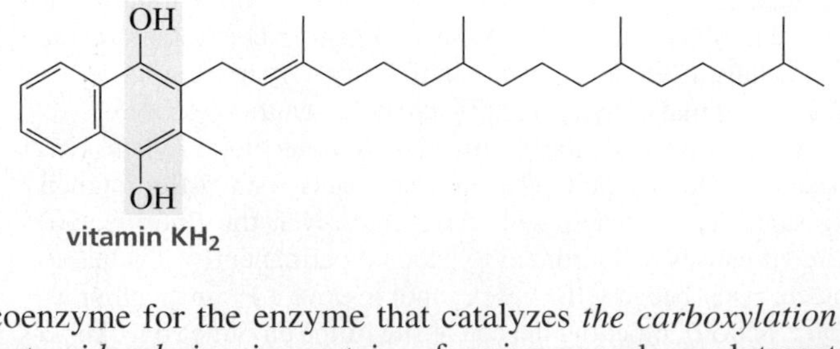

vitamin KH$_2$

Vitamin KH$_2$ is required by the enzyme that catalyzes the carboxylation of the γ-carbon group of a glutamate side chain in a protein.

Vitamin KH$_2$ is the coenzyme for the enzyme that catalyzes *the carboxylation of the γ-carbon of glutamate side chains in proteins*, forming γ-carboxyglutamates. γ-Carboxyglutamates complex Ca^{2+} much more effectively than glutamates do. The enzyme uses CO_2 for the carboxyl group that it puts on glutamate side chains. The proteins involved in blood clotting all have several glutamates near their N-terminal ends. For example, prothrombin, a blood-clotting protein, has glutamates at positions 7, 8, 15, 17, 20, 21, 26, 27, 30, and 33.

glutamate side chain

γ-carboxyglutamate side chain

calcium complex

Warfarin and dicoumerol are used clinically as anticoagulants. They prevent clotting by inhibiting the enzyme that synthesizes vitamin $\overline{K}H_2$ by binding to the enzyme's active site. Warfarin is also a common rat poison; it causes death by internal bleeding.

warfarin

dicoumarol

TOO MUCH BROCCOLI

An article describing two women with diseases characterized by abnormal blood clotting reported that they did not improve when they were given warfarin. When questioned about their diets, one woman said that she ate at least a pound (0.45 kg) of broccoli every day, and the other ate broccoli soup and a broccoli salad every day. When broccoli was removed from their diets, warfarin became effective in preventing the abnormal clotting of their blood. Because broccoli is high in vitamin K, these patients had been getting enough dietary vitamin K to compete with the drug for the enzyme's active site, thereby making the drug ineffective.

Summary

Essentially all organic reactions that occur in biological systems require a catalyst. Most biological catalysts are **enzymes**. The reactant of an enzyme-catalyzed reaction is called a **substrate**. The substrate specifically binds to the **active site** of the enzyme, and all the bond-making and bond-breaking steps of the reaction occur while the substrate is at that site. The remarkable catalytic ability of enzymes is due to reacting groups being brought together at the active site in the proper orientation for reaction and to catalytic groups being in the proper position needed for catalysis.

Cofactors assist enzymes in catalyzing a variety of reactions that cannot be catalyzed solely by their amino acid side chains. Cofactors can be metal ions or organic molecules. Cofactors that are organic molecules are called **coenzymes**; coenzymes are derived from **vitamins**. A vitamin is a substance the body cannot synthesize that is needed in small amounts for normal body function. All the water-soluble vitamins except vitamin C are precursors for coenzymes. Vitamin K is the only water-insoluble vitamin that is a precursor for a coenzyme. The vitamins that are precursors for coenzymes, the coenzymes, and the type of reaction that each catalyzes are summarized in Table 18.1.

The coenzymes used by enzymes to catalyze oxidation reactions are **NAD$^+$** and **FAD**; **NADH** and **FADH$_2$** catalyze reduction reactions. Many enzymes that catalyze oxidation reactions are called **dehydrogenases**. **Thiamine pyrophosphate (TPP)** is the coenzyme required by enzymes that catalyze the transfer of a two-carbon fragment. **Biotin** is the coenzyme required by enzymes that catalyze carboxylation of a carbon adjacent to a carbonyl group. **Pyridoxal phosphate (PLP)** is the coenzyme required by enzymes that catalyze transamination. In a **transamination**

reaction, the amino group is removed from an amino acid and transferred to another molecule, leaving a ketone group in its place.

Coenzyme B$_{12}$ is required by enzymes that catalyze reactions in which a group bonded to one carbon changes places with a hydrogen bonded to an adjacent carbon. **Tetrahydrofolate (THF)** is the coenzyme used by enzymes that catalyze the transfer of a group containing a single carbon—methyl, methylene, or formyl—to their substrates. **Vitamin KH$_2$** is the coenzyme for the enzyme that catalyzes the carboxylation of the γ-carbon of glutamate side chains— a reaction required for blood clotting.

Table 18.1   The Vitamins, Their Coenzymes, and the Reactions They Catalyze

Vitamin	Coenzyme	Reaction catalyzed	Human deficiency disease
Niacin	NAD$^+$	Oxidation	Pellagra
	NADH	Reduction	
Riboflavin (vitamin B$_2$)	FAD	Oxidation	Skin inflammation
	FADH$_2$	Reduction	
Thiamine (vitamin B$_1$)	Thiamine pyrophosphate (TPP)	Two-carbon transfer	Beriberi
Pantothenic acid	Coenzyme A (CoASH)	Activates carboxylic acids for acyl transfer	—
Biotin (vitamin H)	Biotin	Carboxylation	—
Pyridoxine (vitamin B$_6$)	Pyridoxal phosphate (PLP)	Transamination and other reactions of amino acids	Anemia
Vitamin B$_{12}$	Coenzyme B$_{12}$	Isomerization	Pernicious anemia
Folic acid	Tetrahydrofolate (THF)	One-carbon transfer	Megaloblastic anemia
Vitamin K	Vitamin KH$_2$	Carboxylation	Internal bleeding

Problems

17. From what vitamins are the following coenzymes derived?
    a. NAD$^+$ **b.** FAD   **c.** pyridoxal phosphate   **d.** N^5-methylenetetrahydrofolate

18. What is the product of the following reaction?

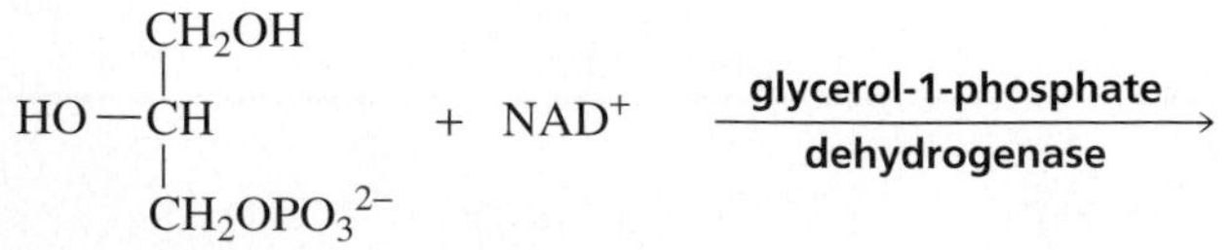

$$\begin{array}{c} CH_2OH \\ | \\ HO-CH \\ | \\ CH_2OPO_3{}^{2-} \end{array} \quad + \quad NAD^+ \quad \xrightarrow{\text{glycerol-1-phosphate dehydrogenase}}$$

19. Name two coenzymes that act as oxidizing agents.

20. Name the three one-carbon groups that various tetrahydrofolates are capable of donating to substrates.

21. What reaction is necessary for proper blood clotting that is catalyzed by vitamin KH$_2$?

22. What two coenzymes are used for carboxylation reactions?

23. For each of the following enzyme-catalyzed reactions, name the required coenzyme:

a. 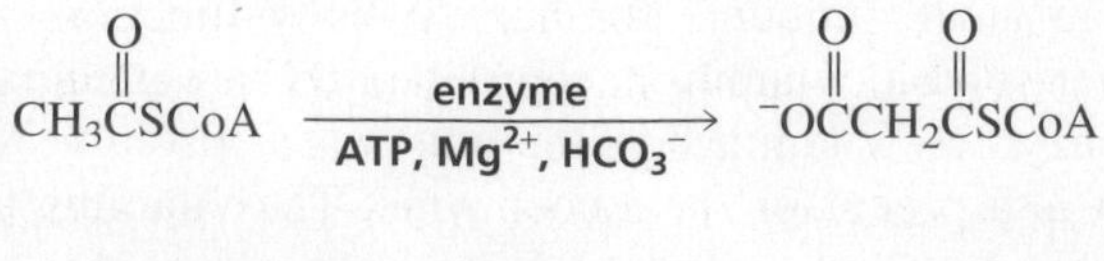 $CH_3\overset{O}{\overset{||}{C}}SCoA \xrightarrow[\text{ATP, Mg}^{2+}\text{, HCO}_3{}^-]{\text{enzyme}} {}^-O\overset{O}{\overset{||}{C}}CH_2\overset{O}{\overset{||}{C}}SCoA$

b. 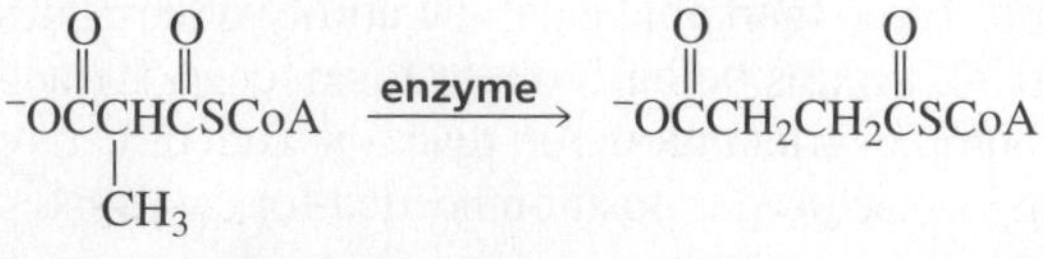 ${}^-O\overset{O}{\overset{||}{C}}\underset{\underset{CH_3}{|}}{C}H\overset{O}{\overset{||}{C}}SCoA \xrightarrow{\text{enzyme}} {}^-O\overset{O}{\overset{||}{C}}CH_2CH_2\overset{O}{\overset{||}{C}}SCoA$

24. What is the product of the following reaction?

$$RCH=CHCSR \ + \ FADH_2 \ \longrightarrow$$

(with C=O shown above the central carbon)

25. S-Adenosylmethionine (SAM) is formed from the reaction between ATP (Section 10.11) and methionine. The other product of the reaction is triphosphate. Propose a mechanism for this reaction. (*Hint:* it is an S_N2 reaction.)

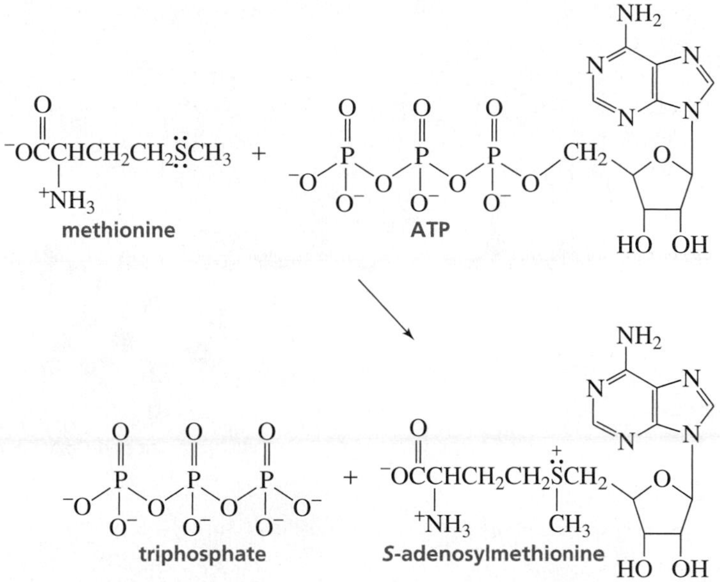

methionine   ATP

triphosphate   *S*-adenosylmethionine

26. For each of the following enzyme-catalyzed reactions, name the required coenzyme:

a. $^-OCCH_2CHCO^- \ + \ ^-OCCH_2CH_2CCO^- \ \xrightarrow{\textbf{enzyme}} \ ^-OCCH_2CCO^- \ + \ ^-OCCH_2CH_2CHCO^-$
(with $^+NH_3$ on first substrate, O on second; products: O on first, $^+NH_3$ on second)

b. $CH_3CH_2CSCoA \ \xrightarrow{\textbf{enzyme}} \ ^-OCCHCSCoA$
(with CH_3 substituent)

27. Which of the following compounds is more likely to loose CO_2?

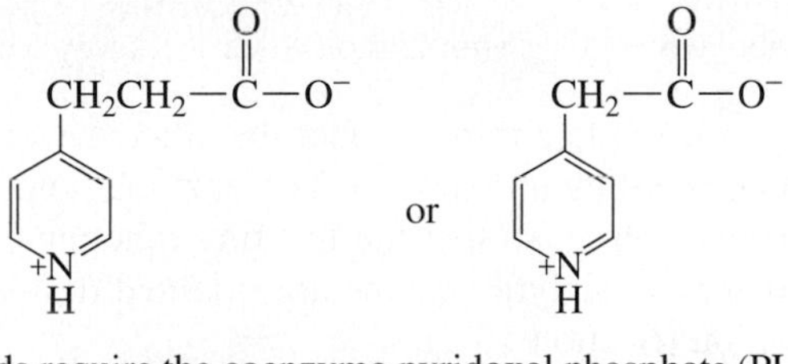

$$CH_2CH_2-C-O^- \qquad or \qquad CH_2-C-O^-$$

(each attached to a pyridinium ring with ^+N–H)

28. Enzymes that decarboxylate amino acids require the coenzyme pyridoxal phosphate (PLP).

$$RCH-C-O^- \ \xrightarrow[\text{PLP}]{\textbf{E}} \ RCH_2\overset{+}{N}H_3 \ + \ CO_2$$
(with $^+NH_3$ on the amino acid)

Show how the decarboxylation reaction occurs using the following information:
a. In the first step of the reaction, the amino acid forms an imine with pyridoxal phosphate.
b. The CO_2 is removed and the electrons it leaves behind are delocalized onto the positively charged nitrogen atom of the pyridine ring.
c. The nitrogen donates these electrons back into the ring, causing electrons to flow toward the α-carbon of the amino acid, which picks up a proton.
d. The imine undergoes acid-catalyzed hydrolysis.

29. Propose a mechanism for the aldolase-catalyzed cleavage of fructose-1,6 diphosphate if it did not form an imine with the substrate. What is the advantage gained by imine formation?

30. In glycolysis, why must glucose-6-phosphate isomerize to fructose-6-phosphate before the cleavage reaction with aldolase occurs?

19

The Chemistry of Metabolism

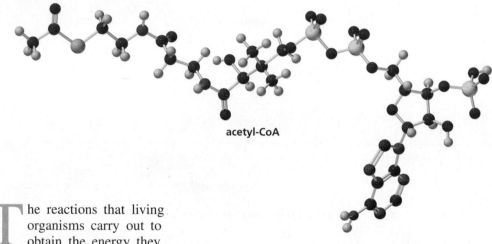

acetyl-CoA

The reactions that living organisms carry out to obtain the energy they need and to synthesize the compounds they require are collectively known as **metabolism**. Metabolism can be divided into two parts: *catabolism* and *anabolism*. **Catabolic reactions** break down complex nutrient molecules to provide energy and simple precursor molecules for synthesis. **Anabolic reactions** require energy and result in the synthesis of complex biomolecules from simpler precursor molecules.

catabolism: complex molecules $\longrightarrow$ simple molecules + energy

anabolism: simple molecules + energy $\longrightarrow$ complex molecules

It is important to remember that almost every reaction that occurs in a living system is catalyzed by an enzyme. The enzyme holds the reactants and any necessary coenzymes in place, so that the reacting functional groups and the amino acid side chains that act as catalytic groups are oriented in a way that will cause a specific chemical reaction (Section 18.1).

Most of the reactions that you will see in this chapter are reactions that you have studied in previous chapters. If you take the time to review these reactions, you will see that the organic reactions done by cells are the same as the organic reactions done by chemists.

DIFFERENCES IN METABOLISM

Humans do not necessarily metabolize compounds in the same way as do other species. This becomes a significant problem when drugs are tested on animals (Section 22.4). For example, chocolate is metabolized to different compounds in humans and in dogs; the metabolites formed by humans are nontoxic, whereas those formed by dogs can be highly toxic. Differences in metabolism have been found even within the same species. For example, isoniazid—an antituberculosis drug—is metabolized by Eskimos much faster than by Egyptians. Current research shows that men and women metabolize certain drugs differently. For example, kappa opioids—a class of painkillers—have been found to be about twice as effective in women than in men.

19.1   Digestion

The reactants required for all life processes ultimately come from our diet. In that sense, we really are what we eat. Catabolism can be divided into four stages (Figure 19.1). The *first stage* is called digestion. In this first stage, fats, carbohydrates, and proteins are hydrolyzed to fatty acids, monosaccharides, and amino acids, respectively. These reactions occur in the mouth, stomach, and small intestine.

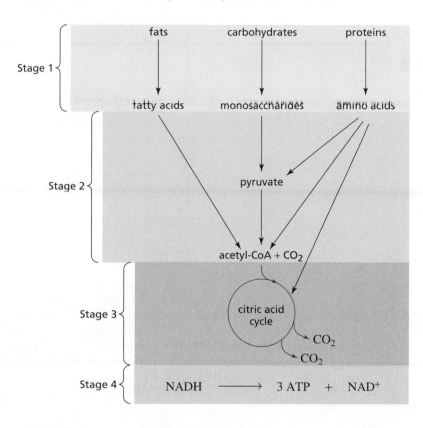

In the first stage of catabolism, fats, carbohydrates, and proteins are hydrolyzed to fatty acids, monosaccharides, and amino acids.

◀ **Figure 19.1**
The four stages of catabolism:
1. digestion
2. conversion of fatty acids, monosaccharides, and amino acids to compounds that can enter the citric acid cycle
3. the citric acid cycle
4. oxidative phosphorylation

You are what you eat.

In the *second stage of catabolism*, the products obtained from the first stage—fatty acids, monosaccharides, and amino acids—are converted to compounds that can enter the citric acid cycle. In order to enter the citric acid cycle, a compound must be either one of the compounds in the cycle itself (called a citric acid cycle intermediate), acetyl-CoA, or pyruvate. Acetyl CoA is the only non–citric acid cycle intermediate that can enter the cycle; pyruvate can enter the citric acid cycle only because it can be converted to acetyl-CoA. (This is the reaction catalyzed by the pyruvate dehydrogenase system that was discussed in Section 18.7).

In the second stage of catabolism, fats, carbohydrates, and proteins are converted to compounds that can enter the citric acid cycle (citric acid cycle intermediates, acetyl-CoA, and pyruvate).

The citric acid cycle constitutes the *third stage of catabolism*. In the citric acid cycle, the two carbons in the acetyl group of each molecule of acetyl-CoA—formed by the catabolism of fats, carbohydrates, and amino acids—are converted to two molecules of CO_2.

The citric acid cycle is the third stage of catabolism.

$$\underset{\textbf{acetyl-CoA}}{CH_3\overset{\overset{\textstyle O}{\|}}{C}SCoA} \longrightarrow 2\ CO_2\ +\ CoASH$$

Metabolic energy is measured in terms of adenosine triphosphate (ATP). Cells get the energy (ATP) they need by using nutrient molecules to make ATP. Only a little ATP is formed in the first three stages of catabolism. Most ATP is formed in the fourth stage of catabolism.

Metabolic energy is measured in terms of adenosine triphosphate (ATP).

We will see that many catabolic reactions are oxidation reactions. In the *fourth stage of catabolism*, every molecule of NADH that is formed as a result of NAD$^+$ being used to carry out an oxidation reaction in one of the earlier stages of catabolism is converted into three molecules of ATP in a process known as *oxidative phosphorylation*. In addition, oxidative phosphorylation converts every molecule of FADH$_2$ formed as a result of FAD carrying out an oxidation reaction into two molecules of ATP. Thus, the bulk of the energy (ATP) provided by fats, carbohydrates, and proteins is obtained in the fourth stage of catabolism.

Oxidative phosphorylation is the fourth stage of catabolism.

19.2 ATP

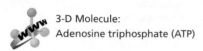

3-D Molecule:

Adenosine triphosphate (ATP)

ATP is enormously important to many biological reactions because these reactions could not occur without it. For example, the reaction of glucose with hydrogen phosphate to form glucose-6-phosphate does not occur because the 6-OH group of glucose would have to displace a very basic HO$^-$ group from hydrogen phosphate.

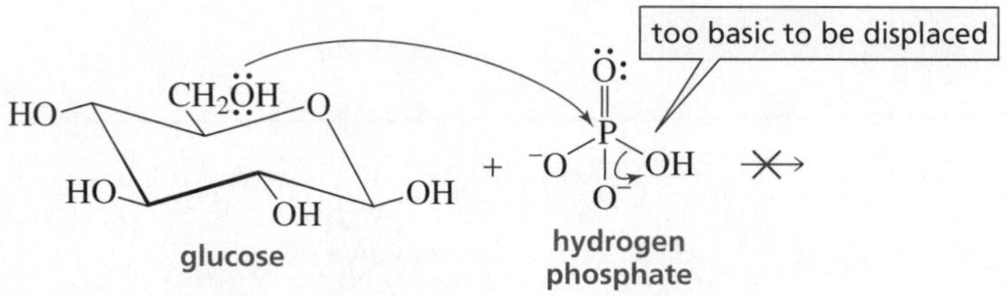

Tutorial:

Phosphoryl transfer reactions

If ATP is present, the 6-OH group of glucose attacks the terminal phosphate of ATP, breaking a **phosphoanhydride bond**. The phosphoanydride bond is a weaker bond than the P = O π bond, so the reaction is essentially an S$_N$2 reaction (Section 10.2). Because the phosphoanydride bond is weak, ADP is a very good leaving group.

ATP provides a reaction pathway involving a good leaving group for a reaction that cannot occur because of a poor leaving group.

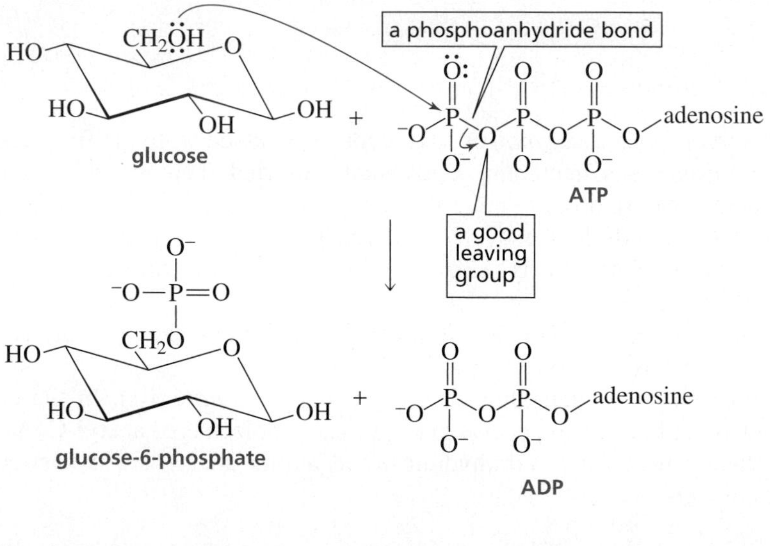

The transfer of a phosphoryl group from ATP to glucose is an example of a **phosphoryl transfer reaction**. There are many phosphoryl transfer reactions in biological systems. This example of a phosphoryl transfer reaction demonstrates the chemical function of ATP—*it provides a reaction pathway involving a good leaving group for a reaction that cannot occur (or would occur very slowly) because of a poor leaving group.*

PROBLEM 1 SOLVED

Why is ADP a weak base and, therefore, a good leaving group?

SOLUTION   When the bond attached to the leaving group breaks, the electrons are not localized on the departing oxygen atom. Instead, they are delocalized—they are shared by two oxygen atoms. Electron delocalization stabilizes the leaving group, and since stable bases are weak bases, they are good leaving groups.

ADP

THE NOBEL PRIZE

The Nobel Prize is considered by many to be the most coveted award a scientist can receive. The awards were established by **Alfred Bernhard Nobel (1833–1896)**. The first prizes were awarded in 1901. Nobel was born in Stockholm, Sweden. When he was nine, he moved with his parents to St. Petersburg, where his father manufactured the torpedoes and submarine mines he had invented for the Russian government. As a young man, Alfred did research on explosives in a factory his father owned near Stockholm. In 1864, an explosion in the factory killed his younger brother, causing Alfred to look for ways to make it safer to handle and transport explosives. The Swedish government would not allow the factory to be rebuilt because of the many accidents that had occurred there. Nobel, therefore, established an explosives factory in Germany, where, in 1867, he discovered that nitroglycerin could be mixed with diatomaceous earth, and the mixture could be molded into sticks that could not be set off without a detonating cap. Thus, Nobel invented dynamite. He also invented blasting gelatin and smokeless powder. Although he was the inventor of explosives used by the military, he was a strong supporter of peace movements.

The 355 patents Nobel held made him a wealthy man. He never married, and when he died, his will stipulated that the bulk of his estate ($9,200,000) be used to establish prizes to be awarded to those who "have conferred the greatest benefit on mankind." He instructed that the money be invested and the interest earned each year be divided into five equal portions "to be awarded to the persons having made the most important contributions in the fields of chemistry, physics, physiology or medicine, literature, and to the one who had done the most toward fostering fraternity among nations, the abolition of standing armies, and the holding and promotion of peace congresses."

Nobel also stipulated that, in awarding the prizes, no consideration be given to the nationality of the candidate, that each prize be shared by no more than three persons, and that no prize be awarded posthumously.

Nobel gave instructions that the prizes for chemistry and physics were to be awarded by the Royal Swedish Academy of Sciences, for physiology or medicine by the Karolinska Institute in Stockholm, for literature by the Swedish Academy, and for peace by a five-person committee selected by the Norwegian Parliament. The deliberations are secret, and the decisions cannot be appealed. In 1969, the Swedish Central Bank established a prize in economics in Nobel's honor. The recipient of this prize is selected by the Royal Swedish Academy of Sciences. On December 10—the anniversary of Nobel's death—the prizes are awarded in Stockholm, except for the peace prize, which is awarded in Oslo.

Alfred Bernhard Nobel

19.3 Catabolism of Fats

In the first stage of fat catabolism, the three ester groups of a fat are hydrolyzed to glycerol and three fatty acid molecules (Section 20.1).

$$
\begin{array}{ccc}
\text{CH}_2\text{O}-\overset{\overset{\displaystyle O}{\|}}{\text{C}}-\text{R}^1 & & \text{CH}_2\text{OH} \quad \text{R}^1-\overset{\overset{\displaystyle O}{\|}}{\text{C}}-\text{OH} \\
\text{CHO}-\overset{\overset{\displaystyle O}{\|}}{\text{C}}-\text{R}^2 \;+\; 3\,\text{H}_2\text{O} \;\longrightarrow\; & & \text{CHOH} \;+\; \text{R}^2-\overset{\overset{\displaystyle O}{\|}}{\text{C}}-\text{OH} \\
\text{CH}_2\text{O}-\overset{\overset{\displaystyle O}{\|}}{\text{C}}-\text{R}^3 & & \text{CH}_2\text{OH} \quad \text{R}^3-\overset{\overset{\displaystyle O}{\|}}{\text{C}}-\text{OH} \\
\textbf{a fat} & & \textbf{glycerol} \qquad \textbf{fatty acids}
\end{array}
$$

Glycerol reacts with ATP, in the same way that glucose reacts with ATP (Section 19.2), to form glycerol-3-phosphate. The enzyme that catalyzes this reaction is called glycerol kinase—a **kinase** is an enzyme that puts a phosphoryl group on its substrate. The secondary alcohol group of glycerol-3-phosphate is then oxidized by NAD^+ to a ketone. The enzyme that catalyzes this reaction is called glycerol phosphate dehydrogenase. Recall that NAD^+ is a common biological oxidizing agent and a **dehydrogenase** is an enzyme that oxidizes its substrate (Section 18.5). The product of the reaction, dihydroxyacetone phosphate, is one of the compounds in the glycolytic pathway, so it can enter that pathway to be broken down further (Section 19.4).

$$
\begin{array}{ccccc}
\text{CH}_2\text{OH} & \textbf{ATP} \quad \textbf{ADP} & \text{CH}_2\text{OH} & \textbf{NAD}^+ \quad \textbf{NADH, H}^+ & \text{CH}_2\text{OH} \\
\text{CHOH} & \xrightarrow{\;\;\textbf{glycerol}\;\;} & \text{CHOH} & \xrightarrow{\quad} & \text{C}=\text{O} \\
\text{CH}_2\text{OH} & \textbf{kinase} & \text{CH}_2\text{OPO}_3{}^{2-} & \begin{array}{c}\textbf{glycerol}\\\textbf{phosphate}\\\textbf{dehydrogenase}\end{array} & \text{CH}_2\text{OPO}_3{}^{2-} \\
\textbf{glycerol} & & \textbf{glycerol-3-phosphate} & & \begin{array}{c}\textbf{dihydroxyacetone}\\\textbf{phosphate}\end{array}
\end{array}
$$

a phosphoryl group

Notice how biochemical reactions are written. The structures of only the primary reactant and the primary product are shown. The structures of other reactants and products are abbreviated and shown on a curved arrow that intersects the reaction arrow.

PROBLEM 2

Give the mechanism for the reaction of glycerol with ATP to form glycerol-3-phosphate.

PROBLEM 3

The asymmetric center of glycerol-3-phosphate has the R configuration. Draw the structure of (R)-glycerol-3-phosphate.

Before a fatty acid can be metabolized, it must be activated. We have seen that one way carboxylic acids are activated in biological systems is by being converted into thioesters (Section 18.7). Thus, the fatty acid is activated by being converted into a fatty acyl-CoA.

$$
\begin{array}{ccc}
& \textbf{CoASH} & \\
& \textbf{ATP} \quad \textbf{ADP, HPO}_4{}^{2-} & \\
\text{RCH}_2\text{CH}_2\text{CH}_2\text{CH}_2\overset{\overset{\displaystyle O}{\|}}{\text{C}}\text{O}^- & \xrightarrow[\textbf{synthetase}]{\textbf{acyl-CoA}} & \text{RCH}_2\text{CH}_2\text{CH}_2\text{CH}_2\overset{\overset{\displaystyle O}{\|}}{\text{C}}\text{SCoA} \\
\textbf{a fatty acid} & & \begin{array}{c}\textbf{a fatty acyl-CoA}\\\textbf{a thioester}\end{array}
\end{array}
$$

The fatty acyl-CoA is converted to acetyl-CoA in a pathway called **β-oxidation**—a repetitive series of four reactions. Each passage through the four reactions removes two carbons from the fatty acyl-CoA by converting them into acetyl-CoA (Figure 19.2). Each of the four reactions is catalyzed by a different enzyme.

1 The first of the four reactions is an oxidation reaction that removes hydrogen from the α- and β-carbons, forming an α,β-unsaturated fatty acyl-CoA. The oxidizing agent is FAD, which is the oxidizing agent used for noncarbonyl oxidations (Section 18.6.). The enzyme that catalyzes this reaction has been found to be deficient in 10% of babies that experience sudden infant death syndrome (SIDS).

2 In the second reaction, water adds to the α,β-unsaturated fatty acyl-CoA (Section 13.8).

3 The third reaction is another oxidation reaction: NAD^+ oxidizes the secondary alcohol to a ketone.

4 The fourth reaction is the reverse of a Claisen condensation (Section 14.7). The final product is acetyl-CoA and a fatty acyl-CoA with *two fewer* carbons than the starting fatty acyl-CoA. The mechanism for this reaction is shown below.

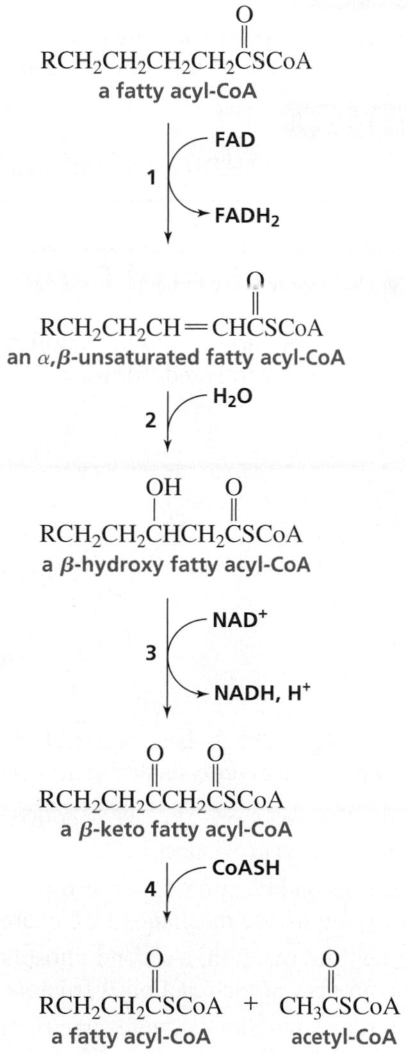

$$RCH_2CH_2CH_2CH_2CSCoA$$
a fatty acyl-CoA

1 $\quad$ FAD $\rightarrow$ FADH$_2$

$$RCH_2CH_2CH\!=\!CHCSCoA$$
an α,β-unsaturated fatty acyl-CoA

2 $\quad$ H$_2$O

$$\overset{OH}{\underset{}{RCH_2CH_2CHCH_2CSCoA}}$$
a β-hydroxy fatty acyl-CoA

3 $\quad$ NAD$^+$ $\rightarrow$ NADH, H$^+$

$$RCH_2CH_2CCH_2CSCoA$$
a β-keto fatty acyl-CoA

4 $\quad$ CoASH

$$RCH_2CH_2CSCoA \quad + \quad CH_3CSCoA$$
a fatty acyl-CoA $\qquad$ acetyl-CoA

◀ **Figure 19.2**
β-oxidation—a series of four enzyme-catalyzed reactions that are repeated until the entire fatty acyl-CoA molecule has been converted to acetyl-CoA molecules. The enzymes that catalyze the reactions are

1 acyl-CoA dehydrogenase
2 enoyl-CoA hydratase
3 3-L-hydroxyacyl-CoA dehydrogenase
4 β-ketoacyl-CoA thiolase

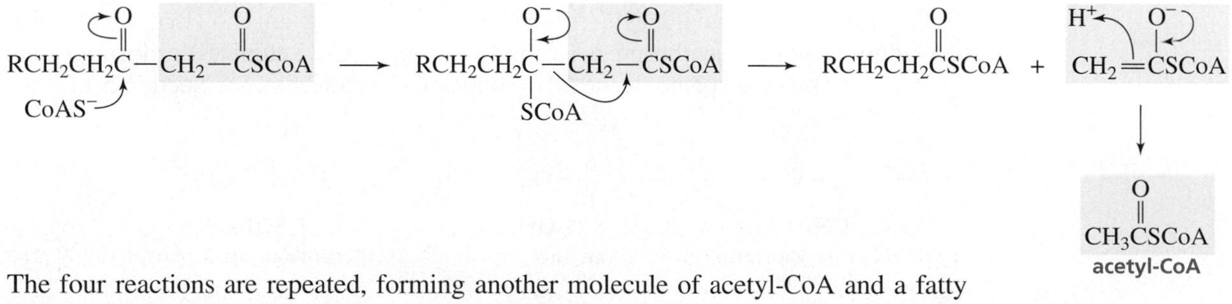

$$RCH_2CH_2\overset{O}{\overset{\|}{C}}-CH_2-\overset{O}{\overset{\|}{C}}SCoA \longrightarrow RCH_2CH_2\overset{O^-}{\underset{SCoA}{\overset{|}{C}}}-CH_2-\overset{O}{\overset{\|}{C}}SCoA \longrightarrow RCH_2CH_2CSCoA \ + \ CH_2=\overset{O^-}{\overset{|}{C}}SCoA$$
$$CoAS^-$$

$$\downarrow$$

$$CH_3CSCoA$$
acetyl-CoA

The four reactions are repeated, forming another molecule of acetyl-CoA and a fatty acyl-CoA that is now four carbons shorter than the original fatty acyl-CoA molecule. Each time the series of four reactions is repeated, two more carbons are removed (as acetyl-CoA) from the fatty acyl-CoA. The series of reactions is repeated until the entire fatty acid has been converted into acetyl-CoA molecules.

We will see that acetyl-CoA enters the citric acid cycle by reacting with oxaloacetate (a citric acid cycle intermediate) to form citrate, another citric acid cycle intermediate (Section 19.7).

Fatty acids are converted to molecules of acetyl-CoA.

PROBLEM 4♦

In the second reaction that occurs in the catabolism of fats, why does the OH group add to the β-carbon rather than to the α-carbon? (Hint: See Section 13.8.)

PROBLEM 5♦

Palmitic acid is a 16-carbon saturated fatty acid. How many moles of acetyl-CoA are formed from the catabolism of 1 mole of palmitic acid?

PROBLEM 6♦

How many moles of NADH are formed from the β-oxidation of 1 mole of palmitic acid?

19.4 Catabolism of Carbohydrates

In the first stage of carbohydrate catabolism, the acetal groups that hold glucose sub-units together are hydrolyzed, forming individual glucose molecules (Section 16.13).

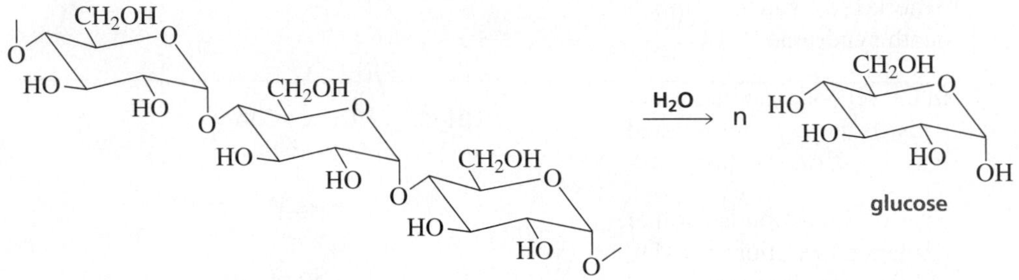

glucose

Glucose is converted to two molecules of pyruvate.

Each glucose molecule is converted to two molecules of pyruvate in a series of 10 reactions known as **glycolysis** or the **glycolytic pathway** (Figure 19.3).

1 In the first reaction, glucose is converted to glucose-6-phosphate, a reaction we just looked at in Section 19.2.

2 In the second reaction, glucose-6-phosphate isomerizes to fructose-6-phosphate, a reaction whose mechanism we examined in Section 18.2.

3 In the third reaction, a second phosphoryl group is put on fructose-6-phosphate. The product of the reaction is fructose-1,6-diphosphate.

4 The fourth reaction is the reverse of an aldol addition reaction. We looked at the mechanism of this reaction in Section 18.3.

5 Dihydroxyacetone phosphate, which is produced in the fourth reaction, is converted into glyceraldehyde-3-phosphate by forming an enol that can reform dihydroxyacetone phosphate (if the OH group at C-2 ketonizes) or form glyceraldehyde-3-phosphate (if the OH group at C-1 ketonizes); see Section 5.11.

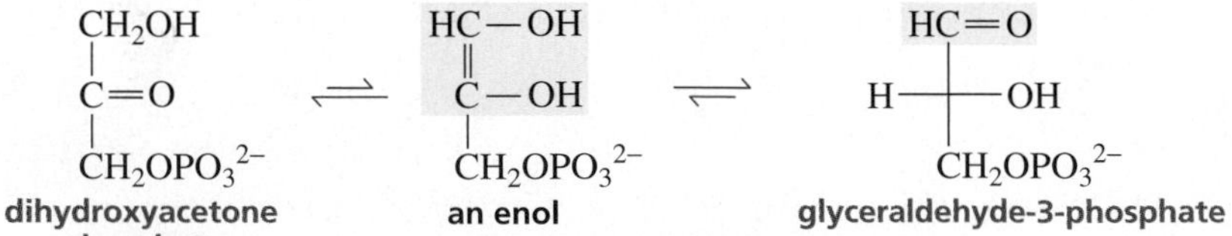

dihydroxyacetone phosphate   an enol   glyceraldehyde-3-phosphate

Overall, therefore, each molecule of D-glucose is converted to two molecules of glyceraldehyde-3-phosphate.

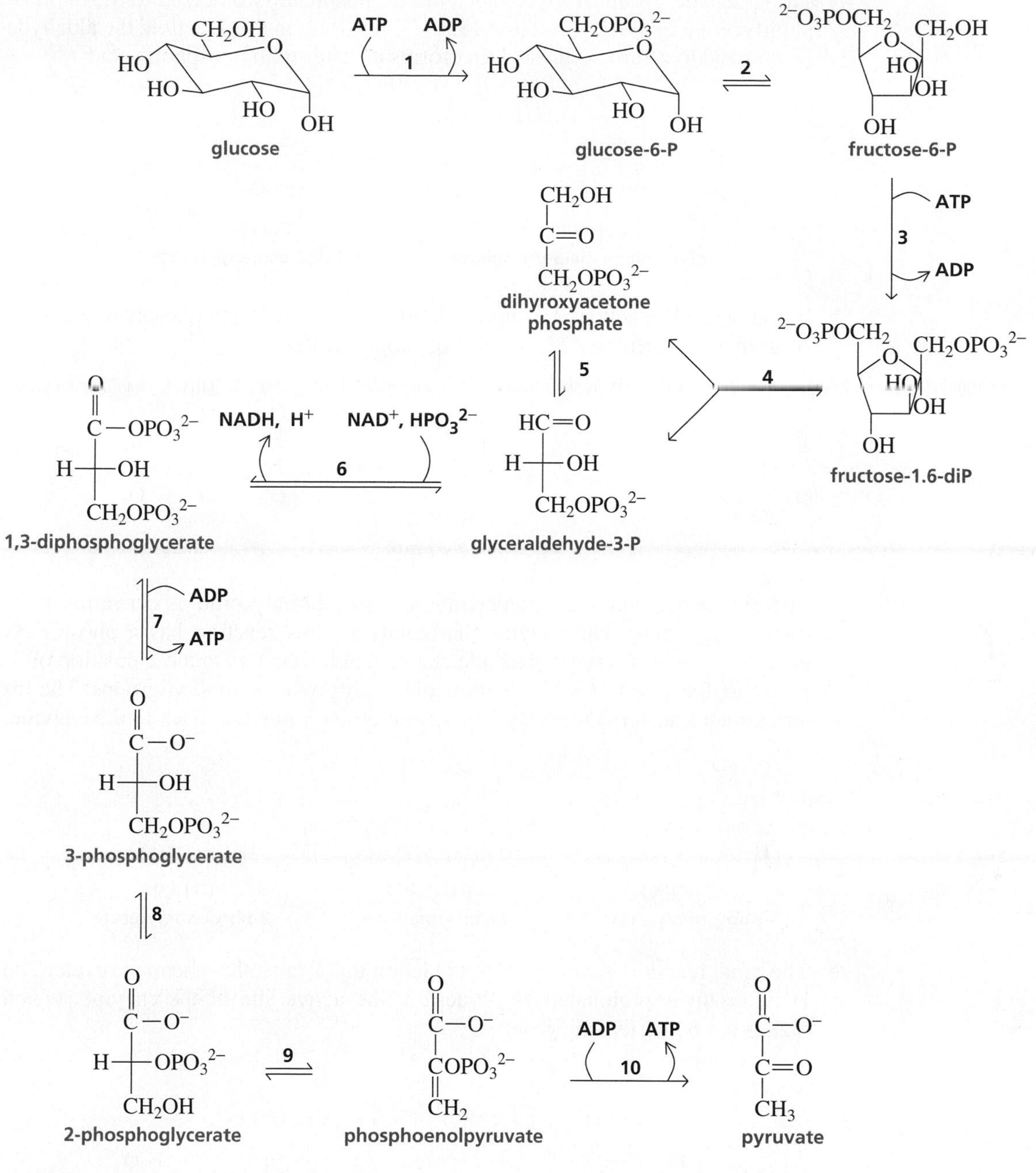

▲ Figure 19.3
Glycolysis—the series of enzyme-catalyzed reactions responsible for the conversion of
1 mole of glucose to 2 moles of pyruvate. The enzymes that catalyze the reactions are

1 hexokinase   **6** glyceraldehyde-3-phosphate dehydrogenase

2 phosphoglucose isomerase   **7** phosphoglycerate kinase

3 phosphofructokinase   **8** phosphoglycerate mutase

4 aldolase   **9** enolase

5 triose phosphate isomerase   **10** pyruvate kinase

6 The aldehyde group of glyceraldehyde-3-phosphate is oxidized to 1,3-diphosphoglycerate by NAD^+ (Section 18.5). Notice that, in this reaction, the aldehyde is oxidized to a carboxylic acid that forms an ester with phosphoric acid.

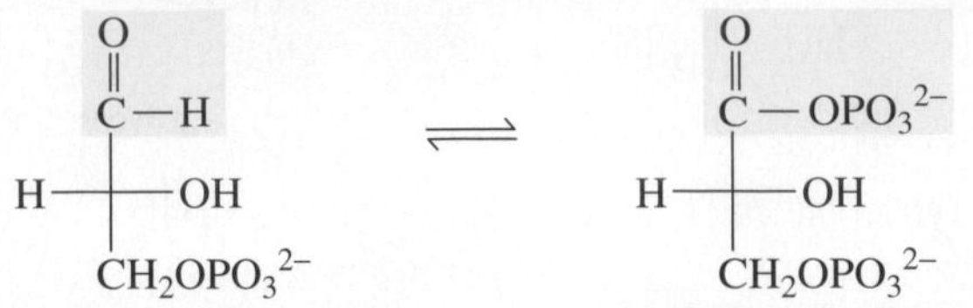

D-glyceraldehyde-3-phosphate   1,3-diphosphoglycerate

7 In the seventh reaction, 1,3-diphosphoglycerate transfers a phosphate group to ADP thereby forming ATP and 3-phosphoglycerate.

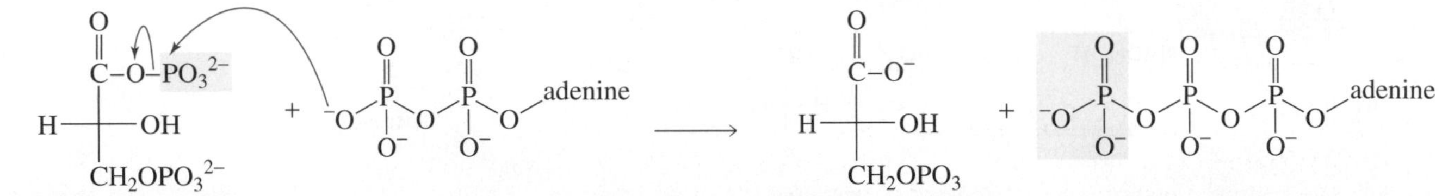

1,3-diphosphoglycerate   ADP   3-phosphoglycerate   ATP

8 The eighth reaction is an isomerization: 3-phosphoglycerate is converted to 2-phosphoglycerate. The enzyme that catalyzes this reaction has a phosphoryl group on one of its amino acid side chains that it transfers to the 2-position of 3-phosphoglycerate to form an intermediate with two phosphoryl groups. The intermediate transfers the phosphoryl group on its 3-position back to the enzyme.

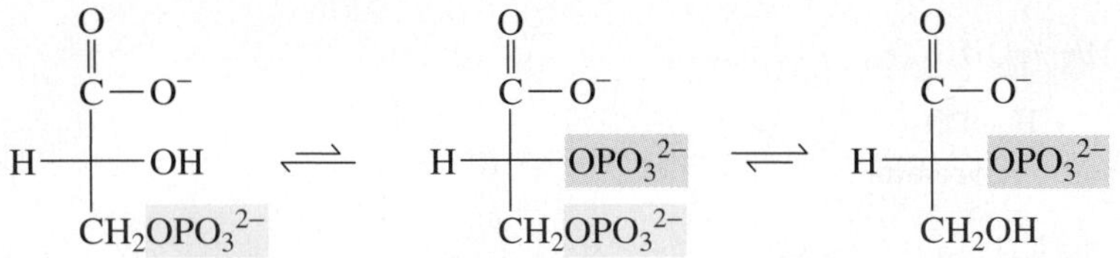

3-phosphoglycerate   an intermediate   2-phosphoglycerate

9 The ninth reaction is a dehydration reaction that forms phosphenolpyruvate. The HO^- group is protonated by an acid at the active site of the enzyme, which makes it a better leaving group.

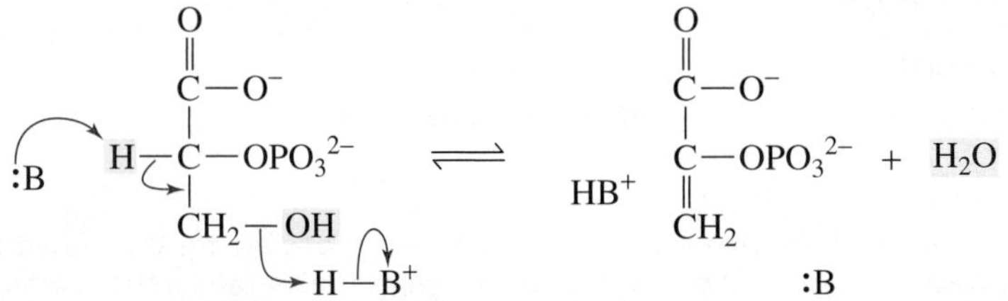

2-phosphoglycerate   2-phosphoenolpyruvate

10 In the last reaction of the glycolytic pathway, phosphenolpyruvate transfers its phosphate group to ADP, forming ATP and pyruvate.

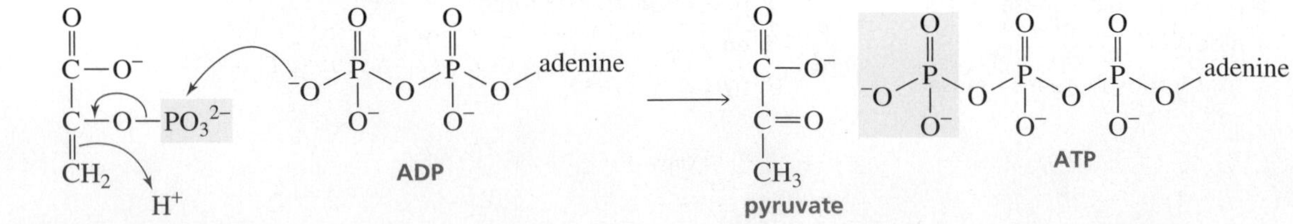

2-phosphoenolpyruvate   ADP   pyruvate   ATP

PROBLEM 7

Give the mechanism for the third reaction in glycolysis—the formation of fructose-1,6-diphosphate from the reaction of fructose-6-diphosphate with ATP.

PROBLEM 8

a. Which steps in glycolysis require ATP?

b. Which steps in glycolysis produce ATP?

PROBLEM-SOLVING STRATEGY

How many molecules of ATP are obtained from each molecule of glucose that is metabolized to pyruvate?

In solving this problem, we first need to count the number of ATPs that are used in the conversion of glucose to pyruvate. We see that there are two: One is used to form glucose-1-phosphate, and the other to form fructose,1,6-diphosphate. Next, we need to see how many ATPs are formed. Each glyceraldehyde-3-phosphate that is metabolized to pyruvate forms two ATPs. Because each glucose molecule forms two molecules of glyceraldehyde-3-phosphate, four ATPs are formed from each molecule of glucose. Therefore, each molecule of glucose that is metabolized to pyruvate forms two molecules of ATP.

Now continue on to Problem 9.

PROBLEM 9◆

How many moles of NAD^+ are required to convert one mole of glucose to pyruvate?

19.5   The Fates of Pyruvate

We have just seen that NAD^+ is used as an oxidizing agent in glycolysis. If glycolysis is to continue, the NADH produced as a result of the oxidation reaction has to be oxidized back to NAD^+ so NAD^+ will continue to be available as an oxidizing agent.

If oxygen is present, oxygen is the oxidizing agent used to oxidize NADH back to NAD^+; this happens *in the fourth stage of catabolism*. If oxygen is not present—for example, in muscle cells when all the oxygen has been depleted—pyruvate (the product of glycolysis) is used to oxidize NADH back to NAD^+. Pyruvate, therefore, is reduced to lactate (lactic acid). The acidic conditions caused by a build up of lactic acid in muscles is responsible for the burning sensation experienced when exercising.

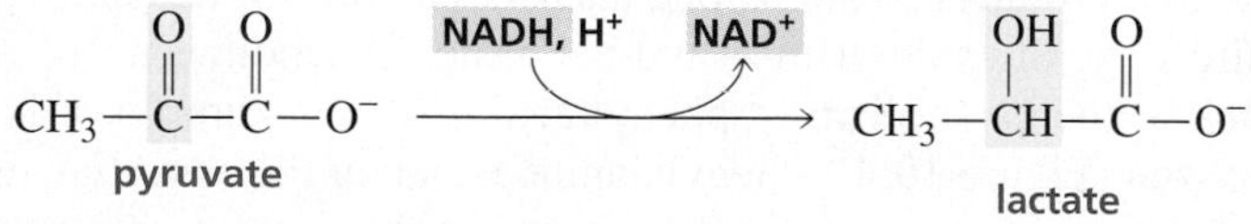

Under normal (aerobic) conditions, when oxygen rather than pyruvate is used to oxidize NADH to NAD^+, pyruvate is converted to acetyl-CoA so it can enter the citric acid cycle. This reaction is catalyzed by the pyruvate dehydrogenase system, a series of reactions that requires three enzymes and five coenzymes. We have seen that one of the coenzymes is thiamine pyrophosphate, the coenzyme required by enzymes that catalyze the transfer of a two-carbon fragment from one species to another (Section 18.7); in this reaction, the two carbons of the acetyl group of pyruvate are transferred to coenzyme A.

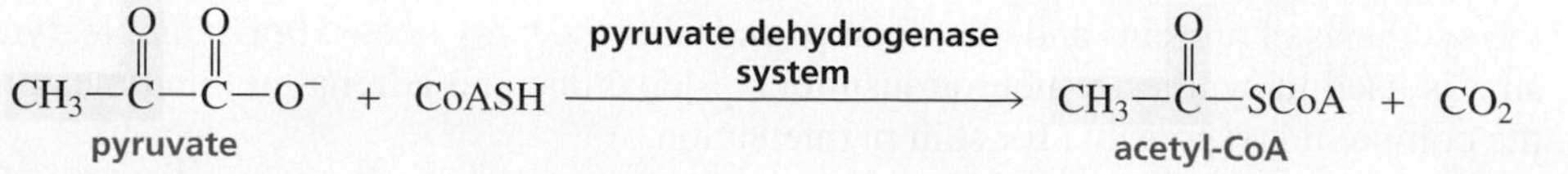

Under anaerobic (no oxygen) conditions, we have just seen that pyruvate is reduced to lactate. Under anaerobic conditions in yeast however, pyruvate has a different fate; it is decarboxylated to acetaldehyde by pyruvate decarboxylase, an enzyme we looked at in Section 18.7. Acetaldehyde oxidizes NADH back to NAD$^+$ Acetaldehyde, therefore, is reduced to ethanol—a reaction that has been used by mankind for thousands of years.

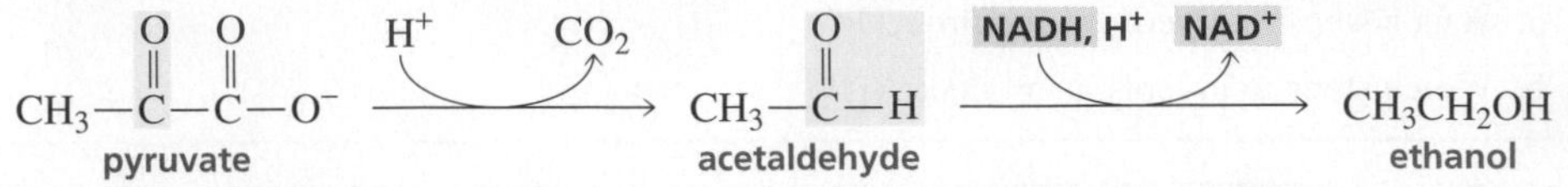

PROBLEM 10◆

What functional group of pyruvate is reduced when it is converted to lactate?

PROBLEM 11◆

What coenzyme is required to convert pyruvate to acetaldehyde?

PROBLEM 12◆

What functional group of acetaldehyde is reduced when it is converted to ethanol?

PROBLEM 13

Give the mechanism for the reduction of acetaldehyde by NADH to ethanol. (*Hint:* See Section 18.5.)

19.6 Catabolism of Proteins

In the first stage of protein catabolism, proteins are hydrolyzed to amino acids.

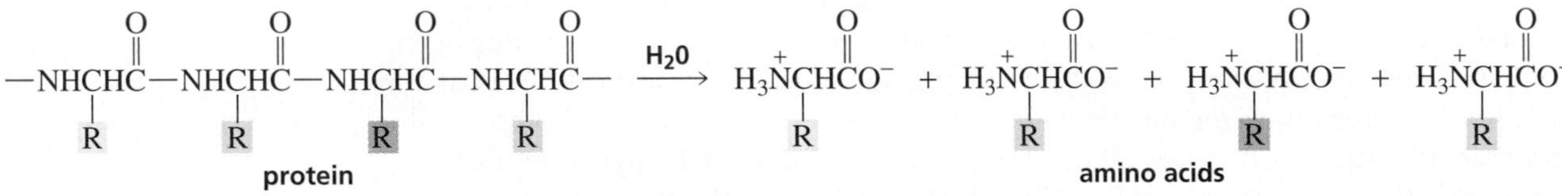

Amino acids are converted to acetyl-CoA, pyruvate, or citric acid cycle intermediates.

The amino acids enter the second stage of catabolism where they are converted to acetyl-CoA, pyruvate, or citric acid cycle intermediates, depending on the amino acid. Therefore, the final products of the second stage of catabolism can enter the citric acid cycle—the third stage of catabolism— and be further metabolized.

We will look at the catabolism of phenylalanine as an example of how an amino acid is metabolized (Figure 19.4). Phenylalanine is one of the essential amino acids; it must be included in our diet (Section 17.1). Tyrosine is not an essential amino acid because the enzyme phenylalanine hydroxylase converts phenylalanine into tyrosine.

We have seen that the first reaction in the catabolism of most amino acids is transamination—replacement of the amino group of the amino acid by a ketone group (Section 18.9). *para*-Hyroxyphenylpyruvate, the product of transamination, is converted by a series of reactions to fumarate and acetyl-CoA. Fumarate is a citric acid cycle intermediate so it can directly enter the citric acid cycle, and acetyl-CoA gets into the cycle by reacting with oxaloacetate to form citrate (Section 19.6). Each of the reactions in this catabolic pathway is catalyzed by a different enzyme.

The amino acids that we ingest are used not only for energy. They also are used for the synthesis of proteins and other compounds that the body needs. For example, tyrosine is used to synthesize neurotransmitters (dopamine and adrenaline) and melanin, the compound responsible for skin pigmentation.

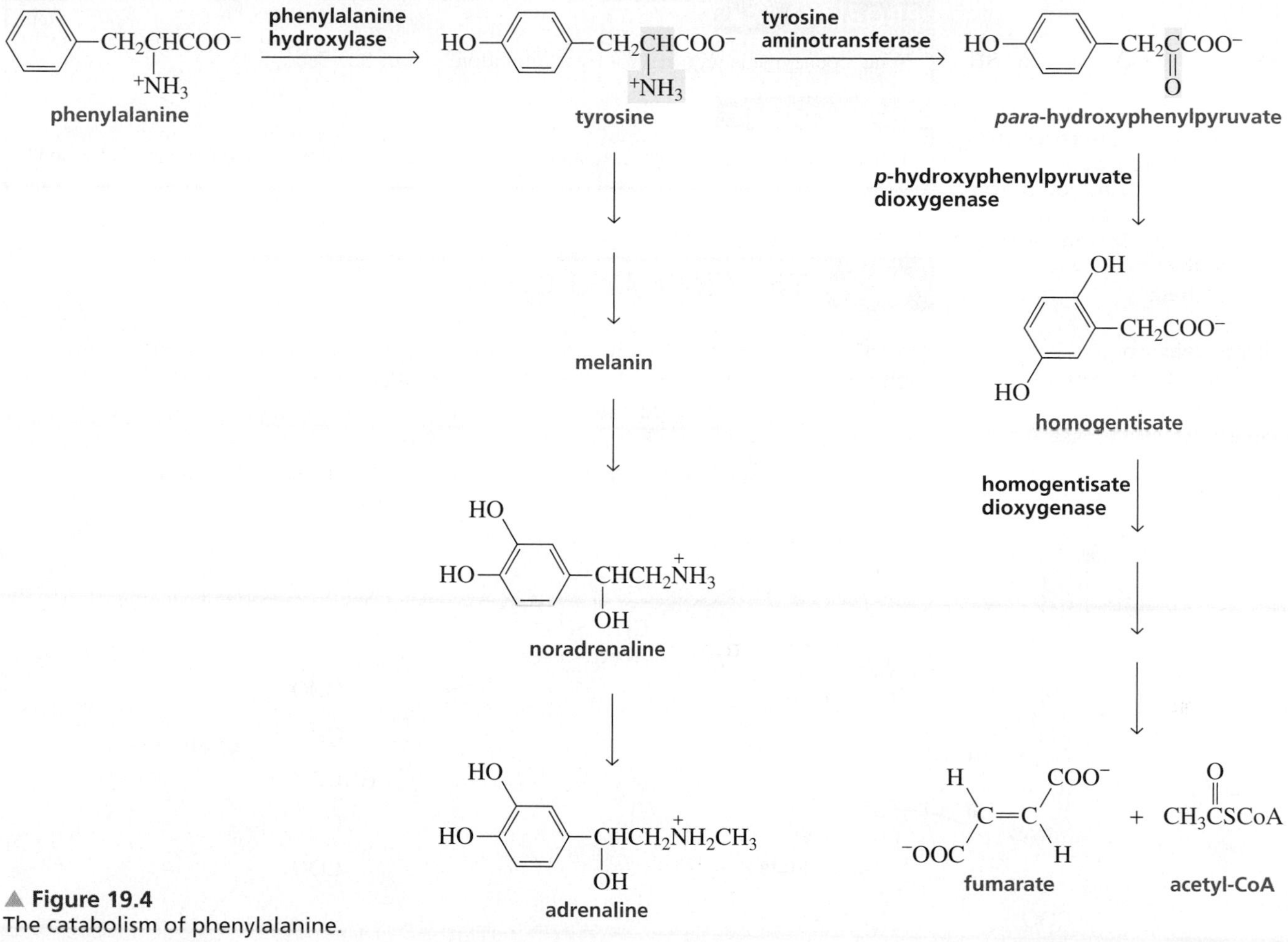

▲ **Figure 19.4**
The catabolism of phenylalanine.

PHENYLKETONURIA: AN INBORN ERROR OF METABOLISM

About 1 in every 20,000 babies is born without phenylalanine hydroxylase, the enzyme that converts phenylalanine into tyrosine. This genetic disease is called phenylketonuria (PKU). Without phenylalanine hydroxylase, the level of phenylalanine builds up; then, when it reaches a high concentration, it is transaminated to phenylpyruvate. The high level of phenylpyruvate found in urine gives the disease its name.

Within 24 hours after birth, all babies born in the United States are tested for high serum phenylalanine levels, which indicate a buildup of phenylalanine caused by an absence of phenylalanine

hydroxylase. Babies with high levels are immediately put on a diet low in phenylalanine and high in tyrosine. As long as the phenylalanine level is kept under careful control for the first 5 to 10 years of life, the baby will experience no adverse effects.

If the diet is not controlled, however, the baby will be severely mentally retarded by the time he or she is a few months old. Untreated children have paler skin and fairer hair than other members of their family because, without tyrosine, they cannot synthesize melanin, the black skin pigment. Half of untreated phenylketonurics are dead by age 20. When a woman with PKU becomes pregnant, she must return to the low phenylalanine diet she had as a child, because a high level of phenylalanine can cause abnormal development of the fetus.

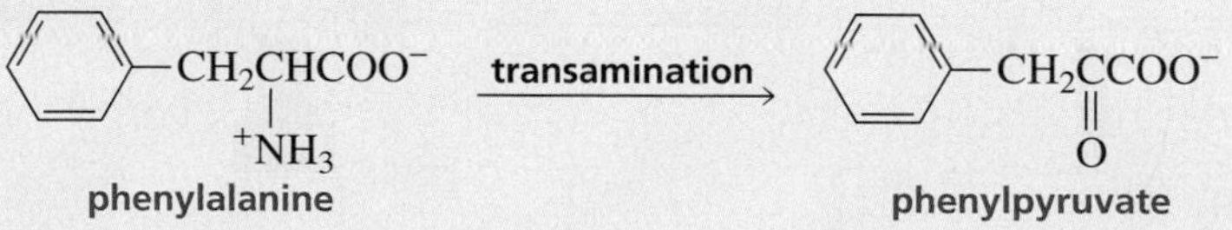

ALCAPTONURIA

Another genetic disease that results from a deficiency of an enzyme in the pathway for phenylalanine degradation is alcaptonuria, which is caused by lack of

homogentisate dioxygenase. The only ill effect of this enzyme deficiency is black urine. The urine of those afflicted with alcaptonuria turns black because the homogentisate they excrete immediately oxidizes in the air, forming a black compound.

$$\underset{\substack{\| \\ O}}{CH_3CSCoA} \longrightarrow 2\,CO_2 + CoASH$$

▲ **Figure 19.5**
The citric acid cycle—the series of enzyme-catalyzed reactions responsible for the oxidation of the acetyl group of acetyl-CoA to 2 molecules of CO_2. The enzymes that catalyze the reactions are:
1 citrate synthase
2 aconitase
3 isocitrate dehydrogenase
4 α-ketoglutarate dehydrogenase
5 succinyl-CoA synthetase
6 succinate dehydrogenase
7 fumarase
8 malate dehydrogenase

PROBLEM 14◆

What coenzyme is required for transamination? (*Hint:* See Section 18.9.)

PROBLEM 15◆

When the amino acid known as alanine undergoes transamination, what compound is formed?

19.7 The Citric Acid Cycle

The **citric acid cycle** is the third stage of catabolism. In this series of eight reactions, each molecule of acetyl-CoA—formed by the catabolism of fats, carbohydrates, and amino acids—is converted into two molecules of CO_2 (Figure 19.5). The series of reactions is called a *cycle* because the eight reactions are in a closed loop in which the product of the last reaction (oxaloacetate) is a reactant for the first reaction.

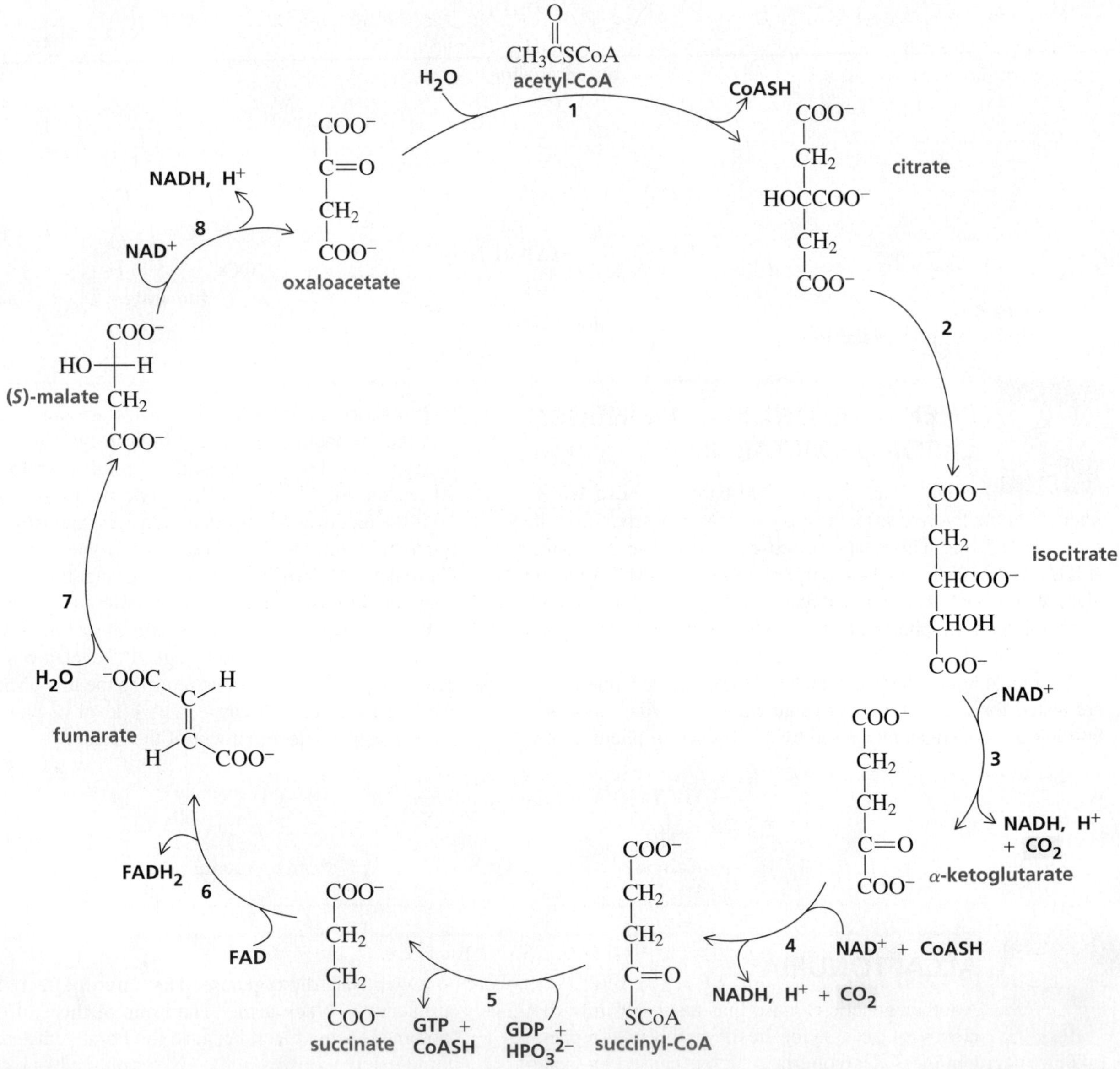

1 In the first reaction, acetyl-CoA reacts with oxaloacetate to form citrate. An aspartate side chain of the enzyme that catalyzes the reaction removes a proton from the α-carbon of acetyl-CoA, creating a nucleophile that attacks the carbonyl carbon of oxaloacetate. The carbonyl oxygen picks up a proton from a histidine side chain of the enzyme. The intermediate that is formed is hydrolyzed to citrate.

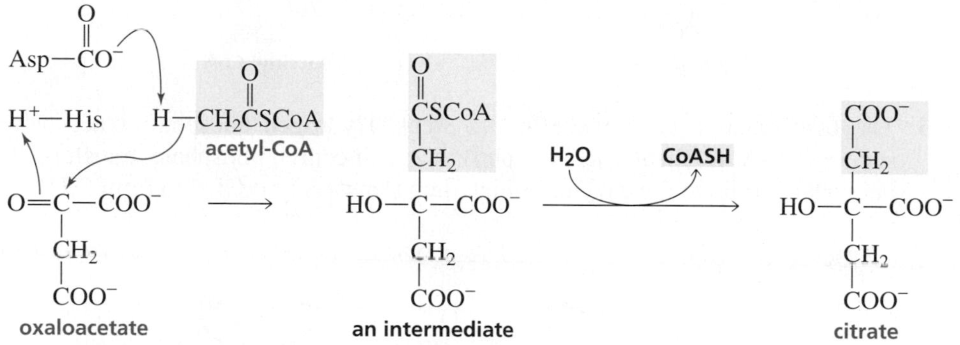

2 In the second reaction, citrate is converted to isocitrate, its isomer. The reaction takes place in two steps. The first step is a dehydration reaction (Section 11.3). In this step, a serine side chain of the enzyme removes a proton, and the OH leaving group is protonated by a histidine side chain to make it a weaker base and, therefore, a better leaving group. In the second step, water is added back to the intermediate (Section 13.8).

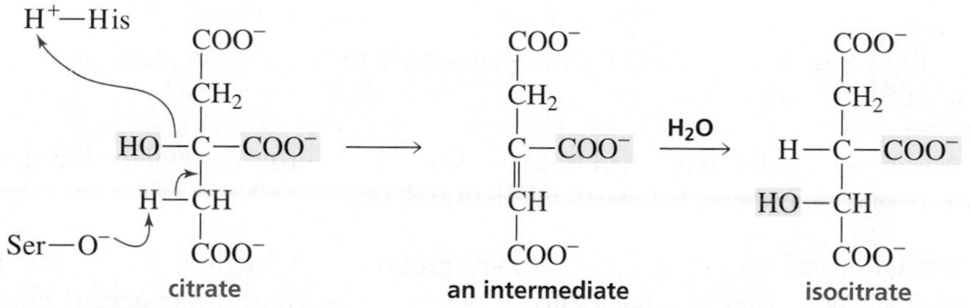

3 The third reaction also has two steps. In the first step, the secondary alcohol is oxidized to a ketone by NAD^+. In the second step, the ketone loses CO_2. We have seen that if a COO^- group is bonded to a carbon that is adjacent to a carbonyl carbon, the CO_2 group can be removed because the electrons left behind can be delocalized onto the carbonyl oxygen (Section 14.8); the enolate ion that is formed immediately tautomerizes to a ketone (Section 14.3).

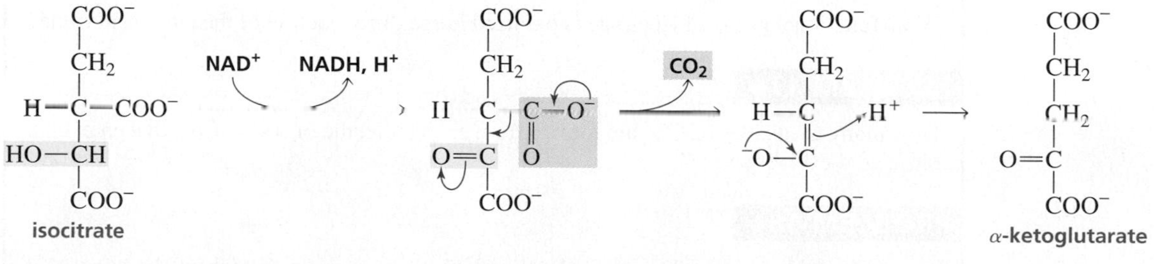

4 NAD^+ is again the oxidizing agent in the fourth reaction. This is the reaction that releases the second molecule of CO_2. This reaction requires a group of enzymes and the same five coenzymes that are required by the pyruvate dehydrogenase system that forms acetyl-CoA (Section 19.5). The product of the reaction is succinyl-CoA.

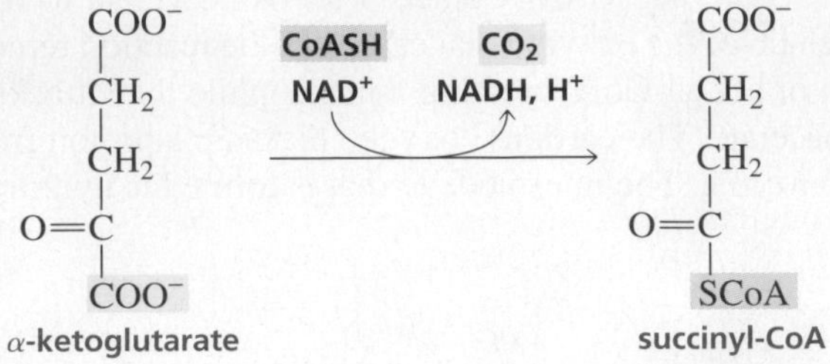

$$
\alpha\text{-ketoglutarate} \xrightarrow[\text{NAD}^+ \quad \text{NADH, H}^+]{\text{CoASH} \quad \text{CO}_2} \text{succinyl-CoA}
$$

α-ketoglutarate   succinyl-CoA

5 The fifth reaction takes place in two steps. Hydrogen phosphate reacts with succinyl-CoA to form succinyl phosphate. Succinyl phosphate transfers its phosphate group to the enzyme, which then transfers it to GDP to form GTP.

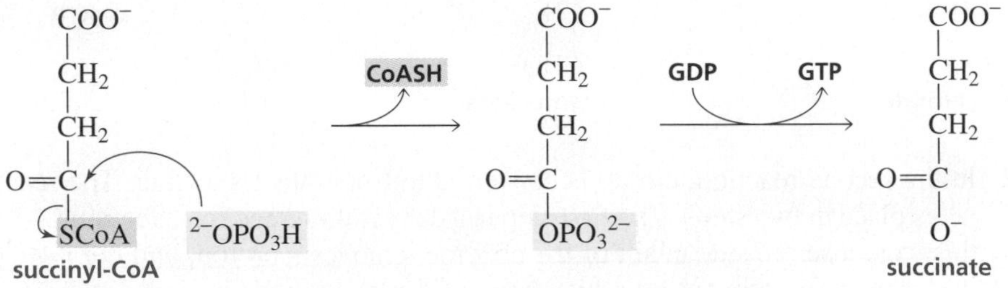

$$
\text{succinyl-CoA} \xrightarrow{\text{CoASH} \quad {}^{2-}\text{OPO}_3\text{H}} \xrightarrow{\text{GDP} \quad \text{GTP}} \text{succinate}
$$

succinyl-CoA   succinate

GTP transfers a phosphate group to ADP to form ATP.

$$
\text{GTP} + \text{ADP} \rightleftharpoons \text{GDP} + \text{ATP}
$$

6 In the sixth reaction, FAD oxidizes succinate to fumarate, a reaction we looked at in Section 18.6.

7 Addition of water to the double bond of fumarate forms (*S*)-malate. We saw why only one enantiomer is formed in Section 8.14.

8 Oxidation of the secondary alcohol group of (*S*)-malate by NAD^+ forms oxaloacetate. The cycle has returned to its starting point. Oxaloacetate now goes around the cycle again, reacting with another molecule of acetyl-CoA and coverting the acetyl group of acetyl-CoA to two molecules of CO_2.

Notice that reactions 6, 7, and 8 in the citric acid cycle are similar to reactions 1, 2, and 3 in β-oxidation.

PROBLEM 16◆

What functional group of isocitrate is oxidized in the third reaction of the citric acid cycle?

PROBLEM 17◆

How many molecules of CO_2 are formed from each molecule of acetyl-CoA that enters the citric acid cycle?

PROBLEM 18◆

The citric acid cycle is also called the tricarboxylic acid cycle. Which of the citric acid cycle intermediates are tricarboxylic acids?

19.8   Oxidative Phosphorylation

Each round of the citric acid cycle forms three molecules of NADH, one molecule of $FADH_2$ and one molecule of ATP. The NADH and $FADH_2$ molecules undergo **oxidative phosphorylation**—the fourth stage of catabolism—where they are oxidized back to NAD^+ and FAD. For each molecule of NADH that undergoes oxidative phosphorylation, three molecules of ATP are formed, and for each molecule of $FADH_2$ that undergoes oxidative phosphorylation, two molecules of ATP are formed.

$$NADH \longrightarrow NAD^+ + 3\ ATP$$

$$FADH_2 \longrightarrow FAD + 2\ ATP$$

Therefore, for every acetyl-CoA molecule that enters the citric acid cycle, 12 molecules of ATP are formed.

$$3\ NADH + FADH_2 + ATP \longrightarrow 3\ NAD^+ + FAD + 12\ ATP$$

> **The acetyl group of every molecule of acetyl-CoA that enters the citric acid cycle is converted to two molecules of CO_2.**
>
> **In oxidative phosphorylation, each NADH is converted to three ATPs and each $FADH_2$ is converted to two ATPs.**

PROBLEM 19◆

How many molecules of ATP are obtained when the NADH and $FADH_2$ formed during the metabolism of a molecule of acetyl-CoA to CO_2 undergo oxidative phosphorylation?

19.9   Anabolism

Anabolism can be thought of as the reverse of catabolism. In anabolism, acetyl-CoA, pyruvate, and citric acid cycle intermediates are the starting materials for the synthesis of fatty acids, monosaccharides, and amino acids. These compounds are then used to form fats, carbohydrates, and proteins. The mechanisms utilized by biological systems to synthesize fats and proteins are discussed in Sections 14.11 and 21.8.

BASAL METABOLIC RATE

The basal metabolic rate (BMR) is the number of calories you would burn if you stayed in bed all day. The BMR is affected by gender, age, and genetics; the BMR is greater for men than for women, it is greater for young people than for old people, and some people are born with a faster metabolic rate than others. BMR is also affected by the percentage of body fat—the higher the percentage, the lower the BMR. For humans, the average BMR is about 1600 kcal/day.

In addition to consuming sufficient calories to meet the needs of basal metabolism, calories also must be consumed to obtain energy for physical activities. The more active the person, the more calories they must consume in order to maintain their current weight. If a person consumes more calories than required by their BMR plus their level of physical activity, they will gain weight. If they consume fewer calories, they will lose weight.

Summary

Metabolism—the set of reactions that living organisms carry out to obtain energy and to synthesize the compounds they require—can be divided into **catabolism** and **anabolism**. **Catabolic reactions** break down complex molecules to provide energy and simple molecules. **Anabolic reactions** require energy and lead to the synthesis of complex biomolecules from simple molecules.

ATP is a cell's most important source of chemical energy; ATP provides a reaction pathway involving a good leaving group for a reaction that would otherwise not occur because of a poor leaving group. This occurs by way of a **phosphoryl transfer reaction** in which a phosphoryl group of ATP is transferred to a nucleophile as a result of breaking a **phosphoanhydride bond**.

Catabolism can be divided into four stages. In the *first stage*, fats, carbohydrates, and proteins are hydrolyzed to fatty acids, monosaccharides, and amino acids, respectively. In the *second stage*, the hydrolysis products obtained from the first stage are all converted to compounds that can enter the citric acid cycle. The citric acid cycle is the *third stage* of catabolism. In order to enter the citric acid cycle, a compound must be either one of the compounds in the cycle itself (called a citric acid cycle intermediate), acetyl-CoA, or pyruvate.

A fatty acyl-CoA is converted to acetyl-CoA in a pathway called **β-oxidation**. The series of four reactions is repeated until the entire fatty acid has been converted to acetyl-CoA molecules.

Glucose is converted to two molecules of pyruvate in a series of 10 reactions known as **glycolysis**. Under normal (aerobic) conditions, pyruvate is converted to acetyl-CoA, which then enters the citric acid cycle.

Amino acids are metabolized to acetyl-CoA, pyruvate, or citric acid cycle intermediates, depending on the amino acid. The amino acids we ingest are used not only for energy but also for the synthesis of proteins and other compounds that the body needs.

The **citric acid cycle** is a series of eight reactions, in which the acetyl group of each molecule of acetyl-CoA that enters the cycle is converted to two molecules of CO_2.

Metabolic energy is measured in terms of ATP. In **oxidative phosphorylation**, each NADH and $FADH_2$ formed as a result of oxidation reactions is converted to three ATPs and two ATPs, respectively.

Problems

20. Indicate whether an anabolic pathway or a catabolic pathway does the following:
   a. produces energy in the form of ATP
   b. involves primarily oxidation reactions

21. Galactose can enter the glycolytic cycle but it must first react with ATP to form galactose-1-phosphate. Give the mechanism for this reaction.

22. Pyruvate is the leaving group in the tenth reaction in glycolysis. Why is it a good leaving group?

23. When pyruvate is reduced by NADH to lactate, which hydrogen in pyruvate comes from NADH?

24. What reactions in the citric acid cycle form a product with one or more new asymmetric centers?

25. If the phosphorus atom in 3-phosphoglycerate is radioactively labeled, where will the radioactive label be when the reaction that forms 2-phosphoglycerate is over?

26. What carbon atoms of glucose end up as a carboxyl group in pyruvate?

glucose

27. What carbon atoms of glucose end up in ethanol under anaerobic conditions in yeast?

28. How many molecules of acetyl-CoA are obtained from the β-oxidation of a molecule of a 16-carbon saturated fatty acyl-CoA?

29. How many molecules of CO_2 are obtained from the complete metabolism of a molecule of a 16-carbon saturated fatty acyl-CoA?

30. How many molecules of ATP are obtained from the β-oxidation of a molecule of a 16-carbon saturated fatty acyl-CoA?

31. How many molecules of NADH and $FADH_2$ are obtained from the β-oxidation of a molecule of a 16-carbon saturated fatty acyl-CoA?

32. How many molecules of ATP are obtained from the NADH and $FADH_2$ formed in the β-oxidation of a molecule of a 16-carbon saturated fatty acyl-CoA?

33. How many molecules of ATP are obtained from the complete (including the fourth stage of catabolism) metabolism of a 16-carbon saturated fatty acyl-CoA?

34. How many molecules of ATP are obtained from the complete (including the fourth stage of catabolism) metabolism of a molecule of glucose? (Note: one molecule of NADH is formed from each molecule of pyruvate that is converted to acetyl-CoA.)

35. Most fatty acids have an even number of carbon atoms and, therefore, are completely metabolized to acetyl CoA. A fatty acid with an odd number of carbon atoms is metabolized to acetyl-CoA and one equivalent of propionyl-CoA. The following two reactions convert propionyl-CoA into succinyl-CoA, a citric acid cycle intermediate. Each of the reactions requires a coenzyme. Identify the coenzyme for each step.

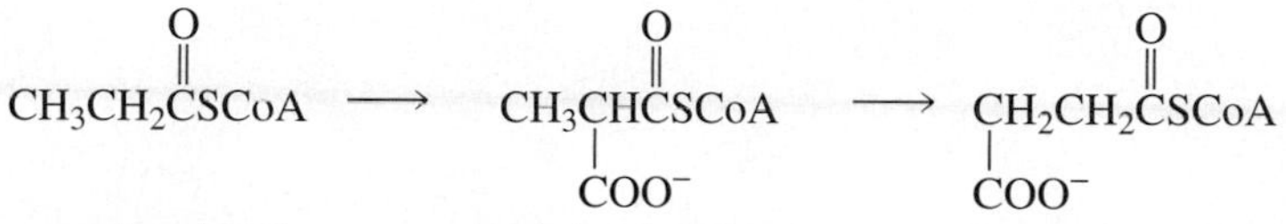

$$
\underset{\text{propionyl-CoA}}{CH_3CH_2\overset{\overset{\displaystyle O}{\|}}{C}SCoA} \longrightarrow \underset{\text{methylmalonyl-CoA}}{CH_3\underset{\underset{\displaystyle COO^-}{|}}{CH}\overset{\overset{\displaystyle O}{\|}}{C}SCoA} \longrightarrow \underset{\text{succinyl-CoA}}{CH_2\underset{\underset{\displaystyle COO^-}{|}}{CH_2}\overset{\overset{\displaystyle O}{\|}}{C}SCoA}
$$

36. Acyl-CoA synthase is the enzyme that activates a fatty acid by converting it to a fatty acyl-CoA (Section 19.2) in a series of two reactions. In the first reaction, the fatty acid reacts with ATP and one of the products formed is ADP. The other product reacts with CoASH to form the fatty acyl-CoA. Give the mechanism for both reactions.

20 Lipids

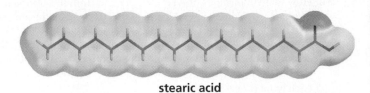

stearic acid

linoleic acid

Lipids are organic compounds that are found in living organisms and that are soluble in nonpolar solvents. Because compounds are classified as lipids on the basis of a physical property—their solubility in a nonpolar solvent—rather than on the basis of their structures, lipids have a variety of structures and functions, as the following examples illustrate:

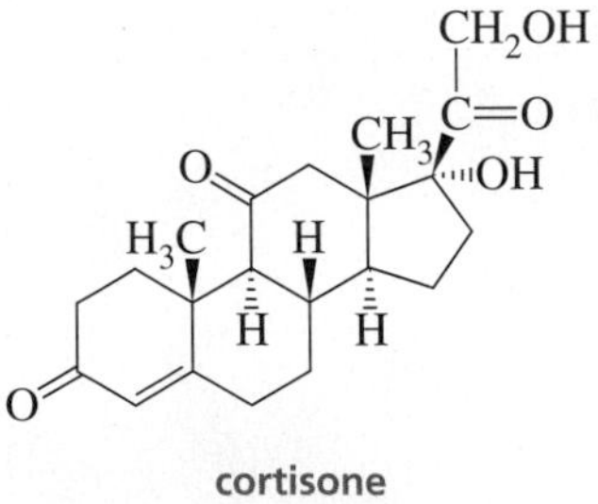

cortisone
a hormone

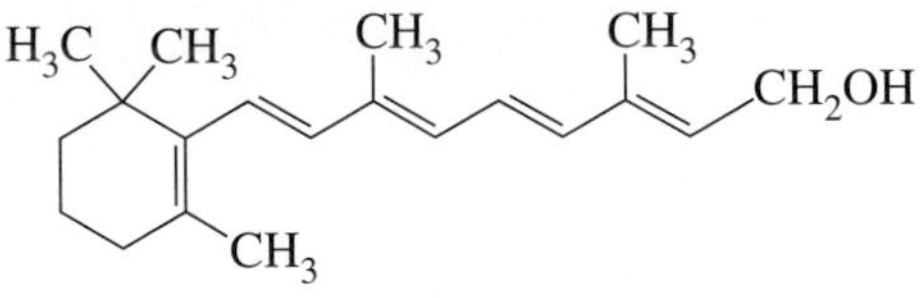

vitamin A
a vitamin

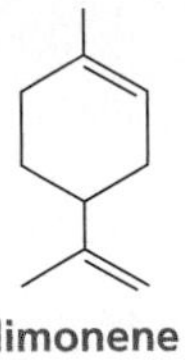

limonene
in orange and
lemon oils

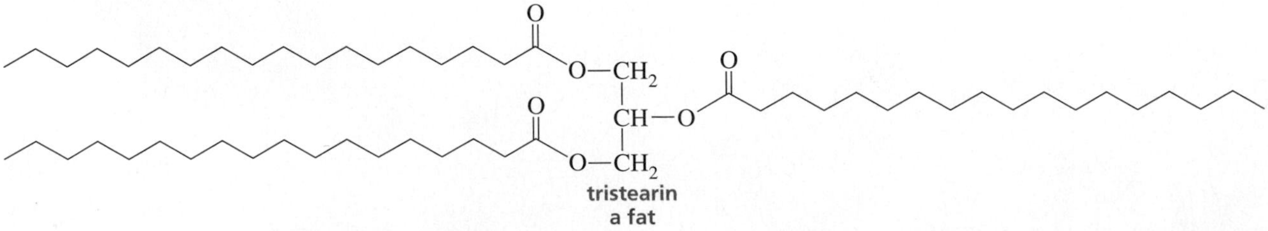

tristearin
a fat

The solubility of lipids in nonpolar solvents results from their significant hydrocarbon component—the part of the molecule that is responsible for its "oiliness" or "fattiness." The word *lipid* comes from the Greek *lipos*, which means "fat."

20.1   Fatty Acids

Fatty acids are carboxylic acids with long hydrocarbon chains. The fatty acids most frequently found in nature are shown in Table 20.1. Most naturally occurring fatty acids contain an even number of carbon atoms and are unbranched, because they are synthesized from acetate, a compound with two carbons (Section 14.11). Fatty acids can be saturated with hydrogen (and therefore have no carbon–carbon double bonds) or unsaturated (and have carbon–carbon double bonds). Fatty acids with more than one double bond are called **polyunsaturated fatty acids**. Double bonds in naturally occurring unsaturated fatty acids have the cis configuration and are always separated by one CH_2 group.

The melting points of saturated fatty acids increase with increasing molecular weight because of increased van der Waals interactions between the molecules (Section 3.7). The melting points of unsaturated fatty acids also increase with increasing molecular weight, but are lower than those of saturated fatty acids with comparable molecular weights (Table 20.1). The cis double bond produces a bend in the

Table 20.1	Common Naturally Occurring Fatty Acids			
Number of carbons	**Common name**	**Systematic name**	**Structure**	**Melting point °C**
Saturated				
12	lauric acid	dodecanoic acid	 COOH	44
14	myristic acid	tetradecanoic acid	 COOH	58
16	palmitic acid	hexadecanoic acid	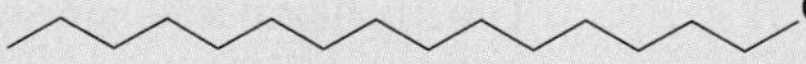 COOH	63
18	stearic acid	octadecanoic acid	 COOH	69
20	arachidic acid	eicosanoic acid	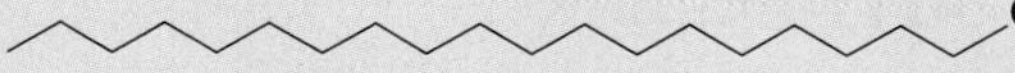 COOH	77
Unsaturated				
16	palmitoleic acid	(9Z)-hexadecenoic acid	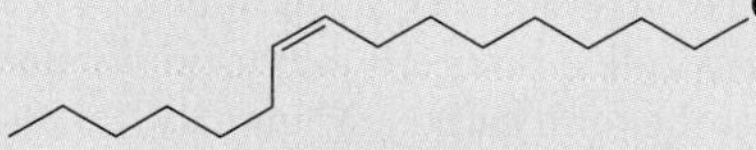 COOH	0
18	oleic acid	(9Z)-octadecenoic acid	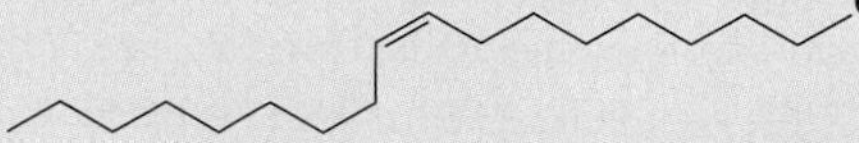 COOH	13
18	linoleic acid	(9Z,12Z)-octadecadienoic acid	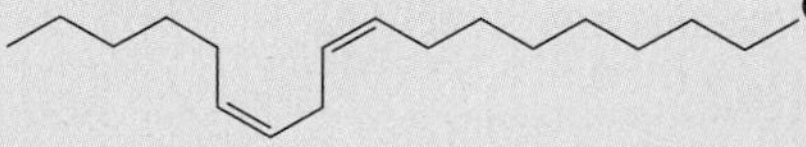 COOH	−5
18	linolenic acid	(9Z,12Z,15Z)-octadecatrienoic acid	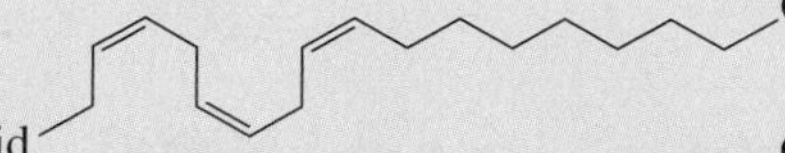 COOH	−11
20	arachidonic acid	(5Z,8Z,11Z,14Z)-eicosatetraenoic acid	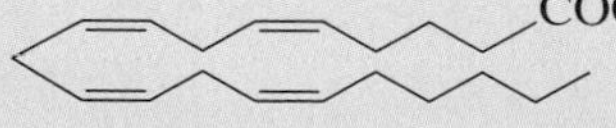 COOH	−50
20	EPA	(5Z,8Z,11Z,14Z,17Z)-eicosapentaenoic acid	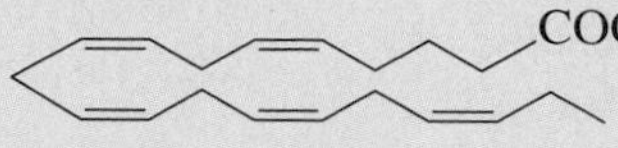 COOH	−50

molecules, which prevents them from packing together as tightly as saturated fatty acids. As a result, unsaturated fatty acids have fewer intermolecular interactions and, therefore, lower melting points.

The melting points of unsaturated fatty acids decrease as the number of double bonds increases. For example, an 18-carbon fatty acid melts at 69 °C if it is saturated, at 13 °C if it has one double bond, at −5 °C if it has two double bonds, and at −11 °C if it has three double bonds.

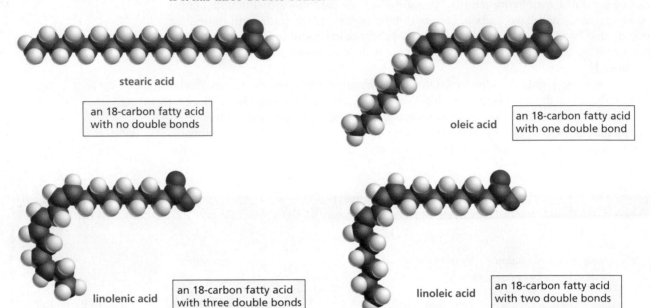

stearic acid

an 18-carbon fatty acid with no double bonds

oleic acid

an 18-carbon fatty acid with one double bond

linolenic acid

an 18-carbon fatty acid with three double bonds

linoleic acid

an 18-carbon fatty acid with two double bonds

3-D Molecules:
Stearic acid;
Oleic acid;
Linoleic acid;
Linolenic acid

PROBLEM 1

Explain the difference in the melting points of the following fatty acids:

a. palmitic acid and stearic acid

b. palmitic acid and palmitoleic acid

c. oleic acid and linoleic acid

OMEGA FATTY ACIDS

Omega is a term used to indicate the position of the first double bond—from the methyl end—in an unsaturated fatty acid. For example, linoleic acid is called omega-6 fatty acid because its first double bond is after the sixth carbon, and linolenic acid is called omega-3 fatty acid because its first double bond is after the third carbon. Mammals lack the enzyme that introduces a double bond beyond C-9 (the carboxyl carbon is C-1). Linoleic acid and linolenic acids, therefore, are essential fatty acids for mammals. In other words, the acids must be included in the diet because, although they cannot be synthesized, they are required for normal body function.

They must have linoleic and linolenic acids in their diets.

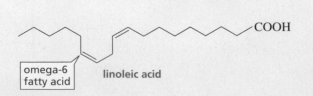

omega-6 fatty acid

linoleic acid

omega-3 fatty acid

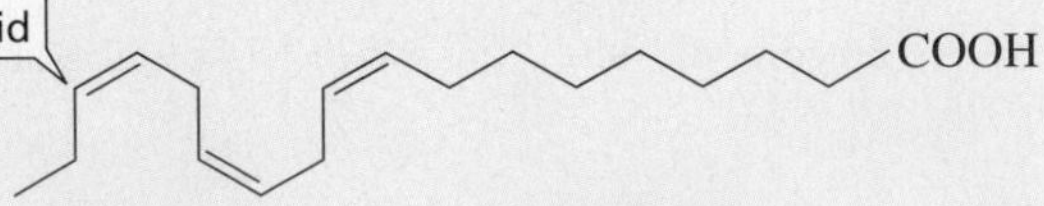

linolenic acid

20.2 Waxes

Waxes are esters formed from long-chain carboxylic acids and long-chain alcohols. For example, beeswax, the structural material of beehives, has a 26-carbon carboxylic acid component and a 30-carbon alcohol component. Carnauba wax is a particularly hard wax because of its relatively high molecular weight; it has a 32-carbon carboxylic acid component and a 34-carbon alcohol component. Carnauba wax is widely used as a car wax and in floor polishes.

Layers of honeycomb in a beehive.

$$CH_3(CH_2)_{24}\overset{O}{\overset{\|}{C}}O(CH_2)_{29}CH_3$$
a major component of
beeswax
structural material
of beehives

$$CH_3(CH_2)_{30}\overset{O}{\overset{\|}{C}}O(CH_2)_{33}CH_3$$
a major component of
carnauba wax
coating on the leaves
of a Brazilian palm

$$CH_3(CH_2)_{14}\overset{O}{\overset{\|}{C}}O(CH_2)_{15}CH_3$$
a major component of
spermaceti wax
from the heads of
sperm whales

Waxes are common in living organisms. The feathers of birds are coated with wax to make them water repellent. Some vertebrates secrete wax in order to keep their fur lubricated and water repellent. Insects secrete a waterproof, waxy layer on the outside of their exoskeletons. Wax is also found on the surfaces of certain leaves and fruits, where it serves as a protectant against parasites and minimizes the evaporation of water.

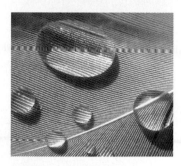
Raindrops on a feather.

20.3 Fats and Oils

Triglycerides, also called **triacylglycerols**, are compounds in which each of the three OH groups of glycerol has formed an ester with a fatty acid (page 504). If the three fatty acid components of a triglyceride are the same, the compound is called a **simple triglyceride**. **Mixed triglycerides**, on the other hand, contain two or three different fatty acid components and are more common than simple triglycerides. Not all triglyceride molecules from a single source are necessarily identical; substances such as lard and olive oil, for example, are mixtures of several different triglycerides (Table 20.2).

Table 20.2 Approximate Percentage of Fatty Acids in Some Common Fats and Oils

		Saturated fatty acids				Unsaturated fatty acids		
	mp (°C)	lauric C_{12}	myristic C_{14}	palmitic C_{16}	stearic C_{18}	oleic C_{18}	linoleic C_{18}	linolenic C_{18}
Animal fats								
Butter	32	2	11	29	9	27	4	—
Lard	30	—	1	28	12	48	6	—
Human fat	15	1	3	25	8	46	10	—
Whale blubber	24	—	8	12	3	35	10	—
Plant oils								
Corn	20	—	1	10	3	50	34	—
Cottonseed	−1	—	1	23	1	23	48	—
Linseed	−24	—	—	6	3	19	24	47
Olive	−6	—	—	7	2	84	5	—
Peanut	3	—	—	8	3	56	26	—
Safflower	−15	—	—	3	3	19	70	3
Sesame	−6	—	—	10	4	45	40	—
Soybean	−16	—	—	10	2	29	51	7

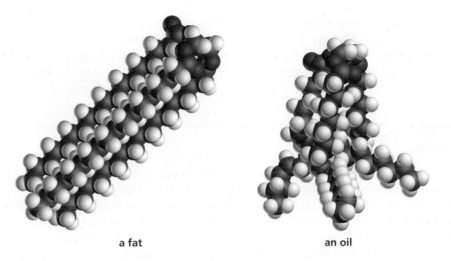

$$
\begin{array}{ccc}
\text{CH}_2\!-\!\text{OH} & \text{R}^1\!-\!\overset{\displaystyle O}{\overset{\|}{\text{C}}}\!-\!\text{OH} & \text{CH}_2\!-\!\text{O}\!-\!\overset{\displaystyle O}{\overset{\|}{\text{C}}}\!-\!\text{R}^1 \\[2ex]
\text{CH}\!-\!\text{OH} & \text{R}^2\!-\!\overset{\displaystyle O}{\overset{\|}{\text{C}}}\!-\!\text{OH} & \text{CH}\!-\!\text{O}\!-\!\overset{\displaystyle O}{\overset{\|}{\text{C}}}\!-\!\text{R}^2 \\[2ex]
\text{CH}_2\!-\!\text{OH} & \text{R}^3\!-\!\overset{\displaystyle O}{\overset{\|}{\text{C}}}\!-\!\text{OH} & \text{CH}_2\!-\!\text{O}\!-\!\overset{\displaystyle O}{\overset{\|}{\text{C}}}\!-\!\text{R}^3 \\[1ex]
\textbf{glycerol} & \textbf{fatty acids} & \textbf{a triacylglycerol}
\end{array}
$$

a fat or an oil

Triglycerides that are solids or semisolids at room temperature are called **fats**. Fats are usually obtained from animals and are composed largely of triglycerides with fatty acid components that are either saturated or have only one double bond. The saturated fatty acid tails pack closely together, giving the triglycerides relatively high melting points—causing them to be solids at room temperature.

a fat   an oil

Liquid triglycerides are called **oils**. Oils typically come from plant products such as corn, soybeans, olives, and peanuts. They are composed primarily of triglycerides with fatty acid components that are unsaturated and therefore cannot pack tightly together. Consequently, they have relatively low melting points—causing them to be liquids at room temperature. The approximate fatty acid compositions of some common fats and oils are shown in Table 20.2.

Some or all of the double bonds of polyunsaturated oils can be reduced by catalytic hydrogenation. Margarine and shortening are prepared by hydrogenating vegetable oils such as safflower oil until they have the desired consistency. Some trans double bonds are created during this process (Section 5.12). The hydrogenation reaction must be carefully controlled, because reducing all the carbon–carbon double bonds would produce a hard fat with the consistency of beef tallow.

$$
\text{RCH}\!=\!\text{CHCH}_2\text{CH}\!=\!\text{CHCH}_2\text{CH}\!=\!\text{CH}\!- \xrightarrow[\text{Pt}]{\text{H}_2} \text{RCH}_2\text{CH}_2\text{CH}_2\text{CH}\!=\!\text{CHCH}_2\text{CH}_2\text{CH}_2\!-
$$

Vegetable oils have become popular in food preparation because some studies have linked the consumption of saturated fats with heart disease. Recent studies have shown that unsaturated fats may also be implicated in heart disease. However, one unsaturated fatty acid—a 20-carbon fatty acid with five double bonds, known as EPA and found in

This puffin's diet is high in fish oil.

high concentrations in fish oils—is thought to lower the chance of developing certain forms of heart disease. Once consumed, dietary fat is hydrolyzed, regenerating glycerol and fatty acids (Section 19.3).

PROBLEM 2◆

Which has a higher melting point, glyceryl tripalmitoleate or glyceryl tripalmitate?

PROBLEM 3◆

Give the structure of an optically inactive fat that, when hydrolyzed, gives glycerol, one equivalent of lauric acid, and two equivalents of stearic acid.

PROBLEM 4◆

Give the structure of an optically active fat that when hydrolyzed gives the same products as the fat in Problem 3.

Animals have a subcutaneous layer of fat cells that serves as both an energy source and an insulator. The fat content of the average man is about 21%, whereas the fat content of the average woman is about 25%. A fat provides about six times as much metabolic energy as an equal weight of hydrated glycogen because fats are less oxidized than carbohydrates and, since fats are nonpolar, they do not bind water. In contrast, two-thirds of the weight of stored glycogen is water (Section 16.13).

Humans can store sufficient fat to provide for the body's metabolic needs for two to three months, but can store only enough carbohydrate to provide for its metabolic needs for less than 24 hours. Carbohydrates, therefore, are used primarily as a quick, short-term energy source.

Polyunsaturated fats and oils are easily oxidized by O_2 by means of a radical chain reaction. In the initiation step, a radical removes a hydrogen from a CH_2 group that is flanked by two double bonds. This hydrogen is the one most easily removed because the resulting radical is resonance stabilized by both π bonds.

$$RCH{=}CH{-}CH{-}CH{=}CH{-} \;+\; X\cdot \xrightarrow{\text{initiation}} RCH{=}CH{-}\overset{\cdot}{C}H{-}CH{=}CH{-} \;+\; HX$$

$$R\overset{\cdot}{C}H{-}CH{=}CH{-}CH{=}CH{-} \;\longleftrightarrow\; RCH{=}CH{-}CH{=}CH{-}\overset{\cdot}{C}H{-}$$

$$\Big\downarrow O_2$$

$$\underset{\textbf{and other short chain}}{\underset{\textbf{carboxylic acids}}{CH_3CH_2CH_2\overset{\displaystyle O}{\overset{\displaystyle \|}{C}}OH}}$$

The reaction of fatty acids with O_2 causes them to become rancid. The unpleasant taste and smell associated with sour milk and rancid butter are due to the products of the oxidation reaction that have strong odors.

PROBLEM 5

Draw the resonance contributors for the radical formed when a hydrogen atom is removed from C-10 of arachidonic acid.

3-D Molecule:
Olestra

OLESTRA: NONFAT WITH FLAVOR

Chemists have been searching for ways to reduce the caloric content of foods without decreasing their flavor. Many people who believe that "no fat" is synonymous with "no flavor" can understand this problem. The Federal Food and Drug Administration (Section 22.10) approved the limited use of Olestra as a substitute for dietary fat in snack foods. Procter and Gamble spent 30 years and more than $2 billion to develop this compound. Its approval was based on the results of more than 150 studies.

Olestra is a semisynthetic compound. That is, Olestra does not exist in nature, but its components do. Developing a compound that can be made from units that are a normal part of our diet decreases the potential toxic effects of the new compound. Olestra is made by esterifying all the OH groups of sucrose with fatty acids obtained from cottonseed oil and soybean oil.

Therefore, its component parts are table sugar and vegetable oil. Olestra's ester linkages are too sterically hindered to be hydrolyzed by digestive enzymes. Therefore, it tastes like fat, but it has no caloric value because it cannot be digested.

Olestra

WHALES AND ECHOLOCATION

Whales have enormous heads, accounting for 33% of their total weight. They have large deposits of fat in their heads and lower jaws. This fat is very different from both the whale's normal body fat and its dietary fat. Because major anatomical modifications were necessary to accommodate this fat, it must have some important function for the animal. It is now believed that the fat is used for echolocation—emitting sounds in pulses and gaining information by analyzing the returning echoes. The fat in the whale's head focuses the emitted sound waves in a directional beam, and the echoes are received by the fat organ in the lower jaw. This organ transmits the sound to the brain for processing and interpretation, providing the whale with information about the depth of the water, changes in the sea floor, and the position of the coastline. The fat deposits in the whale's head and jaw therefore give the animal a unique acoustic sensory system and allow it to compete successfully for survival with the shark, which also has a well-developed sense of sound direction.

Humpback whale in Alaska.

20.4 Soaps, Detergents, and Micelles

Soaps are sodium or potassium salts of fatty acids. Thus, soaps are obtained when fats or oils are hydrolyzed under basic conditions. The hydrolysis of an ester in a basic solution is called **saponification**—the Latin word for "soap" is *sapo*.

$$
\begin{array}{c}
\text{CH}_2\text{O}-\overset{\overset{\text{O}}{\|}}{\text{C}}-\text{R}^1 \\
|\\
\text{CHO}-\overset{\overset{\text{O}}{\|}}{\text{C}}-\text{R}^2 \;+\; \text{H}_2\text{O} \xrightarrow{\ \text{NaOH}\ } \\
|\\
\text{CH}_2\text{O}\ \ \overset{\overset{\text{O}}{\|}}{\text{C}}\ \ \text{R}^3 \\
\textbf{a fat or an oil}
\end{array}
\qquad
\begin{array}{c}
\text{CH}_2\text{OH} \\
|\\
\text{CHOH} \;+ \\
|\\
\text{CH}_2\text{OH} \\
\textbf{glycerol}
\end{array}
\qquad
\begin{array}{c}
\text{R}^1-\overset{\overset{\text{O}}{\|}}{\text{C}}-\text{O}^-\,\text{Na}^+ \\
\text{R}^2-\overset{\overset{\text{O}}{\|}}{\text{C}}-\text{O}^-\,\text{Na}^+ \\
\text{R}^3-\overset{\overset{\text{O}}{\|}}{\text{C}}-\text{O}^-\,\text{Na}^+ \\
\textbf{sodium salts of fatty acids} \\
\textbf{soap}
\end{array}
$$

The following compounds are three of the most common soaps:

$$
\text{CH}_3(\text{CH}_2)_{16}\overset{\overset{\text{O}}{\|}}{\text{C}}\text{O}^-\,\text{Na}^+
\qquad
\text{CH}_3(\text{CH}_2)_7\text{CH}=\text{CH}(\text{CH}_2)_7\overset{\overset{\text{O}}{\|}}{\text{C}}\text{O}^-\,\text{Na}^+
$$

sodium stearate   **sodium oleate**

$$
\text{CH}_3(\text{CH}_2)_4\text{CH}=\text{CHCH}_2\text{CH}=\text{CH}(\text{CH}_2)_7\overset{\overset{\text{O}}{\|}}{\text{C}}\text{O}^-\,\text{Na}^+
$$

sodium linoleate

Long-chain carboxylate ions do not exist as individual ions in aqueous solution. Instead, they arrange themselves in spherical clusters called **micelles**, as shown in Figure 20.1. Each micelle contains 50 to 100 long-chain carboxylate ions. A micelle

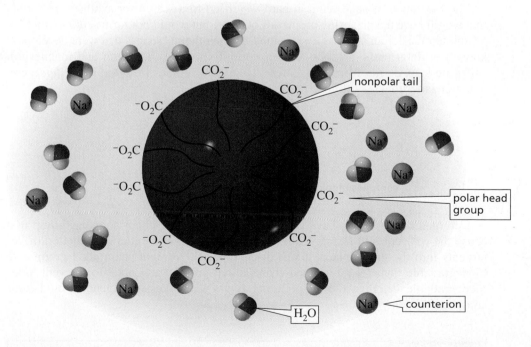

▲ **Figure 20.1**
In aqueous solution, soap molecules form micelles, with the polar heads (carboxylate groups) on the surface and the nonpolar tails (fatty acid R groups) in the interior.

resembles a large ball: The polar head of each carboxylate ion and its counterion are on the outside of the ball because of their attraction for water, and the nonpolar tail is buried in the interior of the ball to minimize its contact with water. Soap has cleansing ability because nonpolar oil molecules, which carry dirt, dissolve in the nonpolar interior of the micelle and are carried away with the soap during rinsing. Because the surface of the micelle is negatively charged, the individual micelles repel each other instead of clustering to form larger aggregates.

MAKING SOAP

For thousands of years, soap was prepared by heating animal fat with wood ashes. Wood ashes contain potassium carbonate, which makes the solution basic. The modern commercial method of making soap involves boiling fats or oils in aqueous sodium hydroxide and adding sodium chloride to precipitate the soap, which is then dried and pressed into bars. Perfume can be added for scented soaps, dyes can be added for colored soaps, sand can be added for scouring soaps, and air can be blown into the soap to make it float in water.

As river water flows over and around rocks, it leaches out calcium and magnesium ions. The concentration of calcium and magnesium ions in water determines its "hardness." Hard water contains high concentrations of these ions; soft water contains few, if any, calcium and magnesium ions. While micelles with sodium and potassium cations are dispersed in water, micelles with calcium and magnesium cations form aggregates. In hard water, therefore, soaps form a precipitate that we know as "bathtub ring" or "soap scum."

The formation of soap scum in hard water led to a search for synthetic materials that would have the cleansing properties of soap, but would not form scum when they encountered calcium and magnesium ions. The synthetic "soaps" that were developed, known as **detergents**, are salts of benzenesulfonic acids. Calcium and magnesium sulfonate salts do not form aggregates. "Detergent" comes from the Latin *detergere*, which means "to wipe off."

a benzenesulfonic acid   **a detergent**

Detergents, like soaps, have a polar group and a long-chain nonpolar group that cause the molecules to form micelles in solution. After the initial introduction of detergents into the marketplace, it was discovered that straight-chain alkyl groups are biodegradable, whereas branched-chain alkyl groups are not. To prevent nonbiodegradable detergents from polluting rivers and lakes, detergents should have straight-chain alkyl groups.

PROBLEM 6

Draw the structure of calcium stearte, a compound in soap scum.

PROBLEM 7 SOLVED

An oil obtained from coconuts is unusual in that all three fatty acid components are identical. The molecular formula of the oil is $C_{45}H_{86}O_6$. What is the molecular formula of the carboxylate ion obtained when the oil is saponified?

SOLUTION When the oil is saponified, it forms glycerol and three equivalents of carboxylate ion. In losing glycerol, the fat loses three carbons and five hydrogens. Thus, the three equivalents of carboxylate ion have a combined molecular formula of $C_{42}H_{81}O_6$. Dividing by three gives a molecular formula of $C_{14}H_{27}O_2^-$ for the carboxylate ion.

20.5 Phospholipids

Lipids that contain a phosphate group are called **phospholipids**. The two main kinds of phospholipids are *phosphoglycerides* and *sphingolipids*.

Phosphoglycerides are the major components of cell membranes. For biological systems to operate, some parts of organisms must be separated from other parts. On a cellular level, the outside of the cell must be separated from the inside. "Greasy" lipid **membranes** serve as the barrier. In addition to isolating the cell's contents, membranes allow the selective transport of ions and organic molecules into and out of the cell.

Phosphoglycerides (also called phosphoacylglycerols) are similar to triglycerides except that a terminal OH group of glycerol forms an ester with phosphoric acid rather than with a fatty acid, forming a phosphatidic acid.

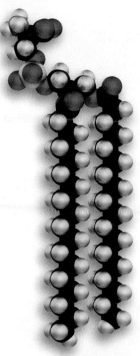

phosphatidylserine

phosphoric acid → an ester of phosphoric acid → **a phosphatidic acid**

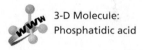

3-D Molecule:
Phosphatidic acid

The most common phosphoglycerides in membranes have a second phosphate ester linkage—they are **phosphodiesters**.

phosphoglycerides

phosphate ester linkages

**a phosphatidylethanolamine
a cephalin**

**a phosphatidylcholine
a lecithin**

a phosphatidylserine

The alcohols most commonly used to form the second ester group are ethanolamine, choline, and serine. Phosphatidylethanolamines are also called *cephalins*, and phosphatidylcholines are called *lecithins*. Lecithins are added to foods such as mayonnaise to prevent the aqueous and fat components from separating—thus, lecithins are emulsifying agents.

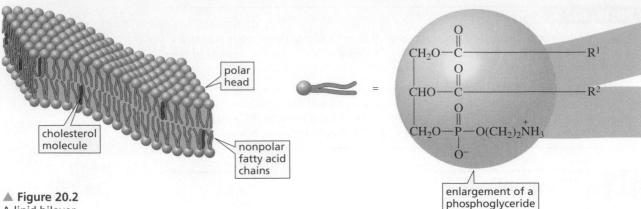

▲ **Figure 20.2**
A lipid bilayer.

3-D Molecule:
Vitamin E

Phosphoglycerides form membranes by arranging themselves in a **lipid bilayer**. The polar heads of the phosphoglycerides are on the outside of the bilayer, and the fatty acid chains form the interior of the bilayer. Cholesterol—a lipid discussed in Section 20.8—is also found in the interior of the bilayer (Figure 20.2).

The fluidity of a membrane is controlled by the fatty acid components of the phosphoglycerides. Saturated fatty acids decrease membrane fluidity because their hydrocarbon chains pack closely together. Unsaturated fatty acids increase fluidity because they pack less closely together. Cholesterol also decreases fluidity because of its rigid ring structure (Section 3.13). Only animal membranes contain cholesterol, so they are more rigid than plant membranes.

The unsaturated fatty acid chains of phosphoglycerides are susceptible to reaction with O_2, similar to the reaction of fats and oils with O_2 (Section 20.3). The oxidation reaction leads to the degradation of membranes. Vitamin E is an important antioxidant that protects fatty acid chains from degradation via oxidation. Vitamin E, also called α-tocopherol, is classified as a lipid because it is soluble in nonpolar solvents. Because vitamin E reacts more rapidly with oxygen than triglycerides do, the vitamin prevents biological membranes from reacting with oxygen (Section 9.6). Because vitamin E also reacts with oxygen more rapidly than fats do, it is added to many foods to prevent spoilage.

PROBLEM 8◆

Membranes contain proteins. Integral membrane proteins extend partly or completely through the membrane, whereas peripheral membrane proteins are found on the inner or outer surfaces of the membrane. What is the likely difference in the amino acid composition of integral and peripheral membrane proteins?

SNAKE VENOM

The venom of some poisonous snakes contains a phospholipase, an enzyme that hydrolyzes an ester

An eastern diamondback rattlesnake.

group of a phosphoglyceride. For example, both the eastern diamondback rattlesnake and the Indian cobra contain a phospholipase that hydrolyzes an ester bond of cephalins, causing the membranes of red blood cells to rupture.

bond hydrolyzed by the phospholipase found in the Indian cobra and the eastern diamondback rattlesnake

PROBLEM 9◆

A colony of bacteria accustomed to an environment at 25 °C was moved to an identical environment at 35 °C. The higher temperature increased the fluidity of the bacterial membranes. What could the bacteria do to regain their original membrane fluidity?

PROBLEM 10

The membrane phospholipids in animals such as deer and elk have a higher degree of unsaturation in cells closer to the hoof than in cells closer to the body. Explain how this trait can be important for survival.

Sphingolipids are also found in membranes. They are the major lipid components in the myelin sheaths of nerve fibers. Sphingolipids contain sphingosine instead of glycerol. In sphingolipids, the amino group of sphingosine is bonded to the acyl group of a fatty acid.

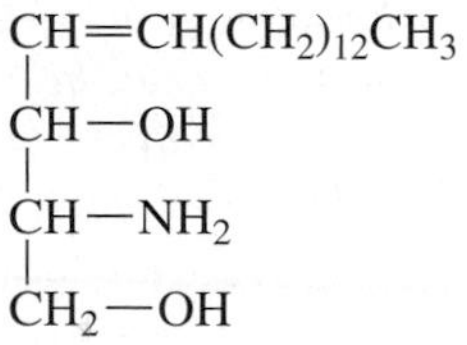

$$CH=CH(CH_2)_{12}CH_3$$
$$CH-OH$$
$$CH-NH_2$$
$$CH_2-OH$$
sphingosine

3-D Molecule: Sphingosine

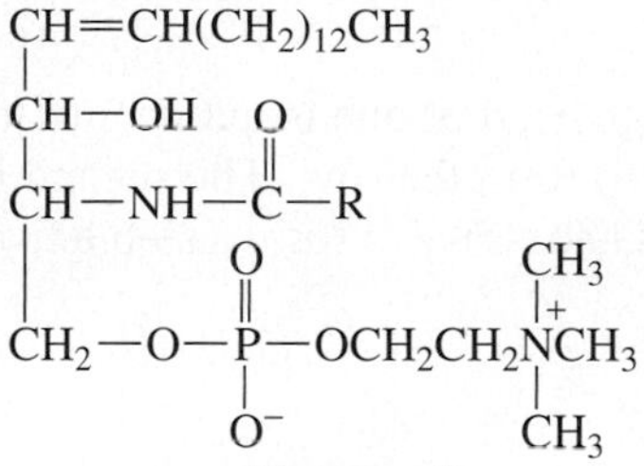

a sphingomyelin

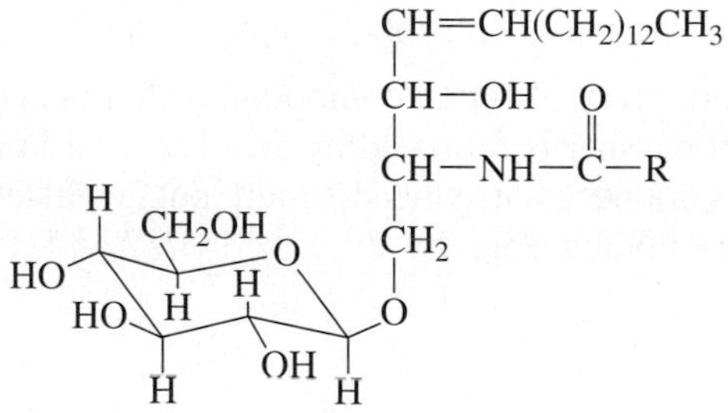

a glucocerebroside

Two of the most common kinds of sphingolipids are *sphingomyelins* and *cerebrosides*. In sphingomyelins, the primary OH group of sphingosine is bonded to phosphocholine or phosphoethanolamine, similar to the bonding in lecithins and cephalins. In cerebrosides, the primary OH group of sphingosine is bonded to a sugar residue through a β-glycosidic linkage (Section 16.10). Sphingomyelins are phospholipids because they contain a phosphate group. Cerebrosides, on the other hand, are not phospholipids.

20.6 Terpenes

Terpenes are a diverse class of lipids. More than 20,000 terpenes are known. They can be hydrocarbons, or they can contain oxygen and be alcohols, ketones, or aldehydes. Certain terpenes have been used as spices, perfumes, and medicines for thousands of years.

menthol peppermint oil	geraniol geranium oil	zingiberene oil of ginger	β-selinene oil of celery

Leopold Stephen Ružička (1887–1976) *was the first to recognize that many organic compounds contain multiples of five carbons. A Croatian, Ružička attended college in Switzerland and became a Swiss citizen in 1917. He was a professor of chemistry at the University of Utrecht in the Netherlands and later at the Federal Institute of Technology in Zürich. For his work on terpenes, he shared the 1939 Nobel Prize in chemistry with Adolph Butenandt (page 516).*

After analyzing a large number of terpenes, organic chemists realized that they contained carbon atoms in multiples of five. These naturally occurring compounds contain 10, 15, 20, 25, 30, and 40 carbon atoms, which suggests that there is a compound with five carbon atoms that serves as their building block. Further investigation showed that their structures are consistent with the assumption that they were made by joining together isoprene units, usually in a head-to-tail fashion. (The branched end of isoprene is called the head, and the unbranched end is called the tail.) Isoprene is the common name for 2-methyl-1,3-butadiene.

head head

tail tail

α-farnesene
a sesquiterpene found in the waxy coating on apple skins

In the case of cyclic compounds, the linkage of the head of one isoprene unit to the tail of another is followed by an additional linkage to form the ring. The second linkage is not necessarily head-to-tail, but is whatever is necessary to form a stable five- or six-membered ring.

head

tail

carvone
spearmint oil
a monoterpene

Terpenes are classified according to the number of carbons they contain. **Monoterpenes** are composed of two isoprene units, so they have 10 carbons. **Sesquiterpenes**, with 15 carbons, are composed of three isoprene units. Many fragrances and flavorings found in plants are monoterpenes and sesquiterpenes. These compounds are known as *essential oils*.

Triterpenes (six isoprene units) and **tetraterpenes** (eight isoprene units) have important biological roles. For example, **squalene**, a triterpene, is the precursor of cholesterol, which is the precursor of all the other steroid molecules (Section 20.8).

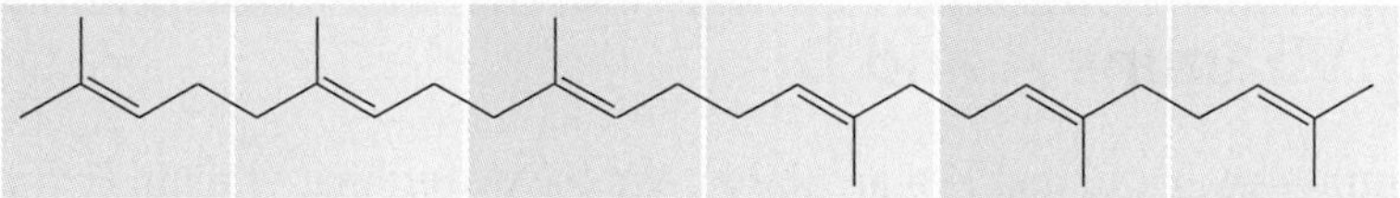

squalene

Carotenoids are tetraterpenes. Lycopene, the compound responsible for the red coloring of tomatoes and watermelon, and β-carotene, the compound that causes carrots and apricots to be orange, are examples of carotenoids. β-carotene is also the coloring agent used in margarine. β-carotene and other colored compounds are found in the leaves of trees, but their characteristic colors are usually obscured by the green color of chlorophyll (Section 6.12). In the fall, when chlorophyll degrades, the colors become apparent. The many conjugated double bonds in lycopene and β-carotene cause the compounds to be colored (Section 6.11).

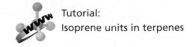

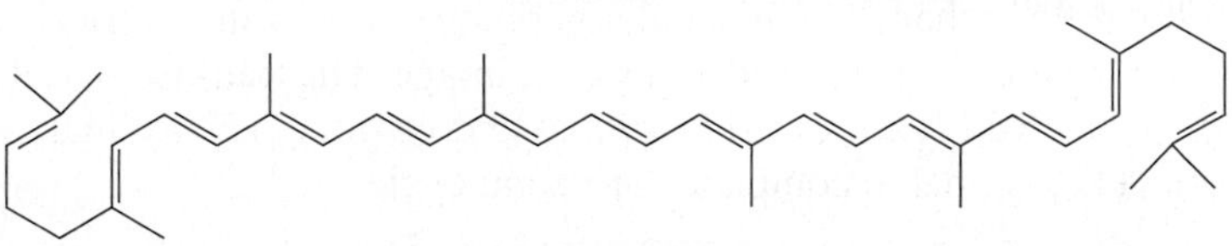

Tutorial:
Isoprene units in terpenes

lycopene

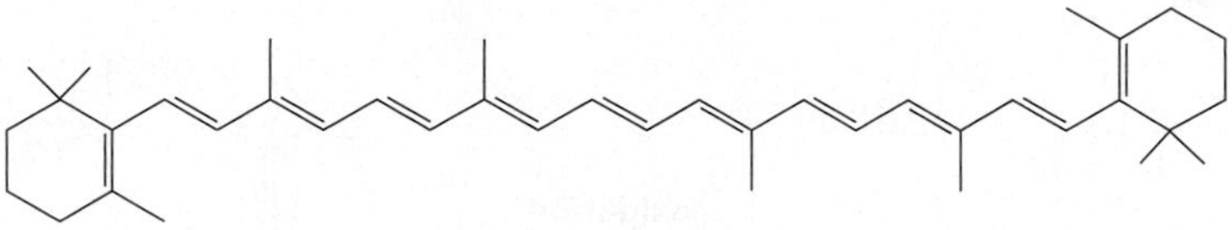

β-carotene

PROBLEM 11◆

Limonene has an asymmetric center so it exists as two different stereoisomers. The *R* isomer is found in oranges and the *S* isomer is found in lemons. Which of the following is found in oranges?

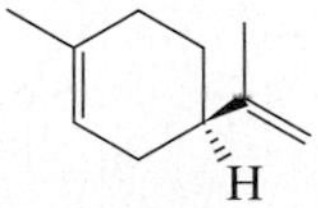

(+)-limonene 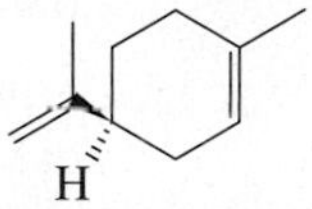   **(−)-limonene**

PROBLEM 12◆   SOLVED

a What class of lipid is α-farnesene?
b What class of lipid is spearmint oil?

SOLUTION TO 12a α-Farnesene has 15 carbons, so it is a sesquiterpene.

PROBLEM 13   SOLVED

Mark off the isoprene units in

a zingiberene   **b** menthol

SOLUTION TO 13a   For zingiberene, we have

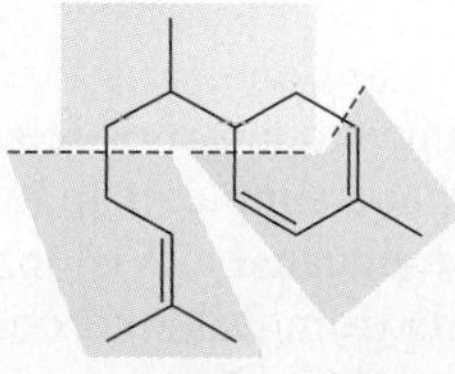

20.7 **Vitamins A and D**

Four vitamins—A, D, E, and K—are lipids. We have seen that vitamin E is an antioxidant (Section 9.6) and vitamin K is a coenzyme (Section 18.12).

Vitamin A, also called retinol, plays an important role in vision. β-Carotene is the major dietary source of the vitamin: Cleaving β-carotene forms two molecules of vitamin A.

The retina of the eye contains cone cells and rod cells. The cone cells are responsible for color vision and for vision in bright light. The rod cells are responsible for vision in dim light. In rod cells, vitamin A is oxidized to an aldehyde and an enzyme catalyzes the isomerization of the trans double bond at C-11 to a cis double bond. The protein *opsin* uses a lysine side chain to form an imine with (11Z)-retinal, forming *rhodopsin*. When rhodopsin absorbs visible light, the cis double bond isomerizes to a trans double bond. This change in molecular geometry causes an electrical signal to be sent to the brain, where it is perceived as a visual image. The trans isomer of rhodopsin is not stable and is hydrolyzed to (11E)-retinal and opsin. (11E)-Retinal is then converted back to (11Z)-retinal to complete the vision cycle.

the chemistry of vision

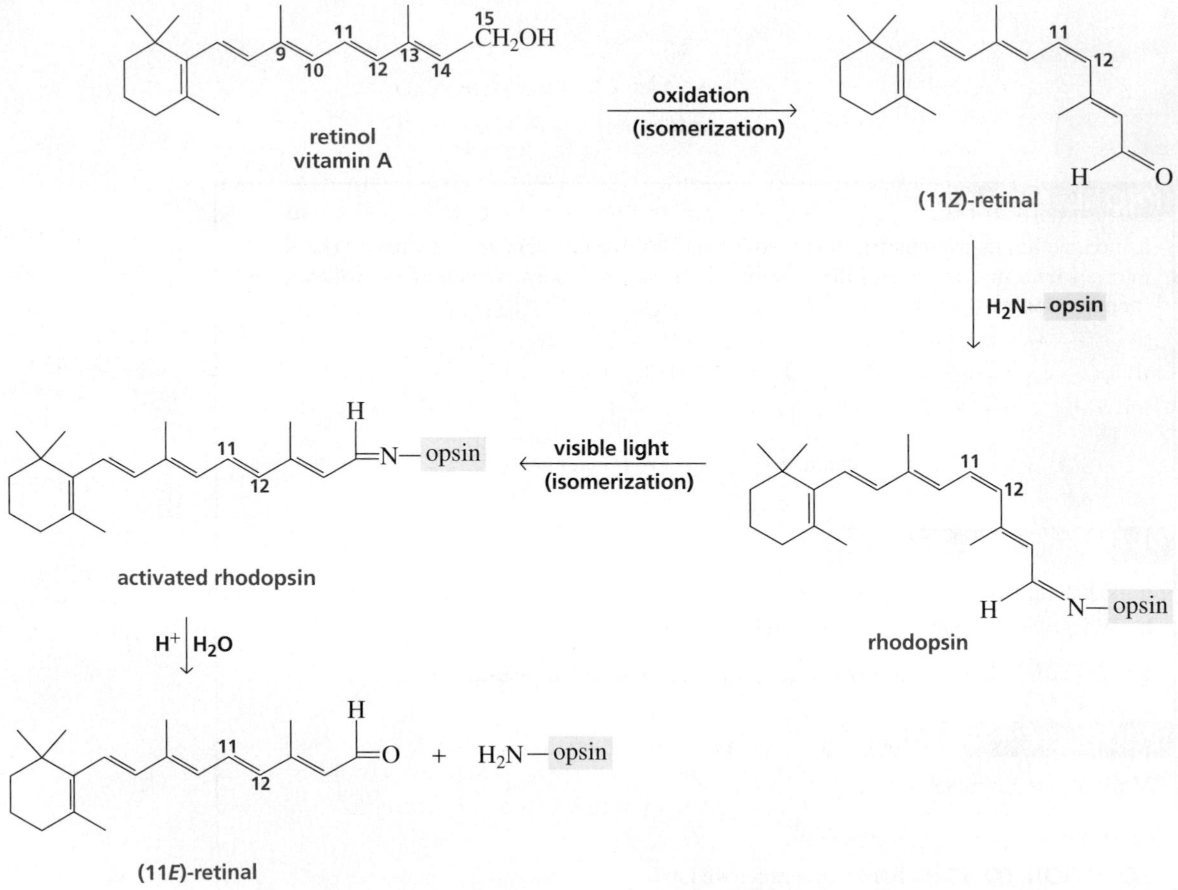

The details of how the foregoing sequence of reactions creates a visual image are not clearly understood. The fact that a simple change from cis to trans can be responsible for initiating a process as complicated as vision, though, is remarkable.

The body synthesizes vitamin D from 7-dehydrocholesterol, a steroid found in the skin, in a series of reactions that requires ultraviolet light.

7-dehydrocholesterol → vitamin D

A deficiency in vitamin D, which can be prevented by getting enough sun, causes a disease known as rickets. Rickets is characterized by deformed bones and stunted growth. Too much vitamin D is also harmful because it causes the calcification of soft tissues. It is thought that skin pigmentation evolved to protect the skin from the sun's ultraviolet rays in order to prevent the synthesis of too much vitamin D. This agrees with the observation that peoples who are indigenous to countries close to the equator have greater skin pigmentation.

20.8 Steroids

Hormones are chemical messengers—delivered by the bloodstream to target tissues in order to stimulate or inhibit some process. Many hormones are **steroids**. All steroids contain a tetracyclic ring system composed of three six-membered rings and one five-membered ring (Section 3.13). Because steroids are nonpolar compounds, they are lipids. Their nonpolar character allows them to cross cell membranes, so they can leave the cells in which they are synthesized and enter their target cells.

The most abundant member of the steroid family in animals is **cholesterol**, the precursor of all other steroids. Cholesterol is biosynthesized from squalene, a triterpene (Section 20.6), and is an important component of cell membranes (Figure 20.2). Because cholesterol has eight asymmetric centers, 256 stereoisomers are possible, but only one exists in nature (Chapter 8, Problem 20). We have seen how cholesterol is related to heart disease and how high cholesterol is treated clinically (Section 3.13).

the steroid ring system

cholesterol

The steroid hormones can be divided into five classes: glucocorticoids, mineralocorticoids, androgens, estrogens, and progestins.

Glucocorticoids, as their name suggests, are involved in glucose metabolism, as well as in the metabolism of proteins and fatty acids. Cortisone is an example of a glucocorticoid. Because of its anti-inflammatory effect, it is used clinically to treat arthritis and other inflammatory conditions.

cortisone

aldosterone

*Two German chemists, **Heinrich Otto Wieland (1877–1957)** and **Adolf Windaus (1876–1959)**, each received a Nobel Prize in chemistry (Wieland in 1927 and Windaus in 1928) for work that led to the determination of the structure of cholesterol.*

Heinrich Wieland, *the son of a chemist, was a professor at the University of Munich, where he showed that bile acids were steroids and determined their individual structures. During World War II, he remained in Germany but was openly anti-Nazi.*

Adolf Windaus *originally intended to be a physician, but the experience of working with Emil Fischer for a year changed his mind. He discovered that vitamin D was a steroid, and he was the first to recognize that vitamin B$_1$ contained sulfur.*

Mineralocorticoids cause increased reabsorption of Na^+, Cl^-, and HCO_3^- by the kidneys, leading to an increase in blood pressure. Aldosterone is an example of a mineralocorticoid.

The male sex hormones, known as *androgens*, are secreted by the testes. They are responsible for the development of male secondary sex characteristics during puberty. They also promote muscle growth. Testosterone and 5α-dihydrotestosterone are androgens.

5 α-dihydrotestosterone   **testosterone**

Adolf Friedrich Johann Butenandt (1903–1995) *was born in Germany. He shared the 1939 Nobel Prize in chemistry (with Ružička see page 512) for isolating and determining the structures of estrone, androsterone, and progesterone. Forced by the Nazi government to refuse the prize, he accepted it after World War II. He was the director of the Kaiser Wilhelm Institute in Berlin and later was a professor at the Universities of Tübingen and Munich.*

Estradiol and estrone are female sex hormones known as *estrogens*. They are secreted by the ovaries and are responsible for the development of female secondary sex characteristics. They also regulate the menstrual cycle. Progesterone is the hormone that prepares the lining of the uterus for implantation of an ovum and is essential for the maintenance of pregnancy. It also prevents ovulation during pregnancy.

estradiol   **estrone**   **progesterone**

Although the various steroid hormones have remarkably different physiological effects, their structures are quite similar. For example, the only difference between testosterone and progesterone is the substituent on the five-membered ring, and the only difference between 5α-dihydrotestosterone and estradiol is one carbon and six hydrogens. However, these compounds make the difference between being male and being female. These examples illustrate the extreme specificity of biochemical reactions.

In addition to being the precursor of all the steroid hormones in animals, cholesterol is the precursor of the *bile acids*. In fact, the word *cholesterol* is derived from the Greek words *chole* meaning "bile" and *stereos* meaning "solid." The bile acids—cholic acid and chenodeoxycholic acid—are synthesized in the liver, stored in the gallbladder, and secreted into the small intestine, where they act as emulsifying agents so that fats and oils can be digested by water-soluble digestive enzymes.

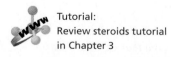

Tutorial:
Review steroids tutorial
in Chapter 3

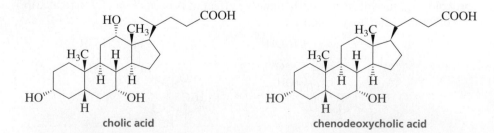

cholic acid   **chenodeoxycholic acid**

LIPOPROTEINS

Lipoproteins are composed of lipids (cholesterol, triglycerides, and phospholipids) and proteins. The density of a lipoprotein depends on the relative amounts of lipid and protein. Proteins are more dense than lipids, so the higher the protein component of a lipoprotein, the more dense it will be. In Section 3.13, we saw that low-density lipoprotein (LDL) is the so-called bad cholesterol, and high density lipoprotein (HDL) is the so-called good cholesterol. LDL is composed of about 75% lipid and 25% protein; HDL is composed of about 50% lipid and 50% protein.

PROBLEM 14◆

What functional groups are in aldosterone?

20.9 Synthetic Steroids

The potent physiological effects of steroids led scientists, in their search for new drugs, to synthesize steroids that are not available in nature and to investigate their physiological effects. Stanozolol and Dianabol are drugs developed in this way. They have the same muscle-building effect as testosterone. Steroids that aid in the development of muscle are called *anabolic steroids*. These drugs are available by prescription and are used to treat people suffering from traumas accompanied by muscle deterioration. The same drugs have been administered to athletes and racehorses to increase their muscle mass. Stanozolol was the drug detected in several athletes in the 1988 Olympics. Anabolic steroids, when taken in relatively high dosages, have been found to cause liver tumors, personality disorders, and testicular atrophy.

stanozolol   **Dianabol®**

Many synthetic steroids have been found to be much more potent than natural steroids. Norethindrone, for example, is better than progesterone in arresting ovulation. Another synthetic steroid terminates pregnancy within the first nine weeks of gestation. Notice that these oral contraceptives have structures similar to that of progesterone.

norethindrone   **RU 486**

PROBLEM 15

How do testosterone, a natural muscle-building steroid, and stanozolol, a synthetic muscle-building steroid, differ in structure?

Summary

Lipids are organic compounds that are found in living organisms and that are soluble in nonpolar solvents. **Fatty acids** are carboxylic acids with long unbranched hydrocarbon chains. Double bonds in fatty acids have the cis configuration. Fatty acids with more than one double bond are called **polyunsaturated fatty acids**. Double bonds in naturally occurring unsaturated fatty acids are separated by one CH_2 group. **Waxes** are esters formed from long-chain carboxylic acids and long-chain alcohols.

Triglycerides are compounds in which the three OH groups of glycerol are esterified with fatty acids. Liquid triglycerides are called **oils**. Those that are solids or semisolids at room temperature are called **fats**. Some or all of the double bonds of polyunsaturated oils can be reduced by catalytic hydrogenation. **Phosphoglycerides** differ from triglycerides in that the terminal **OH** group of glycerol is

esterified with phosphoric acid instead of a fatty acid. Phosphoglycerides form membranes by arranging themselves in a **lipid bilayer**. **Phospholipids** are lipids that contain a phosphate group. **Sphingolipids**, also found in membranes, contain sphingosine instead of glycerol.

Terpenes contain carbon atoms in multiples of five. They are made by joining together five-carbon isoprene units, usually in a head-to-tail fashion. **Monoterpenes**—terpenes with two isoprene units—have 10 carbons; **sesquiterpenes** have 15. **Squalene**, a **triterpene** (a terpene with six isoprene units), is a precursor of steroid molecules. Lycopene and β-carotene are **tetraterpenes** called **carotenoids**.

Hormones are chemical messengers. Many hormones are **steroids**. All steroids contain a tetracyclic ring system. The most abundant member of the steroid family in animals is **cholesterol**, the precursor to all other steroids.

Problems

16. Give the product that would be obtained from the reaction of cholesterol with each of the following reagents:
   a. H_2, Pd/C   **b.** acetyl chloride   **c.** H_2SO_4, Δ **d.** H_2O, H^+

17. Do all triglycerides have the same number of asymmetric centers?

18. Cardiolipins are found in heart muscles. Give the products formed when a cardiolipin undergoes complete acid-catalyzed hydrolysis.

a cardiolipin

19. Nutmeg contains a simple, fully saturated triglyceride with a molecular weight of 722. Draw its structure.

20. The asymmetric center of phospholipids has the R configuration. Which of the following is a lecithin?

21. Draw the structures of three different sphingomyelins.

22. Draw the structure of a galactocerebroside.

23. Mark off the isoprene units in geraniol (page 512).

24. Mark off the isoprene units in squalene. One of the linkages in squalene is tail-to-tail, not head-to-tail. What does this suggest about how squalene is synthesized in nature? (*Hint:* Locate the position of the tail-to-tail linkage.)

25. Mark off the isoprene units in lycopene and β-carotene. Can you detect a similarity in the way in which squalene, lycopene, and β-carotene are biosynthesized?

26. 5-Androstene-3,17-dione is isomerized to 4-androstene-3,17-dione by hydroxide ion. Propose a mechanism for this reaction.

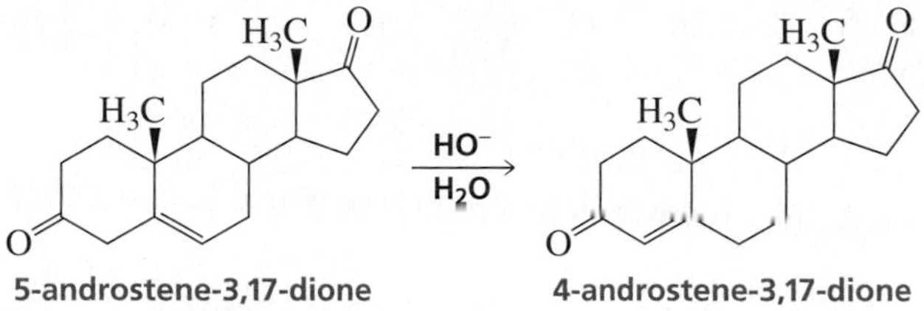

5-androstene-3,17-dione   4-androstene-3,17-dione

27. **a.** How many different triglycerides are there in which one of the fatty acid components is lauric acid and two are myristic acid?
   b. How many different triglycerides are there in which one of the fatty acid components is lauric acid, one is myristic acid, and one is palmitic acid?

28. Diethylstilbestrol (DES) was given to pregnant women to prevent miscarriage, until it was found that the drug caused cancer in both the mothers and their female children. DES has estradiol activity even though it is not a steroid. Draw DES in a way that shows that it is structurally similar to estradiol.

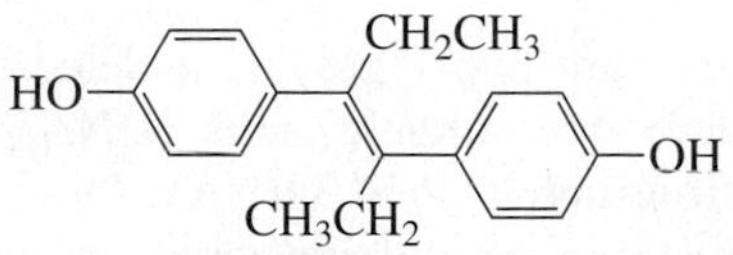

diethylstilbestrol
DES

21 Nucleosides, Nucleotides, and Nucleic Acids

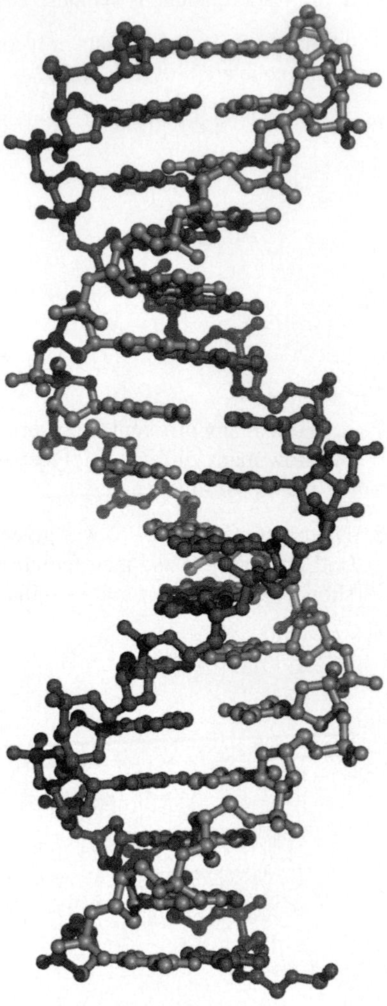

DNA

There are two types of nucleic acids: **deoxyribonucleic acid (DNA)** and **ribonucleic acid (RNA)**. DNA encodes an organism's entire hereditary information and controls the growth and division of cells. The genetic information stored in DNA is transcribed into RNA, and the information in RNA can then be translated for the synthesis of all the proteins needed for cellular structure and function.

In 1869, DNA was first isolated from the nuclei of white blood cells. Because this material was found in the nucleus and was acidic, it was called *nucleic acid*. Eventually, scientists found that the nuclei of all cells contain DNA, but it wasn't until 1944 that they realized that nucleic acids are the carriers of genetic information. In 1953, James Watson and Francis Crick described the three-dimensional structure of DNA—the famed double helix.

21.1 Nucleosides and Nucleotides

Nucleic acids are chains of five-membered-ring sugars linked by forming a **phosphodiester** (Section 20.5) with phosphoric acid (Figure 21.1). The anomeric carbon of each sugar is bonded to a nitrogen of a heterocyclic compound in a β-glycosidic linkage (Section 16.10). Because the heterocyclic compounds are amines, they are commonly referred to as **bases**. In RNA, the five-membered-ring sugar is D-ribose. In DNA, it is 2'-deoxy-D-ribose (D-ribose without an OH group in the 2-position).

The vast differences in heredity among species and among members of the same species are determined by the sequence of the bases in DNA. Surprisingly, there are only four bases in DNA—two are substituted purines (adenine and guanine), and two are substituted pyrimidines (cytosine and thymine).

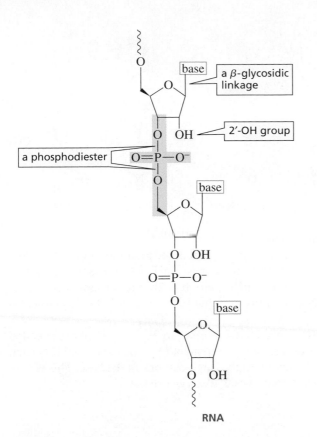

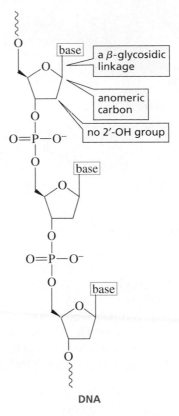

◀ **Figure 21.1**
Nucleic acids consist of a chain of five-membered-ring sugars linked by phosphate groups. Each sugar (D-ribose in RNA, 2'-deoxy-D-ribose in DNA) is bonded to a heterocyclic amine in a β-glycosidic linkage.

RNA   DNA

THE STRUCTURE OF DNA: WATSON, CRICK, FRANKLIN, AND WILKINS

James D. Watson was born in Chicago in 1928. He graduated from the University of Chicago at the age of 19 and received a Ph.D. three years later from Indiana University. In 1951, as a postdoctoral fellow at Cambridge University, Watson worked on determining the three-dimensional structure of DNA.

Francis H. C. Crick (1916–2004) was born in Northampton, England. Originally trained as a physicist, Crick was involved in radar research during World War II. After the war, deciding that the most interesting problem in science was trying to understand the physical basis of life, he entered Cambridge University to study the structure of biological molecules by X-ray analysis. He was a graduate student when he carried out his portion of the work that led to the proposal of the double helical structure of DNA. He received a Ph.D. in chemistry in 1953.

Rosalind Franklin was born in London in 1920. She graduated from Cambridge University and in 1942 went to Paris to study X-ray diffraction techniques. In 1951 she returned to England, accepting a position to develop an X-ray diffraction unit in the biophysics department at King's College. Her X-ray studies showed that DNA was a helix with phosphate groups on the outside of the molecule. Franklin died in 1958 without knowing the significance her work had played in determining the structure of DNA and without being recognized for that work.

Watson and Crick received the 1962 Nobel Prize in medicine or physiology for determining the double helical structure of DNA. They shared the prize with Maurice Wilkins, who contributed X-ray studies that confirmed the double helical structure. Wilkins (1916-2004) was born in New Zealand and moved to England six years later with his parents. During World War II, he joined other British scientists who were working with American scientists on the development of the atomic bomb.

Francis Crick (left) and James Watson (right)

Rosalind Franklin

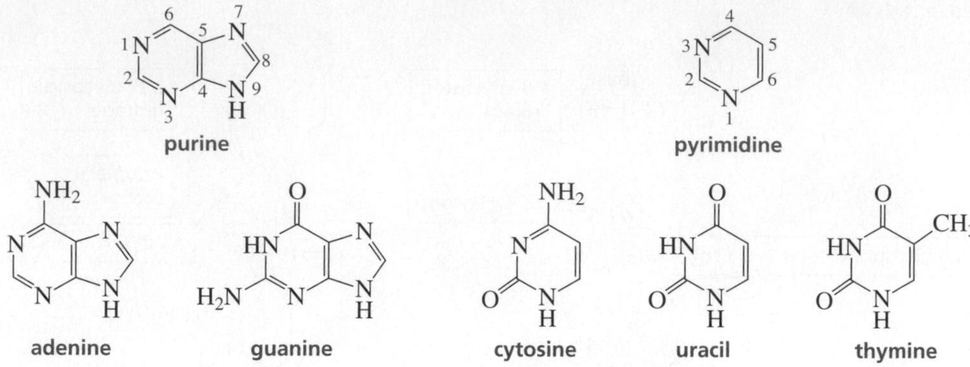

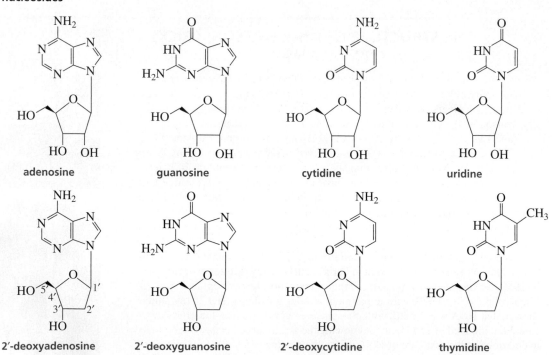

purine

pyrimidine

adenine

guanine

cytosine

uracil

thymine

RNA also contains only four bases. Three—adenine, guanine, and cytosine—are the same as those in DNA, but the fourth base in RNA is uracil instead of thymine. Notice that thymine and uracil differ only by a methyl group—thymine is 5-methyluracil. The reason DNA contains thymine instead of uracil is explained in Section 21.9.

A compound containing a base bonded to D-ribose or to 2′-deoxy-D-ribose is called a **nucleoside**. The ring positions of the sugar component of a nucleoside are indicated by primed numbers to distinguish them from the ring positions of the base. This is why the sugar component of DNA is referred to as 2′-deoxy-D-ribose.

nucleoside = base + sugar

nucleosides

adenosine

guanosine

cytidine

uridine

2′-deoxyadenosine

2′-deoxyguanosine

2′-deoxycytidine

thymidine

A **nucleotide** is a nucleoside with either the 5′- or the 3′-OH group bonded in an ester linkage to phosphoric acid. The nucleotides of RNA—where the sugar is D-ribose—are more precisely called **ribonucleotides**, whereas the nucleotides of DNA—where the sugar is 2′-deoxy-D-ribose—are called **deoxyribonucleotides**.

nucleotide = base + sugar + phosphate

nucleotides

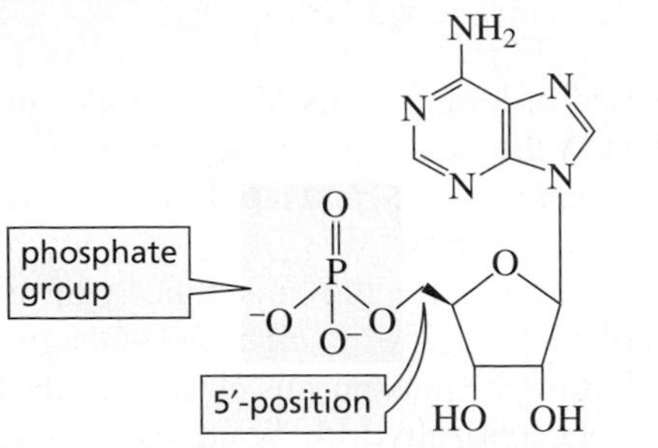

phosphate group

5′-position

adenosine 5′-monophosphate
a ribonucleotide

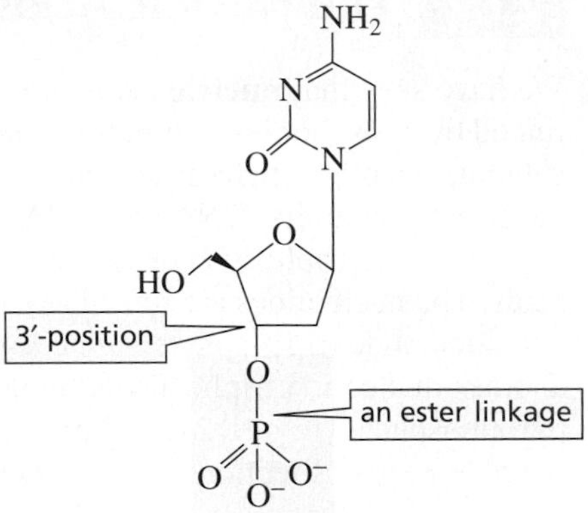

3′-position

an ester linkage

2′-deoxycytidine 3′-monophosphate
a deoxyribonucleotide

3-D Molecules:
Bases; Nucleosides;
Nucleotides

Nucleotides can exist as monophosphates, diphosphates, and triphosphates. They are named by adding *monophosphate* or *diphosphate* or *triphosphate* to the name of the nucleoside.

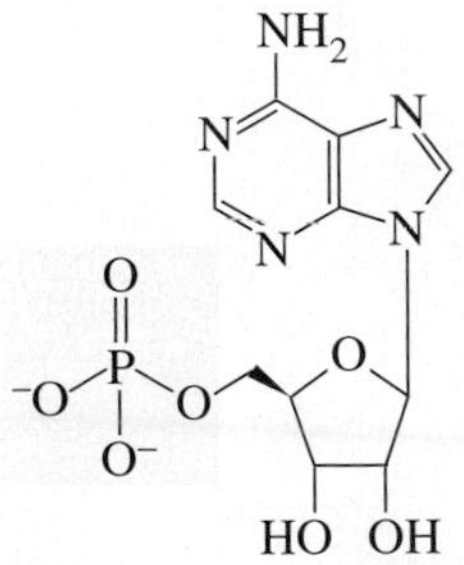

adenosine
5′-monophosphate
AMP

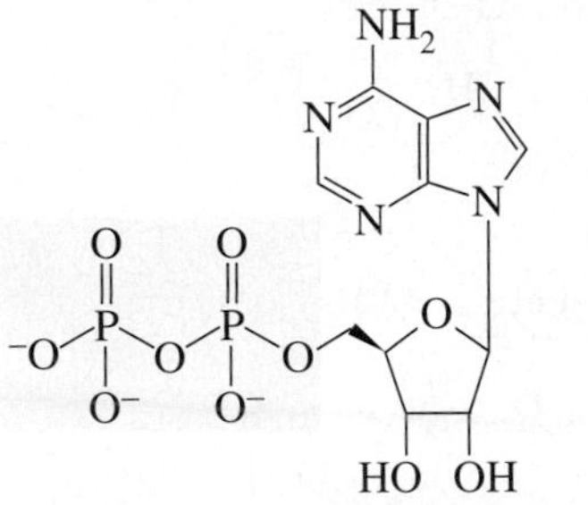

adenosine
5′-diphosphate
ADP

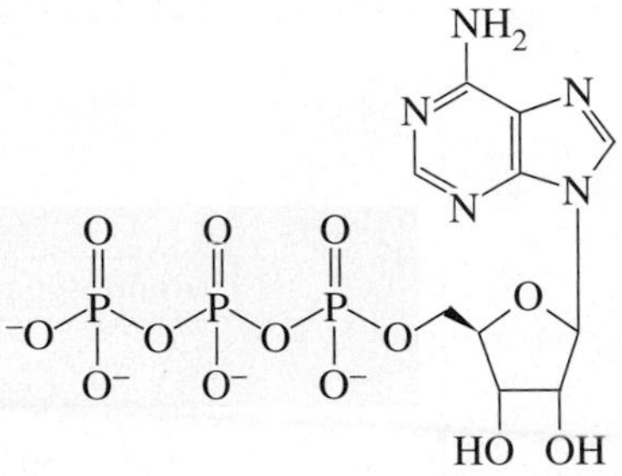

adenosine
5′-triphosphate
ATP

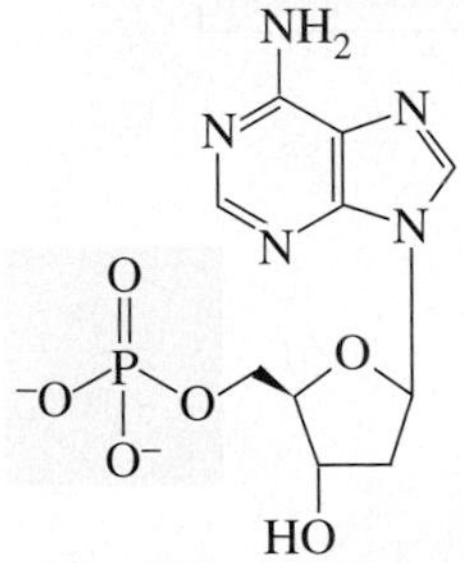

2′-deoxyadenosine
5′-monophosphate
dAMP

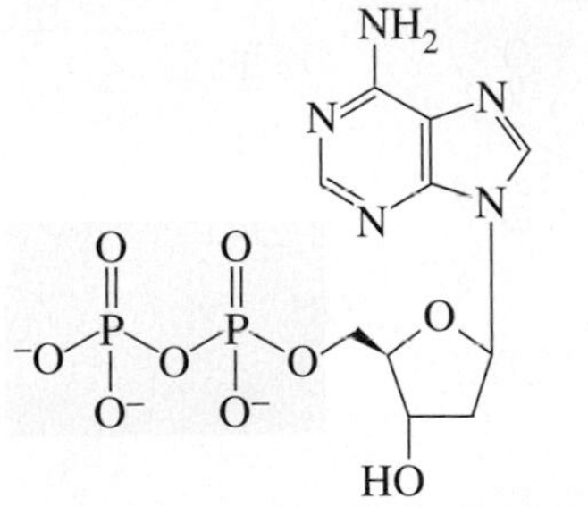

2′-deoxyadenosine
5′-diphosphate
dADP

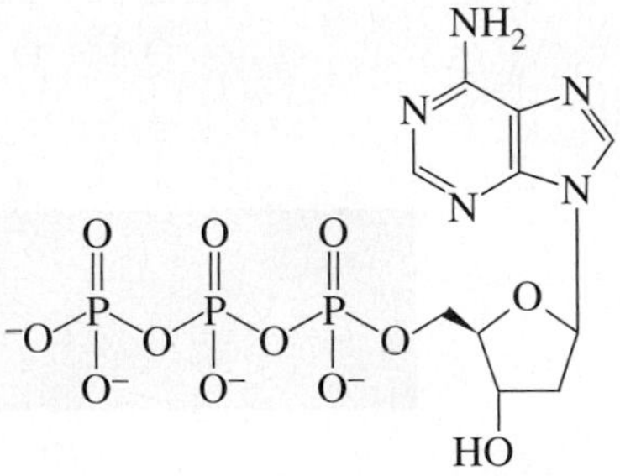

2′-deoxyadenosine
5′-triphosphate
dATP

PROBLEM 1

Draw the structure for each of the following:

a. dCDP

b. dTTP

c. dUMP

d. UDP

e. guanosine 5′-triphosphate

f. adenosine 3′-monophosphate

21.2 The Nucleic Acids

We have seen that **nucleic acids** are composed of long strands of nucleotide subunits linked by phosphodiester bonds (Figure 21.1). A **dinucleotide** contains two nucleotide subunits, an **oligonucleotide** contains three to ten subunits, and a **polynucleotide** contains many subunits. DNA and RNA are polynucleotides.

Nucleotide triphosphates are the starting materials for the biosynthesis of nucleic acids. The nucleotides are linked as a result of nucleophilic attack by a 3'-OH group of one nucleotide triphosphate on the α-phosphorus (the phosphorus closest to ribose) of another nucleotide triphosphate, breaking a phosphoanhydride bond and eliminating pyrophosphate (Figure 21.2). This means that the growing polymer is synthesized in the 5' $\longrightarrow$ 3' direction; in other words, new nucleotides are added to the 3'-end.

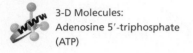

3-D Molecules:
Adenosine 5'-triphosphate
(ATP)

Figure 21.2 ▶
Addition of nucleotides to a growing strand of DNA. Biosynthesis occurs in the 5' $\longrightarrow$ 3' direction.

The **primary structure** of a nucleic acid is the sequence of bases in the strand. By convention, the sequence of bases in a polynucleotide is written in the 5' $\longrightarrow$ 3' direction (the 5'-end is on the left). Notice that the nucleotide at the 5'-end of the strand has an unlinked 5'-triphosphate group, and the nucleotide at the 3'-end of the strand has an unlinked 3'-hydroxyl group.

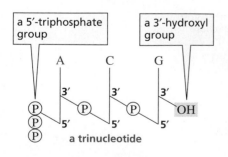

a 5'-triphosphate group

a 3'-hydroxyl group

a trinucleotide

ATGAGCCATGTAGCCTAATCGGC

5'-end

3'-end

DNA consists of two strands of nucleic acids with the sugar–phosphate backbone on the outside and the bases on the inside. The chains are held together by hydrogen bonds between the bases on one strand and the bases on the other strand (Figure 21.3). Adenine (A) always pairs with thymine (T) and guanine (G) always pairs with cytosine (C). This means the two strands are *complementary*—where there is an A in one strand, there is a T in the opposing strand; where there is a G in one strand, there is a C in the other strand (Figure 21.3). Thus, if you know the sequence of bases in one strand, you can figure out the sequence of bases in the other strand.

Why does A pair with T? Why does G pair with C? First of all, the width of the double-stranded molecule is relatively constant, so a purine must pair with a pyrimidine. If the larger purines paired, the strand would bulge; if the smaller pyrimidines paired, the strands would have to contract to bring the two pyrimidines close enough to form hydrogen bonds. What causes A to pair with T rather than with C (the other pyrimidine)? The base pairing is dictated by maximizing the number of hydrogen bonds. Adenine forms two hydrogen bonds with thymine but would form only one hydrogen bond with cytosine. Guanine forms three hydrogen bonds with cytosine but would form only one hydrogen bond with thymine (Figure 21.4).

The two DNA strands run in opposite directions (Figures 21.3 and 21.5). The strands are not linear but are twisted into a helix around a common axis (see Figure 21.6a). The base pairs are planar and parallel to each other on the inside of the helix (Figure 21.6c). The secondary structure is therefore known as a **double helix**. The double helix resembles a circular staircase—the base pairs are the rungs and the sugar–phosphate backbone is the handrail.

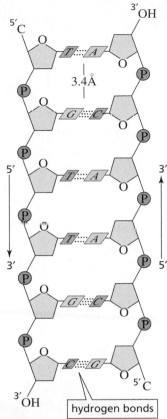

▲ **Figure 21.3**
Complementary base pairing in DNA. Adenine (a purine) always pairs with thymine (a pyrimidine); guanine (a purine) always pairs with cytosine (a pyrimidine).

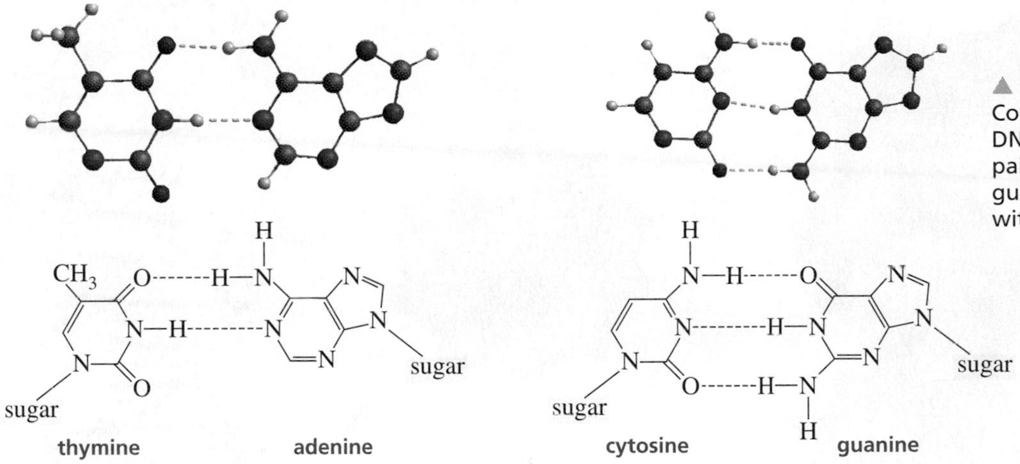

▲ **Figure 21.4**
Base pairing in DNA: Adenine and thymine form two hydrogen bonds; cytosine and guanine form three hydrogen bonds.

3-D Molecules:
Adenine–thymine base pair;
Guanine–cytosine base pair

PROBLEM 2◆

If one of the strands of DNA has the following sequence of bases running in the $5' \longrightarrow 3'$ direction:

$$5'-G-G-A-C-A-A-T-C-T-G-C-3.$$

a. What is the sequence of bases in the complementary strand?

b. What base is closest to the 5'-end in the complementary strand?

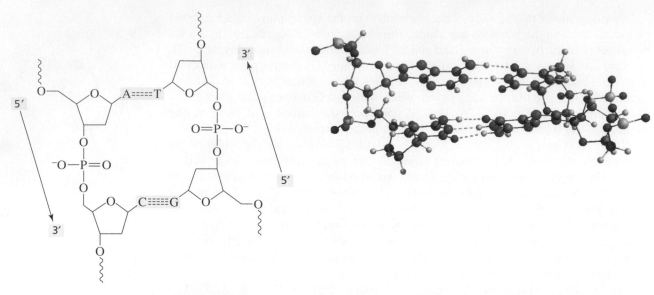

▲ **Figure 21.5**
The sugar–phosphate backbone of DNA is on the outside, and the bases are on the inside—with A's pairing with T's and G's pairing with C's. The two strands are antiparallel—they run in opposite directions.

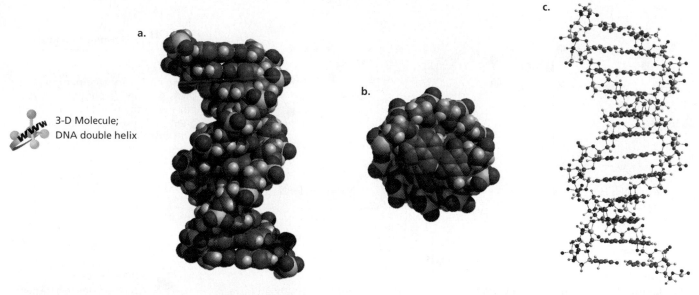

3-D Molecule;
DNA double helix

▲ **Figure 21.6**
(a) The DNA double helix.
(b) The view looking down the long axis of the helix.
(c) The bases are planar and parallel on the inside of the helix.

21.3 DNA is Stable—RNA is Easily Cleaved

RNA is easily cleaved. The 2′-OH group of ribose is the nucleophile that cleaves the strand (Figure 21.7). This explains why the 2′-OH group is absent in DNA. To preserve the genetic information, DNA must remain intact throughout the life span of a cell. Cleavage of DNA would have disastrous consequences for the cell and for life itself. RNA, in contrast, is synthesized as it is needed and is degraded once it has served its purpose.

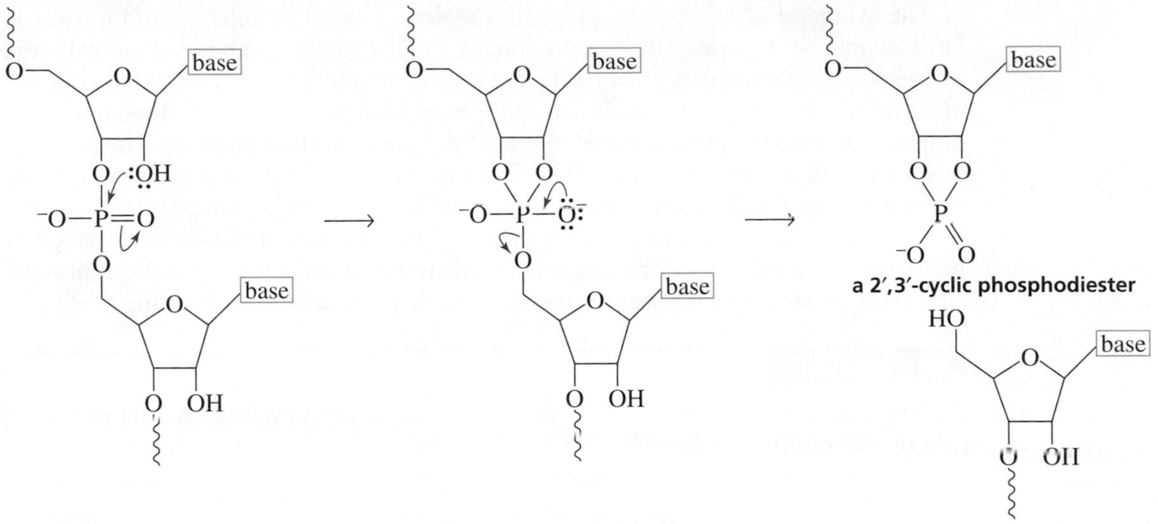

a 2′,3′-cyclic phosphodiester

▲ **Figure 21.7**
Cleavage of RNA by the 2′-OH group. It has been estimated that RNA is cleaved 3 billion times faster than DNA.

21.4 Biosynthesis of DNA: Replication

Each strand of DNA serves as a template for the synthesis of a complementary new strand (Figure 21.8). The new (daughter) DNA molecules are identical to the original (parent) molecule—they contain all the original genetic information. The synthesis of identical copies of DNA is called **replication**.

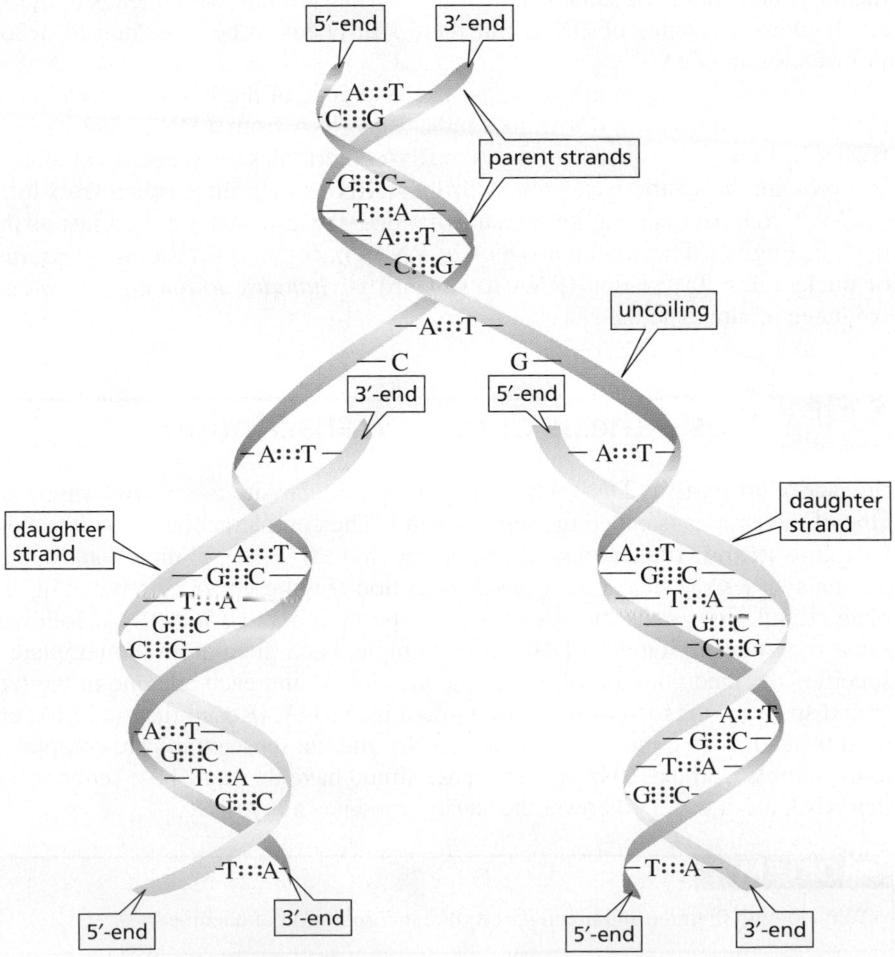

◀ **Figure 21.8**
Replication of DNA. The daughter strand on the left is synthesized continuously in the 5′ ⟶ 3′ direction; the daughter strand on the right is synthesized discontinuously in the 5′ ⟶ 3′ direction.

The synthesis of DNA takes place in a region of the molecule where the strands have started to separate. Because a nucleic acid can be synthesized only in the $5' \longrightarrow 3'$ direction, only the strand on the left in Figure 21.8 is synthesized continuously in a single piece (because it is synthesized in the $5' \longrightarrow 3'$ direction). The other strand needs to grow in the $3' \longrightarrow 5'$ direction, so it is synthesized discontinuously in small pieces. The synthesis of DNA in the $5' \longrightarrow 3'$ direction is catalyzed by an enzyme called DNA polymerase, and the fragments are joined together by an enzyme called DNA ligase. Each of the two new molecules of DNA—called daughter molecules—contain one of the original strands (blue strand) plus a newly synthesized strand (green strand). This process is called **semiconservative replication**.

PROBLEM 3

Using a dark line for parental DNA and a wavy line for DNA synthesized from parental DNA, show what the population of DNA molecules would look like in the fourth generation.

21.5 DNA and Heredity

The genetic information of a human cell is contained in 23 pairs of chromosomes. Each chromosome is composed of several thousand **genes** (segments of DNA). The total DNA of a human cell—the **human genome**—contains 3.1 billion base pairs.

Watson and Crick's proposal for the structure of DNA was an exciting development because the structure immediately suggested how DNA is able to pass on genetic information to succeeding generations. Because the two strands are complementary, both carry the same genetic information and replication leads to the synthesis of identical copies of DNA. But there still needs to be a method to decode the information in DNA.

The decoding is done in two steps. The sequence of the bases in DNA determines the sequence of bases in RNA; the synthesis of RNA from a DNA blueprint is called **transcription**. The sequence of bases in RNA determines the sequence of amino acids in a protein; the synthesis of proteins from an RNA blueprint is called **translation**.

Don't confuse transcription and translation—these words are used just as they are used in English. Transcription (DNA to RNA) is copying *within the same language* of nucleotides. Translation (RNA to protein) is *changing to another language*—the language of amino acids.

Transcription: DNA $\longrightarrow$ RNA

Translation: RNA $\longrightarrow$ protein

21.6 Biosynthesis of RNA: Transcription

Transcription starts by DNA unwinding at a particular site to give two single strands. One of the strands is called the **sense strand**. The complementary strand is called the **template strand**. The template strand is read in the $3' \longrightarrow 5'$ direction, so that RNA can be synthesized in the $5' \longrightarrow 3'$ direction (Figure 21.9). The bases in the template strand specify the bases that need to be incorporated into RNA, following the same base pairing found in DNA. For example, each guanine in the template strand specifies the incorporation of a cytosine into RNA, and each adenine in the template strand specifies the incorporation of a uracil into RNA. (Recall that, in RNA, uracil is used instead of thymine.) Because both RNA and the sense strand are complementary to the template strand, RNA and the sense strand have the same base sequence, except that RNA has a uracil wherever the sense strand has a thymine.

PROBLEM 4

Why do both thymine and uracil specify the incorporation of adenine?

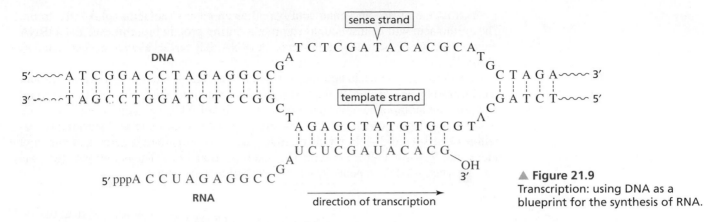

sense strand

DNA

5′ ∼∼∼ A T C G G A C C T A G A G G C C ... G A ... T C T C G A T A C A C G C A ... T G ... C T A G A ∼∼∼ 3′

3′ ∼∼∼ T A G C C T G G A T C T C C G G ... C T ... A G A G C T A T G T G C G T ... C ... G A T C T ∼∼∼ 5′

template strand

U C U C G A U A C A C G ... OH

5′ pppA C C U A G A G G C C ... G A ... 3′

RNA

direction of transcription

▲ **Figure 21.9**
Transcription: using DNA as a blueprint for the synthesis of RNA.

21.7 RNA

RNA is much shorter than DNA and is generally single stranded. There are three kinds of RNA:

- **messenger RNA (mRNA),** whose sequence of bases determines the sequence of amino acids in a protein;
- **ribosomal RNA (rRNA),** which comprises the particles on which the biosynthesis of proteins takes place; and
- **transfer RNA (tRNA),** which carries the amino acid that will be incorporated into a protein.

tRNA is much smaller than mRNA or rRNA. It contains only 70 to 90 nucleotides. The single strand of tRNA is folded into a characteristic cloverleaf structure (Figure 21.10a). All tRNAs have a CCA sequence at the 3′-end. The three bases at the bottom of the loop directly opposite the 5′- and 3′-ends are called an **anticodon** (Figures 21.10a and b).

Elizabeth Keller (1918–1997) *was the first to recognize that tRNA had a cloverleaf structure. She received a B.S. from the University of Chicago in 1940 and a Ph.D. from Cornell University Medical College in 1948. She worked at the Huntington Memorial Laboratory of Massachusetts General Hospital and at the United States Public Health Service. Later, she became a professor at MIT and then at Cornell University.*

◀ **Figure 21.10**
(a) tRNA^Ala: a transfer RNA that carries alanine. Compared with other RNAs, tRNA contains a high percentage of unusual bases (shown as empty circles). These bases result from an enzymatic modification of the four normal bases.
(b) tRNA^Phe: The anticodon is green; the CCA at the 3′-end is red.

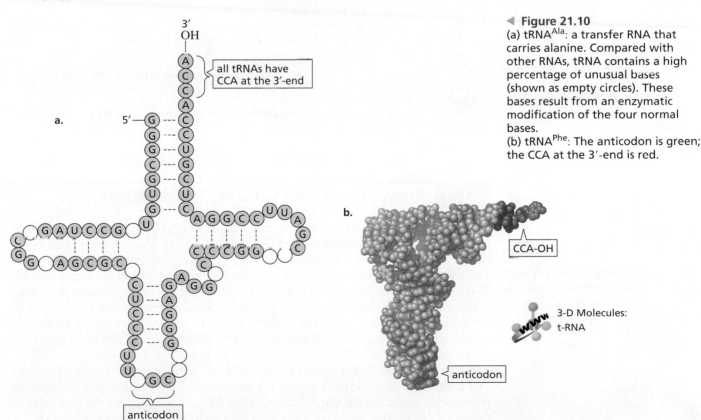

all tRNAs have CCA at the 3′-end

CCA-OH

3-D Molecules:
t-RNA

anticodon

anticodon

Each tRNA carries an amino acid bound as an ester to its terminal 3'-OH group. The amino acid will be inserted into a protein during protein biosynthesis. Each tRNA can carry only one particular amino acid. A tRNA that carries alanine is designated as tRNA^{Ala}.

How does an amino acid become attached to a tRNA? Attachment of the amino acid is catalyzed by an enzyme. In the first step of the reaction (Figure 21.11), the carboxyl group of the amino acid attacks the α-phosphorus of ATP, activating the carboxylate group by putting a good leaving group on it. Then a nucleophilic acyl substitution reaction occurs—the 3'-OH group of tRNA attacks the carbonyl carbon of the acyl adenylate, forming a tetrahedral intermediate (Section 12.5). The amino acyl tRNA is formed when AMP is expelled from the tetrahedral intermediate.

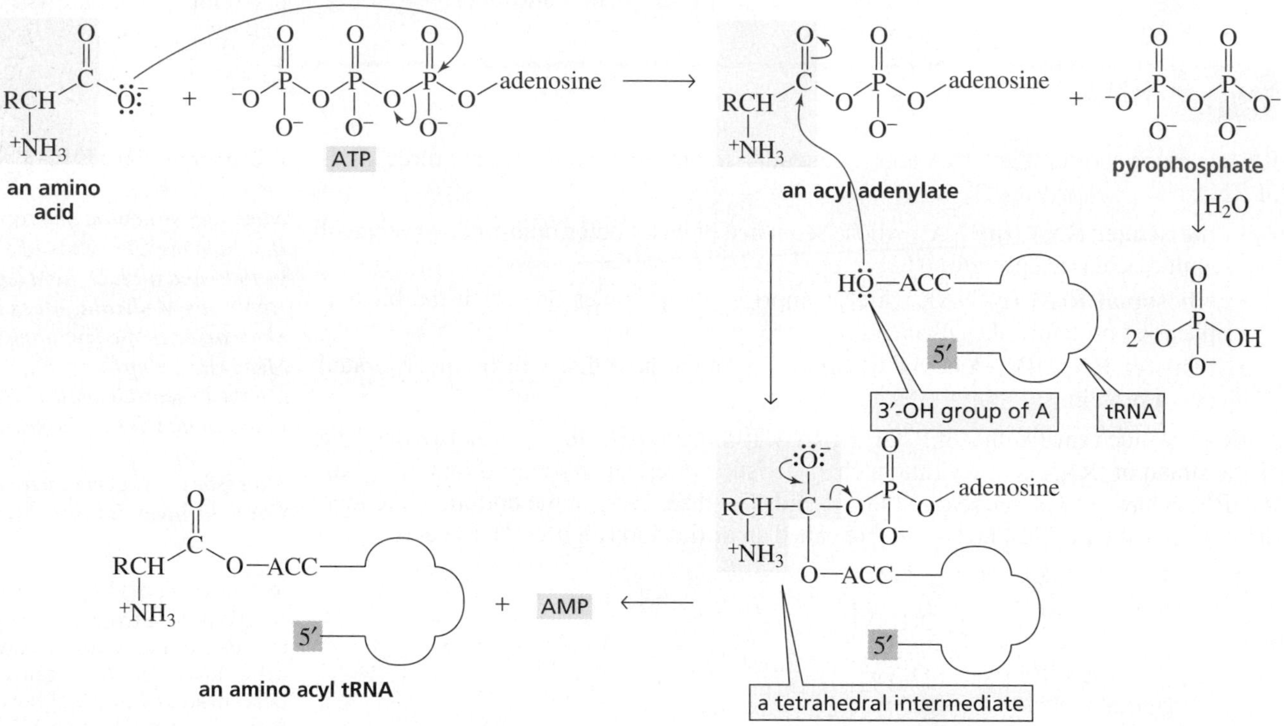

▲ **Figure 21.11**
The proposed mechanism for aminoacyl-tRNA synthetase—the enzyme that catalyzes the attachment of an amino acid to a tRNA.

The genetic code was worked out independently by **Marshall Nirenberg** *and* **Har Gobind Khorana**, *for which they shared the 1968 Nobel Prize in physiology or medicine.*

Marshall Nirenberg *was born in New York in 1927. He received a bachelor's degree from the University of Florida and a Ph.D. from the University of Michigan. He is a scientist at the National Institutes of Health.*

Tutorial:
Translation

21.8 Biosynthesis of Proteins: Translation

A protein is synthesized from its N-terminal end to its C-terminal end by reading the bases along the mRNA strand in the 5' $\longrightarrow$ 3' direction. A sequence of three bases, called a **codon**, specifies the amino acid that is to be incorporated into the protein. The bases are read consecutively and are never skipped. A codon is written with the 5'-nucleotide on the left. Each amino acid is specified by one or more three-base sequences. The three-base sequences and the amino acids they specify are known as the **genetic code** (Table 21.1). For example, UCA on mRNA codes for the amino acid serine, whereas CAG codes for glutamine. **Stop codons** tell the cell to "stop protein synthesis here."

How the information in mRNA is translated into a polypeptide is shown in Figure 21.12. In this figure, serine was the last amino acid incorporated into the growing polypeptide chain. Serine was specified by the AGC codon because the anticodon of the tRNA that carries serine is GCU (3'-UCG-5'). (Remember that a base sequence is

Table 21.1   The Genetic Code

5'-Position	Middle position				3'-Position
	U	C	A	G	
U	Phe	Ser	Tyr	Cys	U
	Phe	Ser	Tyr	Cys	C
	Leu	Ser	Stop	Stop	A
	Leu	Ser	Stop	Trp	G
C	Leu	Pro	His	Arg	U
	Leu	Pro	His	Arg	C
	Leu	Pro	Gln	Arg	A
	Leu	Pro	Gln	Arg	G
A	Ile	Thr	Asn	Ser	U
	Ile	Thr	Asn	Ser	C
	Ile	Thr	Lys	Arg	A
	Met	Thr	Lys	Arg	G
G	Val	Ala	Asp	Gly	U
	Val	Ala	Asp	Gly	C
	Val	Ala	Glu	Gly	A
	Val	Ala	Glu	Gly	G

Har Gobind Khorana *was born in India in 1922. He received a bachelor's and a master's degree from Punjab University and a Ph.D. from the University of Liverpool. In 1960, he joined the faculty at the University of Wisconsin and later became a professor at MIT.*

read in the 5' ⟶ 3' direction, so the sequence of bases in an anticodon must be read from right to left.) The next codon, CUU, signals for a tRNA with an anticodon of AAG (3'-GAA-5'). That particular tRNA carries leucine. The amino group of leucine reacts in an enzyme-catalyzed nucleophilic acyl substitution reaction with the ester on the adjacent tRNA, displacing the tRNA (Section 12.5). The next codon (GCC) brings in a tRNA carrying alanine. The amino group of alanine displaces the tRNA that brought in leucine. Subsequent amino acids are brought in one at a time in the same way, with the codon in mRNA specifying the amino acid to be incorporated by complementary base pairing with the anticodon of the tRNA that carries that amino acid.

Robert W. Holley (1922–1993), *who worked on the structure of tRNA molecules, shared the 1968 Nobel Prize with Nirenberg and Khorana. Holley was born in Illinois and received a bachelor's degree from the University of Illinois and a Ph.D. from Cornell University. During World War II, he worked on the synthesis of penicillin at Cornell Medical School. He was a professor at Cornell and later at the University of California, San Diego. He was also a noted sculptor.*

PROBLEM 5◆

If methionine is the first amino acid incorporated into a oligopeptide, what oligopeptide is encoded for by the following stretch of mRNA?

5'—G—C—A—U—G—G—A—C—C—C—C—G—U—U—A—U—
U—A—A—A—C—A—C—3'

PROBLEM 6◆

Four C's occur in a row in the segment of mRNA in Problem 5. What oligopeptide would be formed from the mRNA if one of the four C's were cut out of the strand?

PROBLEM 7

UAA is a stop codon. Why does the UAA sequence in mRNA in Problem 5 not cause protein synthesis to stop?

PROBLEM 8◆

Write the sequences of bases in the sense strand of DNA that resulted in the mRNA in Problem 5.

A sculpture done by Robert Holley

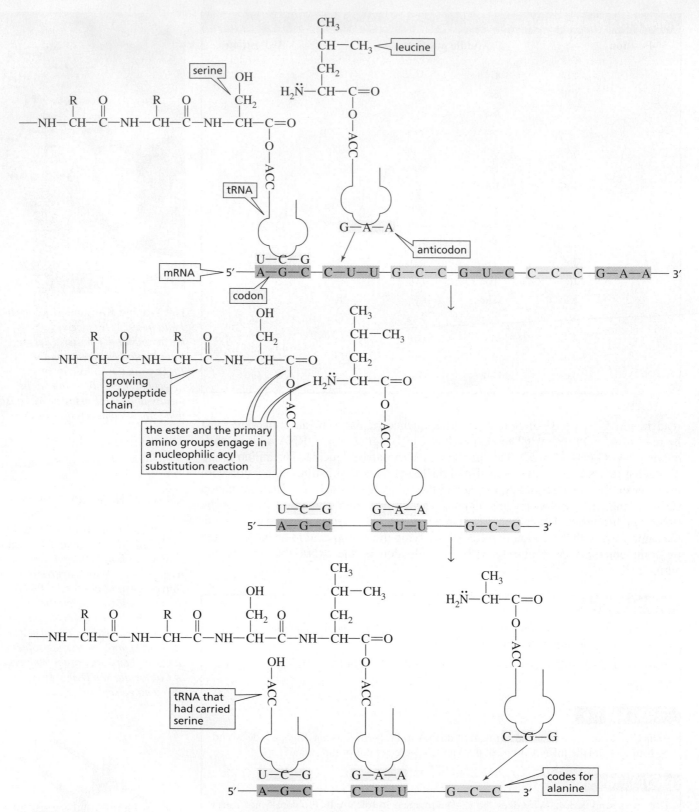

▲ **Figure 21.12**
Translation. The sequence of bases in mRNA determines the sequence of amino acids in a protein.

SICKLE CELL ANEMIA

Sickle cell anemia is an example of the damage that can be caused by a change in a single base of DNA. It is a hereditary disease caused when a GAG triplet becomes a GTG triplet in the sense strand of a section of DNA that codes for the protein component of hemoglobin. As a consequence, the mRNA codon becomes GUG—which signals for the incorporation of valine—rather than GAG—which would have signaled for the incorporation of glutamate. The change from a polar glutamate to a nonpolar valine is sufficient to change the shape of the deoxyhemoglobin molecule and induce aggregation, causing it to precipitate in red blood cells. This stiffens the cells, making it difficult for them to squeeze through a capillary. Blocked capillaries cause severe pain and can be fatal.

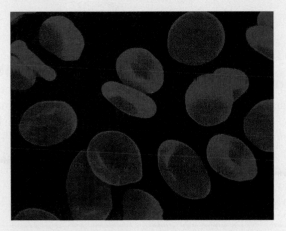

Normal red blood cells

Sickle red blood cells

ANTIBIOTICS THAT ACT BY INHIBITING TRANSLATION

Puromycin is one of several antibiotics that acts by inhibiting translation. Puromycin mimics the 3′-CCA-aminoacyl portion of a tRNA. If, during translation, the enzyme is fooled into transferring the growing peptide chain to the amino group of puromycin rather than to the amino group of the incoming 3′-CCA-aminoacyl tRNA, protein synthesis stops. Because puromycin blocks protein synthesis in eukaryotes as well as in prokaryotes, it is poisonous to humans and therefore is not a clinically useful antibiotic. To be clinically useful, an antibiotic must affect protein synthesis only in prokaryotic cells.

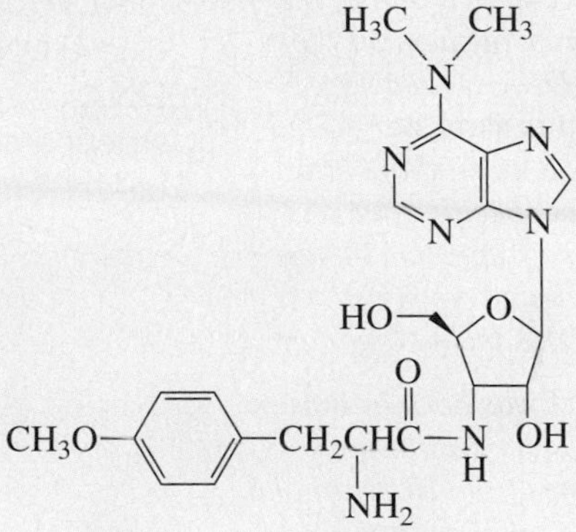

puromycin

Clinically useful antibiotics	Mode of action
Tetracycline	Prevents the aminoacyl-tRNA from binding to the ribosome
Erythromycin	Prevents the incorporation of new amino acids into the protein
Streptomycin	Inhibits the initiation of protein synthesis
Chloramphenicol	Prevents the new peptide bond from being formed

21.9   Why DNA Contains Thymine Instead of Uracil

Thymines are synthesized from uracils in a process that requires considerable energy (Section 18.11). Therefore, there must be a good reason for DNA to contain thymine instead of uracil.

3-D Molecules:
Chloramphenicol complexed to acetyl-transferase;
Tetracycline

The presence of thymine instead of uracil in DNA prevents potentially lethal mutations. Cytosine can tautomerize to form an imine, which can be hydrolyzed to uracil (Section 13.6). The overall reaction is called a **deamination** because it removes an amino group.

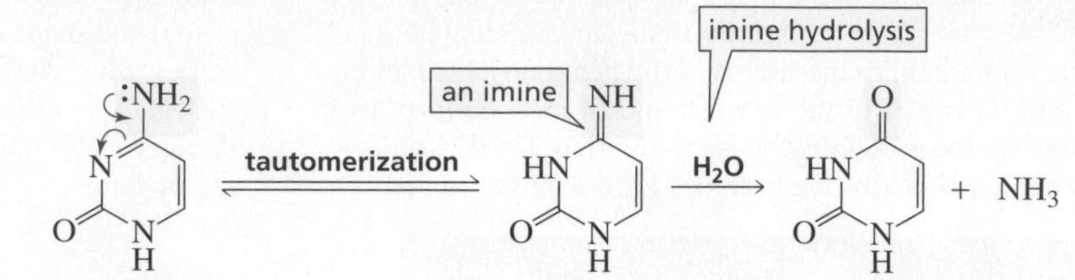

If a C in DNA is deaminated to a U, the U will specify the incorporation of an A into the daughter strand during replication instead of the G that would have been specified by C. Fortunately, a U in DNA is recognized as a "mistake" by an enzyme before an incorrect base can be inserted into the daughter strand. The enzyme cuts out the U and replaces it with a C. If U's were normally found in DNA, the enzyme could not distinguish between a normal U and a U formed by deamination of a C. Having T's in place of U's in DNA allows the U's that are found in DNA to be recognized as mistakes.

Unlike DNA, which replicates itself, any mistake in RNA does not survive for long because RNA is constantly being degraded and then resynthesized from the DNA template. Therefore, it is not worth spending the extra energy to incorporate T's into RNA.

21.10 Determining the Base Sequence of DNA

DNA molecules are too large to sequence as a unit, so DNA is first cleaved at specific base sequences and the resulting DNA fragments are sequenced. The enzymes that cleave DNA at specific base sequences are called **restriction endonucleases**, and the DNA fragments that are formed are called **restriction fragments**. Several hundred restriction endonucleases are now known; a few examples, the base sequence each recognizes, and the point of cleavage in that base sequence are shown here.

restriction enzyme	recognition sequence
*Alu*I	AG|CT TC|GA
*Fnu*DI	GG|CC CC|GG
*Pst*I	CTGCA|G G|ACGTC

The base sequences that most restriction endonucleases recognize are *palindromes*. A palindrome is a word or a group of words that reads the same forward and backward. "Toot" and "race car" are examples of palindromes.[1] A restriction endonuclease recognizes a piece of DNA in which *the template strand is a palindrome of the sense strand*. In other words, the sequence of bases in the template strand (reading from right to left) is identical to the sequence of bases in the sense strand (reading from left to right).

> **PROBLEM 9♦**
>
> Which of the following base sequences would most likely be recognized by a restriction endonuclease?
>
> **a.** ACGCGT   **c.** ACGGCA   **e.** ACATCGT
> **b.** ACGGGT   **d.** ACACGT   **f.** CCAACC

Frederick Sanger *(Section 17.8) and* **Walter Gilbert** *shared half of the 1980 Nobel Prize in chemistry for their work on DNA sequencing. The other half went to* **Paul Berg**, *who had developed a method of cutting nucleic acids at specific sites and recombining the fragments in new ways—a technique known as recombinant DNA technology.*

Walter Gilbert *was born in Boston in 1932. He received a master's degree in physics from Harvard and a Ph.D. in mathematics from Cambridge University. In 1958 he joined the faculty at Harvard, where he became interested in molecular biology.*

Paul Berg *was born in New York in 1926. He received a Ph.D. from Western Reserve University (now Case Western Reserve University). He joined the faculty at Washington University in St. Louis in 1955 and became a professor of biochemistry at Stanford in 1959.*

[1]Some other palindromes are "Mom," "Dad," "Bob," "Lil," "radar," "noon," "wow," "poor Dan in a droop", "a man, a plan, a canal, Panama," "Sex at noon taxes," and "He lived as a devil, eh?"

The restriction fragments can be sequenced using a procedure known as the **dideoxy method**. In this method, a small piece of DNA called a primer, labeled at the 5′-end with ^{32}P, is added to the restriction fragment whose sequence is to be determined. Next, the four 2′-deoxyribonucleoside triphosphates are added, and then DNA polymerase—the enzyme that adds nucleotides to a strand of DNA—is added as well. In addition, a small amount of the 2′,3′-dideoxynucleoside triphosphate of one of the bases is added to the reaction mixture. A 2′,3′-dideoxynucleoside triphosphate has no OH group at either the 2′- or 3′-position.

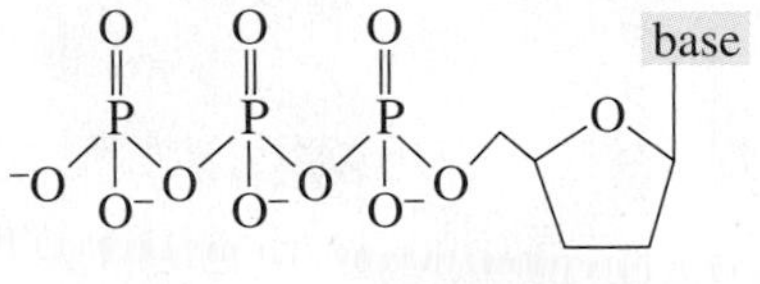

a 2′,3′-dideoxynucleoside triphosphate

Nucleotides will be added to the primer by base pairing with the restriction fragment. Synthesis will stop if the 2′,3′-dideoxy analog of dATP is added instead of dATP, because the 2′,3′-dideoxy analog does not have a 3′-OH to which additional nucleotides can be added. Therefore, three different chain-terminated fragments will be obtained from the DNA restriction fragment shown here.

DNA to be sequenced

3′—AGGCTCCAGTGATCCG—5′

^{32}P—TC

primer

→ **DNA polymerase** / **dATP, dGTP, dCTP, dTTP, 2′,3′-dATP**

^{32}P—TCCGA

^{32}P—TCCGAGGTCA

^{32}P—TCCGAGGTCACTA

The procedure is repeated three more times using a 2′,3′-dideoxy analog of dGTP, then a 2′,3′-dideoxy analog of dCTP, and then a 2′,3′-dideoxy analog of dTTP.

PROBLEM 10

What labeled fragments would be obtained from the segment of DNA shown above if a 2′,3′-dideoxy analog of dGTP had been added to the reaction mixture instead of a 2′,3′-dideoxy analog of dATP?

The chain-terminated fragments obtained from each of the four experiments are loaded onto separate lanes of a gel; the fragments obtained from using a 2′,3′-dideoxy analog of dATP are loaded onto one lane, the fragments obtained from using a 2′,3′-dideoxy analog of dGTP onto another lane, and so on. The smaller fragments fit through the spaces in the gel relatively easily and therefore travel through the gel faster, while the larger fragments pass through the gel more slowly.

After the fragments have been separated, the gel is placed in contact with a photographic plate. Radiation from ^{32}P causes a dark spot to appear on the plate opposite the location of each labeled fragment in the gel. This technique is called autoradiography, and the exposed photographic plate is known as an **autoradiograph** (Figure 21.13).

The sequence of bases in the original restriction fragment can be read directly from the autoradiograph. The identity of each base is determined by noting the column where each successive dark spot (larger piece of labeled fragment) appears, starting at the bottom of the gel to identify the base at the 5′-end. The sequence of the fragment of DNA responsible for the autoradiograph in Figure 21.13a is shown on the left-hand side of the figure.

Figure 21.13 ▶
a. a schematic of an autoradiograph
b. an actual autoradiograph

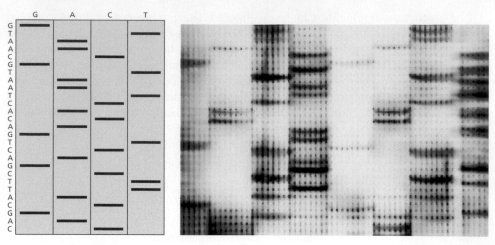

Kary B. Mullis *received the Nobel Prize in chemistry in 1993 for his invention of PCR. He was born in North Carolina in 1944 and received a B.S. from the Georgia Institute of Technology in 1966 and a Ph.D. from the University of California, Berkeley. After several postdoctoral positions, he joined the Cetus Corporation in Emeryville, California. He conceived of the idea of PCR while driving from Berkeley to Mendocino. Currently, he consults and lectures about biotechnology.*

The DNA from a forty-million-year-old leaf preserved in amber was amplified by PCR and then sequenced.

Once the sequence of bases in a restriction fragment is determined, the results can be checked by determining the base sequence of the complementary strand. The base sequence in the original piece of DNA can be determined by repeating the entire procedure with a restriction endonuclease that cleaves at a different site and identifying overlapping fragments.

21.11 Polymerase Chain Reaction (PCR)

PCR (polymerase chain reaction), discovered in 1983, is a technique that allows scientists to amplify DNA—making billions of copies of a segment of DNA in a very short time. With this technique, sufficient DNA for analysis can be obtained from a single hair or sperm.

PCR is done by adding the following to a solution containing the segment of DNA to be amplified (target DNA):

- a large excess of primers (short pieces of DNA) that base pair with (anneal) the DNA that flanks the segment of DNA to be amplified;
- the four deoxyribonucleotide triphosphates (dATP, dGTP, dCTP, dTTP);
- a heat-stable DNA polymerase.

The following three steps are then carried out (Figure 21.14):

- *Strand Separation:* The solution is heated to 95 °C, causing double-stranded DNA to separate into two single strands.
- *Base-Pairing of the Primers:* The solution is cooled to 54 °C, allowing the primers (red and yellow boxes in Figure 21.14) to pair with the bases in the DNA that flank the 3′-end of target DNA.
- *DNA Synthesis:* The solution is heated to 72 °C, a temperature at which DNA polymerase catalyzes the addition of nucleotides to the primer. Notice that because the primers attach to the 3′-end of target DNA, copies of target DNA are synthesized in the required 5′ $\longrightarrow$ 3′ direction.

The solution is then heated to 95 °C to begin a second cycle; the second cycle produces four copies of double-stranded DNA. The third cycle, therefore, starts with eight single strands of DNA and produces sixteen single strands. It takes about an hour to complete 30 cycles at which point the amplification is a billion-fold.

PCR has a wide range of clinical uses. It can be used to detect mutations that lead to cancer, to diagnose genetic diseases, to reveal the presence of HIV that would be missed by an antibody assay, to monitor cancer chemotherapy, and to rapidly identify an infectious disease.

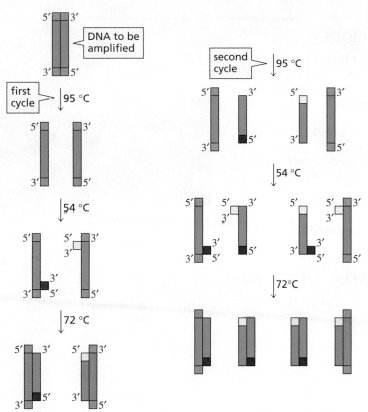

◀ **Figure 21.14**
Two cycles of the polymerase chain reaction.

DNA FINGERPRINTING

PCR is used by forensic chemists to compare DNA samples collected at the scene of a crime with the DNA of the suspected perpetrator. The base sequence of segments of noncoding DNA varies from individual to individual. Forensic laboratories have identified 13 of these segments that are the most accurate for identification purposes. If the sequence of bases from two DNA samples are the same, the chance is about 80 billion to one that they are from the same individual. DNA fingerprinting is also used to establish paternity, accounting for about 100,000 DNA profiles a year.

21.12 | Genetic Engineering

Recombinant DNA molecules are DNA molecules (natural or synthetic) that have been incorporated into the DNA of a compatible host cell. Recombinant DNA technology—also known as genetic engineering—has many practical applications. For example, the DNA for a particular protein can be inserted into a vector where the DNA can replicate; this produces millions of copies, leading to the synthesis of large amounts of protein. Human insulin is made in this way.

Agriculture is benefiting from genetic engineering as crops are being produced with new genes that increase their resistance to drought and insects. For example, genetically engineered cotton crops are resistant to the cotton bollworm, and genetically engineered corn is resistant to the corn rootworm. Therefore, herbicides, which threaten groundwater supplies, no longer have to be used. Genetically modified organisms (GMOs) have been responsible for a nearly 50% reduction in agricultural chemical sales in the United States. On the other hand, there are concerns about the potential consequences of changing the genes of any species.

RESISTING HERBICIDES

Glyphosate, the active ingredient in a well-known herbicide called Roundup, kills weeds by inhibiting an enzyme that plants need to synthesize phenylalanine and tryptophan, amino acids required for growth. Studies are being done with corn and cotton that have been genetically engineered to tolerate the herbicide. Then, when fields are sprayed with glyphosate, the weeds are killed but not the crops. These crops have been given a gene that produces an enzyme that makes glyphosate inactive by acetylating it with acetyl-CoA.

Corn genetically engineered to resist the herbicide glyphosate by acetylating it.

glyphosate
an herbicide

+ CH₃—C(O)—SCoA → (enzyme) →

N-acetylglyphosate
harmless to plants

+ CoASH

Summary

There are two types of nucleic acids—**deoxyribonucleic acid (DNA)** and **ribonucleic acid (RNA)**. The genetic information stored in DNA is **transcribed** into RNA. This information is then **translated** for the synthesis of proteins.

A **nucleoside** contains a base bonded to D-ribose or to 2-deoxy-D-ribose. A **nucleotide** is a nucleoside with either the 5'- or the 3'-OH group bonded to phosphoric acid by an ester linkage. **Nucleic acids** are composed of long strands of nucleotide subunits linked by phosphodiester bonds. These linkages join the 3'-OH group of one nucleotide to the 5'-OH group of the next nucleotide. A **dinucleotide** contains two nucleotide subunits, an **oligonucleotide** contains three to ten subunits, and a **polynucleotide** contains many subunits. DNA contains 2'-deoxy-D-ribose while RNA contains D-ribose. The difference in the sugars causes DNA to be stable and RNA to be easily cleaved.

The **primary structure** of a nucleic acid is the sequence of bases in its strand. DNA contains **A, G, C,** and **T**; RNA contains **A, G, C,** and **U**. DNA is double-stranded; the strands run in opposite directions and are twisted into a helix. The bases are confined to the inside of the helix and the sugar and phosphate groups are on the outside. The strands are held together by hydrogen bonds between bases of opposing strands. The two strands—one is called a **sense strand** and the other a **template strand**—are complementary: **A** pairs with **T**, and **G** pairs with **C**. DNA is synthesized in the 5' ⟶ 3' direction by a process called **semiconservative replication**.

The sequence of bases in DNA provides the blueprint for the synthesis (**transcription**) of RNA. RNA is synthesized in the 5' ⟶ 3' direction by transcribing the DNA template strand in the 3' ⟶ 5' direction. There are three kinds of RNA: messenger RNA, ribosomal RNA, and transfer RNA. Protein synthesis (**translation**) takes place from the N-terminal end to the C-terminal end by reading the bases along the mRNA strand in the 5' ⟶ 3' direction. Each three-base sequence—**a codon**—specifies the particular amino acid to be incorporated into a protein. A tRNA carries the amino acid bound as an ester to its 3'-terminal position. The codons and the amino acids they specify are known as the **genetic code**.

Restriction endonucleases cleave DNA at specific palindromes forming **restriction fragments**. The **dideoxy method** is used to determine the sequence of bases in the restriction fragments. The **PCR method** amplifies segments of DNA.

Problems

11. Name the following compounds:

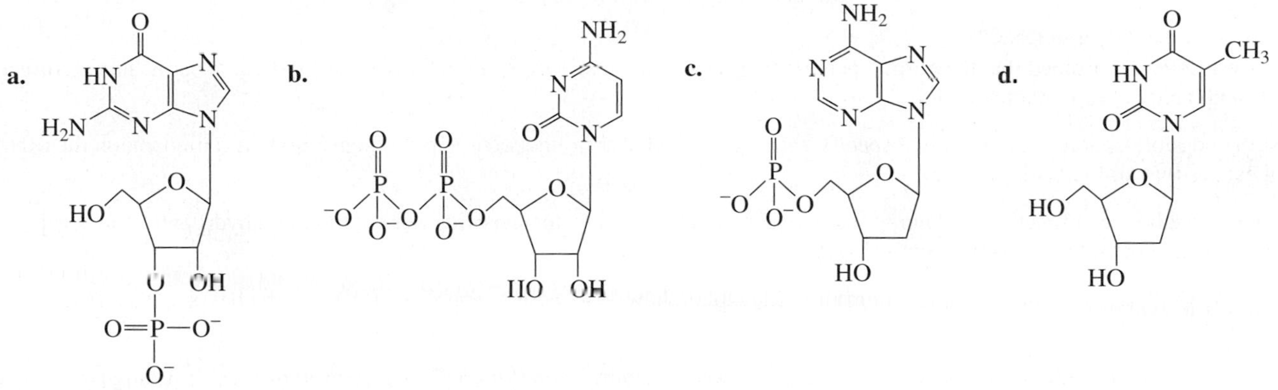

a.

b.

c.

d.

12. What nonapeptide is coded for by the following piece of mRNA?

$$5' - AAA - GUU - GGC - UAC - CCC - GGA - AUG - GUG - GUC - 3'$$

13. What is the sequence of bases in the template strand of DNA that codes for the mRNA in Problem 12?

14. What is the sequence of bases in the sense strand of DNA that codes for the mRNA in Problem 12?

15. What would be the C-terminal amino acid if the codon at the 3'-end of the mRNA in Problem 12 underwent the following mutation:
   a. if the first base was changed to A?
   b. if the second base was changed to A?
   c. if the third base was changed to A?
   d. if the third base was changed to G?

16. What would be the base sequence of the segment of DNA that is responsible for the biosynthesis of the following hexapeptide?

Gly-Ser-Arg-Val-His-Glu

17. Match the codon with the anticodon:

Codon	Anticodon
AAA	ACC
GCA	CCU
CUU	UUU
AGG	AGG
CCU	UGA
GGU	AAG
UCA	GUC
GAC	UGC

18. Using the single-letter abbreviations for the amino acids in Table 17.2, write the sequence of amino acids in a tetrapeptide represented by the first four letters in your first name. Do not use any letter twice. (Because not all letters are assigned to amino acids, you might have to use one or two letters in your last name.) Write the sequence of bases in mRNA that would result in the synthesis of that polypeptide. Write the sequence of bases in the sense strand of DNA that would result in formation of the appropriate mRNA.

19. Which of the following pairs of dinucleotides are present in equal amounts in DNA?
   a. CC and GG
   b. CG and GT
   c. CA and TG
   d. CG and AT
   e. GT and CA
   f. TA and AT

20. The amino acid sequences of peptide fragments obtained from a normal protein and from the same protein synthesized by a defective gene were compared. They were found to differ in only one peptide fragment. The primary sequences of the fragments are shown here:

Normal: Gln-Tyr-Gly-Thr-Arg-Tyr-Val
Mutant: Gln-Ser-Glu-Pro-Gly-Thr

   a. What is the defect in DNA?
   b. It was later determined that the normal peptide fragment is an octapeptide with a C-terminal Val-Leu. What is the C-terminal amino acid of the mutant peptide?

21. List the possible codons on mRNA that specify each amino acid in the oligopeptide in Problem 5 and the anticodon on the tRNA that carries that amino acid.

22. Indicate whether each functional group of the five heterocyclic bases in nucleic acids can function as a hydrogen bond acceptor (A), a hydrogen bond donor (D), or both (D/A).

23. Using the D, A, and D/A designations in Problem 22, explain how base pairing would be affected if the bases existed in the enol form.

24. The 2′,3′-cyclic phosphodiester, which is formed when RNA is hydrolyzed (Figure 21.7), reacts with water, forming a mixture of nucleotide 2′- and 3′-phosphates. Propose a mechanism for this reaction.

25. Adenine can be deaminated to hypoxanthine, and guanine can be deaminated to xanthine. Draw structures for hypoxanthine and xanthine.

26. Explain why thymine cannot be deaminated.

27. In acidic solutions, nucleosides are hydrolyzed to a sugar and a heterocyclic base. Propose a mechanism for this reaction.

28. Why is the codon a triplet rather than a doublet or a quartet?

29. 5-Bromouracil, a highly mutagenic compound, is used in cancer chemotherapy. When administered to a patient, it is converted to the triphosphate and incorporated into DNA in place of thymine, which it resembles sterically. Why does it cause mutations? (*Hint:* The bromo substituent increases the stability of the enol tautomer.)

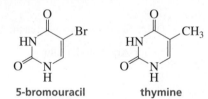

5-bromouracil   thymine

30. Which cytosine in the following sense strand of DNA could cause the most damage to the organism if it were deaminated?

5′—A—T—G—T—C—G—C—T—A—A—T—C—3′

31. The first amino acid incorporated into a polypeptide chain during its biosynthesis in prokaryotes is *N*-formylmethionine. Explain the purpose of the formyl group.

32. Why doesn't DNA unravel completely before replication begins?

22 The Organic Chemistry of Drugs

Discovery and Design

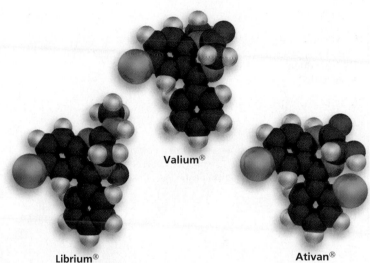

Valium®

Librium®

Ativan®

A **drug** is any absorbed substance that changes or enhances a physical or psychological function in the body. A drug can be a gas, a liquid, or a solid and can have a simple structure or a complicated one. Drugs have been used by humans for thousands of years to alleviate pain and illness. By trial and error, people learned which herbs, berries, roots, and bark could be used for medicinal purposes. The knowledge about natural medicines was passed down from generation to generation without any understanding of how the drugs actually worked. Those who dispensed the drugs—medicine men, shamans, and witch doctors—were important members of every civilization. However, the drugs available to them were just a small fraction of the drugs available to us today.

Even at the beginning of the twentieth century, there were no drugs for the dozens of functional, degenerative, neurological, and psychiatric disorders; no hormone therapies; no vitamins; and—most significantly—no effective drug for the cure of any infectious disease. Local anesthetics had just been discovered, and there were only two analgesics to relieve major pain. One reason that families had many children was because some of the children were bound to succumb to childhood diseases. Life spans were generally short. In 1900, for example, the average life expectancy in the United States was 46 years for a man and 48 years for a woman. In 1920, about 80 of every 100,000 children died before their fifteenth birthday—most as a result of infections in their first year of life. Now there is a drug for treating almost every disease, and this is reflected by current life expectancies: 74 years for a man and 79 years for a woman. Now only about four of every 100,000 children die before the age of 15—mainly from cancer, accidents, and inherited diseases.

The shelves of a typical modern pharmacy are stocked with almost 2000 preparations, most of which contain a single active ingredient, usually an organic compound. These medicines can be swallowed, injected, inhaled, or absorbed through the skin. In 2003, more than 3.1 billion prescriptions were dispensed in the United States.

The most widely *prescribed drugs* are listed in Table 22.1 in order of the number of prescriptions written. Antibiotics are the most widely prescribed class of drugs in the world. In the developed world, heart drugs are the most widely prescribed class, partly because they are generally taken for the remainder of the patient's life. In recent years, prescriptions for psychotropic drugs have decreased as doctors have become more aware of problems associated with addiction, and prescriptions for asthma have increased, reflecting a greater incidence (or awareness) of the disease. The U.S. market accounts for 50% of pharmaceutical sales. The aging U.S. population—nearly 30% is over 50 years old—is leading to an increased demand for treating conditions such as high cholesterol levels, hypertension, diabetes, osteoarthritis, and menopause symptoms.

Table 22.1 The Most Widely Dispensed Drugs in the United States in 2003 in Decreasing Order of Number of Prescriptions Written

Proprietary name	Generic name	Structure	Use
Hydrocodone with APAP	hydrocodone with APAP	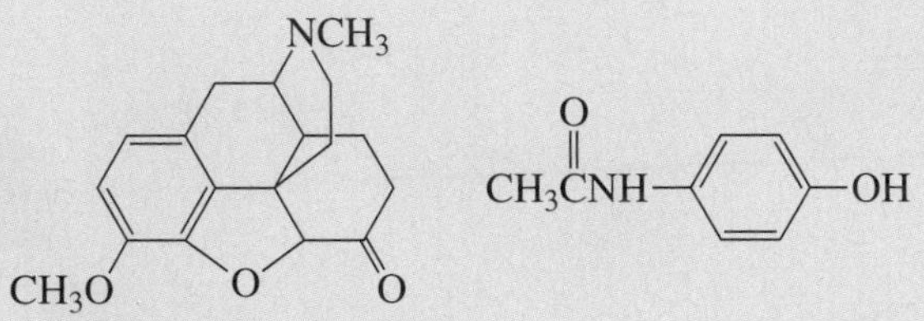	analgesic
Lipitor	atorvastatin	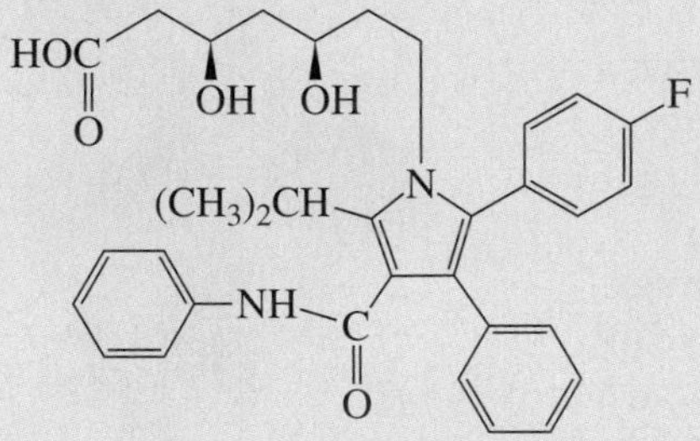	statin (cholesterol-reducing drug)
Synthroid	levothyroxine	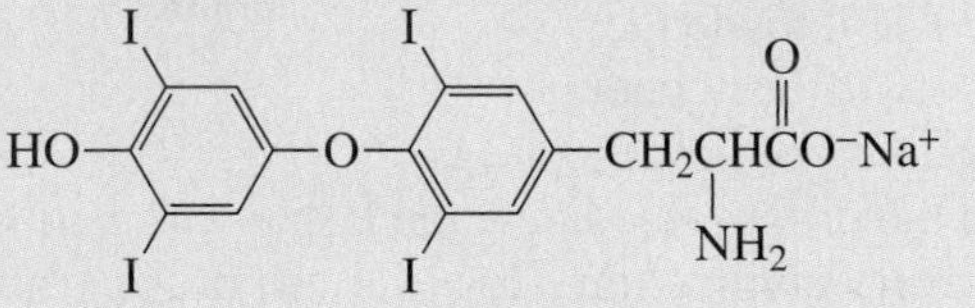	treatment of hypothyroidism
Tenormin	atenolol	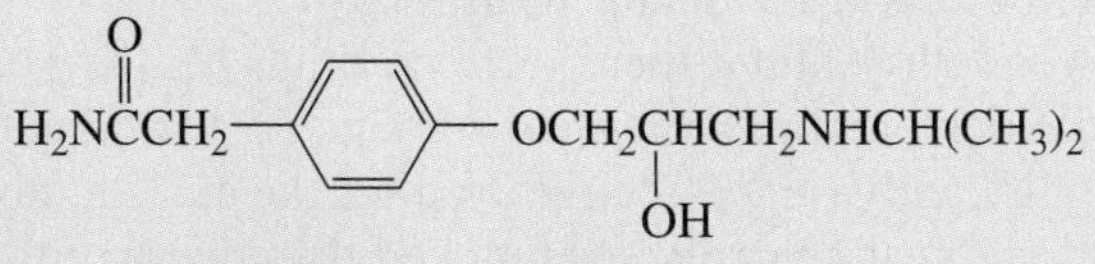	β-adrenergic blocking agent (antiarrhythmic, antihypertensive)
Zithromax	azithromycin	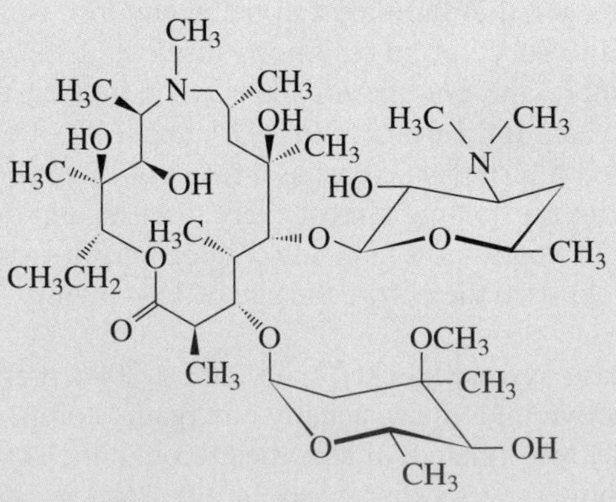	antibiotic

Table 22.1 (continued)

Proprietary name	Generic name	Structure	Use
Trimox	amoxicillin	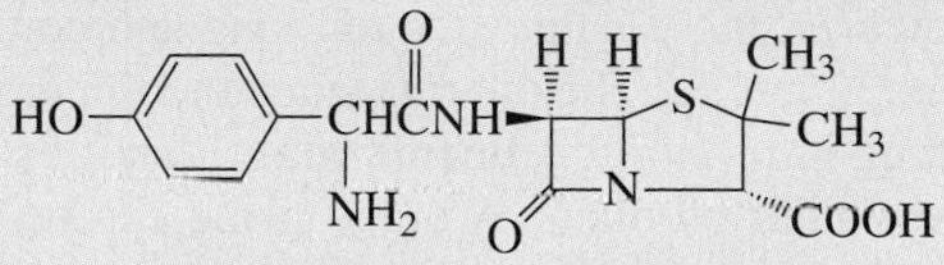 (structure of amoxicillin)	antibiotic
Lasix	furosemide	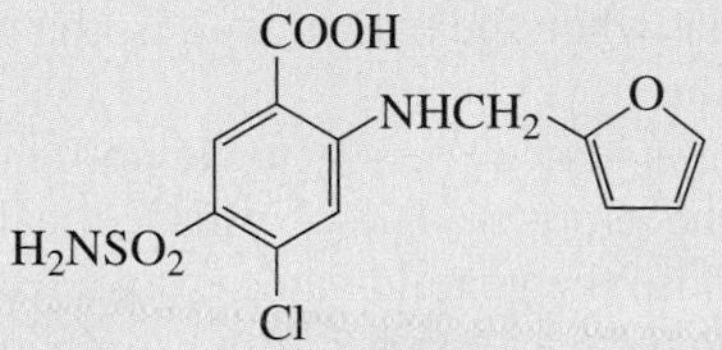 (structure of furosemide)	diuretic
Hydrodiuril	hydrochlorothiazide (HCTZ)	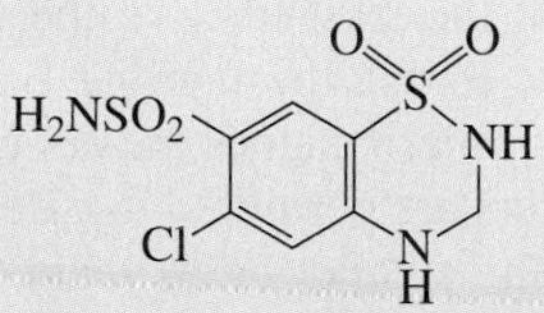 (structure of hydrochlorothiazide)	diuretic
Norvasc	amlodipine	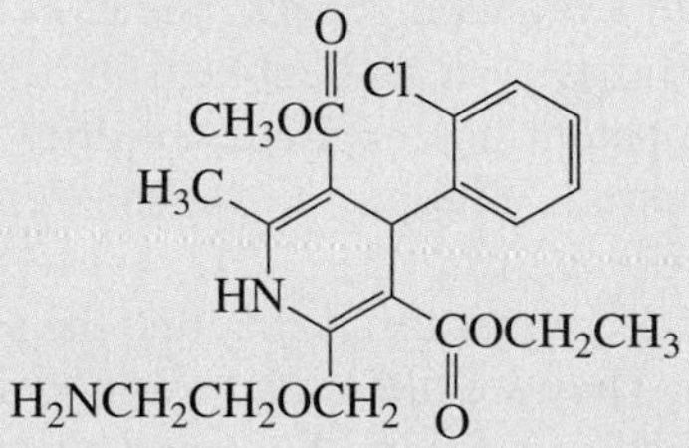 (structure of amlodipine)	calcium channel blocker (antihypertensive)
Lisinopril	lisinopril	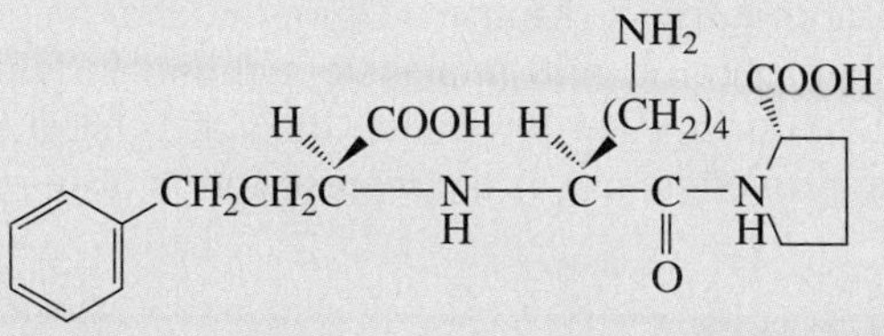 (structure of lisinopril)	antihypertensive
Xanax	alprazolam	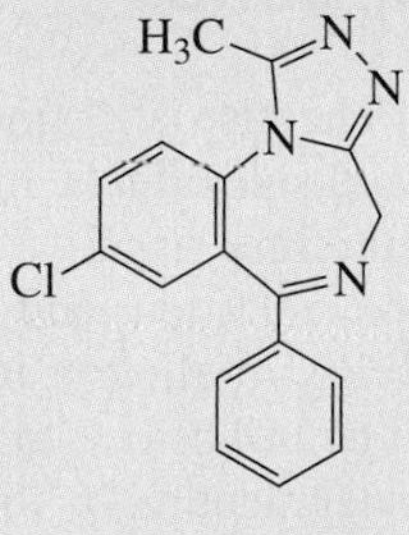 (structure of alprazolam)	tranquilizer
Zoloft	sertraline	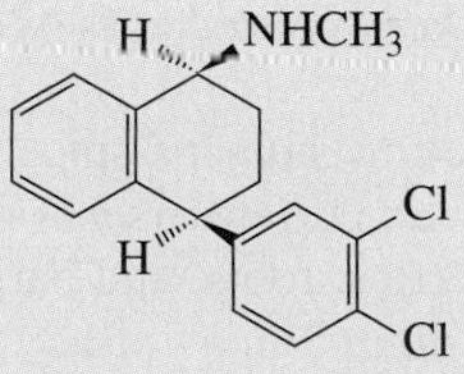 (structure of sertraline)	antidepressant
Albuterol	albuterol	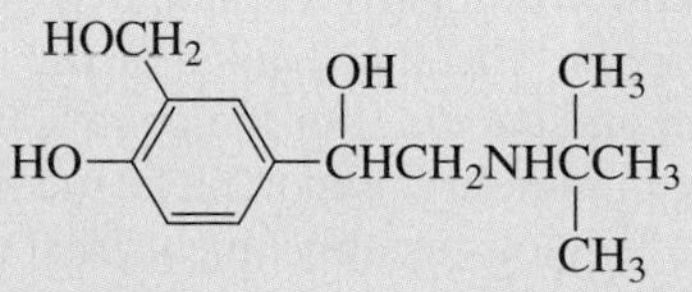 (structure of albuterol)	bronchodilator

22.1 Naming Drugs

The most accurate names of drugs are the chemical names that define their structures. However, these names are too long and complicated to appeal to physicians or to the general public. The pharmaceutical company is allowed to choose the brand name for a drug it develops. A **brand name** identifies a commercial product and distinguishes it from other products. A brand name can be used only by the company that holds the patent. It is in their best interest to choose a name that is easy to remember and pronounce so that when the patent expires, the public will continue to request the drug by its brand name.

Each drug is also given a **generic name** that any pharmaceutical company can use to identify the product. The pharmaceutical company that developed the drug is allowed to choose the generic name from a list of 10 names provided by an independent group. It is in the best interest of the company to choose the generic name that is hardest to pronounce and the least likely to be remembered, so that physicians and consumers will continue to use the familiar brand name. Brand names must always be capitalized, whereas generic names are not capitalized.

Drug manufacturers are permitted to patent and retain exclusive rights to the drugs they develop. A patent is valid for 20 years. Once the patent expires, other drug companies can market the drug under the generic name or under their own brand name, which is called a branded generic name. For example, the antibiotic with the generic name ampicillin is sold as Penbritin by the company that held the original patent. Now that the patent has expired, the drug is sold by other companies as Ampicin, Ampilar, Amplital, Binotal, Nuvapen, Pentrex, Ultrabion, Viccillin, and 30 other branded generic names.

The over-the-counter drugs that line the shelves of drugstores are available without a prescription. They are often mixtures containing one or more active ingredients, plus sweeteners and inert fillers. For example, the preparations called Advil (Whitehall Laboratories), Motrin (Upjohn), and Nuprin (Bristol-Meyers Squibb) all contain ibuprofen, a mild analgesic and anti-inflammatory drug. Ibuprofen was patented in Britain in 1964 by Boots, Inc., and the U.S. Food and Drug Administration (FDA) approved its use as a nonprescription drug in 1984.

22.2 Lead Compounds

The goal of the medicinal chemist is to find compounds that have potent effects on given diseases with minimum side effects. In other words, a drug must react selectively with its target and have minimal negative effects. A drug must get to the right place in the body, at the right concentration, and at the right time. Therefore, a drug must have the appropriate solubility to allow it to be transported to the target cell. If it is taken orally, the drug must be insensitive to the acid conditions of the stomach, and it also must resist enzymatic degradation by the liver before it reaches its target. Finally, it must eventually be either excreted as is or degraded to harmless compounds that can be excreted.

Medicinal agents used by humans since ancient times provided the starting point for the development of our current arsenal of drugs. The active ingredients were isolated from the herbs, berries, roots, and bark used in traditional medicine. Foxglove, for instance, furnished digitoxin, a cardiac stimulant. The bark of the cinchona tree yielded quinine for relief from malaria. Willow bark contains salicylates used to control fever and pain. The sticky juice of the oriental opium poppy provided morphine for severe pain and codeine for the control of a cough. By 1882, more than 50 different herbs were commonly used to make medicines. Many of these herbs were grown in the gardens of religious establishments that treated the sick.

Foxglove

Scientists still search the world for plants and berries and the oceans for flora and fauna that might yield new medicinal compounds. Taxol, a compound isolated from the bark of the Pacific yew tree, is a relatively recently recognized anticancer agent (Section 13.4).

Once a naturally occurring drug is isolated and its structure determined, it can serve as a prototype in a search for other biologically active compounds. The prototype is called a **lead compound** (i.e., a compound that plays a leading role in the search). Analogs of the lead compound are synthesized in order to find one that might have improved therapeutic properties or fewer side effects. The analog may have a different substituent than the lead compound, a branched chain instead of a straight chain, or a different ring system. Changing the structure of the lead compound is called **molecular modification**.

22.3   Molecular Modification

A classic example of molecular modification is the development of synthetic local anesthetics from cocaine, the lead compound. Cocaine comes from the leaves of a bush native to the highlands of the South American Andes. Cocaine is a highly effective local anesthetic, but it produces disturbing effects on the central nervous system (CNS), ranging from initial euphoria to severe depression. By dissecting the cocaine molecule step by step—removing the methoxycarbonyl group and cleaving the seven-membered-ring system—scientists identified the portion of the molecule that carries the local anesthetic activity without the damaging CNS effects. This knowledge gave an improved lead compound—a different ester of benzoic acid, with the alcohol component of the ester having a terminal tertiary amino group.

Coca leaves

cocaine
lead compound

improved lead compound

Hundreds of esters were then synthesized, resulting in esters with substituents on the aromatic ring, esters with alkyl groups bonded to the nitrogen, and esters with the length of the connecting alkyl chain modified. Successful anesthetics obtained through molecular modification were benzocaine, a topical anesthetic, and procaine, commonly known by the brand name Novocain.

Benzocaine®

procaine
Novocain®

lidocaine
Xylocaine®

Because the ester group of procaine is hydrolyzed relatively rapidly by enzymes, procaine has a short half-life. Therefore, compounds with amide groups were synthesized—recall that amides are less reactive than esters (Section 12.6). In this way, lidocaine, one of the most widely used injectable anesthetics, was discovered. The rate at which lidocaine is hydrolyzed is further decreased by its two *ortho*-methyl substituents, which provide steric hindrance to the reactive carbonyl group.

Later, physicians recognized that the action of an anesthetic administered in vivo (in a living organism) could be lengthened considerably if it were administered along with epinephrine. Because epinephrine is a vasoconstrictor, it reduces the blood supply, allowing the drug to remain at its targeted site for a longer period.

In screening the structurally modified compounds for biological activity, scientists were surprised to find that replacing the ester linkage of procaine with an amide linkage led to a compound—procainamide hydrochloride—that had activity as a cardiac depressant as well as activity as a local anesthetic. Procainamide hydrochloride is currently used clinically as an antiarrhythmic.

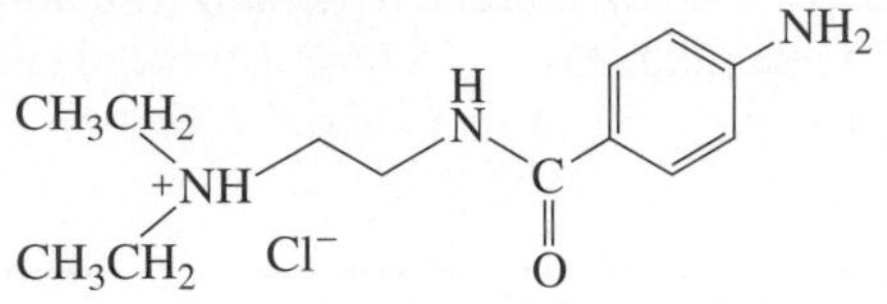

procainamide hydrochloride

Morphine, the most widely used analgesic for severe pain, is the standard by which other painkilling medications are measured. Although scientists have learned how to synthesize morphine, all commercial morphine is obtained from opium, the juice from a species of poppy. Morphine occurs in opium to the extent of 10%. Methylating one of the OH groups of morphine produces codeine, which has one-tenth the analgesic activity of morphine. Codeine profoundly inhibits the cough reflex. Most commercial codeine is obtained by methylating morphine. Putting an acetyl group on (acetylating) one of the OH groups of morphine produces a compound with a similar reduced potency. Acetylating both OH groups forms heroin, which is much more potent than morphine. Because heroin is less polar than morphine, it crosses the blood–brain barrier more rapidly, resulting in a more rapid "high." Heroin has been banned in most countries because it is widely abused. It is synthesized by using acetic anhydride to acetylate morphine (Section 12.10). Therefore, both heroin and acetic acid are formed as products. Drug enforcement agencies use dogs trained to recognize the pungent odor of acetic acid.

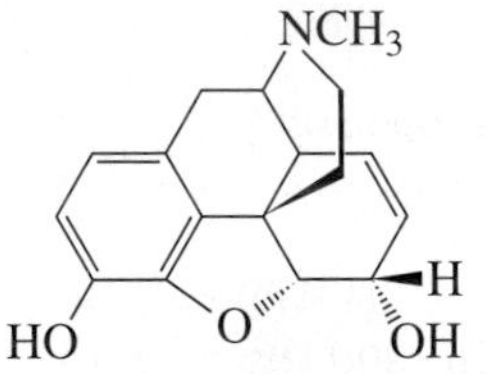

morphine 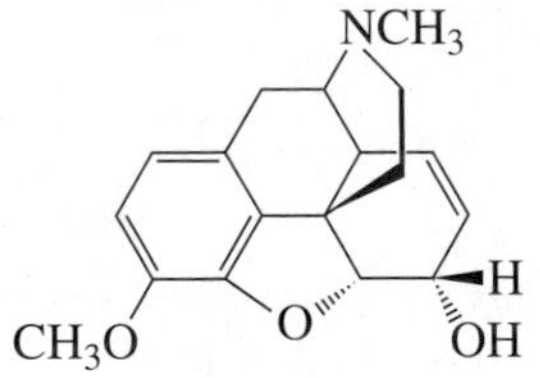   codeine 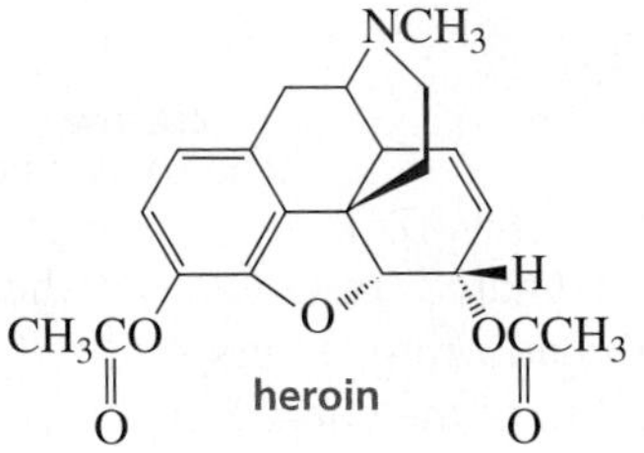   heroin

Molecular modification of codeine led to dextromethorphan, the active ingredient in most cough medicines. Etorphine was synthesized when scientists realized that analgesic potency was related to the ability of a nonpolar segment of the drug to bind to the nonpolar portion of the opiate receptor (Section 22.6). Etorphine is about 2000 times more potent than morphine, but it is not safe for use by humans. It has been used to tranquilize elephants and other large animals. Pentazocine is useful in obstetrics because it dulls the pain of labor, but does not depress the respiration of the infant as morphine does.

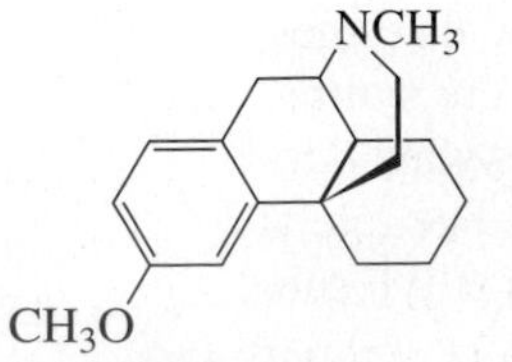

dextromethorphan 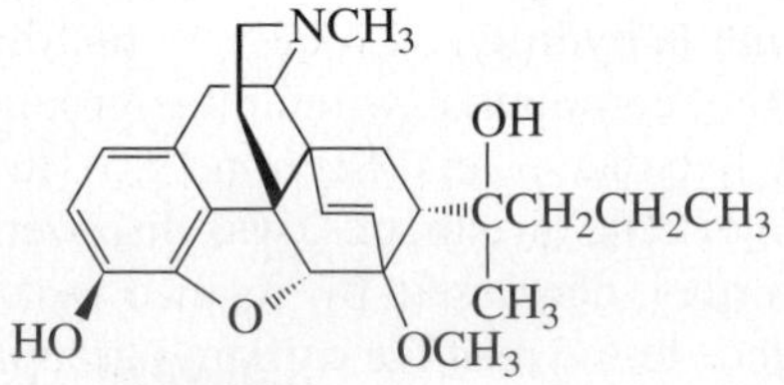   etorphine 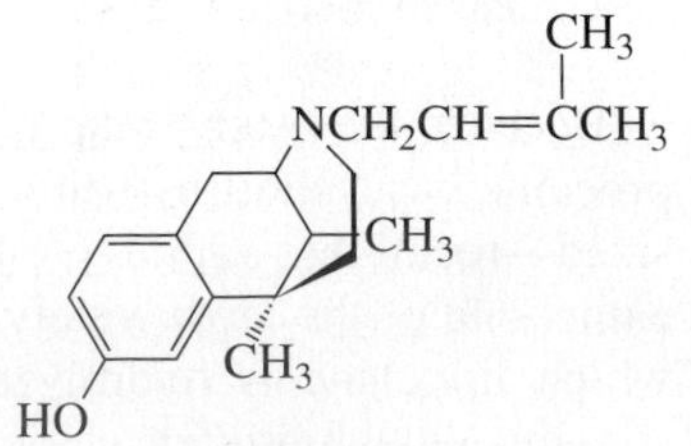   pentazocine

Methadone was synthesized by German scientists in 1944 in an attempt to find a drug to treat muscle spasms. (It was originally called "Adolphile" in honor of Adolph Hitler.) It was not recognized until 10 years later—after building molecular models— that methadone and morphine have similar shapes. In contrast to morphine, methadone can be administered orally. Methadone has a considerably longer half-life (24 to 26 hours) than morphine (2 to 4 hours). Cumulative effects are seen with repeated doses of methadone, so lower doses and longer intervals between doses are possible. Consequently, methadone is used to treat chronic pain and the withdrawal symptoms of heroin addicts. Reducing the carbonyl group of methadone and acetylating it forms α-acetylmethadol. The levorotatory isomer (Section 8.7) of this compound can suppress withdrawal symptoms for 72 hours (Section 8.12). When Darvon (isomethadone) was introduced, it was initially thought to be the long-sought-after nonaddicting painkiller. However, it was later found to have no therapeutic advantage over less toxic and more effective analgesics.

methadone

α-acetylmethadol

isomethadone
Darvon®

Notice that morphine and all the compounds prepared by molecular modification of morphine have a structural feature in common—an aromatic ring attached to a quaternary carbon that is attached to a tertiary amine two carbons away.

a tertiary amine

quaternary carbon

22.4   Random Screening

The lead compound for the development of most drugs is found by screening thousands of compounds randomly. A **random screen**, also known as a **blind screen**, is a search for a pharmacologically active compound without having any information about which chemical structures might show activity. The first blind screen was carried out by Paul Ehrlich, who was searching for a "magic bullet" against trypanosomes—the microorganisms that cause African sleeping sickness. After testing more than 900 compounds against trypanosomes, Ehrlich tested some of his compounds against other bacteria. Compound 606 (salvarsan) was found to be dramatically effective against the microorganisms that cause syphilis.

An important part of random screening is recognizing an effective compound. This requires the development of an assay for the desired biological activity. Some assays can be done in vitro (in glass)—for example, searching for a compound that will inhibit a particular enzyme. Others are done in vivo—for instance, searching for a compound that will save a mouse from a lethal dose of a virus. One problem with in vivo assays is

Paul Ehrlich (1854–1915) *was a German bacteriologist. He received a medical degree from the University of Leipzig and was a professor at the University of Berlin. In 1892, he developed an effective diphtheria antitoxin. For his work on immunity, he received the 1908 Nobel Prize in physiology or medicine, together with Ilya Ilich Mechnikov.*

that drugs can be metabolized differently by different animals (Section 19.0). Thus, an effective drug in a mouse may be less effective or even useless in a human. Another problem involves regulating the dosages of both the virus and the drug. If the dosage of the virus is too high, it might kill the mouse in spite of the presence of a biologically active compound that could save the animal. If the dosage of the potential drug is too high, the drug might kill the mouse whereas a lower dosage would have saved it.

The observation that azo dyes effectively dyed wool fibers (animal protein) gave scientists the idea that such dyes might selectively bind to bacterial proteins, too. Well over 10,000 dyes were screened in vitro in antibacterial tests, but none showed any antibiotic activity. Some scientists argued that the dyes should be screened in vivo because what physicians really needed were antibacterial agents that would cure infections in humans and animals, not in test tubes.

In vivo studies were done in mice that had been infected with a bacterial culture. Now the luck of the investigators improved. Several dyes turned out to counteract gram-positive infections. The least toxic of these, Prontosil (a bright red dye), became the first drug to treat bacterial infections.

Prontosil®

The fact that Prontosil was inactive in vitro but active in vivo should have suggested that the dye was converted to an active compound by the mammalian organism, but this did not occur to the bacteriologists, who were content to have found a useful antibiotic. When scientists at the Pasteur Institute later investigated Prontosil, they noted that mice given the drug did not excrete a red compound. Urine analysis showed that the mice excreted *para*-acetamidobenzenesulfonamide, a colorless compound. Chemists knew that anilines are acetylated in vivo, so they prepared the nonacetylated compound (sulfanilamide). When sulfanilamide was tested in mice infected with streptococcus, all the mice were cured, whereas untreated control mice died. Sulfanilamide was the first antibiotic.

para-**acetamidobenzenesulfonamide**   *para*-**aminobenzenesulfonamide**
sulfanilamide

We have seen that sulfanilamide acts by inhibiting the bacterial enzyme that synthesizes folic acid (Section 18.11). Thus, sulfanilamide is a *bacteriostatic* drug, not a *bactericidal* drug. A **bacteriostatic drug** inhibits the further growth of bacteria, whereas a **bactericidal drug** kills the bacteria. Sulfanilamide inhibits the enzyme because it is similar in size to the carboxylic acid that should be incorporated into folic acid. Many successful drugs have been designed by using similar-size replacements.

Gerhard Domagk (1895–1964) *was a research scientist at I. G. Farbenindustrie, a German manufacturer of dyes and other chemicals. He carried out studies that showed Prontosil to be an effective antibacterial agent. His daughter, who was dying of a streptococcal infection as a result of cutting her finger, was the first patient to receive the drug and be cured by it (1935). Prontosil received wider fame when it was used to save the life of Franklin D. Roosevelt, Jr., son of the U.S. president. Domagk received the Nobel Prize in physiology or medicine in 1939, but Hitler did not allow Germans to accept Nobel Prizes because Carl von Ossietsky, a German who was in a concentration camp, had been awarded the Nobel Prize for peace in 1935. Domagk was eventually able to accept the prize in 1947.*

DRUG SAFETY

In October 1937, patients who had obtained sulfanilamide from a company in Tennessee experienced excruciating abdominal pains before slipping into fatal comas. The FDA asked Eugene Geiling, a pharmacologist at the University of Chicago, and his graduate student, Frances Kelsey, to investigate. They found that this company was dissolving sulfanilamide in diethylene glycol, a sweet-tasting liquid to meet the demand for sulfanilamide that was easy to swallow. However, the safety of diethylene glycol in humans had never been tested. It turned out to be a deadly poison.

At that time, there was no legislation to prevent the sale of lethal medicines. In June 1938, the Federal Food, Drug, and Cosmetic Act was enacted. This legislation required drug companies to prove that their product was safe before they were allowed to market it. Interestingly, Frances Kelsey was the one who prevented thalidomide from being marketed in the United States (Section 8.12).

22.5 Serendipity in Drug Development

Many drugs have been discovered accidentally. Nitroglycerin, the drug used to relieve the symptoms of angina pectoris (heart pain), was discovered when workers handling nitroglycerin in the explosives industry experienced severe headaches. Investigation revealed that the headaches were caused by nitroglycerin's ability to produce a marked dilation of blood vessels. The pain associated with an angina attack results from the inability of the blood vessels to supply the heart adequately with blood. Nitroglycerin relieves the pain by dilating cardiac blood vessels.

$$CH_2—ONO_2$$
$$CH—ONO_2$$
$$CH_2—ONO_2$$
nitroglycerin

Leo H. Sternbach *was born in Austria in 1908. In 1918, after World War I and the dissolution of the Austro–Hungarian Empire, Sternbach's father moved to Krakow in the re-created Poland and obtained a concession to open a pharmacy. As a pharmacist's son, Sternbach was accepted at the Jagiellonian University School of Pharmacy, where he received both a master's degree in pharmacy and a Ph.D. in chemistry. With discrimination against Jewish scientists growing in Eastern Europe in 1937, Sternbach moved to Switzerland to work with Ružička (p. 512) at the ETH (the Swiss Federal Institute of Technology). In 1941, Hoffmann–LaRoche brought Sternbach and several other scientists out of Europe. Sternbach became a research chemist at LaRoche's U.S. headquarters in Nutley, New Jersey, where he later became director of medicinal chemistry.*

The tranquilizer Librium is another drug that was discovered accidentally. Leo Sternbach synthesized a series of quinazoline 3-oxides, but none of them showed any pharmacological activity. One of the compounds was not submitted for testing because it was not the quinazoline 3-oxide he had set out to synthesize. Two years after the project was abandoned, a laboratory worker came across this compound while cleaning up the lab, and Sternbach decided that he might as well submit it for testing before it was thrown away. The compound was shown to have tranquilizing properties and, when its structure was investigated, was found to be a benzodiazepine 4-oxide. Methylamine, instead of displacing the chloro substituent to form a quinazoline 3-oxide, had added to the imine group of the six-membered ring, causing the ring to open and reclose to a seven-membered ring. The compound was given the brand name Librium when it was put into clinical use in 1960.

a quinazoline 3-oxide

CH₃NH₂

an addition reaction occurred

CH₃NH₂ ✗

a substitution reaction did not occur

a benzodiazepine 4-oxide chlordiazepoxide
Librium® (1960)

Librium was structurally modified in an attempt to find other tranquilizers. One successful modification produced Valium, a tranquilizer almost 10 times more potent than Librium. Currently, there are eight benzodiazepines in clinical use as tranquilizers in the United States and some 15 others abroad. Xanax is one of the most widely prescribed medications; Rohypnol is one of the so-called date-rape drugs.

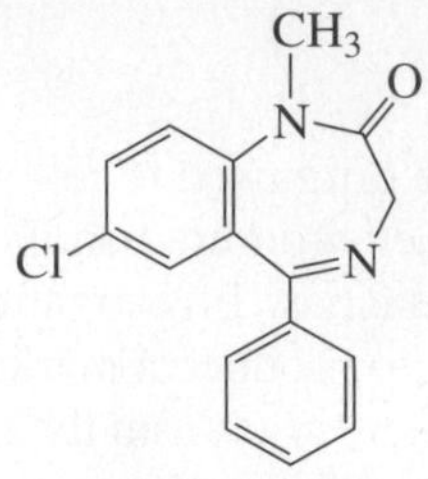

diazepam
Valium® (1963)

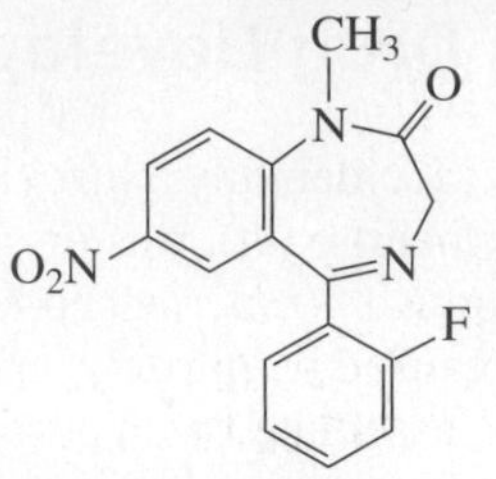

flunitrazepam
Rohypnol® (1963)

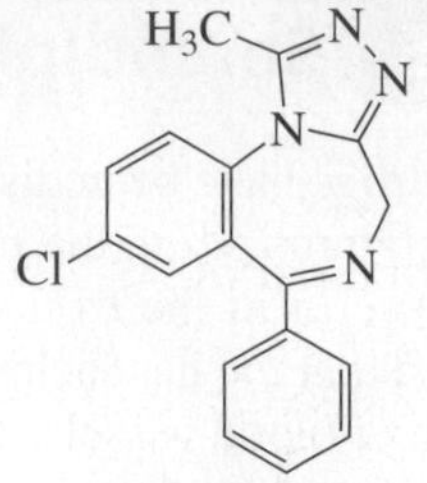

alprazolam
Xanax® (1970)

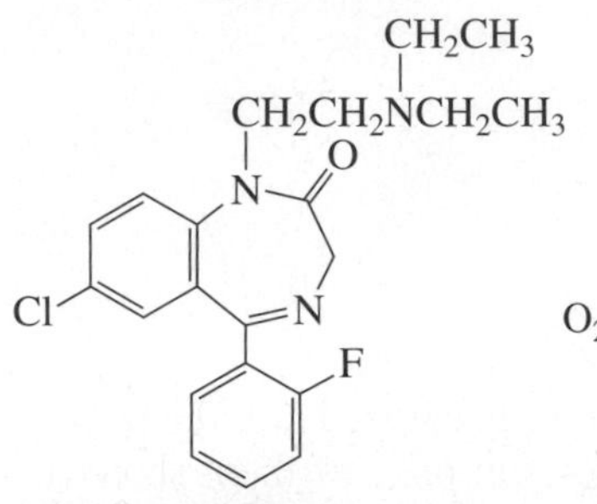

flurazepam
Dalmane® (1970)

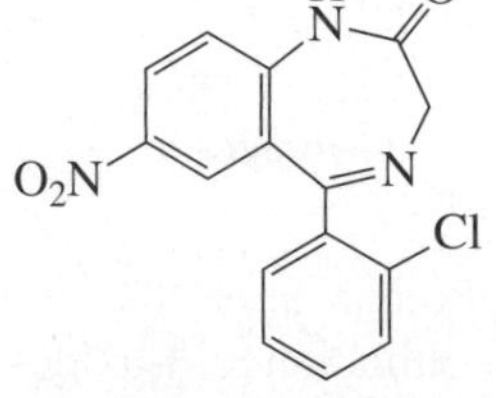

clonazepam
Klonopin® (1975)

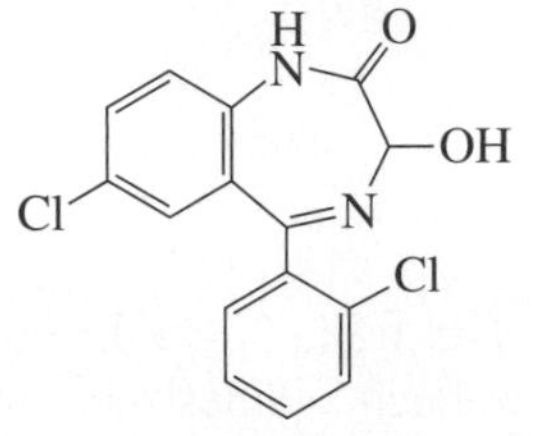

lorazepam
Ativan® (1977)

22.6 Receptors

Many drugs exert their physiological effects by binding to a specific cellular binding site called a **receptor**. That is why a small amount of a drug can bring about a measurable effect. Drug receptors are often lipoproteins or glycoproteins (Section 16.15); because these receptors are chiral, different enantiomers of a drug can have very different physiological effects (Section 8.12). Some receptors are part of cell membranes, whereas others are found in the cytoplasm—the material outside the nucleus. Nucleic acids–particularly DNA—also act as receptors for certain kinds of drugs. Because not all cells have the same receptors, drugs have considerable specificity. For example, epinephrine has intense effects on cardiac muscle, but almost no effect on muscle in other parts of the body.

A drug interacts with its receptor by the same kinds of bonding interactions—hydrogen bonding, electrostatic attractions, and van der Waals interactions—that we encountered in other examples of molecular recognition (Section 18.1). The most important factor in bringing together a drug and a receptor is a snug fit: The greater the affinity of a drug for its binding site, the higher is the drug's potential biological activity. Two drugs for which DNA is a receptor are chloroquine (an antimalarial) and 3,6-diaminoacridine (an antibacterial). These flat cyclic compounds can slide into the DNA double helix between base pairs—like a card being inserted into a deck of playing cards—and interfere with the normal replication of DNA.

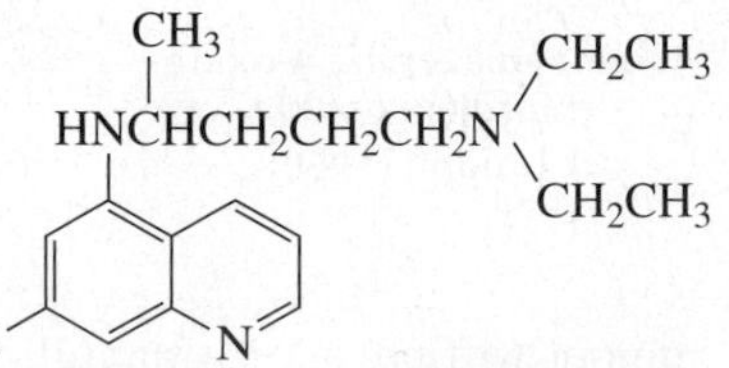

chloroquine

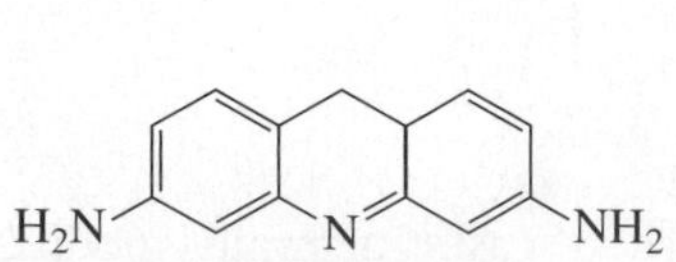

3,6-diaminoacridine

Knowing something about the molecular basis of drug action—such as how a drug interacts with a receptor—allows scientists to design and synthesize compounds that might have a desired biological activity. For example, when excess histamine is produced by the body, it causes the symptoms associated with the common cold and allergic responses. This is thought to be the result of the protonated ethylamino group anchoring the histamine molecule to a negatively charged portion of the histamine receptor.

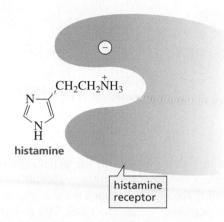

Drugs that interfere with the natural action of histamine—called antihistamines—bind to the histamine receptor but do not trigger the same response as histamine. Like histamine, these drugs have a protonated amino group that binds to the receptor. The drugs also have bulky groups that keep the histamine molecule from approaching the receptor.

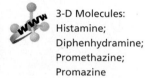

3-D Molecules:
Histamine;
Diphenhydramine;
Promethazine;
Promazine

antihistamines

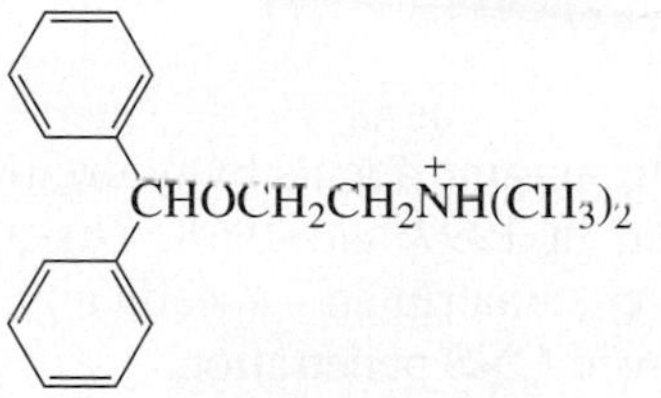

diphenhydramine	promethazine	promazine
Benadryl®	Promine®	Talofen®

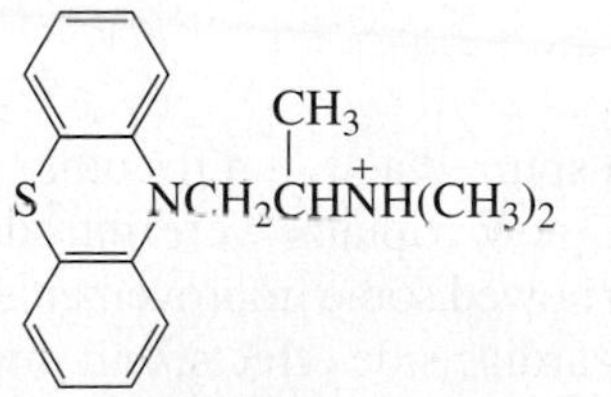

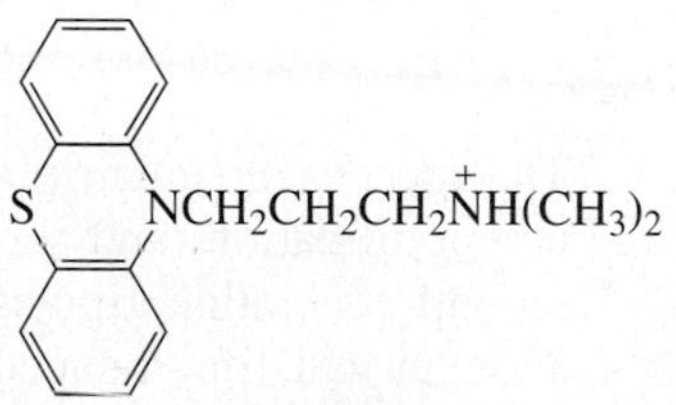

Excess histamine production by the body also causes the hypersecretion of stomach acid by the cells of the stomach lining, leading to the development of ulcers. The antihistamines that block the histamine receptors—thereby preventing the allergic responses associated with excess histamine production—have no effect on HCl production. This fact led scientists to conclude that a second kind of histamine receptor triggers the release of acid into the stomach.

Because 4-methylhistamine was found to cause weak inhibition of HCl secretion, it was used as a lead compound. About 500 molecular modifications were performed over a 10-year period before four clinically useful antiulcer agents were found. Two of these are Tagamet and Zantac. Notice that steric blocking of the receptor site is not a factor in these compounds. Compared with the antihistamines, the effective antiulcer drugs have more polar rings and longer side chains. Tagamet has the same imidazole ring as 4-methylhistidine, but a different side chain. Zantac has a different heterocyclic ring, and its side chain is similar to that of Tagamet.

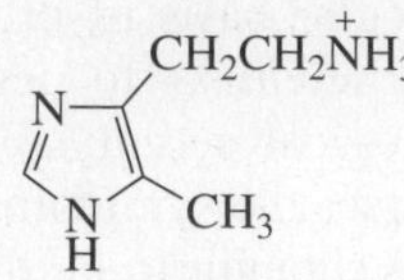

4-methylhistamine

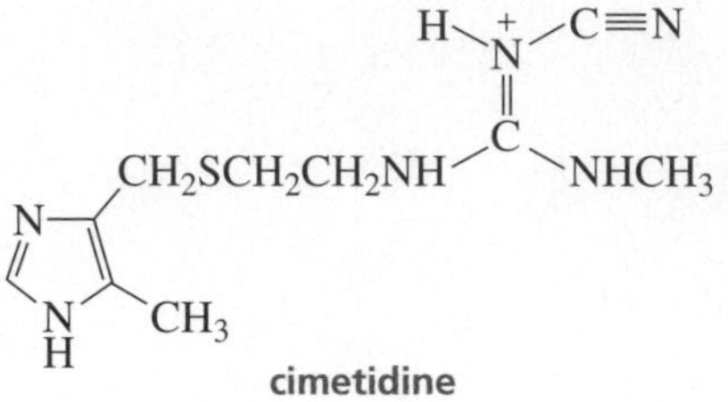

cimetidine
Tagamet®

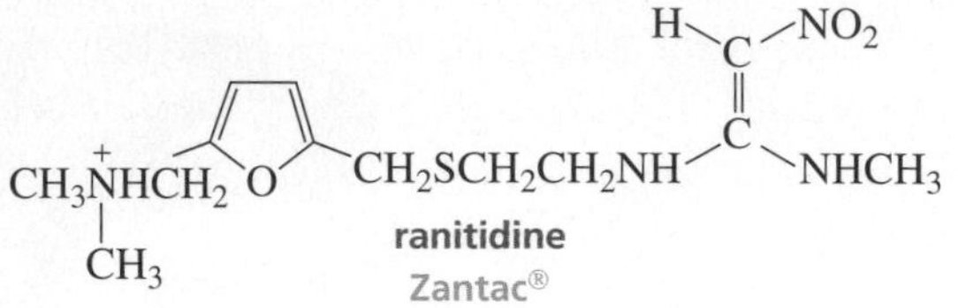

ranitidine
Zantac®

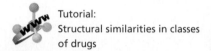

Observations that serotonin was implicated in the generation of migraine attacks led to the development of drugs that bind to serotonin receptors. Sumatriptan, introduced in 1991, relieved not only the pain associated with migraines, but also many of migraine's other symptoms, including nausea and sensitivity to light and sound.

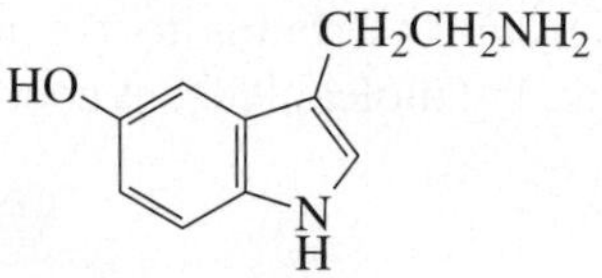

serotonin

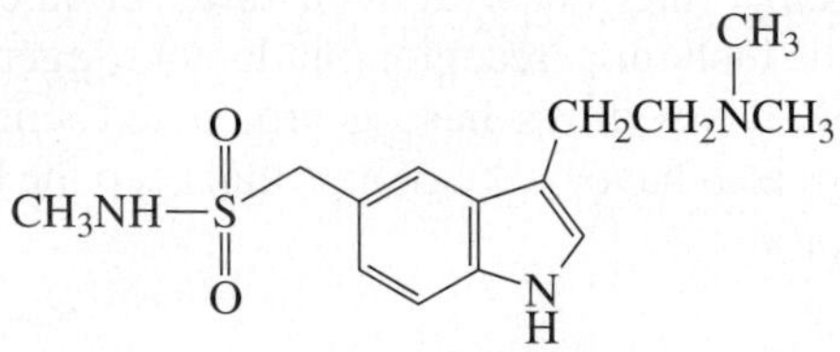

sumatriptan
Imitrex®

The success of sumatriptan spurred a search for other antimigraine agents by molecular modification, and three new triptans were introduced in 1997 and 1998. These second-generation triptans showed some improvements over sumatriptan—specifically, a longer half-life, reduced cardiac side effects, and improved CNS penetration.

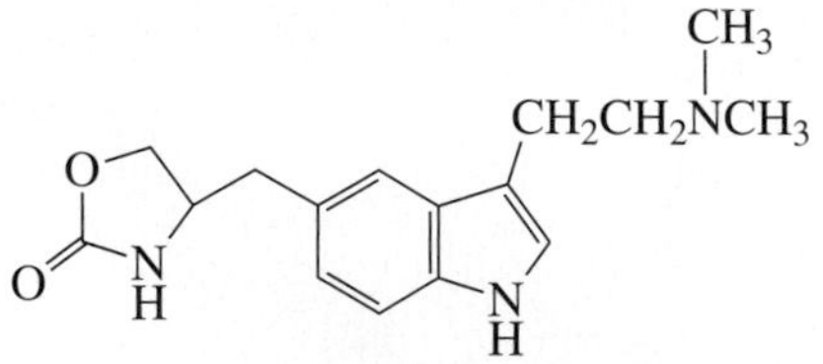

zolmitriptan
Zomig®

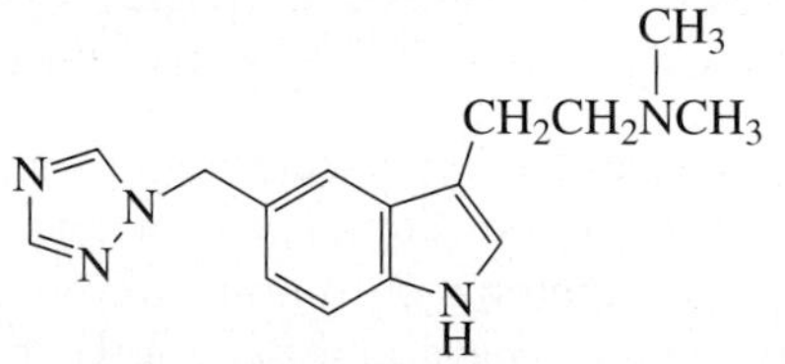

rizatriptan
Maxalt®

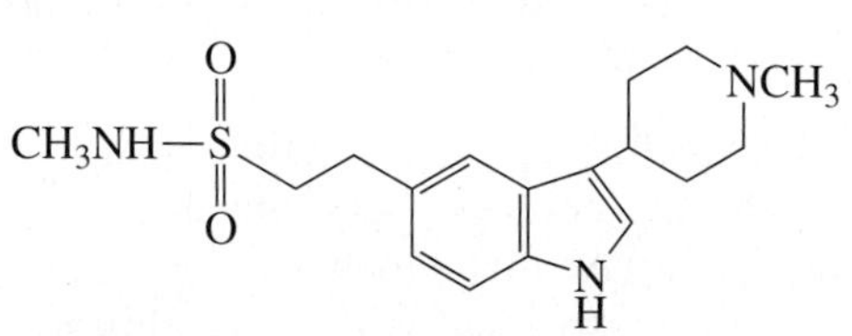

naratriptan
Amerge®

In screening modified compounds, it is not unusual to find a compound with a completely different pharmacological activity than the lead compound. For example, a molecular modification of a sulfonamide, an antibiotic (Section 22.4), led to tolbutamide, a drug with hypoglycemic activity.

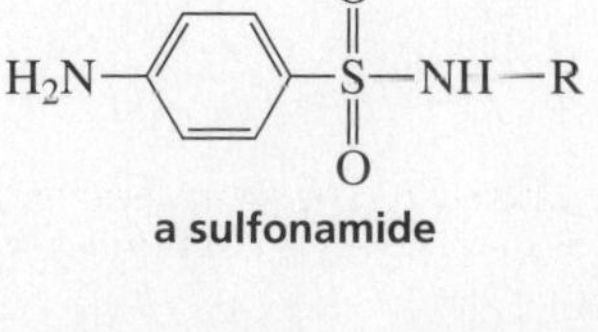

a sulfonamide

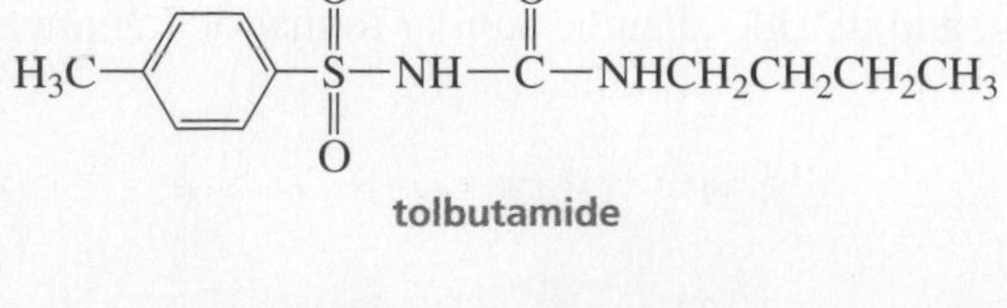

tolbutamide

Molecular modification of promethazine, an antihistamine, led to chlorpromazine, a drug that lacked antihistamine activity but was found to lower body temperature. This drug found clinical use in chest surgery, where patients previously had to be cooled down by wrapping them in cold, wet sheets. Because wrapping in cold, wet sheets had been an old method of calming psychotic patients, a French psychiatrist tried the drug on some of his patients. He found that chlorpromazine was able to suppress psychotic symptoms to the point that the patients assumed almost normal behavioral characteristics. However, they soon developed uncoordinated involuntary movements. After thousands of molecular modifications, thioridazine was found to have the appropriate calming effect with less problematic side effects. It is now in clinical use as an antipsychotic.

chlorpromazine
Thorazine®

thioridazine
Mellaril®

Sometimes a drug initially developed for one purpose is later found to have properties that make it a better drug for a different purpose. Beta-blockers were originally intended to be used to alleviate the pain associated with angina by reducing the amount of work done by the heart. Later, they were found to have antihypertensive properties, so now they are used primarily to manage hypertension.

22.7   Drug Resistance

Typically, it takes a bacterial strain 15 to 20 years to become resistant to an antibiotic. The fluoroquinolones, the last class of antibiotics to be discovered until very recently, were discovered more than 30 years ago, so **drug resistance** has become an increasingly important problem in medicinal chemistry. More and more bacteria have become resistant to all antibiotics—even vancomycin, until recently the antibiotic of last resort.

The antibiotic activity of the fluoroquinolones results from their ability to inhibit DNA gyrase, an enzyme required for transcription (Section 21.6). Fortunately, the bacterial and mammalian forms of the enzyme are sufficiently different that the fluoroquinolones inhibit only the bacterial enzyme.

There are many different fluoroquinolones. All have fluorine substituents, which increase the lipophilicity of the drug to enable it to penetrate into tissues and cells. If either the carboxyl group or the double bond in the 4-pyridinone ring is removed, all activity is lost. By changing the substituents on the piperazine ring, excretion of the drug can be shifted from the liver to the kidney, which is useful to patients with impaired liver function. The substituents on the piperazine ring also affect the half-life of the drug.

ciprofloxacin
Cipro®
active against gram-negative bacteria

sparfloxacin
Zagam®
**active against gram-negative bacteria
and gram-positive bacteria**

The approval of Zyvox by the FDA in April 2000 was met with great relief by the medical community. Zyvox is the first in a new family of antibiotics: the oxazolidinones. In clinical trials, Zyvox was found to cure 75% of the patients infected with bacteria that had become resistant to all other antibiotics.

linezolid
Zyvox®

Zyvox is a synthetic compound designed by scientists to inhibit bacterial growth at a point different from that at which any other antibiotic exerts its effect. Zyvox inhibits the initiation of protein synthesis by preventing the formation of the complex between mRNA, the first amino-acid-bearing tRNA, and the ribosome where protein synthesis takes place (Section 21.8). Because of the drug's new mode of activity, resistance is expected to be rare at first and, hopefully, slow to emerge.

22.8 Molecular Modeling

Because the shape of a molecule determines whether it will be recognized by a receptor, and therefore whether it will exhibit biological activity, compounds with similar biological activity often have similar structures. Because computers can draw molecular models of compounds on a video display and move them around to assume different conformations, computer **molecular modeling** allows more rational drug design. There are computer programs that allow chemists to scan existing collections of thousands of compounds to find those with the appropriate structural and conformational properties. For example, the binding of netropsin, an antibiotic with a wide range of antimicrobial activity, to DNA is shown in Figure 22.1. Retonavir, a drug used to treat the AIDS virus, inactivates HIV protease, an enzyme essential for the maturation of the virus, by binding to its active site (Figure 22.2).

The fit between the compound and the receptor may suggest modifications that can be made to the compound that result in more favorable binding. In this way, the selection of compounds to be synthesized to screen for biological activity can be more rational and will allow pharmacologically active compounds to be discovered more rapidly. The technique will become more valuable as scientists learn more about receptor sites.

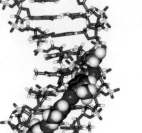

▲ **Figure 22.1**
The antibiotic netropsin bound to DNA.

Figure 22.2 ▶
Retonavir, a drug used to treat the AIDS virus, binds to the active site of HIV protease.

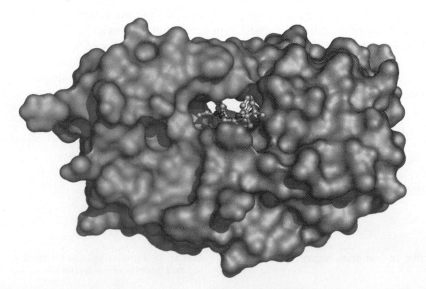

22.9   Antiviral Drugs

Relatively few clinically useful drugs have been developed for viral infections. This slow progress is due to the nature of viruses and the way they replicate. Most **antiviral drugs** are analogs of nucleosides, interfering with DNA or RNA synthesis. In this way, they prevent the virus from replicating. For example, acyclovir, the drug used against herpes viruses, has a three-dimensional shape similar to guanine. Acyclovir can, therefore, fool the virus into incorporating it instead of guanine into the virus's DNA. Once this happens, the DNA strand can no longer grow, because acyclovir lacks a 3′-OH group (Section 21.2).

3-D Molecule:
Rifampin

acyclovir
Aclovir®
used against
herpes simplex infections

cytarabine
Cytosar®
used against acute
myelocytic leukemia

ribavirin
Viramid®
a broad-spectrum
antiviral agent

idoxuridine
Herplex®
approved for topical
ophthalmic use

Cytarabine, used for acute myelocytic leukemia, competes with cytosine for incorporation into viral DNA. Cytarabine contains an arabinose rather than a ribose. The 2′-OH group in the β-position (remember that the 2′-OH group of a ribonucleoside is in the α-position) prevents the bases in DNA from stacking properly.

Ribavirin is a broad-spectrum antiviral agent that interferes with the synthesis of GTP and therefore with the synthesis of all nucleic acid synthesis. It is used to treat children with chronic hepatitis C.

Idoxuridine is approved in the United States only for the topical treatment of ocular infections, although it is used for herpes infections in other countries. Idoxuridine has an iodo group in place of the methyl group of thymine. The drug is incorporated into DNA in place of thymine. Chain elongation can continue because idoxuridine has a 3′-OH group. The DNA that has incorporated idoxuridine, however, is more easily broken and is also not transcribed properly.

22.10   Economics of Drugs •
Governmental Regulations

The average cost of launching a new drug is $100 to $500 million. This cost has to be amortized quickly by the manufacturer because the starting date of a patent is the date the drug is first discovered. A patent is good for 20 years from the date it is applied for, but because it takes an average of 12 years to market a drug after its initial discovery, the patent protects the discoverer of the drug for an average of only 8 years. It is only during the 8 years of patent protection that marketing the drug can provide enough profit so that costs can be recovered and income can be generated to carry out research on new drugs. In addition, the average lifetime of a drug is only 15 to 20 years. After that time, it is generally replaced by a new and improved drug.

Why does it cost so much to develop a new drug? First of all, the Food and Drug Administration (FDA) has very high standards that must be met before it approves a drug for a particular use. Before the U.S. government became involved with the regulation of drugs, it was not uncommon for charlatans to dispense useless and even harmful medicinal preparations. Starting in 1906, Congress passed laws governing the manufacture, distribution, and use of drugs. These laws are amended from time to time to reflect changing situations. The present law requires that all new drugs be thoroughly tested for effectiveness and safety before they are used by physicians.

An important factor leading to the high price of many drugs is the low success rate in progressing from the initial concept to an approved product: Only 1 or 2 of every 100 compounds tested become lead compounds; out of 100 structural modifications of a lead compound, only 1 is worthy of further study; and only 10% of these compounds actually become marketable drugs.

ORPHAN DRUGS

Because of the high cost associated with developing a drug, pharmaceutical companies are reluctant to carry out research on drugs for rare diseases. Even if a company were to find a drug, there would be no way to recoup its expenditure, because of the limited demand. In 1983, the U.S. Congress passed the Orphan Drugs Act. The act creates public subsidies to fund research and provides tax credits for developing and marketing drugs—called **orphan drugs**—for diseases or conditions that affect fewer than 200,000 people. In addition, the company that develops the drug has four years of exclusive marketing rights if the drug is nonpatentable. In the 10 years prior to the passage of this act, fewer than 10 orphan drugs were developed. Today, more than 200 orphan drugs are being used to treat patients, and many hundreds more are in development. Drugs originally developed as orphan drugs include AZT (to treat AIDS), Taxol (to treat ovarian cancer), Exosurf Neonatal (to treat respiratory distress syndrome in infants), and Opticrom (to treat corneal swelling).

Summary

A **drug** is a compound that interacts with a biological molecule, triggering a physiological response. Each drug has a **brand name** that can be used only by the owner of the patent, which is valid for 20 years. Once a patent expires, other drug companies can market the drug under a **generic name** that can be used by any pharmaceutical company. The Food and Drug Administration (FDA) sets high standards that must be met before it approves a drug for a particular use.

The prototype for a new drug is called a **lead compound**. Changing the structure of a lead compound is called **molecular modification**. A **random screen** (or **blind screen**) is a search for a pharmacologically active lead compound without having any information about what structures might show activity.

Many drugs exert their physiological effects by binding to a specific binding site called a **receptor**. Most **antiviral drugs** are analogs of nucleosides, interfering with DNA or RNA synthesis and thereby preventing the virus from replicating.

A **bacteriostatic drug** inhibits the further growth of bacteria; a **bactericidal drug** kills the bacteria. In recent years, many bacteria have become resistant to all antibiotics, so **drug resistance** has become an increasingly important problem in medicinal chemistry.

Problems

1. What is the chemical name of each of the following drugs?
   a. benzocaine (Section 22.3)   **b.** idoxuridine (Section 22.9)

2. Based on the lead compound for the development of procaine and lidocaine, propose structures for other compounds that you would like to see tested for use as anesthetics.

3. Which of the following compounds is more likely to exhibit activity as a tranquilizer?

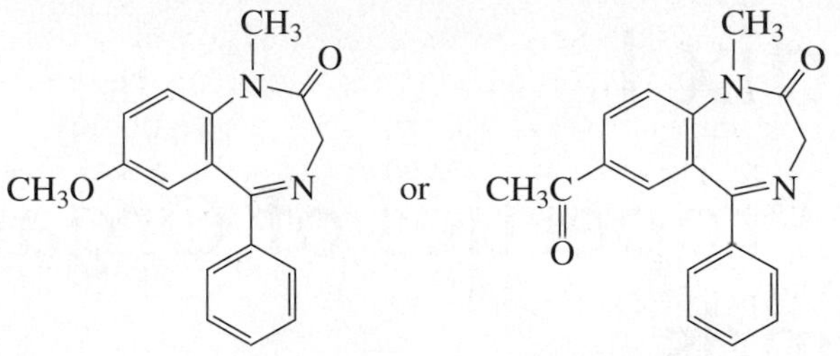

or

4. Which compound is more likely to be a general anesthetic?

$$CH_3CH_2CH_2OH \quad or \quad CH_3OCH_2CH_3$$

5. The therapeutic index of a drug is the ratio of the lethal dose to the therapeutic dose. Therefore, the higher the therapeutic index, the greater is the margin of safety of the drug. The lethal dose of tetrahydrocannabinol in mice is 2.0 g/kg, and the therapeutic dose is 20 mg/kg. The lethal dose of sodium pentathol in mice is 100 mg/kg, and the therapeutic dose is 30 mg/kg. Which is the safer drug?

6. Explain how each of the antiviral drugs shown in Section 22.9 differs from the naturally occurring nucleoside that it most closely resembles.

7. Show a mechanism for the formation of a benzodiazepine 4-oxide from the reaction of a quinazoline 3-oxide with methylamine.

Appendix I

Physical Properties of Organic Compounds

Physical Properties of Alkenes

Name	Structure	mp (°C)	bp (°C)	Density (g/mL)
Ethene	$CH_2{=}CH_2$	−169	−104	
Propene	$CH_2{=}CHCH_3$	−185	−47	
1-Butene	$CH_2{=}CHCH_2CH_3$	−185	−6.3	
1-Pentene	$CH_2{=}CH(CH_2)_2CH_3$	−138	30	0.641
1-Hexene	$CH_2{=}CH(CH_2)_3CH_3$	−140	64	0.673
1-Heptene	$CH_2{=}CH(CH_2)_4CH_3$	−119	94	0.697
1-Octene	$CH_2{=}CH(CH_2)_5CH_3$	−101	122	0.715
1-Nonene	$CH_2{=}CH(CH_2)_6CH_3$	−81	146	0.730
1-Decene	$CH_2{=}CH(CH_2)_7CH_3$	−66	171	0.741
cis-2-Butene	$cis\text{-}CH_3CH{=}CHCH_3$	−180	37	0.650
trans-2-Butene	$trans\text{-}CH_3CH{=}CHCH_3$	−140	37	0.649
Methylpropene	$CH_2{=}C(CH_3)_2$	−140	−6.9	0.594
cis-2-Pentene	$cis\text{-}CH_3CH{=}CHCH_2CH_3$	−180	37	0.650
trans-2-Pentene	$trans\text{-}CH_3CH{=}CHCH_2CH_3$	−140	37	0.649
Cyclohexene		−104	83	0.811

Physical Properties of Alkynes

Name	Structure	mp (°C)	bp (°C)	Density (g/mL)
Ethyne	$HC{\equiv}CH$	−82	−84.0	
Propyne	$HC{\equiv}CCH_3$	−101.5	−23.2	
1-Butyne	$HC{\equiv}CCH_2CH_3$	−122	8.1	
2-Butyne	$CH_3C{\equiv}CCH_3$	−32	27	0.694
1-Pentyne	$HC{\equiv}C(CH_2)_2CH_3$	−98	39.3	0.695
2-Pentyne	$CH_3C{\equiv}CCH_2CH_3$	−101	55.5	0.714
3-Methyl-1-butyne	$HC{\equiv}CCH(CH_3)_2$	−90	29	0.665
1-Hexyne	$HC{\equiv}C(CH_2)_3CH_3$	−132	71	0.715
2-Hexyne	$CH_3C{\equiv}C(CH_2)_2CH_3$	−92	84	0.731
3-Hexyne	$CH_3CH_2C{\equiv}CCH_2CH_3$	−101	81	0.725
1-Heptyne	$HC{\equiv}C(CH_2)_4CH_3$	−81	100	0.733
1-Octyne	$HC{\equiv}C(CH_2)_5CH_3$	−80	127	0.747
1-Nonyne	$HC{\equiv}C(CH_2)_6CH_3$	−50	151	0.757
1-Decyne	$HC{\equiv}C(CH_2)_7CH_3$	−44	174	0.766

Physical Properties of Cyclic Saturated Alkanes

Name	mp (°C)	bp (°C)	Density (g/mL)
Cyclopropane	−128	−33	
Cyclobutane	−80	−12	
Cyclopentane	−94	50	0.751
Cyclohexane	6.5	81	0.779
Cycloheptane	−12	118	0.811
Cyclooctane	14	149	0.834
Methylcyclopentane	−142	72	0.749
Methylcyclohexane	−126	100	0.769
cis-1,2-Dimethylcyclopentane	−62	99	0.772
trans-1,2-Dimethylcyclopentane	−120	92	0.750

Physical Properties of Ethers

Name	Structure	mp (°C)	bp (°C)	Density (g/mL)
Dimethyl ether	CH_3OCH_3	−141	−24.8	
Diethyl ether	$CH_3CH_2OCH_2CH_3$	−116	34.6	0.706
Dipropyl ether	$CH_3(CH_2)_2O(CH_2)_2CH_3$	−123	88	0.736
Diisopropyl ether	$(CH_3)_2CHOCH(CH_3)_2$	−86	69	0.725
Dibutyl ether	$CH_3(CH_2)_3O(CH_2)_3CH_3$	−98	142	0.764
Divinyl ether	$CH_2\text{=}CHOCH\text{=}CH_2$		35	
Diallyl ether	$CH_2\text{=}CHCH_2OCH_2CH\text{=}CH_2$		94	0.830
Tetrahydrofuran		−108	66	0.889
Dioxane	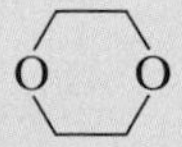	12	101	1.034

Physical Properties of Alcohols

Name	Structure	mp (°C)	bp (°C)	Solubility (g/100 g H$_2$O at 25 °C)
Methanol	CH$_3$OH	−97.8	64	∞
Ethanol	CH$_3$CH$_2$OH	−114.7	78	∞
1-Propanol	CH$_3$(CH$_2$)$_2$OH	−127	97.4	∞
1-Butanol	CH$_3$(CH$_2$)$_3$OH	−90	118	7.9
1-Pentanol	CH$_3$(CH$_2$)$_4$OH	−78	138	2.3
1-Hexanol	CH$_3$(CH$_2$)$_5$OH	−52	157	0.6
1-Heptanol	CH$_3$(CH$_2$)$_6$OH	−36	176	0.2
1-Octanol	CH$_3$(CH$_2$)$_7$OH	−15	196	0.05
2-Propanol	CH$_3$CHOHCH$_3$	−89.5	82	∞
2-Butanol	CH$_3$CHOHCH$_2$CH$_3$	−115	99.5	12.5
2-Methyl-1-propanol	(CH$_3$)$_2$CHCH$_2$OH	−108	108	10.0
2-Methyl-2-propanol	(CH$_3$)$_3$COH	25.5	83	∞
3-Methyl-1-butanol	(CH$_3$)$_2$CH(CH$_2$)$_2$OH	−117	130	2
2-Methyl-2-butanol	(CH$_3$)$_2$COHCH$_2$CH$_3$	−12	102	12.5
2,2-Dimethyl-1-propanol	(CH$_3$)$_3$CCH$_2$OH	55	114	∞
Allyl alcohol	CH$_2$=CHCH$_2$OH	−129	97	∞
Cyclopentanol	C$_5$H$_9$OH	−19	140	s. sol.
Cyclohexanol	C$_6$H$_{11}$OH	24	161	s. sol.
Benzyl alcohol	C$_6$H$_5$CH$_2$OH	−15	205	4

Physical Properties of Alkyl Halides

Name	bp (°C)			
	Fluoride	*Chloride*	*Bromide*	*Iodide*
Methyl	−78.4	−24.2	3.6	42.4
Ethyl	−37.7	12.3	38.4	72.3
Propyl	−2.5	46.6	71.0	102.5
Isopropyl	−9.4	34.8	59.4	89.5
Butyl	32.5	78.4	100	130.5
Isobutyl		68.8	90	120
sec-Butyl		68.3	91.2	120.0
tert-Butyl		50.2	73.1	dec.
Pentyl	62.8	108	130	157.0
Hexyl	92	133	154	179

Physical Properties of Amines

Name	Structure	mp (°C)	bp (°C)	Solubility (g/100 g H₂O at 25 °C)
Primary Amines				
Methylamine	CH_3NH_2	−93	−6.3	v. sol.
Ethylamine	$CH_3CH_2NH_2$	−81	17	∞
Propylamine	$CH_3(CH_2)_2NH_2$	−83	48	∞
Isopropylamine	$(CH_3)_2CHNH_2$	−95	33	∞
Butylamine	$CH_3(CH_2)_3NH_2$	−49	78	v. sol.
Isobutylamine	$(CH_3)_2CHCH_2NH_2$	−85	68	∞
sec-Butylamine	$CH_3CH_2CH(CH_3)NH_2$	−72	63	∞
tert-Butylamine	$(CH_3)_3CNH_2$	−67	46	∞
Cyclohexylamine	$C_6H_{11}NH_2$	−18	134	s. sol.
Secondary Amines				
Dimethylamine	$(CH_3)_2NH$	−93	7.4	v. sol.
Diethylamine	$(CH_3CH_2)_2NH$	−50	55	10.0
Dipropylamine	$(CH_3CH_2CH_2)_2NH$	−63	110	10.0
Dibutylamine	$(CH_3CH_2CH_2CH_2)_2NH$	−62	159	s. sol.
Tertiary Amines				
Trimethylamine	$(CH_3)_3N$	−115	2.9	91
Triethylamine	$(CH_3CH_2)_3N$	−114	89	14
Tripropylamine	$(CH_3CH_2CH_2)_3N$	−93	157	s. sol.

Physical Properties of Benzene and Substituted Benzenes

Name	Structure	mp (°C)	bp (°C)	Solubility (g/100 g H₂O at 25 °C)
Aniline	$C_6H_5NH_2$	−6	184	3.7
Benzene	C_6H_6	5.5	80.1	s. sol.
Benzaldehyde	C_6H_5CHO	−26	178	s. sol.
Benzamide	$C_6H_5CONH_2$	132	290	s. sol.
Benzoic acid	C_6H_5COOH	122	249	0.34
Bromobenzene	C_6H_5Br	−30.8	156	insol.
Chlorobenzene	C_6H_5Cl	−45.6	132	insol.
Nitrobenzene	$C_6H_5NO_2$	5.7	210.8	s. sol.
Phenol	C_6H_5OH	43	182	s. sol.
Styrene	$C_6H_5CH{=}CH_2$	−30.6	145.2	insol.
Toluene	$C_6H_5CH_3$	−95	110.6	insol.

Physical Properties of Carboxylic Acids

Name	Structure	mp (°C)	bp (°C)	Solubility (g/100 g H_2O at 25 °C)
Formic acid	HCOOH	8.4	101	∞
Acetic acid	CH_3COOH	16.6	118	∞
Propionic acid	CH_3CH_2COOH	−21	141	∞
Butanoic acid	$CH_3(CH_2)_2COOH$	−5	162	∞
Pentanoic acid	$CH_3(CH_2)_3COOH$	−34	186	4.97
Hexanoic acid	$CH_3(CH_2)_4COOH$	−4	202	0.97
Heptanoic acid	$CH_3(CH_2)_5COOH$	−8	223	0.24
Octanoic acid	$CH_3(CH_2)_6COOH$	17	237	0.068
Nonanoic acid	$CH_3(CH_2)_7COOH$	15	255	0.026
Decanoic acid	$CH_3(CH_2)_8COOH$	32	270	0.015

Physical Properties of Dicarboxylic Acids

Name	Structure	mp (°C)	Solubility (g/100 g H_2O at 25 °C)
Oxalic acid	HOOCCOOH	189	s
Malonic acid	$HOOCCH_2COOH$	136	v. sol.
Succinic acid	$HOOC(CH_2)_2COOH$	185	s. sol.
Glutaric acid	$HOOC(CH_2)_3COOH$	98	v. sol.
Adipic acid	$HOOC(CH_2)_4COOH$	151	s. sol.
Pimelic acid	$HOOC(CH_2)_5COOH$	106	s. sol.
Phthalic acid	$1,2\text{-}C_6H_4(COOH)_2$	231	s. sol.
Maleic acid	*cis*-HOOCCH=CHCOOH	130.5	v. sol.
Fumaric acid	*trans*-HOOCCH=CHCOOH	302	s. sol.

Physical Properties of Acyl Chlorides and Acid Anhydrides

Name	Structure	mp (°C)	bp (°C)
Acetyl chloride	CH_3COCl	−112	51
Propionyl chloride	CH_3CH_2COCl	−94	80
Butyryl chloride	$CH_3(CH_2)_2COCl$	−89	102
Valeryl chloride	$CH_3(CH_2)_3COCl$	−110	128
Acetic anhydride	$CH_3(CO)O(CO)CH_3$	−73	140
Succinic anhydride	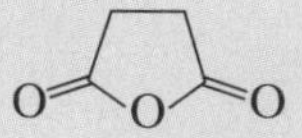	120	261

Physical Properties of Esters

Name	Structure	mp (°C)	bp (°C)
Methyl formate	HCOOCH$_3$	−100	32
Ethyl formate	HCOOCH$_2$CH$_3$	−80	54
Methyl acetate	CH$_3$COOCH$_3$	−98	57.5
Ethyl acetate	CH$_3$COOCH$_2$CH$_3$	−84	77
Propyl acetate	CH$_3$COO(CH$_2$)$_2$CH$_3$	−92	102
Methyl propionate	CH$_3$CH$_2$COOCH$_3$	−87.5	80
Ethyl propionate	CH$_3$CH$_2$COOCH$_2$CH$_3$	−74	99
Methyl butyrate	CH$_3$CH$_2$CH$_2$COOCH$_3$	−84.8	102.3
Ethyl butyrate	CH$_3$CH$_2$CH$_2$COOCH$_2$CH$_3$	−93	121

Physical Properties of Amides

Name	Structure	mp (°C)	bp (°C)
Formamide	HCONH$_2$	3	200 d*
Acetamide	CH$_3$CONH$_2$	82	221
Propanamide	CH$_3$CH$_2$CONH$_2$	80	213
Butanamide	CH$_3$(CH$_2$)$_2$CONH$_2$	116	216
Pentanamide	CH$_3$(CH$_2$)$_3$CONH$_2$	106	232

*d means the substance decomposes.

Physical Properties of Aldehydes

Name	Structure	mp (°C)	bp (°C)	Solubility (g/100 g H$_2$O at 25 °C)
Formaldehyde	HCHO	−92	−21	v. sol.
Acetaldehyde	CH$_3$CHO	−121	21	∞
Propionaldehyde	CH$_3$CH$_2$CHO	−81	49	16
Butyraldehyde	CH$_3$(CH$_2$)$_2$CHO	−96	75	7
Pentanal	CH$_3$(CH$_2$)$_3$CHO	−92	103	s. sol.
Hexanal	CH$_3$(CH$_2$)$_4$CHO	−56	131	s. sol.
Heptanal	CH$_3$(CH$_2$)$_5$CHO	−43	153	0.1
Octanal	CH$_3$(CH$_2$)$_6$CHO		171	insol.
Nonanal	CH$_3$(CH$_2$)$_7$CHO		192	insol.
Decanal	CH$_3$(CH$_2$)$_8$CHO	−5	209	insol.
Benzaldehyde	C$_6$H$_5$CHO	−26	178	0.3

Physical Properties of Ketones

Name	Structure	mp (°C)	bp (°C)	Solubility (g/100 g H$_2$O at 25 °C)
Acetone	CH$_3$COCH$_3$	−95	56	∞
2-Butanone	CH$_3$COCH$_2$CH$_3$	−86	80	25.6
2-Pentanone	CH$_3$CO(CH$_2$)$_2$CH$_3$	−78	102	5.5
2-Hexanone	CH$_3$CO(CH$_2$)$_3$CH$_3$	−57	127	1.6
2-Heptanone	CH$_3$CO(CH$_2$)$_4$CH$_3$	−36	151	0.4
2-Octanone	CH$_3$CO(CH$_2$)$_5$CH$_3$	−16	173	insol.
2-Nonanone	CH$_3$CO(CH$_2$)$_6$CH$_3$	−7	195	insol.
2-Decanone	CH$_3$CO(CH$_2$)$_7$CH$_3$	14	210	insol.
3-Pentanone	CH$_3$CH$_2$COCH$_2$CH$_3$	−40	102	4.8
3-Hexanone	CH$_3$CH$_2$CO(CH$_2$)$_2$CH$_3$		123	1.5
3-Heptanone	CH$_3$CH$_2$CO(CH$_2$)$_3$CH$_3$	−39	149	0.3
Acetophenone	CH$_3$COC$_6$H$_5$	19	202	insol.
Propiophenone	CH$_3$CH$_2$COC$_6$H$_5$	18	218	insol.

Appendix II

pK_a Values

Compound	pK_a	Compound	pK_a	Compound	pK_a
$CH_3C\equiv \overset{+}{N}H$	−10.1	$O_2N-\!\!\bigcirc\!\!-\overset{+}{N}H_3$	1.0	$CH_3-\!\!\bigcirc\!\!-\overset{O}{\overset{\|}{C}}OH$	4.3
HI	−10	(pyrimidinium)	1.0	$CH_3O-\!\!\bigcirc\!\!-\overset{O}{\overset{\|}{C}}OH$	4.5
HBr	−9				
$CH_3\overset{\overset{+}{O}H}{\overset{\|}{C}}H$	−8	$Cl_2CH\overset{O}{\overset{\|}{C}}OH$	1.3	$\bigcirc\!\!-\overset{+}{N}H_3$	4.6
$CH_3\overset{\overset{+}{O}H}{\overset{\|}{C}}CH_3$	−7.3	HSO_4^-	2.0		
HCl	−7	H_3PO_4	2.1	$CH_3\overset{O}{\overset{\|}{C}}OH$	4.8
$\bigcirc\!\!-SO_3H$	−6.5	(purinium)	2.5	(quinolinium)	4.9
$CH_3\overset{\overset{+}{O}H}{\overset{\|}{C}}OCH_3$	−6.5	$FCH_2\overset{O}{\overset{\|}{C}}OH$	2.7	$CH_3-\!\!\bigcirc\!\!-\overset{+}{N}H_3$	5.1
$CH_3\overset{\overset{+}{O}H}{\overset{\|}{C}}OH$	−6.1	$ClCH_2\overset{O}{\overset{\|}{C}}OH$	2.8	(pyridinium)	5.2
H_2SO_4	−5	$BrCH_2\overset{O}{\overset{\|}{C}}OH$	2.9		
(pyrrolium)	−3.8	$ICH_2\overset{O}{\overset{\|}{C}}OH$	3.2	$CH_3O-\!\!\bigcirc\!\!-\overset{+}{N}H_3$	5.3
$CH_3CH_2\overset{\overset{+}{H}}{\overset{\|}{O}}CH_2CH_3$	−3.6	HF	3.2	$CH_3\overset{\overset{+}{N}HCH_3}{\overset{\|}{C}}$ ... CH_3	5.5
$CH_3CH_2\overset{\overset{+}{H}}{\overset{\|}{O}}H$	−2.4	HNO_2	3.4		
$CH_3\overset{\overset{+}{H}}{\overset{\|}{O}}H$	−2.5	$O_2N-\!\!\bigcirc\!\!-\overset{O}{\overset{\|}{C}}OH$	3.4	$CH_3\overset{O}{\overset{\|}{C}}CH_2\overset{O}{\overset{\|}{C}}H$	5.9
H_3O^+	−1.7	$HCOH$	3.8	$HO\overset{+}{N}H_3$	6.0
HNO_3	−1.3			H_2CO_3	6.4
CH_3SO_3H	−1.2	$Br-\!\!\bigcirc\!\!-\overset{+}{N}H_3$	3.9	(imidazolium)	6.8
$CH_3\overset{\overset{+}{O}H}{\overset{\|}{C}}NH_2$	0.0	$Br-\!\!\bigcirc\!\!-\overset{O}{\overset{\|}{C}}OH$	40	H_2S	7.0
$F_3C\overset{O}{\overset{\|}{C}}OH$	0.2	$\bigcirc\!\!-\overset{O}{\overset{\|}{C}}OH$	4.2	$O_2N-\!\!\bigcirc\!\!-OH$	7.1
$Cl_3C\overset{O}{\overset{\|}{C}}OH$	0.64			$H_2PO_4^-$	7.2
(pyridinium $\overset{+}{N}$—OH)	0.79			$\bigcirc\!\!-SH$	7.8

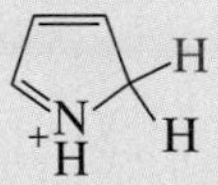

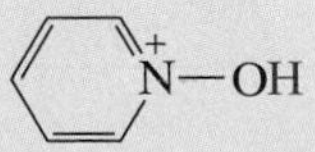

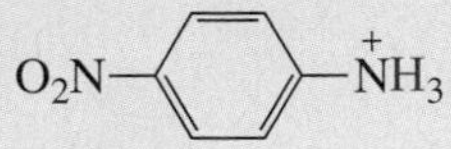

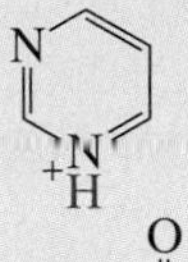

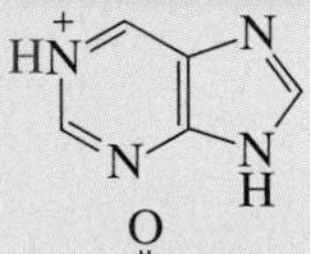

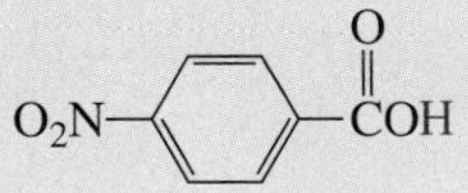

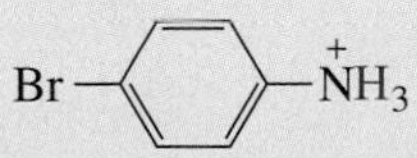

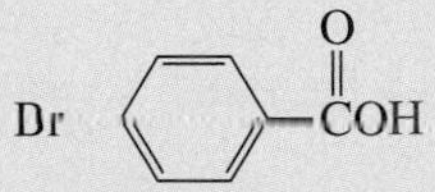

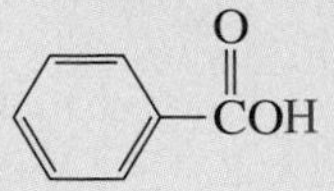

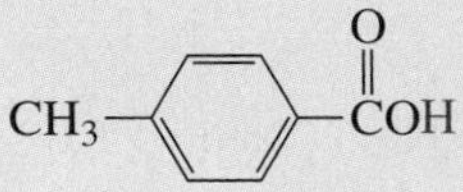

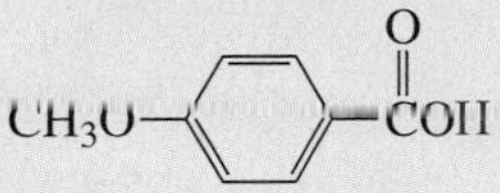

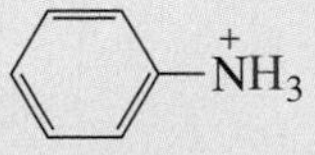

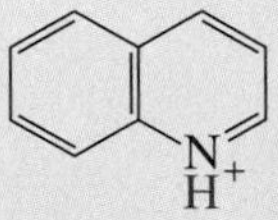

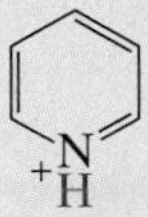

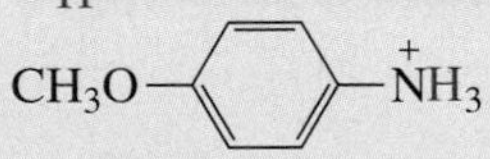

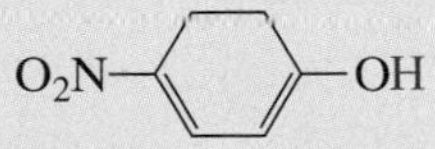

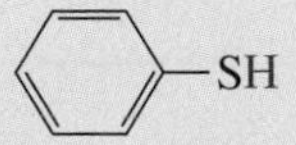

[a] pK_a values are for the red H in each structure

pK_a Values (continued)

Compound	pK_a	Compound	pK_a	Compound	pK_a
aziridinium (N⁺H, H)	8.0	cyclohexyl-NH_3^+	10.7	CH_3CH (O)	17
$H_2N\overset{+}{N}H_3$	8.1	$(CH_3)_2\overset{+}{N}H_2$	10.7	$(CH_3)_3COH$	18
CH_3COOH (O)	8.2	piperidinium (N⁺H, H)	11.1	CH_3CCH_3 (O)	20
$CH_3CH_2NO_2$	8.6	$CH_3CH_2\overset{+}{N}H_3$	11.0	$CH_3COCH_2CH_3$ (O)	24.5
$CH_3CCH_2CCH_3$ (O O)	8.9	pyrrolidinium (N⁺H, H)	11.3	$HC{\equiv}CH$	25
$HC{\equiv}N$	9.1	HPO_4^{2-}	12.3	$CH_3C{\equiv}N$	25
morpholinium (O ring, N⁺H, H)	9.3	CF_3CH_2OH	12.4	$CH_3CN(CH_3)_2$ (O)	30
$Cl{-}C_6H_4{-}OH$	9.4	$CH_3CH_2OCCH_2COCH_2CH_3$ (O O)	13.3	NH_3	36
$\overset{+}{N}H_4$	9.4	$HC{\equiv}CCH_2OH$	13.5	pyrrolidine (N–H)	36
$HOCH_2CH_2\overset{+}{N}H_3$	9.5	H_2NCNH_2 (O)	13.7	CH_3NH_2	40
$H_3\overset{+}{N}CH_2CO^-$ (O)	9.8	$CH_3\overset{+}{N}(CH_3)CH_2CH_2OH$	13.9	$C_6H_5{-}CH_3$	41
$C_6H_5{-}OH$	10.0	imidazole (N, NH)	14.4	benzene	43
$CH_3{-}C_6H_4{-}OH$	10.2	CH_3OH	15.5	$CH_2{=}CHCH_3$	43
HCO_3^-	10.2	H_2O	15.7	$CH_2{=}CH_2$	44
CH_3NO_2	10.2	CH_3CH_2OH	16.0	cyclopropane	46
$H_2N{-}C_6H_4{-}OH$	10.3	CH_3CNH_2 (O)	16	CH_4	60
CH_3CH_2SH	10.5	$C_6H_5CCH_3$ (O)	16.0	CH_3CH_3	>60
$(CH_3)_3\overset{+}{N}H$	10.6	pyrrole (N–H)	~17		
$CH_3CCH_2COCH_2CH_3$ (O O)	10.7				
$CH_3\overset{+}{N}H_3$	10.7				

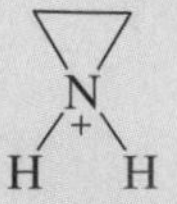

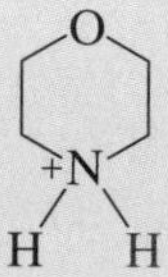

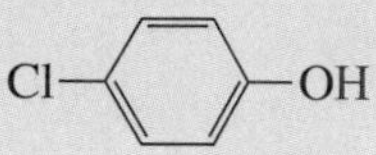

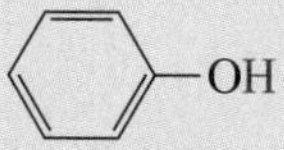

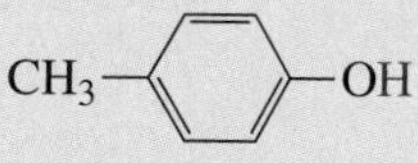

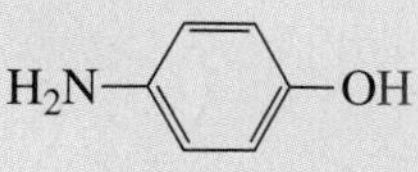

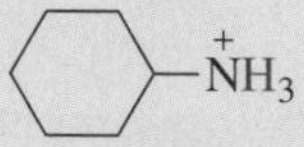

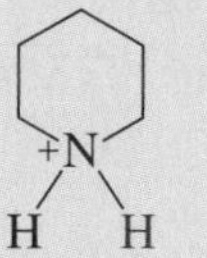

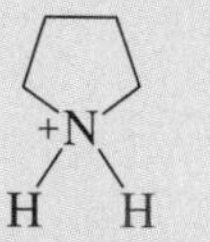

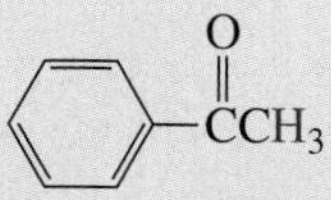

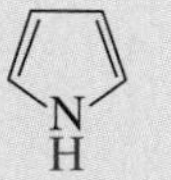

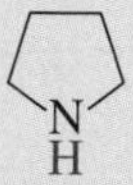

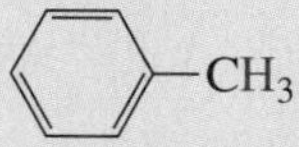

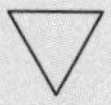

Answers to Selected Problems

CHAPTER 1

1-1. 8 **1-2. a.** 4 **b.** 5 **c.** 6 **d.** 7 **1-3.** 1 **1-4.** 7 **1-5. a.** Cl—CH₃ **b.** H—CH₃
c. H—F **d.** Cl—CH₃ **1-6. a.** KCl **b.** Cl₂ **1-7. a.** LiH and HF **b.** HF

1-8. a. $\overset{\delta-}{H}\overset{}{O}-\overset{\delta+}{H}$ **b.** $\overset{\delta+}{H_3C}-\overset{\delta-}{NH_2}$ **c.** $\overset{\delta-}{HO}-\overset{\delta+}{Br}$ **d.** $\overset{\delta+}{I}-\overset{\delta-}{Cl}$

1-9. a. oxygen **b.** oxygen **c.** oxygen **d.** hydrogen

1-10. a. CH₃—Ö⁺—CH₃ (with H below)
c. CH₃—N⁺—CH₃ (with CH₃ above and below)

b. H—C̈⁻—H (with H above and below)
d. H—N⁺—B⁻—H (with H H above and H H below)

1-11. a. H:C:N:H (with H H above and below)
b. H:C:C:H (with H H above and below)
c. Na⁺ :Ö:H
d. H:N:H :Cl:̈

1-12. a. CH₃CH₂N̈H₂ **c.** CH₃CH₂ÖH **e.** CH₃CH₂C̈l:
b. CH₃N̈HCH₃ **d.** CH₃ÖCH₃ **f.** HÖN̈H₂

1-13. a. CH₃CH₂CH₂Cl
c. CH₃CH₂C̈NCH₂CH₃ (with O double bond above, CH₃ below)
b. CH₃C̈OCH₂CH₃ (with O double bond above)
d. CH₃CH₂C≡N

1-14. a. Cl **b.** O **c.** N **d.** C and H
1-17. 2 C—C bonds from sp^3—sp^3 overlap; 8 C—H bonds from sp^3—s overlap
1-19. greater than 104.5° and less than 109.5° **1-20.** the hydrogens
1-21. Water is the most polar. Methane is the least polar. **1-22.** ~107.3°
1-23. a. relative lengths Br₂ > Cl₂; relative strengths Cl₂ > Br₂ **b.** relative lengths: HBr > HCl > HF; relative strengths: HF > HCl > HBr
1-24. a. 1. C—Br **2.** C—C **3.** H—Cl **b. 1.** C—Cl **2.** C—H **3.** H—H
1-25. σ bond

1-26. a.
CH₃CHCH=CHCH₂C≡CCH₃ (with sp^3, sp^2, sp labels above; CH₃, sp^3 below; sp^3)

1-27. a. 109.5° **b.** 109.5° **c.** 107.3° **d.** 104.5°*

CHAPTER 2

2-1. a. 1. ⁺NH₄ **2.** HCl **3.** H₂O **4.** H₃O⁺ **b. 1.** ⁻NH₂ **2.** Br⁻ **3.** NO₃⁻ **3.** HO⁻
2-3. a. compound with pK_a = 5.2 **b.** compound with dissociation constant = 3.4 × 10⁻³ **2-4.** K_a = 1.51 10⁻⁵; weaker
2-6. a. basic **b.** acidic **c.** basic **2-7. a.** CH₃COO⁻ **b.** ⁻NH₂ **c.** H₂O

2-9. CH₃NH⁻ > CH₃O⁻ > CH₃NH₂ > CH₃C̈O⁻ (with O double bond above) > CH₃OH

2-10. a. HBr **b.** CH₃CH₂CH₂ÖH₂⁺ **2-11. a.** F⁻ **b.** I⁻

2-12. a. oxygen **b.** H₂S **c.** CH₃SH

2-13. a. HO⁻ **b.** NH₃ **c.** CH₃O⁻ **d.** CH₃O⁻
2-14. a. CH₃COO⁻ **b.** CH₃CH₂N̈H₃⁺ **c.** H₂O **d.** Br⁻ **e.** ⁺NH₄ **f.** HC≡N
g. NO₂⁻ **h.** NO₃⁻

2-15. a. 1. neutral **b. 1.** charged **c. 1.** neutral
2. neutral **2.** charged **2.** neutral
3. equal amounts of both **3.** charged **3.** neutral
4. charged **4.** charged **4.** neutral
5. charged **5.** equal amounts of both **5.** neutral
6. charged **6.** neutral **6.** neutral

CHAPTER 3

3-1. a. *n*-propyl alcohol or propyl alcohol **b.** dimethyl ether **c.** *n*-propylamine or propyl amine

3-3. a. CH₃CHOH (with CH₃ below)
d. CH₃CH₂CHI (with CH₃ below)

b. CH₃CHCH₂CH₂F (with CH₃ below)
e. CH₃CNH₂ (with CH₃ above and CH₃ below)

c. CH₃CH₂OCH₂CH₂CH₃ **f.** CH₃CH₂CH₂CH₂CH₂CH₂CH₂CH₂Br

3-4. a. ethyl methyl ether **b.** methyl propyl ether **c.** *sec*-butylamine
d. butyl alcohol **e.** isobutyl bromide **f.** *sec*-butyl chloride

3-5. a. CH₃CH₂CH₂CH₂CH₃
pentane
b. CH₃CCH₃ (with CH₃ above and CH₃ below)
2,2-dimethylpropane

3-6. a. CH₃CHCHCH₂CH₂CH₃ (with CH₃, CH₃ below)
c. CH₃CCH₂CHCH₂CH₂CH₂CH₃ (with CH₃ above, CH₃ CH₂CH₂CH₃ below)

b. CH₃CHCH₂C—CHCH₂CH₃ (with CH₃ CH₃ above, CH(CH₃)₂ below)
d. CH₃CHCH₂CHCHCH₂CH₂CH₃ (with CH₃ CH₃ above, CH₂CH(CH₃)₂ below)

3-8. a. 2,2,4-trimethylhexane **b.** 2,2-dimethylbutane **c.** 3,3-diethylhexane
d. 2,5-dimethylheptane **e.** 4-isopropyloctane **f.** 4-ethyl-2,2,3-trimethylhexane

3-10. a. (skeletal structure with OH) **c.** (skeletal structure with Br)
b. (skeletal structure) **d.** (skeletal structure with O)

3-11. a. 1-ethyl-2-methylcyclopentane **b.** ethylcyclobutane
c. 3,6-dimethyldecane **d.** 5-isopropylnonane **3-12. a.** *sec*-butyl chloride, 2-chlorobutane **b.** isohexyl chloride, 1-chloro-4-methylpentane **c.** cyclohexyl bromide, bromocyclohexane **d.** isopropyl fluoride, 2-fluoropropane
3-13. a. a tertiary alkyl bromide **b.** a tertiary alcohol **c.** a primary amine
3-14. a. methylpropylamine; secondary **b.** trimethylamine; tertiary
c. diethylamine; secondary **d.** butyldimethylamine; tertiary

3-15. a.

b.

c.

3-16. a. ~104.5° **b.** ~107.3° **c.** ~104.5° **3-17.** pentane
3-18. a. O—H hydrogen bond is longer. **b.** O—H covalent bond is stronger.
3-19. a. 1, 4, and 5 **b.** 1, 2, 4, 5, and 6
3-20.

3-22. even number
3-23. a. $HOCH_2CH_2CH_2OH > CH_3CH_2CH_2OH >$
$CH_3CH_2CH_2CH_2OH > CH_3CH_2CH_2CH_2Cl$

b.

3-24. ethanol **3-25.** hexethal **3-29.** isopropylcyclohexane
3-30. a. cis **b.** cis **c.** cis **d.** trans **e.** trans **f.** trans

CHAPTER 4

4-1. a. C_5H_8 **b.** C_4H_6 **c.** $C_{10}H_{16}$ **4-2. a.** 3 **b.** 4 **c.** 1
4-4. a. 4-methyl-2-pentene **b.** 2-chloro-3,4-dimethyl-3-hexene
c. 1-bromo-4-methyl-3-hexene **d.** 1,5-dimethylcyclohexene
4-5. a. 5 **b.** 4 **c.** 4 **d.** 6 **4-6. a.** 1 and 3
4-7. a. —I > —Br > —OH > —CH₃
    b. —OH > —CH₂Cl > —CH=CH₂ > —CH₂CH₂OH
4-10.

4-12. *cis*-3,4-diemthyl-3-hexene > *trans*-3-hexene > *cis*-3-hexene > 1-hexene
4-13. nucleophiles: H⁻ CH₃O⁻ CH₃C≡CH NH₃; electrophiles: CH₃CHCH₃
4-16. a. 2 **b.** B **c.** 3 **d.** the first one **e.** products **f.** C to D **g.** B **h.** yes, since the products are more stable than the reactants

CHAPTER 5

5-1. ethyl cation

5-2.

5-3. a.

5-4. a.

5-5. a. 3 transition states **b.** 2 intermediates **c.** the neutral alcohol **d.** the second and third steps in the forward direction
5-6. a. $CH_3CH_2CH_2CHCH_3$

b.

c. $CH_3CH_2CH_2CH_2CHCH_3$ and $CH_3CH_2CH_2CHCH_2CH_3$

d.

5-9. a. $CH_3CH_2CHCH_3$ and $CH_3CH_2CHCH_3$
    major

b. $CH_3CH_2CHCH_3$ and $CH_3CH_2CHCH_3$
    major

5-11. C_nH_{2n-4} **5-12.** $C_{14}H_{30}(C_nH_{2n+2})$
5-13. a. $ClCH_2CH_2C\equiv CCH_2CH_3$

b. $CH_3C\equiv CCHCH_3$

c. $HC\equiv CCH_2CCH_3$

5-14. a. 2-hexyne **b.** 4-methyl-1-pentyne **5-16. a.** 5-bromo-2-pentyne
b. 6-bromo-2-chloro-4-octyne **c.** 1-methoxy-2-pentyne **d.** 3-ethyl-1-hexyne
5-17. a. sp^2-sp^2 **b.** sp^2-sp^3 **c.** $sp-sp^2$ **d.** $sp-sp^3$ **e.** $sp-sp$ **f.** sp^2-sp^2
g. sp^2-sp^3 **h.** $sp-sp^3$ **i.** sp^2-sp
5-18. a. $CH_2=CCH_3$ **c.** $CH_3CH_2CCH_3$

b. CH_3CCH_3 **d.** $CH_3CCH_2CH_2CH_3$ + $CH_3CH_2CCH_2CH_3$

5-19. $CH_3CH_2CCH_2CH_2CH_2CH_3$ and $CH_3CH_2CH_2CCH_2CH_2CH_3$

5-20. a. $CH_3C\equiv CH$ **b.** $CH_3CH_2C\equiv CCH_2CH_3$ **c.** $HC\equiv C-$

5-21. $CH_3CH=CCH_2CH_2CH_3$ and $CH_3CH_2C=CHCH_2CH_3$

5-23. a. $CH_3CH_2C\equiv CH$ or $CH_3C\equiv CCH_3$
b. $CH_3C\equiv CCH_3$ **c.** $CH_3CH_2CH_2CH_2C\equiv CH$
5-24. The carbanion that would be formed is a stronger base than the amide ion.
5-25. a. $CH_3CH_2CH_2\bar{C}H_2 > CH_3CH_2CH=\bar{C}H > CH_3CH_2C\equiv\bar{C}$
b. $^-NH_2 > CH_3C\equiv C^- > CH_3CH_2O^- > F^-$
5-29. BF₃ does not have an accompanying nucleophile that could act as a chain terminator.
5-30. a. $CH_2=CHCl$ **b.** $CH_2=CCH_3$ **c.** $CF_2=CF_2$

5-32. beach balls

CHAPTER 6

6-1. a. 1, 2, 5, 6, and **8** have delocalized electrons.

6-3. a. $CH_3\overset{\delta+}{C}=\!\!=CH=\!\!=\overset{\delta+}{C}HCH_3$ (with CH_3 below)   **b.** (cyclohexenone with $\delta-$, $\delta-$, O)   **c.** $CH_3\overset{\delta+}{C}H=\!\!=CH=\!\!=\overset{\delta+}{C}HCH_3$

6-4. 6   **6-5.** (carbonate structure $^-O-\overset{O}{\underset{}{C}}-O^-$)

6-6. a. $CH_3CH=\!\!=\overset{+}{C}HCHCH_3$ **b.** (benzene ring)$-\overset{+}{C}CH_3$ (with CH_3 below)

6-7. a. $CH_3-\overset{+NH_2}{\underset{}{C}}-NH_2$

b. $CH_3\overset{O^-}{\underset{}{C}}=CHCH_3$

6-9. a. (cycloheptene ring with CH_3 and Cl)   **b.** $CH_3=\!\!=CHCH_2CH_2CH_2\overset{CH_3}{\underset{Br}{C}}CH_3$

6-10. (structure)   **6-11.** 2,4-heptadiene

6-12. a. $CH_3CH_2-\underset{Cl}{C}H-CH=\!\!=CHCH_3$ + $CH_3CH_2-CH=\!\!=CH-\underset{Cl}{C}HCH_3$

b. $CH_3CH_2\overset{Br}{\underset{CH_3CH_3}{C}}-C=\!\!=CHCH_3$ + $CH_3CH_2\overset{Br}{\underset{CH_3CH_3}{C}}=\!\!=CCHCH_3$

6-13. a. $CH_3CH=\!\!=CHOH$ **c.** $CH_3CH=\!\!=CHOH$

b. $CH_3\overset{O}{\underset{}{C}}-OH$ **d.** $CH_3CH=\!\!=\overset{+}{C}HNH_3$

6-14. a. ethylamine **b.** ethoxide ion **c.** ethoxide ion

6-15. (ring)$-COOH$ > (ring)$-OH$ > (ring)$-CH_2OH$

6-16. (ring)$-CH=\!\!=CH-$(ring) > (biphenyl) >

(ring)$-CH=\!\!=CH_2$ > (benzene)

6-17. a. The compound on the right is blue. **b.** They will be the same color.

CHAPTER 7

7-2. the cycloheptatrienyl cation   **7-3.** only **b**   **7-6. a.** The nitrogen donates electrons by resonance into the ring. **b.** The nitrogen is the most electronegative atom in the molecule. **c.** The nitrogen atom withdraws electrons from the ring.

7-7. a. $CH_3CHCH_2CH_2CH_2CH_3$ (with ring below)   **c.** $CH_3CH_2CHCH_2CH_3$ (with CH_2 + ring below)

b. (ring)$-CH_2OH$

7-10. a. (ring)$-CH_2CH_3$

b. (ring)$-\overset{CH_3}{\underset{}{C}}HCH_2CH_3$

c. (ring)$-CH_2CH=\!\!=CH_2$

7-11. a. *ortho*-ethylphenol or 2-ethylphenol
b. *meta*-bromochlorobenzene or 3-bromochlorobenzene
c. *meta*-bromobenzaldehyde or 3-bromobenzaldehyde
d. *ortho*-ethyltoluene or 2-ethyltoluene

7-12.

a. (phenol with Br para)   **c.** (phenol with Br, I)

b. (aniline with NH_2, NO_2)   **d.** (benzaldehyde $HC=\!\!=O$, NO_2, O_2N)

7-13. a. 1,3,5-tribromobenzene
b. *meta*-nitrophenol or 3-nitrophenol
c. *para*-bromotoluene or 4-bromotoluene
d. *ortho*-dichlorobenzene or 1,2-dichlorobenzene

7-14. a. donates electrons by resonance and withdraws electrons inductively
b. donates electrons inductively
c. withdraws electrons by resonance and withdraws electrons inductively
d. donates electrons by resonance and withdraws electrons inductively
e. donates electrons by resonance and withdraws electrons inductively
f. withdraws electrons inductively

7-15. a. phenol > toluene > benzene > bromobenzene > nitrobenzene
b. toluene > chloromethylbenzene > dichloromethylbenzene > difluoromethylbenzene

7-17. a. (benzene) $\xrightarrow[H_2SO_4]{HNO_3}$ (nitrobenzene NO_2) $\xrightarrow[FeCl_3]{Cl_2}$ (m-chloronitrobenzene NO_2, Cl)

b. (benzene) $\xrightarrow[FeCl_3]{Cl_2}$ (chlorobenzene Cl) $\xrightarrow[H_2SO_4]{HNO_3}$ (p-chloronitrobenzene Cl, O_2N)

7-19. a. $ClCH_2\overset{O}{\underset{}{C}}OH$ **c.** $FCH_2\overset{O}{\underset{}{C}}OH$

b. $H_3\overset{+}{N}CH_2\overset{O}{\underset{}{C}}OH$ **d.** $H\overset{O}{\underset{}{C}}OH$

CHAPTER 8

8-1. a. $CH_3CH_2CH_2OH$ $CH_3\underset{CH_3}{C}HOH$ $CH_3CH_2OCH_3$

b. 7

8-3. a. F, G, J, L, N, P, Q, R, S, Z **b.** A, C, D, H, I, M, O, T, U, V, W, X, Y

8-4. a, c, and **f**   **8-6. a, c,** and **f**

8-7. a.

b.

c.

8-8. a, b, and c

8-9. a. $-CH_2OH$ (1) $-CH_3$ (3) $-CH_2CH_2OH$ (2) $-H$ (4)

b. $-CH=O$ (2) $-OH$ (1) $-CH_3$ (4) $-CH_2OH$ (3)

c. $-CH(CH_3)_2$ (2) $-CH_2CH_2Br$ (3) $-Cl$ (1) $-CH_2CH_2CH_2Br$ (4)

d. $-CH=CH_2$ (2) $-CH_2CH_3$ (3) (1) $-CH_3$ (4)

8-10. a. *R* **b.** *R* **8-11. a.** (*R*)-2-bromobutane **b.** (*R*)-1,3-dichlorobutane
8-12. a. enantiomers **b.** enantiomers **8-14. a.** levorotatory **b.** dextrorotatory
8-15. a. *S* **b.** *R* **c.** *R* **d.** *S* **8-16.** +168° **8-17. a.** −24° **b.** 0° **8-18.** From the data given, you cannot determine the configuration. **8-19. a.** enantiomers **b.** identical compounds (Therefore, they are not isomers.) **c.** diastereomers
8-20. a. 8 **b.** $2^8 = 256$

8-22.

8-23. b

8-26. a.

b.

c.

8-27. a. (*R*)-malate and (*S*)-malate **b.** (*R*)-malate and (*S*)-malate

CHAPTER 9

9-3. a. $CH_3CH_2CHCH_2CCH_2CH_3$ with CH_3 groups **b.** 6 **9-4. a.** 3 **b.** 3 **c.** 1 **d.** 5 **e.** 2 **f.** 1

9-6. a.

$CH_3CH_2CH_2CH_2CH_2Cl$ $CH_3CH_2CH_2CHCH_3$ $CH_3CH_2CH_2CHCH_2CH_3$

21%   53%   26%

b. $ClCH_2CHCH_2CH_2CHCH_3$ $CH_3CCH_2CH_2CHCH_3$ $CH_3CHCHCH_2CHCH_3$

32%   27%   41%

9-7. a. chlorination **b.** bromination **c.** both the same
9-8. a. CH_3CHCH_3 (with CH_3) **b.** $CH_3CH_2CH_2CH_3$

CHAPTER 10

10-1. a. It is tripled. **b.** It is half the original rate.

10-2. $CH_3CH_2CH_2CH_2CH_2Br$ > $CH_3CHCH_2CH_2Br$ (with CH_3) >

$CH_3CH_2CHCH_2Br$ (with CH_3) > CH_3CH_2CBr (with two CH_3)

10-3. b. (*S*)-2-butanol **c.** (*R*)-3-hexanol **d.** 3-pentanol
10-5. a. $CH_3CH_2Br + HO^-$ **c.** $CH_3CH_2Br + I^-$
b. $CH_3CHCH_2Br + HO^-$ (with CH_3)   **d.** $CH_3CH_2CH_2I + HO^-$

10-6. a. $CH_3CH_2CH_2OCH_2CH_3$ **b.** $CH_3C\equiv CCH_2CH_3$
c. $(CH_3)_3NCH_2CH_3$ **d.** $CH_3CH_2SCH_2CH_3$

10-7. CH_3CBr (with two CH_3) > CH_3CHBr (with CH_3) > $CH_3CH_2CH_2Br$ > CH_3Br

10-8. $CH_3CH_2CCH_2CH_3$ (with CH_3, Br) > $CH_3CHCH_2CH_2CH_3$ (with Br) > $CH_3CHCH_2CH_2CH_3$ (with Cl) >

$ClCH_2CH_2CH_2CH_2CH_3$

10-10. a. CH_3CHBr (with CH_3)   **b.** CH_3CHBr (with CH_3)

10-11. a. b is more reactive than a **b.** b is more reactive than a **c.** b is more reactive than a **d.** a is more reactive than b

10-12. b. $CH_3CH_2CH=CCH_3$ (with CH_3)

c. $CH_3CH=CHCHCH_3$ (with CH_3)

d. $CH_3CCH=CH_2$ (with two CH_3)

10-13. $CH_3C=CCH_2CH_3$ (with CH_3) > $CH_3CHC=CHCH_3$ (with CH_3) > $CH_3CHCCH_2CH_3$ (with CH_3, CH_2)

10-18. a. 1. no reaction **b. 1.** primarily substitution
2. no reaction **2.** substitution and elimination
3. substitution and elimination **3.** substitution and elimination
4. substitution and elimination **4.** elimination

10-19. a. $CH_3CH_2CH_2Br$ **c.** $CH_3CH_2CH_2CCH_3$ (with CH_3, Br)

b. (cyclohexyl)—Br   **d.** $CH_3CCH_2CH_2Cl$ (with two CH_3)

CHAPTER 11

11-2. a. 1-pentanol **primary**
    b. 4-methylcyclohexanol **secondary**
    c. 5-chloro-2-methyl-2-pentanol **tertiary**
    d. 2-ethyl-1-pentanol **primary**
    e. 5-methyl-3-hexanol **secondary**
    f. 2,6-dimethyl-4-octanol **secondary**

11-3. $CH_3\overset{CH_3}{\underset{OH}{C}}CH_2CH_2CH_3$ $CH_3CH_2\overset{CH_3}{\underset{OH}{C}}CH_2CH_3$ $CH_3\overset{CH_3}{\underset{OH}{C}}{-}\overset{}{\underset{CH_3}{C}}HCH_3$

      2-methyl-2-pentanol   **3-methyl-3-pentanol**   **2,3-dimethyl-2-butanol**

11-4. Their nucleophilic ability results from the lone pair.

11-6. a. $CH_3CH_2\overset{}{\underset{Br}{C}}HCH_3$ **b.**

11-8. a.   **c.**

    b.

11-10. a. $CH_3CH_2\overset{}{\underset{CH_3}{C}}{=}CCH_3$ **b.**

11-11. a. $CH_3CH_2\overset{CH_3}{C}{=}\overset{CH_3}{\underset{CH_2CH_3}{C}}HCH_3$ **b.**

11-12. a. $CH_3CH_2\overset{O}{\overset{\|}{C}}CH_2CH_3$ **c.**

    b. $CH_3CH_2CH_2CH_2\overset{O}{\overset{\|}{C}}OH$ **d.**

11-14. a. 1. methoxyethane **2.** ethoxyethane **3.** 4-methoxyoctane **4.** 1-propoxybutane
    b. no
    c. 1. ethyl methyl ether **2.** diethyl ether **3.** no common name
      4. butyl propyl ether

11-15.

$CH_3OCH_2CH_2CH_3 \xrightarrow[\Delta]{HI} CH_3I + CH_3CH_2CH_2OH \xrightarrow[\Delta]{HI} CH_3CH_2CH_2I + H_2O$

11-16. b. $HOCH_2CH_2CH_2CH_2CH_2I$

    c. $CH_3\overset{CH_3}{\underset{CH_3}{C}}{-}I + CH_3CH_2OH$

    d. $HOCH_2CH_2CH_2\overset{CH_3}{\underset{I}{C}}CH_3$

11-17. a.   **b.**

11-18. a. $HOCH_2\overset{CH_3}{\underset{OCH_3}{C}}CH_3$ **c.** $HOCH{-}\overset{}{\underset{CH_3\ CH_3}{C}}CH_3$

    b. $CH_3OCH_2\overset{OH}{\underset{CH_3}{C}}CH_3$ **d.** $CH_3OCH{-}\overset{OH}{\underset{CH_3\ CH_3}{C}}CH_3$

11-19. noncyclic ether   **11-20.** The first one is too insoluble; the second is too reactive; the third is too unreactive.

11-21.

CHAPTER 12

12-1. a. propanamide, propionamide
    b. isobutyl butanoate, isobutyl butyrate
    c. potassium butanoate, potassium butyrate
    d. pentanoyl chloride, valeryl chloride
    e. *N,N*-dimethylhexanamide
    f. cyclopentanecarboxylic acid

12-2. a. $CH_3\overset{O}{\overset{\|}{C}}O{-}$ **d.** $CH_3CH_2CH_2\overset{}{\underset{Cl}{C}}H\overset{O}{\overset{\|}{C}}OCH_2CH_3$

    b. $CH_3\overset{O}{\overset{\|}{C}}O^-Na^+$ **e.** $CH_3\overset{}{\underset{Br}{C}}HCH_2\overset{O}{\overset{\|}{C}}NH_2$

    c. $CH_3\overset{O}{\overset{\|}{C}}NHCH_2$ **f.** $\overset{O}{\overset{\|}{C}}CCl$

12-3. The carbon–oxygen bond in an alcohol.

12-4. The bond between oxygen and the methyl group is the longest; The bond between carbon and the carbonyl oxygen is the shortest

12-6. a. a new carboxylic acid derivative
    b. no reaction
    c. a mixture of two carboxylic acid derivatives

12-7. a. acetate ion
    b. no reaction

12-10. a. $CH_3CH_2CH_2OH$ **d.** NH_2

    b. $CH_3CH_2NH_2$ **e.** H_2O

    c. $(CH_3)_2NH$ **f.** $HO{-}\bigcirc{-}NO_2$

12-14. a. $\overset{O}{\overset{\|}{C}}OH + CH_3CH_2OH$

    b. $CH_3CH_2CH_2\overset{O}{\overset{\|}{C}}OH + CH_3OH$

12-15. $HOCH_2CH_2CH_2CH_2\overset{O}{\overset{\|}{C}}OH$

12-16. $CH_3CH_2\overset{O}{\overset{\|}{C}}OCH_2CH_2CH_2CH_3 + CH_3OH$

12-18.

    a. $CH_3CH_2CH_2\overset{O}{\overset{\|}{C}}Cl \xrightarrow{CH_3OH} CH_3CH_2CH_2\overset{O}{\overset{\|}{C}}OCH_3$

    b. $CH_3\overset{O}{\overset{\|}{C}}OCH_2CH_3 \xrightarrow[H^+]{CH_3(CH_2)_7OH} CH_3\overset{O}{\overset{\|}{C}}OCH_2CH_2CH_2CH_2CH_2CH_2CH_2CH_3$

12-20. a. $CH_3CH_2CH_2CCl$ (C=O) $+$ $2\ CH_3CH_2NH_2$

b. CCl $+$ $2\ CH_3NH$ | CH_3

12-21. #2 and #4

12-22. a. 1. $SOCl_2$ **b. 1.** $SOCl_2$
2. $CH_3CH_2NH_2$

2. —OH

12-23. PDS

12-24. a. $CH_3CH_2CH_2Br$ **b.** CH_3CHCH_2Br | CH_3

CHAPTER 13

13-1. If the ketone functional group were anywhere else in these compounds, they would not be ketones and would not have the "one" suffix.

13-2. a. 3-methylpentanal, β-methylvaleraldehyde **b.** 4-heptanone **c.** 2-methyl-4-heptanone **d.** 4-phenylbutanal, γ-phenylbutyraldehyde **e.** 4-ethylhexanal, γ-ethylcaproaldehyde **f.** 6-methyl-3-heptanone **13-3. a.** 2-heptanone **b.** 5-methyl-3-hexanone **13-4. a.** $CH_3CH_3 + HO^-$ **b.** $CH_3CH_3 + CH_3O^-$ **c.** $CH_3CH_3 + CH_3NH$

13-5. a. $CH_3CH_2CH_2CH_2CHCH_3$ (OH)

c. HO CH_3

b. $CH_3CH_2CH_2CCH_3$ (OH) | CH_3

13-6. $CH_3CCH_2CH_3$ (C=O) $+$ $CH_3CH_2CH_2MgBr$ and

$CH_3CH_2CCH_2CH_2CH_3$ (C=O) $+$ CH_3MgBr

13-8. CH_3CHCH_3 and $CH_3CH_2CHCH_2CH_3$
(OH)   (OH)

13-9. A and C

13-10. a. CH_3CHCH_2OH | CH_3

c. —CH_2OH

b. —OH

d. —$CHCH_3$ | OH

13-11. a. —$CNHCH_3$ (C=O)

c. $CH_3CNHCH_2CH_3$ (C=O)

b. CH_3CNH_2 (C=O)

d. CH_3CN (C=O) with CH_2CH_3 / CH_2CH_3

13-12. a. $=NCH_2CH_3$ $+$ H_2O

c. CH_3CH_2 \ $C=N(CH_2)_5CH_3$ $+$ H_2O / CH_3CH_2

b. with N CH_2CH_3 / CH_2CH_3 $+$ H_2O

d. CH_3CH_2 \ $C=N$— $+$ H_2O / CH_3CH_2

13-13. Electron-withdrawing groups decrease the stability of the aldehyde and increase the stability of the hydrate.

13-14. O_2N——C(=O)——NO_2

13-15. a. hemiacetals: 7 **b.** acetals: 2, 3 **c.** hemiketals: 1, 8 **d.** ketals: 5 **e.** hydrates: 4, 6

13-17. a. (Br) **b.** (SCH_3)

CHAPTER 14

14-2. a. the ketone **b.** the ester **14-3.** The electrons left behind when the proton is removed are more readily delocalized onto the oxygen atom. **14-4.** They have a hydrogen bonded to the nitrogen which is more acidic than the hydrogen attached to the α-carbon. **14-5.** It is easier for hydroxide ion to attack the reactive carbonyl group than to remove a hydrogen from an α-carbon.

14-6. a. $CH_3CH=CCH_2CH_3$ (OH) **b.** (OH) $—C=CH_2$ **c.** (OH)

14-7. and

more stable

14-9. a. $CH_3CH_2CCHCH_3$ (C=O) | CH_2CH_3 **b.** CH_2CH_3

14-11. a. $CH_3CH_2CH_2CH$ (C=O) **c.** —CH_2CH (C=O)

b. CH_3CCH_3 (C=O) **d.** $CH_3CH_2CCH_2CH_3$ (C=O)

14-12. (OH) (C=O)

14-14. a. $CH_3CH_2CH_2CCHCOCH_3$ (O O) | CH_2CH_3 **b.** $CH_3CHCH_2CCHCOCH_2CH_3$ (O O) | CH_3 | $CHCH_3$ | CH_3

14-15. "a", "b", and "d"

14-16. $CH_3CH_2CH_2COCH_3$ (C=O) $+$ CH_3O^-

14-17. a and d

14-18. a. methyl bromide **b.** benzyl bromide **c.** isobutyl bromide

14-20. a. ethyl bromide **b.** pentyl bromide **c.** benzyl bromide

14-22. 7

14-23. a. 3 **b.** 7

CHAPTER 15

15-1. a **15-2. a.** 2000 cm^{-1} **b.** 8 μm **c.** 2 μm

15-3. a. C$\equiv$C stretch **b.** C—H stretch **c.** C$\equiv$N stretch **d.** C=O stretch

15-4. a. The carbon–oxygen stretch of phenol. **b.** The carbon–oxygen double-bond stretch of a ketone. **c.** The C—O stretch.

15-5. One bonded to an sp^3 carbon.

15-7. a. a ketone **b.** a tertiary amine

15-10. a. 2 **b.** 1 **c.** 4 **d.** 3 **e.** 3 **f.** 1

15-11. a = 2 signals, b = 1 signal, c = 3 signals.

15-12. a. 2 ppm **b.** 2 ppm **15-13.** 1.5 ppm

15-14. upfield from the TMS peak

15-16. first spectrum = 1-iodopropane

15-17. a. $CH_3CHCHBr$ **c.** $CH_3CH_2CHCH_3$
      | | |
      Br   Br   Cl

      b. CH_3CHOCH_3
             |
             CH_3

15-18. a. $CH_3CH_2CHCH_3$ **b.** $CH_3CH_2CH_2Cl$
              |
              Cl

15-21. 1,4-dimethylbenzene

15-22. a. CH_3CHCOH **b.** $ClCH_2CH_2COH$ (with carbonyl O)
         |
         Cl

15-27. CH_3O—⟨benzene ring⟩—CH_3

15-28. a. 1. 3   **2.** 3   **3.** 4   **4.** 3   **5.** 2   **6.** 3

15-30. $CH_3CH_2CH_2CH_2CH_2CCH_2CH_2CH_2CH_2CH_3$ (with carbonyl O)

CHAPTER 16

16-1. D-Ribose is an aldopentose. D-Sedoheptulose is a ketoheptose. D-Mannose is an aldohexose.   **16-3. a.** D-ribose   **b.** L-talose   **16-4.** D-psicose
16-5. a. 16   **b.** 32   **c.** none   **16-6. a.** D-iditol   **b.** D-iditol and D-gulitol
16-7. L-galactose   **16-8.** D-tagatose   **16-9. a.** L-gulose   **b.** L-gularic acid
c. D-allose and L-allose, D-altrose and D-talose, L-altrose and L-talose, D-galactose and L-galactose   **16-10. a.** D-gulose and D-idose   **b.** L-xylose and L-lyxose   **16-12. a.** none   **b.** C-2, C-3, and C-4   **c.** C-1 and C-3
16-13. b. methyl α-D-galactoside; nonreducing

CHAPTER 17

17-1. (S)-alanine   **17-2.** (2S,3R)-threonine   **17-3.** isoleucine

17-6. b. $H_2NCCH_2CH_2CHCO^-$ **c.** $NH_2CNHCH_2CH_2CH_2CHCO^-$
                 $^+NH_3$ $^+NH_3$
(first with $^+NH_2$ and carbonyl; second with carbonyl groups)

17-8. a. 5.43   **b.** 10.76   **c.** 5.68   **d.** 2.98
17-9. a. Asp   **b.** Arg
17-10. 2-methylpropanal
17-14. A-G-M   A-M-G   M-G-A   M-A-G   G-A-M   G-M-A
17-16. a. glutamate, cysteine, and glycine
17-18. Gly-Arg-Trp-Ala-Glu-Leu-Met-Pro-Val-Asp
17-19. a. His-Lys   Leu-Val-Glu-Pro-Arg   Ala-Gly-Ala
      b. Leu-Gly-Ser-Met-Phe-Pro-Tyr   Gly-Val
17-21. Leu-Tyr-Lys-Arg-Met-Phe-Arg-Ser
17-23. a. cigar-shaped protein   **b.** subunit of a hexamer
17-24. Leu-Val and Val-Val
17-25. 5.8%

CHAPTER 18

18-2. 2, 3, and 4   **18-3.** Only 2 can form an imine with the substrate.

18-4. $^-OCCHCHCH_2CO^-$ + NADH + H^+
     (with COO$^-$ and two carbonyl O)
        |
        OH

18-5. CH_3CH—CO^- + NAD^+
        | ‖
       OH   O

18-6. a. 7   **b.** 3

18-7. (thiolane ring with S–S) $CH_2CH_2CH_2CH_2CO^-$ + $FADH_2$

18-8. a. CH_3C— (with carbonyl O)   **b.** CH_3C— (with carbonyl O)

18-9. a. ^-OC—CH_2C—CO^- (with three carbonyl O)

18-10. 9

18-12. a. CH_3CHCO^- (with carbonyl O)   **b.** $^-OCCH_2CHCO^-$ (with carbonyl O)
         $^+NH_3$ $^+NH_3$

18-13. $CH_3CHCHOH$ (with H and OH arrows)

18-14. By adding four hydrogens to folate.

18-15.

$HSCH_2CH_2CHCO^-$ + N^5-methyl-THF $\longrightarrow$ $CH_3SCH_2CH_2CHCO^-$ + THF
          $^+NH_3$ $^+NH_3$
(both with carbonyl O)

18-16. the methylene group of N^5,N^{10}-methylene-THF

CHAPTER 19

19-4. The β-carbon has a partial positive charge.   **19-5.** eight   **19-6.** seven
19-9. two   **19-10.** a ketone   **19-11.** thiamine pyrophosphate
19-12. an aldehyde   **19-14.** pyridoxal phosphate   **19-15.** pyruvate
19-16. a secondary alcohol   **19-17.** two   **19-18.** citrate and isocitrate
19-19. 11

CHAPTER 20

20-2. glyceryl tripalmitate

20-3.
CH_2—O—C—$(CH_2)_{16}CH_3$ (with carbonyl O)
CH—O—C—$(CH_2)_{14}CH_3$ (with carbonyl O)
CH_2—O—C—$(CH_2)_{16}CH_3$ (with carbonyl O)

20-4.
CH_2—O—C—$(CH_2)_{16}CH_3$ (with carbonyl O)
CH—O—C—$(CH_2)_{16}CH_3$ (with carbonyl O)
CH_2—O—C—$(CH_2)_{14}CH_3$ (with carbonyl O)

20-8. integral proteins

20-9. The bacteria could synthesize phosphoacylglycerols with more saturated fatty acids.

20-11.

(+)-limonene

20-12. b. a monoterpene

20-14. two ketone groups, a double bond, an aldehyde group, a primary alcohol, a tertiary alcohol

CHAPTER 21

21-2. a. 3′——C—C—T—G—T—T—A—G—A—C—G—— 5′
      b. guanine
21-5. Met-Asp-Pro-Val-Ile-Lys-His
21-6. Met-Asp-Pro-Leu-Leu-Asn
21-8.

5′——G—C—A—T—G—G—A—C—C—C—C—G—T—T—A—T—T—A—A—A—C—A—C—— 3′

21-9. a

Glossary

absorption band   a peak in a spectrum that occurs as a result of the absorption of energy.

acetal $R-\overset{\overset{\displaystyle OR}{|}}{\underset{\underset{\displaystyle OR}{|}}{C}}-H$

acetoacetic ester synthesis   synthesis of a methyl ketone, using ethyl acetoacetate as the starting material.

acetylide ion $RC\equiv C^-$

achiral (optically inactive)   an achiral molecule is identical to (i.e., superimposable upon) its mirror image.

acid   a substance that donates a proton.

acid anhydride $R-\overset{\overset{\displaystyle O}{\|}}{C}-O-\overset{\overset{\displaystyle O}{\|}}{C}-R$

acid–base reaction   a reaction in which an acid donates a proton to a base or accepts a share in a base's electrons.

acid catalyst   a catalyst that increases the rate of a reaction by donating a proton.

acid-catalyzed reaction   a reaction catalyzed by an acid.

acid dissociation constant   a measure of the degree to which an acid dissociates in solution.

activating substituent   a substituent that increases the reactivity of an aromatic ring. Electron-donating substituents activate aromatic rings toward electrophilic attack, and electron-withdrawing substituents activate aromatic rings toward nucleophilic attack.

active site   a pocket or cleft in an enzyme where the substrate is bound.

acyl chloride $R-\overset{\overset{\displaystyle O}{\|}}{C}-Cl$

acyl group   a carbonyl group bonded to an alkyl group or to an aryl group.

1,2-addition   addition to the 1- and 2-positions of a conjugated system.

1,4-addition   addition to the 1- and 4-positions of a conjugated system.

addition reaction   a reaction in which atoms or groups are added to the reactant.

alcohol   a compound with an OH group in place of one of the hydrogens of an alkane; (ROH).

alcoholysis   reaction with an alcohol.

aldaric acid   a dicarboxylic acid with an OH group bonded to each carbon. Obtained by oxidizing the aldehyde and primary alcohol groups of an aldose.

aldehyde $R-\overset{\overset{\displaystyle O}{\|}}{C}-H$

alditol   a compound with an OH group bonded to each carbon. Obtained by reducing an aldose or a ketose.

aldol addition   a reaction between two molecules of an aldehyde (or two molecules of a ketone) that connects the α-carbon of one with the carbonyl carbon of the other.

aldol condensation   an aldol addition followed by the elimination of water.

aldonic acid   a carboxylic acid with an OH group bonded to each carbon. Obtained by oxidizing the aldehyde group of an aldose.

aldose   a polyhydroxyaldehyde.

aliphatic   a nonaromatic organic compound.

alkaloid   a natural product, with one or more nitrogen heteroatoms, found in the leaves, bark, or seeds of plants.

alkane   a hydrocarbon that contains only single bonds.

alkene   a hydrocarbon that contains a double bond.

alkyl halide   a compound with a halogen in place of one of the hydrogens of an alkane.

alkyl substituent (alkyl group)   a substituent formed by removing a hydrogen from an alkane.

alkyne   a hydrocarbon that contains a triple bond.

allyl group $CH_2\!=\!CHCH_2-$

allylic carbon   an sp^3 carbon adjacent to a vinylic carbon.

allylic cation   a species with a positive charge on an allylic carbon.

amide $R-\overset{\overset{\displaystyle O}{\|}}{C}-NH_2$, $R-\overset{\overset{\displaystyle O}{\|}}{C}-NHR$, $R-\overset{\overset{\displaystyle O}{\|}}{C}-NR_2$

amine   a compound with a nitrogen in place of one of the hydrogens of an alkane; RNH_2, R_2NH, R_3N

amino acid   an α-aminocarboxylic acid. Naturally occurring amino acids have the L configuration.

amino acid analyzer   an instrument that automates the ion-exchange separation of amino acids.

amino acid residue   a monomeric unit of a peptide or protein.

amino acid side chain   the substituent attached to the α-carbon of an amino acid.

aminolysis   reaction with an amine.

amino sugar   a sugar in which one of the OH groups is replaced by an NH_2 group.

anabolic reaction   a reaction a living organism carries out in order to synthesize complex molecules from simple precursor molecules.

anabolism   reactions living organisms carry out in order to synthesize complex molecules from simple precursor molecules.

angle strain   the strain introduced into a molecule as a result of its bond angles being distorted from their ideal values.

angstrom   unit of length; 100 picometers = 10^{-8} cm = 1 angstrom.

anomeric carbon   the carbon in a cyclic sugar that is the carbonyl carbon in the open-chain form.

anomers   two cyclic sugars that differ in configuration only at the carbon that is the carbonyl carbon in the open-chain form.

antibiotic   a compound that interferes with the growth of a microorganism.

antibodies   compounds that recognize foreign particles in the body.

anticodon   the three bases at the bottom of the middle loop in tRNA.

antigens   compounds that can generate a response from the immune system.

antiviral drug   a drug that interferes with DNA or RNA synthesis in order to prevent a virus from replicating.

applied magnetic field   the externally applied magnetic field.

aprotic solvent   a solvent that does not have a hydrogen bonded to an oxygen or to a nitrogen.

arene oxide   an aromatic compound that has had one of its double bonds converted to an epoxide.

aromatic   a cyclic and planar compound with an uninterrupted ring of p orbital-bearing atoms containing an odd number of pairs of π electrons.

aryl group   a benzene or a substituted-benzene group.

asymmetric center   a carbon bonded to four different atoms or groups.

atomic number   the number of protons (or electrons) that the neutral atom has.

atomic orbital   an orbital associated with an atom.

atomic weight   the average mass of the atoms in the naturally occurring element.

autoradiograph   the exposed photographic plate obtained in autoradiography.

axial bond   a bond of the chair conformation of cyclohexane that is perpendicular to the plane in which the chair is drawn (an up–down bond).

back-side attack   nucleophilic attack on the side of the carbon opposite the side bonded to the leaving group.

bactericidal drug   a drug that kills bacteria.

bacteriostatic drug   a drug that inhibits the further growth of bacteria.

base[1]   a substance that accepts a proton.

base[2]   a heterocyclic compound (a purine or a pyrimidine) in DNA and RNA.

base catalyst   a catalyst that increases the rate of a reaction by removing a proton.

basicity   the tendency of a compound to share its electrons with a proton.

bending vibration   a vibration that does not occur along the line of the bond. It results in changing bond angles.

benzyl group

benzylic carbon   an sp^3 hybridized carbon bonded to a benzene ring.

benzylic cation   a compound with a positive charge on a benzylic carbon.

bimolecular reaction (second-order reaction)   a reaction whose rate depends on the concentration of two reactants.

biochemistry (biological chemistry)   the chemistry of biological systems.

biodegradable polymer   a polymer that can be broken into small segments by an enzyme-catalyzed reaction.

bioorganic compound   an organic compound found in biological systems.

biopolymer   a polymer that is synthesized in nature.

biosynthesis   synthesis in a biological system.

biotin   the coenzyme required by enzymes that catalyze carboxylation of a carbon adjacent to an ester or a keto group.

blind screen (random screen)   the search for a pharmacologically active compound without any information about which chemical structures might show activity.

boiling point   the temperature at which the vapor pressure equals the atmospheric pressure.

bond   an attractive force between two atoms.

bond dissociation energy   the energy required to break a bond, or the amount of energy released when a bond is formed.

bond length   the internuclear distance between two atoms at minimum energy (maximum stability).

bond strength   the energy required to break a bond.

brand name   identifies a commercial product and distinguishes it from other products. It can be used only by the owner of the registered trademark.

buffer   a weak acid and its conjugate base.

carbanion   a compound containing a negatively charged carbon.

carbocation   a species containing a positively charged carbon.

carbohydrate   a sugar or a saccharide. Naturally occurring carbohydrates have the D configuration.

α-carbon   a carbon bonded to a carbonyl carbon or to a leaving group.

β-carbon   a carbon adjacent to an α-carbon carbonyl.

carbonyl carbon   the carbon of a carbonyl group.

carbonyl compound   a compound that contains a carbonyl group.

carbonyl group   a carbon doubly bonded to an oxygen.

carbonyl oxygen   the oxygen of a carbonyl group.

carboxyl group   COOH

carboxylic acid
$$R-\overset{\overset{\displaystyle O}{\|}}{C}-OH$$

carboxylic acid derivative   a compound that is hydrolyzed to a carboxylic acid.

carboxyl oxygen   the single-bonded oxygen of a carboxlic acid or an ester.

carotenoid   a class of compounds (a tetraterpene) responsible for the red and orange colors of fruits, vegetables, and fall leaves.

Catabolic reaction   a reaction a living organism carries out in order to break down complex molecules into simple molecules and energy.

catabolism   reactions living organisms carry out in order to break down complex molecules into simple molecules and energy.

catalyst   a species that increases the rate at which a reaction occurs without being consumed in the reaction. Because it does not change the equilibrium constant of the reaction, it does not change the amount of product that is formed.

catalytic hydrogenation   the addition of hydrogen to a double or a triple bond with the aid of a metal catalyst.

cation-exchange resin   a negatively charged resin used in ion-exchange chromatography.

chain-growth polymer   a polymer made by adding monomers to the growing end of a chain.

chain reaction   a reaction in which propagating steps are repeated over and over.

chair conformation   the conformation of cyclohexane that roughly resembles a chair. It is the most stable conformation of cyclohexane.

chemically equivalent protons   protons with the same connectivity relationship to the rest of the molecule.

chemical shift   the location of a signal in an NMR spectrum. It is measured downfield from a reference compound (most often, TMS).

chiral (optically active)   a chiral molecule has a nonsuperimposable mirror image.

cholesterol   a steroid that is the precursor of all other animal steroids.

chromatography   a separation technique in which the mixture to be separated is dissolved in a solvent and the solvent is passed through a column packed with an absorbent stationary phase.

cis fused   two cyclohexane rings fused together such that if the second ring were considered to be two substituents of the first ring, one substituent would be in an axial position and the other would be in an equatorial position.

cis isomer   the isomer with identical substituents on the same side of the double bond or on the same side of a cyclic structure.

cis–trans isomers   geometric isomers.

citric acid cycle (Krebs cycle)   a series of reactions that converts the acetyl group of acetyl-CoA into two molecules of CO_2.

Claisen condensation   a reaction between two molecules of an ester that connects the α-carbon of one with the carbonyl carbon of the other and eliminates an alkoxide ion.

codon   a sequence of three bases in mRNA that specifies the amino acid to be incorporated into a protein.

coenzyme   a cofactor that is an organic molecule.

coenzyme A   a thiol used by biological organisms to form thioesters.

coenzyme B$_{12}$ the coenzyme required by enzymes that catalyze certain rearrangement reactions.

cofactor   an organic molecule or a metal ion that certain enzymes need to catalyze a reaction.

common name   nonsystematic nomenclature.

complex carbohydrate   a carbohydrate containing two or more sugar molecules linked together.

condensation reaction   a reaction combining two molecules while removing a small molecule (usually water or an alcohol).

configuration   the three-dimensional structure of a particular atom in a compound. The configuration is designated by R or S.

conformation   the three-dimensional shape of a molecule at a given instant that can change as a result of rotations about σ bonds.

conformers   different conformations of a molecule.

conjugate acid   a species accepts a proton to form its conjugate acid.

conjugate addition   1,4-addition.

conjugate base   a species loses a proton to form its conjugate base.

conjugated diene   a hydrocarbon with two conjugated double bonds.

conjugated double bonds   double bonds separated by one single bond.

constitutional isomers   molecules that have the same molecular formula but differ in the way their atoms are connected.

core electrons   electrons in inner shells.

coupled protons   protons that split each other. Coupled protons have the same coupling constant.

coupling constant   the distance (in hertz) between two adjacent peaks of a split NMR signal.

covalent bond   a bond created as a result of sharing electrons.

crown ether   a cyclic molecule that contains several ether linkages.

crown–guest complex   the complex formed when a crown ether binds a substrate.

C-terminal amino acid   the terminal amino acid of a peptide (or protein) that has a free carboxyl group.

cycloalkane   an alkane with its carbon chain arranged in a closed ring.

deactivating substituent a substituent that decreases the reactivity of an aromatic ring. Electron-withdrawing substituents deactivate aromatic rings toward electrophilic attack, and electron-donating substituents deactivate aromatic rings toward nucleophilic attack.

deamination loss of ammonia.

decarboxylation loss of carbon dioxide.

dehydration loss of water.

dehydrogenase an enzyme that carries out an oxidation reaction by removing hydrogen from the substrate.

delocalized electrons electrons that are shared by more than two atoms.

denaturation destruction of the highly organized tertiary structure of a protein.

deoxyribonucleic acid (DNA) a polymer of deoxyribonucleotides.

deoxyribonucleotide a nucleotide in which the sugar component is D-2-deoxyribose.

deoxy sugar a sugar in which one of the OH groups has been replaced by an H.

dextrorotatory the enantiomer that rotates polarized light in a clockwise direction.

diastereomer a configurational stereoisomer that is not an enantiomer.

diene a hydrocarbon with two double bonds.

dinucleotide two nucleotides linked by phosphodiester bonds.

dipeptide two amino acids linked by an amide bond.

dipole–dipole interaction an interaction between the dipole of one molecule and the dipole of another.

disaccharide a compound containing two sugar molecules linked together.

disulfide R—S—S—R

disulfide bridge a disulfide (—S—S—) bond in a peptide or protein.

DNA (deoxyribonucleic acid) a polymer of deoxyribonucleotides.

donation of electrons by resonance donation of electrons through p orbital overlap with neighboring π bonds.

double bond a σ bond and a π bond between two atoms.

double helix the secondary structure of DNA.

doublet an NMR signal split into two peaks.

downfield at a higher frequency in an NMR spectrum.

drug a compound that reacts with a biological molecule, triggering a physiological effect.

drug resistance biological resistance to a particular drug.

eclipsed conformer a conformer in which the bonds on adjacent carbons are aligned as viewed looking down the carbon–carbon bond.

***E* conformation** the conformation of a carboxylic acid or carboxylic acid derivative in which the carbonyl oxygen and the substituent bonded to the carboxyl oxygen or nitrogen are on opposite sides of the single bond.

Edman's reagent phenyl isothiocyanate. A reagent used to determine the N-terminal amino acid of a polypeptide.

effective magnetic field the magnetic field that a proton "senses" through the surrounding cloud of electrons.

***E* isomer** the isomer with the high-priority groups on opposite sides of the double bond.

electronegative element an element that readily acquires an electron.

electronegativity tendency of an atom to pull electrons toward itself.

electronic configuration description of the orbitals that the electrons in an atom occupy.

electrophile an electron-deficient atom or molecule.

electrophilic addition reaction an addition reaction in which the first species that adds to the reactant is an electrophile.

electrophilic aromatic substitution a reaction in which an electrophile substitutes for a hydrogen of an aromatic ring.

electrophoresis a technique that separates amino acids on the basis of their pI values.

electrostatic attraction attractive force between opposite charges.

electrostatic potential map a model that shows the charge distribution in a species.

elimination reaction a reaction that involves the elimination of atoms (or molecules) from the reactant.

enamine an α,β-unsaturated tertiary amine.

enantiomers nonsuperimposable mirror-image molecules.

enkephalins pentapeptides synthesized by the body to control pain.

enol an α,β-unsaturated alcohol.

enolization keto–enol interconversion.

enzyme a protein that is a catalyst.

epimers monosaccharides that differ in configuration at only one carbon.

epoxide an ether in which the oxygen is incorporated into a three-membered ring.

epoxy resin substance formed by mixing a low-molecular-weight prepolymer with a compound that forms a cross-linked polymer.

equatorial bond a bond of the chair conformer of cyclohexane that juts out from the ring in approximately the same plane that contains the chair.

equilibrium constant the ratio of products to reactants at equilibrium or the ratio of the rate constants for the forward and reverse reactions.

E1 reaction a first-order elimination reaction.

E2 reaction a second-order elimination reaction.

essential amino acid an amino acid that humans must obtain from their diet because they cannot synthesize it at all or cannot synthesize it in adequate amounts.

ester
$$\underset{R}{} \overset{\displaystyle O}{\underset{}{\overset{\|}{C}}} \underset{}{OR}$$

ether a compound containing an oxygen bonded to two carbons (ROR).

fat a triester of glycerol that exists as a solid at room temperature.

fatty acid a long-chain carboxylic acid.

favorable reaction a reaction in which the concentration of products is greater than the concentration of reactants at equilibrium.

fibrous protein a water-insoluble protein in which the polypeptide chains are arranged in bundles.

fingerprint region the right-hand third of an IR spectrum where the absorption bands are characteristic of the compound as a whole.

Fischer esterification reaction the reaction of a carboxylic acid with alcohol in the presence of an acid catalyst to form an ester.

Fischer projection a method of representing the spatial arrangement of groups bonded to a chirality center. The chirality center is the point of intersection of two perpendicular lines; the horizontal lines represent bonds that project out of the plane of the paper toward the viewer, and the vertical lines represent bonds that point back from the plane of the paper away from the viewer.

flavin adenine dinucleotide (FAD) a coenzyme required in certain oxidation reactions. It is reduced to $FADH_2$, which can act as a reducing agent in another reaction.

formal charge the number of valence electrons − (the number of nonbonding electrons + 1/2 the number of bonding electrons).

free energy of activation ($\Delta G^{\ddagger}$) the true energy barrier to a reaction.

free radical a species with an unpaired electron.

frequency the velocity of a wave divided by its wavelength (in units of cycles/s).

Friedel–Crafts acylation an electrophilic substitution reaction that puts an acyl group on a benzene ring.

functional group the center of reactivity in a molecule.

functional group interconversion the conversion of one functional group into another functional group.

functional group region the left-hand two-thirds of an IR spectrum where most functional groups show absorption bands.

furanose a five-membered-ring sugar.

furanoside a five-membered-ring glycoside.

gene a segment of DNA.

generic name a commercially nonrestricted name for a drug.

genetic code the amino acid specified by each three-base sequence of mRNA.

geometric isomers cis–trans (or *E,Z*) isomers.

globular protein a water-soluble protein that tends to have a roughly spherical shape.

gluconeogenesis the synthesis of D-glucose from pyruvate.

glycolysis (glycolytic pathway) the sequence of reactions that converts D-glucose into two molecules of pyruvate.

glycoprotein a protein that is covalently bonded to a polysaccharide.

glycoside the acetal of a sugar.

glycosidic bond the bond between the anomeric carbon and the alcohol in a glycoside.

α-1,4-glycosidic linkage a glycosidic linkage between the C-1 oxygen of one sugar and C-4 of a second sugar with the oxygen atom of the glycosidic linkage in the axial position.

α-1,6-glycosidic linkage a glycosidic linkage between the C-1 oxygen of one sugar and C-6 of a second sugar with the oxygen atom of the glycosidic linkage in the axial position.

β-1,4-glycosidic linkage a glycosidic linkage between the C-1 oxygen of one sugar and C-4 of a second sugar with the oxygen atom of the glycosidic linkage in the equatorial position.

Grignard reagent the compound that results when magnesium is inserted between the carbon and halogen of an alkyl halide (RMgBr, RMgCl).

halogenation reaction with halogen (Br_2, Cl_2, I_2).

α-helix the backbone of a polypeptide coiled in a right-handed spiral with hydrogen bonding occurring within the helix.

hemiacetal
$$R-\underset{\underset{OR}{|}}{\overset{\overset{OH}{|}}{C}}-H$$

hemiketal
$$R-\underset{\underset{OR}{|}}{\overset{\overset{OH}{|}}{C}}-R$$

heptose a monosaccharide with seven carbons.

heteroatom an atom other than carbon or hydrogen.

heterocyclic compound (heterocycle) a cyclic compound in which one or more of the atoms of the ring are heteroatoms.

hexose a monosaccharide with six carbons.

hormone an organic compound synthesized in a gland and delivered by the bloodstream to its target tissue.

human genome the total DNA of a human cell.

hybrid orbital an orbital formed by mixing (hybridizing) orbitals.

hydrate (*gem*-diol)
$$R-\underset{\underset{OH}{|}}{\overset{\overset{OH}{|}}{C}}-R \quad (H)$$

hydrated water has been added to a compound.

hydration addition of water to a compound.

hydride ion a negatively charged hydrogen.

hydrocarbon a compound that contains only carbon and hydrogen.

α-hydrogen usually, a hydrogen bonded to the carbon adjacent to a carbonyl carbon.

hydrogenation addition of hydrogen.

hydrogen bond an unusually strong dipole–dipole attraction (5 kcal/mol) between a hydrogen bonded to O, N, or F and the nonbonding electrons of an O, N, or F of another molecule.

hydrogen ion (proton) a positively charged hydrogen.

hydrolysis reaction with water.

hydrophobic interactions interactions between nonpolar groups. These interactions increase stability by decreasing the amount of structured water (increasing entropy).

imine $R_2C=NR$

induced-dipole–induced-dipole interaction an interaction between a temporary dipole in one molecule and the dipole the temporary dipole induces in another molecule.

induced-fit model a model that describes the specificity of an enzyme for its substrate: The shape of the active site does not become completely complementary to the shape of the substrate until after the enzyme binds the substrate.

inductive electron donation donation of electrons through σ bond(s).

inductive electron withdrawal withdrawal of electrons through σ bond(s).

infrared radiation electromagnetic radiation familiar to us as heat.

infrared spectroscopy uses infrared energy to provide a knowledge of the functional groups in a compound.

infrared (IR) spectrum a plot of percent transmission versus wave number (or wavelength) of infrared radiation.

initiation step the step in which radicals are created, or the step in which the radical needed for the first propagation step is created.

intermediate a species formed during a reaction and that is not the final product of the reaction.

intermolecular reaction a reaction that takes place between two molecules.

internal alkyne an alkyne with the triple bond not at the end of the carbon chain.

intramolecular reaction a reaction that takes place within a molecule.

inversion of configuration turning the configuration of a carbon inside out like an umbrella in a windstorm, so that the resulting product has a configuration opposite that of the reactant.

ion–dipole interaction the interaction between an ion and the dipole of a molecule.

ion-exchange chromatography a technique that uses a column packed with an insoluble resin to separate compounds on the basis of their charges and polarities.

ionic bond a bond formed through the attraction of two ions of opposite charges.

isoelectric point (pI) the pH at which there is no net charge on an amino acid.

isolated diene a hydrocarbon containing two isolated double bonds.

isolated double bonds double bonds separated by more than one single bond.

isomers nonidentical compounds with the same molecular formula.

isotopes atoms with the same number of protons, but different numbers of neutrons.

IUPAC nomenclature systematic nomenclature of chemical compounds.

Kekulé structure a model that represents the bonds between atoms as lines.

ketal
$$R-\underset{\underset{OR}{|}}{\overset{\overset{OR}{|}}{C}}-R$$

keto–enol tautomerism (keto–enol interconversion) interconversion of keto and enol tautomers.

keto–enol tautomers a ketone and its isomeric α,β-unsaturated alcohol.

β-keto ester an ester with a second carbonyl group at the β-position.

ketone
$$\underset{R}{\overset{\overset{\displaystyle O}{\|}}{\underset{}{C}}}\!\!-R$$

ketose a polyhydroxyketone.

Kiliani–Fischer synthesis a method used to increase the number of carbons in an aldose by one, resulting in the formation of a pair of C-2 epimers.

λ_{max} the wavelength at which there is maximum UV or Vis absorbance.

lead compound the prototype in a search for other biologically active compounds.

leaving group the group that is displaced in a nucleophilic substitution reaction.

levorotatory the enantiomer that rotates polarized light in a counterclockwise direction.

Lewis acid a substance that accepts an electron pair.

Lewis base a substance that donates an electron pair.

Lewis structure a model that represents the bonds between atoms as lines or dots and the valence electrons as dots.

ligation sharing of nonbonding electrons with a metal ion.

lipid a water-insoluble compound found in a living system.

lipid bilayer two layers of phosphoacylglycerols arranged so that their polar heads are on the outside and their nonpolar fatty acid chains are on the inside.

localized electrons electrons that are restricted to a particular locality.

lone-pair electrons (nonbonding electrons) valence electrons not used in bonding.

magnetic resonance imaging (MRI) NMR used in medicine. The difference in the way water is bound in different tissues produces a variation in signal between organs as well as between healthy and diseased tissue.

major groove the wider and deeper of the two alternating grooves in DNA.

malonic ester synthesis the synthesis of a carboxylic acid, using diethyl malonate as the starting material.

mass number the number of protons plus the number of neutrons in an atom.

mechanism of a reaction a description of the step-by-step process by which reactants are changed into products.

melting point the temperature at which a solid becomes a liquid.

membrane the material that surrounds a cell in order to isolate its contents.

meso compound a compound that contains chirality centers and a plane of symmetry.

metabolism reactions living organisms carry out in order to obtain the energy and to synthesize the compounds they require.

meta director a substituent that directs an incoming substituent meta to an existing substituent.

methine hydrogen a tertiary hydrogen.

methylene group a CH_2 group.

micelle a spherical aggregation of molecules, each with a long hydrophobic tail and a polar head, arranged so that the polar head points to the outside of the sphere.

minor groove the narrower and more shallow of the two alternating grooves in DNA.

mixed triglyceride a triacylglycerol in which the fatty-acid components are different.

molecular modification changing the structure of a lead compound.

molecular recognition the recognition of one molecule by another as a result of specific interactions; for example, the specificity of an enzyme for its substrate.

molecular weight the average weighted mass of the atoms in a molecule.

monomer a repeating unit in a polymer.

monosaccharide (simple carbohydrate) a single sugar molecule.

monoterpene a terpene that contains 10 carbons.

MRI scanner an NMR spectrometer used in medicine for whole-body NMR.

multiplet an NMR signal split into more than seven peaks.

multiplicity the number of peaks in an NMR signal.

multistep synthesis preparation of a compound by a route that requires several steps.

mutarotation a slow change in optical rotation to an equilibrium value.

N + 1 rule an ^1H NMR signal for a hydrogen with N equivalent hydrogens bonded to an adjacent carbon is split into $N + 1$ peaks. A ^{13}C NMR signal for a carbon bonded to N hydrogens is split into $N + 1$ peaks.

natural-abundance atomic weight the average mass of the atoms in the naturally occurring element.

natural product a product synthesized in nature.

nicotinamide adenine dinucleotide (NAD^+) a coenzyme required in certain oxidation reactions. It is reduced to NADH, which can act as a reducing agent in another reaction.

nitration substitution of a nitro group (NO_2) for a hydrogen of a benzene ring.

nitrile a compound that contains a carbon–nitrogen triple bond ($RC \equiv N$).

NMR spectroscopy the absorption of electromagnetic radiation to determine the structural features of an organic compound. In the case of NMR spectroscopy, it determines the carbon–hydrogen framework.

nonbonding electrons (lone-pair electrons) valence electrons not used in bonding.

nonpolar covalent bond a bond formed between two atoms that share the bonding electrons equally.

nonpolar molecule a molecule with no charge or partial charge on any of its atoms.

nonreducing sugar a sugar that cannot be oxidized by reagents such as Ag^+ and Cu^+. Nonreducing sugars are not in equilibrium with the open-chain aldose or ketose.

N-terminal amino acid the terminal amino acid of a peptide (or protein) that has a free amino group.

nucleic acid the two kinds of nucleic acid are DNA and RNA.

nucleophile an electron-rich atom or molecule.

nucleophilic acyl substitution reaction a reaction in which a group bonded to an acyl or aryl group is substituted by another group.

nucleophilic addition reaction a reaction that involves the addition of a nucleophile to a reagent.

nucleophilicity a measure of how readily an atom or a molecule with a pair of nonbonding electrons attacks an atom.

nucleophilic substitution reaction a reaction in which a nucleophile substitutes for an atom or a group.

nucleoside a heterocyclic base (a purine or a pyrimidine) bonded to the anomeric carbon of a sugar (D-ribose or D-2-deoxyribose).

nucleotide a heterocycle attached in the β-position to a phosphorylated ribose or deoxyribose.

observed rotation the amount of rotation observed in a polarimeter.

octet rule states that an atom will give up, accept, or share electrons in order to achieve a filled shell. Because a filled second shell contains eight electrons, this is known as the octet rule.

oil a triester of glycerol that exists as a liquid at room temperature.

oligonucleotide 3 to 10 nucleotides linked by phosphodiester bonds.

oligopeptide 3 to 10 amino acids linked by amide bonds.

oligosaccharide 3 to 10 sugar molecules linked by glycosidic bonds.

operating frequency the frequency at which an NMR spectrometer operates.

optically active rotates the plane of polarized light.

optically inactive does not rotate the plane of polarized light.

orbital the volume of space around the nucleus in which an electron is most likely to be found.

orbital hybridization mixing of orbitals.

organic compound a compound that contains carbon.

organic synthesis preparation of organic compounds from other organic compounds.

organometallic compound a compound containing a carbon–metal bond.

orphan drugs drugs for diseases or conditions that affect fewer than 200,000 people.

ortho-para-director a substituent that directs an incoming substituent ortho and para to an existing substituent.

oxidation reaction a reaction in which the number of $C-H$ bonds decreases.

oxidation–reduction reaction (redox reaction) a reaction that involves the transfer of electrons from one species to another.

oxidative cleavage an oxidation reaction that cuts the reactant into two or more pieces.

packing the fitting of individual molecules into a frozen crystal lattice.

paraffin an alkane.

parent hydrocarbon the longest continuous carbon chain in a molecule.

partial hydrolysis a technique that hydrolyzes only some of the peptide bonds in a polypeptide.

pentose a monosaccharide with five carbons.

peptide polymer of amino acids linked together by amide bonds. A peptide contains fewer amino acid residues than a protein does.

peptide bond the amide bond that links the amino acids in a peptide or protein.

peroxyacid a carboxylic acid with an OOH group instead of an OH group.

perspective formula a method of representing the spatial arrangement of groups bonded to a chirality center. Two bonds are drawn in the plane of the paper; a solid wedge is used to depict a bond that projects out of the plane of the paper toward the viewer, and a hatched wedge is used to represent a bond that projects back from the plane of the paper away from the viewer.

pH the pH scale is used to describe the acidity of a solution (pH $= -\log[H^+]$).

phenyl group

pheromone a compound secreted by an animal that stimulates a physiological or behavioral response from a member of the same species.

phosphoacylglycerol (phosphoglyceride) a compound formed when two OH groups of glycerol form esters with fatty acids and the terminal OH group forms a phosphate ester.

phosphoanhydride bond the bond holding two phosphoric acid molecules together.

phospholipid a lipid that contains a phosphate group.

phosphoryl transfer reaction the transfer of a phosphate group from one compound to another.

pi (π) bond a bond formed as a result of side-to-side overlap of p orbitals.

pK_a describes the tendency of a compound to lose a proton (p$K_a = -\log K_a$, where K_a is the acid dissociation constant).

plane polarized light light that oscillates only in one plane.

plane of symmetry an imaginary plane that bisects a molecule into mirror images.

β-pleated sheet the backbone of a polypeptide that is extended in a zigzag structure with hydrogen bonding between neighboring chains.

polar covalent bond a bond formed by the unequal sharing of electrons.

polarimeter an instrument that measures the rotation of polarized light.

polymer a large molecule made by linking monomers together.

polymerization the process of linking up monomers to form a polymer.

polynucleotide many nucleotides linked by phosphodiester bonds.

polypeptide many amino acids linked by amide bonds.

polysaccharide a compound containing more than 10 sugar molecules linked together.

polyunsaturated fatty acid a fatty acid with more than one double bond.

porphyrin ring system consists of four pyrrole rings joined by one-carbon bridges.

primary alcohol an alcohol in which the OH group is bonded to a primary carbon.

primary alkyl halide an alkyl halide in which the halogen is bonded to a primary carbon.

primary alkyl radical a radical with the unpaired electron on a primary carbon.

primary amine an amine with one alkyl group bonded to the nitrogen.

primary carbocation a carbocation with the positive charge on a primary carbon.

primary carbon a carbon bonded to only one other carbon.

primary hydrogen a hydrogen bonded to a primary carbon.

primary structure (of a nucleic acid) the sequence of bases in a nucleic acid.

primary structure (of a protein) the sequence of amino acids in a protein.

propagating site the reactive end of a chain-growth polymer.

propagation step in the first of a pair of propagation steps, a radical (or an electrophile or a nucleophile) reacts to produce another radical (or an electrophile or a nucleophile) that reacts in the second step to produce the radical (or the electrophile or the nucleophile) that was the reactant in the first propagation step.

protein a polymer containing 40 to 4000 amino acids linked by amide bonds.

proton a positively charged hydrogen (hydrogen ion); a positively charged particle in an atomic nucleus.

proton-coupled ^{13}C NMR spectrum a ^{13}C NMR spectrum in which each signal is split by the hydrogens bonded to the C that produced the spectrum.

pyranose a six-membered-ring sugar.

pyranoside a six-membered-ring glycoside.

pyridoxal phosphate the coenzyme required by enzymes that catalyze certain transformations of amino acids.

quartet an NMR signal split into four peaks.

quaternary structure a description of the way the individual polypeptide chains of a protein are arranged with respect to each other.

racemic mixture (racemate, racemic modification) a mixture of equal amounts of a pair of enantiomers.

radical an atom or a molecule with an unpaired electron.

radical chain reaction a reaction in which radicals are formed and react in repeating propagating steps.

radical inhibitor a compound that traps radicals.

radical substitution reaction a substitution reaction that has a radical intermediate.

random screen (blind screen) the search for a pharmacologically active compound without any information about what chemical structures might show activity.

rate constant a measure of how easy or difficult it is to reach the transition state of a reaction (to get over the energy barrier to the reaction).

rate law the relationship between the rate of a reaction and the concentration of the reactants.

rate-determining step (rate-limiting step) the step in a reaction that has the transition state with the highest energy.

R configuration after assigning relative priorities to the four groups bonded to a chirality center, if the lowest priority group is on a vertical axis in a Fischer projection (or pointing away from the viewer in a perspective formula), an arrow drawn from the highest priority group to the next-highest-priority group goes in a clockwise direction.

reaction coordinate diagram describes the energy changes that take place during the course of a reaction.

reactivity–selectivity principle states that the greater the reactivity of a species, the less selective it will be.

receptor a site on a cell at which a drug binds in order to exert its physiological effect.

reducing sugar a sugar that can be oxidized by reagents such as Ag^+ or Br_2. Reducing sugars are in equilibrium with the open-chain aldose or ketose.

reduction reaction a reaction in which the number of C—H bonds increases.

reference compound a compound added to a sample whose NMR spectrum is to be taken. The positions of the signals in the NMR spectrum are measured from the position of the signal given by the reference compound.

regioselective reaction a reaction that leads to the preferential formation of one constitutional isomer over another.

replication the synthesis of identical copies of DNA.

resonance a compound with delocalized electrons is said to have resonance.

resonance contributor a structure with localized electrons that approxi-

mates the true structure of a compound with delocalized electrons.

resonance electron donation donation of electrons through p orbital overlap with neighboring π bonds.

resonance electron withdrawal withdrawal of electrons through p orbital overlap with neighboring π bonds.

resonance hybrid the actual structure of a compound with delocalized electrons; it is represented by two or more structures with localized electrons.

resonance stabilization the extra stability associated with a compound as a result of its having delocalized electrons.

restriction endonuclease an enzyme that cleaves DNA at a specific base sequence.

restriction fragment a fragment that is formed when DNA is cleaved by a restriction endonuclease.

ribonucleic acid (RNA) a polymer of ribonucleotides.

ribonucleotide a nucleotide in which the sugar component is D-ribose.

RNA (ribonucleic acid) a polymer of ribonucleotides.

saponification hydrolysis of an ester (such as a fat) under basic conditions.

saturated hydrocarbon a hydrocarbon that is completely saturated (i.e., contains no double or triple bonds) with hydrogen.

S configuration after assigning relative priorities to the four groups bonded to a chirality center, if the lowest priority group is on a vertical axis in a Fischer projection (or pointing away from the viewer in a perspective formula), an arrow drawn from the highest priority group to the next-highest priority group goes in a counterclockwise direction.

secondary alcohol an alcohol in which the OH group is bonded to a secondary carbon.

secondary alkyl halide an alkyl halide in which the halogen is bonded to a secondary carbon.

secondary alkyl radical a radical with the unpaired electron on a secondary carbon.

secondary amine an amine with two alkyl groups bonded to the nitrogen.

secondary carbocation a carbocation with the positive charge on a secondary carbon.

secondary carbon a carbon bonded to two other carbons.

secondary hydrogen a hydrogen bonded to a secondary carbon.

secondary structure a description of the conformation of the backbone of a protein.

semiconservative replication the mode of replication that results in a daughter molecule of DNA having one of the original DNA strands plus a newly synthesized strand.

sense strand the strand in DNA that is not read during transcription; it has the same sequence of bases as the synthesized mRNA strand (with a U, T difference).

separated charges a positive and a negative charge that can be neutralized by the movement of electrons.

sesquiterpene a terpene that contains 15 carbons.

shielding phenomenon caused by electron donation to the environment of a proton. The electrons shield the proton from the full effect of the applied magnetic field. The more a proton is shielded, the farther to the right its signal appears in an NMR spectrum.

sigma (σ) bond a bond with a cylindrically symmetrical distribution of electrons.

simple carbohydrate (monosaccharide) a single sugar molecule.

simple triglyceride a triacylglycerol in which the fatty acid components are the same.

single bond a σ bond.

singlet an unsplit NMR signal.

skeletal structure shows the carbon=carbon bonds as lines and does not show the carbon–hydrogen bonds.

S_N1 reaction a unimolecular nucleophilic substitution reaction.

S_N2 reaction a bimolecular nucleophilic substitution reaction.

soap a sodium or potassium salt of a fatty acid.

solvation the interaction between a solvent and another molecule (or ion).

specific rotation the amount of rotation that will be caused by a compound with a concentration of 1.0 g/mL in a sample tube 1.0 dm long.

spectroscopy study of the interaction of matter and electromagnetic radiation.

sphingolipid a lipid that contains sphingosine.

α-spin state nuclei in this spin state have their magnetic moments oriented in the same direction as the applied magnetic field.

β-spin state nuclei in this spin state have their magnetic moments oriented opposite the direction of the applied magnetic field.

squalene a triterpene that is a precursor of steroid molecules.

staggered conformer a conformer in which the bonds on one carbon bisect the bond angle on the adjacent carbon when viewed looking down the carbon–carbon bond.

stereochemistry the field of chemistry that deals with the structures of molecules in three dimensions.

stereoisomers isomers that differ in the way their atoms are arranged in space.

steric effects effects due to the fact that groups occupy a certain volume of space.

steric hindrance refers to bulky groups at the site of a reaction that make it difficult for the reactants to approach each other.

steric strain (van der Waals strain, van der Waals repulsion) the repulsion between the electron cloud of an atom or a group of atoms and the electron cloud of another atom or group of atoms.

steroid a class of compounds that contains a steroid ring system.

straight-chain alkane an alkane in which the carbons form a contiguous chain with no branches.

stretching frequency the frequency at which a stretching vibration occurs.

substitution reaction a reaction in which an atom or group substitutes for another atom or group.

substrate the reactant of an enzyme-catalyzed reaction.

subunit an individual chain of an oligomer.

sulfonation substitution of a hydrogen of a benzene ring by a sulfonic acid group ($-SO_3H$).

synthetic polymer a polymer that is not synthesized in nature.

systematic nomenclature nomenclature based on structure.

tautomerism interconversion of tautomers.

tautomers rapidly equilibrating isomers that differ in the location of their bonding electrons.

template strand (antisense strand) the strand in DNA that is read during transcription.

terminal alkyne an alkyne with the triple bond at the end of the carbon chain.

termination step when two radicals combine to produce a molecule in which all the electrons are paired.

terpene a lipid, isolated from a plant, that contains carbon atoms in multiples of five.

tertiary alcohol an alcohol in which the OH group is bonded to a tertiary carbon.

tertiary alkyl halide an alkyl halide in which the halogen is bonded to a tertiary carbon.

tertiary alkyl radical a radical with the unpaired electron on a tertiary carbon.

tertiary amine an amine with three alkyl groups bonded to the nitrogen.

tertiary carbocation a carbocation with the positive charge on a tertiary carbon.

tertiary carbon a carbon bonded to three other carbons.

tertiary hydrogen a hydrogen bonded to a tertiary carbon.

tertiary structure a description of the three-dimensional arrangement of all the atoms in a protein.

tetrahedral bond angle the bond angle (109.5°) formed by adjacent bonds of an sp^3 hybridized carbon.

tetrahedral carbon an sp^3 hybridized carbon; a carbon that forms covalent bonds by using four sp^3 hybridized orbitals.

tetrahedral intermediate the intermediate formed in a nucleophilic acyl substitution reaction.

tetrahydrofolate (THF) the coenzyme required by enzymes that catalyze reactions that donate a group containing a single carbon to their substrates.

tetraterpene a terpene that contains 40 carbons.

tetrose a monosaccharide with four carbons.

thiamine pyrophosphate (TPP) the coenzyme required by enzymes, which catalyze a reaction that transfers a two-carbon fragment to a substrate.

thioester the sulfur analog of an ester

$$R-\overset{\overset{\displaystyle O}{\|}}{C}-SR$$

thiol the sulfur analog of an alcohol (RSH).

torsional strain the repulsion felt by the bonding electrons of one substituent as they pass close to the bonding electrons of another substituent.

trademark a registered name, symbol, or picture.

transamination a reaction in which an amino group is transferred from one compound to another.

transcription the synthesis of mRNA from a DNA blueprint.

transesterification reaction the reaction of an ester with an alcohol to form a different ester.

trans fused two cyclohexane rings fused together such that if the second ring were considered to be two substituents of the first ring, both substituents would be in equatorial positions.

trans isomer the isomer with identical substituents on opposite sides of the double bond or on opposite sides of a cyclic structure.

transition state the highest point on a hill in a reaction coordinate diagram. In the transition state, bonds in the reactant that will break are partially broken and bonds in the product that will form are partially formed.

translation the synthesis of a protein from an mRNA blueprint.

triglyceride the compound formed when the three OH groups of glycerol are esterified with fatty acids.

triose a monosaccharide with three carbons.

tripeptide three amino acids linked by amide bonds.

triple bond a σ bond plus two π bonds.

triplet an NMR signal split into three peaks.

triterpene a terpene that contains 30 carbons.

ultraviolet light electromagnetic radiation with wavelengths ranging from 180 to 400 nm.

upfield towards lower frequency in an NMR spectrum.

unimolecular reaction (first-order reaction) a reaction whose rate depends on the concentration of one reactant.

unsaturated hydrocarbon a hydrocarbon that contains one or more double or triple bonds.

UV/Vis spectroscopy the absorption of electromagnetic radiation in the ultraviolet and visible regions of the spectrum; used to determine information about conjugated systems.

valence electron an electron in an unfilled shell.

van der Waals forces induced-dipole–induced-dipole interactions.

vinyl group $CH_2=CH-$

vinylic carbon a carbon in a carbon–carbon double bond.

vinylic cation a compound with a positive charge on a vinylic carbon.

vinylic radical a compound with an unpaired electron on a vinylic carbon.

visible light electromagnetic radiation with wavelengths ranging from 400 to 780 nm.

vitamin a substance needed in small amounts for normal body function that the body cannot synthesize at all or cannot synthesize in adequate amounts.

vitamin KH_2 the coenzyme required by the enzyme that catalyzes the carboxylation of glutamate side chains.

wavelength distance from any point on one wave to the corresponding point on the next wave (usually in units of μm or nm).

wavenumber the number of waves in 1 cm.

wax an ester formed from a long-chain carboxylic acid and a long-chain alcohol.

withdraw electrons by resonance withdrawal of electrons through p orbital overlap with neighboring bonds.

Z isomer the isomer with the high-priority groups on the same side of the double bond.

zwitterion a compound with a negative charge and a positive charge on non-adjacent atoms.

Photo Credits

Index

Periodic Table of the Elements

Main groups

1A 1	**2A** 2	**3B** 3	**4B** 4	**5B** 5	**6B** 6	**7B** 7	**8B** 8	**8B** 9	**8B** 10	**1B** 11	**2B** 12	**3A** 13	**4A** 14	**5A** 15	**6A** 16	**7A** 17	**8A** 18
1 **H** 1.00794																	2 **He** 4.002602
3 **Li** 6.941	4 **Be** 9.012182											5 **B** 10.811	6 **C** 12.0107	7 **N** 14.0067	8 **O** 15.9994	9 **F** 18.998403	10 **Ne** 20.1797
11 **Na** 22.989770	12 **Mg** 24.3050											13 **Al** 26.981538	14 **Si** 28.0855	15 **P** 30.973761	16 **S** 32.065	17 **Cl** 35.453	18 **Ar** 39.948
19 **K** 39.0983	20 **Ca** 40.078	21 **Sc** 44.955910	22 **Ti** 47.867	23 **V** 50.9415	24 **Cr** 51.9961	25 **Mn** 54.938049	26 **Fe** 55.845	27 **Co** 58.933200	28 **Ni** 58.6934	29 **Cu** 63.546	30 **Zn** 65.39	31 **Ga** 69.723	32 **Ge** 72.64	33 **As** 74.92160	34 **Se** 78.96	35 **Br** 79.904	36 **Kr** 83.80
37 **Rb** 85.4678	38 **Sr** 87.62	39 **Y** 88.90585	40 **Zr** 91.224	41 **Nb** 92.90638	42 **Mo** 95.94	43 **Tc** [98]	44 **Ru** 101.07	45 **Rh** 102.90550	46 **Pd** 106.42	47 **Ag** 107.8682	48 **Cd** 112.411	49 **In** 114.818	50 **Sn** 118.710	51 **Sb** 121.760	52 **Te** 127.60	53 **I** 126.90447	54 **Xe** 131.293
55 **Cs** 132.90545	56 **Ba** 137.327	71 **Lu** 174.967	72 **Hf** 178.49	73 **Ta** 180.9479	74 **W** 183.84	75 **Re** 186.207	76 **Os** 190.23	77 **Ir** 192.217	78 **Pt** 195.078	79 **Au** 196.96655	80 **Hg** 200.59	81 **Tl** 204.3833	82 **Pb** 207.2	83 **Bi** 208.98038	84 **Po** [208.98]	85 **At** [209.99]	86 **Rn** [222.02]
87 **Fr** [223.02]	88 **Ra** [226.03]	103 **Lr** [262.11]	104 **Rf** [261.11]	105 **Db** [262.11]	106 **Sg** [266.12]	107 **Bh** [264.12]	108 **Hs** [269.13]	109 **Mt** [268.14]	110 [271.15]	111 [272.15]	112 [277]		114 [285]		116 [289]		

Transition metals

*Lanthanide series

57 **La** 138.9055	58 **Ce** 140.116	59 **Pr** 140.90765	60 **Nd** 144.24	61 **Pm** [145]	62 **Sm** 150.36	63 **Eu** 151.964	64 **Gd** 157.25	65 **Tb** 158.92534	66 **Dy** 162.50	67 **Ho** 164.93032	68 **Er** 167.259	69 **Tm** 168.93421	70 **Yb** 173.04

†Actinide series

89 **Ac** [227.03]	90 **Th** 232.0381	91 **Pa** 231.03588	92 **U** 238.02891	93 **Np** [237.05]	94 **Pu** [244.06]	95 **Am** [243.06]	96 **Cm** [247.07]	97 **Bk** [247.07]	98 **Cf** [251.08]	99 **Es** [252.08]	100 **Fm** [257.10]	101 **Md** [258.10]	102 **No** [259.10]

[a]The labels on top (1A, 2A, etc.) are common American usage. The labels below these (1, 2, etc.) are those recommended by the International Union of Pure and Applied Chemistry.

The names and symbols for elements 110 and above have not yet been decided.

Atomic weights in brackets are the masses of the longest-lived or most important isotope of radioactive elements.

Further information is available at *http://www.shef.ac.uk/chemistry/web-elements/*.

The production of element 116 was reported in May 1999 by scientists at Lawrence Berkeley National Laboratory.

Common Functional Groups

Alkane RCH_3

Alkene

$$\overset{\diagdown}{\diagup}C=C\overset{\diagup}{\diagdown} \qquad \overset{\diagdown}{\diagup}C=CH_2$$

internal   terminal

Alkyne $RC\equiv CR$ $RC\equiv CH$

internal   terminal

Nitrile $RC\equiv N$

Ether $R-O-R$

Thiol RCH_2-SH

Disulfide $R-S-S-R$

Epoxide

Aniline

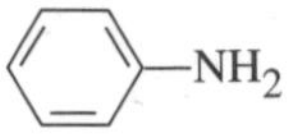

Phenol

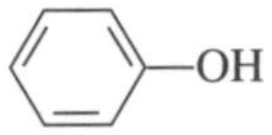

Carboxylic acid
$$R-\overset{\overset{\textstyle O}{\|}}{C}-OH$$

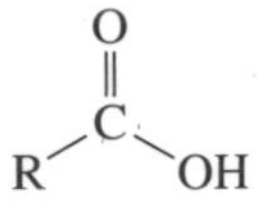

Acyl chloride
$$R-\overset{\overset{\textstyle O}{\|}}{C}-Cl$$

Acid anhydride
$$R-\overset{\overset{\textstyle O}{\|}}{C}-O-\overset{\overset{\textstyle O}{\|}}{C}-R$$

Ester
$$R-\overset{\overset{\textstyle O}{\|}}{C}-OR$$

Amide
$$R-\overset{\overset{\textstyle O}{\|}}{C}-NH_2 \qquad \diagdown NHR \qquad \diagdown NR_2$$

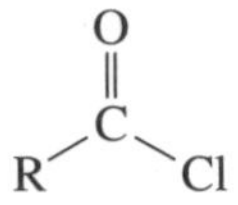

Aldehyde
$$R-\overset{\overset{\textstyle O}{\|}}{C}-H$$

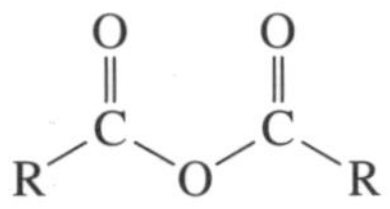

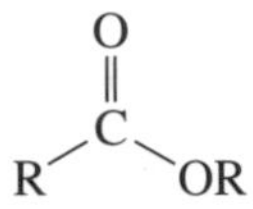

Ketone
$$R-\overset{\overset{\textstyle O}{\|}}{C}-R$$

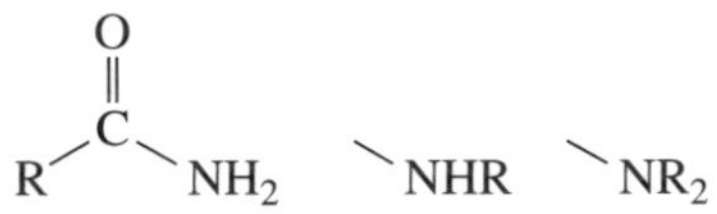

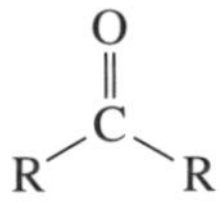

	primary	secondary	tertiary
Alkyl halide	$R-CH_2-X$ X = F, Cl, Br, or I	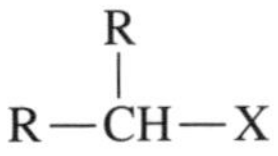 $R-\overset{\overset{\textstyle R}{\|}}{CH}-X$	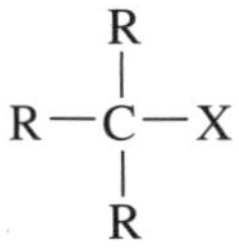 $R-\overset{\overset{\textstyle R}{\|}}{\underset{\underset{\textstyle R}{\|}}{C}}-X$
Alcohol	$R-CH_2-OH$	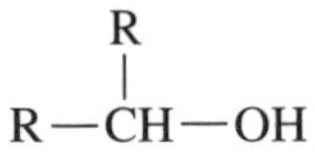 $R-\overset{\overset{\textstyle R}{\|}}{CH}-OH$	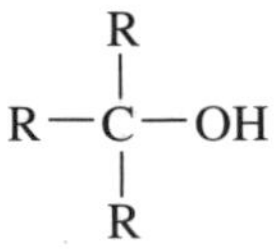 $R-\overset{\overset{\textstyle R}{\|}}{\underset{\underset{\textstyle R}{\|}}{C}}-OH$
Amine	$R-NH_2$	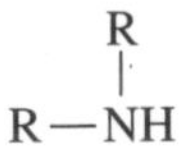 $R-\overset{\overset{\textstyle R}{\|}}{NH}$	 $R-\overset{\overset{\textstyle R}{\|}}{\underset{\underset{\textstyle R}{\|}}{N}}$

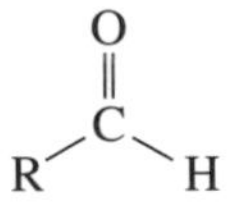